Electrical Machines

Rakesh Kumar Srivastava

Electrical Machines

Third Edition

Springer

Rakesh Kumar Srivastava
Department of Electrical Engineering
Indian Institute of Technology (BHU)
Varanasi
Varanasi, Uttar Pradesh, India

ISBN 978-981-96-5038-5 ISBN 978-981-96-5039-2 (eBook)
https://doi.org/10.1007/978-981-96-5039-2

This Springer imprint is published by the registered company Springer Nature Singapore Pte Ltd.
The registered company address is: 152 Beach Road, #21-01/04 Gateway East, Singapore 189721, Singapore

Dedicated
to
Prof. Som Nath Mahendra

Preface to the Third Edition

Though the solar power is becoming a major source of electrical power, the dependency on rotary machines for power generation will still continue forever. The motive power based on electricity is essentially required in variety of applications. Among them, newer areas are motors for **drones, electric vehicle, linear motor-based levitated** and **wheel on rail high speed ground transportation**, and **hyperloop systems of transport** are some of the major important applications. The electric catapult and launchers used in providing initial propulsion thrust to heavy drone with limited capability to remain air borne and the augmentation of propulsion thrust to fighter aircraft on board of a smaller warship are some of the cutting edge technologies requiring major emphasis on research and development on electrical machines and drives. This book has been written with an emphasis to develop curiosity among students toward electrical unconventional machines—electric motors and generators.

I am deeply indebted to **Prof. (Late) Som Nath Mahendra** for his great capability to teach electrical machines and encourage the students to develop skills to build, test, and analyze newer machines. The cutting edge technology of modern transport system including currently focused hyperloop technology based on linear motors was envisaged in reality in 1980 by Prof. (Late) S. N. Mahendra, when other countries were just thinking to go ahead in these areas of transport.

I am also thankful to my beloved wife Nishi and grand daughters Vidushi and Aanya for their cooperations during the long-extended period of revised book writing. The author is thankful to Prof. R. K. Saket for his encouragements. He is also thankful to research scholars and M.Tech. students who came forward for reading the manuscripts—Mr. Anant Kumar, Mr. Prabhakar Kumar, Mr. Raj Vardhan, Mr. Alok Kumar Tiwari, Mr. Risabh Raj, Mr. Amit Majhi, and Mr. Manish Kumar Pradhan. He is also grateful to Dr. Shailendra Gupta, Dr. Rajaram, and Dr. Ankita Dwivedi for help in revising the manuscripts.

Varanasi, India
(Basant Panchami)
February 2025

Rakesh Kumar Srivastava

Preface to the Second Edition

The most important purpose of the revised book is to provide a coherent, highly motivated, and sound exposition of the theory and characteristics of electrical machines under steady-state conditions. Many topics have been added for meeting various course objectives since the first edition was published, and several examples are added to conceptualize the theoretical contents.

The course contents have been extended to provide appropriate updating as well as to cover topics that furnish the background to gain an easy transition to study electric power system, power electronics, solid-state electric drives, unconventional energy resources utilization of electrical energy, and electrical traction. The treatment of electric machinery is brought up-to-date by the coverage of a number of significant topics. More emphasis is given to create interests among the students for the development of skill in experimental as well as for computer-based analytical techniques. For the overall professional development, students are expected to learn software such as MATLAB/Simulink for steady state and transient studies of electrical machines. References of FEM/BEM software have been mentioned at several places for developing professional expertise in the areas of computer-aided design, analysis, and simulation of electrical apparatus. The engineering institutions in our country now have licensed versions of these software. A few software are also having free versions available on the Internet.

The modern era of computer-based systems and embedded systems has led to the development of modern machinery such as PLC drive, vector-controlled drives, sensor-less control of drives, VFD, VVVF, DSP drives, and dSpace technology.

It is expected that this book will act as a catalyst for dedicated learning and development of professional skills of young budding engineers of our country in areas of electronics, solid-state/PLC/DSP drives, computer simulation, and embedded technology applied to electric machinery.

The course content is aimed for two-semester core courses being offered to undergraduate students of Electrical, and Electrical and Electronics Engineering, and one-semester course being offered to non-electrical undergraduate students of Electronic, Mechanical, Chemical, Ceramics, Metallurgy and Mining, and several other similar branches of engineering studies.

I am thankful to my beloved wife Nishi and son Jayant for their cooperation during the long-extended period of book writing.

The present book is the outcome of my 27 years of teaching electrical machines. I am indebted to my dedicated teachers, Prof. S. N. Mahendra, Prof. M. Bhattacharya, Prof. (Late) R. D. Sen Sharma, Prof. G. Rajsekar, Prof. B. R. M. Gandhi, and Prof. R. S. Gorayan.

I would like to extend my thanks to my classmate Mr. Bimlesh Kumar Singh (AGM), NTPC Singrauli; my teacher and colleague Prof. R. S. Gorayan; and the Ph.D. Scholar Dr. Amrita Sinha for providing their critical suggestions and improvements for the revised course contents. I also acknowledge my colleagues Prof. R. Mahanty, Prof. D. Singh, Prof. R. K. Mishra, and Dr. Kalpana Chaudhary for their encouragement and extended help in the review process.

I am thankful to M.Tech. students (2010–2012) Mr. Prem Kant, Mr. Atul Tiwari, Ms. Ankita Dwivedi, Mr. S. Goladpudi, Mr. Raja Ram, and Ms. Sricha Singh for their extending support in the cross-verification of course contents.

Varanasi, India Rakesh Kumar Srivastava

Preface to the First Edition

Electrical machines have been written with the aim to introduce the principles and techniques of studying electromechanical energy conversion (EMEC) devices. The book has been developed keeping in mind the current trends in the electromechanical energy conversion and the recent research interests in the area of electrical machines. It aims to create interest in the minds of young students for development of newer machines.

A significant component of the many contemporary electrical machines is **power electronics**. The advent of solid-state devices and embedded computing systems has resulted in the development of newer motors of the modern era. These motors are uncommon to the conventional plug-and-play type of motors. They require a logic controller for switching solid-state devices apart from the usual power source. Some of these motors are vector-controlled induction motors, SRM, BLDC motors and WPMG, etc. These machines are opted for because of their drive applications, power conditioning, quality power, and starter-generator technology. Certain machines like SRM and BLDC motors require synchronization of coil switching with the rotor position. BLDC motors are quite ubiquitous. This is because of their presence in personal computer peripherals like SMPS fan drive, CDROM drive, CPU fan drive, and CD/DVD players. Newer machines like doubly excited induction generators have proved to be effective in variable speed wind power generation. The technology depends upon the use of power electronics along with embedded computing for power generation.

Research and development in material science and metallurgical engineering have resulted in commercial production of low-loss sheet steel and ferrites, better conducting materials, insulating materials, powerful permanent magnets, and cooling of electrical machines. Since many years Hall Effect Gauss meter was being used merely for measuring DC flux density. In recent years, commercially available Hall effect devices like voltage and current transducers, Hall effect position sensor and latch, etc. are being used. BLDC motors are utilizing Hall effect-integrated circuits for their operation.

Fabrication of new machines is a difficult task, unless there is a proper workshop facility, materials, infrastructure, and labor for the prototype development. In recent days, newer forms of EMEC devices have been developed. MEMS, linear machines, Switched Reluctance Motor, brush less DC motor, written pole motor-generator are to name just a few. Since last 30 years there has been a radical change in the technology of electrical machines. Applications of personal computer and computer-aided design (CAD) software have grown many folds. It is now possible to predict the performance of any electrical machine based on its design parameters without actually fabricating it. This has been made possible due to the availability of fast computers and commercially available state-of-the-art software like ANSYS, MAGNET, PC-OPERA, and JMAG. These software utilize the solution of Maxwell's equations in and around the geometry of the machine and are normally based on finite element and boundary element methods. However, the discussions of the software are not within the preview of the present book. The license version of some of these software is available in different engineering institutions for academic research. It is expected that the students will learn some of the software and apply to simple electrical machines.

Presently, the use of MATLAB software has become popular in engineering institutions. The software relieves a user of mathematical drudgery, which was normally required in older days. This has resulted in better understanding of the behavior of electrical machines. I hope that the user will learn the software for finding torque speed characteristics of different machines and also transient behavior using MATLAB Simulink.

The book is divided into seven chapters providing a comprehensive coverage of electrical machines.

Chapter 1, **Basic Laws of Electromechanical Energy Conversion**, introduces the basic principles of operation of different electromechanical devices, classifications of electrical machines, duty cycle, conducting, insulating, thermal and mechanical properties of materials used in electrical machines and inductance from different viewpoints. A simple method of solving simultaneous differential equations or state variable equations has been described using MATLAB Simulink software.

Chapter 2, **Forces in an Electromagnetic System**, briefly covers Maxwell's equations from the view point of understanding electrical machines. Simple magnetic circuits along with their relationship with Maxwell's equations, concepts of MMF, magnetic flux and flux density, reluctance, permeance, leakage flux, fringing, saturation, measurement of AC/DC flux density, Hall effect devices and their use in electrical machines are explained.

Chapter 3, **Transformer**, deals with different types of transformers, their classifications, constructional features, cooling methodology, and basic EMF equations. Three-phase transformer connections, Scott-connected transformer and inrush current, etc. have also been explained.

Chapter 4, **Direct Current Machines**, focuses on homopolar and hetro-polar machines, their constructional details, commutation, armature reaction, types of windings, EMF and torque equations, compensating winding, equalizer connection, DC generators, parallel operation, types of DC motors.

Chapter 5, **Induction Machines**, presents three-phase armature winding, MMF distribution, distribution factor, EMF generation, chording, EMF induced and elimination of harmonics, pitch factor, rotating magnetic field, concept of time and space harmonics, cogging and crawling, and Georges's phenomenon.

Chapter 6, **Synchronous Machines**, describes constructional details, generation of three-phase voltages, determination of induced EMF of alternator using general (ZPF), EMF, MMF, and ASA methods, and machines parameters and characteristics.

Chapter 7, **Special Motors and Introduction to Generalized Machine Theory**, discusses special motors like servo motors-printed circuit motor and AC servo motor used in control system, commutator AC motors, stepper motors, PMBLDC motor, Switched Reluctance Motor (SRM), linear motors, written pole motor-generator (WPM, WPG), and eddy current device. The generalized machine theory and symmetrical components have been briefly discussed.

The present book is the outcome of my 24 years of teaching electrical machines. I am indebted to Prof. S. N. Mahendra, who introduced me to the world of linear motors. I pay my sincere gratitude to my teachers, Prof. M. Bhattacharyya Prof. G. Rajsekar, Prof. (late) R. D. Sen Sharma, and Prof. B. R. M. Gandhi.

I acknowledge my colleagues Prof. B. N. Sarkar, Dr. A. K. Tiwari, Dr. R. S. Gorayan, and Mrs. Kalpana Choudhary for their encouragement. I am thankful to my students who came forward for the completion of the present work.

Varanasi, India Rakesh Kumar Srivastava

Competing Interests The author has no competing interests to declare that are relevant to the content of this manuscript.

Contents

About the Author

Rakesh Kumar Srivastava received B.Tech. in electrical engineering, M.Tech. in electrical machines and drives, and Ph.D. in electrical machines (Linear Induction Motor) from the erstwhile Institute of Technology, Banaras Hindu University, Varanasi, India, in 1983, 1985, and 2000, respectively. He is currently working as a Professor in electrical machines and drives. He has published several research papers in international and national journals and conference proceedings. He has authored three books and edited IEEE Conference proceedings of UPCON 2016. His specializations are the electromagnetic field applied to electrical machines, linear induction machines, special purpose electrical machines, permanent magnets motors and generators, eddy current brake and electromagnetic launchers. Prof. Srivastava was Convener of IEEE Uttar Pradesh Section International Conference on Electrical, Computer, and Electronics (UPCON) 2016, IIT (BHU), Varanasi (India). He was an Adviser and Founder of the IEEE Industry Applications Society Student Chapter at IIT (BHU), Varanasi. He is a recipient of Prof. G. K. Dubey Memorial Life-time Achievement Award 2022 of IEEE UP Section.

Chapter 1
Basic Laws of Electromechanical Energy Conversion

Abstract This chapter describes the basic principles of operation of different electromechanical devices, stored energy, counter-EMF and counter-torque/force, types of mechanical loads, hard and soft torque-speed characteristics, AC operation, classifications of electrical machines, duty cycles, IP specifications, material properties of conducting, insulating, thermal and mechanical properties of materials used in electrical machines, inductance from different viewpoints, self, mutual and leakage inductance, A simple method of solving simultaneous differential equations or state variable equation has been described using MATLAB Simulink software.

Keywords Lenz's law · Counter-EMF and counter-torque · Impregnation · Air bearing · Cryogenic magnet · Constant flux theorem · Current and flux doubling · Hall effect sensor · Open magnetic circuit · Power flow · Co-energy

1.1 Introduction

Electrical machines are based on the principles of electromechanical energy conversion. They work on either electromagnetic or electrostatic principles. In the case of machines based on electrostatic principles, the enhancement of stored energy demands an increase in the potential gradient, which is likely to be hazardous for the operator as well as for the insulating materials. Electrostatic voltmeters and recently developed devices like micro-electromechanical systems (MEMS) are some examples of devices working based on electrostatics. Apart from this, all the electrical machines are based on electromagnetic principles. In these machines, the stored energy depends on the square of current. The heat due to resistive losses has to be quickly dissipated to avoid melting of conducting materials and also to safeguard the life of insulating materials.

In this chapter, we will discuss the basic principles of operations of different electromechanical energy conversion devices, and classify electrical machines and material properties.

R. K. Srivastava, *Electrical Machines*, https://doi.org/10.1007/978-981-96-5039-2_1

1.2 Electromagnetic Apparatus

An energy conversion essentially requires storage of energy in the coupling field. Bow-arrow and catapult are two such examples involving the conversions of mechanical energy from one form to another. In electrical machines, electrical energy is stored in capacitors and inductors. In the present context, only inductive storage has been considered. Electrical machines are primarily used in generation of electrical power and converting electrical energy into mechanical energy. These generators and motors are the backbone of the modern civilization.

All electrical machines can be related to the following working principles [1] :

(a) Alignment
(b) Interaction
(c) Induction.

1.2.1 Alignment

Materials having high permeability, like iron, in a low permeability medium, e.g., air, in which magnetic field is established, experience mechanical forces that tend to align them with the field direction in such a way to minimize the reluctance of the system.

In Fig. 1.1a, a ferromagnetic plunger is near a solenoid carrying DC current. The solenoid is energized by a constant voltage source. The magnetized ferromagnetic plunger is attracted towards the solenoid. When the plunger is set suddenly in motion, electrical energy changes into mechanical energy, resulting in sudden movement of the plunger. The current suddenly falls down from its steady-state value, as shown in Fig. 1.2. The inductance of the solenoid is at its maximum, when plunger comes to rest as in Fig. 1.1b. Now, if one attempts to pull the plunger out of the solenoid, an opposing (counter) force is present. The current experiences a sudden rise beyond its steady-state value. A certain amount of mechanical energy is converted into electrical energy and fed back to supply line. The rise and fall in the current can be better explained using Lenz's law. Thus, by using electromagnetic devices, it is possible to convert the energy from electrical to mechanical energy (motoring action) and mechanical to electrical energy (generating action).

Now, we will consider, a rotating variable reluctance machine working on the principle of alignment. Figure 1.3a depicts the forward electromagnetic torque $\mathrm{T_E}$, causing the salient pole rotor to align along the main stator pole. Both stator and rotor are made of ferromagnetic material. Figure 1.3b shows the production of counter-torque $-\ T_{\mathrm{E}}$, when the rotor is forced to move away from main pole by an external torque T_{EX}. In Fig. 1.3c, the reluctance of magnetic circuit is minimum and the rotor is fully aligned with stator pole. In this situation, the coil inductance and normal force between stator and rotor poles are maximum. However, there is no tangential torque. Can you point out the nature of coil current for continuous running in either direction, when multiple poles on stator and rotor are present?

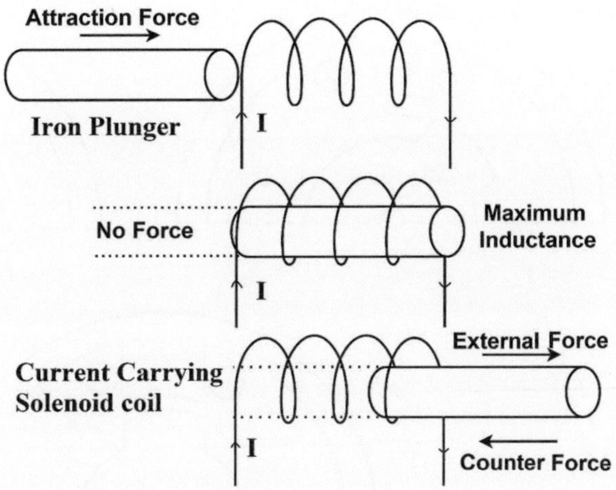

Fig. 1.1 An iron plunger in close proximity to a solenoid carrying DC current

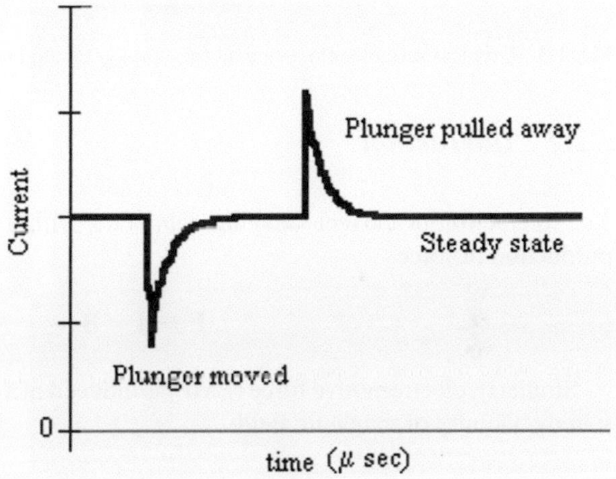

Fig. 1.2 Transient current due to movement of iron plunger

Is it possible for you to draw a curve between coil inductance and rotor position? Is it possible to generate electrical energy using such machines? If yes, how?

Machines based on the principle of alignment require the set of coils to be switched ON or OFF according to the instantaneous position of the rotor. In other words, switching ON/OFF of coils is synchronized with the instantaneous rotor positions. Some examples of such machines are Switched Reluctance Motor (SRM), Variable Reluctance Motor (VRM), permanent magnet brushless DC motor (PMBLDC), stepper motor, multistage coil gun, and reluctance-based electromagnetic thruster and launcher. The operation of these machines essentially requires the use of modern power electronic control circuits.

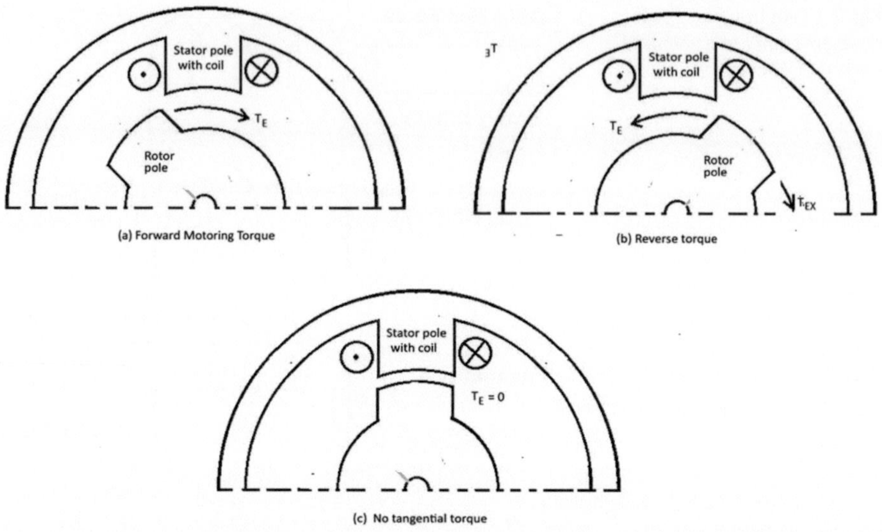

Fig. 1.3 A typical variable reluctance motor working on alignment of a ferromagnetic rotor

1.2.2 Interaction

The interaction of current carrying conductor with a magnetic field results in the production of force.

$$\mathbf{F} = \mathbf{J} \times \mathbf{B} \tag{1.1}$$

Similarly, electromotive force (EMF) is induced in a moving conductor with speed **v** in the vicinity of magnetic field

$$\mathbf{E} = q\mathbf{v} \times \mathbf{B} \tag{1.2}$$

Here **J** is the current density in A/m^2, **B** is the flux density in web/m^2, q is charge in coulomb, and **v** is the speed in m/s.

Can you name these two laws of electromagnetism? Do they satisfy Lenz's law? In a rail gun, as shown in Fig. 1.4, the flux density **B** is directed into the paper. A force **F** (=$I{\cdot}B{\cdot}L$) on armature exists according to Eq. (1.1). The flow of current in moving armature produces a force in the right-hand side of the armature, along with a counter-EMF **E**(=$v{\cdot}B{\cdot}L$) opposite to the applied voltage. Can you point out the reasons for production of counter- or back-EMF?

In Fig. 1.5, here the magnetic field **B** is directed outside paper. An armature (conducting rod) is moved by an external force with a velocity **v** in the right-hand side. A motional-induced EMF in the armature forces a current in a direction downwards in the armature, which produces a counter force $\mathbf{F}_d$ opposite to the direction of external force / velocity. The counter force opposes the external cause of movement. Can

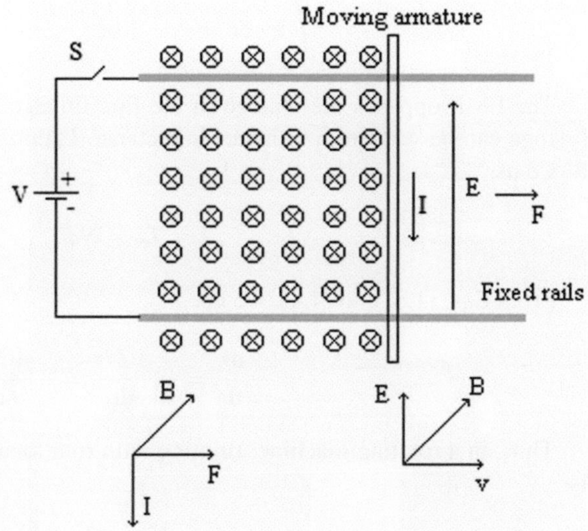

Fig. 1.4 Rail gun excited by a DC source

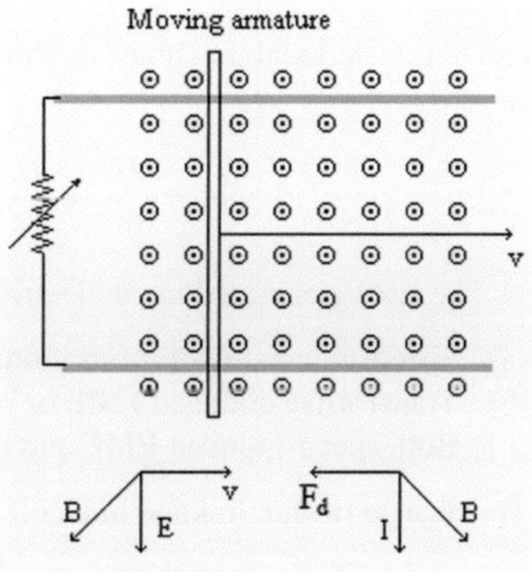

Fig. 1.5 Armature is moved with a Velocity **v** in magnetic field **B**

you give a reason for the counter force or torque in generating action? Is it because of Lenz's law?

Some examples of machines working on the principles of interaction are conventional DC machines, permanent magnet DC machines, and rail guns.

1.2.3 Induction

From Faraday's law, we know that if flux linkage is made to change, an EMF is induced in the electric circuit. This EMF exists only when the flux linkage changes and the magnitude of its time integral is equal to the change in flux linkage [1, 3, 4, 5].

$$\int e \cdot \mathrm{d}t = -\lambda \quad \text{or} \quad e = -\frac{\mathrm{d}\lambda}{\mathrm{d}t} \tag{1.3}$$

The EMF opposes the change in the flux linkage. Can you tell why? The flux linkage can be written in terms of inductance L, current i, number of turns N, and flux ϕ as

$$\lambda = Li = N\phi \tag{1.4}$$

From Eqs. (1.3) and (1.4), we get

$$e = -\frac{\mathrm{d}\lambda}{\mathrm{d}t} = -\frac{\mathrm{d}(Li)}{\mathrm{d}t} = -\frac{\mathrm{d}(N\phi)}{\mathrm{d}t} \tag{1.5}$$

Thus, in a rotating machine, running with rotational speed ω_r, the induced EMF is

$$|e| = \frac{\mathrm{d}\lambda}{\mathrm{d}t} = \frac{\partial\lambda}{\partial t} + \frac{\partial\lambda}{\partial\theta} \cdot \frac{\partial\theta}{\partial t} = \frac{\partial\lambda}{\partial t} + \frac{\partial\lambda}{\partial\theta} \cdot \omega_r \tag{1.6}$$

and in translational machines moving with linear speed v, we have the induced EMF as

$$|e| = \frac{\mathrm{d}\lambda}{\mathrm{d}t} = \frac{\partial\lambda}{\partial t} + \frac{\partial\lambda}{\partial x} \cdot \frac{\partial x}{\partial t} = \frac{\partial\lambda}{\partial t} + \frac{\partial\lambda}{\partial x} \cdot v \tag{1.7}$$

The EMF generated in an electrical machine is of following types:

(a) Speed-induced EMF or motional/rotational EMF.
(b) Transformer-induced EMF or pulsating EMF.
(c) Both speed-induced EMF and transformer-induced EMF.

The change of flux linkage in a coil may occur in the following three ways:

(a) If flux is constant, there is a relative motion between the coil and the constant flux.
(b) If coil is stationary with respect to flux, the flux may vary in magnitude with respect to time.
(c) Both changes may occur together, i.e., there is a relative motion between the coil and a time-varying field is present.

Figure 1.6 shows a conducting (copper or aluminum) plate moving at a linear speed of $\mathbf{v}$. The plate is in proximity to a bar magnet. Stationary magnet with respect to plate ($\mathbf{v} = 0$) does not produce any force. However, when the plate is moving at a high velocity (say, $\mathbf{v} > 5$ m/s), the steady flux produces a speed-induced EMF, causing speed-induced eddy currents to flow on the surface of the plate in a closed path. The direction of these circulating eddy currents opposes the magnetic flux produced by the magnet, thereby producing the following net effects:

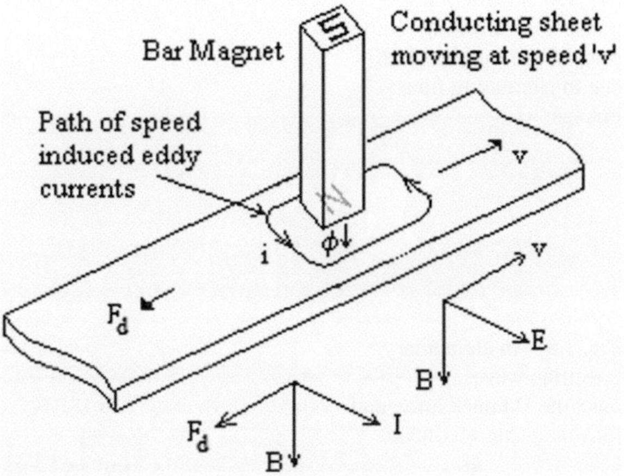

Fig. 1.6 A bar magnet in proximity to a moving conducting plate

(a) A dragging or braking force $\mathbf{F}_d$ appears on the moving sheet in the direction opposite to that of speed.
(b) A normal repulsive force appears between the magnet and the moving sheet, causing the magnet to move away from the sheet.

The above example demonstrates the use of speed-induced EMF in the working of an experimental eddy current brake. The example also shows the production of two-dimensional force in an electromagnetic device. The third force may also appear in the lateral (transverse) direction, if the bar magnet is not placed symmetrically over the width of moving conducting plate. The normal repulsive force is effectively utilized in high-speed levitated trains and also useful in upcoming Hyperloop system of transport for electrodynamic suspension.

Apart from a tangential or longitudinal torque or forces, a radial or normal force is also produced in rotating machines. However, it gets canceled because of the cylindrical structure of machine and has little effect on the machine performance. In cylindrical (round) machines, it is necessary to keep the air-gap uniform; otherwise, the net normal force tends to bend the shaft in one direction. The phenomenon is sometimes referred as *Unbalanced Magnetic Pull (UMP)*. In the case of single sided linear induction motor (SLIM), the normal force does not get canceled. The normal force is around ten times the longitudinal thrust.

A single-phase two-winding transformer is shown in Fig. 1.7, where an alternating current (AC) voltage source excites the primary winding. The alternating flux produced by the primary winding links the secondary winding through mutual coupling. From Eq. (1.5), we know that an alternating EMF of same frequency is induced in the secondary windings. Some examples of electrical machines working on the principles of induction are transformer, reactors, induction machines, synchronous machines, and induction regulators.

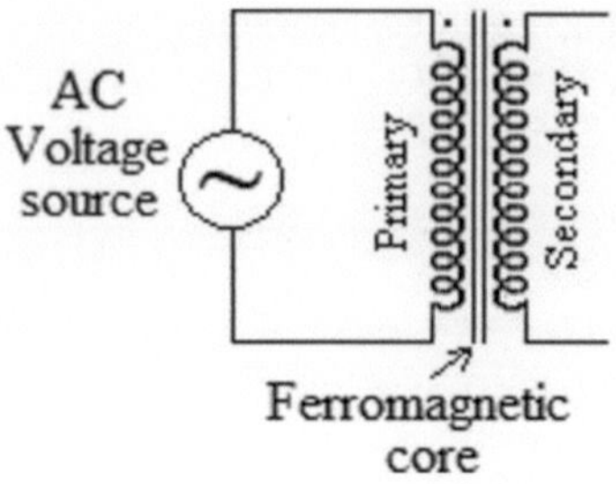

Fig. 1.7 Two-winding transformer working on the principle of induced EMF due to alternating flux linkage

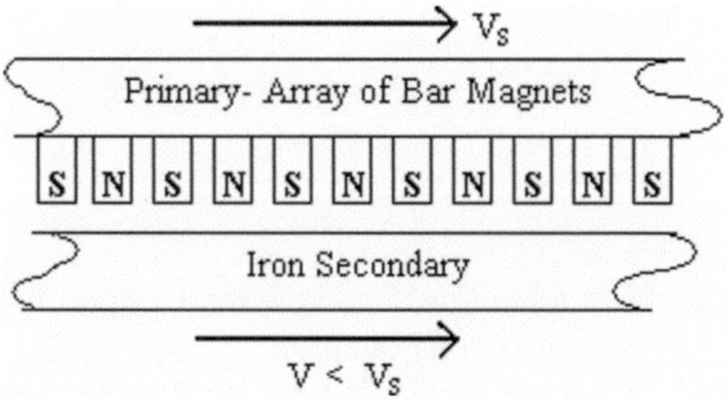

Fig. 1.8 An elementary travelling wave machine-Halbach array and ferromagnetic secondary

A transformer does not work, when a two-winding transformer is excited by a DC supply. Can you give suitable reason? Will there be any transient output voltage, when a transformer is switched ON/OFF with a DC supply?

A simple traveling-wave machine is demonstrated in Fig. 1.8. A traveling magnetic field is created by moving at speed V_s Halbach array as shown in Fig. 1.8. It is assumed that both the members are free to move in longitudinal direction. The iron secondary in proximity to the moving magnets also experiences a tangential (longitudinal) force apart from normal force. This makes it to move in the direction of the movement of magnets. Its speed is always less than the speed of primary Halbach array. Can you point out the reason of tangential (longitudinal) force? If secondary is also having similar number of magnetic poles as that of primary, will there be any movement of secondary when primary is moved?

Alternating current machines, such as induction and synchronous machines, work on the principles of induction. They require a special poly-phase distributed winding energized by a set of poly-phase voltage, which produces rotating magnetic field responsible for their operations.

1.3 Electrical Machine

An electrical machine can work as motor, generator, or brake. In motoring mode, the electrical power supplied to the machine is converted into mechanical power. A counter-EMF opposing the supply voltage becomes inevitable. In generating mode, the machine is driven by a prime mover and converts mechanical power into electrical power. When the generator is loaded, a counter-torque is produced which opposes the driving torque of the prime mover. For maintaining rotational speed, larger mechanical input power is required by the prime mover. The braking mode is a generating action, where the electrical power is either regenerated and fed back to power lines or dissipated within the machine system. It develops a braking action, which acts like a mechanical brake. A generator is said to float, when it neither delivers nor absorbs power from the electrical system, but remains connected to electrical power system as well as to the prime mover.

Conventional electrical machines are based on interaction, induction or alignment principles described. Based on the induced EMF, the machines, in general may be classified as follows:

(a) Constant stator flux and rotor with (or without) commutator winding: These machines are based on speed-induced EMF and the supply is direct current (DC) supply. For examples: DC motors and generators, eddy current coupling and eddy current brake and rail gun.
(b) Constant (stator/rotor) flux and (rotor/stator) with phase winding: Examples are synchronous motor, alternator, permanent magnet synchronous motor (PMSM), permanent magnet synchronous generator (PMSG), permanent magnet brush less DC motor (PMBLDCM).
(c) Alternating stator flux and rotor with commutator winding: The rotor will give both speed-induced and pulsating EMF. Examples of this type are AC commutator motor, Schrage motor, and universal motor
(d) Alternating stator flux and rotor with phase winding: There are both speed- and transformer-induced EMFs present in machine. Example is single-phase induction motor.
(e) Alternating flux primary and stationary secondary having phase winding: There may be slight motion of secondary, only pulsating EMF is present. In some cases, the alignment of secondary is required. Examples are static transformer, induction regulator and inductive reactor.
(f) Traveling-wave machine: Poly-phase stator alternating flux and rotor with phase windings. The machine produces rotating magnetic or traveling wave. Examples are three-phase induction motor and linear induction motors.

Conventional electrical machines can operate on either DC supply or AC supply, or both AC and DC supplies. These machines are used in various applications such as transformer, DC machines, induction machines and synchronous machines. The advent of solid-state devices and embedded computing system has encouraged the development of newer motors of modern era, different from the conventional plug-and-play type of motors. They require a logic controller for switching solid-state devices apart from the usual power source. Some of these motors are

(a) Vector controlled induction motor.
(b) Switched Reluctance Motors.
(c) PMBLDC motors.
(d) Written pole motor and generator.

1.3.1 Alternating Current Versus Direct Current

In the year 1831, Michael Faraday demonstrated all the fundamentals of electromagnetic fields. These included the remarkable inventions of the homopolar disk dynamo for the generation of direct current and the transformer for the voltage conversion of alternating current. Perhaps, he is known to be the greatest experimenter of all time. In the same year, a simple two-pole alternator capable of generating alternating current voltage was built by Hippolyte Pixxi, an instrument maker from Paris. In this device, he rotated a horseshoe magnet mounted on a disk so that the North and South poles passed alternatively beneath a pair of iron core coils arranged on a stationary disk. Later, multipolar alternators were also developed. These machines were elementary forms of axial flux machines.

In the beginning, the only principal uses of electricity were in arc lamps and electroplating. The electric filament lamps and mercury vapor lamps came several decades later.

For transmitting 1 MW power as 100,000 A at 10 V would require the use of very thick copper cables and also the ohmic losses would be a substantial fraction of the power transmitted; but to transmit the same amount of power as 10 A at 100,000 V is far more efficient and economical. However, to use 100,000 V in factory or house is practically impossible because of the danger of electric shock and possible electrocution.

Although, the transformer demonstrated by Michael Faraday in 1831, it was an ideal mechanism for transmission and distribution of bulk electric power, it did not prevent the development of direct current (DC) technology. Ampere invented the commutator, normally used in conventional DC machines. The development of DC motors and generators have helped in developing DC technology. In the beginning, the generation and utilization of direct current electrical energy grew at a fast pace.

Tesla invented the brushless induction motor in 1888. In 1890, in Britain there was an argument concerning whether AC or DC was to be preferred. Great industrialists Ferranti and Tesla were of the firm view to promote AC technology. They advocated the use of not merely AC but three-phase AC, so that best use of remarkable invention of induction motor can be made. Today, we know that over 95% of total power of electrical drives in industrial world is dominated by induction motors. It was induction motor, in addition to the ability to transform voltage level easily for transmission of bulk power at high voltage over long distance and stepping down voltage at lower level for domestic and industrial use, that set the pattern of alternating current distribution networks in the beginning of 1900 century. Alternating current was the obvious right choice for large distribution network situated far from generating stations.

1.3.2 Need for Sinusoidal Supply

The sinusoidal waveform is the only unique waveform whose integration or differentiation is again a sinusoid. In transformer if we apply a sinusoidal waveform, we get a similar waveform on the output side. Here we assume that saturation in the magnetic core is absent. If the waveform is a triangular waveform or a rectangular waveform we get an entirely different waveform on the output of a transformer.

With the use of the Fourier series, it is possible to analyze a non-sinusoidal waveform into several sinusoidal waveforms containing fundamental harmonic and several other harmonics. These harmonics are again sinusoids having different magnitudes and frequencies.

In actual practice, if the supply voltage waveform is deviated from sinusoidal input, the presence of several harmonics in the voltage, current, and magnetic flux are inevitable. The role of these harmonics in current waveform and hence magnetic flux is to cause high frequency eddy current to flow in the ferromagnetic and conducting materials. These harmonics also result in increased eddy current and Hysteresis losses. As such certain amount of power is wasted in this exercise. As far as possible the supply waveform must be close to sinusoid and should have limited number of harmonics present. An alternating current motor working on non-sinusoidal supply gives inferior performance. Here, certain harmonics are responsible for backward rotating magnetic field and negative torque production. Because of increased losses the overall efficiency is decreased. This is responsible for overall de-rating of the machine.

1.3.3 Multiphase

The number of conductors required to supply a single-phase motor with neutral is three, whereas the number of conductors to supply a three-phase motor with neutral is four. Multiphase like two- or three- or higher-phase machines can produce rotating magnetic field. The size or volume and the rated current drawn in a three-phase alternating current machine are smaller in comparison to an equivalent single-phase machine.

With the help of several transformers, a three-phase supply can be converted into multiphase such as 6, 9, 12, 18, 24 phases, etc. Multiphase supply is also preferred in rectification of AC power in to DC power. This helps in reducing the overall ripple contents in the DC output.

1.3.4 Types of Mechanical Loads

An electric drive is defined as a form of electrical machine equipment designed to convert electric energy into mechanical energy and to provide an electrical control of the process.

Electrical motors are used to run several types of mechanical load. Such as household and office equipment (fans, water pumps, hair driers, water coolers, kitchen appliances, vacuum cleaner, air conditioners, fridge, CD players, cooling of personal computer, etc.), public utilities (water pumps, lifts, doors, elevators, tram car, electric vehicle (EV), electric hybrid vehicle (EHV), electric bikes/ scooters, traction locomotives), industrial (cranes, conveyor belts, motorized gates, industrial hammers, tunneling machines, large capacity lifts, pumps, fans and blowers, air compressors, coal pulverizing mills, coal crushers, etc.), machine tools (lathe machines, drilling machines, planar machine, gear cutting machine, blower, etc.). These mechanical loads offer different types of load torque-speed characteristics, i.e., their load torque vary with respect to speed. For driving these loads, the selection of motors is an important task. The electrical motor has its own torque-speed characteristics. The drive motor and driven load runs at the speed at which the two characteristics (developed torque versus speed and load torque versus speed) cross each other. Depending on these, one can decide whether the drive will be stable or unstable. Any discrepancy in this will lead to an uneconomical and unstable drive system. A wrong selection of motor may cause difficulties in its operation.

In a drive system, the driven load offers its own speed torque characteristics. The electrical drive must be chosen in such a way so that the torque requirement of the load must be met at all speeds of operation. Assuming that the friction load torque to be zero, the load torque can be given as

$$T_L \ \alpha \ \left(\frac{\omega}{\omega_0}\right)^x \tag{1.8}$$

where ω is the speed of the operation and ω_0 is the rated speed of machine. The quotient 'x' can take values of 0, 1, 2, and $-$ 1. The load torque versus speed characteristics for different values of 'x' have been shown in Fig. 1.9. Depending upon the quotient 'x', the load torque characteristics is defined as

(a) **Constant load torque, $x = 0$**: Here the load torque remains constant. The examples are frictional torque, hoisting, machine tool feed mechanism, hoist, piston pumps working on a constant pressure head, conveyor belts handling constant weight of material per unit time, and elevator or lift handling constant weight per unit time.
(b) **Torque proportional to speed, $x = 1$**: Here the load torque is linearly rising with speed. It is exhibited by separately excited DC generators connected to a constant resistance.
(c) **Torque proportional to square of speed, $x = 2$**: Here the load torque is nonlinearly rising with speed and is proportional to square of speed. The power is proportional to the cube of speed. Such characteristic is known as fan-type characteristics. It is exhibited by all types of fans, centrifugal pumps, ship propellers, etc.
(d) **Torque inversely proportional to speed, $x = -1$**: Here the load torque is nonlinearly falling with speed and is inversely proportional to speed. However, the power is almost constant. It exhibits large torque at standstill condition and less torque when running. It is offered by certain lathe machines, boring machines, milling machines, metal cutting tools, steel mill coiler machine tools, traction locomotive, and all types of vehicles.

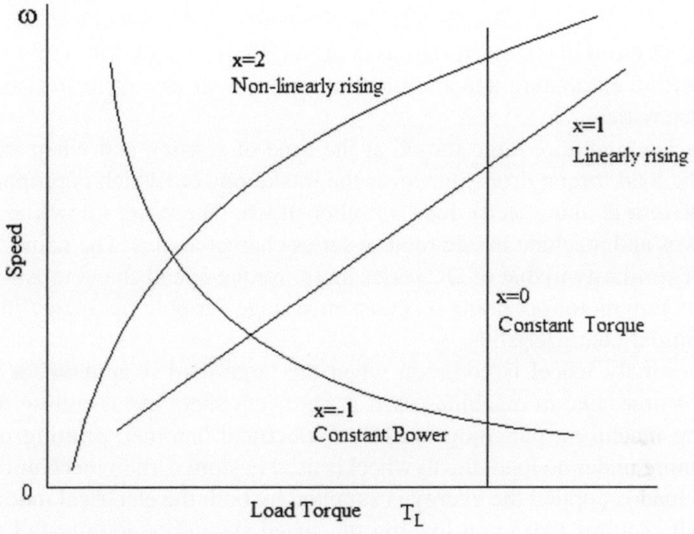

Fig. 1.9 Variation of load torque with respect to speed

If the developed electromagnetic torque in a motor is T and moment of inertia of rotating parts is J, we can write

$$T - T_L = J\frac{d\omega}{dt}$$

$$\begin{aligned}
&\text{if } T > T_L, \frac{d\omega}{dt} > 0, \text{ the drive accelerates.}\\
&\text{if } T < T_L, \frac{d\omega}{dt} < 0, \text{ the drive decelerates.}\\
&\text{if } T = T_L, \frac{d\omega}{dt} = 0, \text{ the drive runs at steady - state speed}
\end{aligned} \tag{1.9}$$

Similar explanations can be given when machine is working as a brake or as a generator.

1.3.5 Hard and Soft Characteristics

The constant speed motors such as synchronous motor, RSM, and hysteresis motor exhibit hard characteristics with no change in speed upon loading. DC shunt motor and an induction motor with low rotor resistance also exhibit the hard torque-speed characteristics. Although a marginal change in speed upon loading is expected in both the motors. The slightly drooping torque-speed characteristics are commonly referred as shunt characteristics, because it is that of DC shunt motor. These motors are also referred to as constant torque motors.

Many electrical motors exhibit soft characteristics, i.e., when motors are mechanically loaded, there is a large variation in speed. Shunt motor with extra resistance in series with the armature and an induction motor with large rotor resistance have soft characteristics.

Certain loads require large torque at the time of starting and when machine is running the load torque drops; however the intake power (which is proportional to product of torque and speed) does not alter much. These are known as constant power drives and machine is said to have series characteristics. The name 'series' is because of similarity to that of DC series motor torque-speed characteristics. Three-phase induction motor operating on constant voltage variable frequency supply also exhibits similar characteristics.

The use of fly wheel is common when the large load is applied for a shorter duration of time, like in machines used in shear cut sheet metal cutting machines, slot cutting machines, punching machines, electrical hammer, printing press, etc. When running under no load, the flywheel is used to store certain mechanical energy and when load is applied the energy is supplied by both the electrical motor and the flywheel. It requires that upon loading the speed should essentially fall to enable the energy transfer from the fly wheel. This minimized the overall size of the motor used. The motor must have a soft speed torque characteristic.

1.3.6 Torque in AC/DC Round Machines

In conventional rotating machines, like DC, induction, and synchronous machines, certain basic principles are common. The geometry of machine is quite complicated and it is quite cumbersome to estimate the force and torque by solving electromagnetic field equations in and around the geometry of machines. Modern state-of-the-art softwares like ANSYS, MSC EMAS, PC OPERA, JMAG, etc. have capability to do so, although it is costly and time-consuming approach. Before the availability of the software, the generalized machine theory or basic principle was used for finding torque in rotary machines [1, 3, 4, 5].

If the torque in round machine at speed ω_r is T, then mechanical power is $P = \omega_r T$, which can be equated to electric power $P = e\, i$. Here $\underline{e}$ and i are the instantaneous speed-induced EMF and instantaneous current, respectively. This gives a simple assessment of torque.

It is assumed that the stator and rotor each have conductors distributed over its periphery to produce a sinusoidal distributed current sheet. A fictitious sheet of thin conductors is assumed to be uniformly distributed over smooth (slot less) active surface, which gives similar fundamental flux density distribution as that of an actual machine. Though, such a machine does not exist physically. However, by doing so, effect of slotting is assumed to be ignored. In Fig. 1.10, one coil in stator and one coil in rotor are shown. The coil axis of stator is due to the currents in stator conductors of the actual machine. The coil axis of rotor is due to the rotor currents of the actual machine. The stator and rotor windings are each having several numbers of

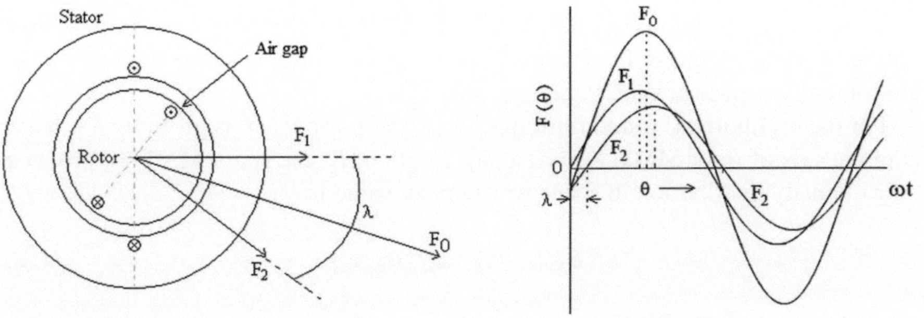

Fig. 1.10 A typical cylindrical rotating machine and distribution of mmf in space

current carrying conductors in their respective slots distributed over periphery. The concept of magneto motive force (mmf or MMF) is described in Sect. 2.3.1. These are responsible for operation of electrical machines.

There are certain simplifying assumptions.

(a) There is a perfect symmetry in windings. In case of three-phase or multiphase winding, each phase is having similar number of turns and the coils are symmetrical displaced in the slots. In other words, windings are perfectly balanced, both electrically and magnetically.
(b) A balanced set of voltage is impressed on the windings.
(c) The effect of slotting is ignored. The winding is replaced by a current sheet assumed to be pasted over the smooth active surface of core.
(d) The MMF due to stator and rotor currents is sinusoidally distributed in the air-gap. The MMF distribution in space is the fundamental component of the actual MMF.
(e) The air-gap or clearance between stator and rotor is uniform.
(f) Magnetic flux crosses the air-gap normally.
(g) Iron parts are unsaturated and are considered to have high magnetic permeability.

In case of an induction motor, the induced currents or MMF in rotor slightly lags the MMF of stator. This is due to *hysteresis effect*. The fundamental stator and rotor MMFs have been shown in Fig. 1.10. The interaction of stator MMF and rotor MMF is responsible for production of net torque in electrical machines.

In case of DC machine, the field winding on stator gives MMF variation in air-gap, and the fundamental of the MMF variation is considered. In case of an induction machine and synchronous machines, the MMF variation due to stator conductor gives a sinusoidal variation of MMF in air-gap. The stator MMF of these machines is represented by F_1. Similarly, in DC machine although the rotor is moving due to the commutator action, the armature currents produces an MMF variation F_2 in air-gap.

In case of induction motor, the induced currents in short-circuited rotor produce an MMF variation F_2 in air-gap almost similar to that of stator MMF. Due to *hysteresis effect* this MMF lags the stator MMF F_1. In case of induction motor both the stator and rotor MMF are rotating in space.

In case of synchronous machines, the field MMF F_2 also gives a sinusoidal distribution of MMF in space. In synchronous machines, both the stator and rotor MMF are rotating in space at synchronous speed.

For the (cylindrical) round rotor machines, the rotor MMF is given by F_2, which is in the axis of rotor MMF. Across a gap length of l_g, the stator MMF F_1 produces a flux density distribution in space whose peak value is

$$B_{1\,\max} = F_{1\,\max}\left(\mu_0/l_g\right) \tag{1.10}$$

At any angle θ, the MMF is

$$F_1(\theta) = F_{1\,\max}\cos\theta \tag{1.11}$$

and the gap flux density,

$$B_1(\theta) = B_{1\,\max}\cos\theta \tag{1.12}$$

In the same manner the peak flux density in rotor due to rotor MMF alone is

$$B_{2\,\max} = F_{2\,\max}\left(\mu_0/l_g\right) \tag{1.13}$$

The rotor flux density distribution in air-gap is

$$B_2(\theta) = B_{2\,\max}\cos(\theta - \lambda) \tag{1.14}$$

whose axis is at angle 'λ'.

The resultant MMF F_0 in air-gap is the vector sum of two MMFs, and is given by

$$F_0^2 = F_1^2 + F_2^2 + 2F_1F_2\cos\lambda \tag{1.15}$$

The resultant peak gap flux density due to this resultant MMF F_0 is assumed to be B_0. The peak gap energy density is

$$W_{\max} = \frac{B_0^2}{2\mu_0} = \frac{\mu_0 F_0^2}{2l_g^2} \tag{1.16}$$

Let, the average diameter of air-gap is D and stack length of machine is L; thus the average active area in air-gap is $\pi.D.L$ and the volume of air-gap is $\pi.D.L.l_g$. The mean gap energy density in active surface in air-gap is one half of the peak value and is given as

$$W_f = \frac{B_0^2}{4\mu_0}\pi DLl_g = \frac{\pi DL\mu_0 F_0^2}{4l_g} \tag{1.17}$$

For a given stator and rotor currents, the torque is given as

$$T = \frac{d}{d\lambda} W_f = -\frac{\pi D L \mu_0}{4 l_g} 2 F_1 F_2 \sin\lambda$$

$$T = -\frac{\pi D L \mu_0}{2 l_g} F_1 F_2 \sin\lambda \tag{1.18}$$

The minus sign indicates that the torque acts to decrease the angle λ or it tries to align the rotor and stator magnetic axes. Thus, the torque in rotating machines depends on the.

(a) Main dimension of the air-gap.
(b) Diameter of machine.
(c) Stack length of machine.
(d) Stator and rotor MMFs.
(e) The angle by which the stator and rotor magnetic axes diverges.

The above expressions are helpful in qualitative analyses of electrical machines.

1.3.7 AC Operation of Magnetic Circuit

Consider a solenoid carrying DC current which is in proximity to a ferromagnetic plunger, as discussed in Sect. 1.2. Under certain position of plunger a net force is produced on the plunger, which attempts to minimize the reluctance. What will happen when solenoid is carrying AC current? Can you answer? The alternating magnetic field produced by solenoid results in alternating magnetization of the ferromagnetic plunger. If only half cycle of AC current is present, (we assume that the solenoid is connected to half wave rectifier), the force of attraction between solenoid and plunger is obvious. However, in presence of alternating current (AC) or direct current (DC) with ripples, the alternating magnetization of ferromagnetic core results in unidirectional force of attraction plus eddy current and hysteresis losses. The DC voltage/current with ripples can be represented by Fourier series. It contains a perfectly DC component and several alternating components of difference frequencies of lower magnitudes.

If the core is excited by non-sinusoidal alternating voltage or excited by DC having ripple contents in voltage waveform, it results in excessive eddy current and hysteresis losses due to harmonic components of the flux densities. The armature of DC machine has rotational magnetizing due to commutator action. The electrical machines working on switched DC supply like Switched Reluctance Motor, Variable Reluctance Motor, and PMBLDC motor magnetizes the core alternatively. Due to repetitive switching of coil currents and hence flux variations, it results in core losses. For limiting core losses either due to alternating magnetization or due to rotational magnetization or due to repetitive switching actions, the magnetic circuits are fabricated using laminated sheet steel (electrical grade) stampings. In case of devices operating at high frequency, ferrite core is preferred. The use of non-electrical grade sheet steel results in large core losses and inferior force/torque production.

1.3.8 Classifications

Electrical machines are built with ratings from milliwatts to mega-watts. They may be classified as either rotating or linear type. For convenient reference, they may also be roughly classified as follows:

1. **Industrial machines**: These machines include large generators and motors for mills, cranes, conveyor belt, electrical hammer, tunnel boring machines, machine tools and industrial pumps and compressor drive, and industrial ventilators of ratings above few kilowatts, etc.
2. **Small machines**: Motors for hand-tools, domestic appliances, main-fed motors, automobile machinery, domestic pumps, etc. come under this category.
3. **Traction machines**: Motors for railways, ships, and battery operated electric vehicle (EV) and hybrid electric vehicle (EHV), and high-power drones.
4. **Miniature machines**: Miniature machines are those motors which do not require precision control like toys, small vibrator, fins controller in coolers, air conditioners, etc. and are used only for intermittent load. Such motors have an output of only a few watts.
5. **Control machines**: Machines associated with open-loop and closed-loop control systems, servo applications, power selsyn, synchro, and selsyn.
6. **Special machines**: Some examples of special machines are linear induction motor (LIM), tubular motor, electromagnetic levitators, controlled/uncontrolled electromagnets, liquid metal pump, electromagnetic guns, coil gun, rail gun and electromagnetic launchers, actuators, eddy current coupling, eddy current brake, and motors with super conducting field windings and unconventional designs.

These electrical machines may be further classified into the following:

(a) Force machines.
(b) Power machines.
(c) Energy machines.

Force machines exert force or torque for a restricted length or angle of travel. Some examples of Force machines are electromagnetic relay, electromagnets, opening and closing of doors and curtains, latches, material transfer machinery, MEMS, etc. *Power machines* require large electrical power for longer period, like pumps, compressors, locomotive, lifts, escalators, etc. *Energy machines* are those applications, which require huge amount of power (pulse power) for a short duration of time. Examples are electropults, rail gun, electromagnetic gun, coil gun and electromagnetic launchers, compulsator, etc.

Electrical machines may also be classified based on duty cycle. The temperature rise limits the time duration for which machine should be operated. An industrial machine is interpreted according to the duty that the machine is called upon to perform. The standard duty cycles designated are as follows:

(a) Continuous running at rated load.
(b) Short-term operation.
(c) Intermittent periodic operation.
(d) Intermittent periodic operation with starting.
(e) Intermittent periodic operation with braking.
(f) Continuous cyclic operation.
(g) Continuous cyclic operation with braking.
(h) Continuous cyclic operation with related load/speed characteristics.

Figure 1.11 describes the continuous, short-term, and intermittent periodic operations of a machine. In continuous duty cycle, the machine is operating continuously (say 24 hours) and delivering rated output. The temperature of the machine will stabilize to well below the permitted temperature as decided based on type of insulation and cooling used. Examples are pumps, compressor, ceiling fans, traction locomotive, etc. In short-time duty cycle the machine is switched ON for a certain period and is switched OFF so that the machine is cooled down to ambient temperature before it is again switched ON. A continuous duty cycle machine can deliver higher output power and torque when used for short-term duty cycle. Examples of short-term duty cycles are domestic food processors, hair drier, electric shaver, domestic vacuum cleaner, etc. In intermittent duty cycle, the motor is loaded intermittently and periodically switched ON and OFF. If the motor is loaded for infinite time in a similar fashion, the final temperature should not cross the permitted temperature rise. Examples of intermittent duty cycle are electrical hammer, slot cutting machine, shear cutting of sheet metal, etc.

1.3.9 Selection of Motor and Environmental Factors

The presence of air in the atmosphere improves the natural heating and cooling cycle of the electrical machine and transformer. Normal electrical appliances can satisfactorily work up to 1000 m above sea level. Beyond this altitude, the temperature of an electrical machine exhibits further rise in temperature even under normal operating conditions. This is because of poor convection of heat through air medium. The air natural type of cooling is seriously affected when the machine is operating at higher altitudes or in presence of low air pressure. Roughly there is an increase of 0.5 °C for per 100 m increase in altitude [5].

Environmental factors do affect the choice of a particular type of machine. For example, the machines used in wet areas such as agricultural farms or rainy atmosphere must have proper protection from the water.

Temperature of environment plays an important role in selection and design of the electrical apparatus. For example, an electrical motor working in day-to-day life may not be useful when used to perform in high temperature zone, say, near a large

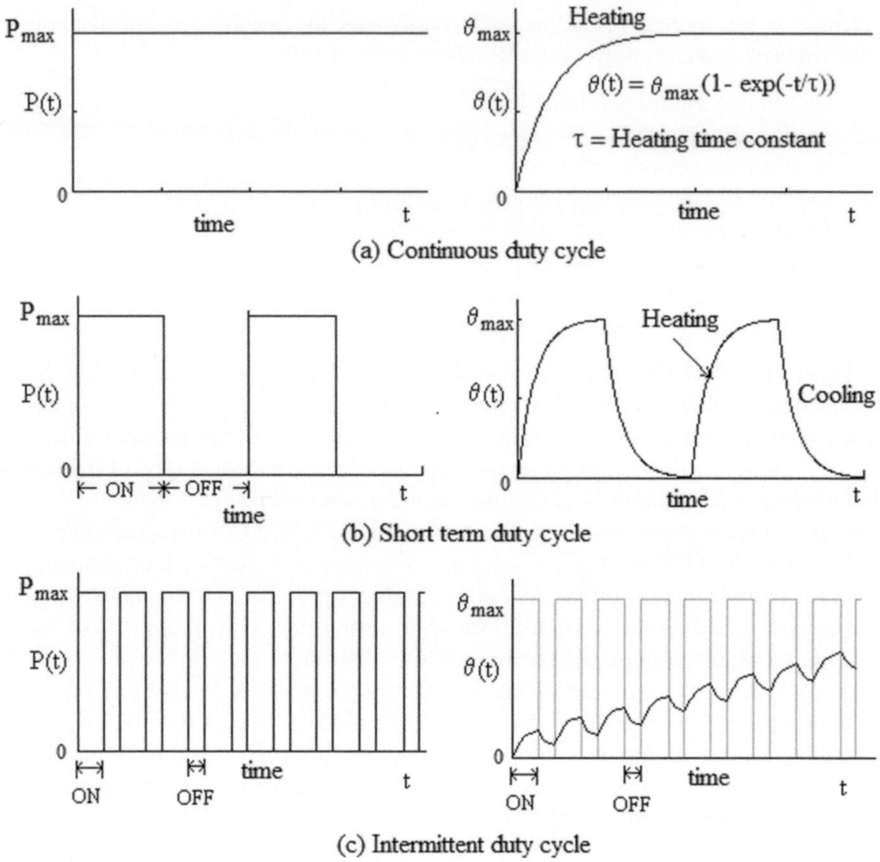

Fig. 1.11 Duty cycle of electrical machines: **a** continuous, **b** short term, **c** intermittent

furnace, pumping hot liquid and blowing hot air and space crafts, etc. A constant heat in proximity to motor may destroy its insulation and may require frequent maintenance. Similarly, the motors used in avionics, space craft or defense areas need to be highly reliable and therefore special attention is required for their design.

All machines are designed to give rated performance at a maximum ambient temperature of 40 °C. For working requirement at higher ambient temperature, a machine needs to be specifically constructed. When a machine is totally enclosed, like compressor motors of fridge, deep freezer, air conditioner, and Hyper loop system, then normally the temperature of environment should not exceed 40 °C. The core of such motor is prepared using low loss silica steel such as cobalt sheet steel and stalloy.

Electric machines and transformers have classifications corresponding to the environmental factors such as

(a) Normal.

(b) Explosive atmosphere as found in mines, storage of oil and gas and other similar material.
(c) Presence of acid vapor.
(d) Other special climate.

The climate also affects the selection of motors. The basic temperate climate is normally referred as *normal technical environment*. Majority of electrical appliances are designed to work in temperate climate. Other climates are humid and dry tropical environment. The tropical climates call for additional measures for protecting the constituent parts of machine and transformer against variations in temperature, humidity, termites/bees, the action of bacteria and corrosion. When a machine is not in use for longer time, sometime beehive inside the machine makes it unusable.

Other environmental factors found in large industries are

(a) Presence of gases, vapors, or explosive powders.
(b) Corrosive and chemically active vapors.

In underground mines, the presence of explosive gases even in small quantity and electrical spark results in explosion. The conventional DC machine where there is continuous sparking between commutator and brush is never recommended for mines. Corrosive vapors are also found in chromium electroplating plants. The fume is said to affect the open electrical contacts, terminals, and electronic circuitry.

Electrical machines and transformers are built with different types of mechanical guards for preventing the entry of align or foreign solid or liquid bodies or water. Measures are taken to prevent the operator from touching the electrically live parts or mechanically rotating parts.

These are described by IP code and are divided into two parts:

(a) Protection against contact.
(b) Protection against penetration of water.

These are described below:

(a) **Degrees of protection against physical contact are**:

   (i) No protection.
   (ii) Protection against solid bodies more than 50 mm thick.
   (iii) Protection against solid bodies more than 12 mm thick.
   (iv) Protection against solid bodies more than 2.5 mm thick.
   (v) Protection against solid bodies more than 1 mm thick.
   (vi) Partial Protection against dust particles.
   (vii) Full protection against dust particles.

(b) **Degrees of protection against penetration of water are as follows**:

   (i) No protection.
   (ii) Protection against drops of condensed moisture.
   (iii) Protection against vertically falling water drops falling at an angle of 15°.

(iv) Protection against water in the form of rain.
(v) Protection against water splash.
(vi) Protection against water jets under pressure.
(vii) Protection against conditions on ships' decks.
(viii) Protection against the effects of immersion in water.
(ix) Submersible equipment.

In IP the first number corresponds to protection against physical contacts and second digit corresponds to the protection from water. For example, in IP 55, the first digit '5' means that any kind of contacts with live or rotating parts is completely prevented. The ingress of dust particles is not totally prevented, but it permits entry of dust only to an extent that does not hamper the operation of machines. The second digit '5' tells that the harmful effects of water projected by hose on to the equipment from any direction are totally averted.

1.4 Material Properties

Electromagnetic energy conversion involves macroscopic electrodynamic interaction between current carrying conductors and electromagnetic fields. There are five important phenomena taking place in an electrical machine, which can be explained using their respective circuit/equations. These circuits/systems are [1, 2, 3]:

(a) Electric circuit.
(b) Magnetic circuit.
(c) Dielectric circuit.
(d) Thermal circuit.
(e) Mechanical system.

The electric current in the coil system (or winding) produces a coupling magnetic field, where the conversion of electrical and mechanical energy takes place. The I^2R loss or ohmic losses in the electrical circuit, iron losses in the magnetic circuit, frictional, and windage losses in rotating mechanical structure, and dielectric losses in insulating materials (though important for high voltage system) contribute to heat generation and hence temperature rise. These are active losses and result in temperature rise. This rise in temperature must be controlled to safeguard the conducting and insulating materials.

1.4.1 Conducting Properties

Conducting material is used in welding electrodes, electrical contactors, cables, bus bars, transmission lines and motor windings, and so on. Copper and aluminum are normally used in electrical engineering applications. However, copper is more

Table 1.1 Conducting materials in electrical engineering

Material	Conductivity (S/m)	Melting point (°C)
Aluminum (with 6% Si)	3.78×10^7	700
Copper	5.96×10^7	1083
Copper zirconium-silver	5.96×10^7	2750
Copper-silver	5.96×10^7	1083

favored because of its electrical and thermal conductivities, inherent corrosion resistance, ease of fabrication, and versatile alloying potential. Moreover, the conductivity of copper is 65% higher than that of aluminum which means that the conductor size of similarly rated machine is proportionally smaller in the case of copper. Copper may be specified into different types based on its conductivity. There is an enormous range of electrical applications for which high conductivity copper is used. There are several different copper metals. Electrolytic Tough Pitch Copper (Cu-ETP) is oxygen bearing high conductivity copper, which is refined using electrolysis and has impurity less than 0.03%. It is an appropriate choice for most of the electrical machines.

The electrical circuit of an electrical machine or winding is made of thin conductors of copper which are insulated from each other by either varnish or paper insulation. The presence of silver (0.05%) traces in the copper-silver alloy gives sufficient strength and easy formability to the thinner winding wire. On the other hand, the presence of zirconium (0.05%) increases the melting point of the winding conductors, making them suitable for defense, avionics, space applications, and atomic energy industries, where the operating temperature is likely to be high. Commercially available winding materials utilize round conductors standard wire gauge (SWG) wire, strip conductors. Hollow conductors are used in winding large capacity water cooled electrical machines. The inner hole of the hollow conductor provides a path for flowing treated water for cooling purpose. Treated water is non-conducting. Untreated water is not used for cooling, as it may cause accumulation of deposits and subsequent blockage of the inner water path. All conductors offer certain resistance, thereby giving rise to ohmic losses. At high frequency, due to the skin effect the current tends to flow on the outer surface of a conductor, thereby causing the variation in the effective resistance.

Table 1.1 shows a list of conducting materials being used in electrical machines. The selection of conducting materials for winding, squirrel cage, and damper bars plays an important role in the characteristics of induction and synchronous motors.

1.4.2 Magnetic Properties

Magnetic properties of materials are due to the interaction between external magnetic fields and moving charges in materials. Atomic scale magnetic fields are produced inside the material due to orbiting electrons. These electrons behave like small magnets or magnetic moments, which gives rise to equivalent current loops. As such, the material produces equivalent current loops, and a certain magnetic moment density arises, thereby giving rise to a net magnetic field inside the material. The internal field is either aligned with the external field to increases the total field or opposes it to decrease the total field. In case of nonmagnetic conducting materials, the internal magnetic moments are randomly oriented, resulting in zero net internal fields, and the material behaves like free space from the magnetic point of view. This has been depicted in Fig. 1.12.

Let us suppose that the total magnetizing vector **M** exists in ferromagnetic material in the presence of an external magnetic field. The magnetic flux density may be written as

$$\mathbf{B} = \mu_0 \mathbf{H} + \mathbf{M} \tag{1.19}$$

where **M** is the magnetization vector and μ_0 is the permeability of free space. Magnetization can be written as

$$\mathbf{M} = \mu_0 \chi_m \mathbf{H} \tag{1.20}$$

where χ_m is the magnetic susceptibility of material and is generally complex

$$\mu = \mu_0(1 + \chi_m) = \mu' + j\mu'' \tag{1.21}$$

$$B = \mu H \tag{1.22}$$

The imaginary part represents losses due to hysteresis, while the real part represents the material without magnetic loss. The permeability of many ferromagnetic

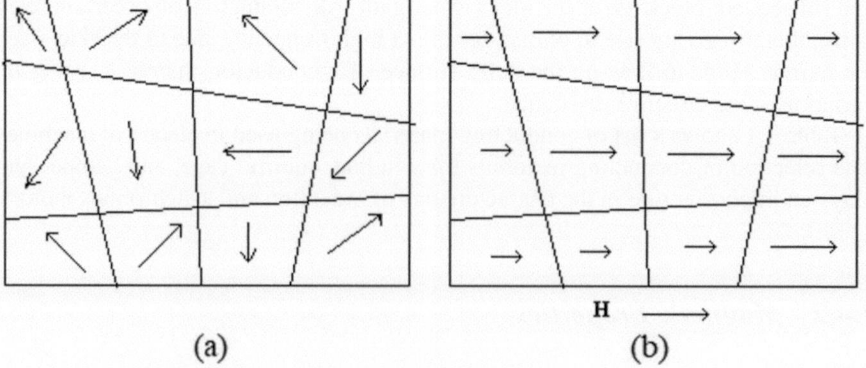

Fig. 1.12 Orientation of magnetic domains in ferromagnetic materials: **a** demagnetized ferromagnetic material, **b** magnetized ferromagnetic material

materials is nonlinear with respect to magnetic flux density or field intensity. The relative permeability is defined using Eq. (1.22).

Magnetic materials are divided into the following three types based on their permeability.

(a) Diamagnetic.
(b) Paramagnetic.
(c) Ferromagnetic.

In *diamagnetic materials,* the internal field is zero under normal conditions. When an external field is applied, a net internal field opposing the external field is induced. The relative permeability is slightly less than 1.0. The group includes important materials used in electrical machines like silver, mercury, lead and water.

Paramagnetic materials are similar to diamagnetic materials. But the internal and external fields are aligned in the same direction, thereby increasing the total field strength. The relative permeability is *slightly* larger than 1.0. The materials in this group are air, vacuum, aluminum, copper, and stainless steels.

Ferromagnetic materials vary from both *diamagnetic* and *paramagnetic materials.* They have very large permeability and the internal structure of their magnetic moments is also different from both *diamagnetic and paramagnetic materials.* When magnetic moments are aligned in a certain direction within a very small region, they form a *magnetic domain.* The neighboring domains have a similar structure and the net magnetic domain is aligned in one direction. These magnetic domains are randomly oriented in demagnetized material and, therefore, the internal fields are either very small or zero (with the exception of natural or artificial permanent magnets). When an external field is applied, the domains that have a net field aligned in the direction of the applied field grow in size while others domains shrink. The total field becomes larger because of the alignment of external and internal fields.

This behavior is related to the *hysteresis curve* of a ferromagnetic material. As the external field is increased, more domains are aligned further increasing the total field. Since beyond a certain field strength, most of the domains are aligned with total field, any further increase in the external field does not increase the internal field. Thus, the magnetic field has reached its saturation. Some examples of ferromagnetic materials are iron, nickel, cobalt, carbon steel, silicon steel, permalloy, and ferrites.

Ferrimagnetic materials are ferromagnetic materials but are essentially insulating materials as far as conductivity is concerned. Ferrite core is manufactured by prolonged heating the powdered ferrites and magnetite under very high pressure and at very high temperature. Since their conductivity is very low, the core losses at high frequency are negligible; this makes the material suitable for high frequency applications.

Table 1.2 lists some typical ferromagnetic and ferrimagnetic materials and their possible maximum permeability, which depends on the composition and treatment of material. It is evident from the *hysteresis curve* that permeability of ferromagnetic material is not constant. Instead, it varies with respect to the magnetic field. Thus, most of the ferromagnetic materials are highly nonlinear. This is exhibited with the

Table 1.2 Maximum values of relative permeability of some ferromagnetic and ferrimagnetic materials

Material	Relative permeability
Cobalt	250
Nickel	600
Iron	6000
Super alloy	1,000,000
Carbon steel (0.9%C)	100
Silicon iron (4%Si)	7000
Permalloy (78.5%Ni)	100,000
Fe_3O_4 (ferrimagnetic)	100
Ferrites (ferrimagnetic)	5000

help of initial magnetization curve, shown in Fig. 1.13. The instantaneous permeability varies with respect to the magnetic field and is shown in Fig. 1.13a. This nonlinearity is given by a relation (1.23),

$$B = \mu(H)H \tag{1.23}$$

Any increase beyond certain magnetic field intensity increases the flux density but not at the same rate as at lowest point of the curve. This is called *saturation region*, where the relative permeability tends to be around 1.0. The level of field at which saturation occurs depends on the material tested and applied magnetic field intensity.

From Fig. 1.14, we can see that if the field is reduced, the curve instead of retracing the initial magnetization curve moves to the left. The flux density is reduced up to point $\mathbf{B}_r$ where the applied field is zero. This residual flux is called *retentivity* or *remanence*. This remanent flux density may exist without a source to sustain. It is very important for a permanent magnet (PM). On applying reverse magnetic field

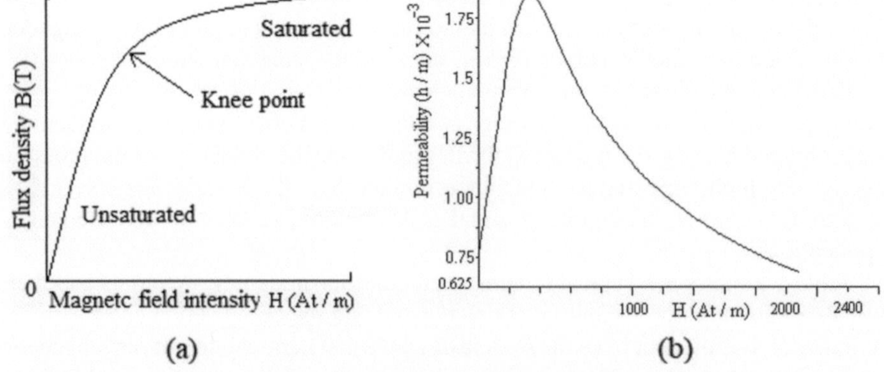

Fig. 1.13 **a** A typical B-H curve of a ferromagnetic material showing non-saturation, knee point, and saturation, **b** typical variation of relative permeability of a magnetic material

intensity up to point $-\mathbf{H}_c$, we find that at this point applied magnetic field intensity exists without an associated flux density. This field intensity is called *Coercivity* or *coercive force* of the material. Any further increase in negative field intensity will trace the magnetization curve up to the saturation point, where both **B** and **H** are negative. Similarly, by decreasing the field a point of $-\mathbf{B}_r$ is reached with zero **H**. Increasing positive **H**, a point is reached when **B** is zero. Permanent magnets manufactured in factories are magnetized using magnetizers, which utilizes a very large value of magnetic field intensity in pulse form, sufficiently above $\mathbf{H}_c$ for a shorter time. The magnitude of current applied may lie near about 1 kA or even more. The intense heat requires the process of magnetization to be completed in liquid nitrogen for certain type of magnets [4, 7].

The magnetization curve is called hysteresis curve and is typical for all magnetic materials, though their coercive force and remanence may be different. Magnetic materials are also referred as *soft* and *hard* magnetic materials. The latter are useful in making permanent magnets. The coercivity and remanence of some of these materials are listed in Table 1.3.

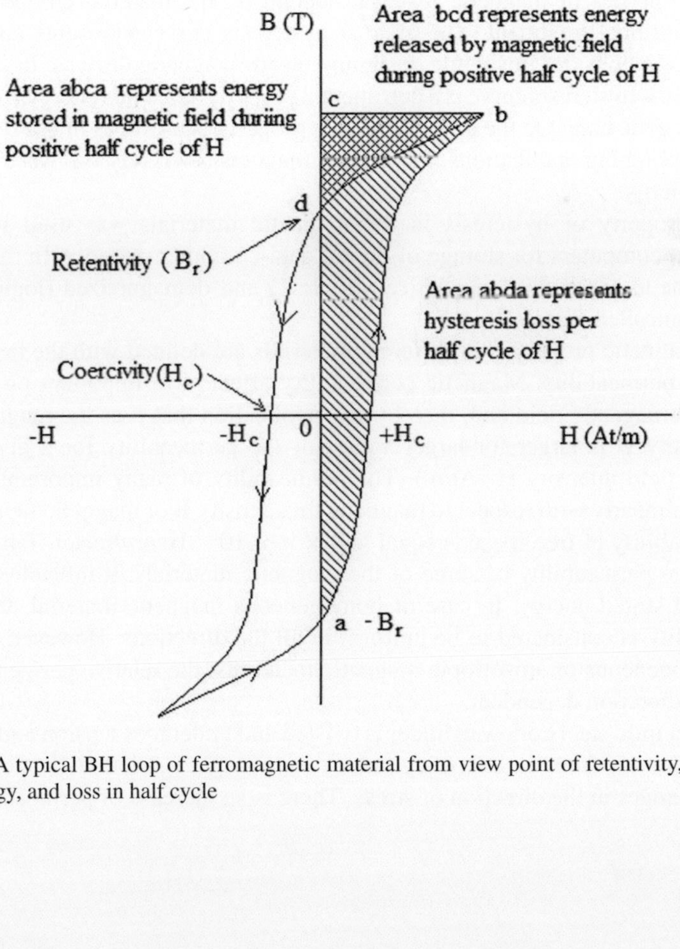

Fig. 1.14 A typical BH loop of ferromagnetic material from view point of retentivity, coercivity, stored energy, and loss in half cycle

Table 1.3 Some soft and hard magnetic materials

Soft				Hard		
Material	$\mathbf{H}_c$ (At/m)	$\mathbf{B}_r$ (T)	$\mathbf{B}_s$ (T)	Material	$\mathbf{H}_c$ (At/m)	$\mathbf{B}_r$ (T)
Pure annealed Iron	100	1.2	2.16	Carbon Steel (0.9%C)	4000	1.0
Super alloy	0.2	0.0001	0.8	Ticonol II	1320	1.18
Steel (0.9%C)	4000	1.0	2.0	Alnico 2	43,000	0.73
Silicon steel (4%Si)	20	0.5	1.95	Alnico 2	126,000	1.04
Ni-Zn ferrites	0	0	0.34	Samarium cobalt	560,000	0.84
				Ferroxdur 2	143,000	0.34
Gyoperm	Usable at cryogenic temperature 4.2 °K			Nd-Fe-B	$6 \times 10^5 - 2 \times 10^6$	$0.6 - 1.4$

The properties of magnetic material depend on the material composition and the metallurgical treatment. The area of hysteresis curve represents energy loss. One needs to consider this while designing electromagnetic devices. In device like transformers, hysteresis curve is a detrimental property since this energy is dissipated in the magnetic core. On the other hand, this property is useful as in use of magnetic core for switching applications. Permanent magnets (PM) rely on hysteresis curve to store energy.

This property of hysteresis in the magnetic materials was used in second-generation computers for storage of digital data-computer memory. In that the tiny ferromagnetic cores were magnetized (logic 1) and demagnetized (logic 0) using certain controlled signals.

The magnetic properties of different materials are defined with the help of their magnetic permeability. Magnetic permeability affects two important properties of the electromagnetic field and, thereby any application that uses the magnetic field. Flux density **B** is larger for larger values of the permeability for a given source magnetic field intensity **H** (At/m). The permeability of many important materials varies nonlinearly with respect to magnetic flux density **B** or magnetic field intensity **H**. Permeability of free space is equal to $4 \times \pi \times 10^{-7}$ Henry/meter. Table 1.2 lists the relative permeability of some of the magnetic materials. It influences the skin depth and stored energy. In case of homogeneous magnetic material, the relative permeability is considered to be uniform in all the directions. However, in case of non-homogeneous or anisotropic magnetic materials, the relative permeability (μ_x, μ_y, μ_z) is direction dependent.

When a mild steel hollow cylinder is twisted and undergoes tension and compression in plane 45° from the axis of cylinder, due to magnetostriction, the permeability of iron changes in the direction of stress. There is an increase of permeability in the direction of tension and decrease of permeability in the direction of compression. The effect is known as Wiedermann effect and used in certain torque transducer.

Permanent magnets

Permanent magnets (PM) are those materials that retain a significant remanant flux density even after external field has been removed. Permanent magnets are also called *hard magnets*. The operating point of a permanent magnet is in the second quadrant of BH loop of the material. Thus the coercive field intensity $\mathbf{H}_c$ and its flux density $\mathbf{B}_r$ are in opposite directions anywhere in the magnet. In other words, permeability of a permanent magnet is negative. The energy associated with the field of a permanent magnet is calculated as

$$W = \frac{1}{2}\int_{v} H \cdot B\mathrm{d}v \tag{1.24}$$

where v is the volume of magnet.

Thus, for a given volume of the magnet, energy depends on the product BH. The B-H loop of the magnet has a point at which this product is maximum, representing the operating point of the magnet at which it can produce maximum energy. At point ($\mathbf{B} = \mathbf{B}_r$, $\mathbf{H} = 0$) and ($\mathbf{B} = 0$, $\mathbf{H} = \mathbf{H}_c$), the product BH is zero and the magnet does not produce any external energy at these points. A good magnet is one, which has a high coercive field intensity $\mathbf{H}_c$ as well as the highest remnant flux density $\mathbf{B}_r$. A high coercive force does not allow the magnet to be demagnetized and a high remanant flux density $\mathbf{B}_r$ is normally associated with the capacity of the magnet to produce high magnetic flux density $\mathbf{B}_r$. The properties of some of the PM materials are given in Table 1.4. High-power injection molded and compression molded permanent magnets are commercially available. These permanent magnets are also available in flexible sheet form for easy manufacturing. This has also been covered in Chap. 2.

The Poynting theorem

Energy in the electromagnetic field is governed by Poynting theorem. The Poynting theorem states that the difference between time rate of change in the stored energy (electric and magnetic) and the power dissipated in a volume is equal to the power leaving that volume. The Poynting vector in terms of **E** and **H** is given as

Table 1.4 Properties of PM materials

	$\mathbf{B}_r$ (T)	$\mathbf{H}_c$ (kA/m)	$\mathbf{(BH)}_{max}$ (kJ/m^3)	μ_r
Alnico	1.25	60	50	$3 - 5$
Ferrite	0.38	240	25	1.1
Sm-Co	0.9	700	150	1.05
Ne-Fe-B	1.15	800	230	1.05

$$\mathbf{P} = \mathbf{E} \times \mathbf{H} \tag{1.25}$$

For periodic time-varying fields, the time average Poynting vector is given as

$$P_{\text{av}} = \frac{1}{T}\int_0^T P(t)\mathrm{d}t \tag{1.26}$$

Complex Poynting vector is defined as

$$\mathbf{P} = \frac{1}{2}\left(\mathbf{E} \times \mathbf{H}^*\right) \tag{1.27}$$

$$\mathbf{P} = \frac{1}{2}\text{Re}\left(\mathbf{E} \times \mathbf{H}^*\right) \tag{1.28}$$

$$\oint_s \mathbf{P} \cdot \mathrm{d}\mathbf{S} = j\omega \int_{\text{Vol}} \left(\frac{1}{2}\varepsilon \text{E} \cdot \text{E}^* - \frac{1}{2}\varepsilon \text{H} \cdot \text{H}^*\right)\mathrm{d}v + \int_{\text{Vol}} \frac{1}{2}E \cdot J_0^* \mathrm{d}v - \int_{\text{Vol}} \frac{1}{2}\sigma|\mathbf{E}|^2 \mathrm{d}v \tag{1.29}$$

The first two terms on right-hand side are the rates of change in the stored magnetic and electric (power) and third term is the dissipated power through ohmic losses. Here, the term containing $\mathbf{J_0}$ or current density represents the losses in electromagnetic system. Other terms on the right-hand side of Eq. (1.29) are the stored energy due to electrostatic field **E** and stored energy due to magnetic field intensity **H**. The Poynting theorem represents any input energy to the electrical system which results in stored energy in electrostatic field, stored energy in electromagnetic field, and ohmic losses. These ohmic losses are due to losses in current carrying conductors and induced eddy currents.

1.4.3 Insulating Materials and Dielectric Losses

Insulators play an important role in electrical engineering. They are commonly known as *dielectric materials*. Electric properties of dielectric materials such as insulating materials are defined by polarization of charges in the material due to an applied external electric field. Since charges in these materials are bound-charges, conduction is negligible, but polarization of charges may be significant. This polarization increases the electric flux density in the material. We can, thus, write

$$\mathbf{D} = \varepsilon_0\mathbf{E} + \mathbf{P} = \varepsilon_0\mathbf{E} + \varepsilon_0\chi_e E = \varepsilon_0(1 + \chi_e)\mathbf{E} \tag{1.30}$$

$$\varepsilon = \varepsilon_0(1 + \chi_e) = \varepsilon' + j\varepsilon'' \tag{1.31}$$

$$\mathbf{D} = \varepsilon \mathbf{E} \tag{1.32}$$

$$\varepsilon = \varepsilon_r \varepsilon_0 \tag{1.33}$$

Permittivity ε does not have to be linear or isotropic. A lossless material has a real permittivity. The imaginary part of permittivity represents dielectric losses. These losses are predominant in high voltage machines, transformer and equipment, and are considered to be absent in low and medium voltage electromagnetic apparatus.

The properties of some insulating materials such as fibers and varnish change with temperature. This is referred to as *aging effect*. The temperature of the machine during steady and transient states should not cross the safe limit, otherwise the insulating materials may get damaged or the life of insulating material is reduced. When a machine gets heated due to ohmic losses, the hidden air in the machine also gets heated up. But, if this heat is not properly dissipated, it creates *hot spot*s (points at which maximum temperature appears) in the machine. Therefore, machines should be cooled properly in order to avoid hot spot creation. The voltage gradient across the insulating materials should be limited in high voltage (HV) machines and transformers for avoiding electrostatic breakdown of insulations, which may result in permanent failure of the machine.

Various insulating materials can be divided in several classes as shown in Table 1.5.

Currently, Kapton tape is also being used. This is made of polymer known as polyimide, which is heat resistance and can sustain − 269 to 400 °C. Another insulating paper in use is Nomex paper. It offers high inherent dielectric strength, mechanical toughness, flexibility, and resilience. It is normally used in separating high voltage coils sides, providing necessary insulation between active conductors and surrounding metallic parts.

The working temperature decides the type of insulating materials. An oil immerged transformer is dipped in transformer oil (mineral oil) for cooling purposes. This protects the windings from insulation failure. In dry type of transformers, a special bonded paper insulation having better capacity to carry away heat is used.

A coating of enamel is provided in standard wire gauge (SWG) winding wires, which insulates the individual conductors in a bunch of several conductors forming a coil.

When an electrical machine is fabricated, the winding is normally dipped in hot (about 75 °C) varnish. In this process, the hidden air bubbles trapped inside the winding or coils comes out and the vacant space is filled by liquid varnish. This process is repeated several times. The varnish solidifies after some times. The winding of the electrical machine is then surrounded by a layer of varnish which becomes a solid structure. This process is known as *impregnation.* It improves the insulating properties of the windings, possibility of hot spot is minimized, the aging effect is mitigated, and the winding gains sufficient mechanical strength to bear the large electromagnetic forces under abnormal operating conditions. For larger machines such as turboalternators, impregnation is done in a vacuum chamber where insulating epoxy resins are poured in windings, which later solidified [1, 3].

Table 1.5 Different types of insulating materials

Class	Maximum permissible temperature	Details	Areas of use
Y	90 °C	Combination or material such as cotton, silk, and paper without impregnation	Hygroscopic in nature, presence of moisture lowers the breakdown strength of the insulating materials Not used for machine insulation
A	105 °C	Cotton, silk, and paper with impregnation or coated with varnish or immersed in dielectric oil	Used for machine insulation
B	130 °C	Mica, glass fiber, asbestos, etc. with suitable bonding, impregnation, or coating of dielectric varnish	Used for machine insulation
H	180 °C	Silicone and combination of materials such as mica, glass fibers, asbestos, etc. with suitable bonding, impregnation, or coating with silicone resins	Used for machine insulation
C	< 180 °C	Mica, porcelain, glass, quartz, and asbestos with or without an organic binder	Brittle natures Not preferred for machine insulation

1.4.4 Thermal Properties

Ohmic losses, magnetic losses, and mechanical losses, such as friction and windage losses, generate heat in machine. As all the active losses appear as heat, the temperature of each part of the machine is raised above that of the medium (surrounding air). The temperature rise is related to

(a) The rate of heat production.
(b) The rate of cooling.
(c) The thermal capacity.

Heat is removed from static transformers or rotating machines through conduction, convection and radiation. Smaller machines are cooled due to natural convection. It has been pointed out earlier that excessive temperature rise in an electrical machine has an adverse effect on the conducting (winding) material and the insulating material. For safeguarding the conducting and the insulating materials, it is imperative to provide sufficient cooling to the machine. The heat can be dissipated with the help of the following:

(a) Natural air (AN).
(b) Natural oil (mineral oil) (ON).
(c) Forced air.

(d) Forcing treated water through hollow conductors.
(e) Hydrogen cooling.
(f) Forced hydrogen cooling.

The type of cooling is marked on the name plate rating of machine. In a transformer, if the cooling is mentioned as (ONAN), then transformer is cooled using natural oil and air natural type of cooling.

In small machines, a fan is attached to the shaft, which aids the natural convection. The outer frame of the machine is made with certain fins for increasing the total area of heat dissipation. In large machines, axial and radial ducts are provided for increasing the dissipation rate. This heat is dissipated in the atmosphere so as to limit the rise in temperature of the machine. The rate of temperature rise in a continuous duty cycle electrical machine operating at rated condition is given as

$$\theta = \theta_m \left(1 - e^{-\left({}^{t}/_{\tau}\right)}\right) \tag{1.34}$$

where θ_{m} is the maximum temperature, which the machine can attain under rated conditions and τ is the heating time constant. When machine is switched OFF, the fall in temperature is exponentially decaying with respect to time.

When machine is overloaded, θ_{m} also rises in proportion to energy loss rate. An electrical machine can work under overloaded conditions, but it will attain permissible maximum temperature more quickly. It has been pointed out earlier in Sect. 1.3.8 and Fig. 1.11. A machine can be re-rated at higher capacity for a short-term duty cycle application. The duty cycle of the motor is then decided based on time taken for acquiring maximum permissible temperature rise.

Is it possible for you to record the temperature of electrical machine or transformer with respect to time? Which transducer will you use for sensing the temperature?

1.4.5 *Mechanical Properties*

The mechanical parts of electrical machines are subjected to mechanical stresses during their operations. The machines' stationary and rotating parts should have proper material strength so as to withstand the harsh operating conditions. The material strength also varies with operating temperature and mechanical stresses. Rotation of electrical motor can be described mathematically as

$$T - T_L = J\frac{\mathrm{d}\omega}{\mathrm{d}t} + B\omega + F \tag{1.35}$$

where T is the electromagnetic torque, T_{L} is the load torque, ω is the angular speed in radians per second, J is the moment of inertia, B is damping coefficient, and F is frictional torque. The moment of inertia J is experimentally determined using retardation test or calculated using known expressions. The retardation test requires variation of speed with respect to time when machine is switched OFF. The solution of Eq. (1.35) helps in understanding variations in transient speed of the rotating system during starting, speed control, braking, load changes, supply side changes, etc.

An electrical machine may have either horizontal shaft or vertical shaft. These rotating shafts are supported by means of bush bearing, ball bearing, roller bearing, or thrust bearing. Bearings are mounted on end covers. A smaller size machine may have bush bearing, which needs frequent lubrication. The research and development of non-contact magnetic bearing for space application is being conducted in certain organizations. The health of the bearing is very important for reliable operation of an electrical machine. In large capacity rotating machinery, the health of bearing is either continuously monitored or monitored at a regular interval of time, using state-of-the-art condition monitoring techniques. A machine exhibits a peculiar audible noise and mechanical vibration patterns in the case of bearing fault.

Is it possible for you to record and compare the noise pattern of different electrical machines in your laboratory? Try recording the noise pattern of a single machine at different load conditions using your personal computer. Can you convert the recorded noise (*.wav) file into alphanumeric values for digital signal analysis?

Which transducer is used for recording mechanical vibration in rotating machines? The vibration signature of high-speed induction motor operating at 400 Hz supply changes when shaft axis deviates from the horizon. In rate gyroscope, the principle is used for detecting angle of tilt of air-borne equipment.

An electrical machine may have a base made of cast iron. These machines are normally grouted or fixed on a concrete foundation and the base is fixed by means of nut and bolt arrangements. Certain smaller capacity motors are axially connected to the frame of another machine. This is referred as *flange mounting*.

The foundation along with the rotating electrical machine has a natural frequency of mechanical vibration. The use of state-of-the-art *air bearings* helps in completely damping the audible noise and mechanical vibrations. The rotating machine exhibits different frequencies of mechanical vibrations at different speeds, loading, supply, and fault conditions. A mechanical resonance is exhibited by the set when the frequency of vibration is matched with the natural frequency of vibration, resulting in large audible noise or vibration at a particular speed. Foundation should be designed to avoid such situations of mechanical resonance.

Is it possible for you to record the vibration pattern of a rotating machine? Which transducer can be used for monitoring the mechanical vibration pattern of an electrical machine?

The requirement of lubrication depends on the ambient temperature. An electrical machine working in the temperature zone of −55 °C to +150 °C (defense use) will have different requirements of lubrication than the machine being used in temperature zone of 0 °C to +120 °C (normal use).

1.5 Inductance

Electrical energy is stored in inductance and capacitance. The electrical machines require the storage of electrical energy in coupling field where the conversion of one form of energy to another form takes place. The electrical circuit in electrical machine is made of several numbers of turns of conducting wire, which offer a finite resistance and finite inductance to the electrical circuit. In high voltage windings interturn capacitance is also sometimes considered. The winding resistance gives rise to ohmic losses. In energy conversion, it is the inductive parameter which is of utmost importance, because of its energy storage capability. It can also be linked to both the electric and the magnetic circuits. The inductance has been summarized in Table 1.6 and dealt in subsequent subsections [4].

1.5.1 Inductance from Circuit Point of View

The instantaneous voltage appearing across an inductance having a changing current with respect to time is given as

$$v_L = L\frac{\mathrm{d}i}{\mathrm{d}t} = N\frac{\mathrm{d}\phi}{\mathrm{d}t} \tag{1.36}$$

$$L = \frac{v_L}{\mathrm{d}i/\mathrm{d}t} \tag{1.37}$$

$$\mathrm{d}i = \frac{1}{L}v_L\mathrm{d}t \tag{1.38}$$

$$\int_{i(0)}^{i(t)} \mathrm{d}i = \frac{1}{L}\int_0^t v_L\mathrm{d}t \tag{1.39}$$

Table 1.6 Inductance from different points of view

Circuit	Energy	Geometrical
$v_L = L\frac{\mathrm{d}i}{\mathrm{d}t} = N\frac{\mathrm{d}\phi}{\mathrm{d}t}$	$W = \int_0^t v_L i\mathrm{d}t = \frac{1}{2}Li^2$	$L = N\frac{\mathrm{d}\phi}{\mathrm{d}i} = N\frac{\phi}{i}$
$L = \frac{V_L}{\mathrm{d}i/\mathrm{d}t}$	$L = \frac{2W}{i^2}$	$L = \frac{\mu N^2 A_m}{\ell} = \frac{N^2}{\mathcal{R}}$
$\int_{i(0)}^{i(t)} \mathrm{d}i = \frac{1}{L}\int_0^t v_L\mathrm{d}t$		
$i(t) = \frac{1}{L}\int_0^t v_L\mathrm{d}t + i(0)$		

$$i(t) = \frac{1}{L}\int_0^t v_L \mathrm{d}t + i(0) \tag{1.40}$$

The current through an inductive circuit cannot be changed instantaneously. When an inductive circuit carrying steady-state DC current is suddenly switched OFF, a transient voltage appears in the circuit. According to Lenz's law, the polarity of induced transient voltage in the coil acts in a direction to oppose the cause. This has been described in Sect. 1.5.4.

1.5.2 Inductance from Energy Point of View

From energy point of view, the stored energy in an inductance is,

$$W = \int_0^t v_L i \mathrm{d}t = \int_0^t \left(L\frac{\mathrm{d}i}{\mathrm{d}t}\right) i \mathrm{d}t = \int_0^t Li \mathrm{d}i = \frac{1}{2}Li^2 \tag{1.41}$$

$$L = \frac{2W}{i^2} \tag{1.42}$$

1.5.3 Inductance from Geometrical Point of View

The inductance in terms of magnetic quantity is given as

$$L = N\frac{\mathrm{d}\phi}{\mathrm{d}i} = N\frac{\phi}{i} \tag{1.43}$$

where N is number of turns; $\phi =$ flux and $i =$ current. Thus,

$$\phi = \frac{\text{MMF}}{\text{Reluctance}} = \frac{Ni}{\mathcal{R}} \tag{1.44}$$

The magneto motive force (MMF) responsible for the creation of flux is given as MMF

$$\mathcal{F} = Ni \tag{1.45}$$

The reluctance $\mathcal{R}$ opposes the flow of flux and is analogous to that of resistance in the electric circuit. If l is the mean length of flux path and A_m is the cross sectional area, the reluctance is given by

$$\mathcal{R} = \frac{\ell}{\mu A_{\mathrm{m}}} \tag{1.46}$$

Thus, inductance in terms of physical parameters is given as

$$L = \frac{\mu N^2 A_{\mathrm{m}}}{\ell} = \frac{N^2}{\mathcal{R}} \tag{1.47}$$

1.5.4 Inductance in DC Circuit

The current through an inductance cannot be suddenly changed to a new value. When an inductive circuit is connected to a DC source, the current cannot be instantaneously changed from one value to the other. The rise of current is given by

$$i(t) = \frac{V}{R}\left(1 - e^{-t/\tau}\right) \tag{1.48}$$

where $\tau = L/R$ is time constant.

If the circuit is short-circuited on its own, a transient decaying current flows. In an inductive circuit, the current cannot be instantaneously brought to zero. The inductance of the circuit stores a sizable amount of energy and unless this energy is dissipated completely the current will continue to flow in same direction. If an inductive circuit is switched OFF, a spark appears between the contactors of the switch. Can you give its reason?

If in RL circuit carrying a DC current is short-circuited, there is no applied voltage but due to current flowing at time $t = 0$, the equation is

$$R \cdot i + L\frac{\mathrm{d}i}{\mathrm{d}t} = 0 \tag{1.49}$$

$$\mathrm{R.i} + L\frac{\mathrm{d}\psi}{\mathrm{dt}} = 0$$

Here current $i(0+) = I = E/R$, solving Eq. (1.49), using Laplace's transform, we get

$$i(t) = I \cdot e^{-\frac{R}{L}t} \tag{1.50}$$

Here, again time constant is $\tau = L/R$

This current flows till all the stored energy in inductance is dissipated in the form of I^2R loss within the circuit. Consider a case of ring of superconducting material. If a strong permanent magnet is thrown at high speed through the ring, it will cause induced current to flow though the ring. If the resistance of the ring-circuit $R \rightarrow 0$, the current through the ring-circuit will continue to flow for time $t \to \infty$ or magnetic flux $\psi = \text{constant}$. This is known as *constant flux theorem* and is useful in understanding transient operation of electrical machines. This is also the operating principle of cryogenic magnets used in high speed ground (levitated) transportation, in which superconducting conductors carrying induced current for time $t \to \infty$, are cooled using liquid nitrogen.

1.5.5 Inductance in AC Circuit

Consider a series RL circuit, excited by an AC source, is switched ON at time t. Let the applied voltage be

$$v(t) = V_{\mathrm{m}} \cdot \sin(\omega \cdot t + \delta) \tag{1.51}$$

where δ is the angle which decides at time $t = 0$ on the sine wave. The circuit equation is

$$v(t) = R \cdot i + L \cdot \frac{\mathrm{d}}{\mathrm{d}t} i \tag{1.52}$$

The steady-state current is

$$I_s(t) = \frac{V_{\mathrm{m}} \cdot \sin(\omega \cdot t + \delta - \theta)}{Z} \tag{1.53}$$

The impedance is $Z = R + j \cdot \omega \cdot L = \sqrt{(R^2 + \omega^2 L^2)} \angle \theta$ and steady-state power factor angle θ is given by

$$\theta = \tan^{-1}\left(\frac{\omega L}{R}\right) = \cos^{-1}\left(\frac{R}{Z}\right) \tag{1.54}$$

Here, angle δ is an arbitrary phase angle permits closing of switch at any instant, or in other words, it controls the points on the voltage wave at which switch in the circuit is closed. Thus when $\delta = 0°$, switch is closed at zero voltage or zero crossing, and when $\delta = 90°$, the switch is closed at voltage peak. At time $t = 0$ and the steady-state current has a value

$$I_{\mathrm{s}}(0) = +\frac{V_{\mathrm{m}} \cdot \sin(\theta)}{Z} \tag{1.55}$$

But current continuity at $t = 0$ requires that the current through the circuit to start at zero. Therefore, a DC counterpart component of current is present,

$$I_c(0) = -\frac{V_m \cdot \sin(\theta)}{Z} \tag{1.56}$$

This decays with the time constant of $\tau = L/R$. The total current in the circuit is given as

$$i(t) = \frac{V_m}{Z}\sin(\omega \cdot t + \delta - \theta) - \frac{V_m}{Z}\exp^{-\left(\frac{t}{\tau}\right)} \cdot \sin(\delta - \theta) \tag{1.57}$$

For $\delta = 0°$, in practice ω L >>> R and $Z \approx \omega L$ θ≈90°. However, when $\delta = 90°$, the current is given as

$$i(t) = \frac{V_m}{Z}\sin(\omega \cdot t - \theta) \tag{1.58}$$

Thus at $\delta = 0°$, maximum DC component is present, and at $\delta = 90°$, no DC component is present. If resistance $R = 0$, time constant is large and the DC component decays very gradually. At the peak of AC current, the total peak current is nearly twice of the AC amplitude. If the switch closes at the instant when $\delta = \theta$, the transient term will be zero and current wave will be symmetrical. If switch closes at $\delta - \theta = +\frac{\pi}{2}$ or $\frac{\pi}{2}$, the transient term attains its maximum amplitude and the first peak of the resulting current wave will approach twice the peak amplitude of the steady-state component. This phenomenon is referred to as *current or flux doubling*. This is also discussed in Chap. 3.

1.5.6 Effect of Saturation on Inductance

The effect of saturation on inductance can be explained using Eq. (1.47). The relative permeability reduces when core is saturated.

1.6 Concept of Self- and Mutual Inductances

In a doubly excited system, two coils having turns N_1 and N_2 are excited by electrical (AC or DC) sources v_1 and v_2, respectively, as shown in Fig. 1.15. We assume that ferromagnetic material has a linear magnetizing curve with no losses. The voltage equations become

$$v_1 = i_1 r_1 + \frac{d\lambda_1}{dt} = i_1 r_1 + L_1\frac{di_1}{dt} \pm M\frac{di_2}{dt} \tag{1.59}$$

$$v_2 = i_2 r_2 + \frac{d\lambda_2}{dt} = i_2 r_2 + L_2\frac{di_2}{dt} \pm M\frac{di_1}{dt} \tag{1.60}$$

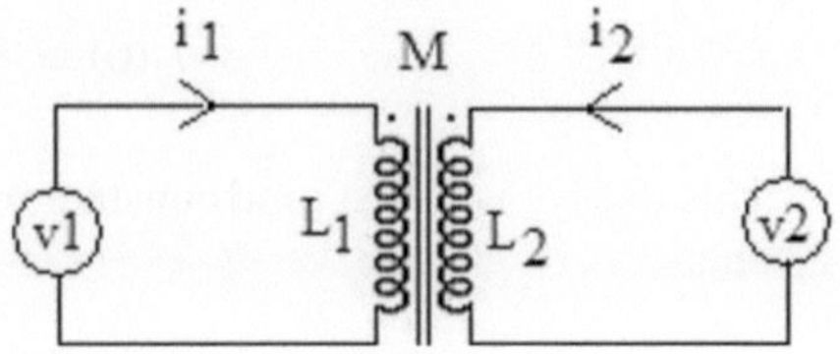

Fig. 1.15 Doubly excited system

Here L_1 and L_2 are the self-inductances of the two coils and M is the mutual inductance. The quantity $L_1 di_1/dt$ refers to the voltage drop associated with total flux produced by current i_1 and linking the primary turns N_1. The quantity $-M di_2/dt$ refers to rise in the voltage associated with the flux produced by current i_2 and linking primary turns. The secondary MMF $N_2 i_2$ produces a flux in counter direction to that of primary MMF $N_1\ i_1$. The + sign denotes a voltage drop, when the flux produced by one coil is aiding the flux produced by another coil.

1.6.1 Concept of Leakage and Mutual Inductance AC Circuit

Consider a case of two-coil system excited by a single AC source. It is a typical two-winding transformer (refer Fig. 1.15). The voltage equation of the first coil is

$$V_1 = I_1 r_1 + j\omega L_1 I_1 - j\omega M I_2 \tag{1.61}$$

If I_2' is the secondary current referred from primary side and transformation ratio, a is the turn ratio (N_1/N_2), then we have

$$I_2 = a I_2' \tag{1.62}$$

Thus, from Eqs. (1.61) and (1.62), we get

$$V_1 = I_1 r_1 + j\omega L_1 I_1 - j\omega a M I_2 \tag{1.63}$$

$$I_2' = I_1 - I_{\mathrm{m}} = I_1 - I_\phi \tag{1.64}$$

Here I_ϕ or I_{m} is the magnetizing current. The magnetizing MMF $N_1 I_{\mathrm{m}}$ is required to magnetize the core. Thus,

$$V_1 = I_1 r_1 + j\omega(L_1 - aM) I_1 - j\omega a M I_m \tag{1.65}$$

The self-inductance of the primary winding is given as

$$L_1 = N_1 \frac{d\phi_1}{di_1} = N_1 \frac{\phi_1}{i_1} = N_1 \frac{\phi_1 + \phi_{\ell 1}}{i_1} \tag{1.66}$$

The magnetizing flux and leakage flux are given as

$$\phi_m = \frac{N_1 i_1}{\mathcal{R}_{\mathrm{m}}} \tag{1.67}$$

$$\phi_{l1} = \frac{N_1 i_1}{\mathcal{R}_{\ell 1}} \tag{1.68}$$

Here $\mathrm{R_m}$ and $\mathrm{R_{l1}}$ are magnetizing reluctance and leakage reluctance respectively. Thus,

$$L_1 = \frac{N_1^2}{\mathcal{R}_{\mathrm{m}}} + \frac{N_1^2}{\mathcal{R}_{\ell 1}} = L_{\mathrm{m}} + L_{\ell 1} \tag{1.69}$$

The mutual inductance between second coil and first coil is

$$M_{21} = N_2 \frac{\mathrm{d}\phi_{\mathrm{m1}}}{\mathrm{d}i_1} = N_2 \frac{\phi_{\mathrm{m1}}}{i_1} \tag{1.70}$$

According to magnetic circuit notation discussed in next section, the magneto motive force MMF is the product of current and number of turns of a coil and is equated as flux flowing through it and reluctance of the path of flux.

$$N_1 i_1 = \phi_{\mathrm{m1}} \mathcal{R}_{\mathrm{m}} \tag{1.71}$$

$$M_{21} = \frac{N_2 N_1}{\mathcal{R}_{\mathrm{m}}} \tag{1.72}$$

Similarly, the mutual inductance between first and second coil is

$$M_{12} = N_1 \frac{\mathrm{d}\phi_{\mathrm{m2}}}{\mathrm{d}i_2} = N_1 \frac{\phi_{\mathrm{m2}}}{i_2} \tag{1.73}$$

$$N_2 i_2 = \phi_{\mathrm{m2}} \mathcal{R}_{\mathrm{m}} \tag{1.74}$$

$$M_{12} = \frac{N_1 N_2}{\mathcal{R}_{\mathrm{m}}} = M_{21} = M \tag{1.75}$$

The leakage inductance of first/primary coil is

$$L_1 - aM = \frac{N_1^2}{\mathcal{R}_{\mathrm{m}}} + \frac{N_1^2}{\mathcal{R}_{\ell 1}} - \frac{N_1}{N_2} \frac{N_2 N_1}{\mathcal{R}_{\mathrm{m}}} = \frac{N_1^2}{\mathcal{R}_{\ell 1}} = L_{\ell 1} \tag{1.76}$$

$$\omega(L_1 - aM) I_1 = \omega L_{\ell 1} I_1 \tag{1.77}$$

$$\omega a M I_{\mathrm{m}} = \omega \frac{N_1}{N_2} \times \frac{N_2 N_1}{R_{\mathrm{m}}} I_{\mathrm{m}} = \omega \frac{N_1^2}{R_{\mathrm{m}}} I_{\mathrm{m}} \tag{1.78}$$

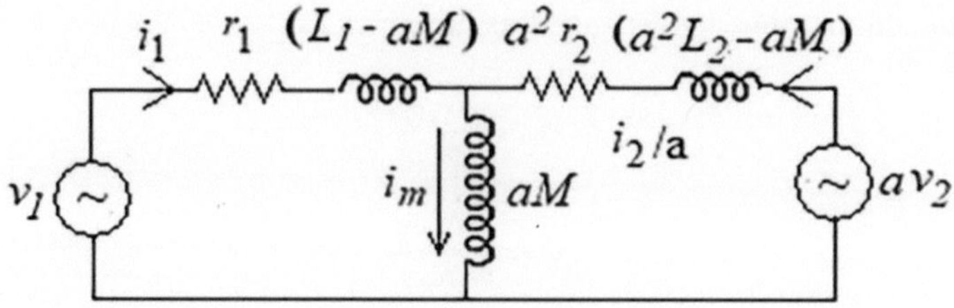

Fig. 1.16 Equivalent *T* network

$$\frac{N_1}{R_{\mathrm{m}}} I_{\mathrm{m}} = \frac{\phi_{\mathrm{m}}}{\sqrt{2}} \tag{1.79}$$

The EMF induced in primary coil is

$$E_1 = \omega a M I_{\mathrm{m}} = \omega N_1 \frac{\phi_{\mathrm{m}}}{\sqrt{2}} = \frac{2\pi}{\sqrt{2}} f N_1 \phi_{\mathrm{m}}. \tag{1.80}$$

A two-winding coupled network can also be expressed in a *T* network as shown in Fig. 1.16. This equation can also be derived using flux changing principle and is discussed in Chap. 2.

The coupled network is also dealt in Chap. 3.

1.6.2 Induced EMF in a Dynamic Circuit

The induced EMF in dynamic circuit is obtained by partial derivative of flux linkage with respect to time. In rotating machines it is given as

$$e = \frac{\mathrm{d}\lambda}{\mathrm{d}t} = \frac{\partial \lambda}{\partial t} + \frac{\partial \lambda}{\partial \theta}\frac{\partial \theta}{\partial t} = \frac{\partial \lambda}{\partial t} + \omega_r \frac{\partial \lambda}{\partial \theta} \tag{1.81}$$

In linear motion,

$$e = \frac{\mathrm{d}\lambda}{\mathrm{d}t} = \frac{\partial \lambda}{\partial t} + \frac{\partial \lambda}{\partial x}\frac{\partial x}{\partial t} = \frac{\partial \lambda}{\partial t} + v \frac{\partial \lambda}{\partial x} \tag{1.82}$$

The EMF induced in a coil can be of the following types:

(a) Transformer-induced EMF or EMF induced due to alternating currents (e_{t}).
(b) Speed-induced EMF (e_{r}).
(c) Both the transformer-induced EMF and speed-induced EMF.

Thus Eqs. (1.81) and (1.82) modify to

$$e = \frac{d\lambda}{dt} = \frac{\partial\lambda}{\partial t} + \frac{\partial\lambda}{\partial\theta}\frac{\partial\theta}{\partial t} = \frac{\partial\lambda}{\partial t} + \omega_r\frac{\partial\lambda}{\partial\theta} = e_r + e_t \quad (1.83)$$

$$e = \frac{d\lambda}{dt} = \frac{\partial\lambda}{\partial t} + \frac{\partial\lambda}{\partial x}\frac{\partial x}{\partial t} = \frac{\partial\lambda}{\partial t} + v\frac{\partial\lambda}{\partial x} = e_r + e_t \quad (1.84)$$

If L_l is the leakage inductance and R is resistance, then circuit equation in presence of transformer (alternating)-induced EMF and speed-induced EMF becomes

$$v = R \cdot i + L_l\frac{\partial i}{\partial t} + e_r + e_t \quad (1.85)$$

where v is the instantaneous applied voltage; i is instantaneous current; e_r is the rotational or speed-induced EMF, and e_t is transformer-induced EMF. On multiplying the Eq. (1.85) with current i, we get the ideal power equation, ignoring the losses in magnetic material.

$$p_e = v \cdot i = R \cdot i^2 + L_l\frac{\partial i}{\partial t}i + e_r \cdot i + e_t \cdot i \quad (1.86)$$

The total input to the machine is $p_e + p_m$, and within the machine, there are three general rates of power absorption:

(a) The rate of loss in I^2R, in the core and in mechanical friction and windage, p_{fw}.
(b) The rate of change of stored magnetic field energy $\frac{dW_f}{dt}$
(c) The rate of change of mechanical energy stored in inertia, deformation, vibration, shaft strain, etc. $\frac{dW_s}{dt}$

The energy rate balance equation is

$$p_e + p_m = \frac{dW_f}{dt} + \frac{dW_s}{dt} + p_{fw} \quad (1.87)$$

The balance of energy has again been discussed in Chap. 2. The behaviors of machine, like synchronous machines under transient conditions are explained using these equations.

1.7 Solution of Differential Equations

The solution of time-varying differential equations is essential in understanding the dynamic behaviors of electrical machines. When machine is running in steady-state conditions, the performance is normally expressed in terms of steady-state equations. The machine operates in transient conditions during starting, speed control, supply and load side changes, braking and fault conditions. These dynamic phenomena are explained using equation containing d/dt terms. These equations can be solved using the followings:

(a) Classical solution of differential equations.
(b) Laplace's transform.
(c) Runge–Kutta method.
(d) MATLAB Simulink.

Classical solutions, Laplace's transform and Runge–Kutta methods are beyond the scope of current book. Readers are expected to refer to books on linear system analysis for these methods.

Earlier, the differential equations for a large dynamic system like aircraft or large system were solved using analog computers. These computers were based on operational amplifiers (OPAMP). In these computers, the solutions were obtained in the form of certain electrical signals. These computers have become obsolete nowadays. Due to developments in computer-aided design and analysis, the analog computation is possible on digital computers using MATLAB Simulink, which utilizes the Runge–Kutta method. Its use totally relieves a user of the mathematical drudgery involved in the solution of large number of differential equations of a system.

Here a simple problem of current in an inductive circuit has been solved using MATLAB Simulink. We assume that a R-L circuit is connected to a DC source of voltage V Volt. The dynamic or time-varying equation is

$$V = Ri + L\frac{\mathrm{d}i}{\mathrm{d}t}$$

$$\frac{\mathrm{d}i}{\mathrm{d}t} = \frac{V}{L} - \frac{R}{L}\cdot i$$

$$i = \int \left[\frac{V}{L} - \frac{R}{L}\cdot i\right]\cdot \mathrm{d}t \tag{1.88}$$

It is very easy to understand the solution of differential equation or state equation using Simulink. A set of large number of such simultaneous equations having large number of state variables can be solved in the same manner. This requires the algebraic determination of $\mathrm{d}/\mathrm{d}t$ of different variables. The patching diagram according to the above equation is given in Fig. 1.17. The expected result $i(t)$ is shown in Fig. 1.18.

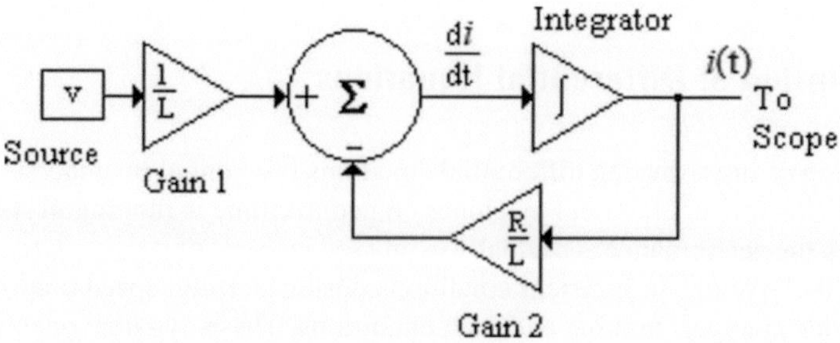

Fig. 1.17 Solution of differential equation using Simulink

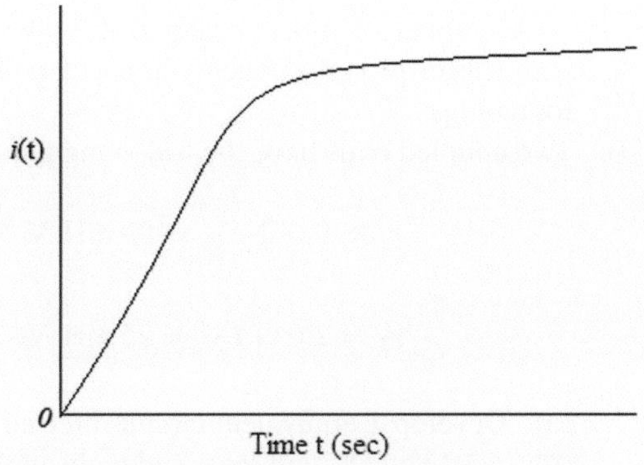

Fig. 1.18 Transient current in the circuit

Problems

1. Compare stored energy densities between (a) 1000 turns uniformly spread over a distance of 1.0 m and carrying 1.0 A (b) Two metal plates, each of 1.0 m × 1.0 m separated by a distance of 1.0 m and each having a potential of + 10 kV and –10 kV, respectively.

   [Hint: H=1000×1/1.0 At/m and E=20,000/1.0 kV/m]
2. What is the importance of ferromagnetic plunger? If plunger is made of copper or aluminum, will it experience any force when placed in proximity to a solenoid?
3. In Problem 2, if supply is AC, will there being any force experienced by iron plunger? Refer Fig. 1.1.
4. Justify sudden fall and rise in current when plunger is attracted toward solenoid coil and drawn by external force away from coil?

   [Hint: When the plunger is suddenly drawn away from coil, this action causes the current in the coil suddenly increase to a value, which maintains the same flux through the coil in spite of the increased reluctance (decreased permeance) of the magnetic circuit and vice versa.]
5. Why there is no tangential force in Fig. 1.3c?
6. In Fig. 1.7, if supply is DC, will there be any secondary induced EMF?
7. A two-turn coil carries a current of 5.0 A. Determine the inductance of the coil if

   (a) the total flux enclosed by coil is 0.002 Wb.
   (b) the total energy stored in the magnetic field is 0.02 J.
8. Why almost all the electrical machines work on the principle of electromagnetic rather than electrostatics?
9. Two coupled networks have the following parameters:

$$R_1 = 1.0\,\Omega \; L_1 = 20.0\,\text{mH}$$

$$R_2 = 1.0\,\Omega \; L_2 = 20.0\,\text{mH} \; M = 10.0\,\text{mH}$$

A number of turns are equal in both the networks. The primary (marked with subscript 1) is suddenly connected to a 10 Volt DC source. Calculate the followings:

10. Two coupled coils have the following parameters:

$$r_1 = 9.0\,\Omega\ L_{11} = 99\,\text{mH}\ N_1 = 99\,\text{turns}\ L_{l1} = 0.1L_{11}$$

$$r_2 = 2.0\,\Omega\ L_{22} = 25\,\text{mH}\ N_2 = 50\,\text{turns}\ L_{l2} = 0.1L_{22}$$

   (a) Develop T equivalent circuit referred from primary and secondary.
   (b) A 10 Volt DC voltage is suddenly applied on primary side. Obtain transient voltage on the secondary terminals.
   (c) If in (b) the secondary is short-circuited. Obtain the transient short circuit currents.

11. What are different principles of force production in an electromagnetic apparatus? Taking an example briefly describe the salient features of force production by principle of alignment? If the method is used in rotary motors, what are the requirements of power supply?

Objective Type Questions

1. In the figure shown, when the coil is excited by DC source.

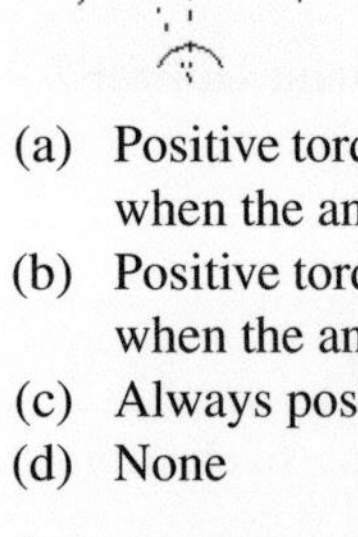

   (a) Positive torque is produced when the angle is negative and negative torque when the angle is positive
   (b) Positive torque is produced when the angle is positive and negative torque when the angle is negative
   (c) Always positive torque
   (d) None

2. A permanent magnet (PM) is in proximity to an aluminum plate moving with a linear speed of 6 m/s. The clearance between PM and moving plate is very small such that the magnetic flux interacts with the moving plate.

   (a) The plate will experience a retarding force
   (b) The plate will experience a repulsive normal force causing the magnet to move away from the moving plate.

(c) (a) & (b) both are true
(d) none

3. A solenoid is excited by a voltage source. A metallic plunger is attracted toward it, if it is (i) DC supply (ii) AC supply (iii) ferromagnetic plunger (iv) aluminum plunger. Following is true.

(a) (i) & (iii)
(b) (i), (ii) & (iii)
(c) (i), (iii) & (iv)
(d) (i), (ii), (iii) & (iv)

4. The noise pattern of an electrical machine can be recorded using a personal computer and

(a) Accelerometer
(b) Thermocouple
(c) Audio pick-up
(d) RTD

5. The vibration pattern of an electrical machine can be recorded using

(a) Accelerometer
(b) Thermocouple
(c) RTD
(d) Magnetic pick-up

6. In a coupled electrical circuit, the voltage equation is

$$v_1 = i_1 r_1 + \frac{\mathrm{d}\lambda_1}{\mathrm{d}t} = i_1 r_1 + L_1 \frac{\mathrm{d}i_1}{\mathrm{d}t} \pm M \frac{\mathrm{d}i_2}{\mathrm{d}t}$$

Here + sign and – sign correspond to the following cases:

(a) +sign means flux produced by one coil is opposite to the flux produced by another coil.
(b) +sign means flux produced by one coil is aiding the flux produced by another coil.
(c) – sign means flux produced by one coil is aiding the flux produced by another coil
(d) None

7. Temperature of an electrical machines can be recorded using following transducers

(a) Accelerometer
(b) Thermometer
(c) RTD
(d) Galvanometer

8. In permanent magnets listed below, the PM having maximum coercive force is

   (a) ALNICO
   (b) Ferrite
   (c) Ni–Fe–B
   (d) Sm-Co

9. Several permanent magnets arranged in an array with alternating poles are known as

   (a) Halbach array
   (b) LIM
   (c) PMBL
   (d) in situ magnetization

10. The relative permeability of a heavily saturated ferromagnetic material is roughly

   (a) 10,000
   (b) 100
   (c) 10
   (d) 1

11. Traces of silver in copper winding wire is responsible for

   (a) Better melting point
   (b) Better formability
   (c) Better shining
   (d) Better texture

12. Traces of zirconium in copper winding wire are used for

   (a) Increasing the boiling point of alloy
   (b) Increasing the melting point of metal
   (c) Increasing the shining
   (d) Improving texture

References

1. M.G. Say, Alternating current Machines, 4th ELBS edition, 1976
2. A. Fransua, R. Magureanu, Electrical Machines and Drive Systems, TP Technical Press, Oxford, 1984
3. Clayton & Hancock, Performance & design of direct current machines, ELBS edition, 1971
4. V Del Toro, Electric machines and power systems, PHI, 1980
5. A.S.Langsdorf, Theory of alternating current machinery, TMH, 2001
6. A. Draper, Electrical Machines, ELBS, 1st Ed, 1956
7. J. R. Hendershot Jr. & THE Miller, Design of Permanent-Magnet Motors, Clarendon Press, Oxford, 1994

Chapter 2
Forces in an Electromagnetic System

Abstract Maxwell's equations have been briefly covered from view point of understanding electrical machines. Simple magnetic circuits along with the Maxwell's equations, concepts of MMF, magnetic flux and flux density, reluctance, permeance, leakage flux, fringing, saturation, measurement of AC/DC flux density, Hall effect devices, and their use in electrical machines. Permanent magnets, magnetic lines of forces, normal force and its use in hyperloop system of transport, relative permeability of iron, Ampere's law applied to an ideal magnetic circuit, analogy between electric and magnetic circuits, magnetic circuit with air-gap, application of Ampere's critical law, losses in magnetic circuits—hysteresis and eddy current, concept of power flow, energy and co-energy, stored energy and forces, λ-I diagram for single excited system, open magnetic circuit and energy in multi-excited system have been dealt.

Keywords Maxwell's equation · Lumped parameter · Magnetic circuit · Lines of forces · Hyperloop · Cryogenic magnet · Shielding · FEM software · Saturation · MMF · Reluctance · Losses · Hysteresis and eddy current · Remanence · Coercivity

2.1 Introduction

In Chap. 1, we discussed that electrical machines follow the basic laws of electromagnetism having three-dimensional variations of fields and forces. Analyzing electrical machine by solving the electromagnetic field equations is quite cumbersome. In present days, the use of FEM and BEM software is in vogue for evaluating the fields and forces in an electrical machine. This requires the use of computer and software. Apart from these more tedious approaches, there are many simple approaches, where machine is regarded as a simple circuit in which output and input have certain established relationships.

An actual machine is quite complex in comparison to simple electromagnetic circuits because several coils are distributed over a large volume of ferromagnetic

R. K. Srivastava, *Electrical Machines*, https://doi.org/10.1007/978-981-96-5039-2_2

core. Certain simplifying assumptions are needed for easy understanding of electrical machines. The first step in reducing the complexity of a practical machine usually consists of assuming the effects, which are actually distributed over a large area and can be supposed to be concentrated in one place without involving serious error. The distributed resistances, inductances, and capacitances of windings are separated as though they were due to individual resistances, inductors, and capacitors connected externally. However, capacitor is rarely considered for analyzing electrical apparatus, except for machines operating at high voltage. The result of this concentration or *lumping*, as it is termed, finally enables the machine to be represented by an electrical network known as the equivalent circuit. In this chapter, we will analyze simple magnetic circuit.

2.2 Review of Maxwell's Equations

The analysis of electric machines and electromechanical devices begins with the adaptation of ***Maxwell's equations***. Here Cartesian coordinate system is assumed. Maxwell's equations which govern the electromagnetic phenomena at any point in space and time are given as following set of equations [1, 2, 4]:

$$\nabla \cdot \mathbf{B} = 0 \tag{2.1}$$

$$\nabla \cdot \mathbf{D} = \rho_v \tag{2.2}$$

$$\nabla \times \mathbf{H} = \mathbf{J} + \frac{\partial \mathbf{D}}{\partial t} \tag{2.3}$$

$$\nabla \times \mathbf{E} = -\frac{\partial \mathbf{B}}{\partial t} \tag{2.4}$$

where

B is the magnetic flux density (T),
H is magnetic field intensity (A/m),
D is electric flux density (C/m^2),
E is electric field intensity (V/m),
ρ_v is volume charge density (C/m^3),
J is current density (A/m^2) and ∇ is the del operator.

The ∇ del operator is given as

$$\nabla = \mathbf{a_x}\frac{\partial}{\partial x} + \mathbf{a_y}\frac{\partial}{\partial y} + \mathbf{a_z}\frac{\partial}{\partial z}$$

For almost all the power frequency (50 Hz) electromagnetic systems, the charge density ρ and the electric flux density **D** are assumed to be negligibly small. By means of ***Stokes's* theorem**, it is possible to transform Eqs. (2.3) and (2.4) into their integral form, applicable over a region of space:

$$\oint \mathbf{H} \cdot \mathbf{dL} = \mathbf{I} \tag{2.5}$$

$$\oint \mathbf{E} \cdot \mathbf{dL} = -\int_{\mathbf{S}} \frac{\partial \mathbf{B}}{\partial \mathbf{t}} \cdot \mathbf{dS} \tag{2.6}$$

Equations (2.5) and (2.6) are known, respectively, as ***Ampere's law*** and ***Faraday's law***. Solving Maxwell's equations in and around the geometry of a machine is a tedious task. For simplification of these equations, it is often assumed that flux density vector is,

$$\mathbf{B} = \nabla \times \mathbf{A} \tag{2.7}$$

where **A** is three-dimensional (complex) magnetic vector potential**.** It can be shown that the **Poisson's equation** for electromagnetic field problems in Cartesian coordinate system is,

$$\nabla^2 \mathbf{A} + \mu \mathbf{J} + \omega^2 \mu \varepsilon \mathbf{A} + j\omega\mu\sigma \mathbf{A} + \mu\sigma \cdot v \frac{\partial \mathbf{A}}{\partial \mathrm{x}} = 0 \tag{2.8}$$

where μ is permeability (H/m); ε is permittivity (F/m); σ is conductivity (S/m); ω is angular frequency (rad/s). **J** is the source current density; $j\omega\mu\sigma\mathbf{A}$ is the eddy currents induced due to transformer action; $\mu\sigma \cdot v\partial A/\partial x$ is the eddy currents induced due to speed **v** in x-direction. For power frequency $f = 50$ Hz, the **displacement current density** is assumed to be zero.

$$\omega^2 \mu \varepsilon \mathbf{A} \approx 0 \tag{2.9}$$

It is important to observe that the directional parameters described by the vector notation in these equations are basis of the left-hand and right-hand **Flemings' rules** that are of use in machine analysis. A significant field relationship is the Lorentz force equation

$$\mathrm{d}\mathbf{F} = (\mathbf{J} \times \mathbf{B})\mathrm{d}V = \mathrm{I}\mathrm{d}\mathbf{L} \times \mathbf{B} \tag{2.10}$$

where I is the current flowing in a differential conductor of length d**L**. One simple application is to integrate the differential force over a volume dV in which the current flows in conductor and flux density **B** is uniform:

$$\mathbf{F} = \oint \mathrm{I}\mathrm{d}\mathbf{L} \times \mathbf{B} \tag{2.11}$$

$$\mathbf{F} = \mathbf{IL} \times \mathbf{B} \tag{2.12}$$

$$= BIL \sin(\theta) \tag{2.13}$$

Here θ is the angle between the direction of current in the conductor and the magnetic field. In many machine configurations, this angle is 90°. Thus, we get

$$F = BLI \tag{2.14}$$

Equation (2.14) is a well-known rule used in machine analysis.

The **Lorentz force equation** is useful in determining the forces on a conducting material having a finite conductivity. The Lorentz force equation can be used for deriving **Maxwell's stress tensor**, which is useful in numerical computation of three-dimensional forces on ferromagnetic parts of an electromagnetic device. In general, it is given as

$$T = \frac{1}{4\pi}\begin{bmatrix} B_x^2 - 0.5B^2 & B_xB_y & B_xB_z \\ B_yB_x & B_y^2 - 0.5B^2 & B_yB_z \\ B_zB_x & B_zB_y & B_z^2 - 0.5B^2 \end{bmatrix} \tag{2.15}$$

The three components of forces can then be evaluated using surface integration, using

$$\mathbf{F} = \int_S T \cdot n \cdot \mathrm{d}\mathbf{S} \tag{2.16}$$

Many machine or device configurations are evaluated using **Biot–Savart law**:

$$\mathrm{d}\mathbf{H} = \frac{I\mathrm{d}\mathbf{L} \times \mathbf{a}_\mathrm{R}}{4\pi R^2} \tag{2.17}$$

which describes the magnetic field intensity d**H** at a point in space due to a differential current element of length d**L** carrying current I at a distance R in the direction (from current element) $\mathbf{a}_\mathrm{R}$. The integral form of Biot–Savart law becomes

$$\mathbf{H} = \oint \frac{I\mathrm{d}\mathbf{L} \times \mathbf{a}_\mathrm{R}}{4\pi R^2} \tag{2.18}$$

2.3 Magnetic Circuit

The magnetic flux in electrical machines is produced using either current carrying coils known as winding or permanent magnets (PM). Due to recent research and development (R&D), the availability of powerful PMs has encouraged the development of high-power PM machines, having higher power to weight ratio. The electrical circuit of current carrying coils (of several conductors insulated from each other) encircles the path of magnetic flux and vice versa as shown in Fig. 2.1. The presence of ferromagnetic material increases the flux density in the core and also inductance of the coil. Figure 2.2 shows an actual magnetic circuit, in which the magnetic core is made of ferromagnetic material and the electric circuit contains a coil encircling the magnetic core. A proper size coil is preformed by winding several turns of insulated conductors on a *former*. Then it is inserted into the limb of core.

It is important to emphasize that magnetic field is a *distributed parameters phenomenon*, i.e., it is distributed over a region of space. However, under proper conditions, it is possible to apply *lumped parameter analysis* to certain class of magnetic field problems, just as it is applied in electrical circuit analysis. The accuracy and precision of such analysis in magnetic field problem is much less, than in electrical circuit problems.

This section describes the lumped circuit analysis as applied to magnetic systems, often called *magnetic circuit analysis*. It follows the approach of simple DC electrical circuit analysis and is applicable to system excited by DC signals or to low frequency AC excitation. Its usefulness is in the sizing of the magnetic components of an electromagnetic device during design stages, in calculating inductances, and

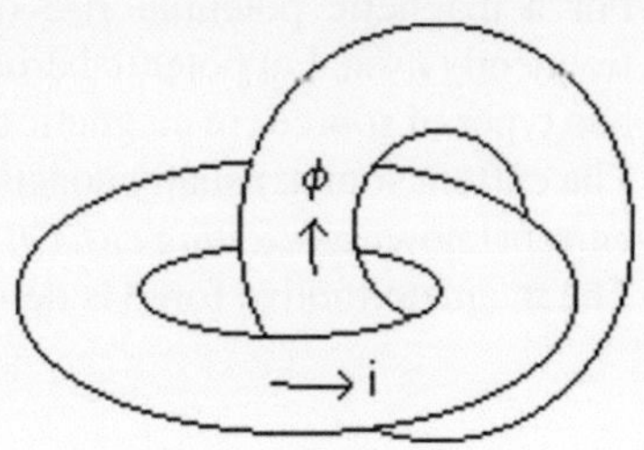

Fig. 2.1 Typical coupling between electric current and magnetic flux

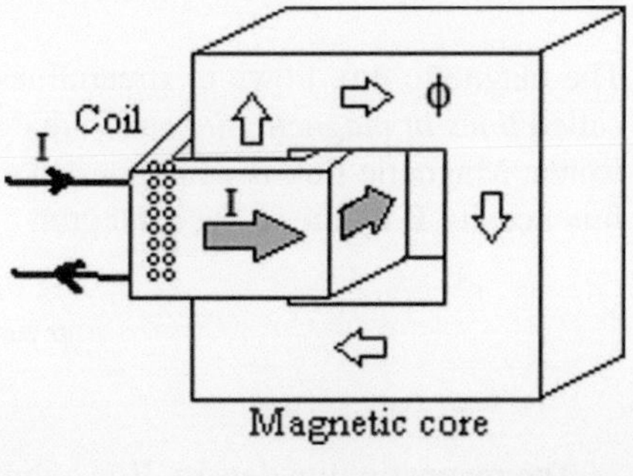

Fig. 2.2 Actual magnetic circuit having current carrying coil of several turns

in determining air-gap flux densities for power and torque calculations. We will now discuss the various quantities used in *magnetic circuit analysis.*

2.3.1 Magneto Motive Force

Magneto motive force, MMF ($\mathcal{F}$) is analogous to electromotive force, EMF (E) in an electric circuit.

In electrostatic field analysis, if V is potential, the electric field intensity is given as

$$\mathbf{E} = -\nabla \cdot V = -\text{grad}\, V \tag{2.19}$$

Similarly, we can define a scalar magnetic potential (MSP) V_m such that.

$$\mathbf{H} = -\nabla . V_\text{m}$$
$$V_\text{m} = \int \mathbf{H} \cdot \text{d}\mathbf{L} = \text{NI} \tag{2.20}$$

The dimension of V_m is ampere turns. It is analogous to electric scalar potential (V). The quantity NI is called magneto motive force (MMF),

$$\mathcal{F} = NI \tag{2.21}$$

For a magnetic potential rise or source of magnetic energy, the term MMF is frequently used. For potential drop, the term 'reluctance drop' is often used. There are two types of sources in magnetic circuits: **electric current** and **permanent magnets**. The current source usually consists of a coil with several number of turns, carrying a current known as *exciting current*. The number of turns is a non-dimensional quantity. The magneto motive force is defined as product of number of turns and current.

2.3.2 Magnetic Flux and Flux Density

The magnetic flux flows in streamlines. These streamlines in a magnetic field are called *lines of magnetic flux*, and are denoted by the symbol ϕ. The SI unit is (Wb) weber. Magnetic flux is analogous to current in an electric circuit and is related to flux density **B** by the surface integral,

$$\phi = \int_S \mathbf{B} \cdot \text{d}\mathbf{S} \tag{2.22}$$

The magnetic flux density **B** is a three-dimensional quantity. The components of flux density (B_x, B_y, and B_z) are defined as flux per unit area in a particular direction.

If we assume that the flux density has only one component, **B**, and is uniform over a cross-sectional area A, which is in perpendicular direction to the direction of **B**, then flux becomes

$$\phi = \mathrm{BA}$$

2.3.3 Reluctance

Reluctance ($\mathcal{R}$) is a component of magnetic impedance. It is analogous to resistance in electric circuit except that the reluctance is not an energy loss component. Reluctance is defined by a relationship similar to that of electric circuit as

$$\phi = \frac{\mathrm{MMF}}{\mathcal{R}} = \frac{\mathrm{NI}}{\mathcal{R}} = \frac{\mathcal{F}}{\mathcal{R}} \tag{2.23}$$

where N is the number of turns and I is the current passing through them. The SI unit of magnetic reluctance is Henry^{-1}. This is used in regions containing magnetic material that is homogeneous and isotropic, where the magnetic field is uniform.

If we assume that magnetic field intensity **H** is not varying along the path of magnetic flux, then

$$\mathcal{R} = \mathrm{MMF}/\phi = \mathrm{HL}/\mathrm{BA} = L/(\mathrm{BA}/H) = L/(\mu.A) \tag{2.24}$$

Equation (2.24) is similar to the geometric relation of resistance of a conductor. Here L is the length of flux path and A is cross-sectional area.

2.3.4 Permeance

The permeance ($\mathcal{P}$) is the reciprocal of the reluctance and has SI unit, Henry. Permeance and reluctance both are used to describe the geometric characteristics of magnetic field, particularly calculating inductance. By inverting Eq. (2.24), we get permeance

$$\mathcal{P} = \frac{\mu \cdot A}{L} \tag{2.25}$$

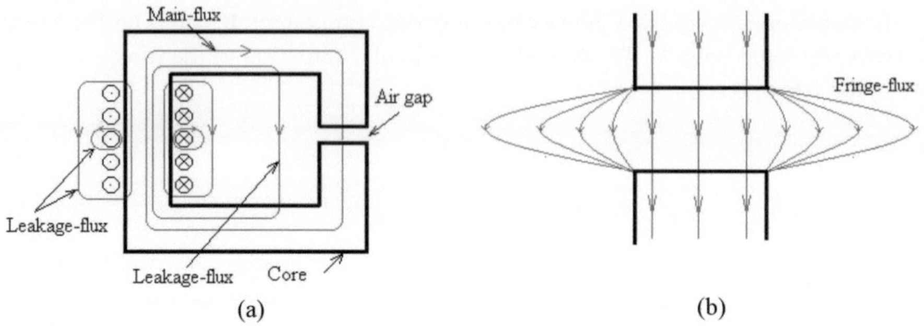

Fig. 2.3 **a** Main flux and leakage flux in magnetic circuit, **b** fringe flux through an air-gap—ballooning effect (enlarged diagram)

2.3.5 Leakage Flux

The magnetic flux produced does not entirely flows through ferromagnetic core, instead a part of this flux leaks through high reluctance path of air. In order to study, it is important to know the behavior of magnetic flux. For an ideal magnetic circuit, we assume that the flux completely flows through the low reluctance path of iron. Although the magnetic flux is analogous to electric current, it behaves in a dissimilar way as that of electric current. Do we have any magnetic insulator? The behavior of magnetic flux in ambience (e.g., air) is analogous to the flow of electric current passing between two electrodes submerged in saline water. The current follows the path of least resistance. The magnetic flux also follows the path of least reluctance of iron, but a small amount of flux produced by a coil or magnet passes through region of space, generally air spaces, electrical insulations, conductor, or structural members of the system—rather than along the main path of the magnetic circuit. These flux lines are termed as *leakage flux lines*. In a coupled circuit with two or more windings, the definition of leakage flux is specific, i.e., flux which links one coil does not link the other. This has been shown in Fig. 2.3a.

Leakage is a specific characteristic of magnetic circuits. Since there is no magnetic insulator, leakage of flux can be never eliminated. Although, at low frequency AC, **magnetic** ***shielding*** can reduce the leakage flux in air up to certain extent. This requires a thin sheet of high permeability material. At high frequency, the ***shielding*** is provided using aluminum foil. The flux penetrating the aluminum foil induces eddy current in the foil. The energy of leakage flux is dissipated in form of eddy current losses.

2.3.6 Fringing

Fringing is similar to leakage. It is used to describe the spreading or ballooning of the flux lines in an air-gap of a magnetic circuit. The reluctance of magnetic path in air is higher than that of iron. It is rather difficult to analytically calculate the effect of fringing unless we solve Maxwell's equation in 2D/3D using FEM or BEM software. Its net effect is to increase the effective area of air-gap. The fringing effect increases relatively with increase in the length of air-gap. This ballooning effect has been shown in Fig. 2.3b.

2.3.7 Saturation

The various microscopic phenomena occurring in ferromagnetic core have been explained in Sect. 1.4.2.

2.3.8 Magnetic Lines of Forces

Permanent magnet, DC electromagnet, and simple solenoid coil carrying DC current are the sources of unidirectional magnetic flux. Permanent magnet (PM) does not require any excitation for magnetic flux. It loses magnetism above Curie temperature. Magnetic flux passes air media in a path starting from N pole to S pole are referred to as lines of forces. In a simple bar magnet, the traces of the locus of tiny magnetic compass from N pole to S poles lines of forces can be roughly traced in 2D as shown in Fig. 2.4a. The lines of forces are disturbed when a ferromagnetic piece is in proximity to permanent magnet. Figure 2.4b shows the lines of forces from N pole falling of the nearby surface of iron piece and the other side of iron piece allows the line of forces closes at S pole. The piece of iron closure to N pole becomes S pole and the surface close to S pole becomes N pole. The iron piece tends to move where the force of attraction is higher than the other. These lines of forces represent direction of magnetic flux and give indication of direction of force. Any ferromagnetic material in these lines of forces alters the original lines of forces. It causes a force on piece of iron to move toward nearest stronger pole. In 3D, several lines of forces representing magnet flux emanate from N pole and terminate at S pole. These lines of forces or magnetic flux can also be computed in 2D or 3D using state-of-the-art FEM / BEM software. Behavior of magnetic flux in presence of another magnet (similar or opposite) can be easily drawn using hand drawing.

For a moving vehicle, the opposing forces are frictional forces between wheel and rail (or tire and road) and air drag (windage) forces. By limiting these opposing (friction and windage) forces, higher speed can be achieved. In case of levitated transport system, it is desired to reduce friction by maintaining a constant air-gap between wheel and rail (for primary suspension) so that higher speed can be achieved.

The vehicle can be lifted in air for few millimeters under dynamic conditions using either force of attraction or force of repulsion. The wheel and rail will be integral part of primary suspension, so that in the event of failure of controlled levitation, the vehicle comes on wheel and rail system under dynamic system. This demands either

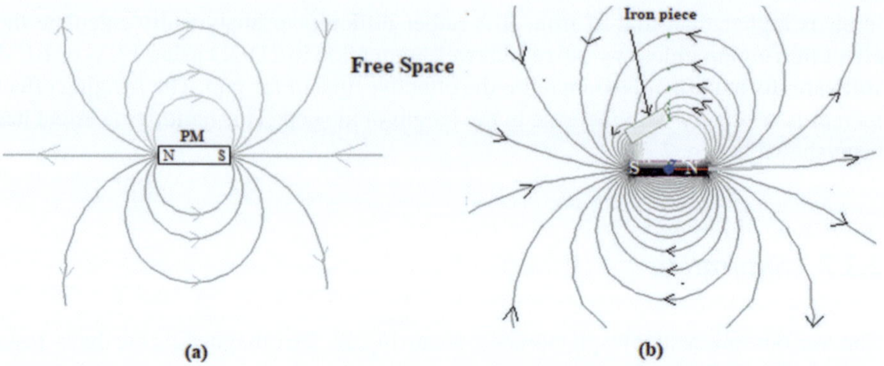

Fig. 2.4 Lines of forces: **a** single bar magnet in free space **b** presence of ferromagnetic piece in lines of forces

controlled electromagnets facing upper ferromagnetic plate on track or a controlled electromagnets facing a lower conducting plate on track. Electromagnets in presence of ferromagnetic material exhibit force of attraction. Such systems with controlled DC current are used in levitated transport system and recently proposed hyperloop system of transport. The force of attraction is almost unchanged under dynamic condition for constant air-gap and controlled current. Repulsive force between electromagnet and aluminum plate moving at high speed is useful for controlled electrodynamic levitation at very high speed.

Figure 2.5 shows an "I" Section Steel beam (ISBM) inside a partially vacuum steel tube, which provides a guideway to the moving pod in a hyperloop system. The top surface of ISBM has aluminum plate backed by iron serving as secondary (or reaction plate) of Single Sided Linear Induction Motor (SLIM). The pod consists of a three-phase SLIM facing the reaction plate, lower portion of pod has E-core or U-core controlled electromagnets facing lower portion of steel beam. Pod has battery and VFD to operate SLIM at high speed and control circuit for lifting pod

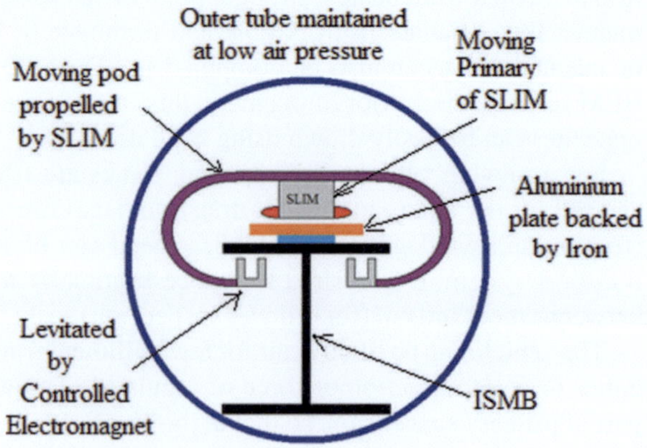

Fig. 2.5 Typical **hyperloop system** of transport (windings not shown for clarity)

by controlling current and force of attraction of electromagnets under dyanamic condition. Many configurations of hyperloop systems using Linear Induction Motor (LIM) are possible.

2.4 Measurement of Flux Density

The measurement of AC/DC flux density is required in understanding the behavior of electrical machines. Hall Effect Gauss meter is used to measure the DC flux density. It is based on the principle of Hall effect. A Hall effect sensor is a very thin wafer of semiconductors such as gallium arsenide (GaAs) or indium-antimonide (InSb) with flux sensitive areas as small as 0.15×0.3 cm and thickness of 0.01 cm. These are used for non-contact speed measurement, non-contact proximity sensing, current/voltage transducers, and magnetic card reader, etc. Hall effect sensors are commercially available and are integral part of several modern motors. Semiconductor Hall effect sensors and latch IC with power conditioners are used in operation of motors like brushless DC (BLDC) motor (a BLDC motor is commonly used in driving cooling fans of a personal computer), Switched Reluctance Motors (SRM) and Variable Reluctance Motors (VRM), etc. In Hall effect AC/DC current and voltage transducers, the measurement is done indirectly by measuring the flux density in air-gap of a magnet circuit. These transducers have several merits over conventional current transformer (CT) and potential transducers (PT). They have smaller size, linearity over wide range and provide perfect isolation.

A typical magnetic circuit made with ferrite with a fine air-gap is shown in Fig. 2.6. The excitation coil has N number of turns and is connected to a DC source. This configuration is being used in a Hall effect voltage transducer, in which a semiconductor Hall effect sensor is placed in the thin air-gap. Since Hall effect probe is non-magnetic, it does not distort the magnetic field. Hall effect integrated circuits (IC) are commercially available. TMAG5213 and DRV5013, A1210LUA-T, and A3144E are some of the Hall effect sensors. Hall effect sensor and latch are integral part of BLDC motors.

The AC flux density is also measured by using search coil. EMF induced in a search coil of very small cross-sectional area. The integration of this time-varying EMF is equal to the flux passing through the cross-sectional area of the coil. Operational

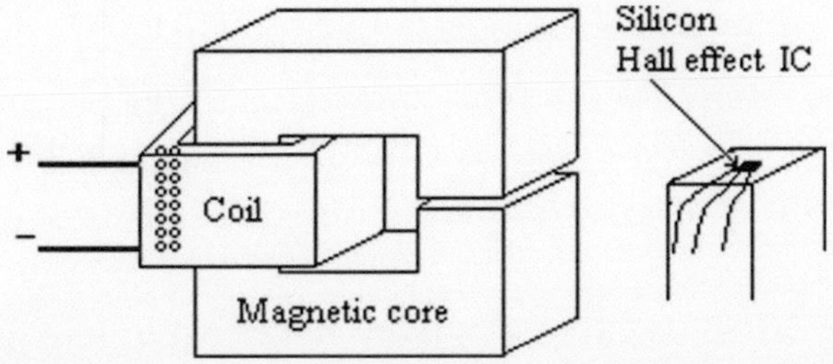

Fig. 2.6 Magnetic circuit having finite air-gap with a Hall effect sensor in its core

amplifier (OP AMP IC-741) may be used for integrating time-varying signal. Some of the power operational amplifiers are LM675 and L272.

2.5 Relative Permeability of Iron

The B-H curve of a ferromagnetic material is shown in Fig. 2.7. The importance of use of Silicon sheet steel in magnetic circuit is clear, as it requires lesser magnetic field intensity for larger flux density. Recall that the magnetic flux density in a ferromagnetic material is

$$B = \mu_0(H + M) \tag{2.26}$$

where M is magnetization vector, given as

$$M = \chi_m H \tag{2.27}$$

Thus,

$$B = \mu H \tag{2.28}$$

But, $\mu = \mu_0 \mu_r$. As such,

$$\mu_r = 1 + \chi_m \tag{2.29}$$

The relative permeability of ferromagnetic material is shown in Table 2.1.

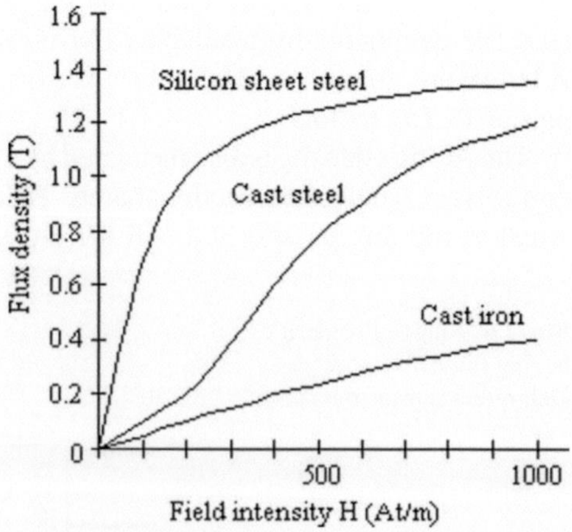

Fig. 2.7 B-H curve of magnetic materials

Table 2.1 Relative permeability of some materials

Material	Relative permeability	Material	Relative permeability
Air	1.0	Cobalt	250
Copper	1.0	Nickel	600
Aluminum	1.0	Iron (0.2% impurity)	5000
Silver	1.0	Iron (0.05% impurity)	200,000

2.6 Ampere's Law Applied to an Ideal Magnetic Circuit

Ampere's law states that the integral of magnetic field intensity around any closed path equals to the electrical current contains in that path. In a simple magnetic circuit MMF,

$$\mathcal{F} = \int_a^b \boldsymbol{H} \cdot \mathrm{d}\boldsymbol{L} \tag{2.30}$$

Here path of integration is from point "a" to point "b". The magnetic flux passing through a surface S is given as

$$\phi = \int_S \mathbf{B} \cdot \mathrm{d}\mathbf{S} \tag{2.31}$$

The MMF now becomes

$$\mathcal{F} = \mathcal{R}\phi \tag{2.32}$$

where $\mathcal{R}$ is reluctance of the path through which the flux flows,

$$\mathcal{R} = \frac{\ell}{\mu A} \tag{2.33}$$

where l is length of flux path, A is area of cross section, and μ is permeability of the medium.

The closed line integral of **H** is the total current enclosed.

$$\oint \boldsymbol{H} \cdot \mathrm{d}\boldsymbol{L} = I \tag{2.34}$$

If there are N numbers of turns, carrying current I, then it is NI.

$$\mathcal{F} = \oint \boldsymbol{H}.\mathbf{d}\boldsymbol{L} = NI \tag{2.35}$$

There is a direct analogy between electric and magnetic circuit as has been given in Table 2.2.

Let us apply Ampere's law to the simple magnetic circuit shown in Fig. 2.8. The mean length of magnetic circuit is L_i. Three coils are with turns N_1, N_2, and N_3 and carrying currents I_1, I_2, and I_3 respectively, are wounded around the core. Coils '1' and '2' are producing magnetic fluxes ϕ_1 and ϕ_2, respectively, in the same direction. These fluxes are additive. Coil '3' produces flux ϕ_3, which opposes the earlier fluxes, ϕ_1 and ϕ_2.

In actual magnetic circuits, the magnetic flux path in air may not be confined to specific path due to fringing. It has been pointed out earlier that it is analogous to the flow of electric current between two electrodes submerged in saline water. In an ideal magnetic circuit, it is assumed that the entire flux flows through the magnetic core and crosses the air-gap normally. It means that there is no leakage of flux and there is no fringing when flux crosses the air-gap. It is also assumed that variation of **H** with respect to **B** is linear, e.g., there is no magnetic saturation. This corresponds to linear portion of B–H curve. The MMF is given by line integral of magnetic field intensity **H**,

Table 2.2 Analogy between electric and magnetic circuit

Electric	Magnetic
EMF E (Volt)	MMF $\mathcal{F}$ (Amp turn)
Current I (Ampere)	Flux ϕ (Weber)
Resistance R (ohm)	Reluctance $\mathcal{R}$ (Amp turn/wb)
Current density (A/m^2)	Flux Density (Weber/m^2)
Conductivity σ (S/m)	Permeability μ (H/m)
Resistivity ρ (ohm m)	Reluctivity $1/\mu$ H^{-1} m
Conductance $G = 1/R$ (mho)	Permeance $\mathcal{P} = 1/\mathcal{R}$ (wb/Amp turn)

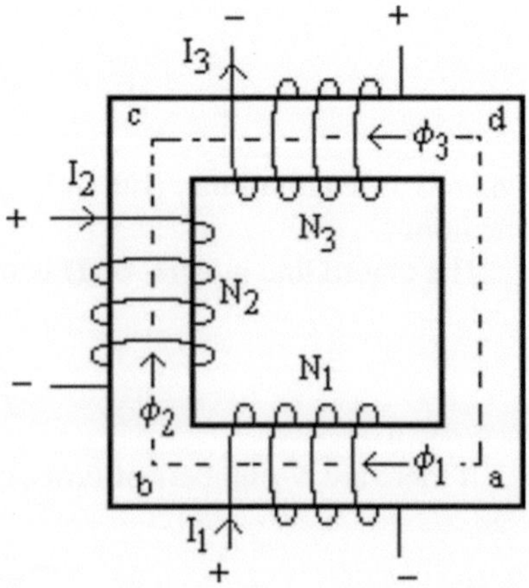

Fig. 2.8 Magnetic circuit with multiple excitation

$$\oint H \cdot \mathrm{d}l = \int_a^b H_m \cdot \mathrm{d}l + \int_b^c H_m \cdot \mathrm{d}l + \int_c^d H_m \cdot \mathrm{d}l + \int_d^a H_m \cdot \mathrm{d}l \tag{2.36}$$

$$\oint H \cdot \mathrm{d}l = H_m \cdot L_{ab} + H_m \cdot L_{bc} + H_m \cdot L_{cd} + H_m \cdot L_{da} = H_m \cdot L_i \tag{2.37}$$

$$H_m L_i = \phi \mathcal{R}_m = \mathcal{F} \tag{2.38}$$

Here $\mathcal{R}_m$ is the reluctance of magnetic path, and $\mathcal{F} = \phi \mathcal{R}_m$ is the magnetic potential or reluctance drop. The MMF is responsible for flow of magnetic flux in the circuit.

This gives,

$$\mathcal{F} = \phi \mathcal{R}_m = N_1 I_1 + N_2 I_2 - N_3 I_3 \tag{2.39}$$

We can say that 'the sum of the magnetic potential around any closed path is equal to zero'. This is analogous to the Kirchhoff's voltage relationship in electrical circuits.

2.6.1 Permanent Magnets

Permanent magnets (PM) are integral parts of many machines. They are the source of unidirectional (DC) magnetic fields in many machines. The permanent magnets are of three types:

(a) Ceramic magnets.
(b) Ferrite magnets.
(c) Rare earth magnets.

These magnets are commercially available in the form of blocks or sheet form. A sheet magnet can be cut in sizes and pasted on to the actual rotor surface. The only demerit of PM is that under certain operating conditions, it may get demagnetized.

A permanent magnet is characterized by second quadrant of B-H curve and by the following quantities:

(a) Residual flux density.
(b) Coercivity.
(c) Maximum energy.

The above quantities are explained in Fig. 2.9.

Fig. 2.9 Characteristics of permanent magnets

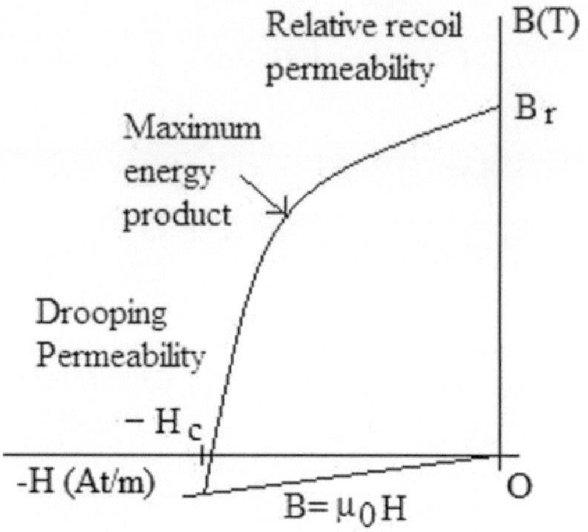

2.6.2 Magnetic Circuit with Permanent Magnet

The main aim of permanent magnets (PM) is to produce unidirectional flux in electrical apparatus. PM is used for

(a) Providing magnetic flux in relay coils, PM stepper motor, BLDC motor, PMSM, and PMSG. In PM motors like BLDC motor, the armature coil currents interact with the PM flux for producing unidirectional torque.
(b) Transducers such as magnetic pick-up, magnetic pick-up RPM meter, non-contact position or speed sensing, mechanical vibration-velocity transducer, proximity or position sensor along with Hall effect sensor. Details of magnetic pick-up and velocity (vibration) transducer are shown in Fig. 2.10.
(c) Audio pick-up and loud speakers, etc. (Students are advised to find the details of audio pick-up and speaker.
(d) Augmenting the DC flux in same direction in case of magnetic speed sensor (magnetic pick-up) and partial controlled levitation, in which a part of DC flux is provided by additional controlled coil current of an electromagnet.

In relay coils, solenoid coils, or ON/OFF switch-type mechanism, normally closed (NC) or normally open (NO) conditions are achieved using PM. In place of PM, mechanical springs are also used as restoring force in electromagnetic contactors.

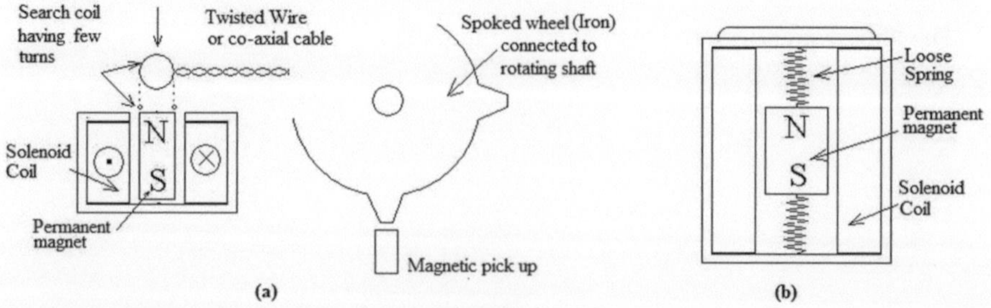

Fig. 2.10 a Magnetic pick-up b velocity transducer

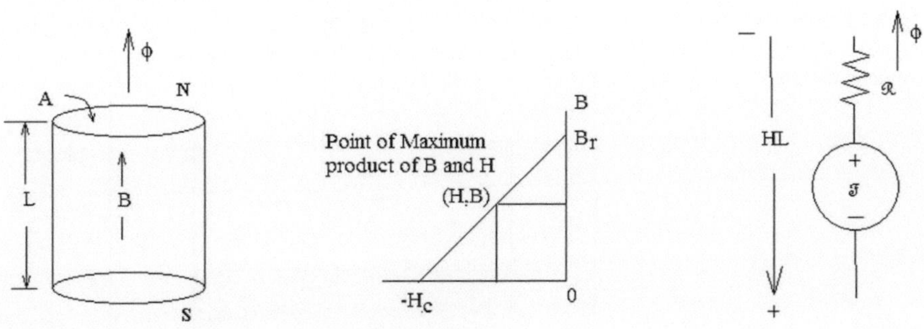

Fig. 2.11 Magnetic circuit with permanent magnet

In absence of coil current, the PM produces force in one direction. However, in presence of coil current, the direction of force reverses. The relay coils are widely used for control and protection circuits based on simple logic or programming logic controllers (PLC), etc. The solenoid (coils) valves are used for controlling supply of gas or fluid.

A PM is described by the second quadrant B–H curve as shown in Figs. 2.9 and 2.11. Considering a linear B–H curve of PM by joining the points $(0, B_r)$ and $(-H_c, 0)$ on the B–H curve. These types of curves are exhibited by PM formed using samarium cobalt, neodymium-iron-boron and ceramic magnets. The PM corresponding to increasing energy products are Alnico, rare earth samarium cobalt (Sm–Co) and rare earth (Nd–Fe–B) neodymium-iron-boron magnets.

2.6.3 Behaviors of Magnet in Case of Relative Speed

Permanent magnets in conjunction with controlled DC electromagnets used in high-speed levitated transport system experience a relative motion with respect to stationary ferromagnetic rails. The magnetic flux interaction with iron rail under dynamic condition develops a negligible or feeble drag force. The normal force of attraction under dynamic condition responsible for controlled levitation also encounters a feeble reduction which is taken care by controlling the coil currents in a close loop control.

However, if permanent magnet/electromagnet is in proximity to conducting material under dynamic condition, a large drag force appears in opposition to the main propulsion force due to eddy current braking action. They produce repulsive normal force, suitable for electrodynamic suspension of levitated vehicle at very high speed. These are special cases where stationary *cryogenic magnets* interact with the switching coil system of moving vehicle.

A magnetic circuit with a permanent magnet is shown in Fig. 2.11. A permanent magnet with air-gap in magnetic circuit is shown in Fig. 2.12. Let the magnet has a cross-sectional area of A and length L. The demagnetization curve of magnet can be expressed as

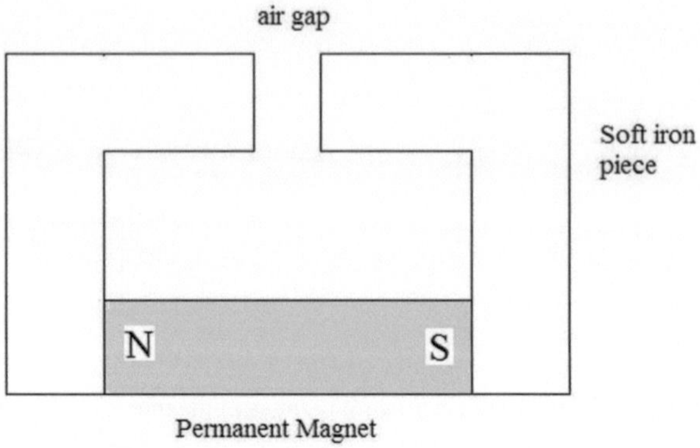

Fig. 2.12 Permanent magnet with air-gap in magnetic circuit

$$B = \frac{B_r}{H_c}(H + H_c) = \mu\,(H + H_c) \tag{2.40}$$

where permeability of magnet,

$$\mu = \frac{B_r}{H_c}$$

Incidentally a similar magnetic flux can be produced by an equivalent current carrying coil.

The magneto motive force (MMF) is given by

$$\left(\frac{B}{\mu} - H_c\right)L = \frac{\phi L}{A\mu} - H_c L = \phi\mathcal{R} - \mathcal{F} \tag{2.41}$$

Thus, a magnet is replaced by a coil having MMF $= \mathcal{F} = H_c L$ and reluctance, $\mathcal{R} = (L/A\mu_0)$.

Here both the flux density and magnetic field intensity are in opposite direction, being in second quadrant of B–H curve. For nonlinear demagnetization curve, having point of maximum B–H product as the magnetic permeability of the magnet becomes

$$\mu = \frac{B}{(H + H_c)} \tag{2.42}$$

The operating point moves closer to the maximum energy or maximum product of B–H as shown in Fig. 2.13. The demagnetization curve for Alnico is shown in Fig. 2.13. It must be noted that for protecting the PM from demagnetization during operation, the magnetic field intensity (H) in the magnet must always be lesser than the coercive force (H_c). Proper care is taken in designing PM machine such that the demagnetization of the PM should not occur under normal over loading or under fault conditions. The normal demagnetization curve is used for determining the amount of flux produced by the magnet. This flux links with the current carrying coil to produce

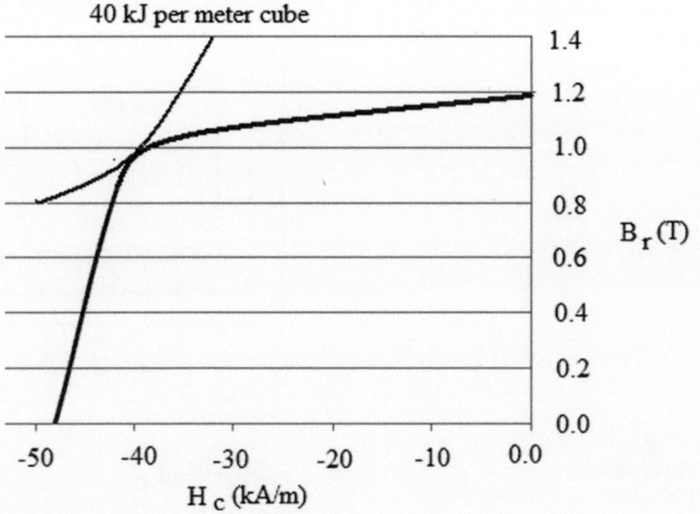

Fig. 2.13 Demagnetization curve for Alnico

unidirectional torque. The intrinsic demagnetization curve is used for determining the maximum allowable stator current before magnet can demagnetize.

2.7 Magnetic Circuit with Air-Gap—Application of Ampere's Circuital Law

The equivalent magnetic circuit with coil having N number of turn carrying current i in a magnetic core (subscript c) and an airgap (subscript g) (Fig. 2.14) is shown in Fig. 2.15. Here MMF is given as

$$N \cdot i = H_c \cdot L_c + H_g L_g \tag{2.43}$$

$$N \cdot i = \frac{B_c \cdot L_c}{\mu_c} + \frac{B_g \cdot L_g}{\mu_g} \tag{2.44}$$

In the case of fringing, the flux is given as

$$\phi = B_c A_c = B_g . A_g \tag{2.45}$$

Here subscript 'c' corresponds to core and 'g' corresponds to air-gap.

In the case of air-gap in the core, the cross-sectional area of core A_c is defined, but the effective area in the air-gap A_g is much larger. Due to ballooning effect, the air-gap flux density $B_g \ll B_c$, and assuming $A_c = A_g = A$, we get

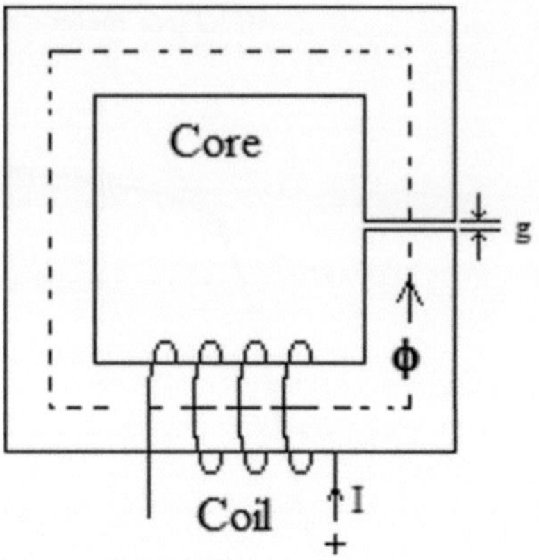

Fig. 2.14 Magnetic circuit with an air-gap

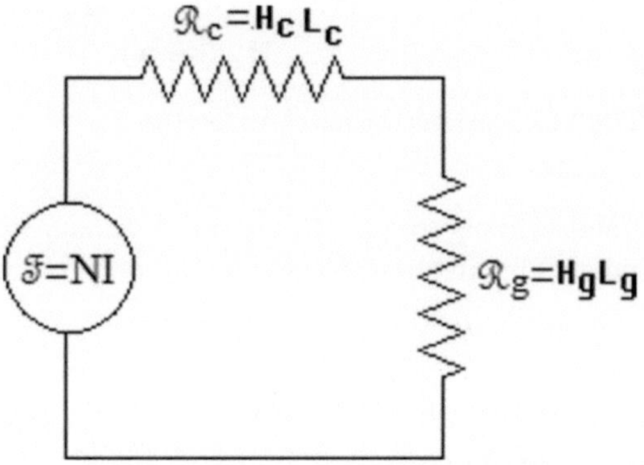

Fig. 2.15 Equivalent magnetic circuit of a magnetic circuit with a gap

$$\phi = B_c A = B_g A \tag{2.46}$$

$$Ni = \mathcal{F} \tag{2.47}$$

Thus,

$$\mathcal{F} = \phi(\mathcal{R}_c + \mathcal{R}_g) = \phi \mathcal{R}_{\text{eq}} \tag{2.48}$$

Here $\mathcal{F}$ is MMF, $\mathcal{R}_c$ is the reluctance of flux path in core, $\mathcal{R}_g$ is the reluctance of flux path in air-gap, and $\mathcal{R}_{\text{eq}}$ is the equivalent reluctance of magnetic circuit.

Now, the magnetic flux in the air-gap can be given as

$$\phi = \frac{\text{MMF}}{\text{Equivalent Reluctance}} \tag{2.49}$$

Since $\mu_r >> 1$, total reluctance of flux path is either equal to $\mathcal{R}_g$ or larger than $\mathcal{R}_c$ ($\mathcal{R}_g \gg> \mathcal{R}_c$) For magnetic circuit with air-gap, the core reluctance may be neglected, without any appreciable loss in accuracy. Similarly, a magnetic equivalent circuit of core magnetic circuit is shown in Fig. 2.15.

2.8 Losses in Magnetic Circuit

The B-H curve of core material is material specific. It is also referred as *magnetization curve*. Figure 2.7 shows the B-H curves for different materials. If we plot the incremental relative permeability, as in Fig. 1.13b, we will observe that when the core is not saturated μ_r is highest at the knee point, and as it gets saturated, the μ_r falls to a very low value. When the core gets saturated, an increment in MMF will not increase the flux density in the same proportion as it did when the core was unsaturated. When core is not saturated, can we assume *infinite permeability*? If a core gets saturated, the inductance as well as the developed force gets relatively reduced. Is magnetic saturation good for electrical machines? Can you explain? In Chap. 4, it is explained that the open-circuit EMF in self-excited DC generator results because of saturation. Similarly, the operation of induction motor (Chap. 5) is due to the hysteresis in B–H curve [1–6].

2.8.1 Magnetic Losses

The core losses occur in the ferromagnetic parts of machine. In AC machines, the magnetic core is subjected to alternating magnetization in which the direction of flux alternates at the supply frequency and due to this certain amount of power is wasted in hysteresis and eddy current losses. The hysteresis loss density P_h depends on the area of the hysteresis loop and is given as

$$P_h = \eta \cdot f \cdot B_m^x \tag{2.50}$$

where η is Steinmetz constant, which is material specific and quotient $x = 1.6$–2.0 for ordinary steel. This is valid for flux density 0.1–1.2 Wb/m^2. The use of proper ferromagnetic material having thinner loop with lesser area of B–H curve reduces the hysteresis loss. The eddy current loss is due to presence of circulating current set-up in conducting materials. These eddy currents are due to rate of change of flux. The rate of change of flux may be due to transformer-induced EMF or due to speed-induced EMF or both. In the case of AC machines, the eddy current loss depends on square of the thickness of stamping t, square of maximum flux density in the core B_m, and square of applied frequency f, and the form factor. The use of thin stamping for the fabrication of the core reduces the eddy current loss. The eddy current loss density is given as

$$P_e = K_e \cdot t^2 \cdot \mathrm{f}^2 \cdot \mathrm{B}_m^2 \tag{2.51}$$

It is possible to reduce this loss by choosing thinner laminations. These stampings while making core must be insulated from each other. In case of p poles DC machine running at speed of n revolution per second, the core is subjected to rotating magnetization. In DC machines the frequency is taken as the frequency of flux reversal, $f_{\mathrm{rm}} = 0.5.\mathrm{p.n.}$ Machines working on switched DC supply or PM machines also have core losses owing to the variations of flux with respect to space and time.

When a magnetic core is subjected to alternating magnetization, it results in microscopic cyclic deformation of the core material at power frequency. This phenomenon is called *magnetostriction.* This results in noise production in large transformers and rotating machines. This humming noise of magnetic core can be limited in variety of ways like clamping of core and use of certain adhesive. This helps in keeping the noise level away from the audible range of human beings (20 Hz to 20 kHz).

2.8.2 Practical Magnetic Circuit

In a practical ferromagnetic material, the functional relationship between MMF $\mathcal{F}$, flux ϕ and the corresponding magnetic field intensity $H = \mathcal{F}/L$ and flux density $B = \phi/\mathrm{A}$ are nonlinear, which indicates that the reluctance $\mathcal{R}$ is dependent on flux density B. Typical B-H curve has been shown in Fig. 2.7. In certain applications the change of reluctance about the operating point is important. This leads to the notion of ***differential reluctance*** $\mathcal{R}_d$, which is defined as

$$\mathcal{R}_d = \frac{\partial \mathcal{F}}{\partial \phi} = \frac{\partial (HL)}{\partial (BA)} = \left(\frac{L}{A}\right)\frac{\partial H}{\partial B} = \frac{1}{\mu_d}\left(\frac{L}{A}\right) \tag{2.52}$$

where $\mu_{\mathrm{d}} =$ differential permeability. The variation of relative permeability of iron with respect to magnetic field intensity H is plotted in Fig. 1.13b.

2.8.3 Open Magnetic Circuit

In most of the electromagnetic devices the path of magnetic flux is inside the core material, like transformer and conventional rotating machines. In certain special purpose machine, the path of magnetic flux is not fully confined within the core material; rather it necessarily has certain flux through the air or conducting material. These open magnetic circuit machines are having lesser efficiency and poor power factor. Such machines are useful in many applications owing to definite advantages offered by them over rotary motors. Examples are linear induction motor, electromagnetic pump capable of pumping hot liquid metal, electromagnetic levitators, and electromagnetic launchers etc.

2.9 General Constructional Features of Static and Rotating Machines

The general construction of machines is of rotating types. The linear machine requires at least one member of machine to be infinitely large so that when motion takes place a continuous force can be imparted to the moving member. Both types of machines have one primary member, the electrical circuit, which is connected to an electrical source. The other member, secondary, is usually on the shaft in rotating machines, and the electromagnetic force brings a continuous rotation to it. The moving member is supported by mechanical bearings, which are fixed in two end covers of stator. The stationary member is known as stator and rotating member is known as rotor. Rotary machines have closed magnetic circuits.

The electromagnets have also been used in variety of places. Their applications are separation of ferromagnetic material from mixed metals, contactors, relay coils, attraction-type electromagnetic levitation, PMBL DC motor, stepper motor, hybrid Switched Reluctance Motor, eddy current devices, non-contact electromagnetic bearings, etc. **Cryogenic electromagnets** (cooled with liquid nitrogen) are used in **electrodynamic suspension** of high-speed levitated vehicles.

2.9.1 Concept of Power Flow

In electromechanical devices, the change of electrical energy to mechanical energy takes place in a coupling field. It has already been mentioned in Chap. 1 that an electromagnetic device consists of an electric circuit, magnetic circuit, dielectric circuit, a cooling system, and a mechanical system. In these subsystems, transfer of energy is associated with certain inherent losses, though primarily the transfers of electrical energy to mechanical energy and vice versa are taking place. These are active losses resulting in rise of temperature. As such, the temperature of various parts of machine rises above the temperature of surrounding air [1, 3].

An ideal electromechanical system is shown in Fig. 2.16. The power transfer may take place from both ways. It is assumed that frequency is sufficiently low such that electromagnetic radiations are not taking place. Here it is assumed that the losses in electrical circuits, the hysteresis and eddy current losses and leakage flux in magnetic circuit, bearing frictional and air frictional losses in mechanical system are all zero. The insulating system is perfect with no dielectric loss. Therefore, there is no generation of heat in the system and temperature deviation is not there.

The material properties of conducting material will offer finite resistance giving ohmic losses unless superconducting materials are used. Low-temperature superconducting materials do require certain energy for cooling. An ideal magnetic material with zero core losses is difficult to achieve. The use of air core may lead to zero core losses, but the machinery is inefficient. Can you give suitable reason? Air friction encountered by moving or rotating bodies leads to marginal losses known as *windage*

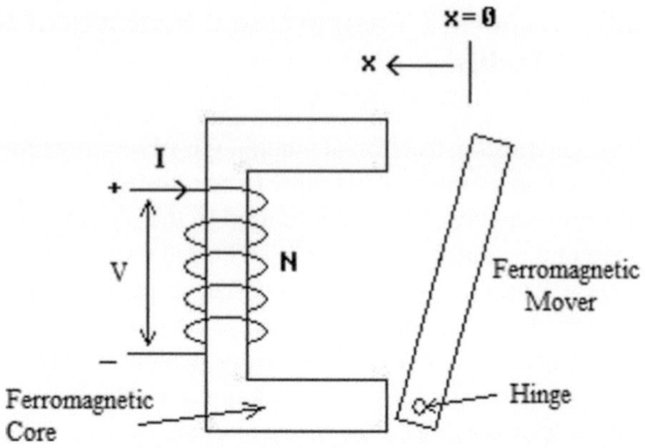

Fig. 2.16 Simple electromagnetic relay consisting of C-core electromagnet having N turns

loss. It is taken as zero if the machine is operating in vacuum, which is not always true. The friction losses arise in the supporting bearing at the two ends of the shaft.

A signal corresponding to torque may be obtained using torque transducer, mechanically coupled between drive motor and driven load. In swinging frame (stator) machine, both the structures of stator and rotor are free to rotate. The shaft and outer frame are supported by minimum four bearings. The outer member has restricted freedom to move. This (dynamometer) arrangement is primarily used for measuring torque, either with the help of spring balances attached to the periphery of outer frame or by means of load cells. The later can give continuous torque signal for recording purpose. In laboratory, usual practice is to couple a separately excited DC generator connected to resistive load box. The output power is taken as generator output and is used for torque calculation. Loss drum with belt and pulley or eddy current brake are also other options for experimentally determining torque of electrical motors or mechanical turbines. The losses in bearing and air friction are termed as *windage and friction loss*. The bearing friction loss can be made zero if the rotating mass is supported by non-contact magnetic bearings. Although, certain amount of controlled electrical power is required for the operation of non-contact bearings.

An ideal electromechanical system shown in Fig. 2.17 is difficult to achieve. The losses are inevitable in any energy conversion process. An actual electromechanical system is shown in Fig. 2.18.

The energy balance diagram for an actual electromechanical system is shown in Fig. 2.18. The total electrical energy W_{TE} is the total energy supplied by the electrical source and W_{TM} is the total energy supplied by the mechanical source.

Fig. 2.17 Ideal electromechanical system

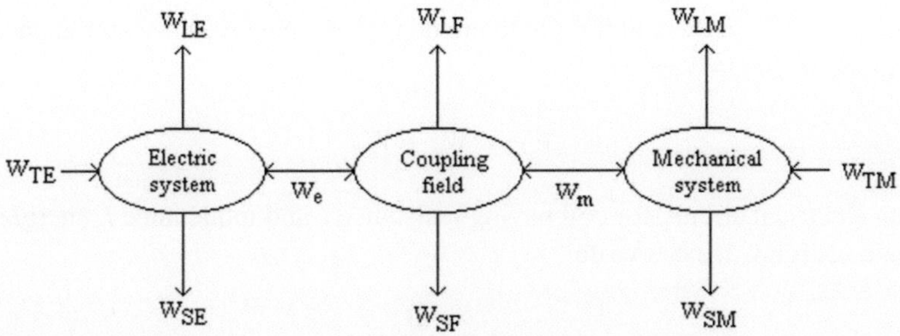

Fig. 2.18 Energy balance in an actual electromechanical system

The total energy supplied by an electrical source is responsible for electrical energy input to the coupling field W_e, losses in electrical system, W_{LE}, and stored energy in electrical system W_{SE}. In other words, we have

$$W_{TE} = W_e + W_{LE} + W_{SE} \tag{2.53}$$

Similarly, the total mechanical energy supplied by a mechanical source is responsible for mechanical energy input to the coupling field W_m, losses in mechanical system (both frictional and windage losses) W_{LM} and stored energy in mechanical system W_{SM}. In other words,

$$W_{TM} = W_m + W_{LM} + W_{SM} \tag{2.54}$$

The stored energy in electric or magnetic fields, which are not coupled with the mechanical system, W_e is the energy transferred to the coupling field. Similarly, W_m is the mechanical energy transferred to coupling field. The energy input may be positive or negative depending on motoring, generating, braking, or floating action. If we define W_F as the total energy transferred to the coupling field, then

$$W_F = W_e + W_m \tag{2.55}$$

$$W_F = (W_{TE} - W_{LE} - W_{SE}) + (W_{TM} - W_{LM} - W_{SM}) \tag{2.56}$$

The total energy transferred to coupling field results in losses within the coupling field, namely, core losses, dielectric losses, and losses due to fringing, leakage, and energy stored in the coupling field. Thus, we have

$$W_F = W_{SF} + W_{LF} \tag{2.57}$$

$$W_{SF} + W_{LF} = W_e + W_m \tag{2.58}$$

If losses (W_{LF}) in the coupling field are negligible, then we have stored energy in coupling field (W_{SF}=W_F)

$$W_F = W_e + W_m \tag{2.59}$$

In an electrical circuit, for coil having resistance r and inductance L energized by a voltage source v, we can write

$$v = ri + L\frac{\mathrm{d}i}{\mathrm{d}t} + e_f \tag{2.60}$$

where e_f is the instantaneous voltage drop or rise due to coupling field.

Consider simple magnetic circuits, as shown in Figs. 2.16, 2.19, and 2.20. When the respective coil is suddenly energized, there is an associated motion along with momentarily rise in flux density, flux linkage, coil inductance, and permeance. An increase in flux linkage causes the induced EMF e_f to oppose applied voltage v. As a result, the current momentarily gets reduced. When the mover is suddenly moved by external force in the opposite direction, there is an increase in air-gap consequently to the decrease in flux density, flux linkage, coil inductance, or permeance. A sudden decrease in flux linkage results in motional-induced EMF to aid the supply voltage, causing the current to momentarily increase. Refer Sect. 1.2.

The dynamic behavior of a rotating mechanical system is given by

$$T - T_L = J\frac{\mathrm{d}^2\theta}{\mathrm{d}t^2} + B.\frac{\mathrm{d}\theta}{\mathrm{d}t} + F\theta \tag{2.61}$$

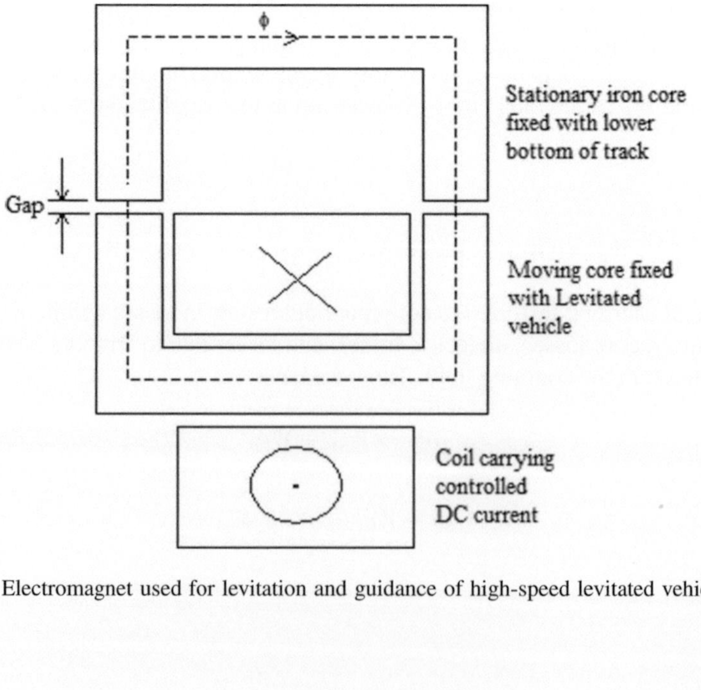

Fig. 2.19 Electromagnet used for levitation and guidance of high-speed levitated vehicle driven by LIM

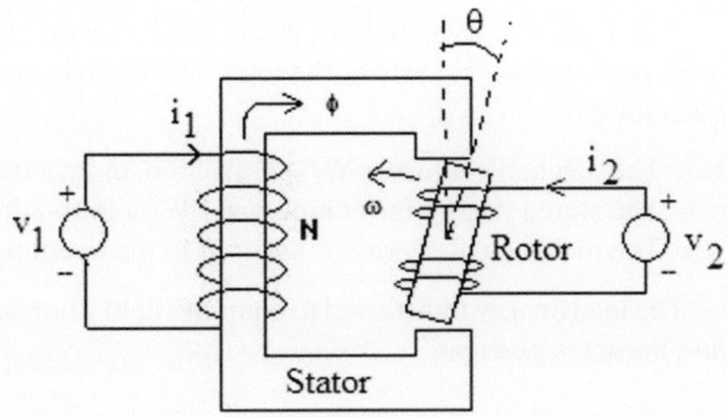

Fig. 2.20 Doubly excited elementary reluctance machine

where T is electromagnetic torque; T_{L} is load torque; J is moment of inertia; B is damping constant, and F is frictional constant. On the same analogy for a linear system we can write

$$f_e - f_L = M\frac{\mathrm{d}^2x}{\mathrm{d}t^2} + D\frac{\mathrm{d}x}{\mathrm{d}t} + Rx \tag{2.62}$$

where $f = f_e$ is the developed force, f_L is the opposing force, M is mass, D damping, and R is the linear frictional constant.

Since power is the time rate of energy transfer, total energy supplied by the electric source is

$$W_{\mathrm{TE}} = \int v \cdot i\mathrm{d}t \tag{2.63}$$

For a rotational or translational system, the total mechanical power supplied by the mechanical source is

$$W_{TM} = \int T \cdot \mathrm{d}\theta = \int f \cdot \mathrm{d}x \tag{2.64}$$

$$W_{TM} = \int T \cdot \frac{\mathrm{d}\theta}{\mathrm{d}t}\mathrm{d}t = \int T \cdot \omega \cdot \mathrm{d}t = \int f \cdot \frac{\mathrm{d}x}{\mathrm{d}t}\mathrm{d}t = \int f \cdot V \cdot \mathrm{d}t$$

Total electrical energy can be given as

$$W_{\mathrm{TE}} = r \cdot \int i^2 \cdot \mathrm{d}t + L\int i \cdot \frac{\mathrm{d}i}{\mathrm{d}t} \cdot \mathrm{d}t + \int e_f \cdot i \cdot \mathrm{d}t \tag{2.65}$$

In the above Eq. (2.65), the total electrical energy is a sum of the following,

(a) The ohmic energy loss, $W_{LE} = r.\int i^2.dt$.
(b) The stored energy in coupling field, $W_{SE} = L\int i.\frac{di}{dt}.dt$.
(c) The energy transferred to the coupling field, $W_e = \int e_f.i.dt$.

This clearly corresponds to the Poynting theorem discussed in Chap. 1. Similarly, for mechanical system, the total mechanical energy W_{TM} is sum of following quantities:

(a) The mechanical losses W_{LM} – friction and windage.
(b) The stored energy in rotating mass W_{SM} in rotating mass.
(c) The mechanical energy transferred to the coupling field, W_m.

The total energy transferred to coupling field from mechanical system (both rotary and linear) is given as

$$W_m = -\int T_e.d\theta = -\int f_e.dx \tag{2.66}$$

A positive force or torque is assumed to be in the same direction as a positive displacement dx or positive rotation dθ. The energy stored in the coupling field is

$$W_f = \int e_f \cdot i \cdot \mathrm{d}t - \int T_e \cdot \mathrm{d}\theta = \int e_f \cdot i \cdot \mathrm{d}t - \int f_e \cdot \mathrm{d}x \tag{2.67}$$

In a typical electromechanical system, having multiple coil system receiving mechanical energy from many sources, we can write,

$$W_f = \int \sum_j e_{fj} \cdot i_j \cdot \mathrm{d}t - \int \sum_k T_{ek} \cdot \mathrm{d}\theta = \int \sum_j e_{fj} \cdot i_j \cdot \mathrm{d}t - \int \sum_k f_{ek} \cdot \mathrm{d}x \tag{2.68}$$

In Eqs. (2.66), (2.67), and (2.68), the second part of equation is for rotary machines and third part of equation is for linear machines.

The electrical machines are normally singly excited or doubly excited. The doubly excited machine may simultaneously have AC as well as DC excitations. Certain large capacity AC machines are even excited by two separate three-phase AC inverters for reducing the power handling capacity of semiconductor power devices. With one mechanical input the energy balance equation becomes

$$\begin{aligned} \mathrm{d}W_f &= \sum_j e_{fj} \cdot i_j \cdot \mathrm{d}t - T_e \cdot \mathrm{d}\theta \quad \text{(rotary)} \\ \mathrm{d}W_f &= \sum_j e_{fj} \cdot i_j \cdot \mathrm{d}t - f_e \cdot \mathrm{d}x \quad \text{(linear)} \end{aligned} \tag{2.69}$$

2.9.2 Concept of Energy and Co-energy in Coupling Field

The energy stored in electrical or mechanical system is functions of the states of the electrical and mechanical variables. It can be assumed that the losses in coupling fields are zero. Nearly all the energy stored in the coupling fields is stored in the air-gap of the electromechanical device. Since air is a conservative medium, all the energy stored can be returned to the electrical or mechanical systems. The energy stored in a conservative field is a function of the state of the system variables and not of the way the variables reached that state. In a simple relay system shown in Fig. 2.16, the change in mover position brings upon consequent changes in the coil inductance and stored energy at the instant of excitation of relay coil, $\mathrm{d}x = 0$. Thus, in the absence of mechanical forces, mechanical stored energy is also zero. Thus,

$$W_m = 0$$

The mechanical stored energy is associated with the consequent displacement. The energy stored in the coupling fields during excitation of electrical circuit is equal to the energy supplied by the coupling field. Therefore, for displacement $\mathrm{d}x = 0$, we can write, the following expressions for energy by the electrical source keeping $\mathrm{d}x = 0$

$$W_f = \int \sum_j e_{\mathrm{fj}} \cdot i_j \cdot \mathrm{d}t \quad \text{for } \mathrm{d}x = 0 \tag{2.70}$$

Here e_{fj} and i_{j} are the instantaneous-induced motional EMF and instantaneous current respectively of the j-th coil in the system.

Initially, a singly excited system is assumed. Here only electromagnetic system is considered. A sudden excitation of coil brings upon changes in the electrical as well as mechanical systems. The flux linkage jumps to a much higher value causing an induced EMF in the coil. The direction of this EMF is opposite to the applied voltage. Can you reason? The induced EMF is given as

$$e_f = \left| -\frac{\mathrm{d}\lambda}{\mathrm{d}t} \right|$$

The energy supplied from electrical system can be written as

$$W_f = \int e_f \cdot i \cdot \mathrm{d}t = \int i \cdot \mathrm{d}\lambda \quad \text{for } \mathrm{d}x = 0 \tag{2.71}$$

Three single excited systems have been shown in Figs. 2.16, 2.19, and 2.20. A typical λ-I diagram for such system is given in Fig. 2.21. It is similar to that of B–H curve or magnetization curve of an electromagnetic system. The quantity W_f corresponds to the **energy** as shown in Fig. 2.21. The area W_{c} is called ***co-energy*** of the system, though it does not have any physical significance. It is given as

$$W_c = \int \lambda \cdot \mathrm{d}i \quad \text{for } \mathrm{d}x = 0 \tag{2.72}$$

For a linear (unsaturated) system,

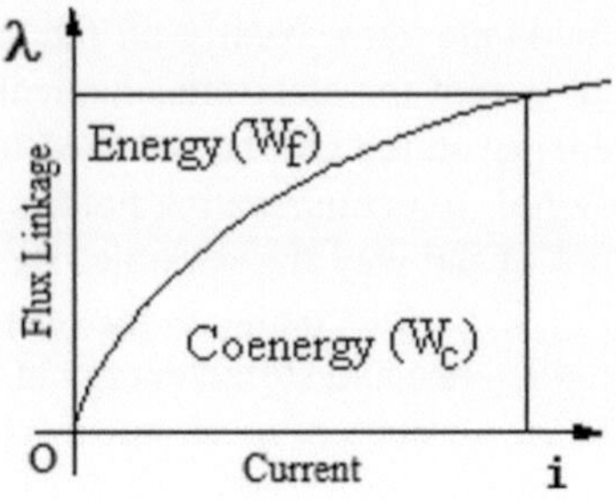

Fig. 2.21 Typical λ-i diagram of a singly excited system at $x = 0$

$$\lambda \cdot i = W_f + W_c \text{ for } \mathrm{d}x = 0 \tag{2.73}$$

It is clear that when the displacement of one of the members is taking place, it brings upon changes in energy and co-energy of the system. Thus, the energy or co-energy is dependent upon the displacement x or the state of the electromechanical system. It is convenient to write that

$$\lambda = \lambda(i, x) \tag{2.74}$$

$$W_f = W_f(i, x) \tag{2.75}$$

Here current i and displacement x are assumed to be independent variables, or in other words,

$$\mathrm{d}\lambda = \frac{\partial\lambda(i, x)}{\partial i}\mathrm{d}i + \frac{\partial\lambda(i, x)}{\partial x}\mathrm{d}x \tag{2.76}$$

Similarly, for a singly excited system, the co-energy is written as

$$W_c(i, x) = \int \lambda(i, x) \cdot \mathrm{d}i \tag{2.77}$$

With λ and x as independent variables, we can write that

$$i = i(\lambda, x) \tag{2.78}$$

$$W_f = W_f(\lambda, x) = \int i(\lambda, x) \cdot \mathrm{d}\lambda \tag{2.79}$$

It can be shown that field energy is

$$W_f = \frac{1}{2}\mathrm{BH} \times \text{Volume} = \frac{1}{2}\mathrm{BH} \times \text{Area} \times \text{Length}$$

or,

$$W_f = \frac{1}{2}\text{BH} \times \text{Volume} = \frac{1}{2}(B \times A) \times (H \times L) = \frac{1}{2}\phi\mathcal{F}$$

Here L is length of flux path and A is cross sectional area. For evaluating the co-energy with λ and x as independent variables, we get

$$\mathrm{d}i = \frac{\partial i(\lambda, x)}{\partial \lambda}\mathrm{d}\lambda + \frac{\partial i(\lambda, x)}{\partial x}\mathrm{d}x \tag{2.80}$$

Since $\mathrm{d}x = 0$, we have

$$W_c(\lambda, x) = \int \lambda \frac{\partial i(\lambda, x)}{\partial \lambda} d\lambda \tag{2.81}$$

2.10 Linear System

In a linear system, λ-I diagram is linear in nature. The electromagnetic device is not saturated or is said to operate in the linear portion of the magnetization curve. Thus, for single excited system, flux linkage can be taken as

$$\lambda(i, x) = i.L(x) \tag{2.82}$$

where $L(x)$ is the inductance of coil which is dependent on the position x, thus

$$\mathrm{d}\lambda = L(x)\mathrm{d}i \tag{2.83}$$

$$W_f(i, x) = \int_0^i i \cdot L(x)\mathrm{d}i = \frac{1}{2}L(x)i^2 \tag{2.84}$$

Similarly, we get co-energy as

$$W_c(i, x) = \frac{1}{2}L(x)i^2 \tag{2.85}$$

This is true for a linear system.

2.11 Stored Energy and Force

The force in an electromagnetic system is given as

$$F_e\mathrm{d}x = -\mathrm{d}W_m + i\mathrm{d}\lambda \tag{2.86}$$

Here Fe is electromagnetic force.

The flux linkage of the coil is

$$\lambda = \lambda(i, x) \tag{2.87}$$

$$\mathrm{d}\lambda = \frac{\partial \lambda(i, x)}{\partial i}\mathrm{d}i + \frac{\partial \lambda(i, x)}{\partial x}\mathrm{d}x \tag{2.88}$$

Considering that the coil is excited by a constant current source, the magnetic stored energy is,

$$W_m = W_m(i, x) \tag{2.89}$$

or,

$$dW_m = \frac{\partial W_m(i, x)}{\partial i}di + \frac{\partial W_m(i, x)}{\partial x}dx$$

Thus, force is given as

$$F_e\mathrm{d}x = -\frac{\partial W_m(i, x)}{\partial i}\mathrm{d}i - \frac{\partial W_m(i, x)}{\partial x}\mathrm{d}x + i\frac{\partial \lambda(i, x)}{\partial i}\mathrm{d}i + i\frac{\partial \lambda(i, x)}{\partial x}\mathrm{d}x$$

$$F_e\mathrm{d}x = \left[-\frac{\partial W_m}{\partial i} + i\frac{\partial \lambda}{\partial i}\right]\mathrm{d}i + \left[-\frac{\partial W_m}{\partial x} + i\frac{\partial \lambda}{\partial x}\right]\mathrm{d}x$$

or,

$$F_e = -\frac{\partial W_m}{\partial x} + i\frac{\partial \lambda}{\partial x} \tag{2.90}$$

Similarly, for rotary machines, the torque is

$$T_e = -\frac{\partial W_m}{\partial \theta} + i\frac{\partial \lambda}{\partial \theta} \tag{2.91}$$

For a voltage-excited system, the stored magnetic energy is

$$W_m = W_m(\lambda, x) \tag{2.92}$$

$$i = i(\lambda, x) \tag{2.93}$$

Taking a partial derivative of W_{m}, we have

$$\mathrm{d}W_m = \frac{\partial W_m(\lambda, x)}{\partial \lambda}\mathrm{d}\lambda + \frac{\partial W_m(\lambda, x)}{\partial x}\mathrm{d}x \tag{2.94}$$

$$F_e \mathrm{d}x = -\frac{\partial W_m(\lambda, x)}{\partial \lambda}\mathrm{d}\lambda - \frac{\partial W_m(\lambda, x)}{\partial x}\mathrm{d}x + i\mathrm{d}\lambda \tag{2.95}$$

$$W_m = \int i\mathrm{d}\lambda \tag{2.96}$$

2.12 Flux Linkage and Current Diagram

In actual system, two types of power supply can be there, viz., **constant current source and constant voltage source**. As far as power sources are concerned, we can have both constant current supply and constant voltage supply. In domestic sector and industrial sector, we have constant voltage supply. Any electrical load is connected to supply terminals with a series switch. A load is switched ON by connecting the switch. In certain areas, such as power distribution in an automobile, the supply may be treated as constant current supply. Each of the series connected loads has a parallel switch or contactor across it. A load is switched ON, by opening the switch and vice versa. Constant current supplies are uncommon in domestic sector and have limited scope. Motors used in dredging operation essentially undergo stalling mode of operation. These motors are operated with constant current generator or Current Source Inverters (CSI). Electric welding and battery charging circuits require constant current generators. What are possible merits and demerits of these supply systems?

In a singly excited relay circuit (Fig. 2.16) a typical B-H curve of λ-I diagram is given in Fig. 2.21. It is evident that magnetic saturation results in larger difference in stored energy or co-energy when motion takes place. It is obvious that when there is a large air-gap in the system, flux linkage–current relationship is almost linear. In absence of air-gap ($x = x_{\max}$), the magnetic circuit tends to saturate. When the coil is excited by DC source, the transition from OFF position (large air-gap or $x = 0$) to ON position (lesser air-gap or $x = x_{\max}$) takes place in a fraction of seconds. In an actual system, the current as well as flux linkage undergoes transient conditions. The coil inductance suddenly rises to a new value, a speed-induced EMF also appears in the coil, satisfying the Lenz's law. This transition may be explained in the following two ways:

(a) **Constant current transition**: As the name suggests, during constant current transition, it is assumed that the current remains constant and only changes in flux linkages are taking place. This has been shown in Fig. 2.22. Although, if coil is excited by a constant current source a momentarily change in current does take place due to the presence of speed-induced EMF in the circuit. In Fig. 2.22, shaded portion depicts the difference in co-energy of the system from armature moves from OFF position to ON position.

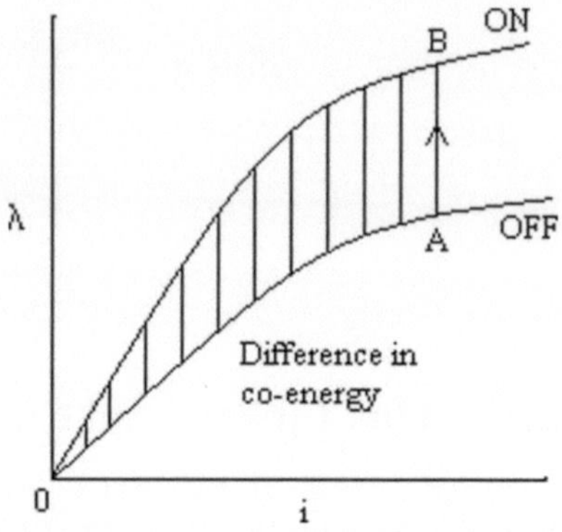

Fig. 2.22 Constant current transition

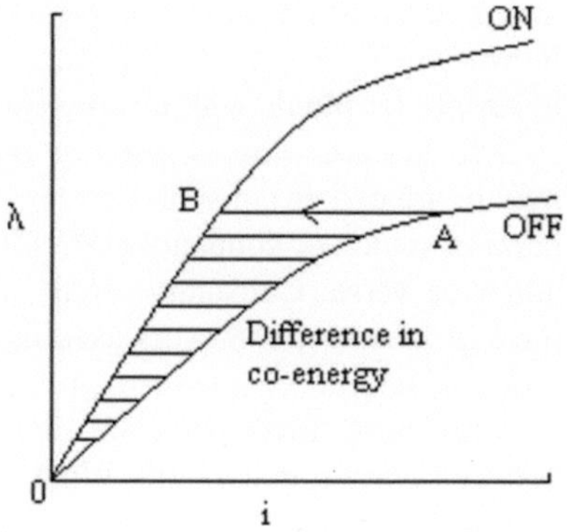

Fig. 2.23 Constant flux transition

(b) **Constant flux transition**: If the coil is excited by an ideal constant voltage source, the flux linkage is likely to remain constant during transition. This has been shown in Fig. 2.23. Here current undergoes a momentarily changes during switching ON of the device. Since supply is DC, the steady-state current remains constant and changing inductance has no effect on the steady-state current. The transient current last for a microseconds and can be visualized using storage oscilloscope.

It is clear that if force in a singly excited system is calculated using constant flux transition, the calculated values will be lesser than that of constant current transition. In actual practice, the transient state incorporates the change in energy, co-energy, inductance, current, and mechanical motion. This has been shown in Fig. 2.24. The variation of inductance with respect to x is depicted in Fig. 2.25.

2.12.1 Stored Energy in Multiply Excited System

In multi-excited system, consisting of several current carrying coils whose fluxes are linking each other, the stored energy is given as

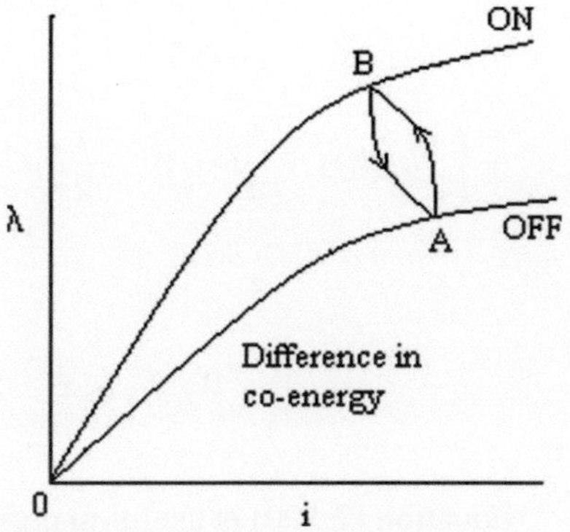

Fig. 2.24 Actual transition in a single excited system

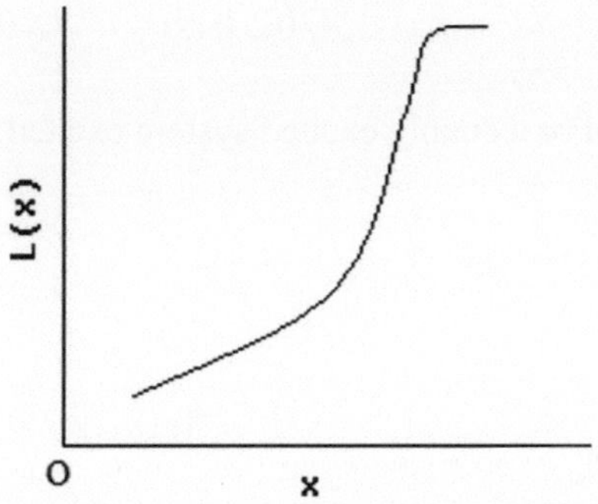

Fig. 2.25 Variation of inductance

$$W_f(i_1, i_2, i_3, \ldots, i_n, \theta) = \int_0^{i_1} \lambda_1 \mathrm{d}i_1 + \int_0^{i_2} \lambda_2 \mathrm{d}i_2 + \int_0^{i_3} \lambda_3 \mathrm{d}i_3 + \ldots + \int_0^{i_n} \lambda_n \mathrm{d}i_n \quad (2.97)$$

where the flux linkages of individual coils are given as

$$\begin{aligned} \lambda_1 &= L_{11}i_1 + L_{12}i_2 + L_{13}i_3 + \cdots + L_{1n}i_n \\ \lambda_2 &= L_{21}i_1 + L_{22}i_2 + L_{23}i_3 + \cdots + L_{2n}i_n \\ \lambda_n &= L_{n1}i_1 + L_{n2}i_2 + L_{n3}i_3 + \cdots + L_{nn}i_n \end{aligned} \quad (2.98)$$

where L_{ii} and L_{jj} are self-inductances of *i*-th and *j*-th coils, respectively, and are not equal. Further L_{ij} and L_{ji} are mutual inductances of between *i*-th and *j*-th coils, respectively, and are equal. In other words, we have

$$\begin{aligned} L_{ij} &= L_{ji} \\ L_{ii} &\neq L_{jj}. \end{aligned} \quad (2.99)$$

Thus, from Eqs. (2.97) and (2.98), we get stored energy.

$$W_f(i_1, i_2, i_3, \ldots, i_n, \theta) = \left[\frac{1}{2}L_{11}i_1^2 + \frac{1}{2}L_{22}i_2^2 + \frac{1}{2}L_{33}i_3^2 + \ldots + \frac{1}{2}L_{nn}i_n^2\right]$$
$$+ \left[\frac{1}{2}L_{12}i_1i_2 + \frac{1}{2}L_{21}i_2i_1 + \frac{1}{2}L_{13}i_1i_3 + \frac{1}{2}L_{31}i_3i_1 + \ldots + \frac{1}{2}L_{1n}i_1i_n + \frac{1}{2}L_{n1}i_ni_1\right]$$

or,

$$W_f(i_1, i_2, i_3, \ldots, i_n, \theta) = \frac{1}{2}\sum_{x,y\to 1}^{n} L_{xy}i_xi_y \tag{2.100}$$

Equation (2.100) is useful in deriving torque as shown previously,

$$T_f(i_1, i_2, i_3, \ldots, i_n, \theta) = -\frac{\partial W_f(i_1, i_2, i_3, \ldots, i_n, \theta)}{\partial \theta} \tag{2.101}$$

For a doubly excited system excited by a voltage source, we have

$$\begin{aligned} \lambda_1 &= L_{11}i_1 + L_{12}i_2 \\ \lambda_2 &= L_{21}i_1 + L_{22}i_2 \end{aligned} \tag{2.102}$$

$$W_f(\lambda_1, \lambda_2, \theta) = \int_0^{\lambda_1} i_1 \mathrm{d}\lambda_1 + \int_0^{\lambda_2} i_2 \mathrm{d}\lambda_2 \tag{2.103}$$

$$W_f(i_1, i_2, \theta) = \int_0^{i_1} \lambda_1 \mathrm{d}i_1 + \int_0^{i_2} \lambda_2 \mathrm{d}i_2 \tag{2.104}$$

Similarly, the stored energy of a current excited system is

$$W_f(i_1, i_2, \theta) = \frac{1}{2}L_{11}i_1^2 + L_{12}i_1i_2 + \frac{1}{2}L_{22}i_2^2 \tag{2.105}$$

The torque can be calculated as

$$T_f = -\frac{\partial W_f(i_1, i_2, \theta)}{\partial \theta} \tag{2.106}$$

Similarly, in a voltage-excited system, we have stored energy

$$W_f(\lambda_1, \lambda_2, \lambda_3, \ldots, \lambda_n, \theta) = \int_0^{\lambda_1} i_1 \mathrm{d}\lambda_1 + \int_0^{\lambda_2} i_2 \mathrm{d}\lambda_2 + \int_0^{\lambda_3} i_3 \mathrm{d}\lambda_3 + \cdots + \int_0^{\lambda_n} i_n \mathrm{d}\lambda_n \tag{2.107}$$

The coil currents are.

$$\begin{aligned} i_1 &= B_{11}\lambda_1 + B_{12}\lambda_2 + B_{13}\lambda_3 + \ldots\ldots + B_{1n}\lambda_n \\ i_2 &= B_{21}\lambda_1 + B_{22}\lambda_2 + B_{23}\lambda_3 + \ldots\ldots + B_{2n}\lambda_n \\ i_n &= B_{n1}\lambda_1 + B_{n2}\lambda_2 + B_{n3}\lambda_3 + \ldots\ldots + B_{nn}\lambda_n \end{aligned} \tag{2.108}$$

The stored energy of system is.

$$\begin{aligned} W_f(\lambda_1, \lambda_2, \lambda_3, \ldots\ldots\lambda_n, \theta) = &\left[\frac{1}{2}B_{11}\lambda_1^2 + \frac{1}{2}B_{22}\lambda_2^2 + \frac{1}{2}B_{33}\lambda_3^2 + \ldots + \frac{1}{2}B_{nn}\lambda_n^2\right] \\ &+ \left[\frac{1}{2}B_{12}\lambda_1\lambda_2 + \frac{1}{2}B_{21}\lambda_2\lambda_1\right. \\ &\left. + \frac{1}{2}B_{13}\lambda_1\lambda_3 + \frac{1}{2}B_{31}\lambda_3\lambda_1 + \ldots + \frac{1}{2}B_{1n}\lambda_1\lambda_n + \frac{1}{2}B_{n1}\lambda_n\lambda_1\right] \end{aligned} \tag{2.109}$$

or,

$$W_f(\lambda_1, \lambda_2, \lambda_3, \ldots, \lambda_n, \theta) = \frac{1}{2}\sum_{x,y \to 1}^{n} B_{xy}\lambda_x\lambda_y \tag{2.110}$$

The torque is

$$T_f(\lambda_1, \lambda_2, \lambda_3, \ldots, \lambda_n, \theta) = -\frac{\partial W_f(\lambda_1, \lambda_2, \lambda_3, \ldots, \lambda_n, \theta)}{\partial\theta} \tag{2.111}$$

It can be shown that for doubly excited system, we have.

$$\begin{aligned} i_1 &= B_{11}\lambda_1 + B_{12}\lambda_2 \\ i_2 &= B_{21}\lambda_1 + B_{22}\lambda_2 \end{aligned} \tag{2.112}$$

The stored energy is

$$W_f(\lambda_1, \lambda_2, \theta) = \frac{1}{2}B_{11}\lambda_1^2 + B_{12}\lambda_1\lambda_2 + \frac{1}{2}B_{22}\lambda_2^2 \tag{2.113}$$

The values of $B_{11}, B_{12}, B_{21}, B_{22}$ are obtained using elimination. The torque developed is then

$$T_f = -\frac{\partial W_f(\lambda_1, \lambda_2, \theta)}{\partial\theta} \tag{2.114}$$

Example 2.1 A magnetic circuit having three coils is shown in Fig. 2.8. It has a cross-sectional area of $1.0 \times 1.0\ \text{cm}^2$. Each limb is 1.0 cm wide and inner window dimension is $2.5 \times 2.5\ \text{cm}^2$. The coils have turns $N_1 = 10$, $N_2 = 20$, and $N_3 = 10$. They are carrying DC currents $I_1 = 1.0$ A, $I_2 = 0.5$ A, and $I_3 = 1.5$ A, respectively. The relative permeability of core is 1000. Determine flux, flux density, and magnetic field intensity in the core.

Solution

Here, the cross-sectional area of core $A = 1.0\ \text{cm}^2 = 0.0001\ \text{m}^2$

Permeability of air, $\mu_0 = 4\pi 10^{-7}$ H/m.

Relative permeability of core, = 1000.

Number of turns, $N_1 = 10$; $N_2 = 20$; $N_3 = 10$.

Currents are $I_1 = 1.0$A, $I_2 = 0.5$A; $I_3 = 1.5$A.

Now, closed line integral is

$$\oint H \cdot \mathrm{d}l = H_m \cdot L_{ab} + H_m \cdot L_{bc} + H_m \cdot L_{cd} + H_m \cdot L_{da} = H_m \cdot L_i$$

The MMF is

$$\mathcal{F} = N_1 I_1 + N_2 I_2 - N_3 I_3$$
$$\mathcal{F} = 10 \times 1.0 + 20 \times 0.5 - 10 \times 1.5$$
$$= 10 + 10 - 15 = 5\ \text{At}$$

Here, third coil produces the flux in opposite direction to that of other two coils. The net flux in core is

$$\phi = \frac{\text{MMF}}{\text{Reluctance}} = \frac{\mathcal{F}}{\mathcal{R}}$$

Length of magnetic path = $2.5 \times 4 + 0.5 \times 2 \times 4 = 10 + 4 = 14\ \text{cm} = 0.14$ m.

The reluctance of magnetic path

$$\mathcal{R} = \frac{L}{\mu_r \mu_0 A} = \frac{0.14}{1000 \times 4 \times 3.14 \times 10^{-7} \times 10^{-4}}$$

Thus, reluctance of magnetic path = 1,114,649.68 MKS unit of reluctance.

The Flux in the core is

$$\phi = \frac{\mathcal{F}}{\mathcal{R}} = \frac{5.0}{1114649.682} = 0.004485\ \text{mWeb}$$

Assuming no leakage, one can apply superposition rule with an assumption that at a time only one coil is carrying current.

Flux density in core, $B = \frac{\phi}{A} = 0.04485$ T or Weber/m^2

Magnetic field intensity in core, $H = \frac{\text{MMF}}{\text{Length}} = \frac{5.0}{0.14} = 35.71$ At per m.

Example 2.2 A toroidal core is made of three ferromagnetic materials and is equipped with a coil having 100 turns. Material ‘a’ is a nickel–iron alloy having mean arc length of 0.3 m. Material ‘b’ is a silicon steel and has a mean arc length of 0.2 m. Material ‘c’ is cast steel and has arc length of 0.1 m. Each material has a cross-sectional area of 0.001m^2.

(a) Find the MMF required to establish a magnetic flux of 6×10^{-4} Wb.

(b) Calculate the coil current.
(c) Compare the relative permeability and reluctance of each ferromagnetic material.
(d) Compute the inductance of coil.

Solution

(a) Applying Ampere's circuital law, we can write for MMF

$\mathcal{F} = \mathcal{F}_a + \mathcal{F}_b + \mathcal{F}_c = H_a L_a + H_b L_b + H_c L_c.$

Here H_{a}, H_{b}, and H_{c} are unknown quantities. These can be obtained from magnetization curves. The flux density in core is

$$B_a = \frac{\phi}{A} = \frac{0.0006}{0.001} = 0.6T = B_b = B_c$$

From the B–H curves of nickel–iron, silicon steel, and cast steel, corresponding to $B = 0.6$ T, we get

$$H_a = 10At/m \; H_b = 77At/m \quad H_c = 270At/m$$

The total MMF required is.

$$\mathcal{F} = H_a L_a + H_b L_b + H_c L_c = 10 \times 0.3 + 77 \times 0.2 + 270 \times 0.1$$
$$\mathcal{F} = 3.0 + 15.4 + 27.0 = 45.4At$$

Here, the MMF required depends on the type of material used. For flowing a specified amount of flux, larger MMF is required for cast steel. Its permeability is lesser as compared to other materials.

(b) Since the number of turns is 100. The MMF required is equal to the number of turns and current. Thus, current in the coil is

$I = \frac{\mathcal{F}}{N} = \frac{45.4}{100} = 0.454A.$

(c) The relative permeability for different core materials is calculated here.

For material 'a':

$$\mu_{ra} = \frac{B_a}{H_a \mu_0} = \frac{0.6}{10 \times 4\pi \times 10^{-7}} = 47771$$

$$\mathcal{R}_a = \frac{\mathcal{F}_a}{\phi} = \frac{3}{6 \times 10^{-4}} = 5000 \text{ MKS unit}$$

For material 'b':

$$\mu_{rb} = \frac{B_b}{H_b \mu_0} = \frac{0.6}{77 \times 4\pi \times 10^{-7}} = 6204$$

$$\mathcal{R}_b = \frac{\mathcal{F}_b}{\phi} = \frac{15.4}{6 \times 10^{-4}} = 25666.7 \text{ MKS unit}$$

$$\mu_{rc} = \frac{B_c}{H_c \mu_0} = \frac{0.6}{270 \times 4\pi \times 10^{-7}} = 1769$$

$$\mathcal{R}_c = \frac{\mathcal{F}_c}{\phi} = \frac{27.0}{6 \times 10^{-4}} = 45000 \text{ MKS unit.}$$

(d) The inductance of coil is given by

$$L = N\frac{\phi}{I} = \frac{100 \times 0.6 \times 10^{-3}}{0.454} = 0.1321 \text{ Henry}$$

Example 2.3 In a toroidal core of earlier example, find the magnetic flux produced by an applied MMF of 35At. Assume that all of the applied MMF appears across the material with the highest reluctance or cast steel having lower permeability.

Solution
Entire MMF is assumed to appear on the cast steel, thus

$$H_c = \frac{\mathcal{F}}{L_c} = \frac{35}{0.1} = 350At/m$$

Looking the B–H curve we get $B_c = 0.65$ T. One can calculate flux in the core as

$$\phi = B_c A_c = 0.65 \times 0.001 = 0.00065 \text{ Wb}$$

Since the cross-sectional area is same, thus flux density in the core will be uniform.

$$B_a = B_b = B_c = 0.65 \text{ T}$$

Example 2.4 A magnetic circuit is shown in Fig. 2.16 and has a cross-sectional area of 1.0 cm^2. The inner window is having an area of 3 × 3 square cm. The outer dimensions are 5X5 square cm. It has a finite air-gap of 2.0 mm for incorporating a semiconductor Hall effect IC for measuring DC flux density. The core is made of ferrite having a relative permeability of 10,000. For a coil having 10 turns and carrying a current of 1.0 A, find the following:

(a) The flux density in air-gap.
(b) If Hall effect sensor generates voltage of 50 mV per 1.0 Tesla. Calculate the output of the sensor.
(c) If there is no air-gap, the ratio of flux density without and with air-gap.

Solution

Cross-sectional area of core $A = 1 \times 1 = 1.0\ \text{cm}^2 = 0.0001\ \text{m}^2$

Length of path abcdef, $L_\text{i} = 4 \times 4.0 - 0.2 = 16 - 0.2 = 15.8\ \text{cm} = 0.158\ \text{m}$.

Length of air-gap, $L_\text{g} = 0.2\ \text{cm} = 0.002\ \text{m}$.

Permeability of air, $\mu_0 = 4 \times 3.14 \times 10^{-7}$ H/m.

Relative permeability of core = 10,000.

Number of turns = 10.

Current, I = 1.0 A.

MMF of coil, $\mathcal{F} = N.I = 10 \times 1.0 = 10$ At.

(a) If the flux flowing through it is ϕ, then the reluctance of core,

$$\mathcal{R}_i = \frac{Li}{\mu_r \mu_0 A} = \frac{0.158}{10000 \times 4\pi 10^{-7} \times 0.0001}.$$
$$= 1.25796 \times 10^{+5}\ \text{MKS unit}$$

The reluctance of air-gap,

$$\mathcal{R}_g = \frac{L_g}{\mu_0 A} = \frac{0.002}{4\pi 10^{-7} \times 0.0001}$$
$$= 1.5923566 \times 10^{+7}\ \text{MKS unit}$$

We observe that the reluctance of air-gap is much higher than that of core. The total reluctance is sum of the two reluctances $\mathcal{R}_i$ and $\mathcal{R}_g$. This is analogous to series connections of two resistances in an electrical circuit. Thus, total reluctance is

$$\mathcal{R} = \mathcal{R}_i + \mathcal{R}_g$$
$$= 1.25796 \times 10^{+5} + 1.592366 \times 10^{+7}$$
$$= 1.604936305 \times 10^{+7}\ \text{MKS unit of Reluctance}$$

The flux in the core is to be given as

$$\phi = \frac{\mathcal{F}}{\mathcal{R}} = \frac{10}{1.604936305 \times 10^{+7}}$$
$$= 6.23 \times 10^{-7}\ \text{Web.}$$

The Flux density in core is,

$$B = \frac{\phi}{A} = \frac{6.23 \times 10^{-7}\ \text{Web}}{0.0001\ \text{m}^2} = 0.00623077\ \text{Tesla}$$

(b) Hall effect sensor gives 50 mV per 1.0 Tesla. Thus, output of Hall effect device is B × 50 mV = 0.00623077 × 50 = 0.31 mV.

(c) In absence of air-gap, length of flux path in ferrite, Li = 0.16 m. Reluctance of magnetic circuit is

$$\mathcal{R} = \frac{L_i}{\mu_r \mu_0 A} = \frac{0.16}{10000 \times 4\pi 10^{-7} \times 0.0001} = 127388.53 \text{ MKS unit}$$

Thus new flux.

$$\phi = \frac{\mathcal{F}}{\mathcal{R}} = \frac{10}{127388.53} = 0.0000785 \text{ Web}$$

The flux density in core

$$B = \frac{\phi}{A} = \frac{0.0000785}{0.0001} = 0.785\text{T}$$

The presence of air-gap in magnetic circuit substantially reduces the flux and flux density in core. The ratio of flux density in two cases, i.e., in the absence and in the presence of airgap, is

$$\text{Ratio} = \frac{0.785}{0.00623077} = 125.98$$

The presence of air-gap in the path of the flux reduces the flux density in the core, increases reluctance and decreases inductance.

Example 2.5 A simple magnetic circuit with an air-gap is shown in Fig. 2.3a. The mean length of the flux path is 55 cm (iron) and 0.2 cm (air-gap). Cross-sectional area of core is 25 cm^2. How many turns should the exciting coil have in order to establish a flux density of 1.0 Wb/m^2 in the air-gap? The coil current is 20 A and core is made of (i) silicon steel 1.0 T at 200 At/m (ii) ferrite core having relative permeability of 20,000.

Solution
Here $B = 1.0$ Wb/m^2. The MMF can be given as

$$\mathcal{F} = \mathcal{F}_{\text{air}} + \mathcal{F}_{\text{steel}}$$

$$\mathcal{R}_{\text{air}} = \frac{2 \times 10^{-3}}{4 \times \pi \times 10^{-7} \times 25 \times 10^{-4}} = 6.36 \times 10^5 \text{ At/wb}$$

The total flux is

$$\phi = BA = 1.0 \times 25 \times 10^{-4} = 2.5 \times 10^{-3} \text{ wb}$$

$$\mathcal{F}_{\text{air}} = \phi \mathcal{R}_{\text{air}} = 6.36 \times 10^5 \times 2.5 \times 10^{-3} = 1590 \text{ At}$$

Given that $B = 1.0$ Wb/m^2 and $H = 200$ At/m. Thus, for iron length of 55 cm, we have MMF of iron path

$$\mathcal{F}_{\text{iron}} = HL_{\text{iron}} = 200 \times 55 \times 10^{-2} = 110\,\text{At}$$

or,

$$NI = 1590 + 110 = 1700\,\text{At}$$

$$N = \frac{1700}{20} = 85\,\text{turns}$$

For a ferrite core of $\mu_r = 20{,}000$, the magnetic field intensity corresponding to 1.0 T is,

$$H = \frac{1.0}{4\pi \times 10^{-7} \times 2 \times 10^4} = 39.8\,\text{At/m}$$

$$\text{MMF} = 39.8 \times 55 \times 10^{-2} = 21.89$$

$$\text{Total MMF} = 1590 + 21.89 = 1611\,\text{At}$$

$$N = \frac{1590 + 22}{20} = 80.6 = 81\,\text{turns}$$

Example 2.6 An elementary reluctance machine is shown in Fig. 2.20. The machine is singly excited with $i_2 = 0$. The rotor is free to rotate. The inductance of the winding with respect to rotor position is given as.

$$L(\theta) = L_1 + L_2 \cos 2\theta$$

Calculate torque on the rotor when rotor is excited by alternating current.

Solution

Let the excitation current be,

$$i = I_m \sin \omega t$$

The stored energy and flux linkage are,

$$W_m = \frac{1}{2} L(\theta) i^2$$

$$\lambda(\theta) = L(\theta) i$$

The developed torque can be given as

$$T_e = -\frac{\partial W_m}{\partial t} + i\frac{\partial \lambda}{\partial \theta}$$

or,

$$T_e = -\frac{1}{2}i^2\frac{\partial L}{\partial \theta} + i^2\frac{\partial L}{\partial \theta}$$

$$T_e = \frac{1}{2}i^2\frac{\partial L}{\partial \theta}$$

$$T_e = -I_m^2 L_2 \sin 2\theta \sin^2 \omega t$$

Let rotor speed be ω'. The rotor position θ is given as

$$\theta = \omega' t - \delta$$

Angle δ is the rotor position at time $t = 0$. Thus, developed torque is

$$T_e = -\frac{1}{2}I_m^2 L_2\left[\sin 2(\omega' t - \delta) - \frac{1}{2}\{\sin 2(\omega' t + \omega t - \delta) + \sin 2(\omega' t - \omega t - \delta)\}\right]$$

If rotor speed ω' is same as angular frequency ω, then for $\omega' = \omega$, the time average value of torque is

$$T_{av} = \frac{1}{4}I_m^2 L_2 \sin 2\delta$$

Example 2.7 In the relay circuit of Fig. 2.26, find the expression for the magnetic force existing on the relay armature. Assume cross-sectional area and air-gap flux. Neglect the reluctance of iron.

Solution
Let the cross-sectional area of core be A, path of flux in iron be L_i and that of air be g, flux be ϕ.

The reluctance of the air-gap is

$$\mathcal{R} = \frac{g}{\mu_0 A}$$

Fig. 2.26 A Relay system

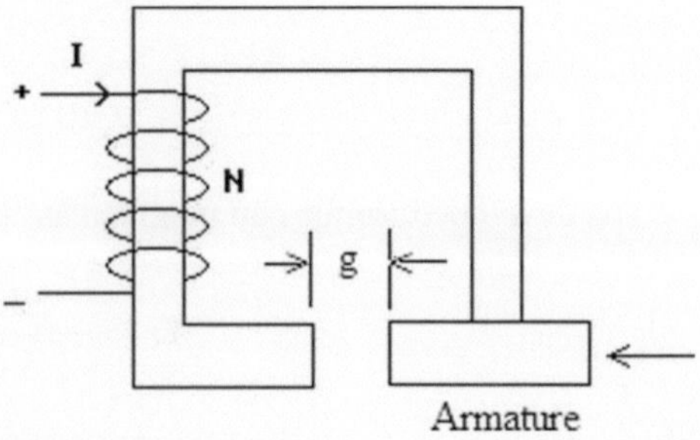

Since force is given by,

$$F = -\frac{1}{2}\phi^2 \frac{d\mathcal{R}}{dx}$$

Thus,

$$\frac{d\mathcal{R}}{dx} = \frac{1}{\mu_0 A}\frac{dg}{dx}$$

Taking, $dg = -dx$, we get,

$$\frac{d\mathcal{R}}{dx} = -\frac{1}{\mu_0 A}$$

Thus, force on the moving armature is

$$F = \frac{1}{2}\frac{\phi^2}{\mu_0 A} = \frac{B^2}{2 \cdot \mu_0}A$$

Example 2.8 In the plunger magnet shown in Fig. 2.27, show that force on it is

$$F = \pi a \mu_0 \frac{\mathcal{F}^2 h^2}{g}\left[\frac{1}{x+h}\right]^2 = 3.94 \times 10^{-6}\frac{a.\mathcal{F}^2 h^2}{g}\left[\frac{1}{x+h}\right]^2$$

Solution
For the plunger magnet, the reluctance of the path of flux is

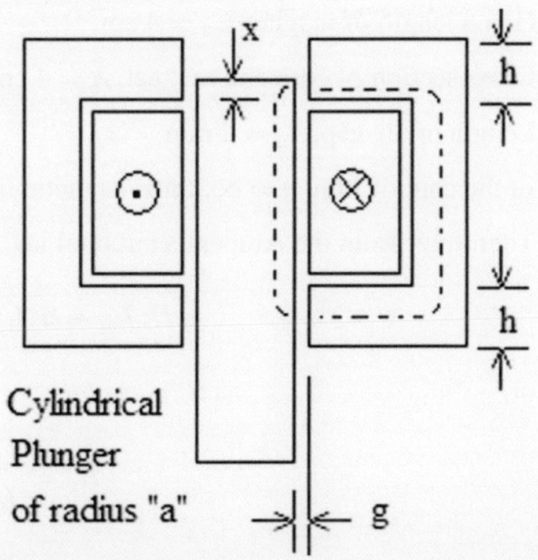

Fig. 2.27 Plunger magnet

$$\mathcal{R} = \frac{g}{2\pi ah\mu_0} + \frac{g}{2\pi ax\mu_0} = \frac{g}{2\pi a\mu_0}\left[\frac{1}{h} + \frac{1}{x}\right]$$

$$\frac{\partial \mathcal{R}}{\partial x} = \frac{g}{2\pi a\mu_0}\left[-\frac{1}{x^2}\right]$$

The flux is

$$\phi = \frac{\mathcal{F}}{\mathcal{R}} = \frac{\mathcal{F}}{g/(2\pi a\mu_0)}\frac{xh}{(h+x)}$$

The force on plunger is

$$F = -\frac{1}{2}\phi^2\frac{\mathrm{d}\mathcal{R}}{\mathrm{d}x}$$

Thus, we get

$$F = \pi a\mu_0\frac{\mathcal{F}^2h^2}{g}\left[\frac{1}{x+h}\right]^2 = 3.94\times 10^{-6}\frac{a\mathcal{F}^2h^2}{g}\left[\frac{1}{x+h}\right]^2$$

Example 2.9 A magnetic circuit is made using high permeability material. A section of magnetic circuit (Refer Fig. 2.12) is made using permanent magnet Alnico $B_\mathrm{m} = 0.96$ T and $H_\mathrm{m} = 42$ kA/m (Refer Fig. 2.9) having cross-sectional area of 2 cm × 2 cm and length 4 cm. The inner window of core is a square of 4 cm × 4 cm. An air-gap of 2 mm is present in the magnetic circuit. Determine the flux density in the air-gap.

Solution
Given length of magnet $L_\mathrm{m} = 4$ cm.

Cross section of core and magnet, $A = 4\ \mathrm{cm}^2$

Length of air-gap, $l_\mathrm{g} = 2$ mm.

In the core of iron, $\mu = \infty$, thus magnetic field intensity in iron core is $H_\mathrm{i} = 0$.

Therefore, from the Ampere's circuital law

$$-H_mL_m + H_gL_g + H_iL_i = 0$$

or,

$$H_g = \frac{L_m}{L_g}H_m$$

$$H_g = \frac{40}{2} \times 42 = 840\text{kAt/m}$$

where H_g and H_m is the magnetic field intensity in air-gap and maximum magnetic field intensity of magnet, respectively.

The flux in the iron core and air-gap, $\phi = B_m A_m = B_g A_g$.

Thus, the flux density in airgap,

$$B_g = \mu_0 H_g = 4\pi 10^{-7} \times 840 \times 10^3 = 1.055T$$

Example 2.10 Design the dimension of permanent magnet which can provide an air-gap flux density $B_g = 0.45$ T for area of air-gap $A_g = 2.5$ cm^2 and length of air-gap $l_g = 0.2$ cm. The magnet is formed using (a) Alnico with operating point 0.96 T at 42 kA/m. (b) Rare earth cobalt, 0.5 T at 300 kA/m.

Solution

Given, $B_g = 0.45$ T, $A_g = 2.5$ cm^2 and $l_g = 0.2$ cm.

(a) For Alnico $B_m = 0.96$ T and $H_m = 42$ kA/m

Assuming linear demagnetization curve and infinite permeability of surrounding medium,

$$B_m l_m = B_g l_g$$

This gives length of magnet,

$$l_m = \frac{B_g l_g}{B_m} = \frac{B_g l_g}{\mu_0 H_m} = \frac{0.45 \times 0.02}{4\pi 10^{-7} \times 42000} = 0.17\text{m}$$

The flux coming from permanent magnet

$$\phi = B_m A_m = B_g A_g$$

Thus

$$A_m = \frac{B_g A_g}{B_m} = \frac{0.45 \times 2.5 \times 10^{-4}}{0.96} = 1.17 \times 10^{-4}\text{m}^2 = 1.17\ \text{cm}^2$$

(b) Rare earth cobalt, 0.5 T at 300 kA/m

The length of magnet,

$$l_m = \frac{B_g l_g}{B_m} = \frac{B_g l_g}{\mu_0 H_m} = \frac{0.45 \times 0.02}{4\pi 10^{-7} \times 300000} = 0.024\ \text{m}$$

The area of magnet,

$$A_m = \frac{B_g A_g}{B_m} = \frac{0.45 \times 2.5 \times 10^{-4}}{0.5} = 2.25 \times 10^{-4}\,\text{m}^2 = 2.25\,\text{cm}^2$$

Example 2.11 An Alnico magnet is manufactured in the form of a toroid with mean diameter of 5 cm and a cross-sectional area of 1 cm^2. An air-gap is provided in the flux path. Determine the length of air-gap so as to maintain maximum energy in the air-gap.

Solution

The point corresponding to maximum energy in Alnico magnet is 0.96 T at 42 kA/ m.

$$\text{Average length of flux path} = l_m + l_g = \pi d = 3.14 \times 0.05 = 0.157\,m \approx l_m$$

Ignoring fringing and leakage, equating MMF we get

$$l_g = \frac{\mu_0 l_m H_m}{B_m} = \frac{\mu_0 (0.157 - l_g) \times 42000}{0.96}$$

$$l_g = \frac{\mu_0 l_m H_m}{B_m} = \frac{\mu_0 0.157 \times 42000}{0.96}$$

This gives the length or air-gap as 8.17 mm or 8.63 mm.

Example 2.12 An Alnico magnet has following details.

$$l_m = 4\,\text{cm};\ A_m = 1.0\,\text{cm}^2;\ l_g = 0.7\,\text{cm};\ A_g = 2.0\,\text{cm}^2$$

Determine the flux density in air-gap considering negligible leakage and fringing.

Solution

The operating point of Alnico is considered as 0.96 T at 42 kA/m.

The flux coming from the magnet, $\phi = B_m A_m = 0.96 \times 10^{-4}$ Wb.

The flux density in air-gap.

$$B_g = \frac{\phi}{A_g} = \frac{0.96 \times 10^{-4}}{2 \times 10^{-4}} = 0.48T$$

Example 2.13 A typical electromagnetic contactor has been shown in Fig. 2.28. Such types of contactors are widely used for control of electrical signals or currents and are made with capability of multiple circuits ON/OFF. The contactors may be normally closed (NC) or normally open (NO). Primary contactors are used for controlling bulk main power with the low voltage low power coil. The auxiliary contactor is used in control, automation, and relying. The cross-sectional area of core is 1.0 cm × 1.0 cm, air-gap is 5 mm. The coil consists of 360 numbers of turns and is carrying 0.5 A. Determine the magnetic flux in air-gaps, assuming the permeability of core to be infinity.

Solution

Given $l_g = 0.005$ m; $A_c = 10^{-4}$ m^2. The equivalent magnetic circuit is shown in Fig. 2.29

The reluctance of air-gap

$$\mathcal{R} = \frac{l_g}{\mu_{0.A_c}}$$

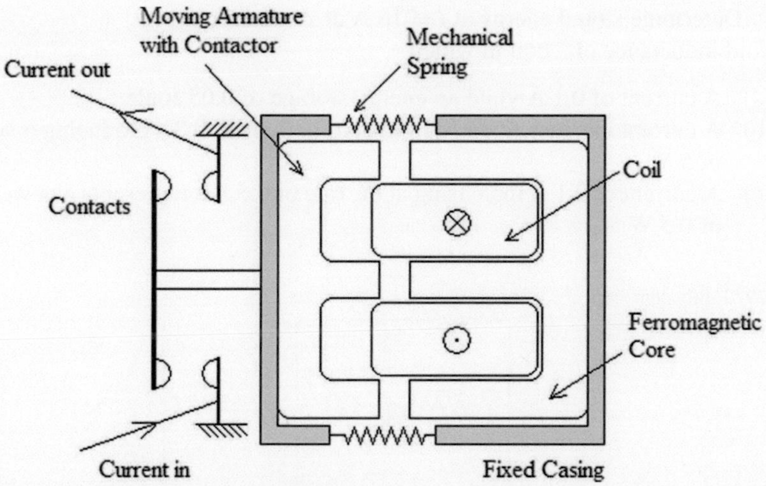

Fig. 2.28 Typical electromagnetic contactor

$$= \frac{0.005}{4\pi 10^{-7} 10^{-4}} = 3.98 \times 10^{7}$$

The equivalent magnetic reluctance

$$\mathcal{R}_{eq} = 3.98 \times 10^{7} \times \frac{1}{3} = 1.327 \times 10^{7} \text{ MKS unit}$$

The MMF

$$\mathcal{F} = NI = \mathcal{R}_{eq}\phi_1$$

Thus, flux in the mid-air-gap

$$\phi_1 = \frac{\text{NI}}{\mathcal{R}_{\text{eq}}} = \frac{360 \times 0.5}{1.327 \times 10^{7}} = 13.6 \times 10^{-6} \text{ wb}$$

The flux density in mid-air-gap, $\mathrm{B}_1 = 13.6 \times 10^{-6}/10^{-4} = 13.6 \times 10^{-2} = 0.136$ T
The flux in outer two air-gaps, $\phi_2 = \phi_1/2 = 6.8 \times 10^{-6}$ Wb
The flux density in outer two air-gaps $B_2 = 0.068$ T
The equivalent magnetic circuit is shown in Fig. 2.29.

Answer

1. An electromagnetic relay has the current and flux linkage, and armature positions

$$i = \lambda^2 + 2\lambda(x - 1)^2$$

   Determine stored energy at $i = 10$ A at $x = 0.25$ cm.
2. Find inductance of a coil in which
   (a) A current of 0.1 A yield an energy storage of 0.05 Joule
   (b) A current increase linearly from zero to 0.1A in 0.2 s producing a voltage of 5 V.
   (c) A current of 0.1 A increasing at the rate of 0.5 A/s represents a power flow of 0.5 W.

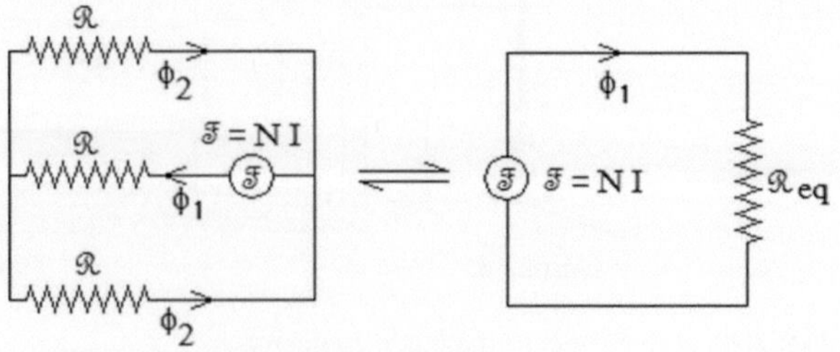

Fig. 2.29 Equivalent magnetic circuit

(d) A current of 100 A yields energy storage of 100 Joule.

3. Show that in a single excited system, force is

$$F = -\frac{1}{2}\phi^2 \frac{d\mathcal{R}}{dx}$$

4. The magnetic circuit shown in Fig. 2.30 is carrying a single current carrying conductor passing through its window area. The core is made of ferrite having relative permeability of 10,000. The cross section of core is $1.0 \times 1.0\ \text{cm}^2$. This arrangement is especially used as Hall effect current sensor. If current flowing through conductor is 50 A, find the core flux density.
5. A DC contactor is made with 1000 turn coil wound on the center limb of E-core shown in Fig. 2.31. The contactor is required to exert a pull of 4.45 N over a gap of 0.635 cm. Find ampere turns required in the coil, flux density in central limb, and force of attraction. The cross-sectional areas of the three pole faces are 6.45, 12.9, and 6.45 cm^2. Choose the central pole face of square cross section and $b = 3a$. Relative permeability of iron 350.
6. An electromagnet used in relay coil having coil resistance of 6 Ω is excited by 20 V. Draw expected current variation when armature is pulled toward core and when armature is drawn away from the coil by external force. What will be steady-state current?
7. A magnetic core shown in Fig. 2.6 is carrying a coil having 10 turns. The arrangement is used for measuring DC current in non-contact manner. A Hall effect IC is inserted in the 2.0 mm slit of the magnetic core, which gives an output of 50 mV per Tesla. If the coil is carrying 100 A of DC current, calculated the output of the Hall effect IC. The B–H curve of the core is given in (a) Fig. 2.32 and (b) Fig. 2.33. Assume that there is no fringing and Hall effect probe is non-magnetic.
8. An E–I core electromagnet is shown in Fig. 2.31 where $a = 1.0$ cm and $b = 3.0$ cm. The central pole face is square. The coil is carrying 100 turns and 1.0 A current. Assume relative permeability of core as 500. Calculate flux density in the central limb and force of attraction experienced by the I-piece.

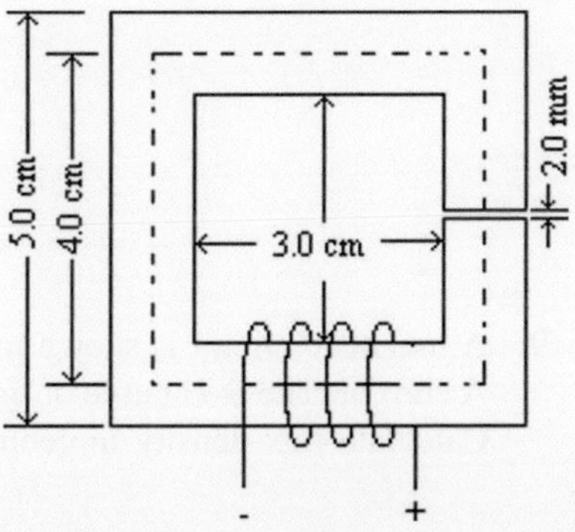

Fig. 2.30 Magnetic circuit with gap

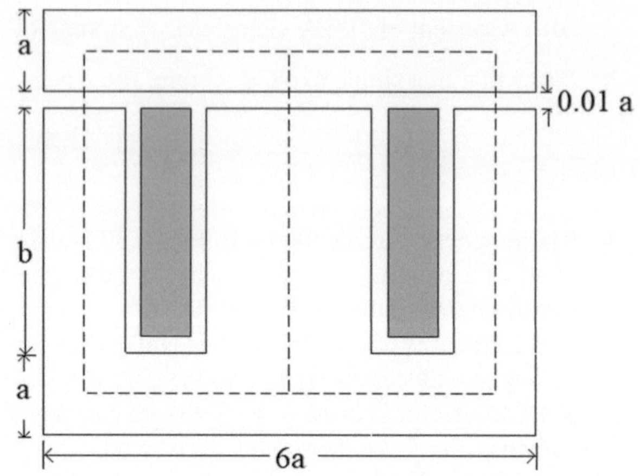

Fig. 2.31 E-core electromagnet

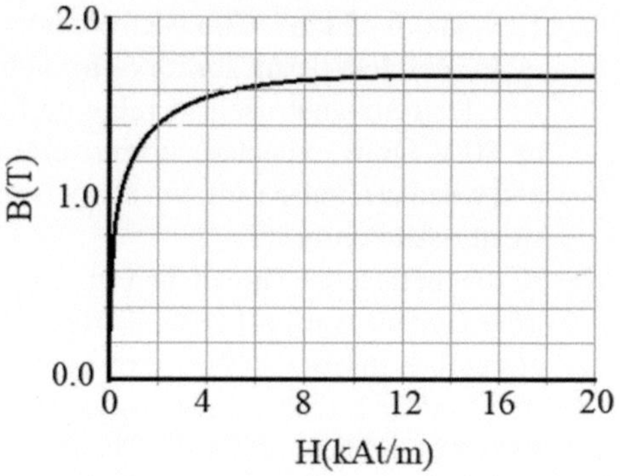

Fig. 2.32 B–H curve for material -1

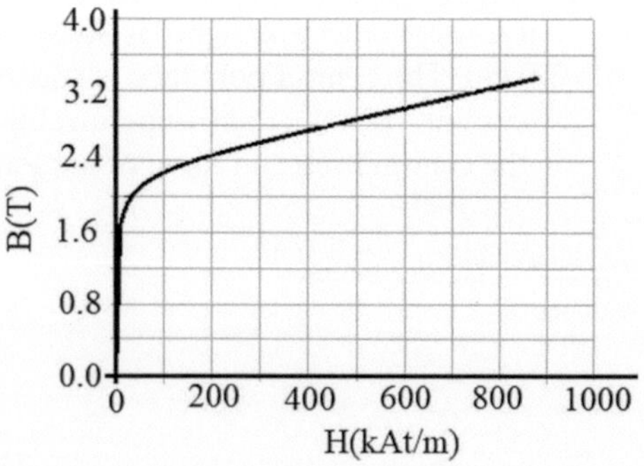

Fig. 2.33 B–H curve for material-2

9. A magnetic circuit is shown in Fig. 2.34. A coil of 1400 turns is carrying 1.0 A current. The B-H curve of the core is given in (a) Fig. 2.32 and (b) Fig. 2.33. Calculate flux density in central limb and force experienced by the plunger.

Assume that main air-gap is of 2.0 mm and side air-gaps are negligible. The cross-sectional area of core is 1.0 cm^2.

10. Experimental controlled magnet-type levitator is shown in Fig. 2.35. The coil has 1400 turns. The B–H curve of the core is given in (a) Fig. 2.32 and (b) Fig. 2.33. If the fringing and leakage allow only 50% of flux in air-gap as compared to that of core, calculate coil current when flux density in air-gap is 0.8 T. The gap between core is 2 mm. Also, compute force of attraction. Assume square cross section and ignore reluctance of flux path in fringing and leakage.
11. The B-H curve of a magnetic material can be expressed as

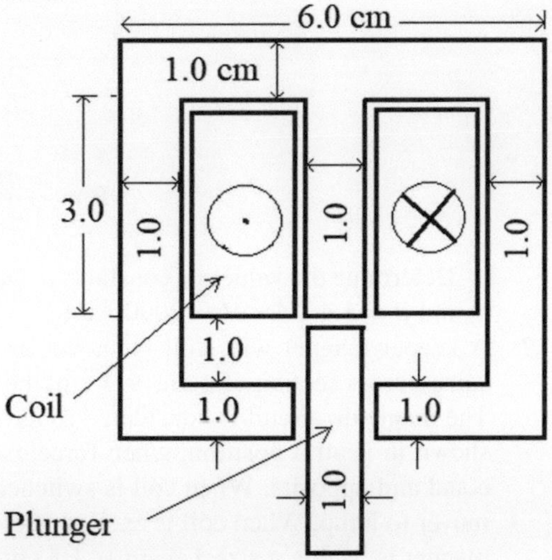

Fig. 2.34 Magnet circuit with plunger

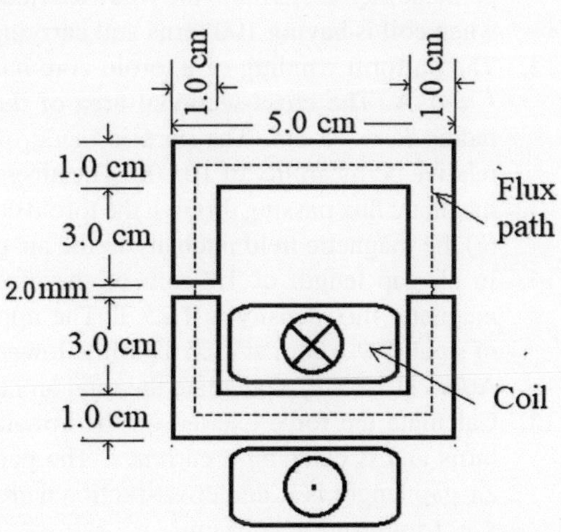

Fig. 2.35 Controlled levitation and guidance magnet

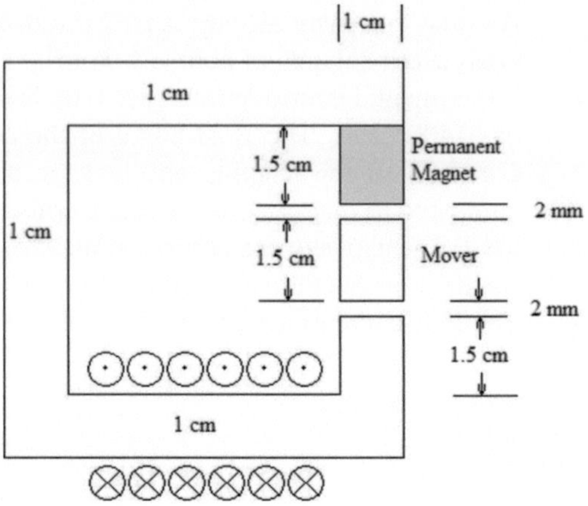

Fig. 2.36 Keeper magnet with coil

$$B = \frac{aH}{1 + bH}$$

Determine the values of constants a and b. Given $B = 1.1$ T for $H = 750$ At/m and $B = 1.4$ T for $H = 2000$ At/m.

12. A keeper magnet with coil is shown in Fig. 2.36. The magnetic circuit has square cross section of 1 cm × 1 cm. the inner window is a square of 4.9 cm. The magnetic circuit is considered to have infinite permeability. The mover is shown in neutral position, when force exerted by coil and that of magnet are equal and opposite. When coil is switched OFF, the magnetic forces make the mover to jump. When coil is excited, the mover is away from the magnet. The magnet is having a size 1.0 cm × 1.5 cm × 1.0 cm. Choose B_r, H_c and recoil permeability. Determine the force exerted on the mover by magnet and by coil when coil is having 100 turns and carrying 1 A current.
13. The uniform winding of a toroid core has 100 turns carrying a direct current $I = 1$ A. The cross-sectional area of the core is 10 cm^2. It has an average radius $R = 15$ cm. The core is made of a ferromagnetic material having a relative permeability of 10,000. The air-gap length is 0.1 mm. Calculate (a) the magnetic flux passing through the toroid core, (b) the inductance of the winding, (c) the magnetic field intensity in the air-gap and ferromagnetic material.
14. In air-gap length of 1.0 mm of the electromagnet shown in Fig. 2.34, the magnetic flux density is 1.25 T. The upper core of magnetic circuit is made of steel (9.9 At/cm at 1.25 T) while lower armature is made of cast steel (14.4 At/cm at 1.25 T). Calculate the total ampere turns, if core flux is 1.0 mWb.
15. Calculate the force exerted on the armature of the electromagnet, coil has N turns and is carrying a current i. The permeability of the core is infinity, the air-gap length is x, and cross-sectional area of the limb is A. $N = 1000$, $i = 1$, $x = 1$ mm, and $A = 25$ cm^2.

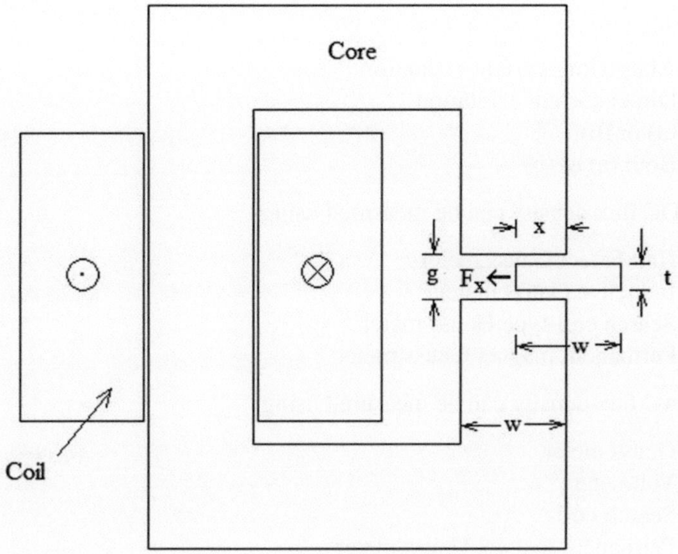

Fig. 2.37 Electromagnet with a gap

16. A magnetic circuit is shown in Fig. 2.37. The coil is carrying N number of turns and carrying dc current i. A ferromagnetic plate of thickness 't' and width 'w' is introduced into the air-gap of length 'g', over a distance x. Show that the force is

$$F_x = \frac{\mu_0 N^2 i^2 tw}{2g(g - t)}$$

Objective Questions

1. In PMBLDC motor, permanent magnet rotor position is judged using
   (a) inductive proximity sensor
   (b) capacitive proximity sensor
   (c) Hall effect sensor
   (d) resistive proximity sensor
2. The motor driving cooling fan of a personal computer is
   (a) Reluctance synchronous motor
   (b) Switched Reluctance Motor
   (c) PMBLDC motor
   (d) Stepper motor

3. An eddy current brake is working on
   (a) Alternating current excitation
   (b) Direct current excitation
   (c) (a) or (b)
   (d) Both (a) & (b)
4. The DC flux density can be measured using
   (a) Hall Effect Gauss meter
   (b) Inductive Gauss meter
   (c) Search coil-type Gauss meter
   (d) Permanent magnet Gauss meter
5. The AC flux density can be measured using
   (a) Gauss meter
   (b) Voltmeter
   (c) Search coil
   (d) Permanent magnet Gauss meter
6. The assumption of linear magnetic circuit is valid when
   (a) Machine is saturated
   (b) Machine is not saturated
   (c) Machine is working at knee point of its magnetization curve
   (d) All the above
7. The hysteresis and eddy current losses in DC machine will be
   (a) Zero, because the supply frequency $f = 0$ Hz
   (b) Machine is subjected to rotating magnetization having frequency $0.5 \times p \times n$, where p is number of poles and n is rotation per second
   (c) Machine is subjected to rotating magnetization having frequency $p \times n$, where p is number of poles and n is rotation per second
   (d) Finite losses independent of frequency and rotation per second
8. Non-contact speed measurement can be done using
   (a) Magnetic pick-up with spoke wheel
   (b) Stroboscope
   (c) PM and Hall effect sensor
   (d) All the above
9. Following permanent magnet having large remanent magnetic flux density (B_r)
   (a) Ferrite
   (b) Alnico
   (c) Sm–Co
   (d) Ni–Fe–B

10. The output of search coil placed in an AC magnetic field is

    (a) Flux
    (b) Flux density
    (c) Rate of change of flux linkage
    (d) Flux linkage

11. Magnetic circuit is analogous to electric circuit except

    (a) There is no magnetic insulator.
    (b) Current cannot flow through air
    (c) Core loss cannot be represented by resistance
    (d) None

12. Storage of electromagnetic energy is in

    (a) Resistance and inductance
    (b) Capacitance and inductance
    (c) Resistance and capacitance
    (d) Resistance, inductance, and capacitance

References

1. Say, M.G.: Alternating Current Machines. 4th ELBS edn (1976)
2. A. Fransua, R. Magureanu, Electrical Machines and Drive Systems, TP Technical Press, Oxford, 1984
3. Clayton & Hancock: Performance & Design of Direct Current Machines, ELBS
4. Del Toro, V.: Electric Machines and Power Systems, PHI (1980)
5. Chapman, S.J.: Electric Machinery Fundamentals, Indian edn, 4e. McGraw Hill Education (Indi) Pvt Ltd (2005)
6. Siskind: Electrical Machines, 2nd edition International Student edn. McGraw Hill International Book Company, 12th printing (1983)

Chapter 3
Transformers

Abstract This chapter on transformer deals with the different types, classifications, constructional features, cooling methodology, and basic EMF equations. The assumption for developing equivalent circuit and phasor diagram from test results, determination of equivalent circuit from self-inductance and mutual inductance and vice versa, hysteresis and eddy current losses, calculation of energy and power efficiency, % regulation, load sharing in parallel-connected transformer, concept of polarity, and auto transformer have been done. Three-phase transformer connections, Scott-connected transformer and inrush current, etc., have been dealt.

Keywords Core type · Shell type · Dry transformer insulation · Magnetic circuit · CRGOS · Cooling · Bucholz relay · Equivalent circuit parameters · Efficiency-power and all day · Polarity · Phase divergence · Multiratio and Multiwinding transformer · Regulation · Parallel operation and load sharing · Heat run test · Tappings · Autotransformer · Inrush current · Three-phase transformer · Scott connection · Instrument transformer

3.1 Introduction

A transformer is an electromagnetic device, consisting of two or more electrical circuits interlinked by a common magnetic circuit for the purpose of transferring electrical energy between them. Transformers do not have any rotating part. Their size ranges from very small one, such as pulse transformer (used in firing of Silicon Controlled Rectifier (SCR)/thyristor), input/output transformers used in radio, TV, and other communication sets (used for impedance matching) to some of the largest single units of electrical equipment ever manufactured, like large power transformers for transmitting several thousand MVA power over larger distances. Based on supply frequency, transformers can be classified into the following groups [1–4]:

(a) Transformers having frequency of 400 Hz and higher are useful in air craft and space applications.

R. K. Srivastava, *Electrical Machines*, https://doi.org/10.1007/978-981-96-5039-2_3

(b) Power transformers having frequency 50–60 Hz are used in industrial & domestic sectors.
(c) Audio transformers having frequency ranging from 20 Hz to 20 kHz.
(d) Radio transformers having frequency ranging from 20 kHz and above.
(e) Pulse transformer.

Alternatively, the transformers can also be classified based on their usage:

(a) **Power Transformer**: These transformers are meant for stepping-up or stepping-down alternating current voltage levels in a large power system for transmission purpose.
(b) **Distribution Transformer**: They convert high-voltage alternating current electricity into lower potentials.
(c) **Testing Transformer**
(d) **Instrument Transformer**: Instrument transformers, like potential transformer (PT) and current transformer (CT) current transformer for measuring high-voltage and high-current respectively.
(e) **Isolation-Two-Winding Transformer**: These transformers have 1:1 turns ratio and are suitable for electrical isolation of input and output.

When the electrical power is transmitted over large distance at higher voltage, it results in lesser voltage drop and lesser Ohmic losses, normally referred to as line-losses. This is explained in Example 3.1.

Example 3.1 It is desired to receive 20 kW at unity power factor at terminal voltage of either 220 V single phase or 1 kV single phase from a generating station 100 km apart. Determine the voltage drop, line losses, and the generating voltages required in both the cases, if the transmission line parameters are $0.05 + j\,0.05$ Ohm/km.

Solution

Here, line current for 220 V is

$$I_1 = \frac{20 \times 1000}{220 \times 1.0} = 90.9 \text{ A}$$

The line current for 1 kV is

$$I_2 = \frac{20 \times 1000}{1000 \times 1.0} = 20 \text{ A}$$

The voltage drop due to currents I_1 and I_2 are

$$I_1.Z = 100 \times (0.05 + j0.05) \times 90.90 = 454.5 + j454.5 = 642.8\angle 45^\circ \text{ V}$$

and

$$I_2.Z = 100 \times (0.05 + j0.05) \times 20 = 100 + j.100 = 141.42\angle 45^\circ \text{ V}$$

Line losses for the two cases are,

$$I_1^2 R = 90.90 \times 90.90 \times 0.05 \times 100 = 4131405 \text{ W} = 41.314 \text{ kW}$$
$$I_2^2 R = 20 \times 20 \times 0.05 \times 100 = 2000 \text{ W} = 2.0 \text{ kW}$$

Thus, input voltages at generator terminals are

$$V_{g1} = 220 + 454.5 + 454.5j = (670.45 + 450.45j) = 813.33\angle 33.97^\circ \text{ V}$$

and

$$V_{g2} = 1000 + 100 + 100j = (1100 + 100j) = 1104.5\angle 5.19^\circ \text{ V}$$

Here one can notice that the voltage drop as well as line losses are higher when power is received at lower voltage. ***A unit of electrical energy saved is equal to a unit of electrical energy generated***. Thus, if electrical power is transmitted over long distances at higher voltage, the line voltage drop and line loss can be limited.

A common solution to reduce voltage drop and line loss is to introduce transformers between the generator and the transmission lines to boost the transmitting voltage to a level where the associated line losses are manageable. This requires step-up transformers. At the receiving end of the transmission lines, transformers are again required to reduce the voltages at the substation to more reasonable levels. A further reduction in the voltage is done for distributing the power. Throughout the process transformers are used in a three-phase mode to provide the required alternating current voltage transformation.

The maximum transmission voltage in our country is three-phase 765 kV AC. Internationally. three-phase 1200 kV transmission voltages are mostly used. HVDC transmission level in our country is ± 500 kV, which is upgradeable to ± 800 kV. HVDC transmission also requires converter transformers, namely, $Y/y - y$ and $Y/y - d$ type respectively for 6-pulse and 12-pulse high-voltage AC to DC conversion and vice versa. The standard frequency used in our country is 50 Hz. The frequency of 60 Hz is also standardized in some other countries. As per **IS12360 Specifications**, the domestic voltage level in our country is 400 V (360–440 V) three-phase and 230 V (207–253 V) with 10% changes in either direction for single phase, the frequency of 50 Hz (48.5–51.5 Hz). In some countries, the voltage used is 220 V for three phase and 110 V for single phase. Different countries have different standardized voltages and frequencies, with standard domestic voltages varying between 200 and 480 V three phase and 100–240 V single phase with marginal differences.

Currently, the standard frequency worldwide in use is either 50 ± 3% Hz or 60 ± 3% Hz. The erstwhile generation of 25 Hz power at Niagara fall (USA) has completely been taken out. Using rotary converters, 25 Hz AC power generated was converted into DC power. Transformers and alternators of 25 Hz are bulkier in comparison with 50 Hz or 60 Hz transformers and alternators, respectively of similar capacities.

As the level of transmission voltage increases, the cost of insulation, cooling, protection requirements, and safety demands also increases. The use of high-frequency transformer or alternator will be much lighter when compared with similar capacity 50 Hz transformer or alternators, respectively.

3.2 Types of Transformers

It has been shown in Fig. 2.1 that magnetic flux and electric current should be tightly coupled reducing leakage flux and hence better performance. The magnetic circuit and electric circuit of a transformer can be arranged in two forms, leading to the following type of transformers [3, 4, 6]:

(a) Core-type transformer.
(b) Shell-type transformer.

In core type transformers, the core encircles the windings as has been shown in Fig. 3.1. In these transformers, the leakage flux is larger than that of shell-type transformers. Core-type construction is also easier to fabricate and repair. In shell-type transformers, as shown in Fig. 3.2, the magnetic and electric circuits are tightly coupled. Here, the coils are usually disk-type. This helps in reducing the leakage flux, but it makes the fabrication and repair difficult. In both types of transformers, the cross section of the core is approximately a circle, which helps in reducing total copper losses, and better utilization of core and conducting materials.

Insulation costs are large in high-voltage transformer. The insulation between different windings and between winding and core are decided based on potential gradients. Insulation coordination is needed to save the overall cost of insulation. Since the core is grounded, the low-voltage windings are placed closer to the core.

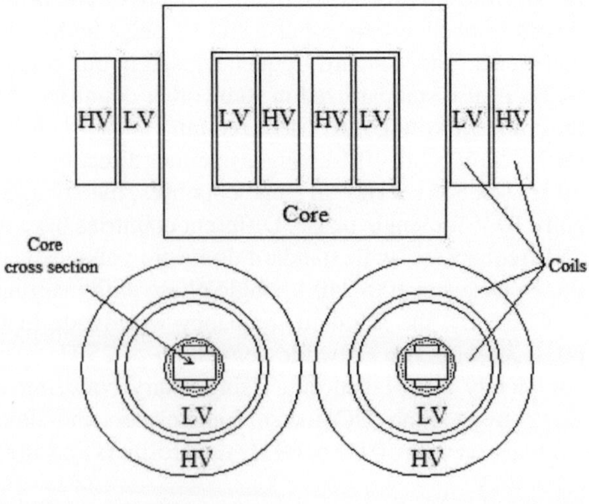

Fig. 3.1 Core-type transformer

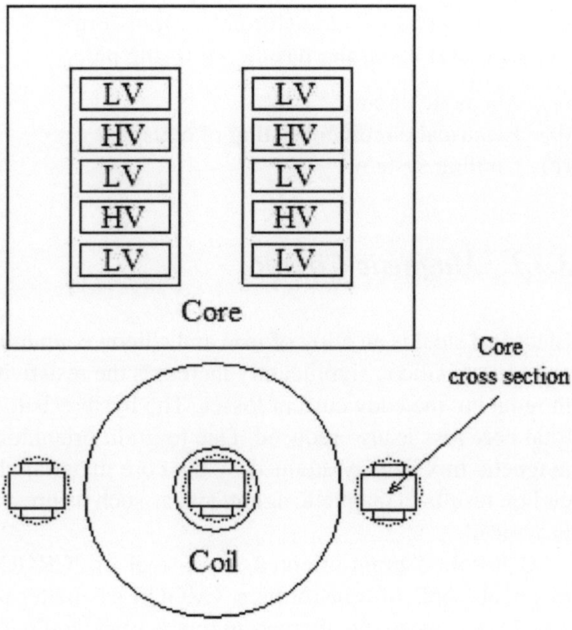

Fig. 3.2 Shell-type transformer

Table 3.1 Comparison between core- and shell-type transformers

Core type transformers	Shell-type transformers
In these transformers, the coils encircle the core	In these transformers, the core encircles the coils
These transformers have high leakage flux and large leakage reactance	These transformers have low leakage flux and smaller leakage reactance
They are easy to fabricate and repair	They are difficult to fabricate and repair
They provide poor regulation	They provide better regulation
Core-type transformers are less costly	Shell-type transformers are costly

Table 3.1 gives a comparison between core-type and shell-type transformers.

3.3 Constructional Details

A transformer is an electromagnetic device which helps in either stepping-up or stepping-down the alternating current (AC) voltage levels. Step-down transformers used in distributing domestic and industrial power are called distribution transformers. The step-up and step-down transformers used in transmission line are referred to as power transformers. The construction of reactors used in power system is similar to that of a transformer. A transformer does not have any rotating parts. Typically, a transformer has the following parts:

(a) Magnetic circuit.
(b) Electrical circuit consisting of coils.
(c) Cooling system.

3.3.1 Magnetic Circuit

Electrical steel is an alloy of iron and silicon containing 0–6.5% of silicon (Si). The presence of silicon significantly increases the resistivity of the steel, thereby helping in reducing the eddy current losses. The hysteresis loop is also narrowed. Thus, the total core loss is also reduced. Due to grain orientation in the rolling direction, the magnetic flux density attained by the core material also increases. The presence of carbon results in magnetic aging, and as such its presence in the electrical steel alloy is avoided.

Cold-rolled grain-oriented sheet-steel or 'CRGOS' is used in manufacturing magnetic cores of transformers. CRGOS has better magnetic properties and lesser core losses, when the alternating flux is made to flow along the direction of rolling. The sheet lamination is coated with tough enamel insulation, which is resistant to hot-transformer-oil (mineral oil) and electromagnetic stresses. In high-frequency transformers, magnetic cores have either simple air core or ferrite core. Ferrite cores are manufactured in multiple pieces to accommodate the preformed coils. Currently, India imports 'CRGOS' sheet steels from China, Japan, Russia, and South Korea.

Smaller capacity transformers are also manufactured using hot-rolled non-grain-oriented sheet-steel, CRGOS, or ferrite. The transformers are designed to operate near knee point of their respective B-H curve.

The magnetic cores shown in Figs. 3.1 and 3.2 are made with several pieces of stampings for smooth placing of preformed coils and insulation. To make stamping, several pieces of appropriate size and shape are shear-cut such that the path of flux is always in the direction of rolling. The laminations are normally cut in rectangular sizes and are assembled in required shape. The joints are overlapped for avoiding a continuous air-gap. In Fig. 3.3a, certain portion of laminations are not in direction of rolling, which causes increased losses. The laminations are assembled in required shape, the corners being interleaved to add to rigidity and to reduce the reluctance of magnetic circuit. Figure 3.3b shows assembly of cores using *mitered joints*.

As already pointed out, the joints of these sheet stampings are mitered and interleaved to avoid a continuous air-gap. Packets of few stampings with different alignments are alternatively placed such that there is no continuous joint in the prepared core. This process of interleaving helps in reducing the leakage reactance and losses of the transformer.

A continuous air-gap in the core increases both the reluctance of magnetic circuit and leakage reactance. This results in poor voltage regulation, when a distribution

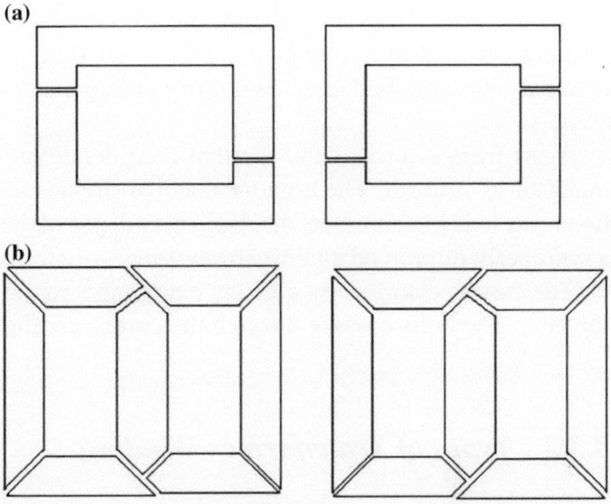

Fig. 3.3 **a** Packing of magnetic core using laminations; **b** interleaving of joints

transformer is loaded. However, in power transformer, it is desired to limit the short-circuit current, an air-gap in the magnetic core increases the leakage reactance of the transformer. The reactors used in power system and welding transformers have similar constructions. However, in the latter case, it is required to limit the short-circuit current at the initiation of arc.

3.3.2 Electric Circuit

The coils are wounded layer-by-layer on former, made of synthetic material, which is inserted into the limbs of the transformer. Copper strip conductors are preferred over round conductors as they give a better copper area inside the window-area of the transformer. Strip conductors have paper insulation. In oil-immersed transformer, the mineral oil provides additional insulation to the winding apart from cooling. Once the winding is made, it is impregnated with varnish or electrical resin. This helps in removal of locked air inside the winding. The varnish insulation or electrical resin on the winding solidifies, this makes the coils stronger enough to withstand large electromagnetic stresses under fault conditions. The constructional details of a particular winding vary with current and voltage ratings. Larger current windings require two or more conductors in parallel. The conductors are transposed at frequent intervals to avoid the circulating currents between the strands. As already stated, these preformed electrical coils are inserted in their respective limbs of magnetic circuit during fabrication.

These windings in a large capacity high-voltage transformer are at high voltages and carry high currents. These coils experience strong electromagnetic forces even under normal operating conditions. The coils are properly reinforced against the strong electromagnetic forces during sudden short circuit of a transformer. The terminals of windings are brought out for connection to outer circuit through the bushings or insulators mounted on the outer portion of the transformer tank. The external cables are bolted to these terminals by means of lugs connected at the cable ends.

Apart from electrical and thermal considerations the windings must have good mechanical strength. The high mechanical stresses set up under short circuit conditions may lead to a complete mechanical collapse of the windings. The inner windings are properly reinforced against strong forces of electromagnetic nature.

The core is clamped by a heavy clamping structure, insulated, and spaced from the tank. The bolts passing through the core is carefully insulated.

3.3.3 *Types of Transformer Winding*

The cost of insulation is an important factor in fabrication of large capacity transformer. It depends on the potential gradient to which the insulating material is subjected to withstand. For insulation coordination, it is a convention to keep low-voltage windings on the inner sides closer to the limb and high-voltage windings are kept on the outer side. This is done in a simpler form of transformer winding, i.e., cylindrical type.

As mentioned earlier, the coils are wounded using either SWG wire or rectangular copper strip. These are standard winding material commercially available in market. The SWG winding wire is having a circular cross section, the outer surface has a thin layer of enamel insulation. The strip conductor is having a rectangular cross section. The outer surface is covered by insulating paper during the process of coil making. The round conductors and rectangular conductors are shown in Fig. 3.4a.

When strip conductor having rectangular cross section is used, the window area is utilized in a better way. With round conductors, certain window area remains unutilized. This has been shown in Fig. 3.4b.

The prefabricated coils are inserted into the limbs of transformer. In large capacity transformers, the low-voltage winding current may be excessively large. For example, in a distribution transformer of 50 kVA, 11 kV/400 V single-phase transformer the rated primary and secondary currents are 4.54 A and 217 A, respectively. The current density in ONAN type of distribution transformer design is kept around 1.8–2.2 A/mm^2. This demands the conductor cross-sectional area in secondary as 98.6 mm^2.

The thickest round conductor of SWG No. 1 is having a cross-sectional area of 45.5 mm^2. Thus when 98.5 mm^2 cross-sectional area is required, one has to use parallel circuits of either SWG wires or strip conductors in parallel. The fabrication of coil with thicker wire is also more difficult than multiple numbers of thinner wires. These parallel conductors are shorted at both ends to share large current. These conductors are transposed so that all strands have equal electrical parameters. This helps in minimizing the possibility of circulating current in certain strands.

For reducing the size of conductor-parallel circuits are formed thus if we go for 8 parallel circuits in secondary—the cross-section area of each conductor is around

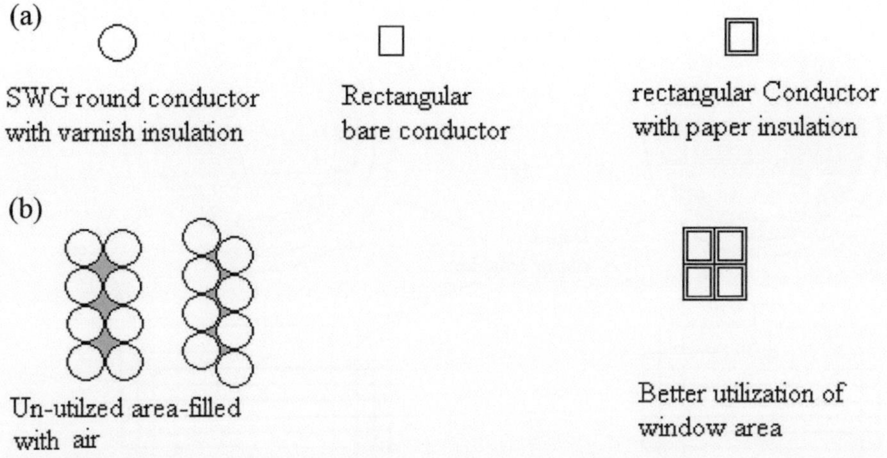

Fig. 3.4 a Cross-sectionals area of different conductors; **b** better utilization of window area using rectangular conductors

12 mm^2. These eight circuits are transposed for avoiding dissimilarity in the winding inductances of conductors.

The types of transformer windings are as follows:

(a) Cylindrical with circular conductors or rectangular conductors.
(b) Continuous disk.
(c) Helical winding.
(d) Cross over windings.

Among these windings, the low-voltage windings are either cylindrical windings or helical windings. The high-voltage windings are either cylindrical winding with round conductors or continuous disk-type windings with rectangular conductors or cross over windings. These are described below.

Cylindrical Windings: These windings are in form of cylindrically wounded coils. There are multiple layers in these windings and each layer consists of several conductors. These are wounded on a Bakelite cylinder as shown in Fig. 3.5a–c. Paper insulation is provided between each layer of conductors.

Figure 3.5b, c shows single-layer/multiple-layer cylindrical windings with single conductor and with four parallel conductors. Cooling ducts for circulation of air (in case of dry transformers) or circulation of mineral oil for cooling (in case of oil immerged transformers are also provided). Figure 3.6 shows the cooling ducts in the cylindrical windings.

The disposition of primary and secondary windings plays an important role in the leakage reactance. In cylindrical windings with one primary coil and one secondary coil of equal length the leakage reactance is more when compared with disk windings or when cylindrical windings are split into multiple windings arranged suitably. This has been highlighted in Fig. 3.7.

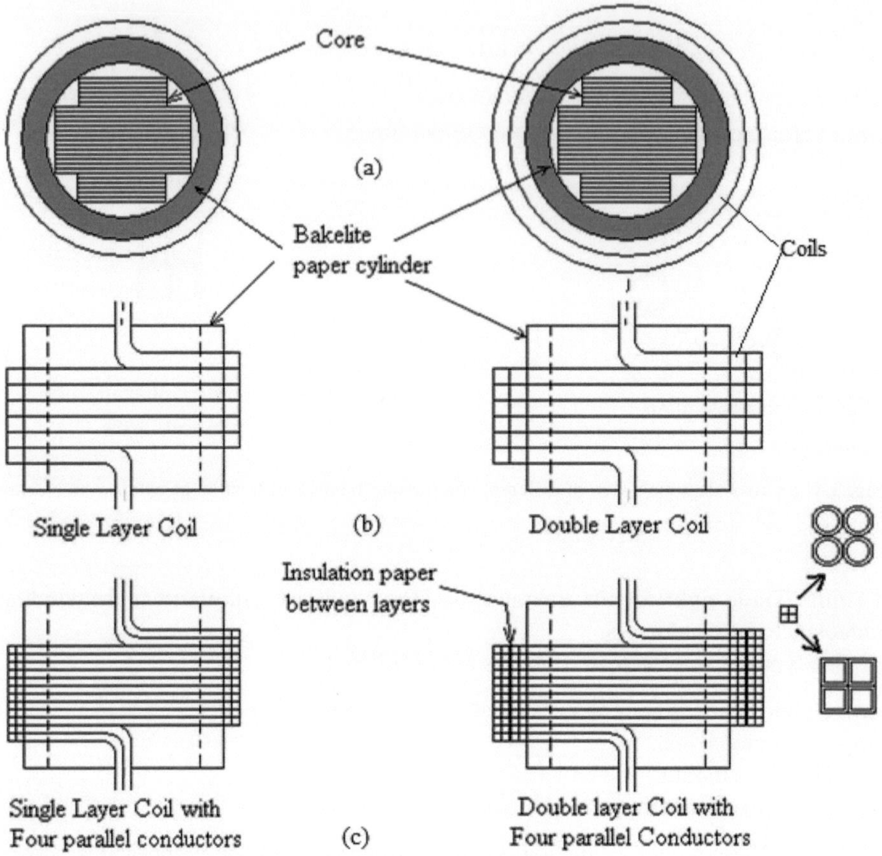

Fig. 3.5 Cylindrical windings—single layer and multiple layer

Fig. 3.6 Cooling ducts in cylindrical windings

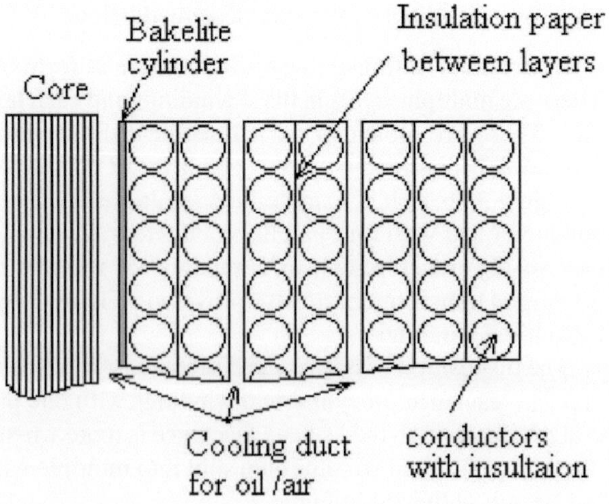

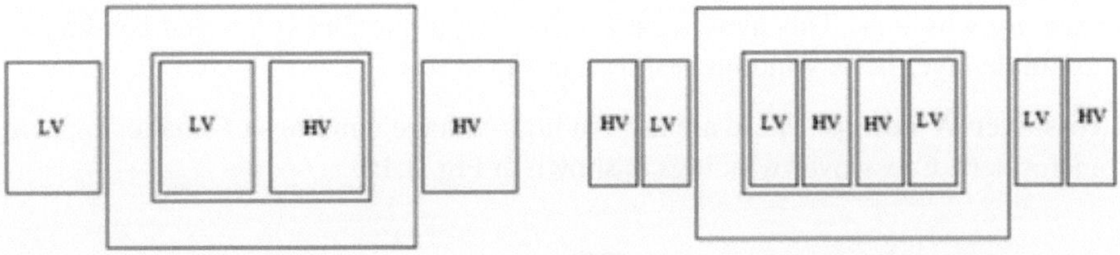

Fig. 3.7 Splitting of windings for reducing the leakage reactance in cylindrical windings

Disk Winding: In shell-type winding, sandwiched disk windings are used as shown in Fig. 3.8. The low-voltage and high-voltage windings are arranged in multiple disk form sandwiching each other. However, on top and bottom sides low-voltage winding of half-length disk is used. Such an arrangement is helpful in insulation coordination. The leakage reactance of sandwiched transformer winding is lesser as compared to shell-type having cylindrical windings. Proper insulation is provided between low-voltage and high-voltage disks and also between the core and disks.

Helical Windings: In helical windings turns are wounded in an axial direction along a screw line with an inclination to the height of the conductor and an oil duct between turns.

Due to high current requirements, the conductors are rectangular strip conductors with paper insulations. The windings are preferred in low-voltage windings of medium- and high-capacity transformers. The parallel connected strips are placed side by side in radial direction. This forms disk helical windings. Helical windings are wounded on Bakelite cylinder with insulating cylinders placed between the conductors. This maintains sufficient gap between two turns. This also maintains high mechanical strength in axial directions. Transpositions of conductors are required for equalizing the conductors' resistance and leakage reactance, so that in parallel circuits there are minimum chances of circulating current to flow in the strands of

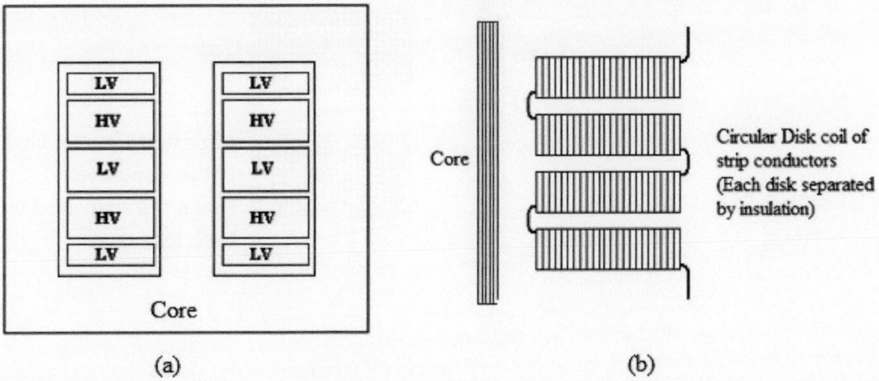

Fig. 3.8 **a** Sandwich disk winding; **b** disk windings—shell-type transformer

secondary windings. This avoids the overloading in parallel connected conductors. A multiple layer helix winding is shown in Fig. 3.9.

Cross-over Windings: These are used in high-voltage windings of smaller capacity transformers. Cross-over windings is shown in Fig. 3.10.

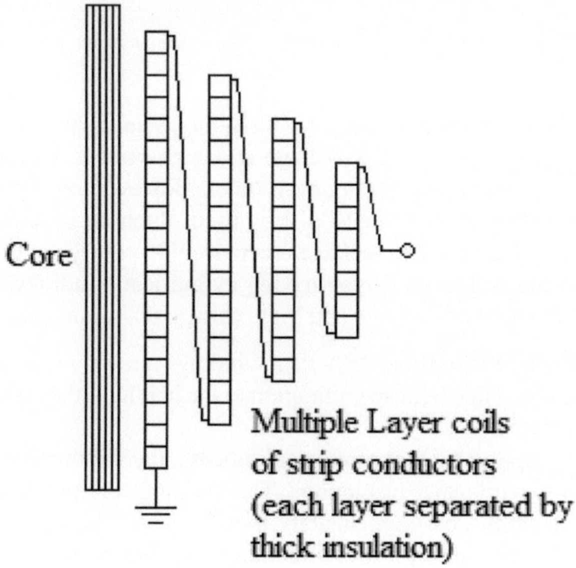

Fig. 3.9 Multilayer helix winding

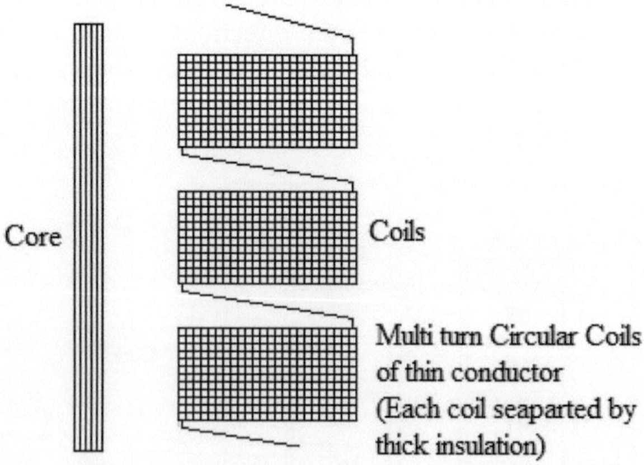

Fig. 3.10 Cross-over winding

3.4 Types of Cooling

The currents flowing in the transformer and AC flux in the core respectively, cause copper loss and core loss. An active power is absorbed, resulting in the generation of heat. If the heat is not properly dissipated, it results in the damage of insulation and electrical faults. The provision of cooling ducts in the core and coils enhances the dissipation of heat. The following are some cooling methods used for dissipating heat in electrical machines and transformers:

(a) Air natural cooling (AN).
(b) Oil-immerged natural cooling (ON).
(c) Oil-immerged forced oil circulation with natural cooling (OFN).
(d) Oil-immerged forced oil circulation with air blast cooling (OFB).
(e) Oil immerged air blast cooling (OB).
(f) Air blast cooling (AB).
(g) Oil-immerged water cooling (OW).
(h) Oil-immerged forced oil circulation with water cooling (OFW).
(i) Hydrogen cooling (for large power rotating machines) (H_2).

Transformers may have mixed type of cooling like, ONAN, OFN, OFB, and OFW. Hydrogen's low-density, high-specific heat, and thermal conductivity make it a superior coolant for higher power electrical machines.

Transformer rating is given in terms of primary and secondary voltage, volt-ampere (or kVA rating), frequency and type of cooling. Oil-immerged natural cooling of a distribution transformer is shown in Fig. 3.11. The heated transformer oil expands in volume, circulates through the tubes attached to the transformer tank, and gets cooled due to natural air flow.

In Fig. 3.12, oil-immersed forced oil air blast cooling of large power transformer of rating 1000-MVA and above is shown. Here heat is exchange from heated oil through water-cooled and air-cooled radiators. The expansion of oil is taken by the conservator. When the transformer oil cools down due to lesser load, the expanded

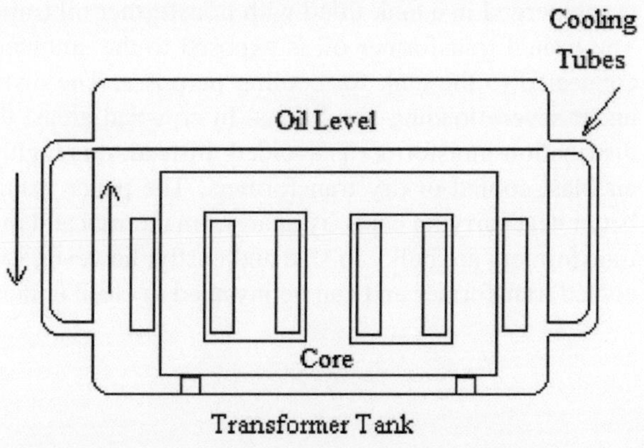

Fig. 3.11 Oil-immerged natural air cooling of distribution transformer

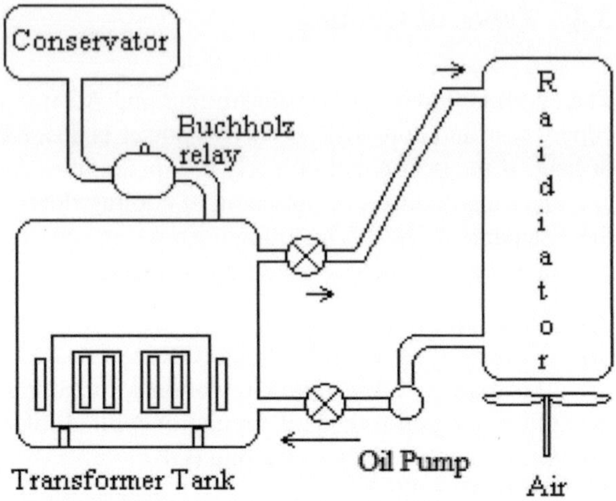

Fig. 3.12 Oil-immersed forced oil-air blast cooling of power transformer

oil shrinks to its original volume causing the outer air to enter the tank. If the air contains the moisture, it distorts the properties of the transformer oil. To avoid this, breather attached to conservator containing silica-gel which absorbs the moisture from outer air entering the transformer tank.

Higher capacity transformer breakdowns are always preceded by more or less violent generation of gas. A broken electrical joint, for example, produces a local arc and vaporizes the transformer oil in the vicinity. An earth fault produces the same result. A sudden short circuit rapidly increases the temperature of the windings, vaporizing the transformer oil in the vicinity. The weakness of insulation also creates electrostatic discharge in a high-voltage transformer. Heat produced due to internal faults in a transformer produces gas-bubbles in transformer oil. If the local temperature goes beyond 150 °C, the transformer oil tends to vaporize. The bubbles of gas thus produced are trapped by the Buchholz relay, causing the activation of trip circuit which ultimately gives an alarm.

In the case of oil-immerged air natural or ONAN type of cooling, the transformer is submerged in a tank filled with transformer oil (mineral oil) for cooling purpose. The heated transformer oil is exposed to the ambience by means of hollow tubes connected to the tank for cooling purposes. The distribution transformer operates under severe loading conditions. In crowded areas, the installation of ON type of distribution transformer is avoided. Instead, it is highly recommended to install dry air blast-cooled or dry transformers. The paper insulation of dry transformer has better heat carrying capacity apart from thermal and insulating properties. These dry transformers are bulky in size and costly; however, they have longer life as ONAN cooled transformer and can be installed in close underground area.

3.5 EMF Equation

A typical two-winding transformer and its schematic representation are shown in Fig. 3.13a, b, respectively. An AC source is connected to the winding which is referred as primary. The electrical load is connected to another winding which is referred to secondary winding. The primary winding has N_1 number of turns, while the secondary winding has N_2 number of turns. A transformer works on the principle of induction (Refer Chap. 1). From Faraday's law, we know that if flux linkage is made to change, an EMF is induced in the electric circuit. The magnitude of this EMF is equal to rate of change in flux linkage.

$$\int e.\mathrm{d}t = -\lambda \quad \text{or} \quad e = -\frac{\mathrm{d}\lambda}{\mathrm{d}t} \tag{3.1}$$

The direction of induced EMF is opposite to the change in flux linkage. Can you tell why?

It assumed here that an AC voltage source v_1 excites the primary winding. This gives a sinusoidal varying flux in the core,

$$\phi = \phi_m \sin \omega t \tag{3.2}$$

where ϕ_m is the maximum value of magnetic flux and $\omega = 2\pi f$ is the angular frequency. The unit of frequency of supply f is Hz. Since primary coil is having N_1 turns, the flux links these turns. The induced EMF in the primary coil which opposes the applied voltage is

$$v_1 - e_1 = 0 \tag{3.3}$$

$$v_1 = e_1 = N_1 \frac{\mathrm{d}\phi}{\mathrm{d}t} = N_1 \phi_m \omega \cos \omega t = -\sqrt{2} E_1 \sin\left(\omega t - \frac{\pi}{2}\right) \tag{3.4}$$

The RMS value of induced EMF in primary winding is,

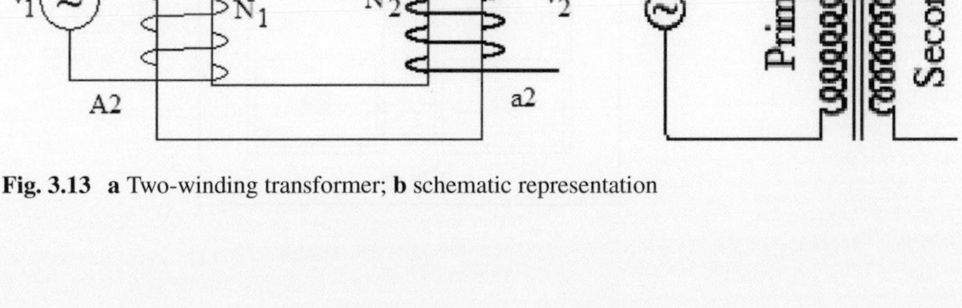

Fig. 3.13 **a** Two-winding transformer; **b** schematic representation

$$E_1 \equiv \frac{N_1.\phi_m.\omega}{\sqrt{2}} = \sqrt{2}.\pi.f.N_1.\phi_m$$

$$E_1 = 4.44 \times f \times \phi_m \times N_1 \tag{3.5}$$

A comparison of Eqs. (3.2) and (3.4) shows that the reaction EMF '$-e_1$' lags the changing flux by angle 90°. The magnetic flux links the secondary having N_2 turns. Thus, the secondary induced EMF is,

$$E_2 \equiv \frac{N_2.\phi_m.\omega}{\sqrt{2}} = \sqrt{2}.\pi.f.N_2.\phi_m \tag{3.6}$$

The operation of a transformer is based on the primary flux linking the secondary number of turns. This flux linkage is automatically changed in constant voltage transformer (CVT) is shown in Fig. 3.14. When transformer is fully loaded the secondary gets slightly moved (attracted) towards primary winding, causing an increase in the total flux linkage and hence voltage. This improves the voltage regulation. Such transformer has limited applications.

Example 3.2 In a single-phase 50 kVA, 6.6 kV/220 V, 50 Hz transformer the maximum flux in the core is not to exceed 0.06 Wb. Determine numbers of turns in each winding and voltage per turn.

Solution

Given voltages are $E_1 = 6.6$ kV $= 6600$ V and $E_2 = 220$ V.

Frequency $f = 50$ Hz and core flux $\phi_m = 0.06$ Wb.

Using EMF equation, we get

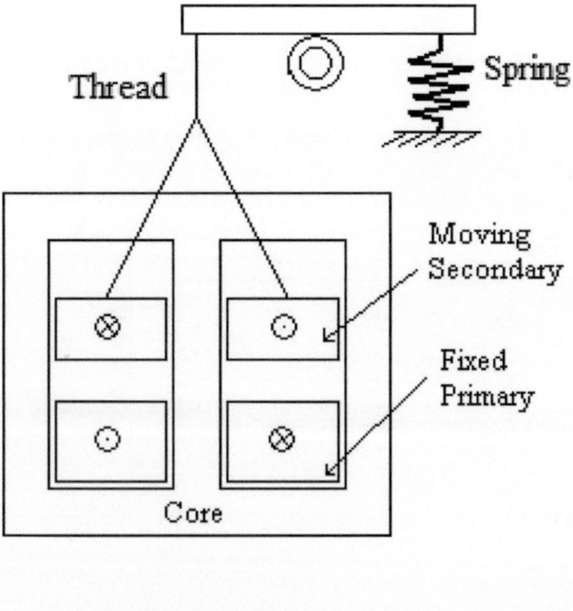

Fig. 3.14 Constant voltage transformer

$$N_1 = \frac{E_1}{4.44 \times f \times \phi_m} = \frac{6600}{4.44 \times 50 \times 0.06} = 495.5 \approx 496 \text{ turns.}$$

Voltage per turn,

$$= \frac{6600}{496} = 13.3 \text{ Volt per turn}$$

$$N_2 = \frac{E_2}{4.44 \times f \times \phi_m} = \frac{220}{4.44 \times 50 \times 0.06} = 16.5 \approx 17 \text{ turns.}$$

Voltage per turn,

$$= \frac{220}{17} = 12.94 \text{ Volt per turn}$$

Example 3.3 The secondary winding of a 50 Hz transformer has 50 turns. If the induced voltage in this winding is 220 V, determine the maximum value of the core flux.

Solution

Given $N = 50$ turns; $f = 50$ Hz; $E = 220$ V.

Applying EMF equation, the maximum value of core flux is,

$$\phi_m = \frac{E}{4.44 \times f \times N} = \frac{220}{4.44 \times 50 \times 50} = 0.0198198 \text{ Wb}$$

Example 3.4 A 220 V/40 V transformer is used to supply high current at low voltage. The rating of transformer is 1.5kVA. Determine the rated currents of the two windings.

Solution

The transformer is capable of transferring power of 1.5 kVA, thus.

Rated HV side current,

$$I_{\text{HV}} = \frac{1500}{220} = 6.81 \text{ A}$$

Rated LV side current,

$$I_{\text{LV}} = \frac{1500}{40} = 37.5 \text{ A}$$

The low-voltage winding has higher rated current.

3.6 Ideal Transformer

An ideal transformer and its schematic representation are shown in Fig. 3.13a, b, respectively. The following assumptions are made for an ideal transformer:

(a) It has zero leakage reactance.
(b) The primary winding is connected to a rated voltage and rated frequency source.
(c) The resistances of both the windings are negligible.
(d) The core loss is zero.
(e) All the magnetic flux generated by one coil links the other coil, and the magnetization curve of the core is linear.

Thus,

$$V_1 \cong E_1 \quad \text{and} \quad V_2 \cong E_2 \tag{3.7}$$

Thus we can say that

$$\frac{E_1}{E_2} \approx \frac{V_1}{V_2} = \frac{N_1}{N_2}$$

A transformer is an energy transfer device. For an ideal transformer, we can assume that apparent powers of primary and secondary are equal, i.e.,

$$E_1 I_1 = E_2 I_2 \quad \text{or} \quad V_1 I_1 = V_2 I_2 \tag{3.8}$$

Alternatively, to magnetize the core a little amount of MMF $\mathcal{F}$ is required irrespective of the fact whether the transformer is on-load or on no-load. Thus, we have

$$N_1 I_1 = N_2 I_2 + \mathcal{F}$$

If, we ignore the magnetizing MMF ($\mathcal{F} = 0$), we get

$$\frac{N_1}{N_2} = \frac{I_2}{I_1}$$

Thus we have,

$$\frac{E_1}{E_2} = \frac{V_1}{V_2} = \frac{N_1}{N_2} = \frac{I_2}{I_1} = a = \text{Transformaton ratio} \tag{3.9}$$

It can be shown that if impedance Z_2 is connected on the secondary terminals, the impedance seen on the primary terminals Z_1 satisfies the following equation.

$$\frac{Z_1}{Z_2} = \left(\frac{N_1}{N_2}\right)^2 = a^2 \tag{3.10}$$

The above expressions (Eqs. 3.8–3.10) imply that a transformer can be used for the following purposes:

(a) Voltage transformation.
(b) Current transformation.
(c) Impedance transformation.

By using different number of turns, one can change the voltage level in the secondary winding. In a typical transformer, the current rating of low-voltage winding is higher than that of current rating of high-voltage winding. The same principle is used for changing the level of high-current to a low measurable value in a current-transformer (CT). The device can be used for changing the referred resistance or referred impedance of certain circuit. This property is utilized in TV, radio and communication networks for impedance matching, leading to maximum power transfer.

3.7 Two-Winding Transformer

A two-winding transformer and its schematic representation have already been shown in Fig. 3.13a, b, respectively. In an ideal transformer, the sinusoidal input voltage results in magnetizing current and core flux. The magnetizing current is in-phase with the mutual flux. The induced EMF e_1 is in-phase opposition to the applied voltage v_1, which satisfies Lenz's law. Their time variations and phasor representation are shown in Fig. 3.15a, b, respectively.

Figure 3.16b shows the determination of magnetizing current waveform using B-H curve, when the flux density is sinusoidal with respect to time. It is observed that

(a) Magnetizing current waveform is non-sinusoidal.
(b) Magnetizing current waveform contains odd-order harmonics, like third, fifth and so on; and
(c) The fundamental Magnetizing current waveform leads the flux waveform by a small angle.

Figure 3.16a shows the experimental visualization of magnetizing current waveform on a Cathode Rays Oscilloscope (CRO). Point N in the figure must be at electrical ground, neutral or zero potential. The auto-transformer set-up can be used for obtaining the B-H curve using Lissajous pattern on a CRO. The resistance R is a low resistance for obtaining the current signal proportional to magnetic field intensity 'H'. The voltage signal proportional to magnetic flux density 'B' is obtained using auto-transformer as shown in figure. The phase difference in the magnetizing current and flux waveform is due to the Hysteresis. It represents the active loss due

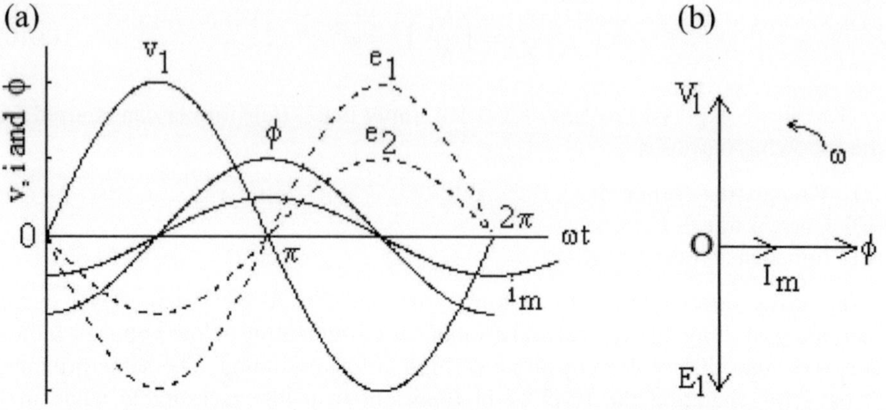

Fig. 3.15 **a** Time variation of applied voltage, induced EMF, flux, and magnetizing current in an ideal transformer; **b** phasor representation of applied voltage, induced EMF, flux, and magnetizing current in an ideal transformer

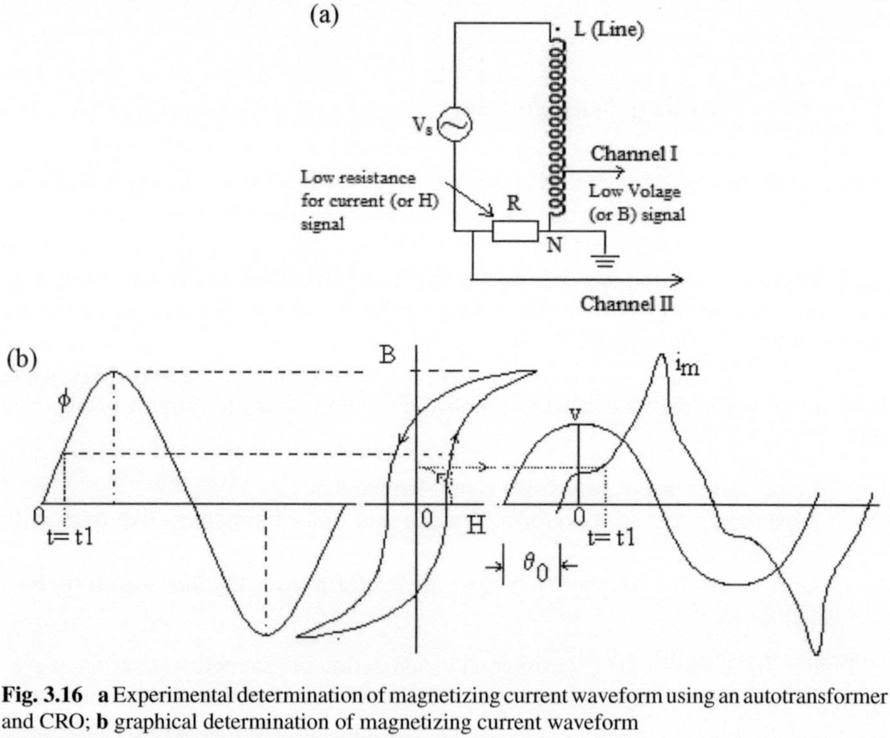

Fig. 3.16 **a** Experimental determination of magnetizing current waveform using an autotransformer and CRO; **b** graphical determination of magnetizing current waveform

to hysteresis and eddy current losses in the transformer. If hysteresis loop is known, then for sinusoidal flux of input voltage, the magnetizing current waveform can be graphically visualized as shown in Fig. 3.16b.

3.7.1 Core Loss and Equivalent Circuit

A Transformer works on the principle of mutual flux coupling the two coils. In order to minimize the leakage flux, the magnetic flux is allowed to flow in the path provided by the ferromagnetic core. This core is subjected to alternating magnetization at the supply frequency. Due to this, a certain amount of power is wasted in hysteresis and eddy current losses. The Hysteresis loss density depends on the area of the Hysteresis loop and is given as

$$P_h = \eta.f.B_m^x \tag{3.11}$$

where η is Steinmetz constant, which is material specific and quotient $x = 1.5$–2.0 for ordinary steel and is valid for flux density 0.1–1.2 wb/m^2, f is the supply frequency and B_m is the maximum flux density. The use of proper ferromagnetic material having thinner or lesser area of hysteresis loop reduces the hysteresis loss.

When a changing or alternating flux penetrates a conducting material, eddy currents are circulated in a closed path on the surface of the conducting material. These eddy current flows in conducting material in 'skin-depth'. The flow of eddy current encounters certain resistance, leading to Ohmic losses. The eddy current loss depends on square of the thickness of stamping t, square of maximum flux density in the core B_m, square of applied frequency f and form factor K_f. The use of thin stamping for the fabrication of the magnetic core reduces the eddy current loss. The eddy current loss density is,

$$P_e = k.K_f^2.t^2.f^2.B_m^2 \tag{3.12}$$

Here k is certain constant. The sum of these two losses, hysteresis loss and eddy current loss, is known as core loss or iron loss (P_c) as they occur in the ferromagnetic core. If we assume that the core loss is dependent on the square of the flux density, then from EMF equation, we get

$$V \approx E = 4.44.f.\phi_m.N = 4.44.f.B_m.A_c N$$

Here, N is the number of turns and A_c is cross sectional area of the core. Thus, the peak value of flux density is proportional to ratio of applied voltage and frequency, i.e.,

$$B_m \quad \alpha \quad \frac{V}{f} \tag{3.13}$$

This technique is adopted for separating different core losses, viz., hysteresis and eddy current losses from the core loss. This implies that if both the applied voltage and frequency are varied but their ratios are held constant, the core loss will remain constant irrespective of the fact that the transformer is at load or at no-load. According to Ohm's law, the power loss in a resistance is given as,

$$I^2R = \frac{V^2}{R} \; \alpha \; \frac{(B_m.f)^2}{R} \tag{3.14}$$

The core loss is a real active power loss leading to heat dissipation and is represented by parallel resistance R_c. Here, it is assumed that hysteresis loss is also proportional to the square of maximum flux density in the core, i.e., ${B_m}^2$. However, this assumption is correct up to a certain extent, as the hysteresis loss is only a fraction of core loss.

As pointed out earlier, in order to magnetize the core, a little amount of MMF is required. This is irrespective of the fact, whether transformer is on-load or on no-load. In this process an imaginary reactive (lagging) power-loss is encountered, which is represented by inductive reactance X_m, a parallel arm of the equivalent circuit.

Based on the above equations, a preliminary equivalent circuit of a two-winding transformer is shown in Fig. 3.17a, b. Here, the primary and secondary windings are representing an ideal transformer having induced EMFs $\boldsymbol{E}_1$ and $\boldsymbol{E}_2$, respectively. Here quantities $\boldsymbol{V}, \boldsymbol{I}$ and $\boldsymbol{E}$ of primary (subscript '1') and secondary (subscript '2') are AC phasor quantities.

The magnetic coupling between $\boldsymbol{E}_1$ and $\boldsymbol{E}_2$ is not needed for developing the equivalent electrical circuit. If these are somehow made equal, then a simple electrical circuit can be developed. If we assume that the secondary number of turns is equal to primary number of turns, i.e., $N_2 = N_1$, then new secondary induced EMF referred to primary is

$$\boldsymbol{E}_2^{/} = a\boldsymbol{E}_2 = \boldsymbol{E}_1$$

Here quantity 'a' is the ratio of numbers of turns of primary and secondary winding or $a = (N_1/N_2)$. This is also known as transformation ratio.

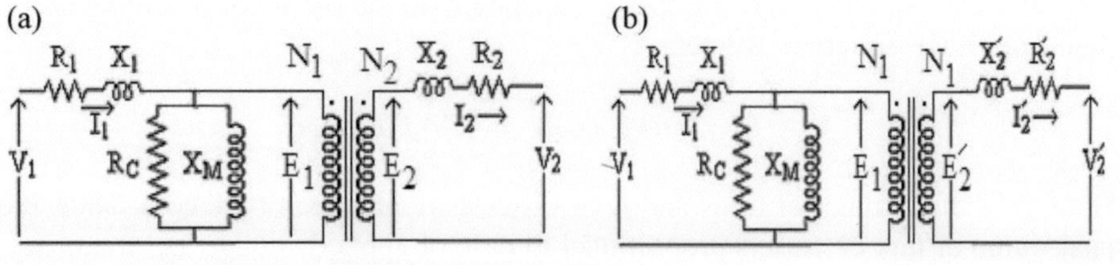

Fig. 3.17 **a** Equivalent circuit of two-winding transformer showing magnetic coupling—unequal number of turns; **b** equivalent circuit of two-winding transformer showing magnetic coupling—equal number of turns

The secondary voltage referred to primary,

$$V_2^{/} = aV_2 = V_1$$

The new secondary current referred to primary is,

$$I_2^{/} = \frac{I_2}{a} = I_1$$

Here quantities with superscripts are the respectively quantities referred to either primary or secondary. For example I_2 is the secondary current, but I'_2 is the secondary current referred to primary. Similarly, I'_1 is the primary current referred to secondary. However, it should be noted that irrespective of the change in the number of turns of secondary, the secondary losses are not changed. Thus, we have

$$I_2^2 R_2 = I_2^{/2} R_2^{/} \tag{3.15}$$

and

$$I_2^2 X_2 = I_2^{/2} X_2^{/} \tag{3.16}$$

where the quantities with superscript '/' are the secondary quantities referred to primary side, i.e.,

$E_2^{/}$ = Secondary EMF referred to primary side.
$V_2^{/}$ = Secondary terminal voltage referred to primary side.
$I_2^{/}$ = Secondary current referred to primary side.

In this process, the loss should remain the same. The secondary resistance referred to primary side is given as

$$R_2^{/} = \frac{I_2^2}{I_2^{/2}} R_2 = \frac{I_2^2}{I_1^2} R_2 = \left(\frac{N_1}{N_2}\right)^2 R_2 = a^2 R_2 \tag{3.17}$$

Similarly secondary leakage reactance referred to primary side,

$$X_2^{/} = \frac{I_2^2}{I_2^{/2}} X_2 = \frac{I_2^2}{I_1^2} X_2 = \left(\frac{N_1}{N_2}\right)^2 X_2 = a^2 X_2 \tag{3.18}$$

An equivalent circuit of a transformer is shown in Fig. 3.18a, when $N_2 = N_1$. Since $E_2^{/} = aE_2 = E_1$, there is no need to show magnetic coupling. As such the exact equivalent circuit of a transformer is derived and is shown in Fig. 3.18a. The parameters of equivalent circuit referred to primary side of a transformer are summarized as follows:

R_1 = Resistance of the primary winding/coil, Ω.

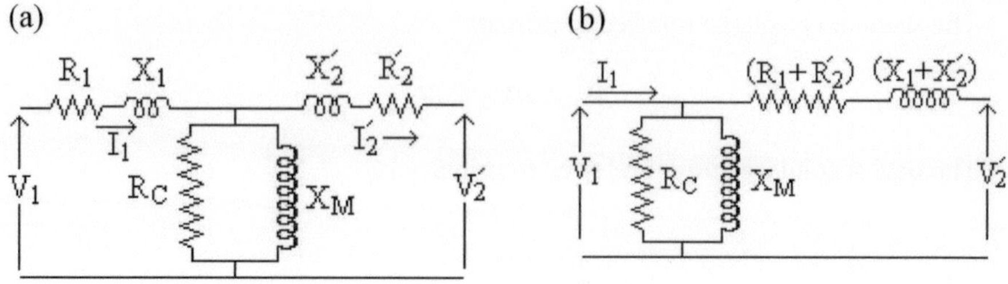

Fig. 3.18 **a** Exact equivalent circuit of a two-winding transformer; **b** approximate equivalent circuit of a two-winding transformer

$X_1 = \omega l_1 =$ Leakage reactance of the primary winding, Ω.
$R_2^/ = \left(\frac{N_1}{N_2}\right)^2 \times R_2 = a^2 R_2 =$ Resistance of secondary winding referred to primary winding, Ω.
$X_2^/ = \left(\frac{N_1}{N_2}\right)^2 \times X_2 = a^2 X_2 =$ Leakage reactance of secondary winding referred to primary winding, Ω.
$R_C =$ Iron loss component of resistance, Ω.
$X_M =$ Magnetizing component of reactance of the transformer, Ω.
$V_1 =$ Primary terminal voltage, V.
$V_2^/ = \left(\frac{N_1}{N_2}\right) V_2 =$ Referred value of secondary terminal voltage, V.
$I_1 =$ Primary winding current, A.
$I_2^/ = \left(\frac{N_2}{N_1}\right) I_2 =$ Secondary winding current referred to primary winding, A.
$I_0 =$ Primary no-load current, A.
$N_1 =$ Primary winding number of turns.
$N_2 =$ Secondary winding number of turns.

The primary leakage inductance is l_1. The secondary leakage inductance referred to primary side $l_2^/ = \left(\frac{N_1}{N_2}\right)^2 l_2$, where l_2 is the secondary leakage inductance.

Now, the primary current is given as

$$\boldsymbol{I}_1 = \boldsymbol{I}_0 + \boldsymbol{I}_2^/$$

But, the no-load current $\boldsymbol{I}_0$ is much less (about 5% or 0.05 p.u. of rated full-load current) than full-loads current of transformer. The parallel branches of the equivalent circuit diagram may be assumed to be connected across the supply terminals as shown in Fig. 3.18b. This approximation does not cause serious error in the calculations, rather it simplifies tedious calculations. The removal of parallel branches further simplifies the equivalent circuit and is shown in Fig. 3.19a. The phasor diagram of a transformer based on the simplified equivalent circuit is shown in Fig. 3.19b. Thus, the resistance of transformer referred to primary,

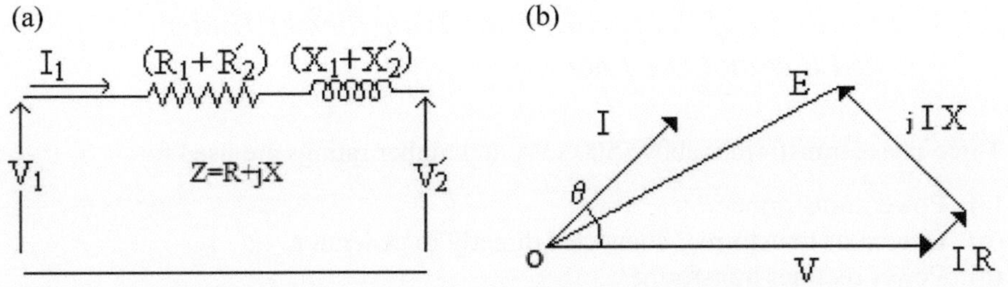

Fig. 3.19 **a** Approximate equivalent circuit; **b** Phasor diagram with leading load

$$R_{\mathrm{EX1}} = R_1 + R_2^{/} = R_1 + a^2 R_2 = R_1 + \left(\frac{N_1}{N_2}\right)^2 R_2 \tag{3.19}$$

Similarly, leakage reactance of transformer referred to primary,

$$X_{\mathrm{EX1}} = X_1 + X_2^{/} = X_1 + a^2 X_2 = X_1 + \left(\frac{N_1}{N_2}\right)^2 X_2 \tag{3.20}$$

3.7.2 Transformer on Load

Since a potential difference is established between the terminals of secondary, we consider that the winding has an induced EMF. This EMF acts as a source in the secondary circuit. Considering that load impedance $\mathbf{Z} = |\mathbf{Z}|\angle\theta°$ across the terminals causes a current given as.

$$\boldsymbol{I}_2 = \frac{V_2\angle 0°}{|\mathbf{Z}|\angle\theta°} = \frac{V_2}{|\mathbf{Z}|}\angle -\theta°$$

where the lagging phasor current has a phase angle of $\angle -\theta°$. This current in secondary flows through N_2 number of turns causes ampere turns 'I_2N_2', which tends to modify the original no-load flux. But the primary flux is due to a constant voltage source, which cannot be altered. This means that total magnetizing ampere turns of both the windings must therefore remain unchanged and, consequently, an additional current is set up in the primary circuit in such a direction so as to cancel the secondary MMF 'I_2N_2'. This additional current is such that the phasor of primary current is obtained by adding $\boldsymbol{I}_o$ and $-\boldsymbol{I}'_2$, and is known as $\boldsymbol{I}_1$. The magnetizing current $\boldsymbol{I}_o$ is relatively small compared to the full-load current $\boldsymbol{I}_1$ and $\boldsymbol{I}_2$. It is approximately 0.05 per unit (5%) of full-load rated current. The phasor diagram of a transformer under load is shown in Fig. 3.19 for leading power factor.

3.7.3 Operation of Practical Power Transformer Under No-Load and On-Load

Three-phase transformer above 500 kVA and higher ratings are used for:

(a) Power transformer.
(b) Generator transformer connected directly to generator.
(c) Power rectifier transformer.
(d) Transformers used in Railways Electrification.
(e) Power Converter transformer Used in HVDC.
(f) Power autotransformer and step voltage regulator.
(g) Regulating transformer.
(h) Grounding transformer.
(i) Transformers used in Electric arc furnaces and electric welding.

The power system is a complex interconnected system of several alternators situated far away, yet electrically connected in a power grid at high voltage level for sharing widely distributed electrical loads. The transformer connected to generator/alternator is referred as generator transformer. Their low voltage rating is normally 11 kV or 15.75 kV or 21 kV and higher voltage rating is normally 400 kV. In a power grid, the output of generator transformer is to be synchronized with the grid. If the transformer is synchronized with the grid and is not connected to any load, it operates under no-load condition for a shorter duration of time. During this period no-load losses of transformer are supplied by the generator. When a power transformer's primary, say, 132 kV/11 kV is connected to the high voltage source, and suppose low-voltage windings are not supplying any load, the magnetizing component of current is drawn from the power system.

The alternator cannot be excited at low RPM, although over-excitation at lower rotational speed (RPM) can result in development of rated terminal voltage. However, due to reduced speed, and hence reduced frequency, the maximum value of core flux rises, causing increase in core losses and hence heating of stator of generator and core of transformer. Reduced frequency rated AC voltage generation cannot be connected to a common grid.

Power transformer rarely operates under no-load conditions. When energized, the magnetic inrush current lasts for about 3–4 cycles during which time it is about 15–20 times rated current. The magnetizing current is almost 1–5% of full-load current. The circuit breaker connected to transformer does not trip and takes this as a healthy condition. If several power transformers in parallel are energized simultaneously, a large inrush current is expected. In such cases, the protection schemes should be reliable to keep the system intact.

When power transformers connected in parallel are switched ON one by one after certain delay, there is a possibility of sustained inrush current because the DC component of the inrush current decays very slowly. When next transformer is switched ON, there is a possibility of presence of decaying DC component. This overloads the main circuit breaker and causes the tripping action.

In case of uninterruptible power supply schemes, the load terminals are disconnected from de-energized power transformer and reconnected to energized power transformer. However, in this process the power outage for certain time should be tolerated.

In case of perfectly uninterrupted power supply schemes supplying sensitive loads, the secondaries are connected in parallel to share a common load, but their high-voltage primary windings are connected with inputs from two different sources. In the event of power failure from one of the source, the other source supplies the total load, though there is a possibility that one of the transformers becomes overloaded.

The power transformer is widely used for transmission and distribution of electrical power in a power system or a power grid. They also enable the transfer of power from one grid to another grid. This requires synchronization of two grids. If grids are operating at different frequencies, they require high-voltage high-current back-to-back solid state circuit for power to flow from one grid to another grid.

When large power is being transmitted to remote locations, far from the generating stations, the voltage of transmission is stepped up in sequence. We already know that overall transmission losses are limited when we transmit power at high voltage. At distribution locations where the power is being utilized, the voltage levels are again stepped down in sequence.

In a power system, the demand of power in 24 h varies from time to time. Several power transformers are connected in parallel. It is recommended that individual power transformer must have their own circuit breaker. When the total transmitted power is less than the capacity of single power transformers, few power transformers are then disconnected from the system. This saves the energy wasted in core losses in the transformer. A power transformer may suffer.

(a) Winding open circuit,
(b) Winding short circuit, and
(c) Overheating.

Normally, open circuit conditions are not harmful and are attended separately. In case of external fault conditions, protection is attended by automatic tripping. When transformer bank is larger than 5 MVA, it is recommended to provide individual circuit breakers in all the windings. The operation of two transformer banks without individual circuit breakers is to be avoided. Differential protection is used for protecting the banks connected in parallel. The CT's (current transformer) are connected either in star or in delta.

Bucholtz relay is useful for detecting incipient faults in transformer. Power transformers are enabled with the remote tripping via power line carrier (PLC) or wireless signals or GPRS.

The rated short-circuit current of power transformer is limited at design level. Short-circuit protection of power transformer is provided using over current protection schemes.

A power transformer may be subjected to over voltage or load rejection when they are operating during disturbance in a power system. Both have the same effect of increasing the real value of maximum flux density in core. The increase in the core loss results in thermal damage of insulation if proper cooling is not provided. These damages occurs even when disturbance occurs for extremely short duration.

Automatic method of detection of over excitation conditions is recommended. The level of excitation for power transformer are specified in ASA standard, which are highlighted here.

At larger excitation or applied voltages, the increased flux strays or leaks, into locations inside the transformer where the presence of normal flux is not even anticipated. The stray flux links transformer windings, leads, nuts and bolts, tank boundary, etc. The stray flux causes the eddy current and the hysteresis losses in ferromagnetic materials to which it is linked and eddy current losses to conducting materials. The heat generated due to stray flux linking cause damage to the insulation material. Under extreme conditions of heat the plastic parts such bobbin (former) used in fabrication of power transformer may also deform.

In case of overheating, alarms are sounded to activate cooling accessories, like bank of fans, coolant pumps, etc.

A power transformer must be capable of operating:

(a) At 110% of rated output voltage at no-load.
(b) At 105% of rated secondary voltage while delivering rated output kVA.

These limitations apply at rated frequency and the rated voltage and kVA at any tap position.

3.7.4 Phasor Diagram

The transformer has two separate windings namely, primary and secondary. Their primary and secondary resistances, respectively, are R_1 and R_2. The flux produced by the primary winding almost links the secondary winding and vice versa. A small amount of flux produced by these windings does not link any of the windings. This leakage is reflected in equivalent circuit as *leakage reactance* of these two windings, i.e., X_1 and X_2, respectively. When an AC voltage of magnitude $\boldsymbol{V}_1$ is applied across the primary winding, it gives rise to an induced EMF, $E_1 = 4.44 f \phi_{\max} N_1$. From Lenz's law, we know that this EMF opposes the applied voltage and results in a flow of current given as

$$\boldsymbol{I}_1 = \frac{\boldsymbol{V}_1 - \boldsymbol{E}_1}{R_1 + jX_1}$$

Thus, for the primary winding we can write

$$\boldsymbol{V}_1 = \boldsymbol{E}_1 + \boldsymbol{I}_1(R_1 + jX_1) \tag{3.21}$$

The secondary induced EMF $E_2 = 4.44 f \phi_{\max} N_2$ is responsible for transfer of power from primary to secondary winding. Thus

$$\boldsymbol{E}_2 = \boldsymbol{V}_2 + \boldsymbol{I}_2(R_2 + jX_2) \tag{3.22a}$$

When referred to primary side, we get

$$\boldsymbol{E}_2^/ = \boldsymbol{V}_2^/ + \boldsymbol{I}_2^/\left(R_2^/ + jX_2^/\right) \tag{3.22b}$$

Here subscript '1' is used for primary and '2' for secondary. The quantities with superscript '/' refers to referred quantities, as explained earlier. Equations (3.21–3.22) can be represented in form of phasor diagrams of a transformer. Figure 3.20a–c show the phasor diagrams of transformer for lagging, unity, and leading power factors on full-load on the secondary-side. Similar phasor diagrams for different power factors such as zero power factor (Z.P.F.) lagging (purely inductive load) and Z.P.F. leading (purely capacitive load) can be developed. Angle θ_1 is the primary power factor and θ_2 is the secondary p.f. angle, and θ_0 is the no-load power factor angle. Here, it must be clearly understood that the quantities X_1 and X_2 are the leakage reactance of primary and secondary windings respectively. Although, convention is to use lower case letters for equivalent circuit parameters of electrical motors and transformer – like resistances and leakage reactance, magnetizing reactance and core loss resistance, in this book upper case letters are used. These quantities should not be confused respectively with the quantity ωL_1 and ωL_2, where L_1 and L_2 are the self-inductances of the respective winding.

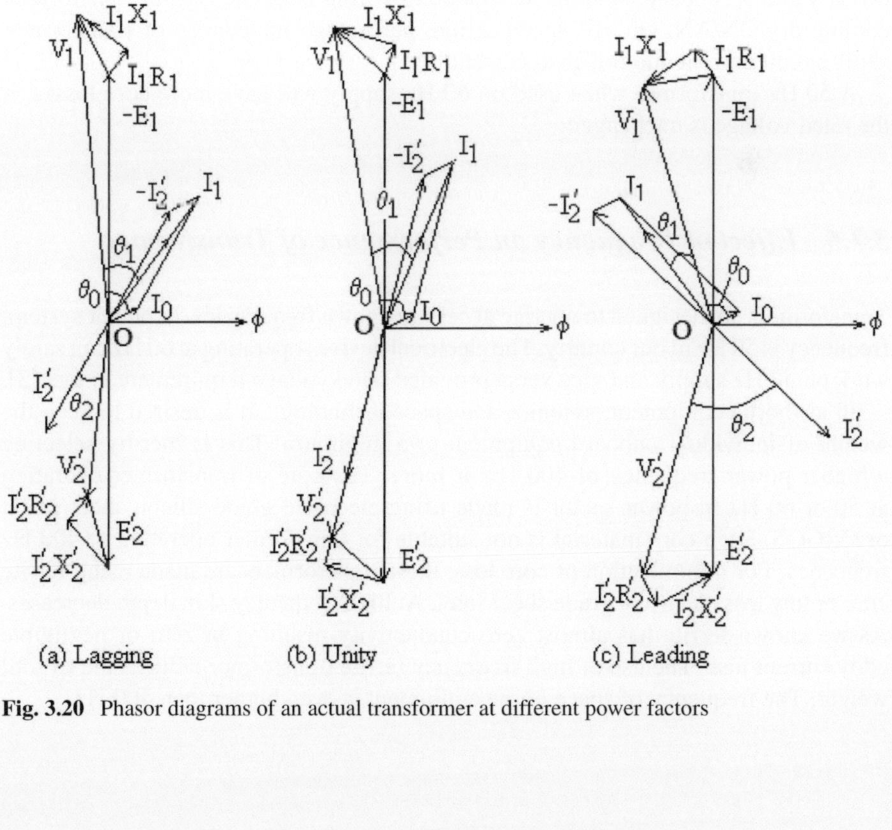

Fig. 3.20 Phasor diagrams of an actual transformer at different power factors

3.7.5 Transformer Labeling and Nomenclature

Labelling of transformer is done as per IS specifications. The high-voltage (HT) terminals are marked as 1 U, 1 V, 1 W and low-voltage (LT) terminals are marked as 2 U, 2 V, 2 W. When individual windings are connected in star or delta, the same should be done step-by-step. The terminals joined together for forming delta connection must have zero potential [3, 6].

The name plate rating of a transformer is given on the outer frame of a transformer. It has the name of manufacturer and its address. Since transformer is an energy transfer device, its rating is in kVA and not in kW. Although, it consumes certain losses, the prime function is the transfer of electrical power at different potential. The current through the windings of transformer is calculated as Volt Ampere/Volt. Due to heating the temperature rises above ambient temperature. The allowable temperature rise is also indicated on the name plate. If the transformer is designed to operate at 110 °C and the maximum allowable temperature rise is 50 °C, the maximum operating temperature can be 160 °C. The voltage ratings of the primary and secondary windings are indicated on the name plate.

The information of number of phases, operating frequency, connection in parallel or series, three-phase star or delta, tertiary winding, etc., are also mentioned.

It also bears the model number, serial number, year, weight, wiring instructions of primary and secondary winding, schematic of wiring diagram, type of transformer, cooling dry/ON/AN, etc., IP specification, percentage impedance of transformer, short-circuit current and full-load current.

A 50 Hz transformer when used on 60 Hz supply will have more core losses, if the rated voltage is unchanged.

3.7.6 Effect of Frequency on Performance of Transformer

Transformers are designed to operate at certain known frequencies. In power system, frequency is 50 Hz in our country. The electrical devices operating at 60 Hz can safely work on 50 Hz supply and vice versa provided rated voltage requirement is met [3].

In air-borne equipment, avionics and space technology, it is desired to keep the weight of individual onboard equipment to a minimum. This is met by selecting a higher power frequency of 400 Hz or more. The core of transformer operating at 50 or 60 Hz in power sector is made using electrical grade silicon sheet metal or CRGOS. Such core material is not suitable for transformer operating at 400 Hz frequency. For minimization of core loss, these transformers are made using ferrite core or low loss electrical grade sheet steel. At high frequency skin depth decreases. As we known ferrite has almost zero conductivity resulting in zero or negligible eddy current loss. The use of high frequency ferrite transformer reduces the overall weight. The frequency of space borne equipment is even higher than 400 Hz.

High frequency ferrite core transformers were also used for obtaining large DC voltage (of the order of 2000–5000 V) required for operation of cathode ray tube (CRT) of earlier models of television and cathode rays oscilloscope (CRO), etc. For this power frequency voltage is converted to high-frequency alternating current voltage, which is stepped up using multiwinding transformer. The individual secondary windings are then connected to a rectifier. The available DC from individual rectifiers is again cascaded to yield large DC voltage sufficient to operate CRT.

Intermediate Frequency Transformer is tiny ferrite or air core high-frequency transformers used in radio frequency circuits for modulation and demodulation. Similarly pulse transformers used in Power electronics circuit transfers high-frequency low-voltage signal are smaller in size and also made using ferrite core.

Consider a case when a normal transformer operating at base frequency f_b = 50 Hz is subjected to varying frequency. The behavior of transformer can be easily explained using per phase equivalent circuit. At any other frequency f, the parameter magnetizing reactance X_M will be modified by a factor of (f/f_b). Similarly, the leakage reactance will also modify by a factor (f/f_b).

At very low frequency, this causes large current to flow in primary winding. However, the voltage available across X_M reduces significantly. The core loss resistance depends on the core loss-hysteresis and eddy current losses. The output voltage drops considerably. For DC input voltage, the current is limited by the winding resistance alone with no-voltage on the secondary side. Is there any back EMF on primary winding? For a medium range of frequency, the transformer operates very well, as such voltage gain is almost unaffected. For high frequency, magnetizing reactance increases manifold, the core loss also increases, the core loss resistance 'R_c' reduces considerably, the winding resistance increases due to severe skin effect and also there

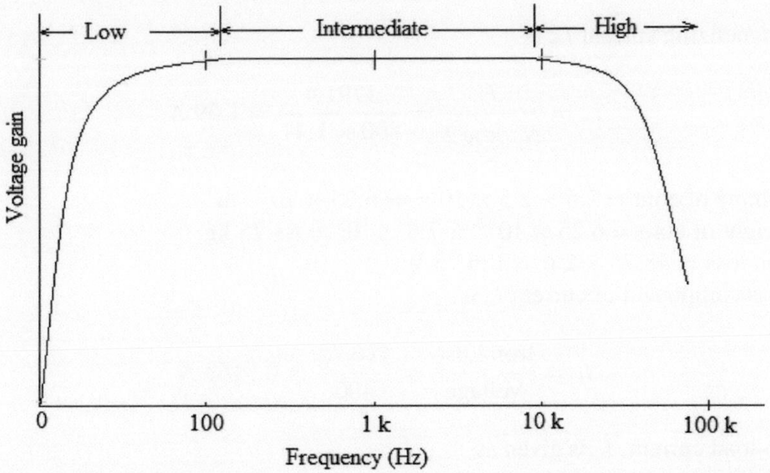

Fig. 3.21 Voltage gain (ratio of output voltage and input voltage) with respect to frequency

is presence of inter turn capacitance. Due to these reasons the voltage gain reduces drastically. These have been shown in Fig. 3.21.

Example 3.5 A single-phase, 400 V, 50 Hz, transformer is built from stamping having a relative permeability of 1000. The length of the flux path is 2.5 m, the area of cross section of core is 2.5×10^{-3} m^2 and the primary winding has 800 turns. Estimate the maximum flux and no-load current and no-load power factor of the transformer. The iron loss at the working flux density is 2.6 W/kg. Density of iron is 7.8×10^3 kg/m^3.

Solution

Cross-sectional area of core $A_i = 2.5 \times 10^{-3}$ m^2. From EMF equation, we have

$$E = 4.44f\,\phi_{\max} N_p = 4.44 f B_m A_i N_p$$

Thus,

$$B_m = \frac{400}{4.44 \times 50 \times 2.5 \times 10^{-3} \times 800} = 0.9\ \text{T}$$

Flux in core is

$$\phi_m = B_m \times A_i = 0.9 \times 2.5 \times 10^{-3} = 2.25 \times 10^{-3}\ \text{Wb}$$

Magnetizing MMF is given as

$$\mathcal{F} = \mathcal{R} \times \phi_m = \frac{L_i \times B_m}{\mu_0 \mu_r} = \frac{2.5 \times 0.9}{4 \times 3.14 \times 10^{-7} \times 1000} = 1791.4\ \text{At}$$

Magnetizing current I_m is

$$I_m = \frac{\mathcal{F}}{N_p \times \sqrt{2}} = \frac{1791.4}{800 \times 1.41} = 1.59\ \text{A}$$

Volume of core $= 2.5 \times 2.5 \times 10^{-3} = 6.25 \times 10^{-3}$ m^3.
Weight of core $= 6.25 \times 10^{-3} \times 7.8 \times 10^3 = 48.75$ kg.
Iron loss $= 48.75 \times 2.6 = 126.75$ W.
Loss component of current I_c is

$$I_c = \frac{\text{Iron Loss}}{\text{Voltage}} = \frac{126.75}{400} = 0.3169\ \text{A}$$

No-load current, I_0 is given as

$$I_0 = \sqrt{I_c^2 + I_m^2} = \sqrt{1.76^2 + 0.3169^2} = 1.788\ \text{A}$$

No-load power factor angle

$$\theta_0 = \tan^{-1}\left(\frac{1.76}{0.3168}\right) = 79.796^\circ$$

Example 3.6 A 220/110 V, 50 Hz ideal transformer is rated at 5 kVA. Calculate

(a) the turn ratio,
(b) the primary and secondary currents (i) at full-load unity power factor and (ii) at 4 kW load at 0.8 lagging power factor.

Solution

Given VA rating, $S = 5 \times 1000$ VA; $V_1{:}V_2$: 220:110; $f = 50$ Hz.

(a) Turn ratio,

$$a = \frac{V_1}{V_2} = \frac{220}{110} = 2$$

(b) (i) At full-load unity power factor,

The primary current at full-load, $I_1 = \frac{S}{V_1} = \frac{5000}{220} = 22.73$ A.
The secondary current at full-load, $I_2 = \frac{S}{V_2} = \frac{5000}{110} = 45.45$ A.

(ii) At 4 kW 0.8 lagging power factor load,

Primary current, $I_1 = \frac{P}{V_1 \times PF} = \frac{4000}{220 \times 0.8} = 22.72$ A.
Secondary current, $I_1 = \frac{P}{V_1 \times PF} = \frac{4000}{110 \times 0.8} = 45.45$ A.

Example 3.7 In an ideal transformer of turns ratio 100/200, a resistance of 1.0 Ω is connected across secondary having 200 turns. Obtain the referred value of resistance from primary side.

Solution

Here, $R = 1.0\ \Omega$; Turn ratio $a = 100/200 = 0.5$.
The referred value of 1.0 Ω resistance from primary side is,

$$R^{/} = a^2 R = 0.5^2 \times 1.0 = 0.25\ \Omega$$

Example 3.8 A 60 kVA, 50 Hz, 6.6 kV/220 V transformer has primary and secondary resistances of 7.8 Ω and 0.0085 Ω, respectively. Calculate the equivalent resistances as referred to primary and secondary respectively. Also, calculate the full-load copper loss.

Solution

Given $S = 60$ kVA; $V_1{:}V_2$:: 6600:220; $R_1 = 7.8\ \Omega$; $R_2 = 0.0085\ \Omega$.
Transformation ratio, $a = \frac{V_1}{V_2} = \frac{6600}{220} = 30$.

The resistance referred to primary side,

$$R_p = R_1 + a^2 R_2 = 7.8 + 30^2 \times 0.0085 = 15.45\ \Omega$$

The resistance referred to secondary side,

$$R_s = \frac{R_1}{a^2} + R_2 = \frac{7.8}{30^2} + 0.0085 = 0.01716\ \Omega$$

Total copper loss

$$R_p I_p^2 = 15.45 \times \left(\frac{60 \times 1000}{6600}\right)^2 = 1276.84\ \text{W}$$

Similarly, total copper loss from secondary can also be calculated,

$$R_s I_s^2 = 0.01716 \times \left(\frac{60 \times 1000}{220}\right)^2 = 1276.36\ \text{W}$$

Example 3.9 In an ideal 11 kV/440 V transformer a resistance of 1.0 Ω is connected on secondary side. Obtain its referred value from primary side.

Solution

The transformation ratio, $a = \frac{V_1}{V_2} = \frac{11000}{440} = 25$.

The referred value of resistance from primary side,

$$R_2' = a^2 R_2 = 25^2 \times 1.0 = 625\ \Omega$$

3.8 Determination of Equivalent Circuit Parameters

The equivalent circuit parameters are experimentally determined using open-circuit and short-circuit tests [2, 3, 6].

3.8.1 Open-Circuit Test

A simple circuit diagram for conducting open-circuit test is shown in Fig. 3.22a. This test is conventionally performed from low-voltage side. In a laboratory, transformers of different voltage ratings such as 11 kV/440 V, 6.6 kV/440 V can be easily tested. This requires a voltage of 440 V, which is easily available in laboratories. The open-circuit parameters are referred to low-voltage side. The connected voltmeter should be able to read rated voltage of LT winding. The magnetizing current being 1–5% of full-load current, as such the ammeter and wattmeter should be chosen accordingly.

The phasor diagram under open circuit condition is shown in Fig. 3.22b. The voltages $\boldsymbol{V}_1$ and $\boldsymbol{V}_2$ and flux ϕ under no-load conditions are shown in the phasor diagram. The flux ϕ lags behind the voltage $\boldsymbol{V}_1$ or $-\boldsymbol{E}_1$ by 90°.

In open-circuit condition, open-circuit current $\boldsymbol{I}_o$ is assumed to flow in the parallel combination of R_c and X_m. The no-load current is

$$\boldsymbol{I}_0 = \boldsymbol{I}_c + \boldsymbol{I}_m \tag{3.23}$$

The real component of no-load current is given as

$$I_c = \frac{V}{R_c} = I_0 \cos\theta_0 \tag{3.24}$$

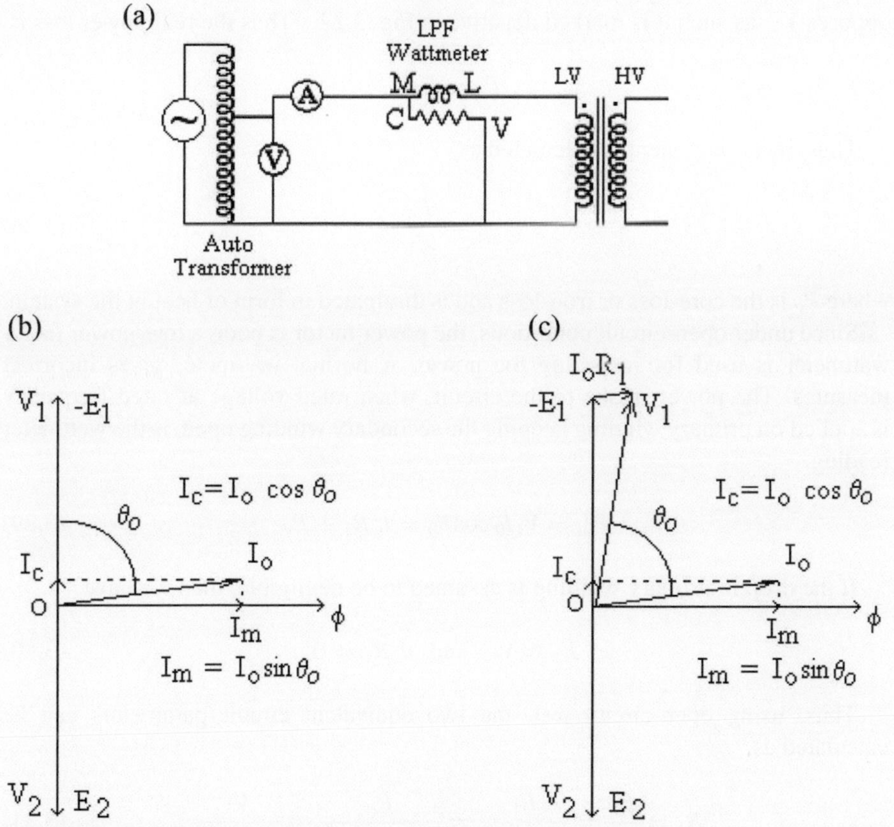

Fig. 3.22 **a** Open circuit test on two-winding transformer; **b** phasor diagram in open circuit conditions and **c** open-circuit phasor diagram with correction for resistive drop

where θ_0 is no-load power factor angle. The imaginary component of open-circuit current is

$$I_m = \frac{V}{jX_m} = I_0 \sin\theta_0 \tag{3.25}$$

In a two-winding transformer, the transformer-induced EMF in primary winding is,

$$\boldsymbol{E}_1 = \boldsymbol{V}_1 - \boldsymbol{I}_0(R_1 + jX_1) \tag{3.26}$$

See Fig. 3.22c, if the winding resistance R_1 is known, then the EMF $\boldsymbol{E}_1$

$$\boldsymbol{E}_1 \approx \boldsymbol{V}_1 - \boldsymbol{I}_0 R_1$$

I_1R_1 drop is usually less and is ignored. It is assumed that $\boldsymbol{E}_1 \approx \boldsymbol{V}_1$. Since $\boldsymbol{E}_1$ opposes $\boldsymbol{V}_1$, as such it is marked negative in Fig. 3.22c. Thus the real power loss is

$$P_c = E_1 I_c \tag{3.27}$$

Thus, real component is calculated as,

$$I_c = \frac{P_c}{E_1} \tag{3.28}$$

where P_c is the core-loss or iron-loss and is dissipated in form of heat in the system.

Since under open-circuit conditions, the power factor is poor, a low-power factor wattmeter is used for recording the power. A normal wattmeter gives incorrect measures. The power intake to the circuit, when rated voltage at rated frequency is applied on primary winding keeping the secondary winding open, is the wattmeter reading.

$$P_{\text{in}} = V_1 I_0 \cos\theta_0 = I_0^2 R_1 + P_c \tag{3.29}$$

If the drop in primary winding is assumed to be negligible, then we have

$$E_1 \approx V_1 \quad \text{and} \quad I_0 R_1 = 0 \tag{3.30}$$

Thus, using open-circuit test, the two equivalent circuit parameters can be calculated as,

$$R_c = \frac{E_1}{I_c} = \frac{E_1}{I_0 \cos\theta_0} = \frac{P_c}{(I_0 \cos\theta_0)^2} \approx \frac{V_1}{I_0 \cos\theta_0} \tag{3.31}$$

$$X_m = \frac{E_1}{I_m} = \frac{E_1}{I_0 \sin\theta_0} \approx \frac{V_1}{I_0 \sin\theta_0} \tag{3.32}$$

These values of R_c and X_m are referred to low-voltage side. Their values referred from high-voltage sides can be calculated as,

$$R_c^/ = a^2 R_c$$

$$X_m^/ = a^2 X_m$$

Here transformation ratio,

$$a = \left(\frac{V_{\mathrm{HV}}}{V_{\mathrm{LV}}}\right)$$

3.8.2 Short-Circuit Test

A simple circuit diagram for conducting a short-circuit test is shown in Fig. 3.23a. As a convention, the short-circuit test is performed from high-voltage side. In a single 100 kVA, 11 kV/440 V transformer, the rated high-voltage side and low-voltage side currents are,

$$I_{\mathrm{HV}} = \frac{100 \times 1000}{11 \times 1000} = 9.09\ \mathrm{A}$$

$$I_{\mathrm{LV}} = \frac{100 \times 1000}{440} = 227.3\ \mathrm{A}$$

In this test, one of the winding (low-voltage winding) is short-circuited and a voltage of about 5–10% is applied across high-voltage winding terminals, causing the rated current to flow. Since the core loss depends on the applied voltage, the losses in short-circuit test is entirely taken as copper loss. The measurement of lower current (say, 9.09 A) on high-voltage winding is easier, than that of low-voltage winding (say, 227.3 A) current. Thus, as a convention, the short-circuit test is conducted from high-voltage side keeping low voltage terminals short-circuited. The ratings of the instruments used are accordingly selected.

The phasor diagram is shown in Fig. 3.23b. The secondary EMF is entirely absorbed by circulating current in the secondary impedance $Z_2 = R_2 + jX_2$. The current is likely to be high. The applied voltage is very less, around 10% of rated voltage, which is absorbed by primary EMF $\boldsymbol{E}_1$ and voltage drop in primary leakage impedance $Z_1 = R_1 + jX_1$. If $\boldsymbol{E}_1$ is about 0.05 per unit (p.u.) or (5%) of rated voltage, the flux drops to about 0.0025 of original value and is well below the saturation level. As such, the core loss is ignored and, $E_1 >> E_2$ and $I_2 >> I_1$.

In short-circuit test, if W_{sc} is the wattmeter reading or rated copper loss, V_{sc} is applied voltage, I_{sc} (rated high-voltage side current) is short-circuit current. Total

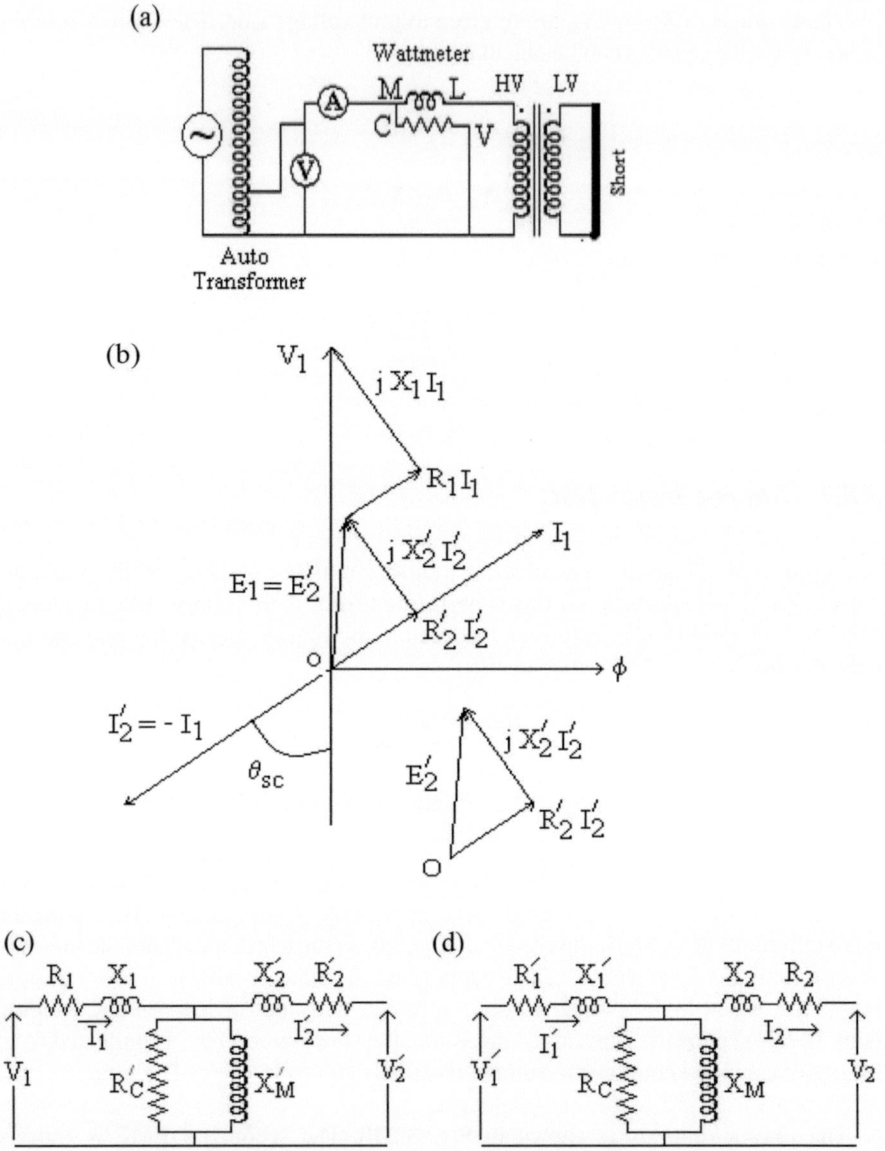

Fig. 3.23 **a** Short-circuit test on two-winding transformer; **b** phasor diagram under short-circuit conditions; **c** equivalent circuit referred to primary side and **d** equivalent circuit referred to secondary side

resistances and leakage reactance of transformer referred to primary (HV) are

$$R_{\text{EX1}} = R_1 + R_2^{/} = R_1 + a^2 R_2 \tag{3.33}$$

$$X_{\text{EX1}} = X_1 + X_2^{/} = X_1 + a^2 X_2 \tag{3.34}$$

From simple circuit analysis, we have

$$R_{\text{EX1}} = \frac{W_{\text{sc}}}{I_{\text{sc}}^2} \tag{3.35}$$

$$Z_{\text{EX1}} = \frac{V_{\text{sc}}}{I_{\text{sc}}} \tag{3.36}$$

$$R_{\text{EX1}} = \sqrt{\left(Z_{\text{EX1}}^2 - R_{\text{EX1}}^2\right)} \tag{3.37}$$

These quantities are referred from high-voltage side. Their values referred to low-voltage side are calculated as,

$$R_{\text{EX2}} = \frac{1}{a^2} R_{\text{EX1}}$$

$$X_{\text{EX2}} = \frac{1}{a^2} X_{\text{EX1}}$$

Here a is the transformation ratio. The equivalent circuit referred to primary and secondary sides are given in Fig. 3.23c, d, respectively.

Example 3.10 A single-phase 440/220 V two-winding transformer takes a no-load current of 5 A at 0.2 power factor lagging. If the secondary supplies a current of 120 A at 0.8 power factor lagging, calculate the current taken by the primary.

Solution

Here transformation ratio, $a = \frac{440}{220} = 2$.

No-load power factor angle, $\theta_0 = \cos^{-1} 0.2 = 78.46^\circ$ lagging,

$$\sin\theta_0 = 0.979$$

No-load current, $\boldsymbol{I}_0 = 5\angle - 78.46^\circ = 5 \times (0.2 - j0.979)$A.

The power factor angle at load $\theta = \cos^{-1} 0.8 = 36.87^\circ$, $\sin\theta = 0.6$.

The secondary current is, $\boldsymbol{I}_2 = 120\angle - 36.87^\circ = 120 \times (0.8 - j0.6)$A.

The secondary current referred to primary side is,

$$\frac{I_2}{a} = \frac{120(0.8 - j0.6)}{2} = 60(0.8 - j0.6)\text{A}$$

The primary current is,

$$I_1 = I_0 + \frac{I_2}{a} = 5(0.2 - j0.979) + 60(0.8 - j0.6) = 49 - j40.895 \text{ A}$$
$$|I_1| = \sqrt{49^2 + 40.895^2} = 63.82 \text{ A}$$

and $\theta_1 = \tan^{-1}\left(\frac{40.895}{49}\right) = 39.84°$.

Thus, primary current is, $\boldsymbol{I}_1 = 63.82\angle - 39.84°$ A.

Students are expected to draw *to-the scale* phasor diagram and verify the above primary current's magnitude and phase angle.

Example 3.11 A 50 kVA, 2300/230 V, 50 Hz is connected to 230 V mains. The high-voltage winding is kept open. The meters indicate the following readings:

$$W_0 = 190 \text{ W}, \; V_0 = 230 \text{ V}, \; I_0 = 6.5 \text{ A}$$

The resistance of low-voltage winding is 0.06 Ω. Find

(a) Core loss.
(b) No-load power factor angle and power factor.
(c) Current through core loss component and core loss resistance.
(d) Current through magnetizing component X_m.
(e) % of exciting current.

Solution

Given $S = 50$ kVA; $f = 50$ Hz; transformation ratio $a = 10$;
Test data $W_0 = 190$ W, $V_0 = 230$ V, $I_0 = 6.5$ A, $R_2 = 0.06\ \Omega$.

(a) Core loss $= W_c = W_0 - I_0^2 R_2 = 190 - 6.5^2 \times 0.06 = 187.46$ W.
(b) No-load power factor,

$$\cos\theta_0 = \frac{W_c}{V_0 I_0} = \frac{187.46}{230 \times 6.5} = 0.1253$$

The no-load power factor angle $\theta_0 = 82.79°$, $\sin\theta_0 = 0.9921$.

(c) The current through R_c is

$$I_c = I_0 \cos\theta_0 = 6.5 \times 0.1253 = 0.8125 \text{ A}$$

$$E = 230\angle 0° - 6.5 \times (0.1253 - j0.9921) = 229.19\angle 1.61° \text{ V}.$$

The core loss resistance is

$$R_c = \frac{W_c}{I_c^2} = \frac{187.46}{0.8125^2} = 283.96\ \Omega$$

(d) The current through the magnetizing branch is given as

$$I_m = I_0 \sin\theta_0 = 6.5 \times 0.9921 = 6.488\ \text{A}$$

The magnetizing reactance,

$$X_m = \frac{E}{I_m} = \frac{229.18}{6.488} = 35.22\ \Omega$$

(e) Full-load current is

$$I_{fl} = 50 \times 1000/230 = 217.39A$$

Percentage of exciting current

$$= \frac{6.5}{217.39} \times 100 = 2.99\%$$

3.9 Separation of Hysteresis and Eddy Current Losses

As pointed out in Eqs. (3.11) and (3.12), the hysteresis loss and eddy current losses are given as,

$$P_h \ \alpha \ fB_m^x$$
$$P_e \ \alpha \ f^2B_m^2$$

We know that, the EMF equation is given as

$$E = 4.44f\phi_m N$$

The maximum value of flux density or flux can be kept constant by keeping the ratio of voltage and frequency constant. The variable frequency source in laboratories is generally obtained using the following:

(a) Synchronous generator/alternator with speed control using prime mover.
(b) Slip ring induction machine connected to three phase with open slip rings run using prime mover in direction opposite to rotating magnetic field.
(c) Three-phase static inverter.

The frequency of alternator output can be changed by changing rotational speed. Voltage can be controlled by using either field control or using an autotransformer. If the maximum value of flux density in core is to be kept indirectly constant, then the ratio of voltage and frequency is maintained constant. Then we have

$$P_h = Af$$
$$P_e = Bf^2$$

where A and B are the constants. Thus, the core loss P_c is

$$P_c = Af + Bf^2$$

The core loss is obtained by no-load test of the transformer under rated voltage conditions. If the resistance of winding R is known then

$$\text{Core loss} = \text{No - load active power}(W) - I_0^2 R = P_c$$

The measurement is taken at two different frequencies f_1 and f_2, keeping applied voltage accordingly so as to maintain the ratio of voltage and frequency a constant. The no-load corrected loss are:

$$P_{c1} = Af_1 + Bf_1^2 \tag{3.38}$$

$$P_{c2} = Af_2 + Bf_2^2 \tag{3.39}$$

Here constants A and B can be separated for computing the hysteresis and eddy current losses at any supply frequency.

Example 3.12 A six-pole single-phase alternator is used to obtain variable frequency supply for separating the core loss of an autotransformer. A two-winding transformer is tested at no-load condition at two different frequencies. The winding has 200-turns and has resistance of 0.66 Ω. The test data are

$$N_1 = 1000 \text{ RPM}; \ V_0 = 250 \text{ V}, \ I_0 = 0.5 \text{ A}, \ W_0 = 230 \text{ W};$$
$$N_2 = 900 \text{ RPM}; \ V_0 = 225 \text{ V}, \ I_0 = 0.5 \text{ A}, \ W_0 = 200 \text{ W}$$

Calculate hysteresis and eddy current losses at 50 Hz.

Solution

Given pole = 6;

$$N_1 = 1000 \text{ RPM}; \ V_0 = 250 \text{ V}, \ I_0 = 0.5 \text{ A}, \ W_0 = 230 \text{ W};$$
$$N_2 = 900 \text{ RPM}; \ V_0 = 225 \text{ V}, \ I_0 = 0.5 \text{ A}, \ W_0 = 200 \text{ W}$$

The supply frequency is $f_1 = N_1 p/120 = 50$ Hz;
Ratio of voltage and frequency is 250/50 = 5 V/Hz.
We know that,

$$A + Bf = \frac{W_0 - I_0^2 R}{f}$$

$$A + 50B = \frac{230 - 0.5^2 \times 0.66}{50} = 4.5967$$

The supply frequency is $f_2 = N_2 p/120 = 45$ Hz;
Ratio of voltage and frequency is 225/45 = 5.
We know that,

$$A + 45B = \frac{200 - 0.5^2 \times 0.66}{45} = 4.441$$

Thus, we get

$$B = 0.03114 \text{ W/Hz}^2$$

$$A = 4.5967 - 50 \times 0.03114 = 3.0397 \text{ W/Hz}$$

At 50 Hz, hysteresis loss is

$$P_h = 50 \text{ A} = 151.985 \text{ W}$$

Eddy current loss is

$$P_e = Bf^2 = 77.85 \text{ W}$$

Example 3.13 A transformer is tested at two different voltages and frequencies under open circuit conditions. The variable frequency supply was obtained using four-pole single-phase alternator. The test data are:

$$V_{01} = 250 \text{ V} \quad W_{01} = 55 \text{ W} \quad N_1 = 1500 \text{ RPM}$$
$$V_{02} = 200 \text{ V} \quad W_{02} = 40 \text{ W} \quad N_2 = 1200 \text{ RPM}$$

Calculate hysteresis and eddy current losses at these two frequencies? Ignore the primary resistance.

Solution

Given

$$V_1 = 250 \text{ V} \quad W_1 = 55 \text{ W}$$
$$V_2 = 200 \text{ V} \quad W_2 = 40 \text{ W}$$

The supply frequencies are $f_1 = N_1 p/120 = 50$ Hz and $f_2 = N_2 p/120 = 40$ Hz. Since the ratios of voltage and alternator speed (or frequency) are same, we have

$$Af + Bf^2 = W_0$$

$$A + 50B = \frac{55}{50} = 1.1$$

$$A + 40B = \frac{40}{40} = 1.0$$

Solving, we get

$$10B = 0.1 \quad \text{or} \quad B = 0.01 \text{ W/Hz}^2$$

$$A = 1.0 - 40 \times 0.01 = 0.6 \text{ W/Hz}$$

Hysteresis loss at 50 Hz is $P_h = Af = 0.6 \times 50 = 30$ W.
Eddy current loss at 50 Hz, $P_e = Bf^2 = 0.01 \times 50^2 = 25$ W.
Hysteresis loss at 40 Hz is $P_h = Af = 0.6 \times 40 = 24$ W.
Eddy current loss at 40 Hz, $P_e = Bf^2 = 0.01 \times 40^2 = 16$ W.

3.10 Difference Between Circuit Parameters and Equivalent Circuit Parameters

It is important to note that self-inductance (L) of a transformer winding is not equal to its leakage inductance l. Also, the mutual inductance of the transformer coils is not equal to the magnetizing inductance. This has already been discussed in Chap. 1. In other words,

$$L_1 \neq l_1 \tag{3.40}$$

$$L_2 \neq l_2 \tag{3.41}$$

$$M \neq \frac{X_m}{\omega} \tag{3.42}$$

where

L_1 = Self-inductance of primary.
L_2 = Self-inductance of secondary.
l_1 = Primary leakage inductance, $l_1 = \frac{X_1}{\omega}$.
l_2 = Secondary leakage inductance, $l_2 = \frac{X_2}{\omega}$.
ωM = Mutual reactance of the transformer.
X_m = The magnetizing reactance of the transformer.

A suitable relationship between the coil parameters (L_1, L_2, M) and the equivalent circuit parameters ($l_1, l_2, X_m/\omega$) can be developed by comparing Fig. 1.15 of Chap. 1.

Now, we have

$$\text{Impedance of the primary winding alone} = R_1 + j\omega L_1 \tag{3.43}$$

$$\text{Impedance of secondary winding alone} = R_2 + j\omega L_2 \tag{3.44}$$

$$\text{Mutual reactance} = j\omega \text{M} \tag{3.45}$$

The transformation ratio, $a = N_1/N_2$.
The angular frequency of the supply $= \omega$.
It can be shown that primary leakage inductance,

$$jX_1 = j(\omega L_1 - a\omega M)$$

or, primary leakage inductance,

$$l_1 = L_1 - aM \tag{3.46}$$

Also secondary leakage reactance is

$$j.X_2 = j\left(\omega L_2 - \frac{\omega M}{a}\right)$$

or, the secondary leakage inductance

$$l_2 = L_2 - \frac{M}{a} \tag{3.47}$$

The magnetizing reactance referred to primary side

$$X_m = a\omega M \tag{3.48}$$

We have already assumed the value of R_c as

$$R_c = \frac{E_1}{I_c} \tag{3.49}$$

The equivalent circuit referred to primary as derived from the circuit parameters is given (Fig. 1.16) in Fig. 3.24.

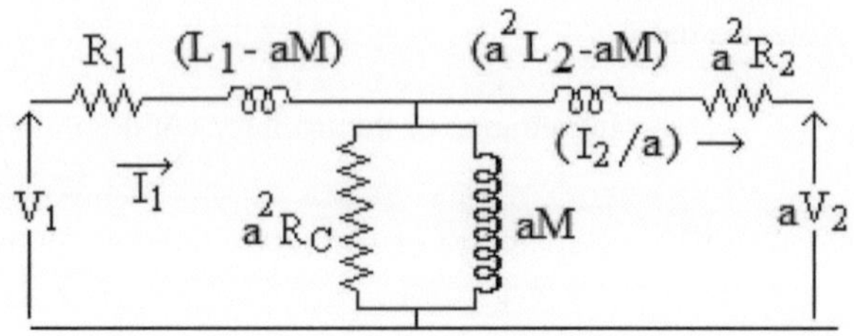

Fig. 3.24 Equivalent circuit of a two-winding transformer derived from circuit parameters

3.11 Power Efficiency

In a transformer, a certain fraction of power is wasted in heat as copper and core losses. Amongst these losses, the copper loss (I^2R or Ohmic loss) depends on the current flowing through secondary terminals, and a replica (of referred value) of this current also flows in the primary windings. The core-loss depends on the square of the voltage. Since the supply voltage does not change, it is taken as a constant. It is assumed that volt ampere or VA rating of transformer is S, rated copper loss or the wattmeter reading under short-circuit test is P_{CU}, and rated core-loss or wattmeter reading under open-circuit test is $P_I = P_o$. The % power efficiency, η, of a two-winding transformer operating at kth fraction of load and at a power factor $\cos\theta$ is,

$$\eta = \left[\frac{\text{Output}}{\text{Input}}\right]100 = \left[\frac{\text{Input} - \text{Losses}}{\text{Input}}\right]100$$

$$\eta = \left[\frac{kS\cos\theta}{kS\cos\theta + P_I + k^2 P_{\text{CU}}}\right]100$$

$$\eta = \left[1 - \frac{P_I + k^2 P_{\text{CU}}}{kS\cos\theta + P_I + k^2 P_{\text{CU}}}\right]100 \tag{3.50}$$

If I_{fl} is the rated current then current at kth fraction of load is $I(k) = k\, I_{fl}$. A typical variation of power efficiency is shown in Fig. 3.25. It has a maximum value at certain fraction of load k and has a lesser efficiency if the load current is either increased or decreased. Mathematically, the maxima of Eq. (3.50) can be determined. Differentiating the equation with respect to k and putting it to zero, we get

$$\frac{\mathrm{d}\eta}{\mathrm{d}k} = 0 \tag{3.51}$$

Solving this we, get

$$P_I = k^2 P_{\text{CU}} \tag{3.52}$$

The power efficiency is highest at a fraction of load k at which the variable copper loss is equal to rated core loss. The fraction k for maximum power efficiency is,

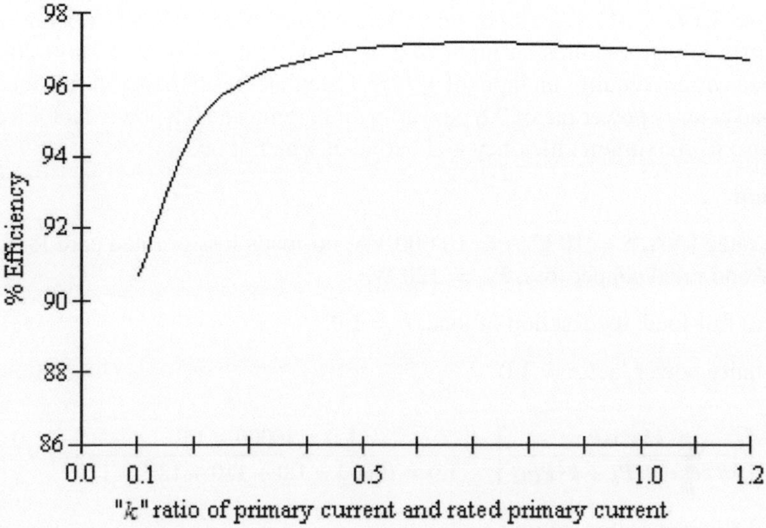

Fig. 3.25 Variation of power efficiency with respect to fraction of load k

$$k = \sqrt{\frac{P_I}{P_{CU}}} = \sqrt{\frac{\text{Rated iron or core loss}}{\text{Rated copper loss}}} \tag{3.53}$$

If supply voltage changes by δth fraction. The core loss is proportional to square of applied voltage. The power efficiency is given as,

$$\%\eta = \left[1 \quad \frac{\delta^2 P_I + k^2 P_{CU}}{k\delta S\cos\theta + \delta^2 P_I + k^2 P_{CU}}\right]100 \tag{3.54}$$

where

$$k = \frac{\text{New Current}}{\text{Rated Current}} \quad \text{and} \quad \delta = \frac{\text{New Voltage}}{\text{Rated Voltage}}$$

However, since rated core loss is much less than that of rated copper loss, the % power efficiency is not largely affected if supply voltage marginally changes. Distribution transformers are designed to have maximum power efficiency near 90–99%, whereas large power transformers are designed to have maximum efficiency at about 50% of rated load. Several power transformers operate in parallel. Some of parallel-connected power transformers are switched OFF, if total shared load is less. This saves the core losses of power transformers (not in circuit) otherwise being wasted in lightly loaded power transformers. High-frequency ferrite transformers have core loss about 1% of their output power.

Example 3.14 A 10 kVA, 50 Hz single-phase transformer, when tested on no-load at normal voltage requires an input of 110 W and when tested on short-circuit at full-load current requires an input of 120 W. Calculate its efficiency at full-load, half full-load at unity power factor 0.8 power factor lagging and 0.8 power factor leading. Find also its maximum efficiency and output at which it occurs.

Solution

Given rated kVA, $S = 10$ kVA $= 10{,}000$ VA, no-loads loss or rated core loss, $P_c = 110$ W and rated copper loss, $P_{\mathrm{sc}} = 120$ W.

(a) For full-load, the fraction of load, $k = 1.0$.

At unity power factor = 1.0

$$\eta = \left[\frac{kS\cos\theta}{kS\cos\theta + P_I + k^2 P_{\mathrm{CU}}}\right] = \frac{1.0 \times 10000 \times 1.0}{1.0 \times 10000 \times 1.0 + 110 + 1.0^2 \times 120} = 97.75\%$$

At 0.8 power factor lagging,

$$\eta = \left[\frac{kS\cos\theta}{kS\cos\theta + P_I + k^2 P_{\mathrm{CU}}}\right] = \frac{1.0 \times 10000 \times 0.8}{1.0 \times 10000 \times 0.8 + 110 + 1.0^2 \times 120} = 97.20\%$$

The efficiency at 0.8 power factor leading will also be 97.20%.

(b) For half full-load, the fraction of load, $k = 0.5$.

At unity power factor = 1.0

$$\eta = \left[\frac{kS\cos\theta}{kS\cos\theta + P_I + k^2 P_{\mathrm{CU}}}\right] = \frac{0.5 \times 10000 \times 1.0}{0.5 \times 10000 \times 1.0 + 110 + 0.5^2 \times 120} = 97.28\%$$

At 0.8 power factor lagging,

$$\eta = \left[\frac{kS\cos\theta}{kS\cos\theta + P_I + k^2 P_{\mathrm{CU}}}\right] = \frac{0.5 \times 10000 \times 0.8}{0.5 \times 10000 \times 0.8 + 110 + 0.5^2 \times 120} = 96.62\%$$

The efficiency at 0.8 power factor leading will also be 96.62%.
The maximum power efficiency will occur at

$$k = \sqrt{\frac{P_c}{P_{\mathrm{SC}}}} = \sqrt{\frac{110}{120}} = 95.74\%$$

$$\eta = \left[\frac{kS\cos\theta}{kS\cos\theta + P_I + k^2 P_{\mathrm{CU}}}\right] = \frac{0.9574 \times 10000 \times 1.0}{0.9574 \times 10000 \times 1.0 + 110 + 0.9574^2 \times 120} = 97.77\%$$

The maximum power efficiency occurs at 95.74% of rated power output. The maximum efficiency at unity power factor is 97.77%.

3.12 Energy Efficiency

Power transformers face load variations 24 hours a day. Sometimes, the load is high, while at other times, it is low. In a power system, a regional load dispatch center is responsible for strategic distribution planning of the bulk power generated from power generating stations for safeguarding the power grid. A typical load curve in 24 hours is shown in Fig. 3.26. Here, the load varies with respect to time in a haphazard manner, the figure shows a normalized curve. The core loss remains constant but copper loss changes with the square of load current. This information is normally shown as *all day efficiency*, which is, in fact, the ratio of output energy to input energy.

$$\begin{aligned}\text{All Day Efficiency} &= \frac{\sum \text{Output Power} \times \text{Duration}}{\sum \text{Input Power} \times \text{Duration}} \\ &= \frac{\sum \text{Output Power} \times \text{Duration}}{\sum (\text{Output Power} + \text{Losses}) \times \text{Duration}}\end{aligned} \tag{3.55}$$

Example 3.15 The load variation in a 500 MVA 400 kV/131 kV power transformer is shown in Fig. 3.26. Rated copper loss and core loss are 5% and 1% respectively, of output power. Calculate all-day efficiency.

Solution

Given rated output power, $S = 500 \times 10^6$ VA, rated copper loss $P_{\text{CU}} = 0.05 \times S = 0.05 \times 500 \times 10^6 \text{ VA} = 25 \times 10^6$ W, rated core loss $P_I = 0.01 \times 500 \times 10^6 \text{ VA} = 5 \times 10^6$ W.

$$\text{Output Energy} = S \times \sum k \times \Delta t$$

$$\begin{aligned}&\text{Output Energy} \\ &= S \times [0.25 \times 3 + 0.75 \times 3 + 1 \times 3 + 0.5 \times 3 + 1.0 \times 3 + 0.25 \times 6 + 1.0 \times 3] \\ &= 7500 \times 1000 \text{ kWH}\end{aligned}$$

$$\begin{aligned}&\text{Energy dissipated in losses} \\ &= 24P_I + \sum (\text{copper Loss} \times \text{duration}) \\ &= 24 \times P_I + P_{\text{CU}} \times \sum k^2 \times \Delta t \\ &= 24 \times 5 \times 10^6 + 25 \times 10^6 \\ &\quad \times \left[0.25^2 \times 9 + 0.75^2 \times 3 + 1^2 \times 9 + 0.5^2 \times 3\right] \\ &= 420{,}000 \text{ kWH}\end{aligned}$$

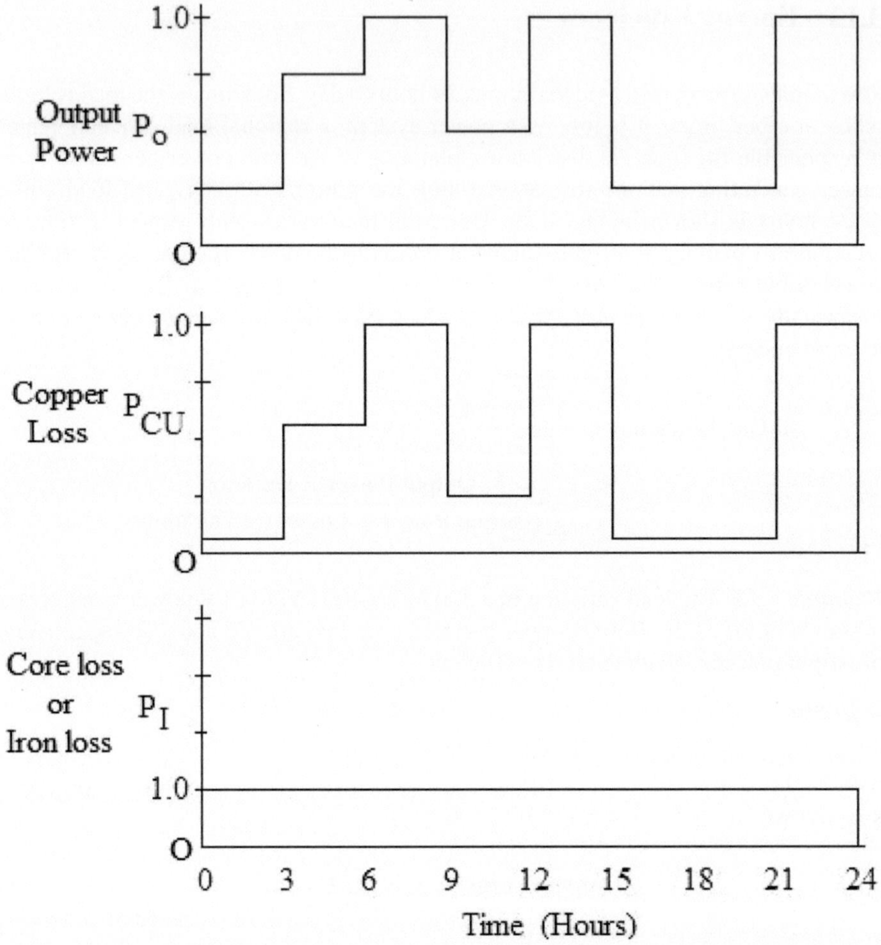

Fig. 3.26 A typical variations of load, copper, and core losses of a power transformer in 24 h

$$\% \text{ All Day Efficiency} = \frac{\text{Output Energy in 24 h}}{\text{Output Energy} + \text{Energy Lost in 24 h}} \times 100$$

$$= \frac{75{,}00{,}000}{75{,}00{,}000 + 420{,}000} = 0.9469 \text{ or } 94.69\%$$

The all-day-efficiency will be different for different load pattern in one day.

3.13 Non-sinusoidal Waveform

The magnetizing current in a transformer will not vary sinusoidally if the flux wave is sinusoidal. When an iron core is subjected to an alternating magnetization MMF, the magnetic core operates in nonlinear portion of B-H curve. If we assume that there is no hysteresis loss in the core, then B-H curve is simply a straight line. A B-H curve having hysteresis is shown in Fig. 3.16b, where on *y*-axis, we have total flux ϕ or flux density B and on *x*-axis, we have magnetic field intensity $H = NI_m/L$, or magnetizing current I_m. Let us assume that flux ϕ varies sinusoidally with respect to time, as depicted in the Fig. 3.15a. When flux has instantaneous value $\phi_1(t)$, the current should have instantaneous value $i_1(t)$ as fixed by the B-H curve and so on. The current waveform thus obtained is no longer varying in sinusoidal manner, unless we assume a linear B-H relationship. If we resolve the curve using Fourier series, we get fundamental frequency components and odd harmonic components. The thirds harmonic is more dominant. It can be shown that if magnetizing current is sinusoidal; the flux waveform will become flattened or peaked, due to third order time harmonics. Here *n*th order flux produces *n*th order secondary EMF. The effective values of secondary EMF in presence of harmonics is

$$E = \sqrt{E_1^2 + E_3^2 + E_5^2 + E_7^2 + \dots}$$

3.14 Concept of Polarity

The concept of polarity is not significant with reference to a single coil when used in an AC circuit. However, polarity assumes great importance when more than one coils are used together in a coupled circuit (electric and magnetic) with suitable interconnections between the coils. **In a coupled AC network, two terminals are said to have same polarity if the instantaneous rise and fall of the alternating voltages are identical.** This can be realized from the following discussions.

Let 'AB' and 'CD' be two coils as shown in Fig. 3.27a, b. If these coils are to be connected in series, we will find that there are two alternative schemes of connections as illustrated in Fig. 3.27a. If these coils are connected to DC supply, each will offer its resistance alone. In either case, the net resistance appearing due to the combination will be $R_1 + R_2$. If **flux is not a consideration**, therefore either form of connection is acceptable when working in steady state DC system.

With an **AC system**, however, each coil will present self-inductance effects in addition to resistance. Moreover, the nature of coupling between these coils will decide the extent of mutual inductance effect that will be present. This is because of the fact that the alternating flux produced along the axis of a coil due to current in it is linked by the coil in its vicinity, which induces a voltage. The nature of this induced voltage decides whether the net inductance offered by the combination will

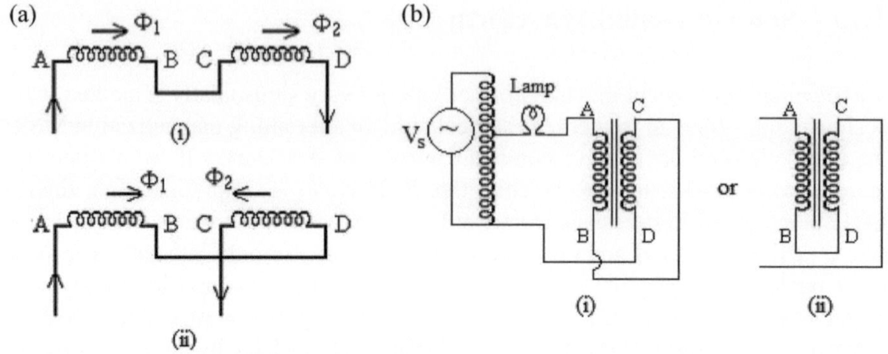

Fig. 3.27 **a** Series addition and series opposition connections of two windings and **b** testing of polarity using lamp

increase or decrease on connecting the coils in series. To further elaborate this point, let us assume that both coils are wound in the same way, say, clockwise when looked at from left hand side. Referring to the Fig. 3.27a, we can see that for connection (i), the current entering terminal '*A*' will cause both coils to produce flux in the same direction so that two fluxes, ϕ_1 and ϕ_2, add. Such a series connection is called *series addition.* In series addition, *B* and *D* are at same polarity. In other words, *A* and *C* are also at same polarity. In connection (ii), the resultant flux is subtractive and the combination is identified as *series opposition or subtractive.* In series opposition, *B* and *C* are at same polarity.

For series addition, the net impedance of the system is given as

$$Z_{eq} = Z_1 + Z_2 + \text{Mutual Reactance Component}$$
$$Z_{eq} = R_1 + R_2 + j\omega(L_1 + L_2 + 2M) \tag{3.56}$$

In series opposition, the net impedance of the system is

$$Z_{eq} = Z_1 + Z_2 - \text{Mutual Reactance Component}$$
$$Z_{eq} = R_1 + R_2 + j\omega(L_1 + L_2 - 2M) \tag{3.57}$$

It is clear that the inductive reactance for the series addition (Eq. 3.56) is more than series opposition (Eq. 3.57). In practice, it is necessary to connect coils in series such that fluxes are added. This requires the identification of correct polarities of the second coil relative to the first.

Polarity can be experimentally checked using any of the following methods:

(a) Test lamps connected in series with the series connected windings (Fig. 3.27b).
(b) Measuring coil voltage (Fig. 3.28a) and supply voltage in series-connected coils
(c) Using battery (Fig. 3.29).

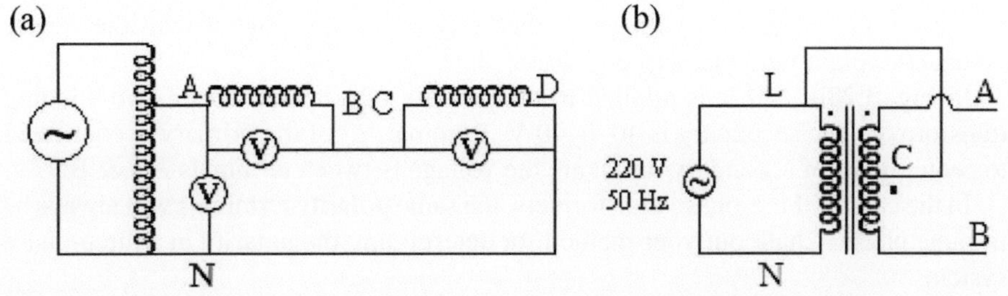

Fig. 3.28 **a** Polarity check by measuring coil voltages; **b** transformer connections for getting different voltages

Fig. 3.29 Checking of polarity using battery

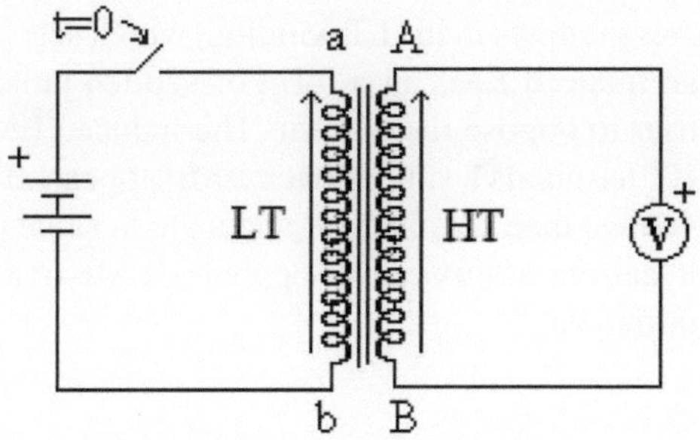

Method (a) is self-explanatory. The windings of a two-winding transformer are connected in two different fashion as shown in Fig. 3.27b. If in case (i) the lamp glows brighter than other case (ii). Can you determine polarity of the windings? In case of series addition, the equivalent impedance is higher than that of series opposition connection. The lamp glows brighter in case of series opposition for same input voltage.

In Fig. 3.28a, let the applied voltage be V_S and the voltages appearing across coils 'AB' & 'CD' be V_{AB} and V_{CD} respectively. For series additive connection, we have

$$V_{AN} = |V_{AB} + V_{CD}| \tag{3.58}$$

If Eq. (3.58) is satisfied, then terminals '*A*' and '*C*' are said to have same polarity. Can we say that in series additive connection as discussed above, terminals *B* and *D* are also at same polarity?

For series subtractive (opposing) connection, we have

$$V_{AN} = |V_{AB} - V_{CD}| \tag{3.59}$$

If Eq. (3.59) is satisfied, then terminals, '*A*' and '*D*' are said to have same polarity. Can we say that in series opposing or subtracting connection, as discussed above, terminals '*B*' and '*C*' are at same polarity? In the case of series opposing connection, why does one of the coil voltages (either V_{AB} or V_{CD}) is higher than the supply voltage? Explain with the help of phasor.

In Fig. 3.28b, 220 V is applied to the primary of a center-tapped two-winding transformer. The secondary is 40–0–40 V. Terminal '*L*' of the primary is connected to center point of secondary. What are the voltage between terminals AN & BN?

In the case of three-phase transformers, the same polarity terminals will always be in same phase. Chalk out your method for determining the polarity in a three-phase system.

The testing of polarity in field needs a simpler arrangement than mentioned earlier. A DC battery is connected to low-voltage winding and a DC voltmeter is connected to the HT terminals as shown in Fig. 3.29. The flow to transient DC current is resisted by an opposing transient EMF. At time $t = 0$ when switch S is closed, a transient DC current flows in the LT winding, which sets up DC flux in the core. As a consequence, an induced EMF in a form of sudden pulse appears on the primary terminals that tries to oppose the current. The induced EMF vanishes within few seconds. On the HT terminals a similar back EMF appears. If the voltmeter on HT side reads positive voltage then terminals A and a are at same polarity. When switch is opened, the fall in current or flux causes opposing EMF or a negative voltage appears on the winding terminals.

3.14.1 *Multiratio and Multiple Winding Transformers*

While manufacturing a transformer, extra electrical connections or tapping are normally provided at different number of turns on its windings. Such a transformer is useful for operating at different output voltages, because of different voltage ratios. Multiratio transformer is shown in Fig. 3.30. Autotransformers used in starting of induction motor are also wounded with multiratios; this makes them suitable for limitation of starting current by selection of starting voltage.

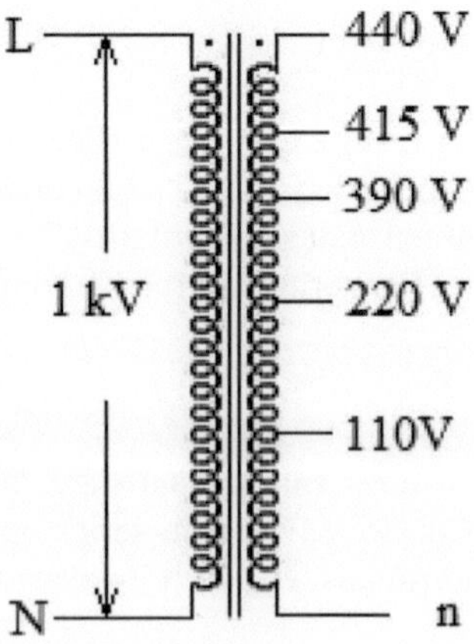

Fig. 3.30 Typical multiratio single-phase transformer

When the secondary consists of multiple windings, it is referred to as multiple winding transformer or multiwinding transformer. These transformers are useful for supplying load rated at different voltages. Such a 11 kV/440 V/390/220 V three-phase transformer can be used for supplying electrical power to three-phase motors or instruments rated at 440, 390, and 220 V. Multiratio three-phase transformers also have similar features.

A typical multiwinding transformer having separate secondary winding of 30, 40 and 50 V for an input voltage of 220 V is shown in Fig. 3.31. As compared with the multiratio transformer, in multiwinding transformers the secondary (or primary, if applicable) can be connected in different way taking care of their respective polarity. For example, in the above transformer, 10 V output can also be obtained by connecting 40 and 50 V windings in series phase opposition. Similar results can also be obtained by connecting 30 and 40 V is series phase opposition. How will you get 20, 70, 80, 90 V output from the same multiwinding transformer? Can you answer? Do you find any other difficulty when load is connected as in Fig. 3.32a, b? In these cases, can you tell the potential of each terminal of load with respect to electrical ground or neutral?

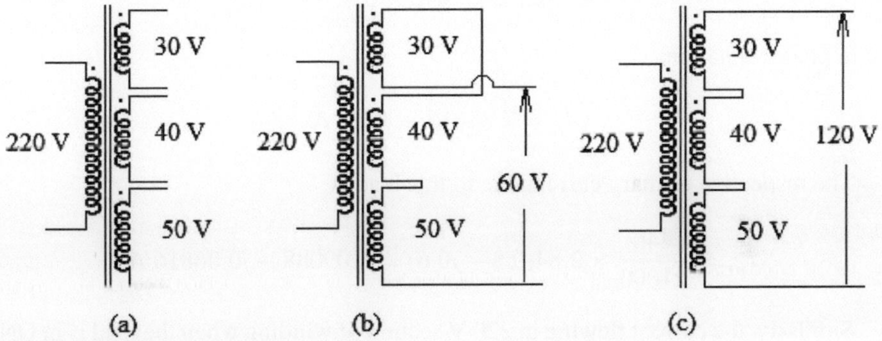

Fig. 3.31 Typical multiwinding transformer

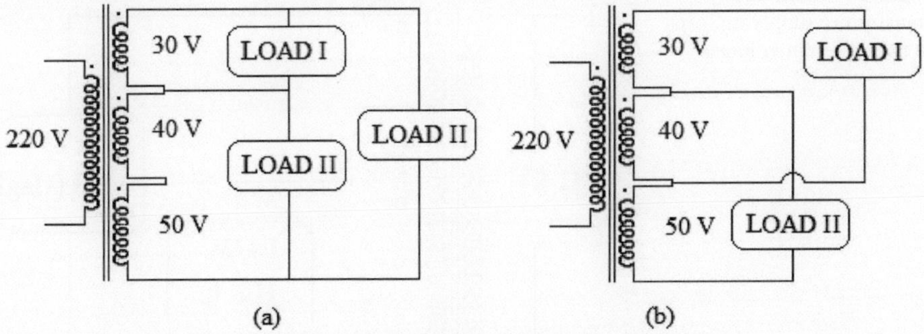

Fig. 3.32 Connection of electrical load at different voltages in multiwinding transformer

As has been shown in Fig. 3.31b, in case of multiwinding transformers, secondary or primary can be connected in a different ways taking into account their polarity. However, in case of multiratio transformers, one of the segments of secondary winding cannot be reconnected in a reverse fashion. Mark the voltage across different loads in Fig. 3.32a, b.

Example 3.16 A single-phase 11 kV/440 V/220 V multiple ratio transformer is used to supply power to 1.0 kW,440 V, 0.8 lag p.f. Load and 1.0 kW, 220 V, unity power factor load. Determine the currents drawn by different sections of secondary winding and net primary current.

Solution

Given single-phase 11 kV/440 V/220 V multiple ratio transformer.

Loads are 1.0 kW, 440 V, 0.8 lag p.f. load and 1.0 kW, 220 V, unity power factor as shown in Fig. 3.33.

The current flowing in 440 V section of winding is.

$$I_a = \frac{P \times 1000}{V \times p.f.} = \frac{1.0 \times 1000}{440 \times 0.8} = 2.84 \text{ A}$$

or in phasor notation,

$$\overline{I}_a = 2.84(0.8 - j0.6) \text{ A}$$

The respective primary current due to this load is

$$\overline{I}_{p1} = \frac{440}{11000} \times 2.84(0.8 - j0.6) = 0.09088 - j0.06816 \text{ A}$$

Similarly, the current flowing in 220 V section of winding when the load is in ON condition.

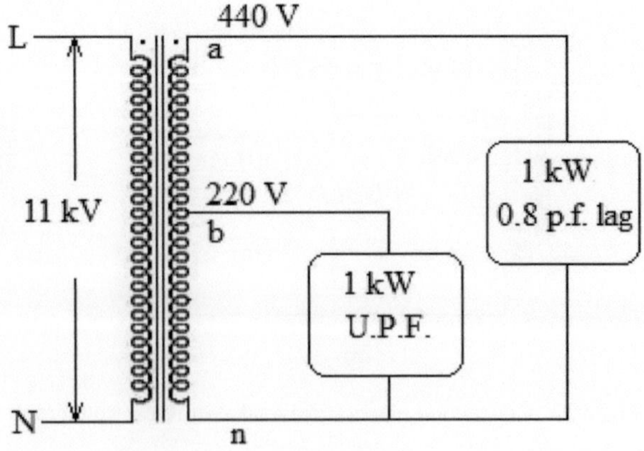

Fig. 3.33 Multiratio transformer supplying loads at two different voltages

$$I_b = \frac{P \times 1000}{V \times p.f.} = \frac{1.0 \times 1000}{220 \times 1.0} = 4.55 \text{ A}$$

or in phasor notation,

$$\bar{I}_b = 4.55(1.0 - j0.0)\text{A}$$

When both the loads are ON, the current in 'bn' section of winding carries the current of both the loads. The current in this section is obtained by phasor summation of two currents.

$$\bar{I}_{\text{bn}} = \bar{I}_a + \bar{I}_b$$
$$\bar{I}_{\text{bn}} = 2.84(0.8 - j0.6) + 4.55(1.0 + j0.0)$$
$$\bar{I}_{\text{bn}} = 6.82 - j1.704 = 7.029\angle - 14.2^\circ \text{ A}$$

The respective primary current due to this load is

$$\bar{I}_{p2} = \frac{220}{11000}(6.82 - j1.704) = 0.1364 - j0.03408\text{A}$$

Total primary current

$$\bar{I}_p = I_{p1} + \bar{I}_{p2} = 0.09088 - j0.06816 + 0.1364 - j0.03408$$
$$= (0.227 - j0.10224)\text{A}$$

Example 3.17 A single-phase 11 kV/440 V/390 V/220 V multiple secondary transformer is supplying power from each secondary 1.0 kW at 440 V, 0.8 lag, 1.0 kW AT 390 V, U.P.F. and 1.0 Kw at 220 V, 0.8 leading power factor. Determine winding currents and the net primary current.

Solution

There are three independent secondary windings providing electrical isolation between primary and secondary windings. Such a transformer will be costly than multiratio transformer. However there will be complete electrical isolation among different windings.

Given single-phase transformer 11 kV/440 V/390 V/220 V multiple secondary. The loads are.

(i) 1.0 kW, 440 V, 0.8 p.f. lag
(ii) 1.0 kW, 390 V, U.P.F
(iii) 1.0 kW, 220 V, 0.8 p.f. leading.

The current flowing in 440 V section of winding is

$$I_{L1} = \frac{P \times 1000}{V \times p.f.} = \frac{1.0 \times 1000}{440 \times 0.8} = 2.84 \text{ A}$$

or in phasor notation,

$$\bar{I}_{L1} = 2.84(0.8 - j0.6) \text{ A}$$

The respective primary winding current due to this load is

$$\bar{I}_{p1} = \frac{440}{11000} 2.84(0.8 - j0.6) = 0.09088 - j0.06816 \text{ A}$$

Similarly, the current flowing in 390 V section of winding is

$$I_{L2} = \frac{P \times 1000}{V \times p.f.} = \frac{1.01000}{390} = 2.564 \text{ A}$$

or in phasor notation,

$$\bar{I}_{L2} = 2.564(1.0 - j0.0) \text{ A}$$

The respective primary winding current due to this load is

$$\bar{I}_{p2} = \frac{390}{11000} 2.564(1.0 - j0.0) = 0.090905 - j0.0 \text{ A}$$

Further, the current flowing in 220 V section of winding is

$$I_{L3} = \frac{P \times 1000}{V \times p.f.} = \frac{1.0 \times 1000}{220 \times 0.8} = 5.682 \text{ A}$$

or in phasor notation,

$$\bar{I}_{L3} = 5.682(0.8 + j0.6) \text{ A}$$

The respective primary winding current due to this load is

$$\bar{I}_{p3} = \frac{220}{11000} \times 5.682(0.8 + j0.6) = 0.090909 + j0.06818 \text{ A}$$

The net primary current in multiwinding transformer is the phasor sum of currents flowing on primary due to different secondary loads. Thus

$$\bar{I}_p = \bar{I}_{p1} + \bar{I}_{p2} + \bar{I}_{p3} \text{A}$$
$$\bar{I}_p = 0.09088 - j0.06816 + 0.090905 + 0.090909 + j0.06818$$
$$\bar{I}_p = 0.272694\angle 0^\circ \text{ Ampere}$$

3.15 Transformer Regulation

The magnitude of voltage drop corresponding to full-load is often expressed as a fraction of the rated voltage of the transformer. In an 11 kV/440 V transformer, the significance of the fall of the primary voltage, say, by 110 V, cannot be appreciated unless it is compared with the rated voltage. The ratio of the fall in primary voltage and the input voltage at no-load of the transformer is called the per unit drop and indicates directly the relative importance of the voltage drop as compared to input voltage. In this example, the per unit drop is 110/11000 or 0.01.

Figure 3.34 shows the approximate phasor diagram of transformer with lagging power factor. From Fig. 3.34, we get angle $\angle$ BAD $= \theta = \angle$KCG and angles $\angle$DKB $= \angle$CKG $= 90 - \theta$. Also, AD $=$ IR cos θ, BH $=$ DG $=$ IX sin θ, BD $=$ GH $=$ IR sin θ and CH $=$ IX cos θ.

From ΔOCG, we get

$$\text{OG} = \text{OA} + \text{AD} + \text{BH} = \text{V} + \text{IR}\cos\theta + \text{IX}\sin\theta.$$

$$\text{CG} = \text{CH} - \text{DB} = \text{IX}\cos\theta - \text{IR}\sin\theta$$

Thus, for a lagging power factor angle, no-load terminal voltage E can be written as,

$$E = \sqrt{\left[(V + \text{IR}\cos\theta + \text{IX}\sin\theta)^2 + (\text{IX}\cos\theta - \text{IR}\sin\theta)^2\right]} \tag{3.60}$$

Similarly, the expression for E can be derived for both unity and leading power factors. Thus, the % regulation (UP) & (DOWN) can now be calculated using Eqs. (3.61) and (3.62):

$$\%\text{Regulation(UP)} = \left(\frac{E - V}{V}\right)100 \tag{3.61}$$

$$\%\text{Regulation(DOWN)} = \left(\frac{E - V}{E}\right)100 \tag{3.62}$$

The per unit resistance or per unit reactance is defined as,

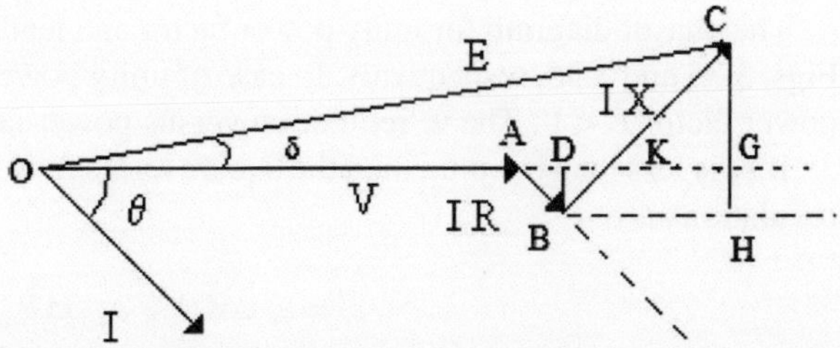

Fig. 3.34 Phasor diagram for lagging power factor

$$\varepsilon_r = \frac{I_1^2 R_{\mathrm{EX1}}}{V_1 I_1} = \frac{I_2^2 R_{\mathrm{EX2}}}{V_2 I_2} = \frac{P_{\mathrm{CU}}}{S} \tag{3.63}$$

Similarly, per unit leakage reactance is defined as,

$$\varepsilon_x = \frac{I_1^2 X_{\mathrm{EX1}}}{V_1 I_1} = \frac{I_1 X_{\mathrm{EX1}}}{V_1} = \frac{I_2 X_{\mathrm{EX2}}}{V_2} \tag{3.64}$$

Here '$R_{\mathrm{EX1}} = R_1 + R_2^{/}$' is the transformer resistance referred to primary side. Transformer resistance referred to secondary side is '$R_{\mathrm{EX2}} = R_1^{/} + R_2$'. Similarly, leakage reactance referred to primary side is '$X_{\mathrm{EX1}} = X_1 + X_2^{/}$' and leakage reactance referred to secondary side is '$X_{\mathrm{EX2}} = X_1^{/} + X_2$'. The rated copper loss is P_{cu}, which can be taken as wattmeter reading of short-circuit test for rated current and volt-ampere rating of transformer is S. The per unit leakage impedance is given as,

$$\varepsilon_z = \sqrt{\left(\varepsilon_r^2 + \varepsilon_x^2\right)} \tag{3.65}$$

If we assume that transformer's no-load voltage (E) is 1.0 per unit and it is delivering current I of 1.0 per unit, then from Fig. 3.34, we have

$$\mathrm{AG} = \varepsilon_r \cos\theta + \varepsilon_x \sin\theta \quad \text{and} \quad \mathrm{CG} = \varepsilon_x \cos\theta - \varepsilon_r \sin\theta$$

The angle between

$$\sin\delta = \angle\mathrm{COA} = \delta \cong \frac{\mathrm{CG}}{E} = \mathrm{CG}$$

$$\cos\delta = \frac{\mathrm{OG}}{E} = \mathrm{OA} + \mathrm{AG}$$

$$V + \mathrm{AG} = \cos\delta = 1 - \frac{1}{2}\mathrm{CG}^2 + \ldots\ldots\ldots$$

Per unit regulation given as scalar expression is

$$\varepsilon = E - V = 1 - V = \varepsilon_r.\cos\theta + \varepsilon_x.\sin\theta + \frac{1}{2}(\varepsilon_x.\cos\theta - \varepsilon_r.\sin\theta)^2 - \ldots. \tag{3.66}$$

The phasor diagram for unity power factor and leading power factor are shown in Figs. 3.35 and 3.36, respectively. In case of unity power factor $E > V$ and for leading power factor $E < V$. The % regulation versus power factor is shown in Fig. 3.37.

If ε_z is less than 0.1 p.u., then the square terms may be ignored. Thus per unit (pu) regulations is

$$\varepsilon = \varepsilon_r \cos\theta + \varepsilon_x \sin\theta \tag{3.67}$$

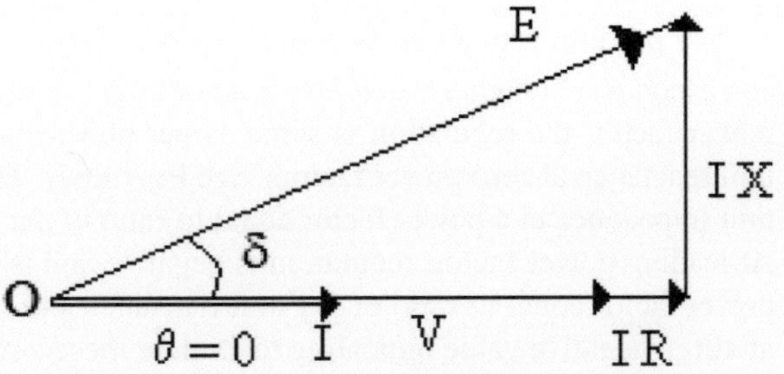

Fig. 3.35 Phasor diagram for unity pf

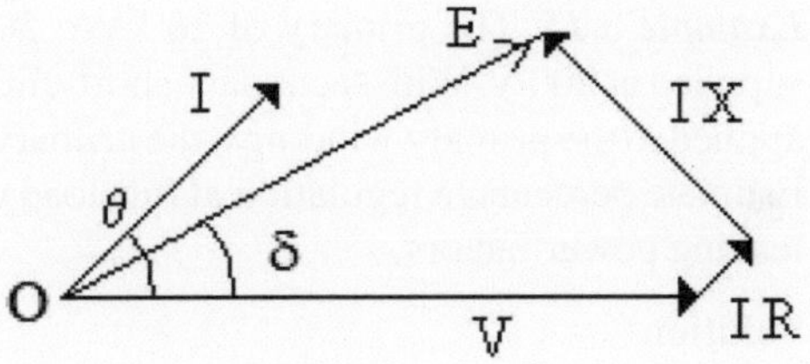

Fig. 3.36 Phasor diagram for leading pf

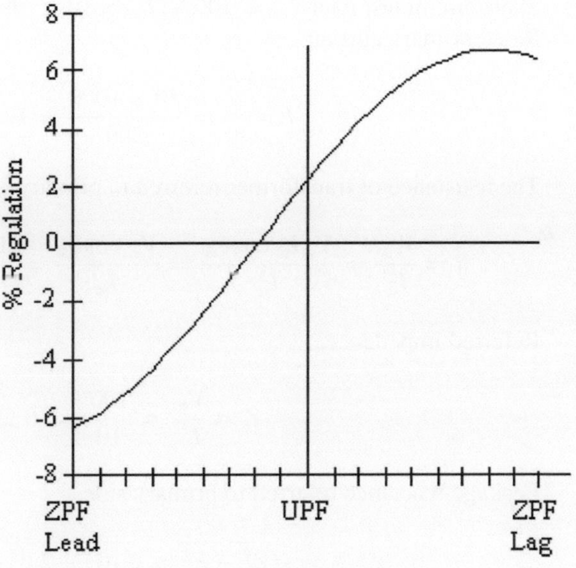

Fig. 3.37 Variation of percentage regulation with power factor

The full-load regulation is maximum when power factor angle is

$$\theta = \tan^{-1}\left(\frac{\varepsilon_x}{\varepsilon_r}\right) = \tan^{-1}\left(\frac{X_{\text{EX1}}}{R_{\text{EX1}}}\right) = \tan^{-1}\left(\frac{X_{\text{EX2}}}{R_{\text{EX2}}}\right)$$

This corresponds to a very low power factor.

The per unit regulation is zero for power factor angle of $\theta = -\tan^{-1}\left(\frac{\varepsilon_r}{\varepsilon_x}\right)$. It corresponds to a higher-capacitive load or high leading power factor load. For a unit power factor, the regulation is same as per unit resistance, which increases to per unit reactance at zero power factors. See Eq. (3.67). The maximum values being per unit impedance at a power factor equal to ratio of per unit resistance and reactance. At leading power factor, regulation is negative and is reduced to zero when leading power factor equal to ratio of per unit reactance and impedance. After this, it again attains a negative value indicating that under these circumstances the output voltage rises to a higher value, as capacitive load is increased.

Example 3.18 The primary of 20 kVA, 50 Hz step down transformer is normally supplied at 200 V. With secondary short-circuited and a potential difference of 100 V applied to the primary windings, the primary current is 10 A and power factor is 0.28. Estimate percentage regulation at full-load unity power factor and at 0.6 lagging and leading power factors.

Solution

Given $S = 20$ kVA; $V_1 = 200$ V;

Short-circuit test data $V_{sc} = 100$ V; $I_{sc} = 10$ A; short circuit pf 0.28.

Rated primary current,

$$I = \frac{S}{V} = \frac{20 \times 1000}{200} = 100 \text{ A}$$

The resistance of transformer referred to primary side,

$$R = \frac{W_{sc}}{I_{sc}^2} = \frac{V_{sc} I_{sc} \cos\theta_{sc}}{I_{sc}^2} = \frac{V_{sc} \cos\theta_{sc}}{I_{sc}} = \frac{100 \times 0.28}{10} = 2.8\ \Omega$$

Referred impedance,

$$Z = \frac{V_{sc}}{I_{sc}} = \frac{100}{10} = 10\ \Omega$$

Leakage reactance referred to primary side,

$$X = \sqrt{Z^2 - R^2} = \sqrt{10^2 - 2.8^2} = 9.6\ \Omega$$

Per unit resistance, $\varepsilon_r = \frac{IR}{V} = \frac{100 \times 2.8}{200} = 1.4$

Similarly per unit reactance, $\varepsilon_x = \frac{IX}{V} = \frac{100 \times 9.6}{200} = 4.8$

The per unit regulation is given as, $\varepsilon = \varepsilon_r \cos\theta + \varepsilon_x \sin\theta$.

At power factor = 1.0, per unit regulation is,

$$\varepsilon = \varepsilon_r \cos\theta + \varepsilon_x \sin\theta = 1.4 \times 1 + 4.8 \times 0 = 1.4$$

At 0.6 lagging power factor, per unit regulation is,

$$\varepsilon = \varepsilon_r \cos\theta + \varepsilon_x \sin\theta = 1.4 \times 0.6 + 4.8 \times 0.8 = 4.68$$

At 0.6 leading, per unit regulation,

$$\varepsilon = \varepsilon_r \cos\theta - \varepsilon_x \sin\theta = 1.4 \times 0.6 - 4.8 \times 0.8 = -3.00$$

Example 3.19 The maximum efficiency of a three-phase 11 kV/400 V, 500 kVA, 50 Hz transformer is 98% and occurs at 80% full-load unity power factor. Its percentage impedance is 4.5%. Load power factor is varied while the load current and the supply voltage are held constant at rated values. Determine the load power factor at which the secondary terminal voltage is minimum.

Solution

The output at 80% full-load unity power factor = 500 × 0.8 × 1.0 = 400 kW.
The input at full-loads unity power factor = 500 × 0.8 × 1.0/0.98 = 408.16 kW.
Total losses at full-load = Input – Output = 408.16 − 400 = 8.16 kW.
At 80% of full-load fixed core loss = variable copper loss.
Copper loss at 80% full-load = 8.16/2 = 4.08 kW.

Here fraction of load k = 0.8, full-load copper loss is

$$P_{CU} = \frac{\text{Copper loss at 80\% load}}{k^2} = \frac{4.08}{0.8^2} = 6.375 \text{ kW}$$

Per unit resistance

$$\varepsilon_r = \frac{P_{CU}}{S} = 0.01275 \text{ per unit}$$

Per unit impedance = 0.045 per unit. Thus $\varepsilon_x = \sqrt{(\varepsilon_z^2 - \varepsilon_r^2)} = 0.04315$ per unit
The per unit regulation is $\varepsilon = \varepsilon_r \cos\theta + \varepsilon_x \sin\theta$.

For terminal voltage to be minimum, the drop or percentage regulation will be maximum, i.e.,

$$\frac{d\varepsilon}{d\theta} = 0$$

$$\theta = \tan^{-1}\left(\frac{\varepsilon_x}{\varepsilon_r}\right) = 73.54^\circ$$

Or,
Load power factor for minimum voltage cos 73.54° = 0.2833 lagging.

3.16 Parallel Operation

The electricity is generated at generating stations, that are located at faraway places from the place of utility (consumption). Power is transmitted to long distances at higher voltage. The voltage from generating station is increased to much higher value for transmitting bulk power which is subsequently reduced to lower level of voltages suitable for domestic and industrial use. However, a transformer has limited capacity to transfer power. Bulk consumers such as large industries receive power at a level higher than three phase, 440 V. The reason being the tariff for receiving power at higher voltages is less. Moreover, a transformer's capacity to transfer electrical power is limited. In case of overloading, there are chances of breakdown due to several faults. To avoid this, both the primary and secondary sides of several transformers are connected in parallel to share the electrical load. The generated power is transmitted to long distances by means of parallel connected power transformers. In the event of light loading, some power transformers are switched OFF. While distributing power at different localities, the distribution transformers, which convert higher voltages at 440 V are remain connected in parallel. For parallel operation of transformers, the following conditions are to be met.

(a) Terminals with similar polarity must be joined together as shown in Fig. 3.38. A wrong connection will result in a dead short circuit, causing extensive damage to the transformer.
(b) Terminals joined together must have same phase sequence and same phase displacement, when poly phase transformers are connected in parallel. Connecting wrong terminals with different phase sequence and/or different phase displacements will cause circulating currents. The secondary voltage of

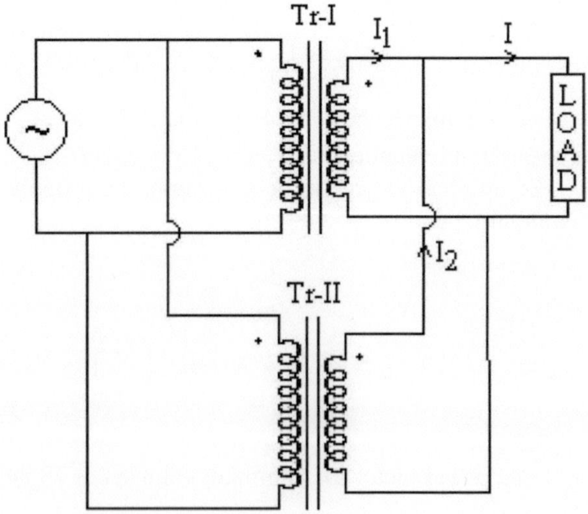

Fig. 3.38 Parallel-connected transformers

star/star and star/delta transformers have a phase difference of 30° and, therefore, these transformers cannot be connected in parallel even when their input and output voltages are matching. In three-phase connection, voltage and phase displacement may have different magnitudes are possible. The magnitude of voltage can be adjusted using tapping, but phase divergence cannot be adjusted.

(c) Terminals joined together must have near-identical voltages. A large difference between the open circuit terminal voltages will cause circulating currents even when the parallel connected transformers are at no-load.

(d) The phase sequence must be identical for two parallel-connected transformers. If the polarity and voltage ratio are correct, then secondary terminals 'a_1-b_1-c_1' of transformer-I are connected to secondary terminals 'a_2-b_2-c_2' of transformer-II. When their primary windings are connected in parallel and energized. Terminals 'a_1-a_2' should show zero voltage and should be joined. Similar is the case for terminals 'b_1-b_2' and 'c_1-c_2'.

Three-phase connections have been explained here and in Sect. 3.21.1. In three-phase transformers, terminals having similar phase sequence and phase divergence can be joined in parallel. Thus all transformers of same phase group can be connected in parallel. These are given in Table 3.2.

The transformers of + 30° and − 30° angle, like Dy1 and Dy11 can be joined together in parallel by reversing the primary and secondary terminals phase sequence by one, provided other conditions are satisfied.

Two single-phase transformers are connected in parallel, as shown in Fig. 3.38. It is assumed that their equivalent circuit series parameters or leakage impedances are $Z_1 = R_1 + jX_1$ and $Z_2 = R_2 + jX_2$, and open circuit EMFs are $\boldsymbol{E}_1$ and $\boldsymbol{E}_2$. It is further assumed that their open-circuit EMFs are in same phase, but their magnitudes are different. When the terminal are joined together, a circulating current will arise within the transformer secondary and primary windings. The magnitude of this current on secondary is given as

$$\boldsymbol{I}_{cir} = \frac{\boldsymbol{E}_1 - \boldsymbol{E}_2}{\boldsymbol{Z}_1 + \boldsymbol{Z}_2} \tag{3.68}$$

Equation (3.68) is valid when magnitudes are same but phase differences are there. The circulating current causes Ohmic losses resulting in a higher operating temperature than expected. The actual operating temperatures are much higher when transformers are loaded. It may result in overheating of the transformer. Thus, equal voltage ratio is necessary to avoid any circulating current. When leakage reactance is low, even a small voltage difference will cause considerable circulating current on both the primary and secondary sides. When parallel connected transformers are

Table 3.2 Types of transformer connections

Transformer	Connections		
Transformer-I	Star/star	Star/delta	Star/delta
Transformer-II	Delta/delta	Delta/star	Star/zigzag

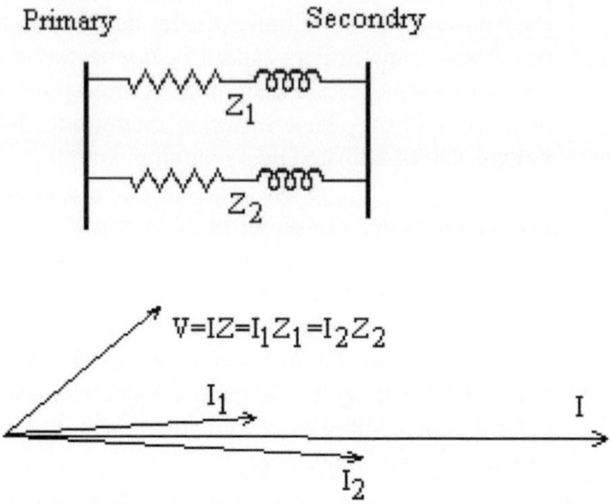

Fig. 3.39 Equivalent representation and phasor diagram-unequal load sharing

fully loaded, overcurrent may occur in one of the transformers. A typical unequal load sharing in parallel connected transformers is shown in Fig. 3.39. We will now discuss load sharing under different conditions.

3.16.1 Sharing of Load-Case A: Load Current is Known

Let $\boldsymbol{V}_1$ and $\boldsymbol{V}_2$ be the common primary and secondary terminal voltages of the transformers operating in parallel. The primary voltages referred to secondary sides are $(\boldsymbol{V}_1/a_1)$, $(\boldsymbol{V}_1/a_2)$ and $(\boldsymbol{V}_1/a_3)$, and so on. Here a_1, a_2 or a_3 are the respective transformation ratios. Thus the impedance drops, which, in each case, is the phasor difference between $\boldsymbol{V}_1/a_n$ and $\boldsymbol{V}_2$ ($n = 1, 2, 3, \ldots$ etc.) for different transformers. Thus, we get

$$
\begin{aligned}
\frac{\mathbf{V}_1}{\mathrm{a}_1} - \mathbf{V}_2 &= \mathbf{I}_1\mathbf{Z}_1 = \frac{\mathbf{I}_1}{\mathbf{Y}_1} \\
\frac{\mathbf{V}_1}{\mathrm{a}_2} - \mathbf{V}_2 &= \mathbf{I}_2\mathbf{Z}_2 = \frac{\mathbf{I}_2}{\mathbf{Y}_2} \\
&\vdots \\
\frac{\mathbf{V}_1}{\mathrm{a}_n} - \mathbf{V}_2 &= \mathbf{I}_n\mathbf{Z}_n = \frac{\mathbf{I}_n}{\mathbf{Y}_n}
\end{aligned}
\tag{3.69}
$$

Here $\mathbf{Z}_1$, $\mathbf{Z}_2$ and $\mathbf{Z}_{3,}$ etc. are the equivalent impedance referred to secondary and $\boldsymbol{Y}_1$, $\boldsymbol{Y}_2$ $\boldsymbol{Y}_3$ .. etc. are the corresponding admittance. The output current is the phasor sum of secondary currents. Therefore, we get

$$\mathbf{I} = \mathbf{I}_1 + \mathbf{I}_2 + \mathbf{I}_3 + \ldots\ldots\ldots\ldots + \mathbf{I_n}$$

$$\mathbf{I} = \mathbf{I}_1 + \mathbf{I}_2 + \mathbf{I}_3 + \ldots\ldots\ldots\ldots + \mathbf{I_n} = \mathbf{V}_1 \sum \frac{\mathbf{Y_n}}{\mathrm{a_n}} - \mathbf{V}_2 \sum \mathbf{Y_n} \tag{3.70}$$

$$\mathbf{V}_2 = \frac{\mathbf{V}_1 \sum \frac{\mathbf{Y_n}}{a_n} - \mathbf{I}}{\sum \mathbf{Y_n}} \tag{3.71}$$

Thus the individual transformer currents can be obtained

$$\mathbf{I}_1 = \frac{1}{\mathbf{Z}_1}\left(\frac{\mathbf{V}_1}{a_1} - \mathbf{V}_2\right) \tag{3.72}$$

$$\mathbf{I}_n = \frac{\mathbf{I} + \mathbf{V}_1\left[\sum \frac{\mathbf{Y_n}}{\mathbf{a}_1} - \sum \frac{\mathbf{Y_n}}{\mathbf{a_n}}\right]}{\mathbf{Z_n} \sum \mathbf{Y_n}} \tag{3.73}$$

3.16.2 *Sharing of Load-Case B: When Load Impedance is Known*

Let the load impedance be $\mathbf{Z}_{\mathrm{L}}$. Since the secondary voltage is known, the load current is given as,

$$\mathbf{I} = \frac{\mathbf{V}_2}{\mathbf{Z_L}} = \mathbf{V}_2 \mathbf{Y_L} \tag{3.74}$$

$$\mathbf{V}_2 = \frac{\mathbf{V}_1 \sum \frac{\mathbf{Y_n}}{a_n}}{\mathbf{Y_L} + \sum \mathbf{Y_n}} \tag{3.75}$$

$$\mathbf{I}_1 = \frac{1}{\mathbf{Z}_1}\left(\frac{\mathbf{V}_1}{a_1} - \mathbf{V}_2\right) \tag{3.76}$$

$$\mathbf{I}_1 = \frac{\mathbf{V}_1\left[\frac{(\mathbf{Y_L} + \sum \mathbf{Y_n})}{\mathbf{a}_1} - \sum \frac{\mathbf{Y_n}}{\mathbf{a_n}}\right]}{\mathbf{Z}_1(\mathbf{Y_L} + \sum \mathbf{Y_n})} \tag{3.77}$$

Let there be only two parallel-connected transformers, then

$$\sum \mathbf{Y_n} = \frac{1}{\mathbf{Z}_1} + \frac{1}{\mathbf{Z}_2} \tag{3.78}$$

$$\sum \frac{\mathbf{Y_n}}{a_n} = \frac{1}{a_1 \mathbf{Z}_1} + \frac{1}{a_2 \mathbf{Z}_2} \tag{3.79}$$

Solving these equations we get,

$$\mathbf{I}_1 = \frac{\mathbf{Z}_2}{\mathbf{Z}_1 + \mathbf{Z}_2}\mathbf{I} + \frac{\left(\frac{\mathbf{V}_1}{\mathbf{a}_1} - \frac{\mathbf{V}_1}{\mathbf{a}_2}\right)}{\mathbf{Z}_1 + \mathbf{Z}_2} \tag{3.80}$$

$$\mathbf{I}_2 = \frac{\mathbf{Z}_1}{\mathbf{Z}_1 + \mathbf{Z}_2}\mathbf{I} - \frac{\left(\frac{\mathbf{V}_1}{\mathbf{a}_1} - \frac{\mathbf{V}_1}{\mathbf{a}_2}\right)}{\mathbf{Z}_1 + \mathbf{Z}_2} \tag{3.81}$$

The second term in Eq. (3.81) is circulating current which flows in the direction of $\boldsymbol{I}_1$ even when load current $\boldsymbol{I}$ is zero. The circulating currents are avoided by selecting equal turn ratio or $a_1 = a_2 = a_3 = \ldots = a_{n.}$

When the circulating currents are zero for two parallel-connected transformers, their individual currents can be given as

$$\mathbf{I}_1 = \frac{\mathbf{Z}_2}{\mathbf{Z}_1 + \mathbf{Z}_2}\mathbf{I} = \frac{\mathbf{I}}{\mathbf{Z}_1 \sum \mathbf{Y}_\mathbf{n}} \tag{3.82}$$

$$\mathbf{I}_2 = \frac{\mathbf{Z}_1}{\mathbf{Z}_1 + \mathbf{Z}_2}\mathbf{I} = \frac{\mathbf{I}}{\mathbf{Z}_2 \sum \mathbf{Y}_\mathbf{n}} \tag{3.83}$$

Similarly, the kVA shared by individual transformer will be

$$S_1 = \frac{\mathbf{Z}_2}{\mathbf{Z}_1 + \mathbf{Z}_2}S$$

$$S_2 = \frac{\mathbf{Z}_1}{\mathbf{Z}_1 + \mathbf{Z}_2}S$$

The current in transformer windings is inversely proportional to its leakage impedance. If transformers with same voltage ratings and same transformation ratios have to be loaded in proportional to their kVA ratings, then their impedances must be inversely proportional to their respective kVA ratings. At the same time, their per unit impedances should be same for equal sharing of load. If it is assumed that three transformer turn ratios and voltage ratios are identical, then the currents can be given as

$$\mathbf{I}_1 = \frac{\left(\frac{\mathbf{V}_1}{\mathbf{a}_1} - \mathbf{V}_2\right)}{\mathbf{Z}_1} = \mathbf{V}_\mathbf{d}\mathbf{Y}_1$$

$$\mathbf{I}_2 = \mathbf{V}_\mathbf{d}\mathbf{Y}_2 \tag{3.84}$$

$$\mathbf{I}_3 = \mathbf{V}_\mathbf{d}\mathbf{Y}_3 \tag{3.85}$$

Here voltage difference

$$\mathbf{V_d} = \left(\frac{\mathbf{V}_1}{a_1} - \mathbf{V}_2\right)$$

However, the current supplied by the three transformers are not be in phase with one other. Therefore, we have

$$\mathbf{I} = \mathbf{I}_1 + \mathbf{I}_2 + \mathbf{I}_3 \tag{3.86}$$

$$\mathbf{I} = \mathbf{V_d}(\mathbf{Y}_1 + \mathbf{Y}_2 + \mathbf{Y}_3) \tag{3.87}$$

$$\mathbf{I} = \mathbf{V_d}(\mathbf{Y}_1 + \mathbf{Y}_2 + \mathbf{Y}_3) = \mathbf{V_d}\mathbf{Y_S} \tag{3.88}$$

where $\mathbf{Y_S}$ is the total admittance or $\mathbf{Y_S} = (\mathbf{Y}_1 + \mathbf{Y}_2 + \mathbf{Y}_3)$. The individual currents are

$$\begin{aligned} \mathbf{I}_1 &= \frac{\mathbf{Y}_1}{\mathbf{Y_S}}\mathbf{I} \\ \mathbf{I}_2 &= \frac{\mathbf{Y}_2}{\mathbf{Y_S}}\mathbf{I} \\ \mathbf{I}_3 &= \frac{\mathbf{Y}_3}{\mathbf{Y_S}}\mathbf{I} \end{aligned} \tag{3.89}$$

If the currents are in phase, both the magnitude of total current and the output kVA will be maximum. The currents should divide between individual transformers in such a way that the currents are not only in direct proportion to the respective kVA ratings, but also in phase with one other and with the total current. For identical phase currents one can prove that

$$\frac{X_1}{R_1} = \frac{X_2}{R_2} = \frac{X_3}{R_3} = \ldots.. = \frac{X_n}{R_n} \tag{3.90}$$

If we ignore the saturation, we will find that both the current and flux are sinusoidal. The transformers are designed to operate near knee point of the magnetization curve. The magnetizing current and the flux produced in the magnetic core are non-sinusoidal. The ampere turns required to produce a sinusoidal flux can be obtained by point-by-point interpolating from the B-H curve. The sinusoidal flux wave is drawn on the left of the loop to the time axis. Flux values are projected on the loop point-by-point, and corresponding values of current are plotted on the time base as shown in Fig. 3.16b. The current waveform is re-plotted where it contains fundamental frequency component in phase with the flux, and a fundamental frequency component in quadrature with the flux.

In the case of saturation, if the flux is sinusoidal, the magnetizing current will contain harmonics and vice versa. The value of no-loaded current is $\mathbf{I}_0$ and $\mathbf{I}_0 N_1$ is the RMS value of the magnetizing ampere turns in phase with the flux.

Example 3.20 Show that when two transformers having unequal secondary EMF are connected in parallel to supply power to load impedance Z_L, the load current shared by them is given by.

$$I_A = \frac{\mathbf{E_A Z_B} + (\mathbf{E_A} - \mathbf{E_B})\mathbf{Z_L}}{\mathbf{Z_A Z_B} + \mathbf{Z_L}(\mathbf{Z_A} + \mathbf{Z_B})} \quad \text{and} \quad I_B = \frac{\mathbf{E_B Z_A} - (\mathbf{E_A} - \mathbf{E_B})\mathbf{Z_L}}{\mathbf{Z_A Z_B} + \mathbf{Z_L}(\mathbf{Z_A} + \mathbf{Z_B})}$$

Solution

Let V = supply voltage on primary side.

Z_A = equivalent impedance of transformer referred to secondary.
Z_B = equivalent impedance of transformer referred to secondary.
I_A = output current of transformer A.
I_B = output current of transformer B.
V_2 = secondary voltage = $\mathbf{Z_L.I}$
Induced EMF in secondary winding, $\mathbf{E_A}$, is given as

$$\mathbf{E_A} = \frac{V_1}{a_A} = \mathbf{V_2} + \mathbf{I_A Z_A}$$

Similarly, $\mathbf{E_B}$ is

$$\mathbf{E_B} = \frac{V_1}{a_B} = \mathbf{V_2} + \mathbf{I_B Z_B}$$

The terminal voltage is

$$\mathbf{V_2} = \mathbf{Z_L I} = \mathbf{Z_L}(\mathbf{I}_A + \mathbf{I}_B)$$

Thus, we get

$$\mathbf{E_A} = \mathbf{Z_L}(\mathbf{I}_A + \mathbf{I}_B) + \mathbf{I_A Z_A}$$
$$\mathbf{E_B} = \mathbf{Z_L}(\mathbf{I}_A + \mathbf{I}_B) + \mathbf{I_B Z_B}$$

On subtracting above two equations, we get

$$\mathbf{E_A} - \mathbf{E_B} = \mathbf{I_A Z_A} - \mathbf{I_B Z_B}$$

Solving we get,

$$\mathbf{I_A} = \frac{(\mathbf{E_A} - \mathbf{E_B}) + \mathbf{I_B Z_B}}{\mathbf{Z_A}}$$

On subtracting the value of $\mathbf{I_A}$ in $\mathbf{E_A}$ above we get

$$\mathbf{I_B} = \frac{\mathbf{E_B Z_A} - (\mathbf{E_A} - \mathbf{E_B})\mathbf{Z_L}}{\mathbf{Z_A Z_B} + \mathbf{Z_L}(\mathbf{Z}_A + \mathbf{Z}_B)}$$

and

$$\mathbf{I_A} = \frac{\mathbf{E_A Z_B} + (\mathbf{E_A} - \mathbf{E_B})\mathbf{Z_L}}{\mathbf{Z_A Z_B} + \mathbf{Z_L}(\mathbf{Z}_A + \mathbf{Z}_B)}$$

Example 3.21 Two 200 kVA, 50 Hz single-phase transformers having equal transformation ratio. One transformer has equivalent impedance of $0.005 + j0.08$ pu and the other has equivalent impedance of $0.0075 + j0.04$ pu. Their rated voltages are equal. How will the two parallel-connected transformer share the total load of 400 kW at unity power factor?

Solution

Given $\mathbf{Z}_1 = 0.005 + j\,0.08$ pu and $\mathbf{Z}_2 = 0.0075 + \mathrm{j}\,0.04$ pu.

Since the transformers have equal transformation ratio, there will be no circulating currents. The currents shared by individual transformers and kVA shared will be I_1, S_1, I_2 and S_2, respectively.

$$\mathbf{I}_1 = \frac{\mathbf{Z}_2}{\mathbf{Z}_1 + \mathbf{Z}_2}\mathbf{I} \quad S_1 = \frac{\mathbf{Z}_2}{\mathbf{Z}_1 + \mathbf{Z}_2}S$$

$$\mathbf{I}_2 = \frac{\mathbf{Z}_1}{\mathbf{Z}_1 + \mathbf{Z}_2}\mathbf{I} \quad \mathrm{S}_2 = \frac{\mathbf{Z}_1}{\mathbf{Z}_1 + \mathbf{Z}_2}\mathrm{S}$$

The load kW, P = 400 kW power factor is unity power factor;

$$\mathrm{kVA} = \frac{400}{1.0} = 400\ \mathrm{kVA}$$

$$S = P + jQ = 400 + j0\ \mathrm{kVA}$$

$$S_1 = \frac{0.0075 + j0.04}{0.005 + j0.08 + 0.0075 + j0.04} \times 400$$

$$= \frac{0.0075 + j0.04}{0.0125 + j0.12} \times 400 = 134.47 - j11\ \mathrm{kVA}$$

$$\text{Or } S_2 = S - S_1 = 400 - 0j + 134.47 + 11j = (265.52 + 11j)\ \mathrm{kVA}$$

Thus the transformer having higher R/X ratio shares larger kVA.

3.17 Back-to-Back Test or Sumpner's Test

Sometimes, it becomes necessary to conduct *heat run* test on a transformer for experimentally checking the temperature rise due to loading. A smaller capacity transformer can be conveniently tested for temperature rise by either direct load or short-circuit connection with rated currents in the windings. If heat run test is to be performed on single transformer, the iron loss and copper losses can be provided to the transformer by connecting it under short-circuit conditions. The input voltage is adjusted such that the input power is sum of rated iron and copper losses of the transformer. It is very difficult to conduct similar test on a large capacity transformer. Further, two transformers can be connected back-to-back so that each transformer has its full-load core loss as well as copper losses. The test is useful for finding temperature versus time curve as well as for obtaining equivalent circuit parameters. Can you suggest the name of transducers required for continuous monitoring of temperature? Are you familiar with computer aided data-logging techniques?

The current flowing in two high-voltage (HV) secondary windings induces currents in the respective low-voltage (LV) primary windings. The actual current flowing in the individual primary is $\boldsymbol{I}_0 \pm \boldsymbol{I}'_2$. Since secondary windings are in phase opposition, the induced current $\mathbf{I}'_2$ on primary side keeps circulating. The current and power in circuit connected to parallel-connected LV primary windings do not vary, unless the two transformers have different transformation ratios. Although the individual primary windings are carrying full-load currents, the source current on LV side is only twice the no-load magnetizing current. The low power factor wattmeter on LV side reads twice the core loss of individual transformers. On HV side, rated HT currents are flowing. The voltmeter and wattmeter on HV side read respectively, twice the short-circuit voltage and twice the copper loss.

The essential condition in this test is that the high-voltage (HV) secondary winding should be in phase opposition. If this condition is not satisfied, there will be a dead short circuit, causing extensive damage to the transformer as well as to the supply system. The low-voltage primary of the two transformers are connected in parallel and maintained at rated voltage. The power supplied from this side is basically twice the core loss of individual transformer plus small Ohmic losses. Similarly, the current is twice the no-load current of individual transformers. The secondary are connected in series and its open-circuit voltage is observed with a voltmeter having rating of more than twice the rating of HV (high-voltage) windings. Depending on the polarities of secondary terminals, this voltmeter reads either zero voltage or around twice the high-voltage rating of the transformer. If voltmeter reads zero, then secondary (HV) windings are in phase opposition, and polarities are marked as in Fig. 3.40. When the (HV) secondary windings are in phase opposition, switch is closed and a reduced voltage on series-connected secondary windings is applied such that the rated HV winding current can flow.

The back-to-back test is used to find temperature rise, various losses—rated core loss, rated copper loss, equivalent circuit parameters, and rise in temperature with

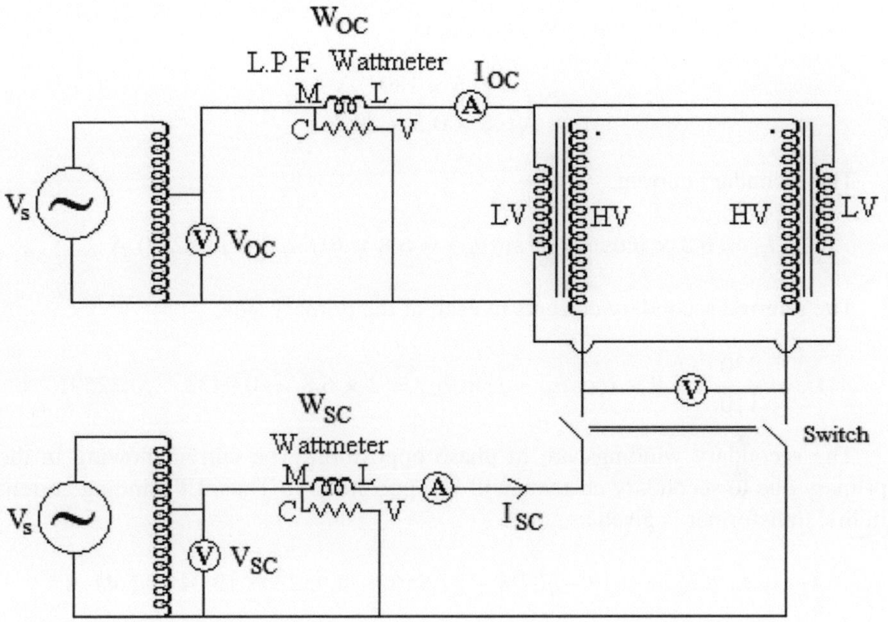

Fig. 3.40 Sumpner's or back-to-back test

respect to time. Back-to-back test on three-phase transformer has been dealt in Sect. 3.21.6.

In heat–run test, the temperature of transformer oil is recorded using either mercury thermometer or a RTD or thermocouple. When signal corresponding to the temperature of transformer oil is available, computer aided data acquisition and data logging can be done.

Example 3.22 Two single-phase transformers are connected for back-to-back test. The two primaries are connected in parallel across a 110 V supply. The primary current and power intake are 2.0 A and 40 W. The secondary are connected in phase opposition across a source of 28 V for circulating a current of 6.8 A. The power intake is 180 W. The turn ratio is 220:110. Calculate the current flowing in primary windings.

Solution

The primary power factor angle θ_0 is

$$\theta_0 = \cos^{-1}\left(\frac{0.5 \times 40}{0.5 \times 2 \times 110}\right) = 79.52^\circ$$

The no-load current in individual transformer is given as,

$$I_0 = (0.5 \times 2.0) \times (\cos\theta_0 - j\sin\theta_0) = (0.18 - j0.98)\ \text{A}$$

The pf angle of secondary currents θ_{SC} is

$$\theta_{sc} = \cos^{-1}\left(\frac{0.5 \times 180}{6.8 \times 0.5 \times 28}\right) = 19.02^{\circ}$$

The secondary current,

$$I_{sc} = 6.8 \times (\cos\theta_{sc} - j\sin\theta_{sc}) = 6.8 \times (0.9433 - j0.3259)\ \text{A}$$

The referred secondary currents in each of the primary side,

$$I_{sc}^{/} = \frac{220}{110} \times 6.8 \times (\cos\theta_{sc} - j\sin\theta_{sc}) = 2 \times 6.8 \times (0.9433 - j0.3259)$$

The secondary windings are in phase opposition. The current flowing in the primary due to secondary currents will be opposite to it. Thus, LT winding current in first transformer is given as,

$$\mathbf{I}_{\mathbf{T1}} = \mathbf{I}_0 + \mathbf{I}_{\mathbf{sc}}^{/} = 0.18 - j0.98 + 12.856 - j4.4322 = 13.04 - j5.41\ \text{A}$$

The LT winding current in the second transformer is,

$$\mathbf{I}_{\mathbf{T2}} = \mathbf{I}_0 - \mathbf{I}_{sc}^{/} = 0.18 - j0.98 - 12.856 + j4.4322 = -12.68 + 3.45j\ \text{A}$$

Thus, although the current supplied to parallel-connected primary is almost twice that of the no-load current of the individual transformers, the actual current flowing in each of the transformer's primary winding is approximately full-load current. The copper losses are supplied by tertiary supply connected on HT terminals. The LT windings are connected in parallel. The current flowing on parallel-connected LT windings is the sum of $\mathbf{I}_{\mathbf{T1}}$ & $\mathbf{I}_{\mathbf{T2}}$. It can be cross-verified that

$$\begin{aligned}\mathbf{I}_0 = \mathbf{I}_{\mathbf{T1}} + \mathbf{I}_{\mathbf{T2}} &= (13.063 - 5.51224j) + (-12.676 + 3.4522j) \\ &= 0.387 - 2.06004j = 2.096\angle -79.36^{\circ}\ \text{A}\end{aligned}$$

This is same as no-load current.

3.18 Tapping in Transformer

Tapping in a transformer are extra electrical terminals connected at different number of turns either on primary winding or on secondary winding such that by selective changing of tap, voltage correction can be done. For example, in a two-winding 220/110 V transformer if we want an additional voltage at 55 V such that the output is either 55 V or 110 V for an AC input voltage of 220 V, then the mid-point of the secondary winding is available for connection. In other words, if number of turns in secondary winding is N_2, then an extra electrical terminal is also provided at $0.5 \times N_2$ turns. This additional terminal is referred as tap. There can be multiple tapping in a transformer. In the present case since the tap is at 50% of secondary winding, the transformer is known as center tap transformer.

A center tap transformer has output voltage terminals for 50% and 100% voltage. These transformers have widespread applications in rectification conversion of AC voltage into DC voltage using solid state devices like diodes or power diodes. Transformer in rectifier circuit provides inherent electrical isolation.

Tapping in transformer is provided for connection of load operating at slightly different voltages. They are also used for minor corrections of input or output voltages. Frequent changing of tap in a transformer, or tap changing, is required in many applications. These are:

(a) Rated voltages of 440 V three phase and 220 V single phase with variation of ± 5% is supplied to the electrical consumers in our country. Due to various reasons, during 24 hours in day, there can be large variations in the HV input voltage, causing variation of more than ± 5% in the voltages available to the utility. This requires the correction of voltages supplied to domestic or industrial users. The tap in a distribution transformer normally provided on low-voltage sides are changed after switching OFF the transformer. It is referred as OFF load tap changing. This type of tap changing is also useful in supplying electrical furnaces operating at different voltages.
(b) When an electric locomotive moves on rail track from one substation to another substation situated at about 150 km (say), variation in line voltage from 25 to 18 kV is observed by the locomotive due to line voltage drop. Both the substations are supplying 25 kV power to the overhead line. The input voltage drop cannot be avoided due to the large power requirement of locomotive. The onboard voltage control is also required for speed control and starting of traction motors. Since onboard transformer is having tapping on high voltage side, these are changed without disconnecting the transformer from supply and load. This type of tap changing is referred as on-load tap changing. When traction motor supply is disconnected momentarily, it causes re-switching transient current. These currents are absent when on-load tap changer is used. On board, the voltage in traction locomotive is not affected by location related line voltage drop. Using online tap changer in traction transformer, the voltage is regulated.

As mentioned above, there are two types of tap changing methods in transformer. These are:

(a) OFF load tap changing
(b) ON load tap changing.

3.18.1 OFF Load Tap Changing

The control of voltage in a distribution supply network is essential due to the reasons mentioned below:

(a) Due to loading of the transformer or due to lesser input voltage on the primary of transformer, the output voltage on load side may fall or rise beyond statutory limit. This requires adjustment of tap on the transformer.
(b) The taps are corrected for controlling the KW or KVAR flow over a parallel connected supply line or grid.
(c) In substations, a seasonal, daily or short period correction of voltage is required for taking care of cyclic changes of the load.

If we consider a three-phase distribution transformer, say 100 kVA, 11 kV/440 V, there is always possibility that the local voltage might be either more than 11 kV or even less than 11 kV. As such the line-to-line voltage will be either more or less than 440 V respectively. There is a convention to provide extra tapping on the low-voltage side for adjustment of voltage as per requirements. Using low-voltage or LV tapping one can correct 5% higher voltage and 10% lower voltage. In distribution transformer, taps are sometimes provided on high voltage or HV terminals also for voltage correction. However, the use of HV side OFF load tap changing is avoided due to operational difficulties. If the local high voltage is 11.55 kV or 9.9 kV, then by using suitable tap on secondary winding, the output line voltage can be corrected to nearly 440 V. In these distribution transformers, the multiple terminals on the three-phase LT secondary are provided by the manufacturer for local adjustment of the distribution voltage. Such transformer may have multiple tapping for 390 V, 400 V, 420 V, 460 V, 480 V for rated 11 kV input voltage. If the input voltage differs from 11 kV, the secondary tapping is adjusted in offline conditions. For example, in 11 kV/440 V transformer, if the voltage drops to 10 kV, the output voltage drops to 440 × (10/11) or 400 V. This can be again corrected to 440 V by changing the transformation ratio. The transformation ratio is increased by reducing the number of turns on LV secondary side. Similarly, if the input voltage is 13 kV, the secondary voltage will shoot to 440 × (13/11) or 520 V. This is a higher voltage giving 300 V per phase. Under such conditions, the supply is switched OFF. The voltage is corrected by reducing the transformation ratio or by increasing the secondary number of turns. This has been explained in Fig. 3.41. Why do we have off line or OFF-load tapping on low-voltage side? Can you answer?

3.18.2 On Load Tap Changing

Let us consider the single-phase voltage of 25 kV of electrical locomotive as mentioned earlier. Due to large current taken by the locomotive and hence subsequent line loss, the available voltage may vary from (say) 25 kV to 18 kV in between two

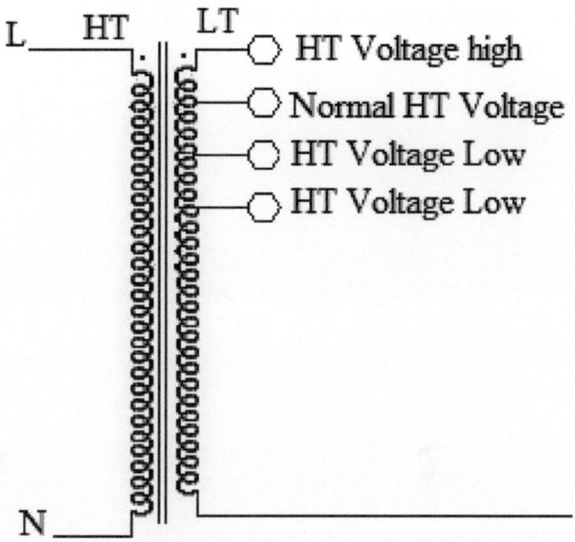

Fig. 3.41 OFF load or offline tap changing transformer

feeder sub-stations as the train moves. Momentary disconnection and re-connection of supply may cause severe damages to the solid state converters supplying power to traction motors, supply system and traction motors due to re-switching transient current. As such voltage control requires on-line or on-load tap changing as shown in Fig. 3.42. For maintaining a constant voltage, the tapping is provided on the high-voltage side of the transformer on board of locomotive. These tap changing may also be required for speed control of traction motors. These taps are changed online, using tap changer switch, without switching off the main supply. Since the high-voltage rated current is quite less, it causes lesser sparking when taps are changed online. However, the insulation requirement is stringent. These contacts are dipped in transformer oil. In Fig. 3.43a, line terminal is connected to a certain tap when it is required to join the line to different tap without disconnecting the line. When transition from one voltage to another voltage is required, then two subsequent taps are short-circuited, as shown in Fig. 3.43b, during which a local short-circuit current flows. The local short-circuit current is limited using either a center tapped inductor as shown in figure or using a resistance. The opening of the upper contactor in Fig. 3.43c joins the line to different tap on HT (HV) side. The arc is quenched rapidly. These tap changers are oil immersed and the contactors for changing the tap are operated in a closed loop manner using sophisticated automated control. Online tap changers frequently require maintenance due to insulation breakdown and sparking.

If ON-load or online tap changers are provided on low-voltage side, the secondary current is much higher than high-voltage side currents, causing severe sparking and, large local short-circuit current flows between two taps (nearby turns) when taps are being changed.

Example 3.23 In certain rural area, it is found that when 11 kV/440 V three-phase transformer is installed, but the single-phase voltage is 280 V. What is the turn ratio

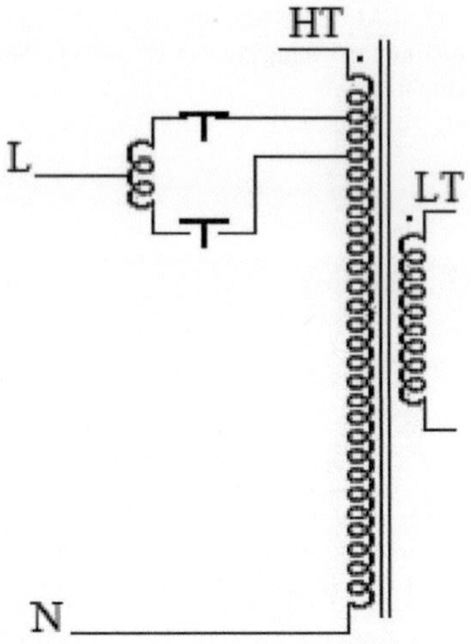

Fig. 3.42 ON load or online tap changer

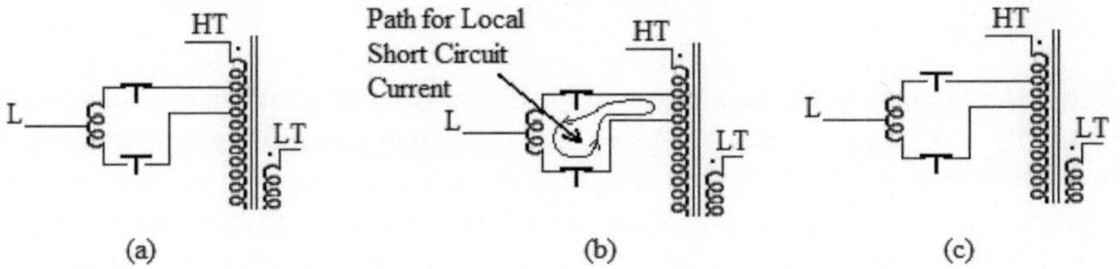

Fig. 3.43 Transition in ON load tap changing **a** Line connected to one tap, **b** turns between two tap are short-circuited giving local short-circuit current, and **c** line connected to another tap

and input voltage and how the voltage can be corrected? What is turn ratio for correcting the voltage to 440 V three phase? What is the output line-to-line voltage when 11 kV is applied for this turn ratio?

Solution

Considering the transformer to be delta/star.

Here $V_1 = 11$ kV and $V_2 = 440/1.7321 = 254$ V

Turn ratio of transformer,

$$N_1/N_2 = V_1/V_2 = 11,000/254 = 43.307$$

Here single-phase voltage is V_2 (new) = 280 V.

Thus, corresponding input voltage is

$$V_1(\text{new}) = V_2(\text{new}) \times (V_1/V_2) = 280 \times (11/254) = 12.124 \text{ kV}$$

Input side voltage is higher than 11 kV. If we change tap on low-voltage side such that for input voltage 12.124 kV, we get three-phase 440 V voltage or 254 V single-phase voltage on secondary side. The new turn ratio is

$$N_1(\text{new})/N_2(\text{new}) = V_1(\text{new})/V_2 = 12,124/254 = 47.73$$

The new turn ratio is 47.73 which is higher than 43.307. It means, tap on the low voltage side is to be used for reducing the voltage from 280 to 254 V. However, if 11 kV is the input voltage, the output voltage for this connection is,

$$V_1/V_2(\text{new}) = 11,000/V_2(\text{new}) = 47.73$$
$$V_2(\text{new}) = 11.000/47.73 = 230.45 \text{ V}$$

The line to line voltage,

$$V_{2L}(\text{new}) = 230.45\sqrt{3} = 399.2V$$

Example 3.24 The single-phase voltage on traction locomotive near substation is 25 kV. When locomotive is 150 km away from any nearby substation, the voltage drops to 18 kV. The on board transformer rating is 25 kV/600 V single phase. Calculate (a) the output voltage if turn ratio is not changed; (b) the effective number of turns when tap changer is operating for maintaining the rated output voltage; and (c) the output voltage for new turn ratio for 25 kV input voltage.

Solution

Here $V_1 = 25$ kV and $V_2 = 600$ V and turn ratio is

$$N_1/N_2 = V_1/V_2 = 25,000/600 = 41.67$$

Here input voltage is V_1 (new) = 18 kV.

Thus corresponding output voltage when turn ratio is not altered is.

$$V_2(\text{new})/V_1(\text{new}) = (V_2/V_1) = N_1/N_2$$

$$V_2(\text{new}) = 18(600/25) = 432 \text{ V}$$

The effective number of turns for maintaining the 600 V output voltage at 18 kV input voltage is.

$$N_1(\text{new})/N_2(\text{new}) = V_1(\text{new})/V_2(\text{new}) = 18,000/600 = 30$$

For this turn ratio, the output voltage for 25 kV input is,

$$V_2 = 25,000/30 = 833.33 \text{ V}$$

3.19 Autotransformer

Autotransformer is normally single-winding transformer from which variable voltages can be obtained. An autotransformer with fixed output voltages utilizes the magnetic core similar to that of transformer with common winding having multiple tapings brought out for connection. Similar three-phase autotransformers are used for starting three-phase induction motors used in submersible pumps. For smooth variation of output voltage as used in lab electrics, common winding is wound on a toroidal core, the variable voltage is obtained using carbon brush rocking over turns of the common winding. An autotransformer with its winding terminals connected to supply terminals and also to load is shown in Fig. 3.44a, b. The number of turns on primary and secondary differs, causing the variation of voltage, thereby giving step-up or step-down autotransformers. The transfer of energy takes place by means of both induction and conduction. Multiphase or three-phase autotransformers are also common. These transformers have separate phase windings, which are normally connected in star. There is no magnetic coupling between different phases.

The common winding has large current to supply by means of induction, whereas the uncommon winding will supply current through conduction. These two currents are superimposed.

Since the turns links the same flux in the core, we have

$$\frac{V_1}{V_2} = \frac{N_1}{N_2} = a = \text{transformation ratio}$$

If the number of turns of secondary is less than the number of primary turns, voltage steps down. The ampere turns of different segments of the windings are given as.

For uncommon windings,

$$\mathcal{F}_u = (N_1 - N_2)I_1 \tag{3.91}$$

For common windings,

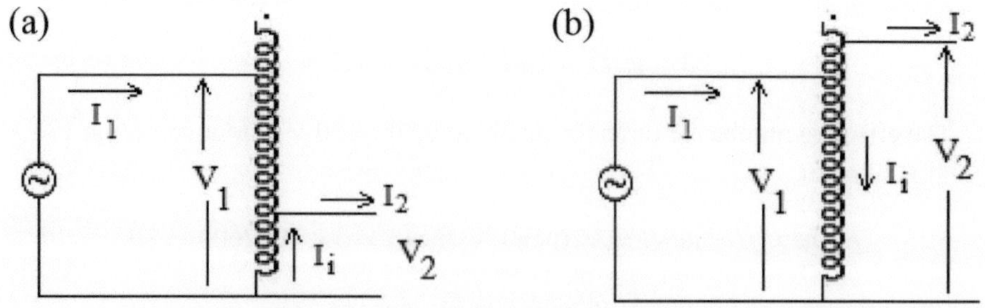

Fig. 3.44 **a** Step-down autotransformer; **b** Step-up autotransformer

$$\mathcal{F}_c = N_2(I_2 - I_1) = \frac{N_1}{a}(I_2 - I_1) = \frac{N_1}{a} I_i \tag{3.92}$$

$$(N_1 - N_2)I_1 = \frac{N_1}{a}(I_2 - I_1) \tag{3.93}$$

Ignoring the magnetizing amperes-turns, the two MMF are equal. Thus, we get

$$\left(1 - \frac{1}{a}\right)N_1 I_1 = \frac{N_1}{a}(I_2 - I_1) \tag{3.94}$$

Solving Eq. (3.94), we get

$$\frac{I_2}{I_1} = a \tag{3.95}$$

Equation (3.95) is same as for two-winding transformer. Here current $\boldsymbol{I}_1$ is supplied by means of conduction and current $(\boldsymbol{I}_2 - \boldsymbol{I}_1)$ is due to induction, $\mathbf{I}_\mathrm{i}$ as marked in Fig. 3.44. Now we have:

The transformed output power $= V_1 I_1 = V_2(I_2 - I_1) = V_2\, I_i$.

The conducted output power $= V_2 I_1$.

The total power available in auto transformer $= V_2 I_2$.

Let us compare the output power of the auto transformer to the output power of the two-winding transformer having similar current and voltage ratings.

The output VA of autotransformer $= V_2 I_2$.

The output power of the equivalent two-winding transformer $= V_2(I_2 - I_1)$.

Thus, the ratio of output power of autotransformer to that of a two-winding transformer is

$$\frac{V_2 I_2}{V_2(I_2 - I_1)} = \frac{a}{a - 1} \tag{3.96}$$

Thus $\frac{a-1}{a}$ times the output power is transferred by transformer (induction) action and the remainder $\frac{1}{a}$ times the output power is transferred through electrical conduction.

A considerable amount of copper can be saved in an autotransformer. If we assume that the current density in the windings and core flux remain same for an autotransformer and a two-winding transformer having same input and output power, then we get ratio of conducting material in

$$\begin{aligned}\frac{\text{Autotransformer}}{\text{Two windng transformer}} &= \frac{I_1(N_1 - N_2) + (I_2 - I_1)N_2}{I_1 N_1 + I_2 N_2} \\ &= 1 - \frac{I_1}{I_2} = 1 - \frac{V_2}{V_1} = \frac{a-1}{a}\end{aligned} \tag{3.97}$$

A considerable amount of copper is saved when V_2 approaches V_1. For a voltage ratio of 2:1, approximately 50% copper can be saved. The core in autotransformer is large as compared to that of two-winding transformer and, therefore, overall saving is about 30–40%. A step down autotransformer is shown in Fig. 3.45. The equivalent circuit of step-down autotransformer, based on approximate equivalence for a power factor angle of θ, is shown in Fig. 3.46.

If R_1 and X_1 are the parameters of uncommon windings, whereas R_2 and X_2 are the parameters of common windings. N_1 is the total number of turns and N_2 is common winding number of turns, then voltage equation can be given as

$$\begin{aligned}\mathbf{V}_1 &= \mathbf{E}_1 + \mathbf{I}_1(R_1\cos\theta + X_1\sin\theta) + (\mathbf{I}_1 - \mathbf{I}_2)(R_2\cos\theta + X_2\sin\theta)\\ \mathbf{V}_2 &= \mathbf{E}_2 - (\mathbf{I}_2 - \mathbf{I}_1)(R_2\cos\theta + X_2\sin\theta)\end{aligned} \tag{3.98}$$

We know

$$E_2 = E_1\left(\frac{N_2}{N_1}\right) = \frac{E_1}{a}$$
$$I_2 = I_1\left(\frac{N_1}{N_2}\right) = aI_1$$

Thus we get,

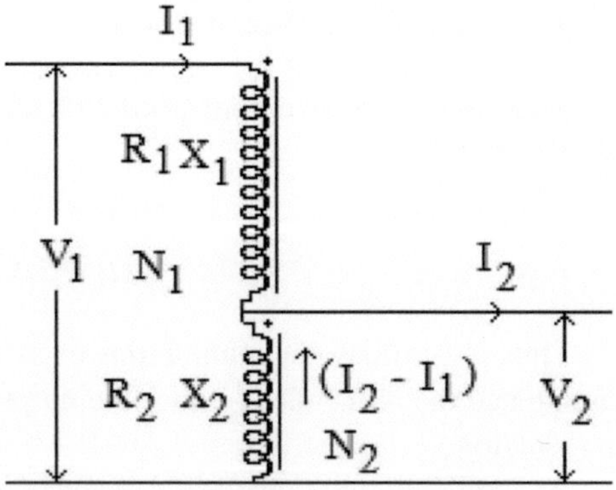

Fig. 3.45 Step-down autotransformer

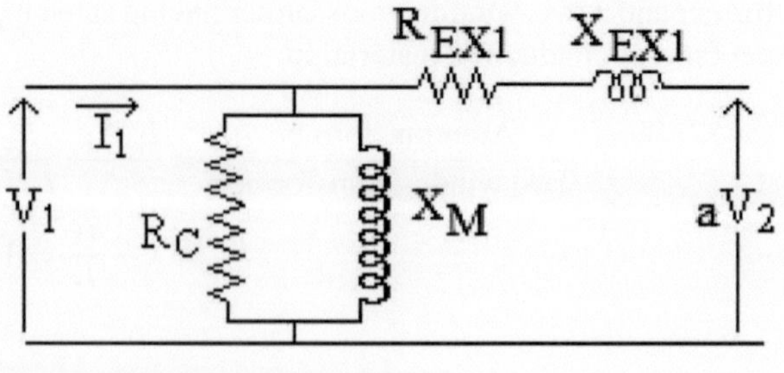

Fig. 3.46 Equivalent circuit of autotransformer

$$\mathbf{V}_1 = \text{a}\,\mathbf{V}_2 + \mathbf{I}_1\left(R_1 + R_2^{/}\right)\cos\theta + \mathbf{I}_1\left(X_1 + X_2^{/}\right)\sin\theta$$

Combining the above two expressions, we get,

$$\mathbf{V}_1 = \text{a}\,\mathbf{V}_2 + \mathbf{I}_1 R_{EX1}\cos\theta + \mathbf{I}_1 \text{X}_{\text{EX1}}\sin\theta \quad (3.99)$$

The equivalent resistance and leakage reactance referred to primary (HT) are

$$R_{\text{EX1}} = \left(R_1 + R_2^{/}\right) \text{ and } X_{\text{EX1}} = \left(X_1 + X_2^{/}\right)$$

where $\text{R}_2^{/} = \text{R}_2(\text{a} - 1)^2$ and $\text{X}_2^{/} = \text{X}_2(\text{a} - 1)^2$

Using above derivations, show that the.

(i) % Full-load losses of auto/two-winding transformer is $\frac{a-1}{a}$.
(ii) % Full-load drop of auto/two-winding transformer is $\frac{a-1}{a}$
(iii) % Regulation as auto/two-winding transformer is $\frac{a-1}{a}$
(iv) % Short-circuit current in auto/two-winding transformer is $\frac{a}{a-1}$

The autotransformers have the following advantages:

(a) They have lower leakage reactance.
(b) The copper losses are lowered in autotransformers.
(c) They have lower magnetizing current.
(d) The kVA ratings are increased in autotransformers because the power is also transferred by means of conduction.
(e) A variable voltage from zero to slightly higher than supply voltage can be obtained when sliding contacts are used.
(f) A saving in copper is achieved.
(g) Autotransformer may be connected as star or delta or even in Scott-connection.

However, autotransformer have some demerits as well. They are as follows:

(a) The low-voltage secondary is connected to primary terminal by conductive links. This may result in appearance of high voltage on secondary in the event of open-circuit fault in the common winding. The turn ratio rarely goes beyond 3:1, with secondary voltage around one-third of primary voltage and secondary rated current around three-times to that of the primary rated current.
(b) Stepped-up autotransformers require more insulation due to high voltage.

Autotransformers have the following applications:

(a) They serve as variable voltage AC source in laboratories. Also known as variac.
(b) They are required for starting three-phase induction motor.
(c) Autotransformers are needed to connect 380 V instruments with 440 V supply.
(d) They are also required for interconnecting two power grids operating at different voltages.

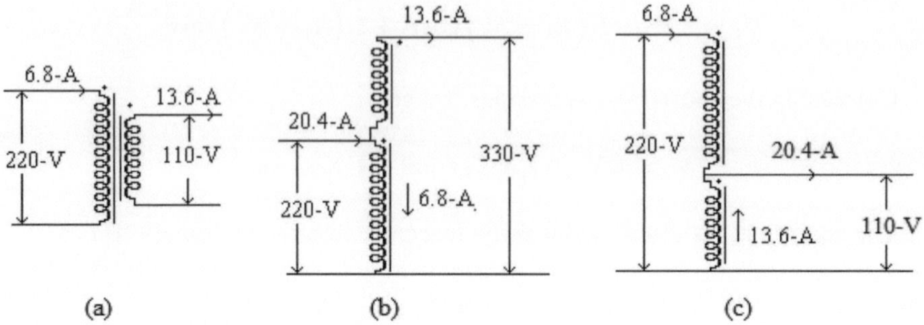

Fig. 3.47 **a** Two-winding 1.5 kVA, 220/110 transformer as, **b** Step-up autotransformer and **c** Step-down autotransformer

Example 3.25 If two-winding 220 V/110-V, 50 Hz, 1.5 kVA transformer is reconnected as step-up and step-down autotransformers, calculate the kVA rating and currents of autotransformer shown in Fig. 3.47b, c.

Solution

The rated currents of windings are

$$I_1 = \frac{1500}{220} = 6.8 \text{ A}$$

$$I_2 = \frac{1500}{110} = 13.6 \text{ A}$$

The step-up autotransformer produces 330 V and input voltage is 220 V. Therefore, for full-load operation of the autotransformer, the terminal currents are.

$I_H = 13.6$ A and $I_L = 13.6 + 6.8 = 20.4$ A.

$V_L = 220$ V and $V_H = 330$ V

$$\text{KVA}_L = \frac{220 \times 20.4}{1000} = 4.488 \text{ KVA}$$

$$\text{KVA}_H = \frac{330 \times 13.6}{1000} = 4.488 \text{ KVA}$$

Thus, we see that a 1.5 kVA two-winding transformer, when reconnected as step-up autotransformer can deliver 4.888 kVA power. The common 220 V winding will carry a current of 6.8 A. See Fig. 3.47b.

If the transformer is reconnected as step-down autotransformer to produce 110 V, then the rated currents of two-winding transformer are, $I_2 = 13.6$ A and $I_1 = 6.8$ A.

Thus for full-load operation, we have

$$I_H = 6.8 \text{ A}$$

$$I_L = 13.6 + 6.8 = 20.4\text{ A}$$

Here 13.6 A current is supplied by the common winding. Thus, we get

$$\text{KVA}_L = \frac{110 \times 20.4}{1000} = 2.244\text{ KVA}$$
$$\text{KVA}_H = \frac{330 \times 6.8}{1000} = 2.244\text{ KVA}$$

A voltage on secondary with open-circuit fault in the common winding will be 220 V. Thus 1.5 KVA, 220/110 V two-winding transformer when connected as step-up autotransformer is re-rated as 4.488 kVA and when connected as step-down autotransformer is re-rated as 2.244 kVA without exceeding the currents in the winding. In step-down autotransformer, a voltage on secondary with open fault in the common winding will be 220 V.

Example 3.26 A single-phase, 10 KVA, 50 Hz, 2000/200 V two-winding transformer is connected as an autotransformer to produce (a) 2200 V and (b) 200 V. Calculate the following:

(a) The current supplied by the common windings
(b) kVA rating output of autotransformer
(c) Voltage on secondary, if common windings are open.

Solution

(a) For step-up autotransformer, the rated currents of windings are
$I_2 = 10{,}000/200 = 50$ A
$I_1 - 10{,}000/2000 = 5$ A

Therefore, for full-load operation of the autotransformer, the currents are. $I_H = 50A$ and $I_L = 50 + 5 = 55A$.
Now, we have

$$V_L = 2000\text{ V and } V_H = 2000 + 200 = 2200\text{ V}$$
$$\text{KVA}_L = \frac{2000 \times 55}{1000} = 110\text{ KVA}$$
$$\text{KVA}_H = \frac{2200 \times 50}{1000} = 110\text{ KVA.}$$

Thus we see that a 10 kVA transformer as an autotransformer can deliver 110-kVA. If common windings are open due to any reason, 2000 Volt will appear on the secondary side.

(b) The input voltage is 2200 V and output voltage is 200 V.

The rated currents are

$$I_L = 50\text{ A}$$

$$I_H = 5\text{ A}$$

Thus, we have

$$I_H = 5\text{ A}$$

$$I_L = 50 + 5 = 55\text{ A}$$

Here, 50A current is supplied by common winding. Thus, we get

$$\text{KVA}_L = \frac{200 \times 55}{1000} = 11\text{ KVA}$$
$$\text{KVA}_H = \frac{2200 \times 5}{1000} = 11\text{ KVA}$$

(c) A voltage on secondary with open-circuit fault (opening of winding conductor) in common winding will be 2000 V, which is very high voltage and can damage the insulation of the common winding. This is a significant demerit of auto transformers. However, this property is used in three-phase autotransformer starting of a three-phase induction motor. Here the induction motor is supplied a reduced three-phase voltage during starting. When the motor is started, full voltage is given to the motor by opening the star point of the three-phase autotransformer as shown in Fig. 5.42. During this process, there is no electrical disconnection, which reduces the transient current due to re-switching from low voltage to full voltage.

3.20 Inrush Current

A transformer behaves like a resistive-inductive (R-L) circuit at power frequency. We know that when an RL circuit is switched ON at voltage peak, a transient current appears, as shown in Fig. 3.48c. Consider an RL circuit which is connected to an AC voltage $V(t) = E_m \sin \omega t$. The switching transient in RL circuit has been dealt in Sect. 1.5.5.

When the circuit is highly inductive giving $\theta \approx 90°$, the current lags the voltage by an angle θ. The nature of current when switching at zero crossing voltage is explained as:

(a) A decaying transient component of direct current having a time constant of $\tau = L/R$.
(b) A steady state current which is lagging the voltage.

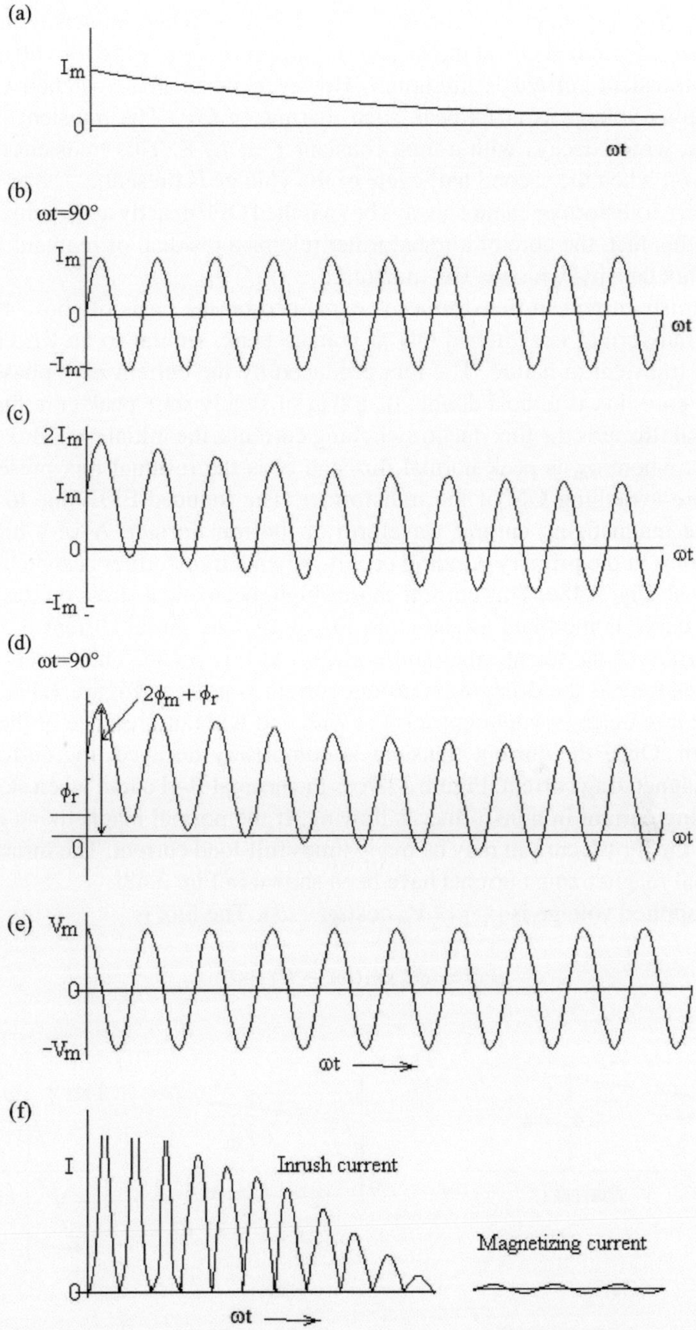

Fig. 3.48 **a** Decaying DC component of current when switch is closed at instant $\omega t = 90°$, **b** steady state component of current, **c** transient current in inductive circuit when switch is closed at instant $\omega t = 90°$, **d** effect of residual flux and switching transients–flux-doubling phenomenon, **e** switching at maximum voltage condition, and **f** inrush current and normal magnetizing current

These two components are shown in Fig. 3.48a, b. When inductive circuit or a transformer is switched ON at the instant of current zero crossing (or voltage peak), the peak transient current is minimum. However, when it is switched ON when instantaneous voltage is at its peak such that $\omega t \approx 90°$. The transient current is maximum, which decays with a time constant $\tau = L/R$. This transient current is present even when the second half cycle of the voltage is present.

Further, a transformer cannot always be switched OFF exactly at zero flux crossing owing to this fact, the core of a transformer retains a residual or remnant flux ϕ_r or residual flux density, which is DC in nature.

The inrush current in transformer is explained on the basis of above two facts. When a transformer is switched ON at voltage peak, similar to an RL circuit the current is transient in nature. The flux produced by the current is in phase with it. The peak core flux is almost double than that of steady state peak core flux. When the residual flux aids the flux due to switching currents, the initial peak flux is almost $2\phi_m + \phi_r$, where ϕ_m is peak normal flux and ϕ_r is the residual flux present in the core before switching ON of the transformer. The induced EMF due to this flux affects the magnetizing current waveform of the transformer. A very high initial surge current in the primary winding occurs, when a transformer is switched ON at voltage peak Fig. 3.48e. This current causes high mechanical stress on the winding. The B-H curve is modified as shown in Fig. 3.49. The initial current is known as inrush current of the transformer and is shown in Fig. 3.48d. The current is unidirectional as long as the decaying transient current is present. Figure 3.49a, b shows the B-H curve during switching transient with and without presence of the residual magnetism. Once the current transient is completely decayed, the current is the normal magnetizing current. Figure 3.49c is the normal B-H curve when steady state magnetizing current in transformer is flowing. If the normal magnetizing current is 0.05 p.u., the inrush current may be many times full-load current. The inrush current and normal magnetizing currents have been shown in Fig. 3.48f.

If the applied voltage is $v(t) = V_m \cos(\omega t - \theta)$. The flux is

$$\phi = \phi_m \sin(\omega t - \theta) + C$$

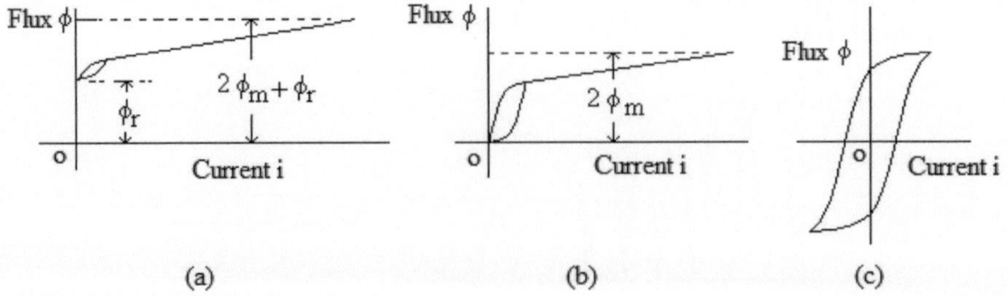

Fig. 3.49 Modified BH or flux-current diagram due to **a** Switching and residual flux, **b** Only switching, **c** Normal magnetizing current in a transformer

where

$$\phi_m = \frac{V_m}{\omega N_1}$$

At $t = 0 \quad \phi = \phi_r$

$$\phi = \phi_m \sin(\omega t - \theta) + \phi_r - \phi_m \sin\theta$$

when it is switched ON at zero crossing, $\theta = 0°$

$$\phi = \phi_m \sin(\omega t - \theta) + \phi_r$$

when switch is closed at voltage peak, $\theta = -90°$

$$\phi = -\phi_m \cos \omega t + \phi_r + \phi_m$$

3.21 Three-Phase Transformer

Single-phase system is rarely used for generating and transmitting electrical power, except in remote generation and utilization of power. The three-phase generation and transmission is cost effective over single-phase system. A simple single-phase transmission corridor will require three-wire transmission lines, whereas a three-phase system will require four-wire transmission line. For the same voltage and current, the transmitted power in three-phase system will be about 73% more than that of single-phase transmission lines. The overall sizes and cost of transformer as well as that of an alternator will also reduce. The transmission of power with polyphase AC system (more than three-phases) increases complexities and is not preferred.

Power transmitted either through three-phase AC or bipolar High Voltage Direct Current (HVDC), both the systems require polyphase transformers for their functioning. As already mentioned, polyphase transformers, where numbers of phases are more than three, are used for solid state power conversions from AC to HVDC at sending-end and HVDC to AC at receiving-end in HVDC system. These transformers are referred to as *converter-transformers*. The use of polyphase converter transformers reduces ripple contents in rectified DC.

The concept and determination of polarity in AC circuits have already been discussed. Determination of polarity is essential for connecting two or more transformers to similar power supply and load. When primary/secondary of a three-phase transformer is connected in delta a wrong connection may result a dead short circuit.

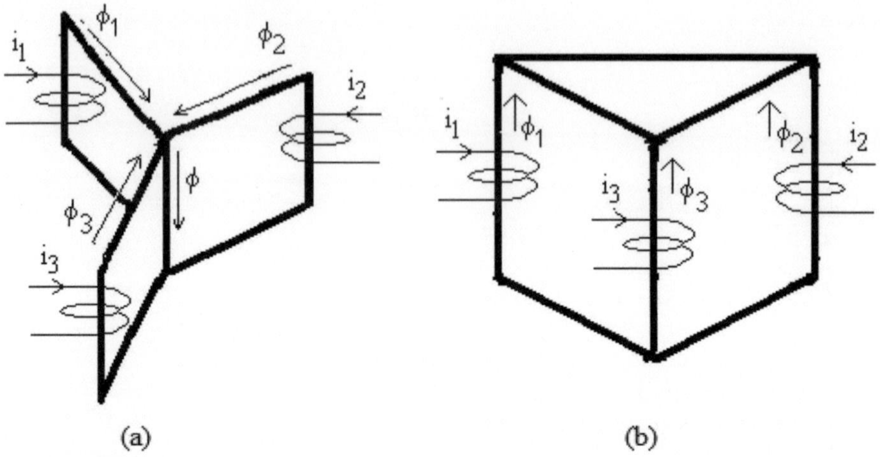

Fig. 3.50 **a** Typical three-phase core-type transformer, **b** path of magnetic flux in a typical three-phase shell-type transformer

Bulk electrical energy is transmitted using three-phase power transformers. These are required to either step-up or step-down the voltage level, and are of the following two types:

(a) Single unit of three-phase transformer
(b) Bank of three single-phase transformers connected in three phase.

A large single unit will be quite bulky and needs special arrangements for transportation from the factory to the installation site. Moreover, in the event of a fault or breakdown, the whole unit needs to be shut down for repair. Further, for maintaining regular supply, it is required to keep at least one spare unit for immediate replacement. On the other hand, when bank connections are used, in the case of a fault, only faulty single-phase transformer needs to be removed. Also, the cost of spare is less in one number single-phase transformers connected in three phases than single unit of three-phase transformer.

Three-phase transformers (with flux path in core) are shown in Fig. 3.50a, b. In core-type transformer, middle limb is avoided since summation of fluxes are zero at any instant of time. Shell-type three-phase transformers are more compact. They are also more difficult to assemble and repair as compared to core-type three-phase transformers. The primary and secondary windings, are connected in different ways.

3.21.1 Three-Phase Connections

Three-phase transformers are either single-unit transformer or assembled using three units of single-phase transformers. The primary and secondary windings are symmetrically connected for the transfer of three-phase electrical energy at different voltage levels. The windings of either single unit of three-phase or a bank of three single-phase transformers are connected in star or delta or zigzag forms. The latter requires multiple segments in the windings of each phase of the secondary.

The three-phase transformer connections are symmetrical. It needs three transformers having identical characteristics and ratings. If primary is supplied with a balanced three-phase voltage, then the secondary voltages will be balanced, provided the loads connected to individual phases are balanced. The three-phase transformers have following connections:

(a) **Primary in Y and Secondary in Y**: Also known as Y-y (or Star-star) connection, it is rarely used because of problems associated with exciting current and induced voltages.
(b) **Primary in Δ and Secondary in Δ**: Also known as Δ –Δ (Dd) (or Delta-delta) connection, it can be used with a provision to use V-V connection in the event of fault in one of the winding.
(c) **Primary in Y and Secondary in Δ**: Also known as Y-Δ (Yd) (or Star-delta) connection, it is commonly used for stepping down a higher voltage to lower voltage.
(d) **Primary in Δ and Secondary in Y**: Also known as Δ –y (Dy) (or Delta-star) connection, it is commonly used for stepping-up and stepping-down a voltage.

The primary and secondary windings are connected in either star or delta configurations. In the case of star-connected windings, the voltage per turn is lesser than that of delta-connected windings. In the case of high-voltage transformers, the use of star-connected windings requires lesser insulation and hence, will be less costly when compared with that of delta-connected (HT) windings. In star-connected transformers, the line current is same as the load current as such, thicker conductors are required.

If the primary voltages are balanced, then secondary will also be balanced, provided the load on secondary side is same on each phase. The operational characteristics under unbalanced supply, unbalanced load and fault conditions require special treatment. Polyphase transformers are grouped under four groups as given in Table 3.3. Here ‘Y’ and ‘y’ stand for star connections; and ‘D’ & ‘d’ stand for delta connection; ‘z’ stands for zigzag connection. The capital letters, ‘Y’ & ‘D’ correspond to primary and lower case letters, ‘y’ & ‘d’ correspond to secondary. In zigzag connection, the secondary has multiple winding segments which are connected symmetrically to form complete three-phase secondary. The nomenclature is according to the clock position. If the primary line voltages are at 12’O clock, the secondary line voltage is at 12 O’clock [Group-I (0°)], at 6 O’clock [Group-II (180°)], at 1 O’clock [Group-III (−30°)] and at 11 O’clock [Group-IV (+30°)].

This has been shown in Table 3.3.

A converter transformer having multiple three-phase secondary can have mixed nomenclature. For example, Yy0y6 transformer, which converts three-phases into six-phases, has two secondaries connected in star configuration of Group-I and Group-II.

Table 3.3 Transformer grouping

Group I (0°)	Group II (180°)	Group III (−30°)	Group IV (+30°)
Yy0	Yy6	Dy1	Dy11
Dd0	Dd6	Yd1	Yd11
Dz0	Dz6	Yz1	Yz11

The primary (HT) terminals are marked as A1-A2, B1-B2, C1-C2 and secondary (LT) terminals are marked as a1-a2, b1-b2 and c1-c2. Accordingly, the polarity of terminals A1 & a1, B1 & b1 and C1 & c1 are identical. The phase voltages on primary sides are OA, OB, OC, while on secondary sides are oa, ob, oc, etc.

(a) **Star/Star-Connected Transformers**: The Y-y transformer connection can have either 0° or 180° phase difference between primary and secondary set of voltages. The primary neutral point is formed by joining A2-B2-C2. The secondary neutral point is formed by joining a2-b2-c2 and phase terminals are a1-b1-c1 as shown in Yy0 transformer in Fig. 3.51. The phase difference between input and output is then 0°.

If the terminals of secondary are interchanged making a1-b1-c1 as neutral and a2-b2-c2 as line terminals, then the secondary voltage phasor are changed as shown in Yy6 transformer in Fig. 3.52. The line voltage and phase voltages of primary and secondary windings differ by an angle 180°. The phasor diagrams are shown in Fig. 3.54.

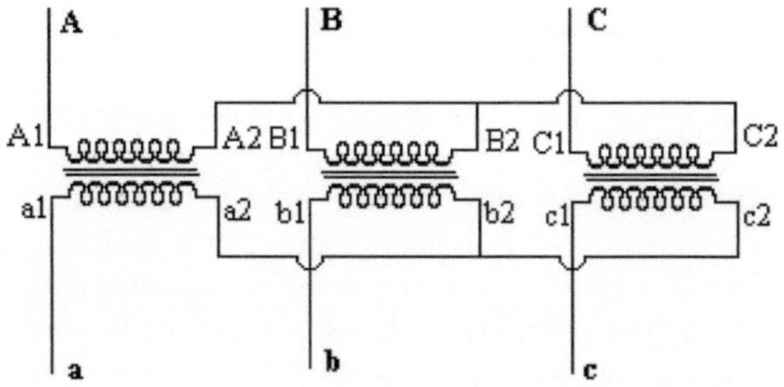

Fig. 3.51 Star/star-connected three-phase transformers Yy0 (0° phase displacement)

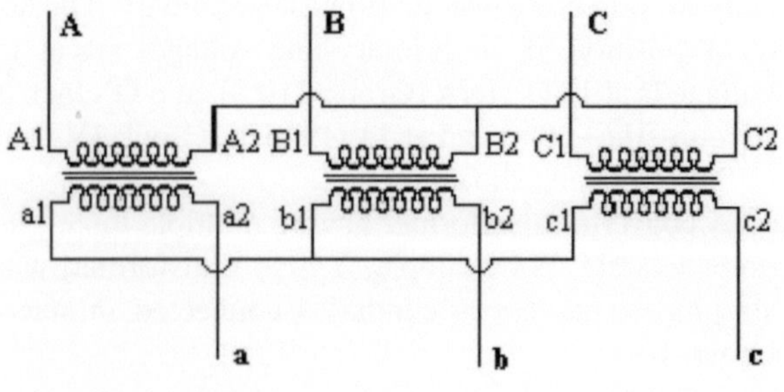

Fig. 3.52 Star/star-connected three-phase transformers Yy6 (180° phase displacement)

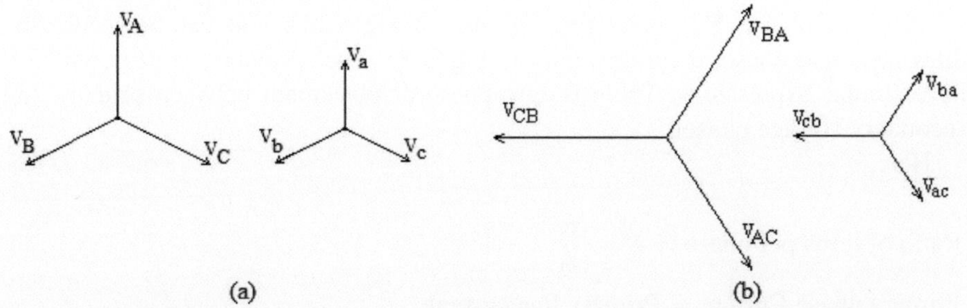

Fig. 3.53 **a** Primary and secondary phase voltage phasor Yy0 (0° phase displacement), **b** primary and secondary line voltage phasor Yy0 (0° phase displacement)

Fig. 3.54 Primary and secondary phase voltage phasor Yy6 (180° phase displacement)

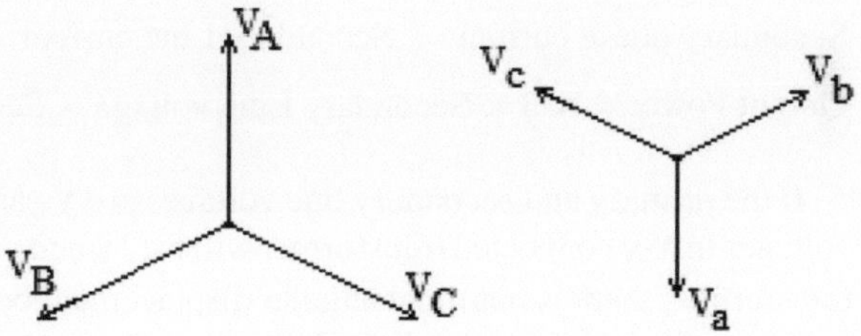

The phasors of Yy0 are shown in Fig. 3.53a, b. The primary and secondary line voltages of Yy6 are shown in Fig. 3.54. Such transformer requires fewer numbers of turns of thicker wire. As such, the windings are mechanically strong and can withstand stresses due to short circuit. Also, the voltage per turn is low, which reduces the dielectric stress on the insulation.

In Yy0 transformer, the HT terminals are connected at A1-B1-C1 and LT terminals are connected at a1-b1-c1. The neutral terminals are A2-B2-C2 and on secondary side a2-b2-c2. The line voltage is obtained by adding $-\mathbf{V}_{\mathrm{OB}}$ with $\mathbf{V}_{\mathrm{OA}}$ for getting $\mathbf{V}_{\mathrm{BA}}$. In other words, we have

$$\boldsymbol{V}_{\mathrm{BA}} = \boldsymbol{V}_{\mathrm{OA}} + \boldsymbol{V}_{\mathrm{BO}}$$

Here,

$$\begin{aligned} \boldsymbol{V}_{\mathrm{OA}} &= V\angle 0^\circ = V + j0 \\ \boldsymbol{V}_{\mathrm{OB}} &= V\angle 120^\circ = V(-0.5 + j0.866) \\ \boldsymbol{V}_{\mathrm{OC}} &= V\angle -120^\circ = V(-0.5 - j0.866) \end{aligned}$$

On substituting these values, we get

$$\boldsymbol{V}_{\mathrm{BA}} = \boldsymbol{V}_{\mathrm{OA}} + \boldsymbol{V}_{\mathrm{BO}} = V(0.5 - 0.866j + 1) = V\angle -30^\circ$$

Thus, the phasor $\mathbf{V}_{BA}$ is lagging $\mathbf{V}_{OA}$ by an angle 30°. This can be verified by drawing actual phasor diagram shown in Fig. 3.53. The secondary voltage will also have similar expressions. There is zero-phase displacement between primary and secondary voltage phasor.

Here, if

$$\text{Ratio of turns per phase} = a = \frac{N_1}{N_2}$$

$$\text{Primary phase Current} = \text{Primary line current}$$

$$\text{Secondary phase voltage} = \frac{\text{Primary phase voltage}}{a}$$

$$\text{Secondary Line voltage} = \text{Secondary phase voltage} \times \sqrt{3}$$

$$\text{Secondary phase current} = \text{Secondary Line current} = a \times \text{Primary phase current}$$

$$\text{Output Power} = \sqrt{3} \times \text{Secondary Line voltage} \times \text{Secondary Line current}$$

If the primary and secondary line voltages are V_1 and V_2 respectively, their phase voltages in Y-y connected transformer will be $\frac{V_1}{\sqrt{3}}$ and $\frac{V_2}{\sqrt{3}}$, respectively. In three-phase transformer, there is marginal angular displacement between primary and secondary voltages, due to the leakage impedances and magnetizing current. This displacement is considered to be negligible.

In Y-y-connected transformer, if only phase ‘a’ is shorted-circuited, the drop across this terminal will be only the small amount due to the flow of current through leakage impedance. The potential of neutral point becomes same as phase ‘a’. When only phase ‘a’ is loaded, the current flowing in the phase ‘a’ is the geometrical sum of load current of phase ‘a’ and the sum of exciting currents of other phases. The unbalanced loading in Y-y connections, either due to switching OFF of one phase (blowing of fuse) or due to load, the neutral of the secondary as well as primary oscillates.

The shifting of neutral is prevented by connecting the primary neutral to the generator neutral. However, if there is third harmonic present in the generator phase voltage, it results in third harmonic currents in secondary circuit. In the neutral wire, all the triplen (third-order) harmonic currents are in phase with each other as such these currents instead of neutralizing add. If secondary neutral is grounded, the triplen currents will be set up in the secondary circuit. Since all the three triplen currents are in phase with each other and, these currents, instead of neutralizing, add up. If secondary neutral is grounded, the triplen currents will be set up the secondary circuit. Since all the triplen currents are in phase, they add up and tend to produce inductive interference with telephone lines and other electronic circuits. The triple frequency currents in the secondary windings take the place of the missing triple harmonic current of primary and restore harmonic variation to the fluxes.

In the absence of primary neutral connections, there is no return path for the third harmonic components of exciting current. Hence, the flux wave is distorted and a third harmonic appears in the phase voltage of each transformer between neutral and line. This is true for both primary and secondary.

If the neutral of both the primary and secondary sides are insulated, there will be triple harmonic effect in the voltage wave. However, the voltage regulation of all the phases will be poor if load is unbalanced or if one of the line voltage is missing. If primary neutral is connected to generator, a third harmonic in the generator phase voltage will appear in the secondary phase voltage. If secondary neutral is grounded, objectionable triple frequency currents will flow causing inductive interference.

The triple harmonic secondary current can be avoided using a tertiary winding connected in closed delta. The fundamental induced EMF will be 120° apart and, therefore, will be balanced, but the EMF due to triple frequency will be in time phase around the closed delta of tertiary winding. The resultant triple frequency current will supply the small magnetizing current, which cannot flow in the primary winding. The rating of tertiary winding is about one-third of the rating of transformer.

In AC power system, there are chances of unbalanced load as well as unbalanced power supply, causing the neutral to oscillate. This oscillation of neutral gives rise to unequal voltages in each of the phase. Therefore, star/star transformers are rarely used in AC power system.

In HVDC, the star/star-star transformers in which two sets of secondary are connected in star, with each set having phase difference 0° and 180° for six pulse power (AC to DC and DC/AC) conversion are used.

Example 3.27 A three-phase 100 kVA, 50 Hz, 11 kV/400 V, Y-y ideal transformer is connected to an unbalanced load. The LT terminals are connected to 0.6 Ω, 0.6(0.8 + j 0.6) Ω and 0.6(0.5 − j 0.866) Ω. Determine current and phase voltage and the voltage across neutral assuming an ideal transformer.

Solution

Let the voltage is

$$V_{ph} = \frac{400}{\sqrt{3}} = 231 \text{ V}$$

Let the voltages be

$$\begin{aligned}
\boldsymbol{V}_{\text{AB}} &= 400\angle 0^\circ \text{ Volt} \\
\boldsymbol{V}_{\text{BC}} &= 400\angle -120^\circ \text{ Volt} \\
\boldsymbol{V}_{\text{CA}} &= 400\angle 120^\circ \text{ Volt}
\end{aligned}$$

and, $V_{\text{AN}} = 231\angle -30^\circ$ Volt

The phase voltage lags the corresponding reference line voltage by 30°. Similarly, we have

$$\begin{aligned}
V_{\text{BN}} &= \frac{400}{\sqrt{3}}\angle -30^\circ - 120^\circ = 231\angle -150^\circ \text{ Volt} \\
V_{\text{CN}} &= 231\angle -30^\circ + 120^\circ = 231\angle 90^\circ \text{ Volt}
\end{aligned}$$

$$Z_A = 0.6\ \Omega$$
$$Z_B = 0.6(0.8 + 0.6j) = 0.6\angle 36.86^\circ\ \Omega$$
$$Z_C = 0.6(0.5 - 0.866j) = 0.6\angle -69.9^\circ\ \Omega$$

Thus, individual currents are

$$V_{\mathrm{AN}} Y_A = \frac{231\angle -30^\circ}{0.6\angle 0^\circ} = 385\angle -30^\circ\ \mathrm{A}$$
$$V_{\mathrm{BN}} Y_B = \frac{231\angle -150^\circ}{0.6\angle 36.86^\circ} = 385\angle -186.86^\circ\ \mathrm{A}$$
$$V_{\mathrm{BN}} Y_B = \frac{231\angle 90^\circ}{0.6\angle -59.9^\circ} = 385\angle 149.9^\circ\ \mathrm{A}$$

The current in neutral is

$$\mathbf{I}_{\mathrm{A}} + \mathbf{I}_{\mathrm{B}} + \mathbf{I}_{\mathrm{C}} = 385(\angle -30^\circ + \angle -186.86^\circ + \angle 149.9^\circ)$$
$$= 385\angle -186.86^\circ\ \mathrm{A}$$

The net admittance

$$\mathbf{Y} = \mathbf{Y}_{\mathrm{A}} + \mathbf{Y}_{\mathrm{B}} + \mathbf{Y}_{\mathrm{C}} = \frac{1}{0.6} + \frac{1}{0.6\angle 36.86^\circ} + \frac{1}{0.6\angle -59.9^\circ}$$
$$= 3.86\angle 6.57^\circ\ \mathrm{mho}$$

The neutral potential is

$$\mathbf{V}_{\mathrm{N}} = \frac{\mathbf{I}_{\mathbf{N}}}{\mathbf{Y}} = \frac{385\angle -186.86^\circ}{3.86\angle 6.59^\circ} = 230.5\angle -193.45^\circ V$$

The voltage drop across Z_A,

$$\mathbf{V}_{\mathbf{dA}} = \mathbf{V}_{\mathbf{AN}} - \mathbf{V}_{\mathbf{N}} = 231\angle -30^\circ - 230.5\angle -193.45^\circ = 327.3\angle -25.08^\circ\ \mathrm{Volt}$$

Similarly, the voltage drop across Z_B and Z_C,

$$\mathbf{V}_{\mathbf{dA}} = \mathbf{V}_{\mathbf{BN}} - \mathbf{V}_{\mathbf{N}} = 231\angle -150^\circ - 230.5\angle -193.45^\circ = 172.65\angle -126.52^\circ\ \mathrm{Volt}$$
$$\mathbf{V}_{\mathbf{dC}} = \mathbf{V}_{\mathbf{CN}} - \mathbf{V}_{\mathbf{N}} = 231\angle 90^\circ - 230.5\angle -193.45^\circ = 229.38\angle 64.91^\circ\ \mathrm{Volt}$$

The potential across neutral is 230.5 Volt, which is quite high. Ideally this voltage must be zero volt. The voltages across other impedances are 327.3 V, 172.65 V and 229.38 V with their respective phase angles are 78.17°, − 65.85° & 84.1°. The unbalanced load on Y-y connected transformer results in.

(a) Unbalanced magnitude of phase-voltages
(b) Unbalanced phase differences between different phase voltages
(c) Large neutral voltage, the neutral voltage changes with unbalanced load.

The example demonstrates that in case of unbalanced loading, the neutral of the Yy-connected transformer shifts.

(b) **Delta/Delta-Connected Transformers**: The primary and secondary winding must be designed to withstand full line voltage. The Dd0 and Dd6 transformers, and their typical phasor have been shown in Fig. 3.55, 3.56, 3.57 and 3.58, respectively. The cost of insulation is more in these transformers. Proper care must be taken to connect secondary into delta. In the case of fault, the unit of two transformers can be used for maintaining three-phase supplies using open delta connection. The following can be easily proved for Delta/delta-connected transformers:

$$\frac{\text{Primary phase voltage}}{\text{Secondary phase voltage}} = \frac{\text{Primary line voltage}}{\text{Secondary line voltage}} = a$$

$$\frac{\text{Primary phase current}}{\text{Secondary phase current}} = \frac{\text{Primary line current}}{\text{Secondary line current}} = \frac{1}{a}$$

The transformer provides the path for triplen harmonics.

(c) **Open Delta Connected Transformers**: Open-delta transformers are also known as V-V connections or V connections. Here, only two transformers are required. This has been shown in Fig. 3.59. The V-V connection is normally used to maintain the three-phase supply in the event of fault in Delta/delta connected transformers. This can be done by removing both the primary and secondary connections of the faulty unit. In the case of balanced load, it can be shown that one of the units operates at a power factor of *cos* (30 + θ) and the other at

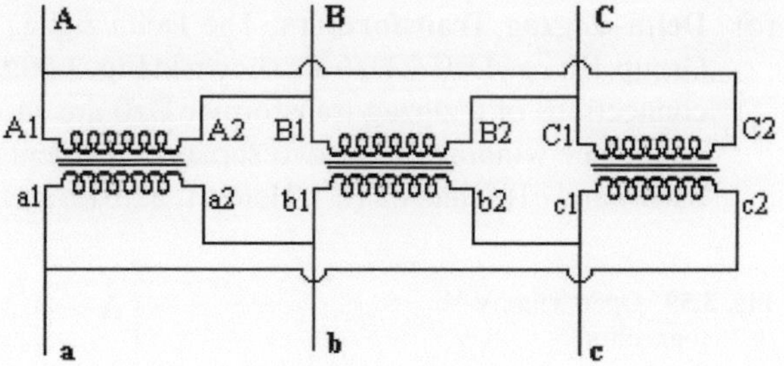

Fig. 3.55 Delta/delta-connected three-phase transformers Dd0 (0°)

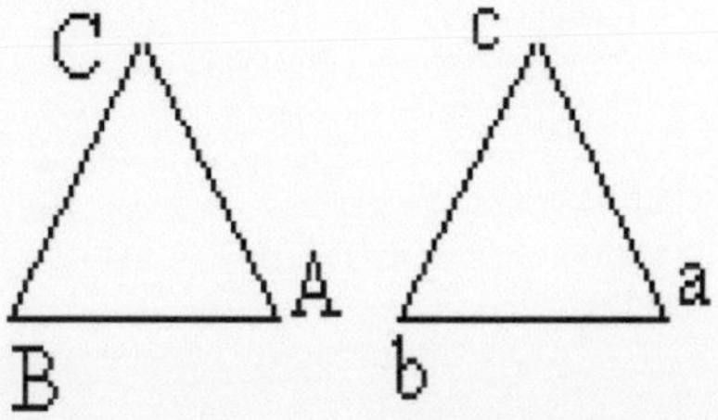

Fig. 3.56 Phasor diagram of Dd0 (0°) connected transformers

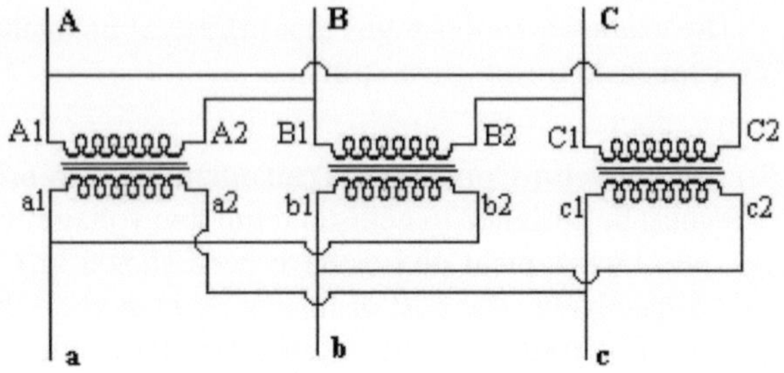

Fig. 3.57 Delta/delta connected three phase transformers Dd6 (180°)

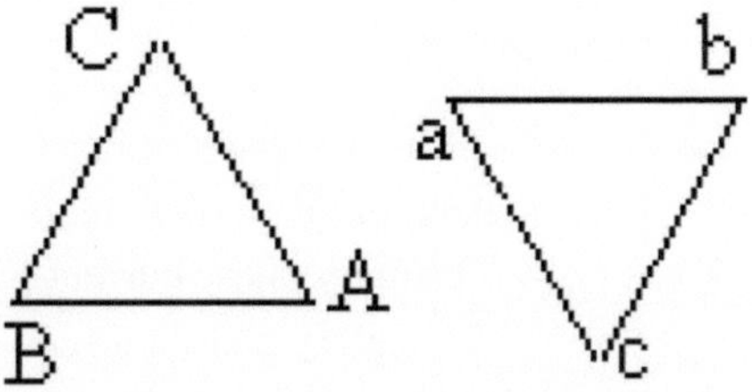

Fig. 3.58 Phasor diagram of Dd6 (180°) connected transformers

power factor of *cos* (30 − θ). The combined capacity of the two transformers in V-V connection, i.e., two-third of the three-phase output of Dd connection is not available. If V_L and I_L be the rated line voltage and line current then the rated output of the Delta/delta connection is, $\sqrt{3}V_L I_L$. Since the line current becomes the transformer winding current in case of open delta, the winding current should not exceed its rated value, i.e., $I_L/\sqrt{3} = 58.8\%\ \ of\ I_L$. Thus, only $\sqrt{3}\big/2$ or 86.6% capacity of single-phase transformer is available in open delta. This factor is referred as *utilization factor*.

(c) **Delta–Zigzag Transformers**: The Delta/zigzag transformers of Group-I and Group-III, i.e., Dz0 & Dz6 are shown in Fig. 3.60a, b respectively. The secondary connections of a zigzag transformer Dz0 are shown in Fig. 3.61a. Each of the secondary windings have two separate segments a1(a) and a2 (a'), b1(b) and b2(b') and c1(c) and c2 (c'). Here a1, a2,b1,b2,c1 and c2 are segments of phase

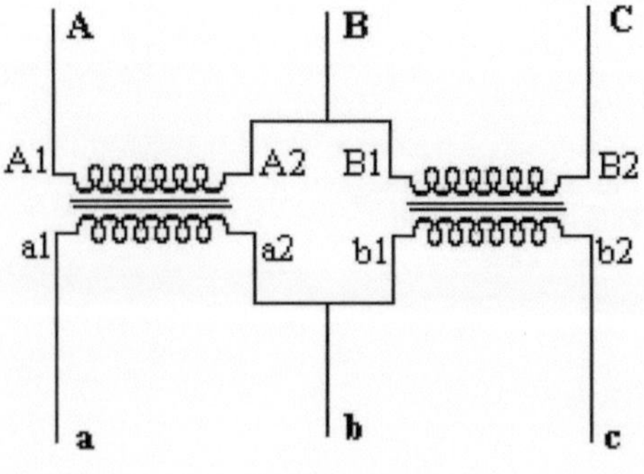

Fig. 3.59 Open delta, V-V or V-connection

winding. These segments are symmetrically connected. The phasor sum of $\mathbf{V}_{c1}$ and $\mathbf{V}_{a2}$ gives the summation of phasor 'a-phase'. Phasor $\mathbf{V}_{a2}$ is in opposition to phasor $\mathbf{V}_{a1}$, as shown in Fig. 3.54b.

The secondary 'a-phase' consists of part 'c-phase' and part 'a-phase'. Let the voltage across the each secondary segment be V_2. The resultant voltage across zigzag secondary 'a-phase' is

$$V_a = V_{c1} + V_{a2}$$
$$V_{c1} = V_2\angle 60^\circ = V_2(0.5 + j0.866)$$
$$V_{a2} = -V_2\angle -120^\circ = V_2\angle -120^\circ = V_2(-0.5 + j0.866)$$

Thus, we get.

$$V_a = 1.732j \quad \text{or} \quad V_a = V_2\sqrt{3}$$

If the transformation ratio between primary and secondary segment is 'a', then.

$$V_2 = V_1/a$$

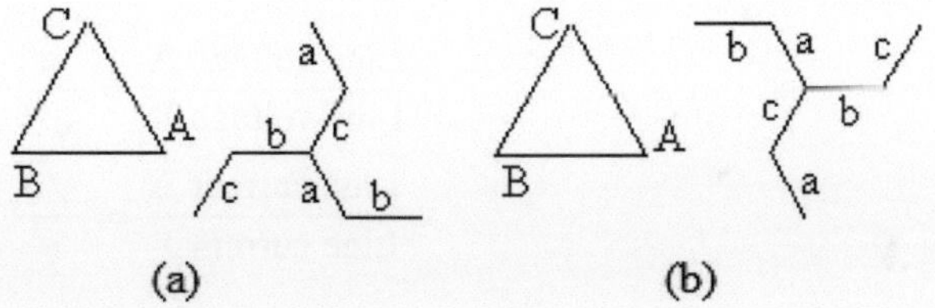

Fig. 3.60 **a** Phasor diagram for Dz0, **b** Phasor diagram for Dz6

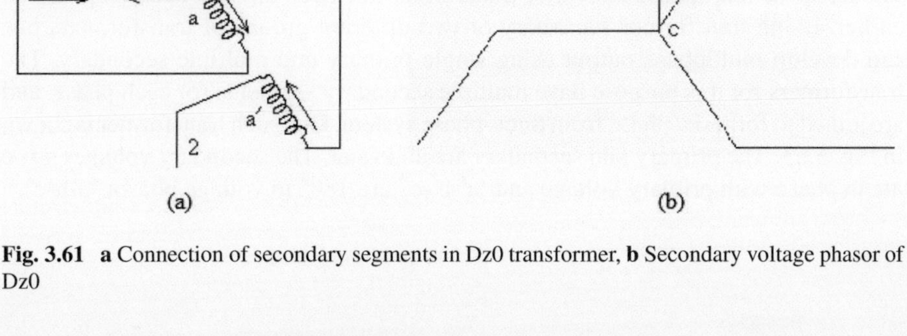

Fig. 3.61 **a** Connection of secondary segments in Dz0 transformer, **b** Secondary voltage phasor of Dz0

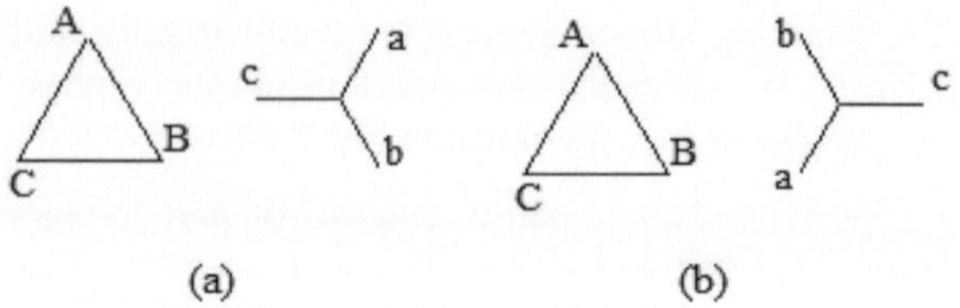

Fig. 3.62 **a** Delta star Dy1 (−30°), **b** Delta star Dy11 (+30°)

Thus, secondary a-phase voltage of zigzag transformer is '$V_1\sqrt{3}/a$' and line-to-line voltage is '$3\ V_1/a$'.

(d) **Star/Zigzag Transformers**: These transformers have phase difference of either –30° or + 30° between primary and secondary line phasor. They are classified as Yz1 or Yz11 (Table 3.2).

(e) **Delta/Star Transformers**: These transformers can have phase difference of either –30° or + 30° between primary and secondary line voltages and are classified as Dy1 & Dy11, respectively. The delta winding provides the path for triplen (third-order) harmonics. These transformers are at both the transmitting end and the distribution ends of transmission lines. The phasors are shown in Fig. 3.62a, b, respectively.

It can be shown that

$$\frac{\text{Line voltage } \Delta}{\text{Line voltage } Y} = \frac{a}{\sqrt{3}}$$
$$\frac{\text{Line current } \Delta}{\text{Line current } Y} = \frac{\sqrt{3}}{a}$$

(g) **Star/Delta Transformers**: These transformers are found at receiving end of transmission lines. The neutral of primary is grounded to avoid difficulties during unbalanced load.

3.21.2 Multiphase Transformers

The need for multiphase converter transformer has been already been emphasized earlier. Using transformer secondary of two different groups of transformers, one can develop multiphase output using single primary and multiple secondary. The transformers for this purpose have multiple secondary segments for each phase, and are joined to form six-phase from three-phase system. One such transformer is shown in Fig. 3.63. The primary and secondary are all in star. The secondary voltages a-b-c are in phase with primary voltage and 'a'-b'-c'' are 180° to voltage phasor 'a-b-c'.

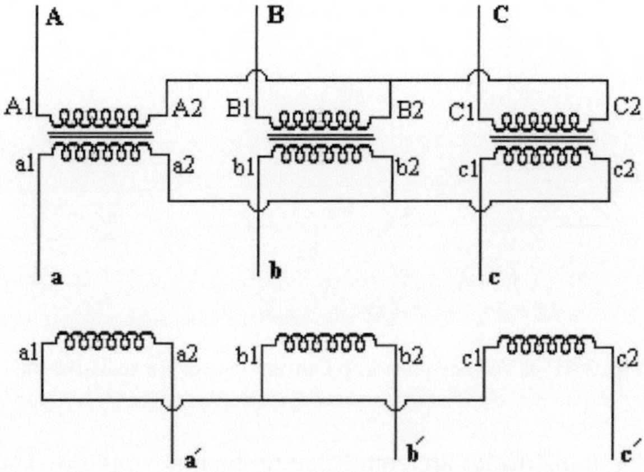

Fig. 3.63 Multiphase converter transformer having two sets of secondary of group-I and group-III (three-phase to six-phase transformer)

3.21.3 Scott-Connected

Scott-connected transformers are used for conversion of three-phase voltage into two-phase voltage and vice versa. They are used in supplying power to two-phase servo motor, induction furnace and instruments operating on two-phase supply. If two-phase supply is available, it can be converted into balanced three-phase supply. This is also used in obtaining precision three-phase balanced supply in laboratory. A Scott connected transformer is shown in Fig. 3.64a. If the two secondaries are connected in series, as in Fig. 3.64b, the secondary no-load RMS voltage will $V\sqrt{2}$.

The voltage phasors are arranged in a triangular form as shown in Fig. 3.65a. Secondary voltage phasor 'a1-a2' is in phase with the phase voltage of primary of 'A-phase'. The secondary voltage 'b1-b2' is in phase with the line voltage 'CB'. The secondary voltages are in phase quadrature. For equal number of turns in secondary, let us assume that main transformer has 50% tapping at point 'S'. The terminals of

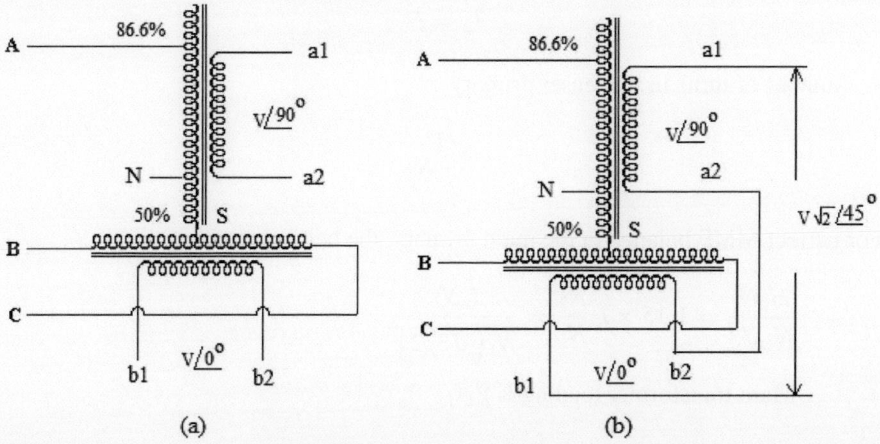

Fig. 3.64 **a** Scott-connected transformer, **b** Scott-connected transformer series-connected secondary

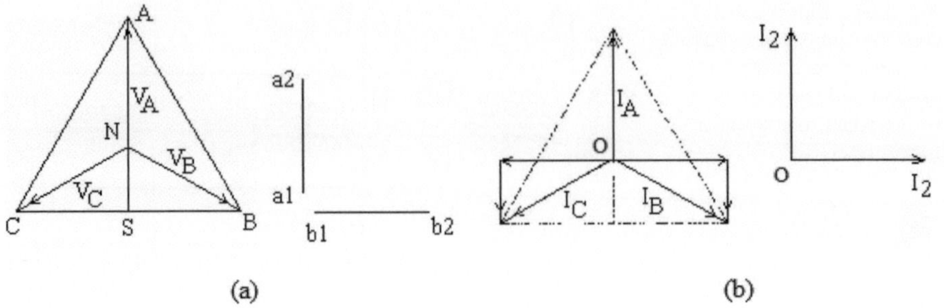

Fig. 3.65 **a** Voltage phasor, **b** Current phasor for resistive load

the transformer are connected to the line voltages. The voltage between 'AS' is $\sqrt{3}/2$ times of line voltage. If the teaser primary has $\sqrt{3}/2$ times number of turns of the main primary, the voltage per turn of both the main and the teaser transformer will be the same. The secondary voltages will be in quadrature. The midpoint of 'BC' is 'S'. The three-phase neutral 'N' is located on the teaser primary at one-third of the number of turns of winding from point 'S'. The phase voltages are

$$V_{AN} = \frac{V}{\sqrt{3}} = 0.58\ \text{V} = 58\%\ \text{of V}$$

$$V_{SN} = \frac{V_{AS}}{3} = \frac{1}{3} \times \frac{3}{2} \times V_{AN} = 29\%\ \text{of V}$$

If the secondary is carrying a balanced current of unity power factor, then we have

$$\frac{I_1}{I_2} = \frac{T_2}{T_1}$$

Number of turns in the primary (main) T_1 is

$$T_1 = N_1$$

Number of turns in the teaser primary

$$\frac{\sqrt{3}}{2} N_1$$

For perfect MMF balance in the main primary, the balancing current is,

$$\frac{2I_2N_2}{N_1\sqrt{3}} = 1.15 \times I_2\frac{N_2}{N_1} = \frac{I_2N_2}{N_1\left(\sqrt{3}/2\right)}$$

Main transformer loading is V_1I_1

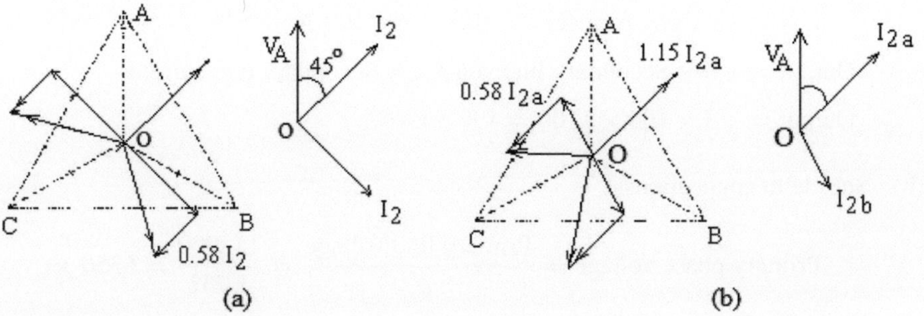

Fig. 3.66 **a** Current phasor for lagging load, **b** Current phasor for unbalanced lagging load

$$\text{Teaser transformer loading} = \frac{\sqrt{3}}{2} V_1 I_1$$

$$\text{The total current in primary is one} - \text{half of teaser primary} = I_2 \frac{N_2}{N_1}$$

Choosing the number of turns $N_1 = N_2$, the voltage and current phasors for resistive load, lagging load and unbalanced loads are plotted in Fig. 3.65a, b, and Fig. 3.66a, b, respectively.

Example 3.28 A bank of three-identical single-phase transformer is connected to 11,000 V supply. The line current is 10 A. If the ratio of turn per phase is 10, calculate secondary line voltage and current, primary & secondary phase current and, output for the following connections:

(a) Star-star.
(b) Star-delta.
(c) Delta-delta.
(d) Delta-star.

Solution

Given line-to-line voltage = 11000 V and input line current = 10 A and ratio of transformation, a = 10.

(a) Star-star connection:

$$\text{Primary phase voltage} = \frac{\text{Primary line voltage}}{\sqrt{3}} = \frac{11,000}{1.732} = 6350\text{ V}$$

$$\text{Primary phase current} = \text{Primary line current} = 10\text{ A}$$

$$\text{Secondary phase voltage} = \frac{\text{Primary phase voltage}}{a} = \frac{6350}{10} = 635\text{ V}$$

$$\text{Secondary line voltage} = \text{Secondary phase voltage}\sqrt{3} = 635\sqrt{3} = 1100\text{ V}$$

$$\text{Secondary phase current} = \text{Secondary line current}$$

$$= a \times \text{Prmary phase current} = 10 \times 10 = 100\ \text{A}$$

$$\text{Output} = \sqrt{3} \times \text{Secondary line voltage} \times \text{Secondary line current}$$

$$\text{Output} = \sqrt{3} \times 100 \times 1100 = 190.5\ \text{kVA}$$

(b) Star-delta connections:

$$\text{Primary phase voltage} = \frac{\text{Primary line voltage}}{\sqrt{3}} = \frac{11,000}{1.732} = 6350\ \text{V}$$

$$\text{Primary phase current} = \text{Primary line current} = 10\ \text{A}$$

$$\text{Secondary phase voltage} = \frac{\text{Primary phase voltage}}{a} = \frac{6350}{10} = 635\ \text{V}$$

$$\text{Secondary line voltage} = \text{Secondary phase voltage} = 635\ \text{V}$$

$$\text{Secondary phase current} = 10 \times 10 = 100\ \text{A}$$

$$\text{Secondary line current} = 100\sqrt{3} = 173\ \text{A}$$

$$\text{Output} = \sqrt{3} \times \text{Secondary line voltage} \times \text{Secondary line current}$$

$$= \sqrt{3} \times 635 \times 173 = 190.2\ \text{kVA}$$

(c) Delta-delta connections:

$$\text{Primary phase voltage} = \text{Primary line voltage} = 11,000\ \text{V}$$

$$\text{Primary phase current} = \frac{\text{Primary line current}}{\sqrt{3}} = \frac{10}{\sqrt{3}} = 5.77\ \text{A}$$

$$\text{Secondary phase voltage} = \frac{\text{Primary phase voltage}}{a} = \frac{11,000}{10} = 1100\ \text{V}$$

$$\text{Secondary line voltage} = \text{Secondary phase voltage} = 1100\ \text{V}$$

$$\text{Secondary phase current} = a \times \text{Primary phase current} = 10 \times 5.77 = 57.7\ \text{A}$$

$$\text{Secondary line current} = \sqrt{3} \times \text{Secondary phase current} = \sqrt{3} \times 57.7 = 99.94\ \text{A}$$

$$\text{Output} = \sqrt{3} \times \text{Secondary line voltage} \times \text{Secondary line current}$$

$$= \sqrt{3} \times 1100 \times 99.94 = 190.4\ \text{kVA}$$

(d) Delta-star connections:

$$\text{Primary phase voltage} = \text{Primary line voltage} = 11000\ \text{V}$$

$$\text{Primary phase current} = \frac{\text{Primary line current}}{\sqrt{3}} = \frac{10}{\sqrt{3}} = 5.77\ \text{A}$$

$$\text{Secondary phase voltage} = \frac{\text{Primary phase voltage}}{\text{a}} = \frac{11,000}{10} = 1100\ \text{V}$$

$$\text{Secondary line voltage} = \sqrt{3} \times \text{Secondary phase voltage} = 1100\sqrt{3}\ \text{V}$$

Secondary phase current = $a \times$ Primary phase current = $10 \times 5.77 = 57.7$ A

Secondary line current = Secondary phase current = 57.7 A

Output = $\sqrt{3} \times$ Secondary line voltage $\times$ Secondary line current

$= \sqrt{3} \times \sqrt{3} \times 1100 \times 57.7 = 190.4$ kVA

3.21.4 T-Connected Transformer

In T-connected transformer, one of the transformers has center-tapped primary and center-tapped secondary which is known as main transformer. The other transformer has tapping at 86.6% both on primary and secondary and is known as teaser transformer. This is similar to the teaser primary of Scott-connected transformer. One end of both the primary and the secondary of the teaser transformer are joined to the center-tap of the primary and secondary of the main transformer, respectively. The other end of primary of teaser transformers and both the ends of primary of main transformer are connected to balanced three-phase supply as shown in Fig. 3.67. Similarly the end terminal of secondary of teaser and end terminals of secondary of main transformer are connected to three-phase load. Figure shows the connection of two single-phase transformers in T-T configuration with three-phase four-wire system.

The T-connected transformer, star/delta or star/star have been reportedly used in STATCOM (Static Compensator) in power system for reactive power compensation. The scheme utilizes Insulated Gate Bipolar Transistor (IGBT)-based inverter for controlling lagging or leading reactive power. It absorbs or injects lagging or leading current in the power system.

The T-connected windings of transformer not only provides a path for zero sequence fundamental current and harmonic currents but it also offer a path for neutral current when connected in shunt at Point of Common Coupling (PCC) in power system. Under unbalanced load, the zero sequence load-neutral current divides equally into three currents and takes a path through the T-connected windings of the transformer.

Merits of T-Connected Transformer

(1) It has low manufacturing costs.
(2) The T-connected transformer has a physical neutral that allows the transformer to be grounded.
(3) It provides supply to three-phase or two-phase or single-phase loads simultaneously.
(4) The neutral point is used either for grounding or for loading purpose.
(5) It requires only two transformers.
(6) It reduces the total KVA rating requirements.

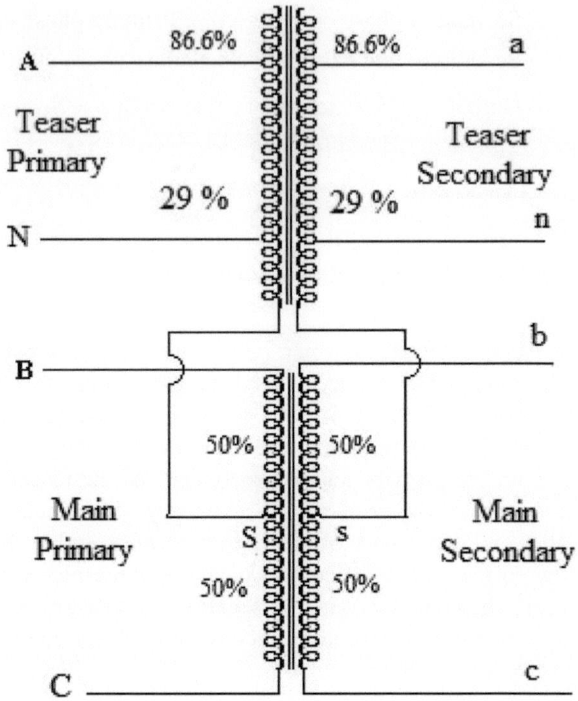

Fig. 3.67 T-connected transformer

3.21.5 Three-Phase Autotransformer

Three-phase autotransformer works on the same principle of changing the turn ratio in the winding. For each phase individual winding is provided. In fact, three (or multiple) single-phase autotransformer can be suitably connected to work as three- (or multi-) phase autotransformer. Mutual fluxes between all the phases are not required for their operation. The windings of the three-phase autotransformer are not space distributed. The schematic representation of three- phase autotransformer is shown in Figure 3.68a, b. The windings can be connected in star or delta. The application of three-phase autotransformer for starting three-phase induction motor has been reported in subsequent chapter.

In three-phase transformer, either fixed voltage tap are provided or the output voltage can be smoothly controlled from zero voltage to slightly more than maximum input voltage. The latter requires the simultaneous rocking of a set of three carbon brushes over the three-phase toroidal core windings. It is also known as three-phase variac. In laboratory, three-phase variac or autotransformer is used for controlled voltage manual starting of three-phase induction motor for limiting starting current. They are also useful for (a) connection of three-phase 380 V induction motor on 440 V three-phase supply, (b) connection of 220 V three-phase instruments or motors on 440 V three-phase supply; (c) controlling DC voltage of three-phase

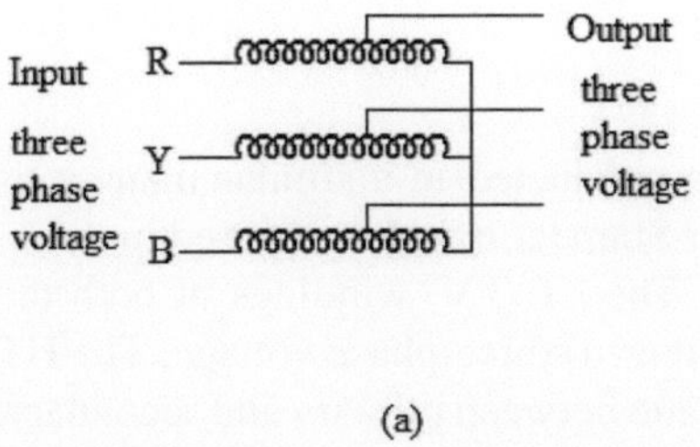

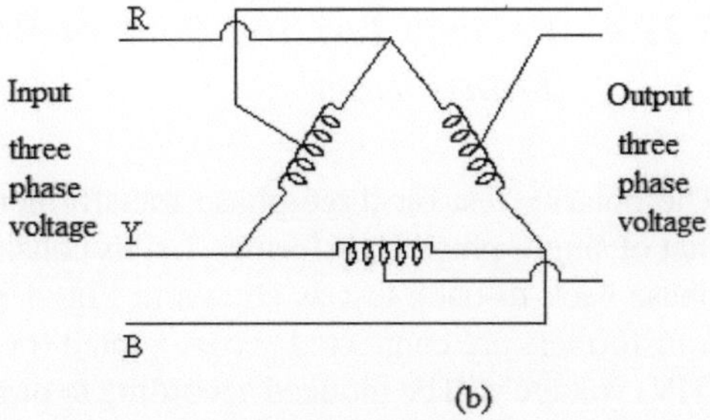

Fig. 3.68 **a** Star-connected autotransformer, **b** Delta-connected autotransformer

rectifier by controlling AC side voltage; (d) starting of medium capacity induction or synchronous motors.

There is no autotransformer action in star-connected three-phase autotransformer (almost full voltage appears on output side), if common neutral point making star is not connected. Although this is a demerit of autotransformer, this property is used in starting three-phase induction motor without disconnecting the motor terminals from supply.

Figure 3.68a, b shows star-connected and delta-connected three-phase autotransformer, respectively. Autotransformer can also be connected in T connection or Scott connection as shown in Fig. 3.69 for converting three-phase power into two-phase power or vice versa. However, in such connection, electrical isolation is missing.

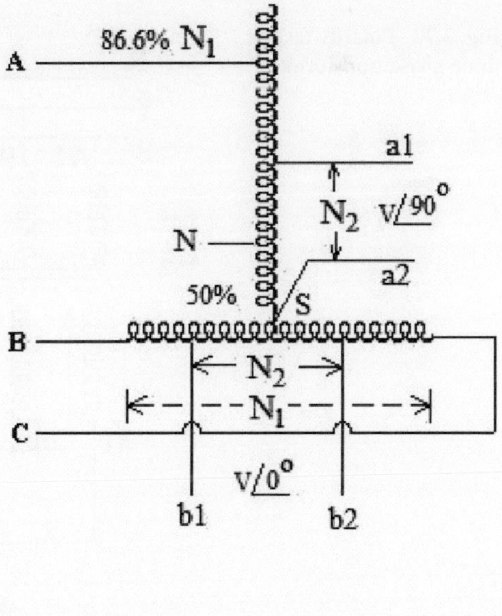

Fig. 3.69 Scott-connected autotransformer

3.21.6 Polarity Test and Back-to-Back Test on Three-Phase Transformer

The polarity test on three-phase transformer is conducted in a similar manner as that of single-phase transformer. Let us consider Star/star transformer used in three-phase back-to-back test as shown in Fig. 3.70. The LT (LV) windings of both the transformers are connected across a rated (or reduced) three-phase voltage. The HT (HV) voltage will be induced according to turn ratio between primary and secondary. We require a voltmeter of sufficient rating which can read roughly two times the HT-induced voltage. The voltages between one terminal of the first transformer (a1 or b1 or c1) and one terminal of second transformer (a'1 or b'1 or c'1) are measured. The terminals such as a1-a'1, b1-b'1 and c1-c'1 will be at same polarity respectively, if the voltmeter across them reads zero voltage or very low voltage. The latter is due to slightly different transformation ratio of individual transformer. The voltage across a1-b'1 or a1-c'1 will be about $\sqrt{3}$ times HT phase voltage.

Polarity test is conducted for parallel connection of transformers, connection of three-phase transformer, connection of secondary in delta, etc. Chalk out your procedure for connecting three single-phase transformers in different groups of transformer.

Sumpner's test or back-to-back test on three-phase Star/star-connected transformer requires that the HT windings are connected in phase opposition and both the LT windings are connected as shown in Fig. 3.71a. This requires the polarity test on the transformer. A wrong connection will lead to completely dead short-circuit

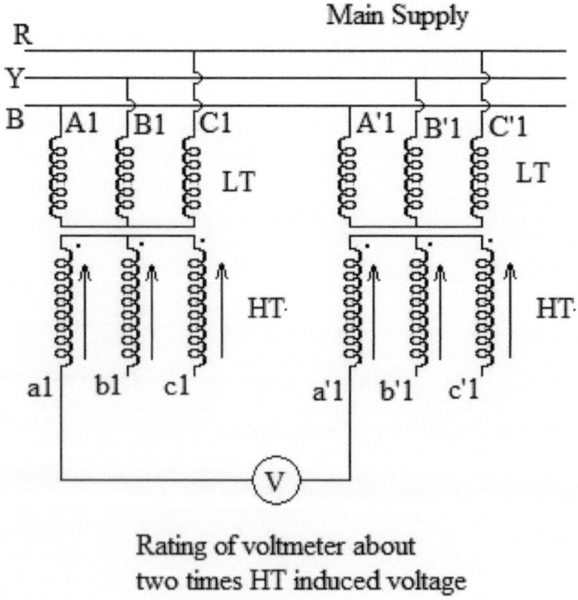

Fig. 3.70 Polarity test on three-phase transformer

damaging both the LT and HT windings if rated voltage on LT windings is applied. When the HT windings are in perfect phase opposition, no current flows in the closed circuit formed by HT windings connected in phase opposition. Using auxiliary three-phase supply, another set of three-phase voltage of reduced magnitude is injected in series with one of the LT winding. This in turn induces a three-phase voltage on HT side causing currents to flow in all the short-circuited windings. If the secondary winding is provided with the tapping, there is no need of auxiliary power supply and auxiliary transformer. In that case one of the three-phase LT is connected to different tapping as shown in Fig. 3.71b. The difference in the three-phase LT voltages causes an induced voltage on the closed HT circuit, which is responsible for the flow of short-circuit currents in HT windings.

This test is sometimes conducted for heat run-test of identical three-phase transformers. In heat-run test, the delta-connected secondary is also sometimes connected to a single-phase source by opening one of its delta points, when the primary remained connected to a three-phase source. Heat–run test on a single three-phase transformer is done by conducting short-circuit test and by providing either total loss (rated copper plus rate core loss) or only rated winding currents under short-circuit conditions. Develop schemes of recording temperature with RTD and your personal computer.

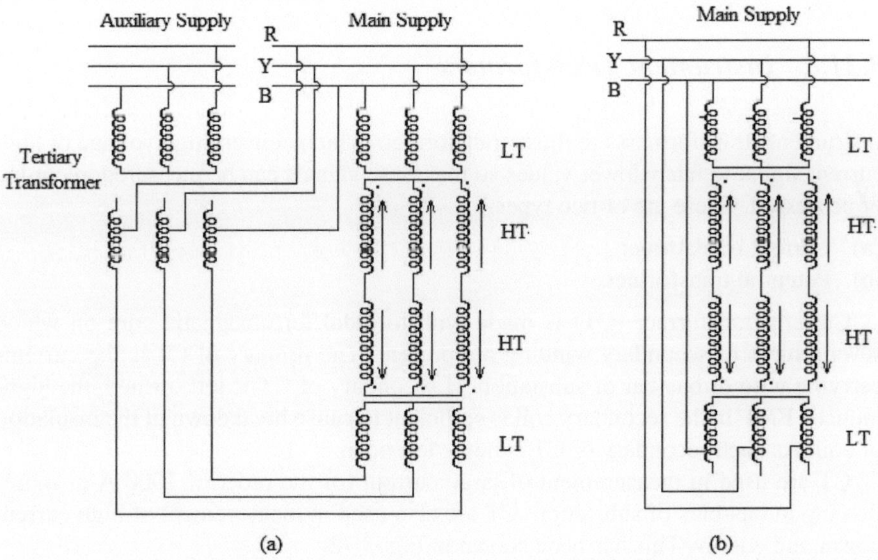

Fig. 3.71 **a** Back-to-back test on three-phase transformer, **b** back-to-back test on three-phase transformer when tapping are present in LT windings

3.21.7 Transformer Used in Welding

The electric welding requires a special type of transformers. For initiating the arc, welding electrodes are short-circuited momentarily for the initiation of arc and then they are separated by a certain gap for conduction of electric current through the arc. If we supply power to arc welding apparatus using normal transformer, it will cause short circuit of winding before the initiation of arc. Welding transformers are special transformers having larger leakage reactance. When welding electrodes are short-circuited, maximum voltage drop occurs in leakage reactance, which limits the short-circuit current required for the initiation of arc. We also recall that in power transformer the current during accidental short circuit is limited by increasing the leakage reactance at design level. In the construction of such transformers having large leakage reactance, an air-gap is introduced in the core of the transformer in the direction of flow of flux.

In absence of proper welding transformers, inductive reactors are provided in the electrical circuit for limiting the current. The leakage reactance is controlled by changing the air-gap between the two magnetic cores. Similar reactive loads are required for conducting ZPF load test on three-phase alternator.

3.21.8 Instrument Transformers

Instruments transformers are those transformers which convert high voltage or high current signal to fairly lower values so that these signals can be measured, recorded or processed. These are of two types.

(a) Current transformer
(b) Potential transformer.

Current transformer (CT) is made with toroidal ferromagnetic core on which several turns of secondary winding are present. The primary of CT is the current-carrying wire or bus-bar of substation. If secondary of CT is left opened, the high-induced EMF in the secondary coil is sufficient to cause breakdown of the insulation of coil. As such secondary of CT is never left open.

CT are used in measurement of large current (of the order of 1000 A or more) flowing in bus-bars of substation. CT are also used in measurement of high current power and energy. This has been shown in Fig. 3.72.

The CT output is also used for recording current. A tong tester which measures current in a non-contact manner, and Hall effect current transducer are also based on similar technique. In modern era, the state-of-the-art Hall effect current sensors are being preferred.

The potential transformer (PT) is step-down transformer providing electrical isolation between high-voltage and low-voltage windings. Using this voltage 132 kV (say) is stepped down to 250 V so that it can be safely measured without putting the life of operator in danger.

PT and CT are used in the measurement of high-voltage high-current power and energy in power system. They are also used to record the energy in high-voltage

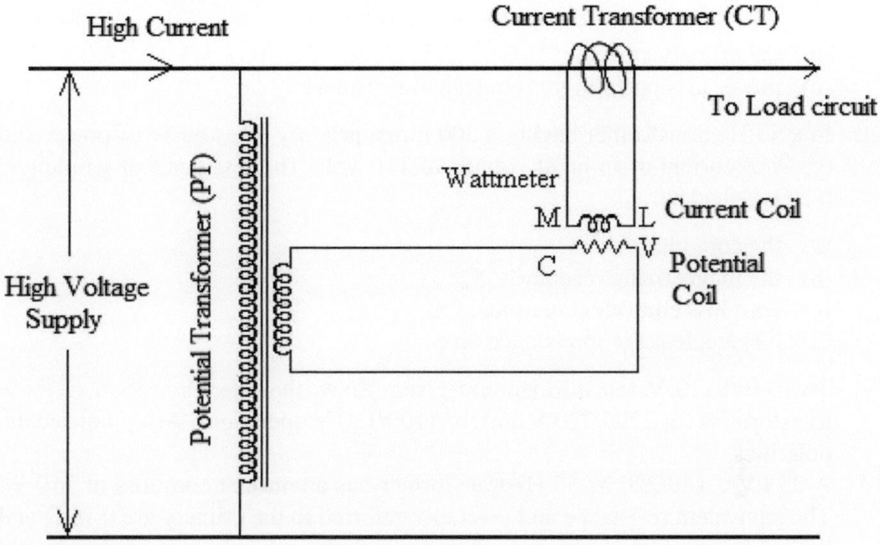

Fig. 3.72 Application of CT and PT in measurement of high-voltage, high-current alternating current power

and high-current power system. The schematic diagram for measurement of power at high voltage and high current is shown in Fig. 3.72.

Answer the Followings

1. Why self-inductances and leakage inductances of a transformer are not equal?
2. What is the significance of turn ratio?
3. A transformer is having turn ratio 200/10, is it possible to change the effective value of a load impedance using transformer ? If, Yes, how?
4. Is it possible to obtain 260 Volt and 180 Volt, in a two-winding transformer of rating 220 V/40 V, 1 KVA, 50 Hz. If yes, how? Give suitable connections.
5. A 110 V/220 V, 50 Hz transformer is operating at frequency ranging from 10 to 100 Hz, discuss its response using an equivalent circuit.
6. In a three-phase transformer, how polarity can be tested ? Draw a schematic circuit diagram for the polarity test of two three-phase transformer.
7. Explain briefly, how you will connect primary and secondary of Delta/delta transformer.
8. Three banks of single-phase transformers are to be connected in Delta/star and Star/delta configurations. Discuss the tests which need to be performed for a symmetrical balanced connection.
9. Show how a 220 V/40-0-40 V, 50 Hz transformer can be used to give 260 V and 180 V.
10. A 1.0 Ω Resistance is Connected Across the Secondary of a 220/110 Volt, 50 Hz Ideal Transformer. The Secondary Voltage is 110 Volt. Calculate

(a) the equivalent resistance referred to primary.
(b) the primary current.
(c) power dissipated in this equivalent resistance.

11. In a 50 Hz transformer having a 300 turns, primary takes 60 W of power and 0.8 A of current at an input voltage of 110 Volt. The resistance of winding is 0.2 Ω. Calculate.

    (a) the core loss.
    (b) the magnetizing reactance, X_m.
    (c) core loss equivalent resistance R_c.
    (d) Neglect leakage impedance drop.

12. Two 1100/110 V transformers are given. Show the interconnection of these transformers for 2200/220 V and (b) 1100/110 V operations. Also, indicate the polarities.
13. A 25 kVA, 440/220 V, 50 Hz transformer has a constant core loss of 710 W. The equivalent resistance and reactance referred to the primary are 0.16 Ω and 0.66 Ω, respectively. Calculate efficiency and per unit regulation of transformer at full-load unity, 0.8 lagging and leading power factors.
14. Show that regulation of a transformer is given as

$$\varepsilon = \varepsilon_r \cos\theta + \varepsilon_x \sin\theta + \frac{1}{2}(\varepsilon_x \cos\theta - \varepsilon_r \sin\theta)^2$$

15. A 6600 V, 50 Hz single-phase transformer has sheet steel core. The net cross sectional area of core is 22.6×10^{-3} m^2, the mean length is 2.23 m. There are four interleaved joints, each joint taking as much reactive MMF as is require per meter of core. If $B_m = 1.1$ wb/m^2, determine.

    (a) the number of turns on the 6600 V winding.
    (b) the no-load current.
    (c) For given flux density, the MMF/m is 232 At/m; specific loss = 1.76 W/kg specific density 7800 kg/m^3.

16. A single-phase transformer has 400 primary and 100 secondary turns. The net cross sectional area of core is 60 cm^2. If the primary is connected to a 50 Hz, 500 V source, calculate

    (a) the peak value of flux density in the core.
    (b) the voltage induced in the secondary winding.

17. In a two-winding isolation transformer having turn ratio 1:1, the winding parameters of both the coils are $R_1 = 1.0$ Ω and $L_1 = 1.5$ mH. Determine equivalent circuit of the transformer. The core loss resistance is 700 Ω and magnetizing reactance at 50 Hz is 5000 Ω. Saturation is assumed to be negligible.
18. A 75 kVA transformer has an iron loss of 1.0 kW and a full load copper loss of 1.0 kW. Calculate the transformer efficiency at unity power factor at

(a) 25% load (b) 50% load (c) 75% load (d) 100% load and (e) 125% load.

19. The transformer of Problem 18 operates on full-load at unity power factor for 8 hours, on no-load for 8 hours and on one half load at unity power factor for 8 hours. Determine the all-day efficiency.
20. The maximum efficiency of a 100 kVA transformer is 98.4% and occurs at 90% full-load unity power factor. Calculate the efficiency of the transformer at unity power factor at (a) full-load and (b) 50% of full-load.
21. The results of open circuit and short-circuit tests on a 10 kVA, 440/220 V, 50 Hz transformer are as follows:
    Open-circuit test: HV side open: $V_0 = 220$ V; $I_0 = 1.2$ A and $W_0 = 150$ W.
    Short-circuit test: LV side shorted $V_{SC} = 20.5$ V; $I_{SC} = 42$ A and $W_S = 140$ W.
    (a) Determine the parameters of the approximate equivalent circuit referred from HV side.
    (b) Choosing approximate equivalent circuit, draw to-the-scale phasor diagram at full-load U.P.F, 0.8 lag and lead power factors and, Z.P.F. lag and lead power factors.
22. Using equivalent circuit parameters of Problem 21, calculate the primary voltage of the transformer, when the secondary is supplying full-load at 0.8 lagging power factor. What is the efficiency of the transformer at this load ?
23. Show that magnetizing reactance is

$$X_m = R_c / \tan\theta_0$$

24. A two-winding transformer is on no-load. If $Z_0 = jR_0X_0/(R_0 + jX_0)$ is its exciting impedance, $Y_0 - G_0 + j\,B_0$ is admittance, G_0 is conductance and B_0 is susceptance. Show that
    (a) $R_0 = G_0/[G_0^2 + B^2]$
    (b) $G_0 = R_0/[R_0^2 + X_0^2]$
    (c) $X_0 = B_0/[G_0^2 + B^2]$
    (d) $B_0 = X_0/[R_0^2 + X_0^2]$
25. Two 500 kVA, 50 Hz, 11,000/6600 Volt transformers are operating in parallel. Short-circuit test at rated current of these transformer yields

Transformer	Voltage (Volt)	Power (W)
A	105	1500
B	100	1600

What is the largest load at unity power factor can be carried out by the two transformers without overloading ? Assume that terminal voltage on the load side is 6600 V.

26. Two single-phase transformers rated at 100 kVA and 75 kVA, 11 kV/220 V are connected in parallel. They supply a load of 150 kVA at 0.8 lagging power factor. The per unit values of their parameters are

Transformer	Resistance	Reactance
100 kVA	0.016	0.07
75 kVA	0.025	0.06

The supply voltage being held constant at 220 V, calculate the kVA loading and the power factor of each transformer.

27. Two single-phase transformers rated at 100 kVA each, 11 kV/250 V are connected in parallel. They supply a load of 200 kVA at 0.8 lagging power factor. The per unit values of their parameters are

Transformer	Resistance	Reactance
A	0.02	0.02
B	0.025	0.025

The supply voltage being held constant at 250 V, calculate the kVA loading and the power factor of each transformer.

28. A two-winding transformer is connected to a load of '$R + jX$', Its secondary parameters are '$R_2 + jX_2$'. Show that secondary current differs in phase from the secondary-induced EMF by an angle φ_2,where, $tan\ \varphi_2 = (X_2 + X)/(R_2 + R)$.

29. The test data of 150 kVA, 2400/240 V transformer are as follows:

    Open circuit Test: $V_{oc} = 240$ V; $I_{oc} = 15.67$ A $W_{oc} = 576$ W.

    Short-Circuit Test: $V_{sc} = 61.56$ V; $I_{sc} = 62.5$ A and $W_{sc} = 1562.5$ W.

    Determine equivalent circuit parameters and draw the equivalent circuit referred to primary side. Also determine.

    (a) voltage regulation
    (b) ratio of no-load current and rated current
    (c) efficiency of transformer operating at rated load with 0.8 lagging power factor, unity power factor and 0.8 leading power factor.

30. Show that

    (a) In open-circuit test, $\tan \theta_0 = B_0/G_0$
    (b) In short-circuit test, $(X_1 + X_2) = (R_1 + R_2) \tan \theta_{SC}$

31. A 1000 kVA transformer has a 94% efficiency at full-load and at 50% full-load. The power factor is unity in both the cases. (a) Separate the losses (b) Determine efficiency at unity power factor and 75% full-load.

32. The primary and secondary voltages of an autotransformer are 440 V and 360 V respectively. If the secondary current is 80 A, determine the primary current.

33. A 5 kVA, 400/220 V, 50 Hz two-winding transformer has a full-load efficiency of 96% at unity power factor. The iron loss at 50 Hz is 60 W. This transformer is used as an autotransformer and delivers a 5 kVA unity power factor load at 220 V. Determine (a) the primary current and (b) the efficiency of the autotransformer.
34. A 100 kVA, 42 kV/2400 V, 50 Hz transformer is connected in parallel with a 75 kVA, 42 kV/2400 V transformer. The respective impedances referred to the secondary are $(0.5 + j4)\ \Omega$ and $(0.8 + j5)\ \Omega$. The total load on the transformer is 150 kVA at 0.8 lagging power factor. Calculate the load on each transformer. Also, calculate the primary and secondary currents of the two transformers.
35. Three single-phase transformers are connected to make a three-phase delta-star bank. The line voltages of the bank are 11 kV and 440 V on the primary and secondary, respectively. The load on the transformer is 440 V, 15 kW, 0.8 power factor lagging, star-connected. Determine the primary and secondary line and phase currents.
36. A 100 kVA, 2200/220 V star-star three-phase 50 Hz transformer bank has an iron loss of 1500 W and a copper loss of 3000 W at full-load. Determine the efficiency of the transformer for 75% full-load at 0.8 power factor lagging.
37. Two transformers, each rated at 50 kVA, are connected in open delta to form a three phase system. What total kVA which can be obtained from the open-delta bank without over loading the transformers?
38. Draw the voltage phasor diagram of a star/star and delta/star connected transformers. Should these two banks be operated in parallel?
39. A three-phase 75 kVA, 440 V, 0.8 lagging power factor load to be supplied by single-phase transformers having a turn ratio of 10. Calculate the currents in the windings of each transformers and the corresponding power factor if the transformers are connected in

    (a) open-delta
    (b) delta-delta.

40. Two Scott-connected transformers are supplied from a 440 V three-phase source. The two-phase secondary supplies a 100 kVA load at 220 V. Determine the currents in the windings of the two transformers.
41. A 15 kVA single-phase two-winding transformer takes a current of 2.0 A at 0.2 power factor. When primary winding is connected to a 3300 V supply, the corresponding voltage at secondary terminals is 240 V. The resistances of the primary and secondary windings are $1.1\ \Omega$ and $0.005\ \Omega$, respectively. The equivalent reactance referred to the secondary side is $0.02\ \Omega$. Determine the efficiency and secondary terminal voltage when the input to the primary is 15 kVA at 3300 V, 0.8 power factor lagging.
42. A 10 kVA, 220/110 V transformer has maximum efficiency of 96.97% at power factor lagging. Its core loss is 100 W and full-load regulation at 0.8 power factor lagging is 5%. Find the efficiency and regulation at full-load 0.8 power factor leading. Determine the power factor at full-load for which the regulation will be maximum, minimum and zero. Also, calculate the maximum and minimum regulations.

43. A 20 kVA, 2000/200 V, two-winding transformer is to be used as step-down autotransformer, with a constant source voltage of 2000 V. At full-load of unity power factor, calculate the power output, power transformed and power conducted. If the efficiency of the two-winding transformer at 0.7 power factor lagging is 97%, find the efficiency of the autotransformer.
44. A three-phase transformer, 33/6.6 kV, Δ/Y, 1.0 MVA has a primary resistance of 8.0 Ω per phase and a secondary resistance of 0.08 Ω per phase. The percentage impedance is 7%. Find the secondary voltage with rated primary voltage and hence regulation at full-load, 0.8 power factor lagging.
45. A single-phase 6.6 kV/440 V/220 V multiple ratio transformer is used to supply power to 1.5 kW, 440 V, 0.8 lagging power factor load and 2 kW, 220 V, unity power factor load. Determine the currents drawn by different sections of secondary winding and net primary current.
46. A single-phase 11 kV/440 V/220 V/110 V multiple secondary transformer is supplying power from each secondary 10.0 kW at 0.8 power factor lagging, U.P.F., 0.8 power factor leading respectively. Determine winding currents and net primary current. How the overall power factor will change if loads are 10.0 kW 0.8 lead, U.P.F. and 0.8 lagging power factors, respectively.
47. In certain rural area, it is found that when 132 kV/440 V three-phase Delta/star connected transformer is installed and the single-phase voltage is 180 V. What is the turn-ratio and input voltage and how the voltage can be corrected? What is turn ratio for correcting the voltage to 440 V three-phase? What is the output line-to-line voltage when 132 kV is applied for this turn ratio?
48. The single-phase on traction locomotive near feeding substation is 25 kV. When the locomotive is 150 km away from any nearby substation, the voltage drops to 20 kV. The transformer rating is 25 kV/600 V single phase. Calculate (a) the output voltage if turn ratio is not changed; (b) the effective number of turns when tap changer is operating for maintaining the rated output voltage; (c) the output voltage for new turn ratio for 25 kV input voltage.
49. Develop algorithm for on-load tap changer switch for controlling several electromagnetic switches in a flawless manner.
50. In multiwinding 220 V, 50 Hz transformer having three identical isolated windings which can be used as primary or secondary. A lagging load of 1.2 kW, 0.8 power factor is connected across one of the secondary. Calculate the leading load to be connected across another secondary which can correct the overall power factor on primary to unity. What is the effect of turn ratio of the second secondary winding on the overall power factor.

Objective Type Questions

1. The magnetizing current of a transformer will be minimum for

   (a) Core having no air-gap but interleaved joints
   (b) Core having perfectly no air-gap
   (c) Core having an air-gap

(d) Air core

2. With on-load tap changer, the tapping on transformer are provided on

(a) Low-voltage side
(b) High-voltage side
(c) Both on low- and high-voltage sides
(d) None.

3. In which of the following types of transformers the leakage reactance is kept large?

(a) Distribution transformer and power transformer
(b) Power transformer and welding transformerhigh-voltage sides
(c) Welding transformer and transformer used in Neon lighting
(d) Ferrite transformers

4. Using Scott-connected transformer, it is possible to convert

(a) Three phase into two phase
(b) Two phase into balanced three phase
(c) Both (a) and (b)
(d) None

5. In back-to-back connection of two number single-phase transformers and back-to-back connection of two number Three-phase Y/y transformers, power in LT is transferred by ______________ and by ______________ respectively

(a) Conduction, conduction
(b) Conduction, induction
(c) Induction, induction
(d) None

6. A single-phase 220 V/110 V, 50 Hz transformer will satisfactorily work on

(a) 110 V/220 V, 60 Hz
(b) 22 V/11 V, 0.1 Hz
(c) 220 V/110 V, 1 MHz
(d) 2.2 V/1.1 V, 1 kHz.

7. In multiwinding transformer shown in figure, it is possible to get set of isolated voltages with rated input voltages

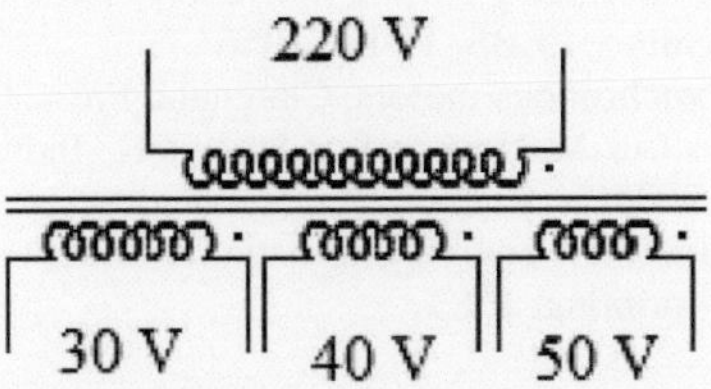

(a) 10 V
(b) 20 V
(c) 60 V
(d) All the above.

8. In which of the following transformer, the segments of secondary windings can be connected in phase addition or phase opposition?
   (a) Multiratio transformers having rating of 220 V/110 V, 70 V, 40 V
   (b) Multiwinding transformer having ratings 220 V/110 V, 70 V,40 V
   (c) Both (a) and (b)
   (d) None.
9. While measuring high-voltage and high-current power using CT and PT and single-phase wattmeter,
   (a) One has to short one terminal of current coil with another terminal of potential coil of wattmeter
   (b) Short as in (a) for very small duration
   (c) Without shorting as in (a), the wattmeter will give zero deflection
   (d) There is no need of it.
10. We require only two transformers for following type of connections with three-phase input.
   (a) V-V connection
   (b) T-T connection
   (c) Scott connection
   (d) All (a), (b) and (c).

Bibliography

1. Philip L. Alger, Induction Machines, Gordon & Breach Science Pub. N.Y. 2nd Ed. 1965
2. M.G.Say, Alternating current Machines, 4th ELBS edition, 1976
3. A. Fransua, R. Magureanu, Electrical Machines and Drive Systems, TP Technical Press, Oxford, 1984
4. V Del Toro, Electric machines and power systems, PHI, 1980
5. S.A.Nasar, Electromagnetic energy conversion & Systems, Prentice-Hall, Inc., N.J., 1970
6. A.S.Langsdorf, Theory of alternating current machinery, TMH
7. A. Draper, Electrical Machines, ELBS, 1st Ed, 1956
8. TJE Miller, Reluctance Synchronous motors, Clarendon Press, Oxford, 1993
9. Stephen J Chapman, Electric Machinery Fundamentals, Indian Edition, 4e, McGraw hill education (Indi) Pvt Ltd, 2005
10. Siskind, Electrical Machines, 2nd edition International student edition, McGraw hill international book company,12th printing, 1983

Chapter 4
Direct Current Machines

Abstract Homo-polar and hetro-polar machines; constructional details; commutation; armature reaction; types of windings-lap and -wave; EMF and torque equations; compensating winding; equalizer connection; DC generators-self, separately excited and compound generators; parallel operation; Types of DC motors; starting, speed control-conventional and solid state and braking; characteristics and control, testing, cross-field machines.

Keywords Homopolar machine · Armature · Field poles · Inter pole · Commutation · Torque · Speed control · Starting · Braking

4.1 Introduction

Direct current motors and generators are used in many industrial applications. These machines had widespread applications in chemical and metallurgical plants, electric traction, automobiles, airplane, submarines, electric vehicle (EV), train, lighting and battery charging, rolling mills, bucket conveyors, synchronous machine excitation, Unmanned Aerial Vehicle (UAV), air borne devices, drones, and automatic control systems etc [1–5].

The principle of operation of machines has already been described in Chap. 1. We know that a current carrying conductor in a magnetic field produces torque (force) in a motor. When the motor is rotating, a back EMF opposing the supply voltage appears. Similarly, when a DC generator is electrically loaded a counter torque opposing the driving torque (of prime mover) is produced. These facts satisfy the Lenz's law. A DC motor is based on the interaction of magnetic field with current carrying conductors. One of the members of an electrical machine is called *stator* and the other is called *rotor*. The latter is free to rotate due to interaction torque. A DC machine has the following two parts:

(a) Field
(b) Armature.

R. K. Srivastava, *Electrical Machines*, https://doi.org/10.1007/978-981-96-5039-2_4

The field provides a steady DC flux either through field coils or permanent magnets. The armature is an arrangement of current carrying conductors oriented in such a way to develop unidirectional interaction force and EMF.

Certain machines operating on pulse DC supply, such as Switched Reluctance Motor (SRM), Variable Reluctance Motor (VRM) and Permanent Magnet Brushless DC (PMBLDC) motor work on the *principle of alignment*. These machines are shaped to give rise to an *alignment or reluctance torque*. In these machines the coils are switched ON and OFF according to the instantaneous position of the rotor.

A simple Faraday disk generator is shown in Fig. 4.1a, in which a homopolar permanent magnet (PM) provides unidirectional magnetic flux. In actual homopolar generator, two annular shape PM are in use. The surface of the conducting disk, which is being rotated, is experiencing a unidirectional flux. The DC current is collected from the rim of the disk and the shaft. When used as a motor, Faraday disk requires very large DC current that too at a considerably lower voltage to produce a sufficient torque. A homopolar disk generator running at very high speed (of the order of 20,000 RPM) can provide large DC current of the order of 1.0 kA or more at a reduced voltage level of about 5 V. Such types of homopolar generators are useful in chemical industries, extraction of metals, metal refinement, defense sector and in applications where there is a demand of very large DC current at a lower voltage level. A cylindrical homopolar DC machine may have a slotted rotor carrying unidirectional currents in each of its slots. However, there are enormous fabrication and operational difficulties in the physical realization of cylindrical homopolar DC generator, like collection of current, arrangement of multiple conductors in slots with a condition that the currents in all the conductors of each slot are unidirectional. Can you point some other difficulties, like collection of large DC current? Note that the ohmic loss is proportional to the square of current. The use of homopolar machine is for special purpose DC generator only.

A heteropolar machine is shown in Fig. 4.2. A conductor in a slot faces North and South poles alternatively. For continuous unidirectional toque production, it is

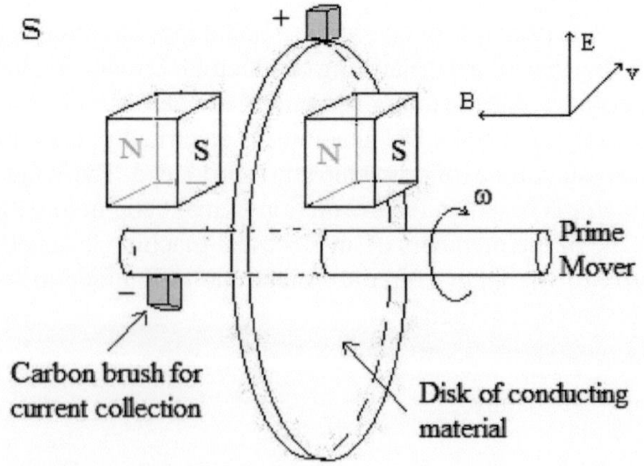

Fig. 4.1 A typical homopolar disk generator—faraday disk generator

desired that the conductors should undergo a successive reversal of current after a regular travel of one pole pitch. Successive reversal of current in the conductors is achieved by means of commutator and brush arrangement. Here, each slot can have multiple conductors in series carrying the same current. The machine is supplied by considerably higher voltage and lower current as compared to that of homopolar machine. Commutation (current reversal) of coils and associated sparking in the trailing edge of the brushes restricted the use of these DC machines in explosive environment, like underground mines where inflammable gases are present. However, these difficulties have been overcome to a large extent by considerable researches in the field of high-power permanent magnets, electronic switching of coils, Hall-effect sensors, sensor-less control and new developments in PMBLDC motors of modern era.

A conventional DC motor and generator both are having similar constructional details. However, a DC generator should not be used as a DC motor and vice versa because of poor commutation. A typical hetero polar DC motor is shown in Fig. 4.2.

Direct Current (DC) generators are sometimes made with a swinging frame for dynamometer application, in which both the rotor-shaft and the stator frame are free to rotate on the shaft axis. The construction requires four bearing supports. The movement of outer frame or stator fame is restricted by using spring balances or load cells. The latter gives an electrical signal according to the torque experienced by the

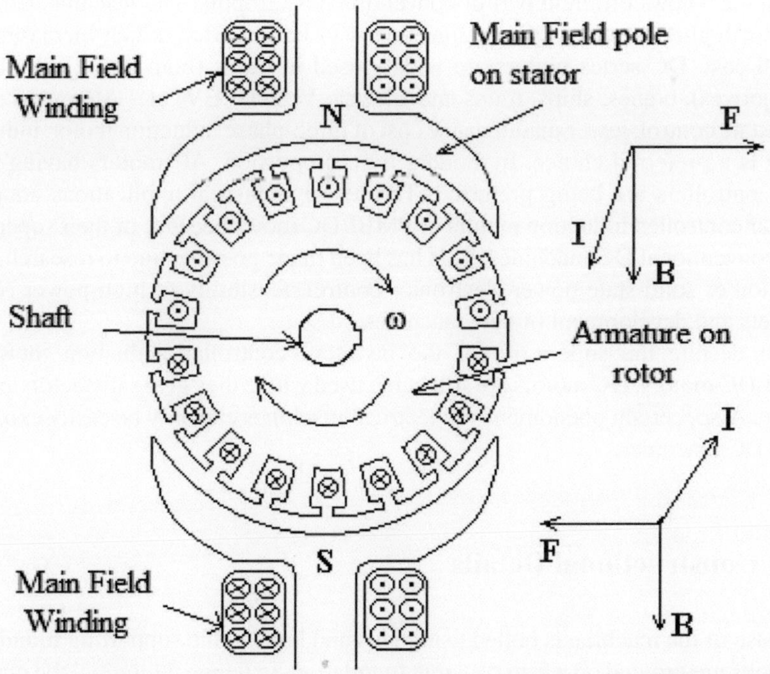

Fig. 4.2 Development of torque in heteropolar DC motor

stator. Such an arrangement is useful in testing small and medium capacity motors of any type. The eddy current type of loading device is also used for testing small and medium capacity motors.

Nowadays, the bulk DC power generation and distribution have been stopped, DC power is now generated only in smaller capacity in certain control system applications, charging of batteries, DC arc welding, pilot generator for smaller and medium size synchronous motors and alternators, lab electric and also for testing different types of motors. In a laboratory, a DC power supply is maintained using either MG (Motor-generator) set or three-phase solid-state rectifier. In older days, Mercury Arc rectifier, motor-generator sets or rotary converters were used. The latter converted three-phase AC into DC. The three-phase power is normally converted into three wires DC, capable of supplying either 220 V and 440 V DC loads. Certain motors operate on 440 V DC supply or even higher DC voltage. The unconventional modern energy resources like solar-cell and fuel- cell also give DC power. These schemes do not require any prime mover.

The DC motors exhibit very good speed control characteristics. They have been adopted for many industrial purposes. However, due to the development of computer-controlled vector drives of three-phase induction-motors, conventional DC motors use in newer applications is being avoided, because of maintenance and cost factors. A DC machine exhibits sparking in case of poor commutation. Also, the costs of procurement and maintenance of a DC machine are higher than newer alternatives. Figure 4.2 shows different part of conventional heteropolar DC machine. Commutator for higher capacity DC machines need to be imported which increases their overall cost. DC series motors are widely used in trains (both diesel and electric locomotives), cranes, ships, trams and Electric Vehicle (EV) etc. Although, due to solid-state control, lesser maintenance cost of three-phase induction motor, induction motor is a preferred choice. In modern train propulsion, AC motors having solid-state controllers are being preferred. The newer industrial applications adopt the inverter controlled induction motors or PMBLDC motor because of their superiority over conventional DC machines. This has been made possible due to research in the direction of solid-state power electronics control, feasibility of high-power NdFeB magnets and development of new machines.

But despite the superiority of the inverter –controlled induction motors or PMBLDC motors, DC motors are still being used where they give satisfactory performance. Also, certain phenomena of electrical machines can only be easily explained using DC machines.

4.2 Constructional Details

The base of the machine is bolted using nuts and bolts to the supporting foundation. The bolts are grouted on a firm concrete foundation. In flange mounting, the machine is coupled to the other shaft and its frame is fixed by means of nuts and bolts to the

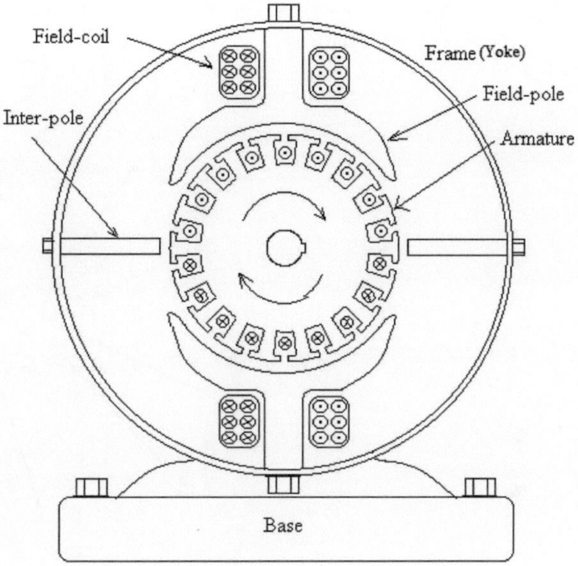

Fig. 4.3 Details of DC machine *(inter-pole windings, commutator and brushes not shown for clarity)*

axial side of the frame of another machine. Some of the details of a DC machine are shown in Fig. 4.3. A DC machine has following main electrical parts [6–8]:

(a) Stator having field poles and field windings or PM poles, carbon brush gear
(b) Rotor having shaft, armature, armature windings and commutator

Various parts of a DC machine have been shown in Fig. 4.4. The field poles are made with ferromagnetic material and are properly camphored—all sharp edges are removed by machining process. The field poles carry preformed DC coils having several numbers of turns of insulated standard wire gauge (SWG) wire or of insulated rectangular-strip conductors. These coils are inserted in the field poles with proper insulation. The field poles along with the field windings or coils are then bolted inside the cylindrical frame of the machine. Even number of poles are energized by DC, such that magnetic flux of alternative (N-S) magnetic poles appears in air-gap.

The rotating shaft carries armature, armature windings and commutator. The carbon brushes are stationary and are mounted on separate brush holders attached to the inner stator. The rotating shaft is supported by two bearings attached to the end covers, which are fixed on both the sides of the stator. The armature is made using slotted stamping as shown in Fig. 4.3. The bundles of stamping are fixed to the shaft by means of a 'key-slot' as shown in Fig. 4.4. It also carries a commutator, to which the end terminals of each of the coils of the armature winding are brazed to the riser of the commutator. Inter-poles are also provided in the inter-polar or brush axis. The purpose of inter-poles is to improve commutation. A typical commutator has been shown in Fig. 4.5. A schematic representation of DC machine is shown in Fig. 4.6.

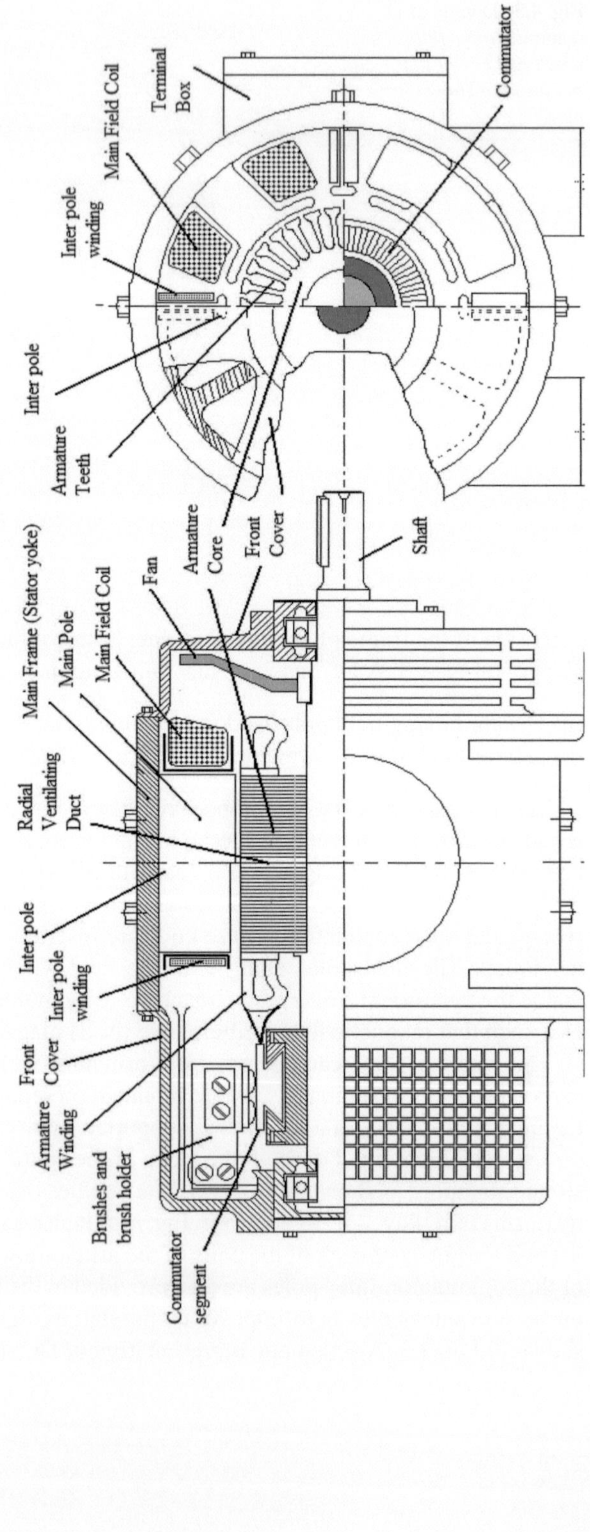

Fig. 4.4 Various parts of direct current (DC machine) [2]

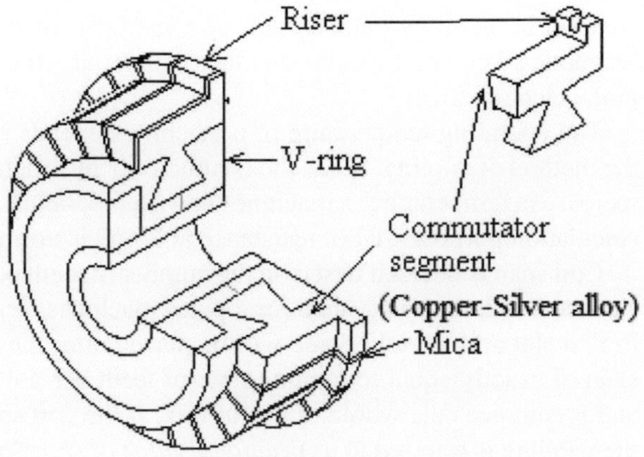

Fig. 4.5 A typical commutator and commutator segment

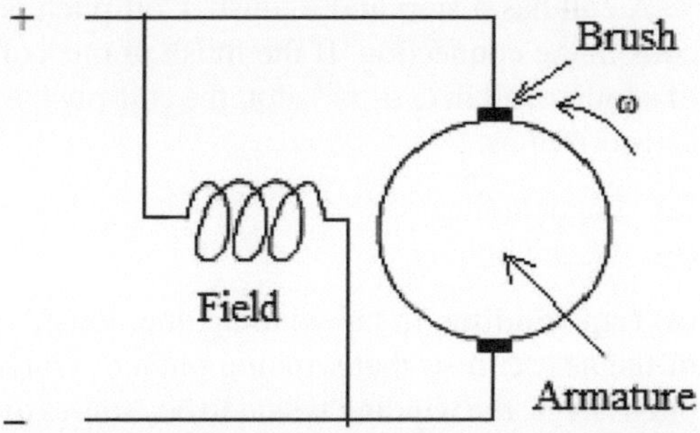

Fig. 4.6 Schematic representation of DC machine

4.2.1 Windings

Windings are essentially the heart of an electromagnetic device through which the energies change. Electrical windings are a source of magneto motive forces (MMF) responsible for production of torque in a motor and development of electromotive force (EMF) in case of a generator. The current carrying conductors are embedded in the slots of ferromagnetic armature core for minimizing the leakage flux. In an electromechanical energy conversion device, the magnetic and electric circuits are coupled. The magnetic circuit is formed by the ferromagnetic core, while the electric circuit is essentially the windings [1, 3].

Windings are made of several coils having multiple turn of insulated fine wire. Each conductor is insulated from the other. The shape of coil is decided based on axial length of stator, coil throw and slot area. One of the coil side of a preformed coil is kept in one of the slots and the other is placed in another slot almost one pole pitch apart. The winding wire is made of copper, and has sufficient strength to bear mechanical creep and thermal resistance. SWG (Standard Wire gauge) is commercially available winding wire. Copper strip conductors are also used; however, they need to be insulated. These windings carry electrical current, resulting in ohmic losses and, hence, rise in temperature. An insulating material can withstand heat up to a certain temperature only. At high temperature, the insulation may break down, resulting in creation of hotspots, short circuit of some turns of the insulated coil embedded in magnetic core, melting of the conductor material, grounding of windings and a break in the electric circuit.

The operating temperature of a machine depends on the insulating material and the method of cooling. Since the conductivity of a metallic conductor increases with increase in temperature, a machine shows temperature variation of its resistance. For calculation purposes, a hot resistance at 75 °C is normally taken.

Coil span is decided based on the number of teeth per pole. If there are 24-slots or slot-teeth in a stamping, then for 4 poles machines- one side of a coil will be placed in first slot and the other side will be placed after the fifth slot-teeth. A coil having span of exactly equal to total number of teeth per pole is referred as *full-pitch coil*, and it embraces the whole of useful flux. If the coil span is less than the pole pitch, the winding is referred to as *fractional-pitch* or *short-pitch* winding. Such a winding is said to be chorded. Chording reduces certain space harmonics.

A coil has a start and a finish. Coil-pitch is referred in terms as total number of coils in the connection. If the finish of one coil in the first slot is connected to start of another coil in $(z + 1)^{\text{th}}$ slot, the coil-pitch is z. The DC commutator windings are categorized as:

(a) Lap winding
(b) Wave winding.

(a) Lap winding: In lap winding, the 'finish' of each coil is connected to the 'start' of the next coil so that winding pitch or commutator pitch is unity. If commutator pitch is + 1, the winding is said to be *progressive*, and when commutator pitch is − 1, the winding pitch is *retrogressive*. In latter case, the order of connection is reversed. In case of a 'p pole machine' with lap windings, we have

The number of commutator brushes, p

The number of parallel paths, $a = p$.

The conductors per parallel path, $Z_c = \frac{Z_a}{p}$,

where Z_a is total number of armature conductors.

A simple lap winding has been shown in Fig. 4.7. There are 12 slots on armature and number of poles are two. The darker lines show the upper-coil side's conductor and lighter lines show lower-coil side's conductors in a slot. Only two coils are shown for clarity. It is assumed that each slot has two coil sides- upper coil side and lower coil side. First coil is placed in $1 - 7'$ slots. The second is placed in $2 - 8'$ slots. The finish of first coil is connected to beginning of next coil.

A complete winding diagram for four pole, 12-slots lap winding is shown in Fig. 4.8.

These details are also provided in Table 4.1.

(b) Wave winding: In wave winding, all the coils carrying current in the clockwise direction is connected in series and those coils carrying current in opposite direction is connected in series. Here the 'finish' of first coil is connected to the 'start' of another coil, one pole pitch apart. In these windings, we have

The number of commutator brushes is 2.

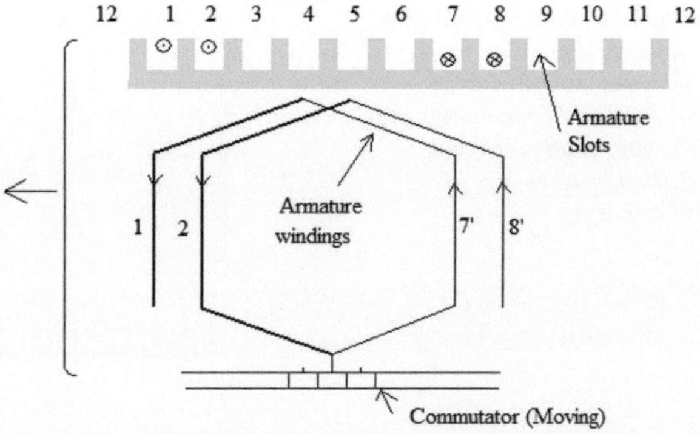

Fig. 4.7 Simple Lap winding 12 slots, two poles (*slots on the periphery of armature are represented in linear form*)

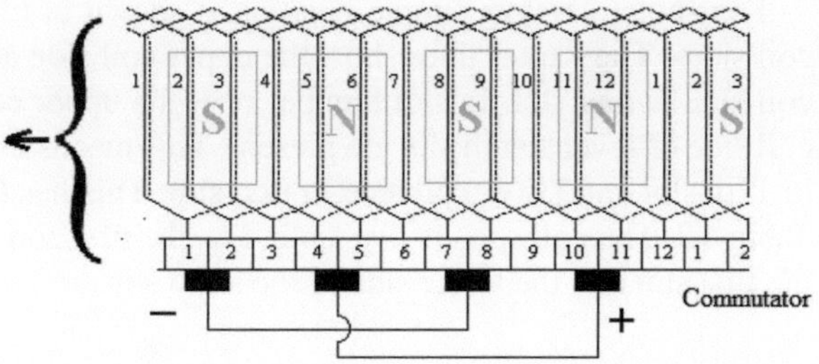

Fig. 4.8 Lap winding 4 pole 12 slots, 2 coil side per slot

Table 4.1 Winding details for lap winding 12 slots, four poles

Slot No	1	2	3	4	5	6	7	8	9	10	11	12
UC*	1	2	3	4	5	6	7	8	9	10	11	12
LC**	10′	11′	12′	1′	2′	3′	4′	5′	6′	7′	8′	9′

*UC—Upper coil side in a slot and **LC—lower coil side in a slot. Coils are marked 1 − 1′, similarly for other coils

The number of parallel path is a = 2.

The conductors per parallel path, $Z_c = \frac{Z_a}{2}$.

The number of parallel paths is independent of number of poles. In our example, the number of parallel paths is two. The windings are useful in comparatively high voltage and low current machines. A typical wave winding is shown in Fig. 4.9.

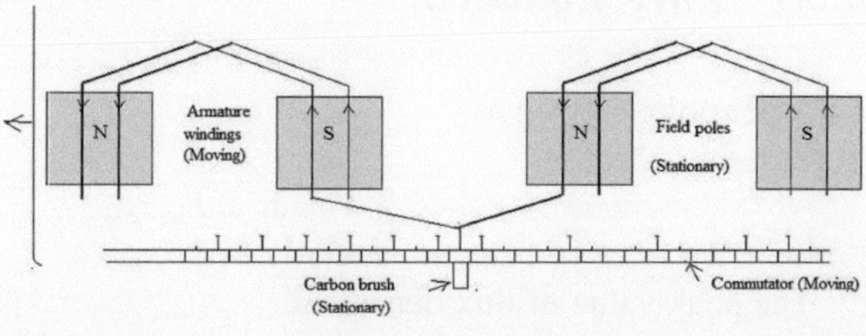

Fig. 4.9 Simple wave windings

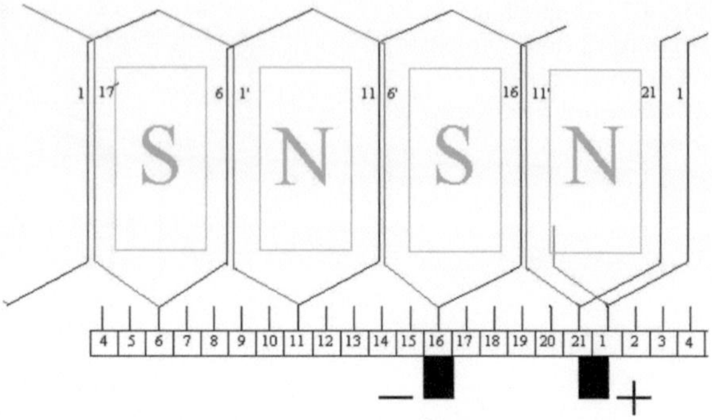

Fig. 4.10 Four pole, 21 slots wave winding, with two coil sides per slot (*armature winding and commutator are moving. Field poles and carbon brushes are stationary*)

Four-pole, 21 slots simple winding is shown in Fig. 4.10. Each slot carries two coil sides. The darker lines show the upper coil side and lighter line show the lower coil side in any slot. In slot number one, the upper coil side of first slot and lower coil side of seventeenth slot are present. This means that a coil has its upper coil side in 17th slot and lower coil side in first slot. This has been shown in tabular form in Table 4.2. The coil span is five teeth, i.e., the first coil side is placed in upper side of the first slot and the lower side of the sixth slot.

4.3 Production of EMF, Torque and Losses in DC Machine

The flux distribution in airgap will be discussed in subsequent sections. Due to the projected or salient poles, the flux density in interpolar region falls to a considerably lower value. The distribution of flux density over the air-gap is non-sinusoidal in nature. In a DC machine, the flux produced by main field poles is concentrated in air-gap. Some flux leaks through the surrounding air, and is not utilized in force or torque production. The form factor of the flux density distribution is defined as,

$$K_f = \frac{\text{Mean air gap flux density}}{\text{Maximum flux density}} \tag{4.1}$$

4.3.1 EMF Equation

Flux per pole is given as,

$$\phi = B_{\text{av}} \tau_p L \tag{4.2}$$

The peak value of flux density is,

Table 4.2 Winding details for wave winding

Slot	1	2	3	4	5	6	7	8	9	10	11	12	13	14	15	16	17	18	19	20	21
UC	1	2	3	4	5	6	7	8	9	10	11	12	13	14	15	16	17	18	19	20	21
LC	17′	18′	19′	20′	21′	1′	2′	3′	4′	5′	6′	7′	8′	9′	10′	11′	12′	13′	14′	15′	16′

UC—Upper coil side in a slot & LC—lower coil side in a slot. Coil is marked 1–1′

$$B_m = \frac{\phi}{K_f \tau_p L} = \frac{B_{\text{av}}}{K_f} = \frac{\phi p}{K_f \pi DL} \tag{4.3}$$

where B_{av} = Average value of flux density in air-gap (Tesla or Weber per sq. m)
τ_p = Pole pith = $\frac{\pi D}{p}$(m)
L = axial length of the machine (m).
p = number of poles (even number 2, 4,…).
D = diameter of armature (m).
The electromotive force (EMF) induced in one armature conductor is

$$E_C = B_{\text{av}} vL = K_f B_m vL = \frac{1}{\alpha} \times (B_1 + B_2 + B_3 + \cdots B_\alpha) \times v \times L \tag{4.4}$$

where α = Slots in one pole pitch
v = Peripheral speed (m/s).
E_C = Total flux cut per second.
= Flux per pole × Number of poles × Revolution per second
$E_C = p\phi n$
$n = \frac{N}{60}$ revolutions or rotation per second (rps)
N = Revolution per minute (RPM).
$\omega = 2\pi \text{n} = 2\pi \frac{N}{60}$ Radians per second.
B_α = maximum flux density over αth slot.

It is obvious that when conductors are in polar axis, the EMF and torque productions are at their highest than when the conductor is in proximity to the interpolar axis. If 'Z_C' conductors are in series per circuit or if 'Z_a' is the number of total conductors in the armature circuit and number of parallel paths is 'a', then we have

$$Z_c = \frac{Z_a}{a} \tag{4.5}$$

The total EMF induced due to rotation in a DC machine is,

$$E = E_C Z_C = p\phi n Z_C = \frac{p}{a}\phi n Z_a = K_f Z_C B_m vL = k_e^{/}\phi n = k_e \phi \omega \tag{4.6}$$

where $k_e = \frac{p}{2\pi a} Z_a$ = EMF constant (volt per Weber per radians per second)
a = mp for lap winding.
a = 2m for wave winding.
Here 'm' is the multiplicity of windings.

The induced EMF is referred to as back EMF. In case of motor it acts in opposition to the applied voltage. In other words, if the armature circuit resistance is R_a and the armature current is I_a, then back EMF is

$$E = V - I_a R_a$$

In the case of generator, the induced EMF is the source of terminal voltage and the terminal voltage is given

$$V = E - I_a R_a \tag{4.7}$$

In a DC generator, the EMF induced is in such a direction so as to establish a current which gives rise to a counter torque opposing the torque of the prime mover. On the other hand, in a motor, the direction of induced EMF is in such direction so as to oppose the flow of the current.

4.3.2 Torque Production

Force on a current carrying conductor under the influence of magnetic field is given as

$$F = (\mathbf{J} \times \mathbf{B})dV = I_a BL \tag{4.8}$$

where $\mathbf{B}$ = the air-gap flux density (T)
$\mathbf{J}$ = current density (A/sq m).
I_a = the armature current, (A).
L = the active length of conductor (m).
= axial length of machine – number of axial cooling ducts × length of ducts.
dV = differential volume.

If a current carrying conductor moves through a distance 'δx' in a uniform magnetic field, then work done is given as

$$W = F.dx = I_a BL\delta x$$

The conductor cuts through a magnetic flux '$\delta\phi$', or

$$\delta\phi = BL\delta x \tag{4.9}$$

Further, the work done is expressed as

$$W = \phi I_a \tag{4.10}$$

Equation (4.10) is valid for a uniform flux density only. In one complete revolution, each conductor moves through a total flux of '$p\phi$'. When a current carrying armature conductor passes from the influence of one field pole to that of the next field pole, the current in the conductor is reversed by commutator action. Thus, the total work done in one complete revolution for each armature conductor is,

$$W_C = p\phi I_a \tag{4.11}$$

Thus, total work done by all the conductors,

$$W = p\phi I_a Z_a$$

$$W = (p\phi)(I_a Z_a)$$

or,

$$W = (\text{Total magnetic loading}) \times (\text{Total Electric loading}) \quad (4.12)$$

Power developed by the armature is

$$P = (p\phi)(I_a Z_a)n \quad (4.13)$$

Torque is given by work done per revolution per conductor

$$W_C = 2\pi T \quad (4.14)$$

$$T = \frac{p.\phi.I_a.Z_a}{2\pi a} = k_t.\phi.I_a \quad \text{Nm} \quad (4.15)$$

Here, $k_t = \frac{p.Z_a}{2\pi a}$ torque constant, Nm per Weber per Amp

The average force on $\underline{Z_a}$ conductor,

$$F_{\text{av}} = B_{\text{av}} I_a L Z_a \quad (4.16)$$

This force is acting on radius (D/2), and therefore we have torque

$$T = \frac{B_{\text{av}} I_a L Z_a D}{2} = \frac{Z_a B_{\text{av}} I_a L D}{2} \text{ Nm} \quad (4.17)$$

Total flux is,

$$p\phi = B_{\text{av}} \pi D L \quad (4.18)$$

The power developed is

$$P = \frac{2\pi N T}{60} \quad (4.19)$$

4.3.3 Ratio of Back EMF and Developed Torque

EMF is given as

$$E = k_e \phi \omega$$

where EMF constant is

$$k_e = \frac{pZ_a}{2\pi a}$$

The torque can also be expressed as

$$T = k_t \phi I_a$$

Torque constant k_t can be given as

$$k_t = \frac{pZ_a}{2\pi a}$$

The ratio of torque and back EMF is

$$\frac{T}{E} = \frac{I_a}{\omega} \tag{4.20}$$

Or,

$$T\omega = EI_a$$

.

Example 4.1 Calculate the maximum value of EMF induced in the armature conductors of a machine driven at 600 RPM. Diameter of machine is 200 cm and length is 30 cm. The maximum value of flux density is 1.0 Weber /m^2.

Solution: Given that $N = 600$ RPM, $D = 200$ cm $= 2.0$ m, $L = 0.30$ m, $B_m = 1.0$ Wb/m^2

Revolution per seconds $= n = \frac{N}{60} = \frac{600}{60} = 10$ rps.

Peripheral velocity, $v = \pi Dn = 3.14 \times 2.0 \times 10 = 62.8$ m/s.

The maximum EMF induced in a conductor, $E = B_m \times v \times L = 1.0 \times 62.8 \times 0.3 = 18.84$ Volt.

Example 4.2 In a DC machine, if number of poles are 8, average flux in air-gap is 0.095Weber per pole, total number of conductors in armature is 1000, number of parallel paths is 8 and rotation per minutes is 500 RPM. Calculate the EMF induced.

Solution: Given $p = 8$; $a = 8$; $\phi = 0.095$ wb; $N = 500$ RPM

$$E = \frac{p}{a}\phi n Z_a = \frac{8}{8} \times 0.095 \times \frac{500}{60} \times 1000 = 791.7\,V$$

Example 4.3 Calculate the number of conductors in parallel path, if total number of poles is four. EMF induced in armature is 420 Volt, the rotational speed is 900 RPM and flux per pole is 0.06 Web /pole.

Solution: Given $2p = 4$; $E = 420$ V; $N = 900$ RPM; flux per pole $= 0.06$ Wb/m^2

$$E = p\phi n Z_C$$

$$Z_C = \frac{420 \times 60}{4 \times 0.06 \times 900} \approx 117 \text{ Conductors}$$

Example 4.4 A two-pole PMBLDC motor is operating on average flux density of 0.6 T. There are 24 slots in the machine, and each slot has 10 conductors carrying 10 A current. The length of the machine is 0.3 m, obtain torque if the machine diameter 0.1 m.

Solution

Given $L = 0.30$ m; I = 10 A; Conductors per slot = 10; B = 0.6 T.

Total Ampere conductors, NI = 24 × 10 × 10.

Force due to single slot is

$$F = \text{NIBL} = 10 \times 10 \times 0.6 \times 0.3 = 18.0 \text{ N}$$

Torque

$$T = \frac{B_{av}.I_a.L.Z_a.D}{2} = \frac{24 \times 10 \times 10 \times 0.6 \times 0.1 \times 0.3}{2} = 22.6\,\text{Nm}$$

Example 4.5 The armature of a pole DC machine has 100 turns and runs at 600RPM. The EMF generated in open circuit is 220 V. Find the useful flux per pole when the armature is

(i) lap connected and
(ii) wave connected.

Solution: The EMF induced is given as

$$E = \frac{p}{a}\phi Z \frac{N}{60}$$

(a) For lap winding, $a = 4$, Thus flux is

$$\phi = \frac{E \times 60 \times a}{p \times Z \times N} = \frac{220 \times 60 \times 4}{4 \times 200 \times 600} = 0.1\,\text{Web}$$

(ii) Wave windings $a = 2$. Flux is

$$\phi = \frac{E \times 60 \times a}{p \times Z \times N} = \frac{220 \times 60 \times 2}{4 \times 200 \times 600} = 0.05\,\text{Web}$$

Example 4.6 Calculate the torque developed in a six-pole DC motor. The armature is carrying a current of 20 A and has 1000 conductors connected in lap. The flux per pole is 25.0 m Web.

Solution: Given Poles $p = 6$, number of parallel path $a = 6$, flux per pole = 0.25 mWeb, total conductors $Z = 1000$ and current $I = 20$ A.

The developed torque is

$$T = \frac{p}{a}\frac{ZI_a\phi}{2\pi} = \frac{6 \times 1000 \times 20 \times 25 \times 0.001}{6 \times 2 \times \pi} = 79.6\,\text{Nm}$$

4.3.4 Core Loss and Magnetization

There can be two cases when ferromagnetic materials are subjected to changing magnetization. These are

(a) Alternating magnetization
(b) Rotating magnetization.

A ferromagnetic material is subjected to alternating magnetization when stationary or rotating windings are energized by alternating current, like transformers and other alternating current machines. In case of transformer and stator of alternating current machines like induction and synchronous motors, the frequency of alternating magnetization is supply frequency. The rotors of alternating current machines are subjected to alternating current magnetization at slip-frequency.

A magnetic core is subjected to rotating magnetization when in presence of a DC flux there is an associated change in the relative speed. In a conventional DC machine, the field winding is stationary and armature is rotating. The ferromagnetic core of the armature is cut by a DC magnetic field of alternate polarities of N-S poles, when rotor or armature is rotating. In such cases, the armature is subjected to rotating magnetization. In a 'p pole' machine running at 'N RPM' (n rps), the frequency of rotating magnetizing is given by,

$$f_R = \frac{np}{2} = \frac{Np}{120} \tag{4.21}$$

The ferromagnetic core of DC machine armature gives rise to core losses – hysteresis and eddy current losses. The field pole-faces are also subjected to varying DC flux due to slotting. These slot harmonics give rise to additional losses. In order to limit the eddy current losses, the field poles and armature are made using laminated stampings.

In high-speed PMBLDC motors, the ferromagnetic parts of machines are made using low core-loss electrical steel stampings. At high flux density, ferric saturation occurs- as such hysteresis loss due to alternating current magnetization is constant but hysteresis loss due to rotating magnetization increases.

4.4 Commutation

In DC machine, a current carrying armature coil essentially undergoes a current reversal or commutation after one pole pitch. When the magnetic poles become S from N (due to rotation), the conductors under the influence of reverse flux should carry a current in opposite direction, to produce a unidirectional torque.

The purpose of commutator is to continuously substitute a new circuit for EMF generation in a generator or torque production in case of a motor. A commutator is a device for short circuiting each coil of the armature in sequence as soon as the coil has turned to a position of minimum induced voltage (or maximum flux), and then reconnecting the coil into parallel circuit in the reverse direction. Thus, all the coils with decreasing flux are continuously connected in parallel reversed with all the coils having increasing flux, and the voltage of the combined circuit is sustained in one direction. The commutator acts as a mechanical switch or sometimes referred to as *mechanical rectifier.* If the coil current is changed from ' + I' to '–I' in an inductive circuit having inductance L, in time 'T_c', the reactance voltage is given as

$$E_r = L\frac{2I}{T_c} \tag{4.22}$$

The coil undergoing commutation has rotational induced EMF due to poor commutation and reactance voltage. Both cause poor commutation and result in sparking in DC machine.

When the stationary carbon brush is positioned fully on one of the commutator segments, as shown in Fig. 4.11a, the current is equally divided into two different paths. Suppose at that instant of time, coil 'a–b–c–d' is carrying current '0.5 × I' in the clockwise direction. When the two commutator segments are short-circuited, as in Fig. 4.11b, then the coil 'a–b–c–d' gets short-circuited. The current in this coil satisfies the following relationship:

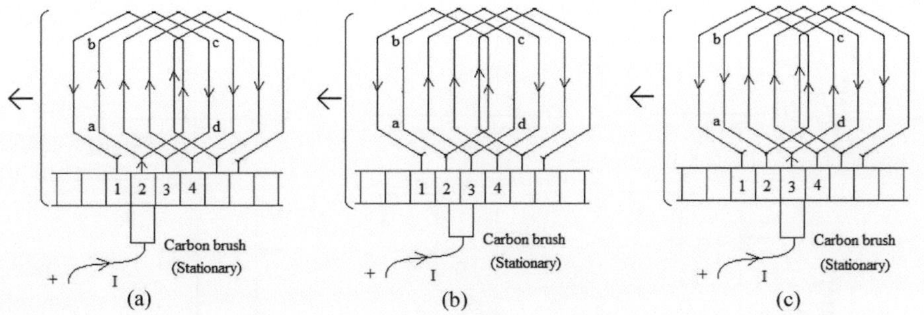

Fig. 4.11 **a** Coil undergoing commutation coil is carrying clockwise current. **b** The coil is short-circuited. **c** Coil is commutated and is carrying anti-clockwise current

$$i(t) = \frac{I}{2} e^{-\frac{t}{\tau}} \tag{4.23}$$

where τ is the time constant of the inductive circuit of the coil. If the resistance and inductance of the individual coil is R and L, respectively, then time constant is $\tau = \frac{L}{R}$ The short-circuited coil is carrying a decaying transient current.

After commutation time t_c, as shown in Fig. 4.11c, the carbon brush is now positioned fully on another commutator segment. The direction of coil 'a–b–c–d' current is changed to anticlockwise direction. If the peripheral speed of commutator is v (m/s) and width of each commutator segment is w, then time t_c is given as,

$$t_c = \frac{w}{v} \tag{4.24}$$

During the commutation time t_c the current may linearly come to '$-$ 0.5I' from '$+$ 0.5I' in the coil. This is referred to as *straight line* or *linear commutation.* The brush current density is uniformly distributed over the brush surface, as shown in Fig. 4.12a. If this change is too fast, the current density over the brush surface rises near the leading edge, as shown in Fig. 4.12b. In the case of under commutation, the reversal is incomplete, and when the brush leaves the commutator segment, a spark extends the effective time of commutation. During this time, the stored energy in the commutating coil is dissipated in the form of a spark. Minor sparking in DC machine is common. A poor commutation causes huge sparking in DC machines, resulting in a *ring-of-fire* over the periphery of the commutator. It leads to non-uniform current density resulting in higher ohmic losses, temperature rise and brush wears. The sparking damages the surface of commutator, leading to still greater wearing of brush. The commutation is also inferior in case of a universal motor operating on AC supply and rectified DC having ripple contents. As already mentioned, sparking restricts the use of conventional DC machines in explosive environments.

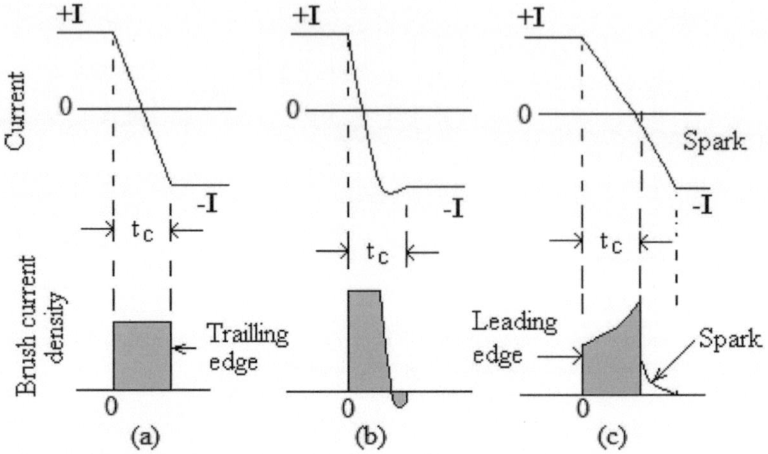

Fig. 4.12 **a** Linear commutation **b** over commutation and **c** under commutation of armature conductors

4.4.1 Brushless Commutation

In motors such as Switched Reluctance Motor or PMBLDC motors operating on DC supply, the commutation is achieved using electronic devices like power transistors, IGBT and MOSFET. There is no sparking in brushless commutation. Such types of electronic commutation of the coil are based on position sensing of the coil either by means of Hall-effect sensors or optical sensors or resolvers or sensor-less control. The switching ON/OFF of coils is synchronized with rotor position. If a coil is commutated before or after its required position, it will lead to faulty operation-leading to damage of switching devices and demagnetization of permanent magnet. The name brushless is due to the replacement of mechanical commutator and carbon brushes by solid-state devices.

Example 4.7 A two-pole lap wound armature running at 1000 RPM delivers a current of 20 A and has 36 commutator segments. The brush width is 1.4 times the commutator segments and inductance of each armature coil is 0.09 mH. Calculate the value of reactance voltage. Assume linear commutation.

Solution: Reactance voltage,

$$E_r = L\frac{2I}{T_c}$$

Here, $L = 0.09$ mH, $I = 20$ A, Number of parallel path = 2

Current through each armature conductor = 20 / 2 = 10 A.

Brush width = 1.4 × Commutator width.

Peripheral velocity of commutator = $\frac{1000 \times 36}{60}$ commutator segments/s.

Time of commutation,

$$T_c = \frac{1.4 \times 60}{1000 \times 36} = 0.0023\,\text{s}$$

Reactance voltage,

$$E_r = L\frac{2I}{T_c} = 0.09 \times 0.001 \times \frac{2 \times 10}{0.0023} = 0.7826\,\text{V}$$

Example 4.8 A commutator rotates at 800 RPM and has a diameter of 0.5 m. Find the approximate time of commutation, if the width of brush is 1.5 cm.

Solution: Diameter of the commutator = 0.5 m = 50 cm.
RPM = 800 RPM.
Peripheral linear speed $v = \omega r = 2 \times 3.14 \times \frac{800}{60} \times \frac{50}{2} = 2093.3$ cm/s.
Brush width = 1.5 cm.
Time of commutation, $T_c = \frac{1.5}{2093.3} = 0.7166$ milli-second.

4.5 Armature Reaction

Figures 4.13 and 4.14 show a linear representation of a DC machine. Here, a sheet of conductors carrying current in + z direction and in –z direction replaces slotted armature structure and winding [3].

The main field poles are having pole-shoe, which concentrate the magnetic flux in the polar axis of the air-gap. Figure 4.14b shows the distribution of flux density in air-gap, when only field poles are excited. One can experimentally obtain such

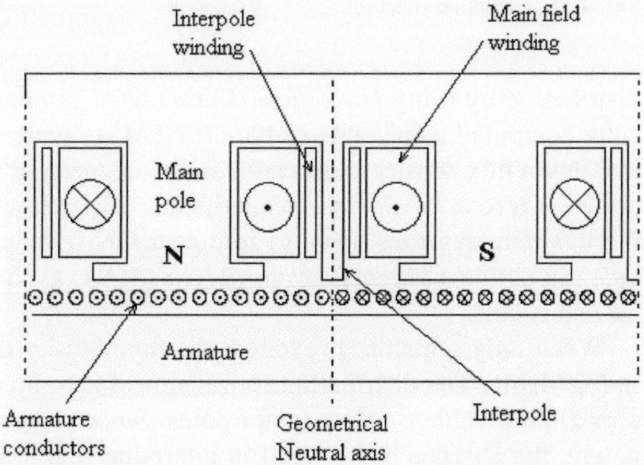

Fig. 4.13 Two-dimensional linear representation of DC machine (slotted armature is replaced by sheet of current carrying conductors)

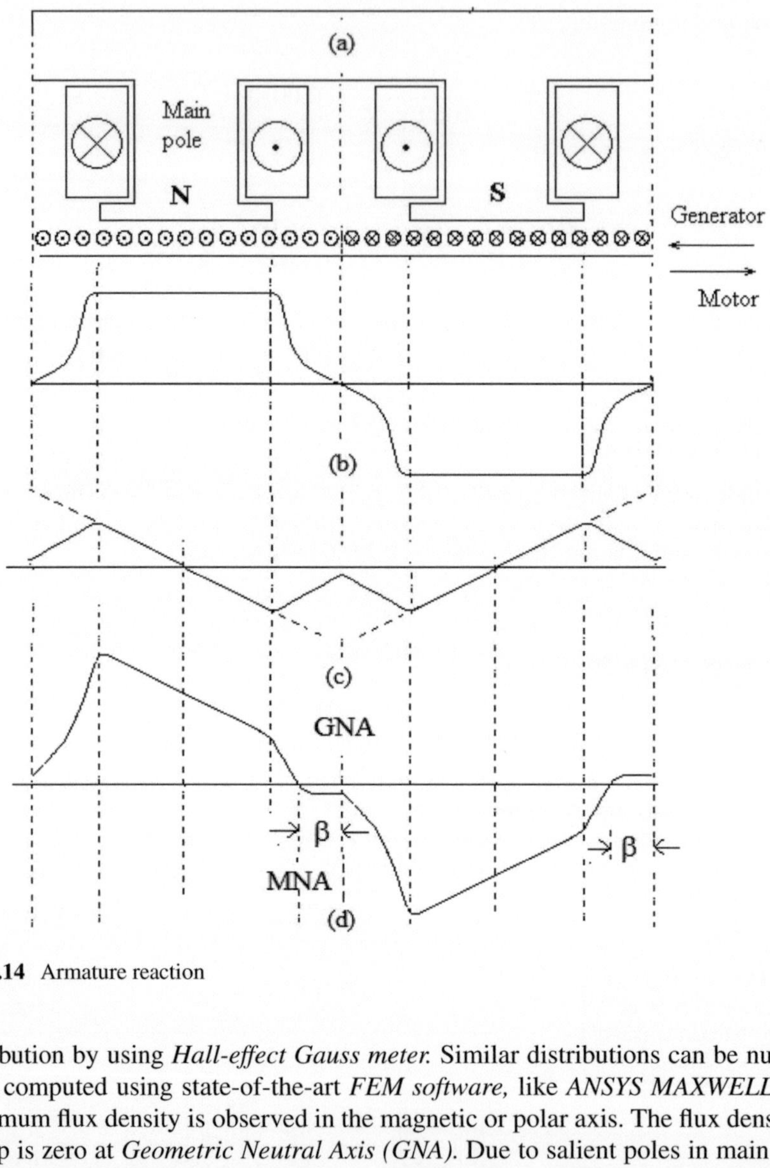

Fig. 4.14 Armature reaction

distribution by using *Hall-effect Gauss meter.* Similar distributions can be numerically computed using state-of-the-art *FEM software,* like *ANSYS MAXWELL.* The maximum flux density is observed in the magnetic or polar axis. The flux density in airgap is zero at *Geometric Neutral Axis (GNA).* Due to salient poles in main field, the flux density drops down to zero near GNA, when only field poles are excited. One can define a *Magnetic Neutral Axis (MNA)* as the axis in which flux density in air-gap is null.

When only armature is excited, the flux density distribution in air-gap is given in Fig. 4.14c. The distribution shows almost zero flux density in the polar-axis and a maximum value on the edge of poles. Since the poles are salient or projected in nature, the absence of iron part in interpolar region leads the flux density to falls. The dotted lines show an ideal flux density distribution due to armature currents in the inter-polar region for a cylindrical stator or presence of iron in interpolar region. When both the armature and field windings are excited, the resultant distribution will be as shown in Fig. 4.14d, which is obtained by the superposition of Fig. 4.14b, c. We observe that due to the operation of DC machine, when both the armature and field windings are excited, the distribution of air-gap flux density is markedly changed. This reaction of armature MMF on the resultant flux density distribution in air-gap is called as **armature reaction**. The following are the net effects of armature reaction:

(a) The MNA shifts from the GNA.
(b) The peak value of flux density rises.
(c) The magnetic field distribution in air-gap is distorted.

A finite flux density is present in GNA. This flux density induces a rotational EMF in the coil undergoing commutation (current reversal). It causes sparking at the trailing edge of the brushes and results in poor commutation. As already explained, the axis at which resultant magnetic flux is null or zero is referred at Magnetic Neutral Axis (MNA or NMA). When only field poles are excited, both MNA and GNA are on the same axis. However, due to armature reaction MNA shifts from GNA. Now, if brushes are shifted to the NMA, the rotational induced EMF in the said coil is minimized. As such, the commutation can be improved. We will discuss this in later sections. Note that MNA shifts in the direction of rotation in case of generators, and opposite to the direction of rotation in the case of motors.

Due to distortion of magnetic flux distribution in air-gap, the flux density rises at one corner of a pole-face and falls on the other corner of the same pole-face. The core losses, i.e., hysteresis and eddy current losses depend on the maximum flux density B_m in air-gap. Thus, *the core loss increases due to the armature reaction.*

The armature conductors experiencing the higher flux density will momentarily induce sufficiently higher value of speed induced EMF, which is hazardous for the operation of large DC machines. This higher induced voltage may cause flashover between two commutator segments and also between two brushes.

We will now discuss the effects and remedies of the armature reaction in conventional DC machines.

4.5.1 Brush Shift

Armature reaction results in poor commutation which leads to sparking on the trailing edge of the brushes. To improve the commutation, if the brush axis is now placed in MNA, the rotational induced EMF in the coil undergoing commutation becomes negligibly small. Following changes are done in smaller capacity DC machines:

(a) **Forward shift**: In case of the DC generators, the MNA shifts in the direction of rotation by few commutator segments or angle 'β'. This is referred to as forward shift of the brush axis.

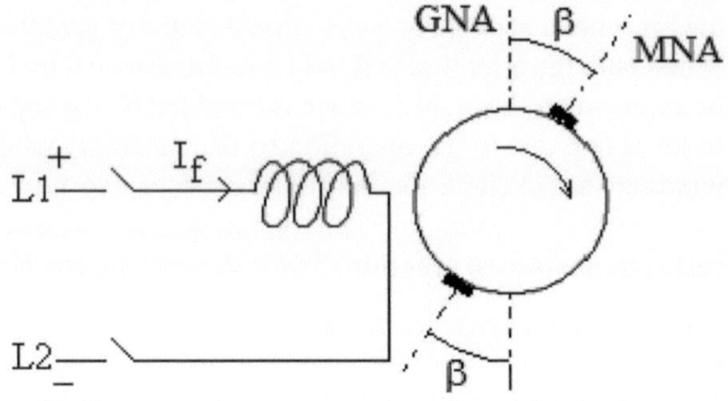

Fig. 4.15 Shifting of brushes in magnetic neutral axis for DC generator

(b) **Backward shift**: In case of DC motor, the MNA shifts in the opposite direction of rotation by few commutator segments or angle 'β'. We call this as *backward shift* of brush axis.

Figure 4.15 depicts a forward shift for DC generator

When brushes are shifted by an angle 'β', then all the conductors within the angle '2β' at both top and bottom of the armature carry current in such a direction as to produce a flux toward the field poles or through the armature from right to left. These conductors act in direct opposition to the main field and are known as demagnetizing armature conductors. From Ampere's law, we know that their role in the polar axis is demagnetizing MMF per pair of poles in a two-pole machine. The demagnetizing MMF per pole is given as,

$$\begin{aligned} AT_d &= \frac{1}{2} \times \text{MMF in } 2\beta \\ AT_d &= \frac{1}{2} \times \frac{2\beta}{2\pi} \times Z_a \times \frac{I_a}{a} \end{aligned} \tag{4.25}$$

The flux produced by the conductors not within the twice the angle of the brush shift is perpendicular to the polar axis and hence cross-magnetizes the field. The MMF per pole for a two-pole machine is

$$\begin{aligned} AT_c &= \frac{1}{2} \times \text{Cross - magnetizing MMF in}(\pi - 2\beta) \\ AT_c &= \frac{1}{2} \times \frac{(\pi - 2\beta)}{2\pi} \times Z_a \times \frac{I_a}{a} = \left(\frac{1}{2} - \frac{\beta}{\pi}\right)\frac{Z_a I_a}{2a} \end{aligned} \tag{4.26}$$

The number of demagnetizing amperes turns per pole is given as,

$$AT_d = \frac{\beta Z_a I_a}{2\pi a} \tag{4.27}$$

The Brush-axis is shifted to have ideal commutation at about three-fourth rated full load or full-load conditions. Once the brush axis is fixed, it cannot be changed.

The commutation is likely to be poor if the machine is not operated at the set load. Similar difficulties are observed when a DC generator is operated as a DC motor or vice versa. If the DC supply has larger ripple contents, commutation is expected to be poor. To improve commutation, brush shifting is adopted in smaller capacity motors, having no interpoles.

4.5.2 Interpoles

Flux density in GNA can also be nullified by using another magnetic pole. The additional poles are known as *interpoles,* and are oriented toward the (GNA) Geometrical Neutral Axis. Interpoles have no pole-shoes and nor do they undergo magnetic saturation. The airgap between interpole and armature is larger than airgap between main pole and armature. The flux produced by it is concentrated toward GNA and neutralizes the local magnetic flux in GNA due to armature reaction as shown in Fig. 4.16. Here Fig. 4.16a–c are self-explanatory. These additional poles help in improving commutation. The interpoles are connected in series with the armature. Though the interpoles have lesser number of turns of thicker wire, they are sufficient to produce an opposing field (Refer Fig. 4.16d) at GNA to neutralize the presence of magnetic field due to the armature reaction. The effect of interpoles on the flux density distribution in air-gap is depicted in Fig. 4.16e.

Due to saturation, the peak flux density decreases according to the magnetization characteristics of ferromagnetic material, as shown in Fig. 4.16e.

4.5.3 Compensating Windings

The armature reaction effect is nullified in interpolar region using interpoles. The resultant flux distortion has no effect due to presence of interpoles. The resultant flux distribution becomes extremely distorted if the machine is heavily loaded or when the field flux is extremely low. The coils under the influence of larger magnetic field will develop excessive rotational EMF, sufficient to cause sparking among the commutator segments connected to the coil. This may ultimately cause flashover between positive and negative brushes. These electric arcs may cause sufficient damage to the supply and to the commutator surface.

The arcing or flashover can be overcome by using pole face winding referred to as compensating winding. The winding is embedded on the slots on the pole face. The compensating winding is connected in series with the armature winding. The armature and compensating windings MMF oppose each other. The direction of current just below the pole face on the armature should be opposite for decreasing the flux density. Half the compensating winding conductors are connected in series with the half of the conductors of compensating winding of neighboring pole (refer Fig. 4.17a). The resultant flux waveform below the pole face is independent of loads

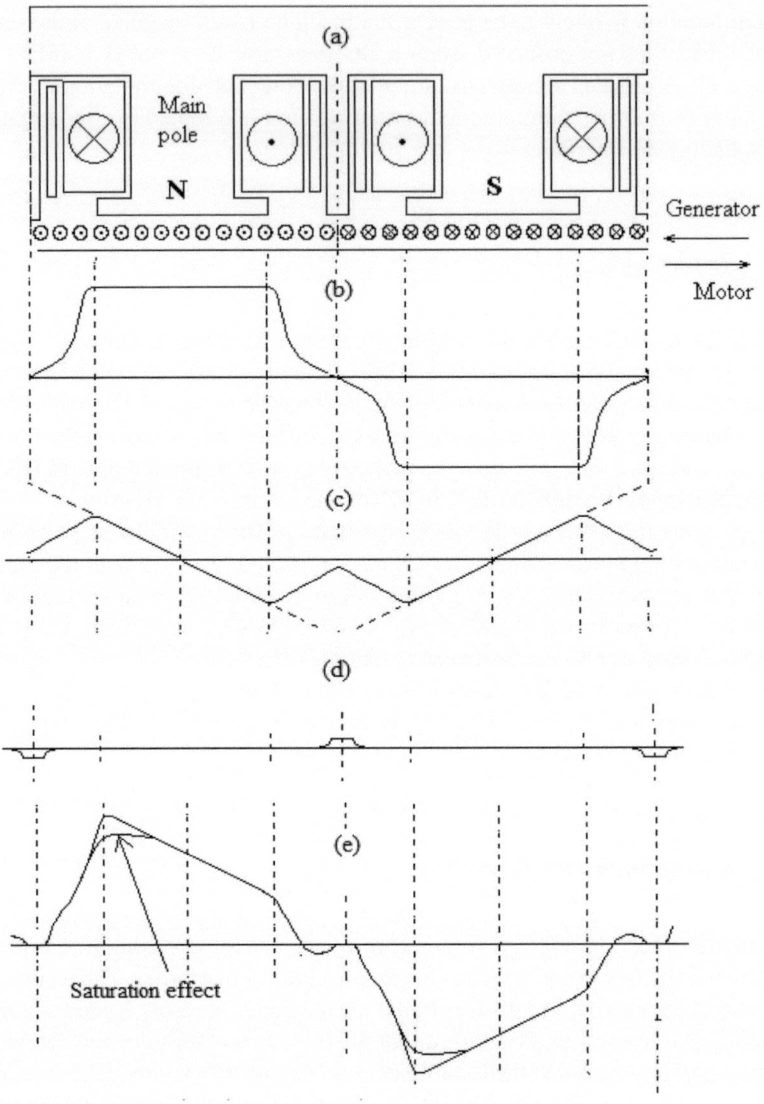

Fig. 4.16 Armature reaction due to Interpoles

current and is very much similar to the flux waveform of original main pole, as shown in Fig. 4.17f. The commutation is also more reliable. Since the compensating winding MMF neutralizes the armature reaction MMF only under the main pole-face, the compensating MMF per pole is

$$AT_w = \left(\frac{\text{pole arc}}{\text{pole pitch}}\right) \times \text{MMF of main field pole} \tag{4.28}$$

where pole-arc = angle subtended by one pole- face (radians or degree)

$$\text{pole pitch} = \frac{2\pi}{\text{number of poles}} \text{ (radians or degree)}$$

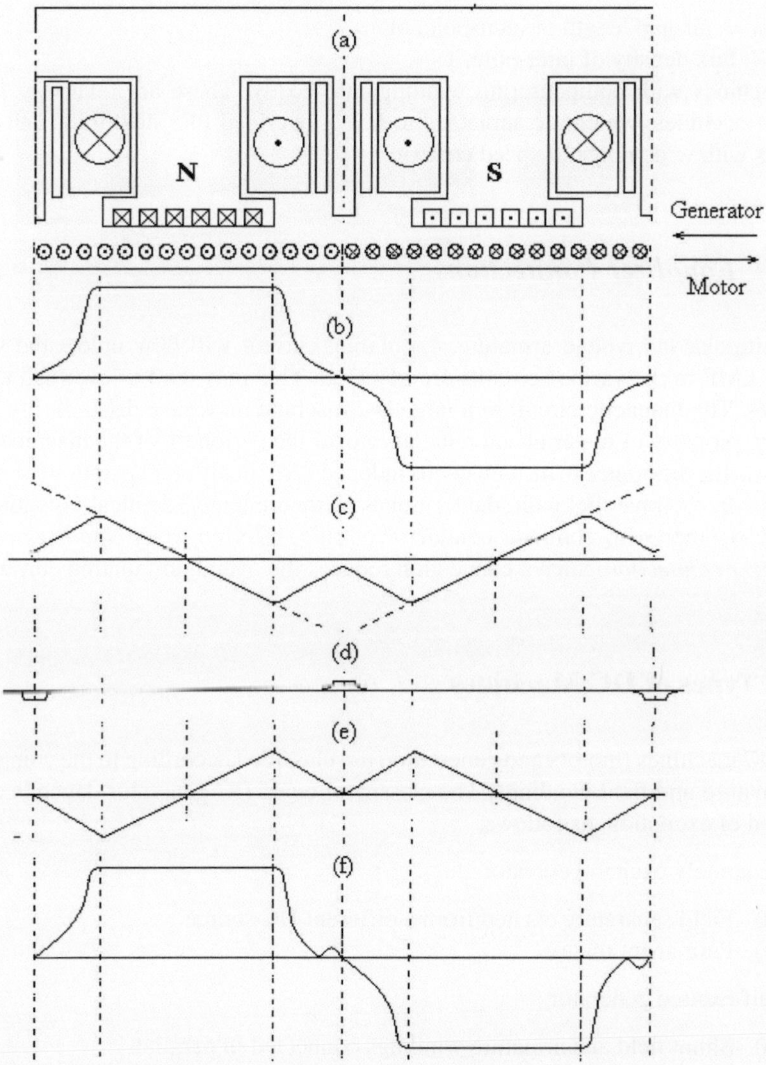

Fig. 4.17 Effect of Compensating windings on the flux density distribution. **a** Schematic representation. **b** Flux density distribution in air-gap due to main field poles only. **c** Flux density distribution in air-gap due to armature currents only. **d** Flux density distribution due to Inter-poles only. **e** Flux density due to compensating windings. **f** Resultant flux density distribution when both field poles and armature are excited.

When compensating windings are used, the MMF requirement of interpoles also gets reduced. The interpoles MMF required is given as,

$$AT_i = \frac{Z_a I}{2ap} + \frac{B_c l_g}{\mu_0} \tag{4.29}$$

Here $l_g =$ air-gap length of inter-pole, m

$B_c =$ flux density of inter-pole, T.

Machines with compensating windings are costly. These are relatively higher power machines, which are subjected to heavy overload like steel mills and shunt motors with wide range of speed control.

4.5.4 Equalizer Connections

In multipolar lap wound armature, circulating current will flow unless the values of the EMF in the various circuits are identical. This may tend to overload certain brushes. The magnetic circuit of a large DC machine may have dissimilarity in the air-gap, property of material and reluctance over the periphery of the machine. This results in the difference in instantaneous induced EMF in different parallel-connected circuits. In each parallel path, the terminals of theoretically identical potentials are joined together with common conductor or ring, referred to as *equalizer ring* or *equalizer connection.* Such a connection reduces the overall circulating current.

4.6 Types of DC Machines

The DC machines (motors and generators) are classified according to the connection of armature and field windings. The classification of DC generator depends on the method of excitation, as follows:

1. Separately excited generator
   (a) Field separately excited from a different DC source
   (b) Permanent magnet.
2. Self-excited generator
   (a) Shunt field and armature windings connected in parallel
   (b) Series field, armature winding and field winding plus load all are in series
   (c) Compound generator having both the series and shunt fields connected in different manner with armature.

The separately excited, self-excited and series generators have been shown, respectively, in Fig. 4.18a–c. A generator receives its motive power through shaft

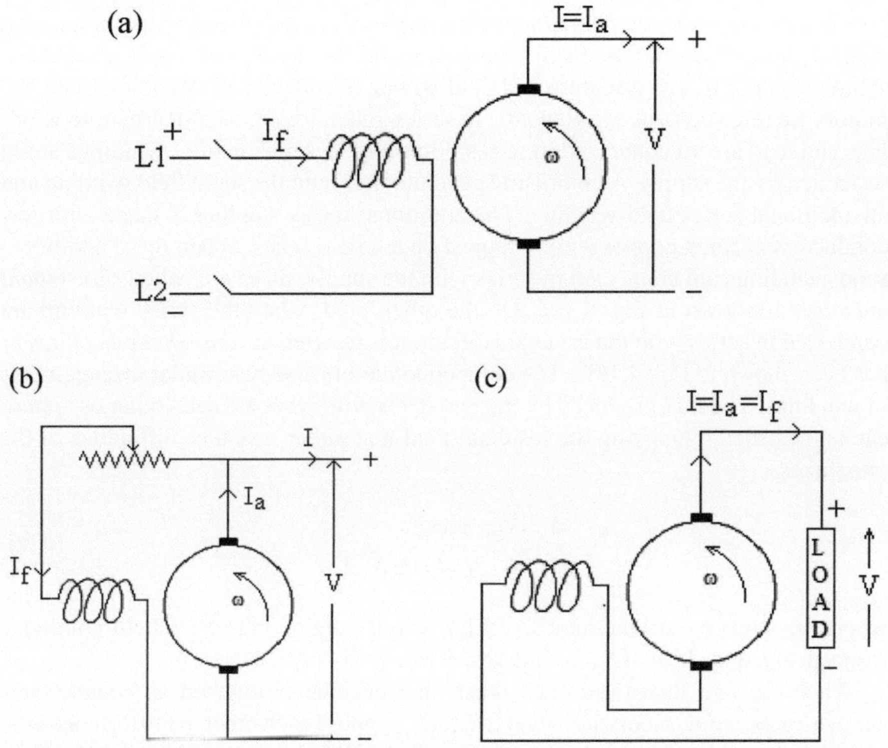

Fig. 4.18 **a** Separately excited generator. **b** Long shunt generator, **c** Series generator (Prime mover not shown)

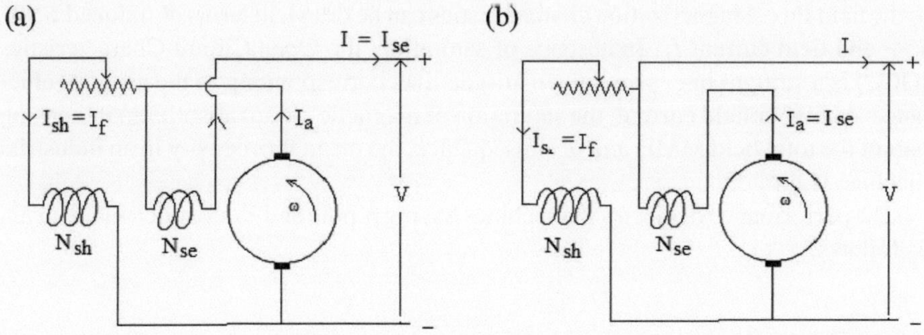

Fig. 4.19 **a** Short shunt DC generator. **b** Long shunt DC generator

connected to a prime mover. A prime mover may be like, any electrical motor, petrol/diesel/gas engine, hydro turbine, steam turbine and wind turbine etc. In case of a separately exited DC machine, the field is either excited by a separate DC source or with the help of permanent magnets. The field winding or permanent magnet pole provides the necessary field flux for operation. However, the air-gap flux in permanent magnate machine cannot be easily changed. As the name suggests, a self-excited DC shunt generator has its armature and field windings connected in parallel. Such DC motors are referred to as *shunt motor*. In series generators, the field, armature windings and load are all connected in series. Similarly, in series motor , windings are in series across the supply. A compound generator has both the shunt field winding and an additional series field winding. The additional series winding is made with few conductors of thick copper wire, wrapped on the field poles. When these additional series windings are connected in series with the supply, these are called *short-shunt machine*, as shown in Fig. 4.19a. On the other hand, when the series winding are connected in series with the armature circuit, it is referred as *long-shunt machine*, as has been shown in Fig. 4.19b. The compound motors also has similar arrangements of windings. The flux produced by the two fields will either aid each other or oppose each other, thereby making the resultant field flux either a sum or difference of the two fluxes.

$$\begin{aligned} \phi &= \phi_{sh} \pm \phi_{se} \\ \phi \quad &\alpha \quad N_{sh}I_{sh} \pm N_{se}I_{se}, \end{aligned} \tag{4.30}$$

where N_{sh} and N_{se} are the number of turns in the shunt and series field windings, respectively, and, I_{sh} and I_{se} are the respective field currents.

When the two fluxes aid each other, the machine is referred as *cumulatively compound machine*, otherwise when two fluxes oppose each other, it is a *differentially compound machine*. The addition and opposition of DC fluxes are shown in Fig. 4.20a, b.

If the MMF produced by the field windings carrying a current I_f and machine is driven at a constant speed, the armature rotation voltage EMF is directly proportional to the field flux. Magnetization characteristics can be drawn in terms of induced EMF, V_{OC} and field current I_f. In absence of saturation, the Open Circuit Characteristics (OCC) is a straight line, also known air-gap line, corresponding to the air-gap reluctance. At higher field current, the saturation of magnetic circuit absorbs an increasing part of the total field MMF, and in consequence, the mean flux density in an industrial machine is limited.

The performance of electrical machines having p-pole and Z_a conductors are given as follows:

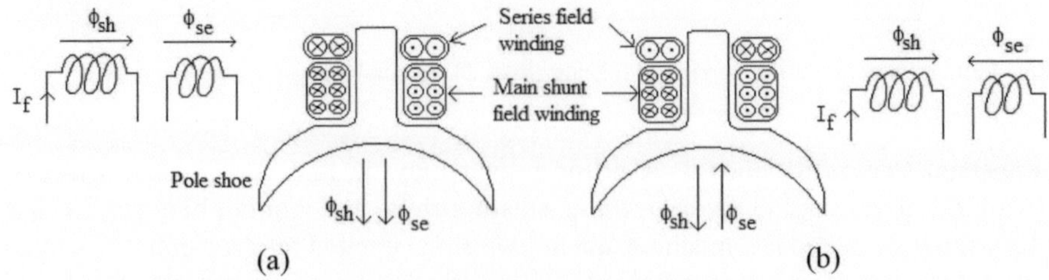

Fig. 4.20 Flux addition and subtraction. **a** Cumulatively compound machine b Differential compound machine

$$\begin{aligned}
\text{Armature voltage,}\quad & V_a = E + I_a r_a \quad \text{(for motor)} \\
& V_a = E - I_a r_a \quad \text{(for generator)} \\
\text{Back - EMF,}\quad & E = k\phi\omega \\
\text{Torque,}\quad & T = k\phi I_a \\
\text{Speed,}\quad & \omega \\
\text{Power,}\quad & P = EI_a = T\omega
\end{aligned} \tag{4.31}$$

4.6.1 Separately Excited Generator

The field winding is made of several turns of thin conductors wounded on a former. Permanent magnet can also replace the field winding. In such cases, field flux cannot be altered. The effective resistance of shunt field winding is much higher than that of armature resistance. Under steady state condition, one can write,

$$\begin{aligned}
V_f &= I_f R_f \\
V_a &= k_e\phi\omega - I_a r_a
\end{aligned} \tag{4.32}$$

where

V_f = Voltage across field circuit (V)
R_f = Resistance of field circuit (Ω)
I_f = Field current (A)
V_a = Voltage across armature (V)
I_a = Armature current (A)
k_e = EMF constant (V/weber/ radians per second).
When armature current $I_a = 0$, the armature no-load or open-circuit voltage is

$$V_{nL} = V_{\text{oc}} = V_{\text{aoc}} = k_e\phi\omega \tag{4.33}$$

The flux ϕ is a nonlinear function of field current I_f. The armature reaction results in distortion of the gap flux. When poles are unsaturated, there is no change in the total gap flux. Under full load condition, a reduction of about 4–8% of total gap flux results. The effect of nonlinearity due to magnetic saturation is difficult to account for in the formulation. The machine is being rotated at constant speed, ω, by a prime mover. When the field current is zero, the field flux is due to the residual magnetism and the generator develops very small voltage of the order of a few volts.

The variable field resistance 'R_f' is normally used to vary the DC field current. Single-phase or three phase controlled rectifier can also be used for the variation of field current. For a constant speed ω, the variation of no-load armature voltage and field current is shown in Fig. 4.21a. This is referred to as *magnetization* characteristics or no-load or (open circuit characteristics) OCC. The OCC is similar to the magnetization characteristics of a ferromagnetic material. Since back EMF is proportional to speed, similar characteristics at different rotational speeds can be plotted. Due to saturation, increasing and decreasing the field current over full range, have little difference in the generated induced voltage is observed, due to the hysteresis. The machine is designed such that the operating point corresponds to the knee point of the magnetization curve or OCC. A good regulation cannot be achieved if the machine is designed to operate in saturation zone. The voltage is unstable, when machine operates in the linear portion of OCC.

The reversal of field winding current results in the reversal of direction of rotation of a DC motor and reversal of polarity of the generated voltage in case of a generator.

When load current $I_a \neq 0$, the voltage across armature drops to a slightly lower value due to '$I_a r_a$' drop or ohmic drop and also due to demagnetizing effect of armature reaction, as depicted in Fig. 4.21b. The terminal voltage drops to about 5–10%. The internal characteristic is obtained by subtracting '$I_a r_a$' drop from open circuit EMF. The brush contact resistance is not constant. Depending on the brush material, the voltage drop varies by about 2 V. If OCC is known, the terminal or external characteristics can be obtained. The demagnetizing field current is given as,

$$I_f^/ = \frac{AT_d}{N_f} \tag{4.34}$$

where N_f = number of turns in field windings

Separately excited generator is also used for amplifying DC signal or control voltage. In laboratory, the machine is used for obtaining torque-speed characteristics of any electrical motor or a mechanical engine, or a turbine. A separately excited

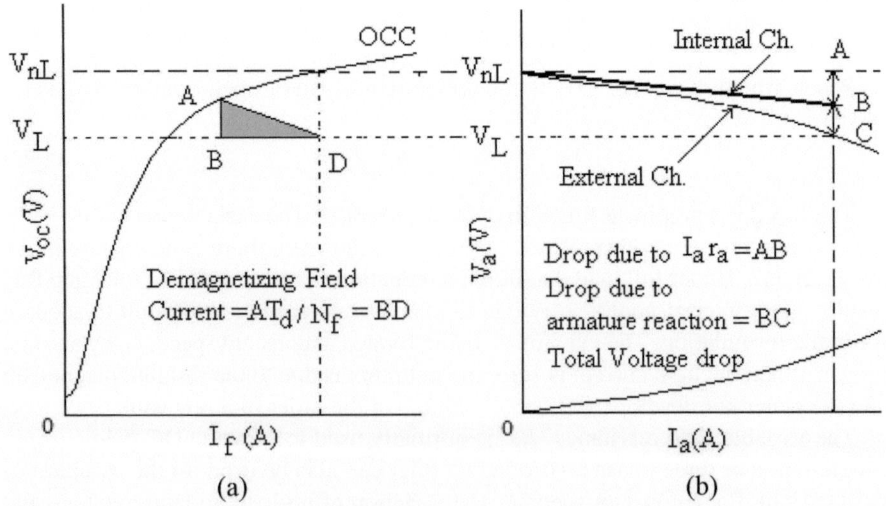

Fig. 4.21 **a** Open circuit or No-load or internal characteristics. **b** Terminal or External characteristics

DC generator connected to constant load resistance R_L exhibits linearly rising (load) torque-speed characteristics.

A separately excited DC generator can be calibrated to offer different load-torque characteristics (wind turbine or vehicle or fan load) for testing electrical motors with different load-torque versus speed characteristics. This requires the sensing of V_a, I_a and ω, thereby changing the field current I_f accordingly. A knowledge of programming microcontroller and solid-state devices are desirable.

4.6.2 Self-excited DC Generator

As already pointed out, the armature and field windings are connected in parallel. A separately excited DC generator with shunt field winding (without permanent magnet) can be connected as a self-excited DC generator. The generator is being rotated using a prime mover at constant speed, ω. If the field poles have residual magnetism, a small developed armature voltage causes the field current to flow. If this field current sets up in a direction to aid the residual magnetism, a large voltage on the armature terminal develops. The generator fails to self-excite, if certain conditions are not satisfied. The following are the essential conditions for self-excitation:

(a) Field poles must have residual magnetism.
(b) Field winding must be connected in a direction to aid the residual magnetism. The net effect is to raise the air-gap flux, field current and armature voltage.
(c) The field circuit resistance R_f should be less than the critical field resistance. The critical field resistance is obtained from the slope of linear portion of magnetization characteristics or OCC Fig. 4.22a.
(d) The speed of the prime mover is higher than the critical speed. The critical speed is defined as the speed at which the machine just becomes self-excited.

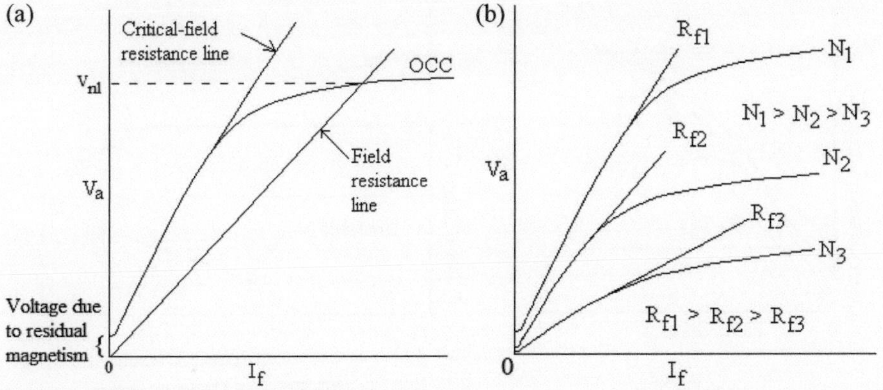

Fig. 4.22 **a** No-load magnetization characteristic of a self-excited DC generator. **b**. OCC of self-excited DC generator at different speeds N_1, N_2 and N_3

In OCC, the terminal voltage corresponds to the voltage where OCC is cutting the field resistance line, as shown in Fig. 4.22a. A self-excited DC generator cannot operate in the linear portion of OCC. Can you give a suitable explanation? Do you find magnetic saturation useful here?

For checking the residual magnetism, the generator is rotated with its field winding disconnected. A voltmeter is connected across the armature. If the field poles have residual magnetism, the voltmeter shows a very small voltage of the order of 5–10 Volt. In absence of residual magnetism, if the field winding is connected to a DC source, it will retain the residual magnetism. In certain cases, the use of magnetizer or discharging of charged capacitors (at high voltage) to the field windings is fruitful. Similar test is useful in checking the residual magnetism of an induction generator or a pilot-generator of a synchronous motor and alternator.

If the field circuit is connected in parallel with armature, and the machine fails to self-excite and generating a DC voltage of merely 5–10 V. Either of the following remedies may be useful:

(a) Check residual magnetism
(b) Interchange the terminals of field winding
(c) Interchange the terminals of armature winding
(d) Change the direction of rotation
(e) Decrease the field resistances
(f) Increase the speed.

In certain cases, the terminal voltage will change its polarity. No-load magnetization characteristic of self-excited DC generator at different speed are shown in Fig. 4.22b. It can be shown that

$$\frac{N_1}{N_2} = \frac{R_{f1}}{R_{f2}} \tag{4.35}$$

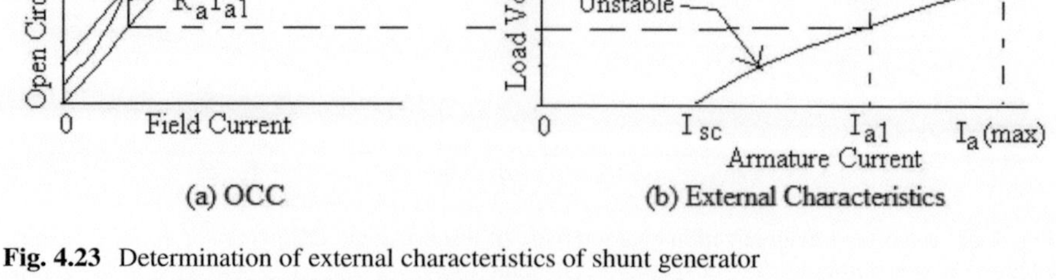

Fig. 4.23 Determination of external characteristics of shunt generator

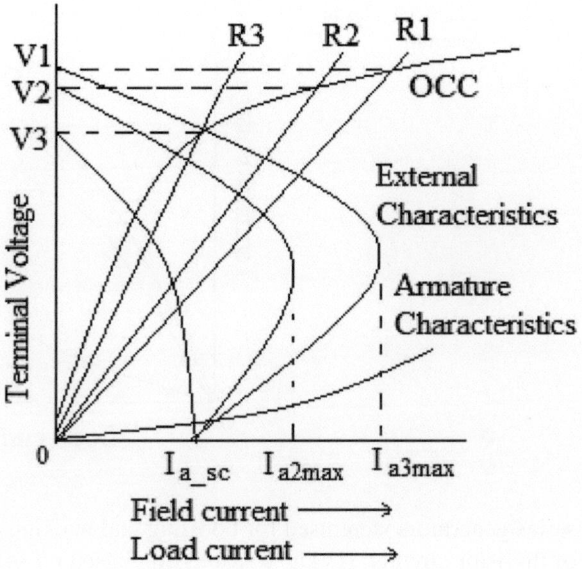

Fig. 4.24 External characteristics of self-excited shunt generator for different values of field resistances

The critical field resistance depends on the operating speed.

The external characteristics of a shunt generator can be derived using OCC. Figure 4.23a shows the OCC characteristics a shunt generator. It cuts the field resistance line at no- load voltage V_{nL}. Lines parallel to the field resistance line cut the OCC at two points. The terminal voltage is given by generated EMF less 'I_aR_a' drop. Ignoring the effect of armature reaction, the external characteristics are shown in Fig. 4.23b. Negative slope of the characteristics forms the stable part, and extends from no-load current to maximum possible armature current '$I_{a(max)}$'. The positive slope of the characteristics is the unstable portion; generator does not operate in this region. When generator is initially short circuited, (prior to running) at rated speed, the steady state short circuit current 'I_{sc}' flows through the generator due to residual magnetism when rotating at rated speed.

Effects of different field resistances on the external characteristics are shown in Fig. 4.24. Here field resistance lines R1, R2, and R3 cut the OCC at different points of OCC giving open circuit voltages V1, V2, and V3, respectively. Here I_{a_sc} is common short circuit current and I_{a2max} and I_{a3max} are the maximum current point of two external characteristics.

4.6.3 DC Series Generator

René Thury (1860–1938), a Swiss electrical engineer developed high voltage direct current (HVDC) electricity transmission system using DC series generators. Such system of HVDC were in use from 1906 to 1936 in foreign countries. These DC

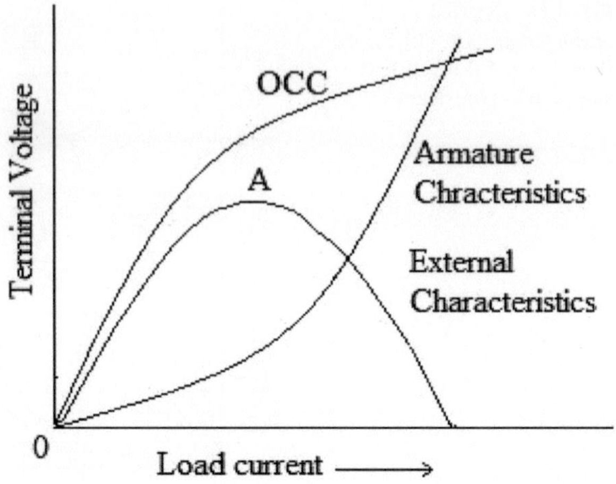

Fig. 4.25 Characteristics of DC series generator

series generators were used for boosting and bucking of DC voltage. Due to advent of thyristor circuits, HVDC transmission based on solid-state circuit grew at larger pace [6, 9, 10].

DC series generators can be used for boosting or bucking the DC battery voltage in electric vehicle (EV). The electrical circuit should be closed for self-excitation. The generation is unstable if the magnetic core is saturated. In other words, if the current flowing through the generator crosses a certain limit, the developed voltage falls to zero. The OCC of the DC series generator is shown in Fig. 4.25, without external battery source. After point A, when load current increases beyond certain limit, the generated voltage falls to zero. Figure 4.26 shows a simple circuit of boosting/bucking of DC voltage using series generator. Here battery and load are all in series with the series generator. The generated voltage aids or subtracts to the battery voltage. This requires the speed control of prime mover for voltage regulation.

4.6.4 Compound Generator

The cumulative compound generator exhibits a small increase in the voltage on loading. Depending on the number of turns in the series field winding, these generators have better voltage regulation. The characteristics of compound generators are shown in Fig. 4.27.

When a differential compound generator is loaded, the terminal voltage falls to very low value. These generators are also referred as constant current generator. The terminal characteristics are suitable for DC arc welding. The operation of DC motor, which undergoes into stalling mode during its duty cycle, also requires constant current generator.

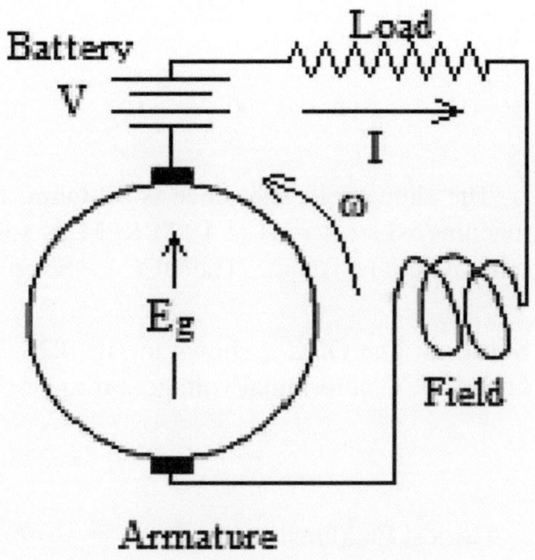

Fig. 4.26 DC series generator for boosting (prime mover not shown)

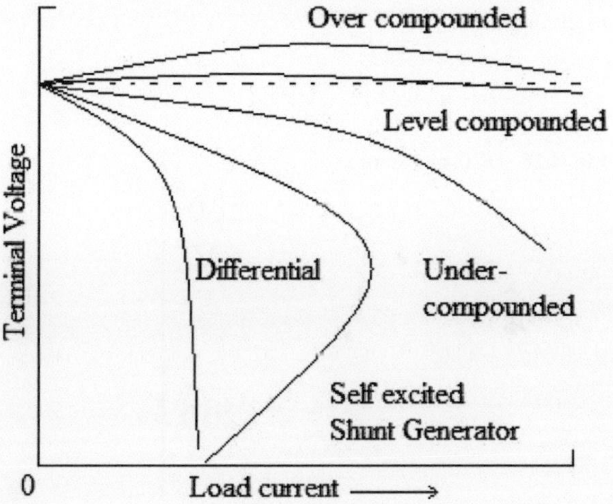

Fig. 4.27 Characteristics of compound generators

Example 4.9 The relationship between the field current and open circuit EMF generated by a DC shunt wound generator running on open circuit at 1000 RPM is

I_f (A)	0	1	2	3	4	5	6
V_{oc}(V)	10	90	170	217.5	251	272.5	281

The shunt field resistance is 50 Ohm. Find the voltage at the terminals of the machine when it runs at 1000 RPM as self-excited generator. Also, calculate the critical field resistance. The OCC is shown in Fig. 4.28.

Solution: The OCC is shown in Fig. 4.28. The slope of the resistance line $R_f = 50$ Ω cuts OCC at terminal voltage. From graph, we have

$$V_t = 277.17 \text{ Volt}$$

Critical Field resistance, $R_C = \frac{90}{1.0} = 90\,\Omega$.

Example 4.10 The open circuit magnetization curve at 600RPM of a DC machine is given as:

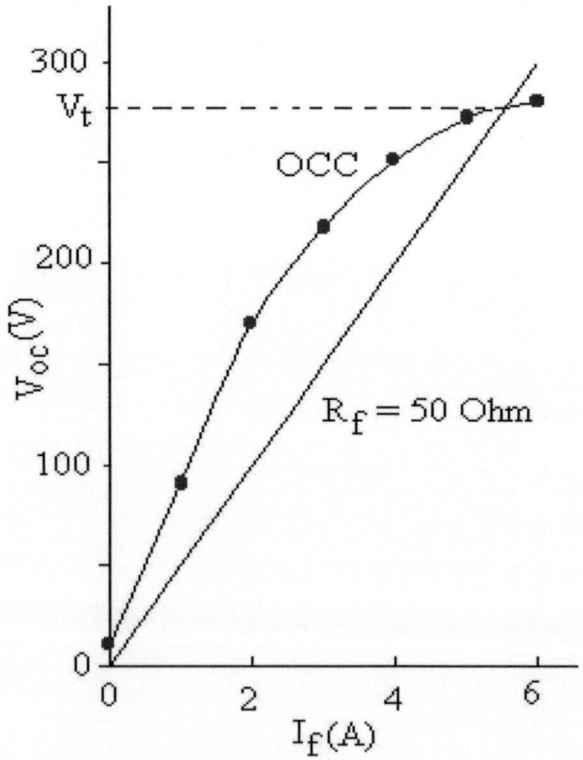

Fig. 4.28 OCC of Example 4.11

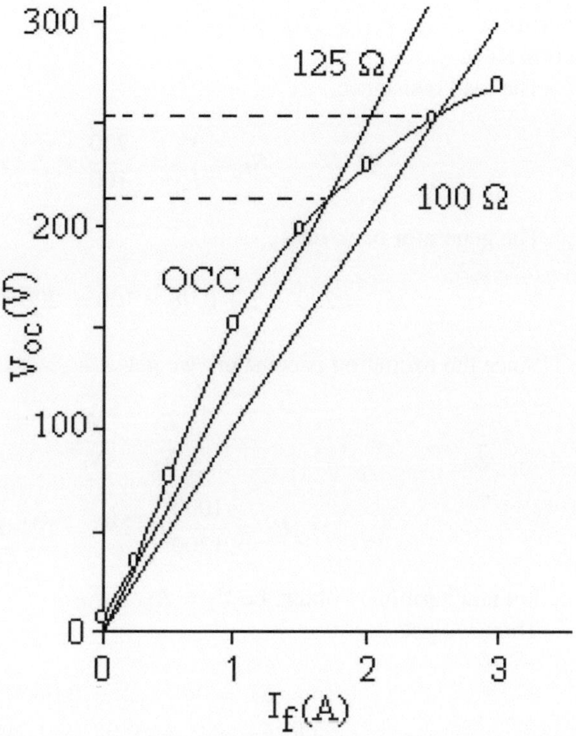

Fig. 4.29 OCC of Example 4.12

V_{oc} (V)	8	43	77	151	198	229	253	269
I_f (A)	0	0.25	0.5	1.0	1.5	2.0	2.5	3.0

Find the critical field resistance. Determine the terminal voltage when the field resistance are 100 and 125 Ω. The OCC is given in Fig. 4.29.

Solution: The OCC is plotted in Fig. 4.29. Approximate critical field resistance is

$$R_C = \frac{151}{1.0} = 151\,\Omega$$

From the graph, we get the following:
The terminal voltage corresponding to field resistance of 125 Ohm is 213.33 V
The terminal voltage corresponding to field resistance of 100 Ohm is 253.33.

Example 4.11 A separately excited generator, when running at 1200 RPM supplies 100 A at 220 V to a fixed resistance. If the field resistance is unaltered, calculate the current when speed drops to 1000 RPM. Armature resistance is 0.08 Ω and total drop at brush is 2.0 V. Ignore effect of armature reaction.

Solution: Given $N_1 = 1200$ RPM, $V_1 = 220$ V, $I_{al} = 100$ A, $R_a = 0.08\ \Omega$, $N_2 = 1000$ RPM.

The load resistance,

$$R_L = \frac{V_1}{I_{a1}} = \frac{220}{100} = 2.2\,\Omega$$

The generator back EMF,

$$E_1 = 220 + 2 + 0.08 \times 100 = 220 + 10 = 230\ \text{V}$$

Since the excitation is constant, we get

$$\frac{E_2}{E_1} = \frac{N_2}{N_1}$$

$$E_2 = \frac{1000}{1200} \times 230 = 191.67V$$

Let new terminal voltage be $V_2 = R_L. \times I_{a2}$.

Thus, we get

$$E_2 = V_2 + 2 + 0.08 \times I_{a2}$$

$$\text{Or, } E_2 = 2 + (0.08 + 2.2) \times I_{a2}$$

New load current is

$$I_{a2} = \frac{E_2 - 2}{0.08 + 2.2} = \frac{191.67 - 2}{2.28} = 83.19\ \text{A}$$

Example 4.12 A separately excited DC generator supplies 50 A to a DC bus bar of 250 V. The useful flux is 0.03 Weber. If the flux is changed to 0.029 Weber, calculate the current flowing immediately after the change. Also, find the percentage change in the speed of the machine that is required to restore the original armature current. The armature resistance of generator is 0.5 Ω.

Solution: Given: Flux $\phi_1 = 0.03$ Wb, $I = 50$ A, $V = 250$ V, $R_a = 0.5\ \Omega$, $\phi_2 = 0.029$Wb.

Voltage drop in armature,

$$I_{a1}R_a = 50 \times 0.5 = 25V$$

The EMF generated,

$$\text{E}_1 = 250 + 25 = 275\,\text{V}$$

Now the flux changes to $\phi_2 = 0.029\,\text{Wb}$. Immediately after the change in the flux, the speed remains same, thus EMF is proportional to the flux, the speed being constant, i.e.,

$$\frac{E_2}{E_1} = \frac{\phi_2}{\phi_1}$$

$$E_2 = \frac{0.029}{0.03} \times 275 = 265.8\,\text{V}$$

Armature current immediately after the change of flux is

$$I_a = \frac{265.8 - 250}{0.5} = 31.6\,\text{A}$$

For making the current equal to the original armature current speed need to be increased, the generated EMF should be raised to the original value. The EMF is proportional to flux and speed. Thus, we have

$$E = \phi_2 N_2 = \phi_1 N_1$$

$$\frac{N_2 - N_1}{N_1} = \frac{\phi_1 - \phi_2}{\phi_2} = \frac{0.03 - 0.029}{0.029} = 3.45\%$$

Thus, percentage change in the speed is 3.45%.

4.7 Parallel Operation of DC Generator

The electrical power is normally generated at several distant locations for meeting a common bulk power demand. When a generator is loaded, its speed falls because of the counter torque. The fall in the speed is checked in a power plant by the regulating the prime mover speed. In hydraulic turbines, the speed governors regulate the speed by increasing or decreasing the flow of water. A balance between electrical power generated and mechanical power input is maintained. A likely fall of $2 - 3\%$ in speed from no load to full load is obvious. Speed governors are also used in the case of steam turbines. In the event of fall in speed, the steam input is increased. Though, currently the bulk DC power generation has been stopped, a study answers some of the basic facts of parallel connections of generators, which is difficult to explain in the case of AC generators. The parallel operation of DC generator requires following essential conditions:

(a) The terminals with identical polarity should be joined together. The 'positive' of generator should be joined with 'positive' of bus bar. Negative terminals should be similarly connected.

(b) The no-load EMF of the generators connected in parallel must be identical.

If the terminals joined together are having different polarities, there will be a dead short-circuit, causing extensive damage.

If no-load EMFs are identical, but the speed regulation is different, the generators will share unequal load. This is shown in Fig. 4.30. The generator having more drooping characteristics will share lesser load current.

Consider a case when the no-load EMFs are different and their speed governors' characteristics are also not identical, as shown in Fig. 4.31b.

In Fig. 4.31, at operating point A, we have.

$I_1 = 0$ (Generator-I floats) and $I_2 =$ Negative (Generator-II acts as motor).

At point B, we have.

$I_1 =$ Positive (Generator-I acts as a generator) and $I_2 = 0$ (Generator-II floats).

At operating point between B & C,

$I_1 > I_2$ (Generator-I shares a larger load than Generator-II.)

At operating Point C,

$I_1 = I_2$ (Both the Generator-I & -II act as generator and share equal load.)

At Point D, we have.

$I_1 < I_2$ (Generator-II shares larger load current than Generator-I).

If it neither delivers nor receives power from the bus bar, though it remains connected to the bus bar and receives mechanical power from turbine. The machine is said to 'float' on the bus-bar or power system. Here, terminal voltage is equal to its open circuit EMF.

Fig. 4.30 Parallel-connected generators having identical open circuit EMF and different speed regulations

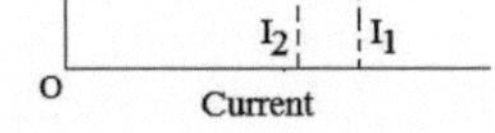

Fig. 4.31 Parallel-connected generators having different open circuit EMF and different speed regulations

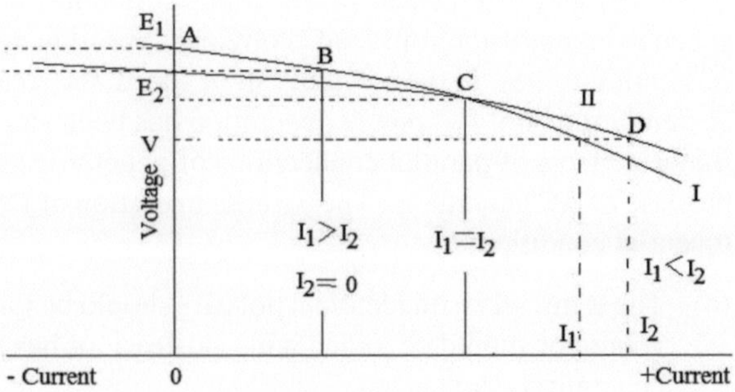

A circulating current flows if the operating point lies between A and B. The above example explains the behaviors of generators in parallel connection. The AC alternators operate in a similar way. In the event of grid failure, one of the alternators connected to the grid starts working as a motor, thereby absorbing all the power generated by other parallel-connected alternators. All the AC alternators connected to the grid may lose synchronism and are subsequently removed from the grid. In the event of sudden load withdrawal due to fault in transmission lines in a large interconnected power grid, the interconnected generators may start over speeding, if not checked on real time, it results in the failure.

Consider a case, when the EMFs of two parallel-connected generators are unequal, but they have identical speed regulation of the prime mover. The generator having higher no-load EMF will share larger load than the other.

The behaviors of machine for different cases as mentioned below may be similarly explained.

(a) Machines have different no-load EMF as well as speed regulation
(b) Machines have same no-load EMF as well as speed regulation. (Both the machines are identical.)

When a generator is loaded, a marginal fall in the speed occurs, causing the no-load EMF to drop. The generators have a drooping characteristic. It helps in the stability of the power system. Consider a system in which generator's speed (and EMF) increases on loading. Such an uncontrolled system is likely to be unstable.

4.7.1 Parallel Operation of Shunt, Compound and Series Generators

From the discussions of parallel connection of DC generators, it is clear that the load sharing of generator depends on following factors.

(a) Open circuit EMF
(b) Parameters of armature and main field circuit
(c) Compensation of field flux-current in the series compensating field windings of compound generators
(d) Mechanical power input to the prime mover
(e) Governor speed regulation.

The open circuit EMF is controlled by controlling the shunt or main field current. In parallel-connected generators, the sharing of load current of individual generator has been explained earlier. It is required that all the parallel-connected generators must share equal load current, there is no circulating current in the parallel-connected generators, and the generator should not work as motor or 'float' on the parallel-connected system.

Cumulatively compound generators have inherent capability to regulate the output voltage. As such when generator is loaded the field weakening is compensated by

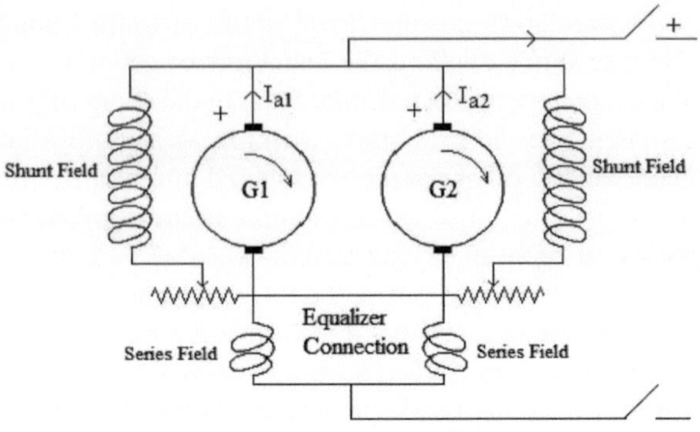

Fig. 4.32 Compound generators in parallel-equalizer connection. (both the generators are driven by separate prime movers)

series field which is carrying the load current. In parallel-connected generators, there is a great possibility that the parameters such as armature resistance, field resistance, armature inductance, field inductance and brush contact drop of the two generators may not be identical even when the two generators are from the same manufacturer. These generators are being driven by separate prime movers. The governor of the individual prime mover also has different drooping characteristics. As such there is a possibility that one generator may share a lesser or higher load than the other. The input power to the prime mover may also differ. The generator sharing comparatively lesser load current will have a lesser compensation of field flux. This causes further reduction in the developed voltage, thereby causing further imbalance in the sharing of the total load current. Similar explanations can be given for generator sharing larger load current. However, by connecting equalizer connection as shown in Fig. 4.32, both the series compensating fields of parallel-connected generators are having equal current. This ensures that both the generators to have equal compensation of field flux. It helps in equalization of the shared load currents of parallel-connected compound generators.

The series generators were useful for voltage boosting and bucking. They are connected in series with the supply and load circuits. If the supply voltage is other than nominal voltage the regulation of DC voltage requires that a certain amount of voltage should be added or subtracted in series with the supply. As mentioned earlier, these series generators were in use for boosting or bucking the DC supply voltage. The parallel-connected DC series generators are used when the load current is much higher than the individual rated current of DC series machine. A DC series generator exhibits unstable terminal characteristics when overloaded and magnetic saturation is present.

The parallel-connected series generators are driven by separate prime movers having different governor characteristics. The series windings are having a low resistance and low inductance. the two generators like armature resistance, field resistance, armature inductance, field inductance and brush contact drop etc. may not be identical even when the two generators are from the same manufacturer. The series windings are having a low resistance and low inductance.

If one series generator is sharing lesser load current (due to any reason) than the other generator, ideally, it will cause further imbalance in the shared load current. There is a possibility of circulating current in the parallel-connected series generators. The generator having lesser field current will develop lesser induced EMF than the other, which causes circulation currents in the local circuit. This also causes further imbalance in the shared load current. By connecting the equalizer connection, as shown in Fig. 4.33a, both the series field winding carry equal load current as such the effect of unbalanced load sharing is mitigated.

If we use cross connection as shown in Fig. 4.33b, the generator sharing larger load current strengthen the field flux of other generator and generator sharing lesser load current weakens the field flux of other generator. This also helps in equalization of shared load current.

The DC power supply is uncommon in the present days. We know that the computer industry demands regulated DC power supply of different level of regulated voltages. A DC grid may be thought for industry where thousands of computers are in use in closed premises. The necessities of DC link voltage in renewable energy sector and on board requirements of battery voltage in electric vehicle (EV) and electric hybrid vehicle (EHV) for boosting and bucking may find an alternative back up method using DC series generators. Solar PV and Fuel Cell, both give DC power output which can be connected in parallel for forming local DC grid.

In traction applications, series motors were frequently arranged to operate as generator during braking. Though, currently these traction motors have been replaced by induction motors operating on VFD (Variable Frequency Drive), their torque-speed characteristics are shaped as that of DC series motor for traction purpose. Self-exited series generators were used in tram car for braking purposes. For satisfactory operation, equalizer connections are used in parallel-connected generators.

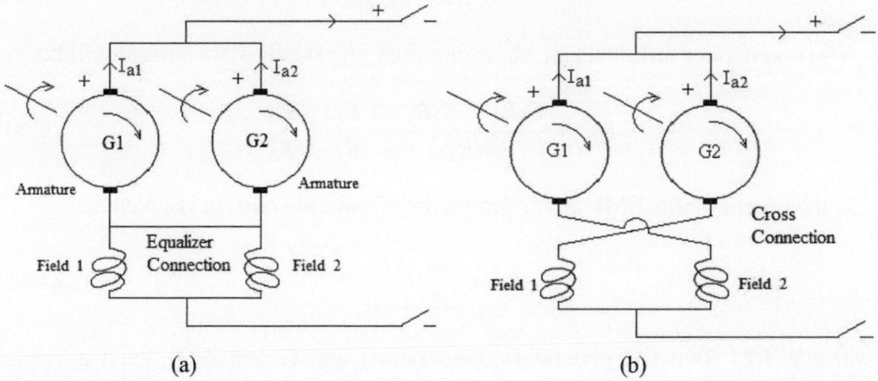

Fig. 4.33 Parallel-connected DC series generator **a** with equalizer connection and **b** Cross connection

4.7.2 Sharing of Load Current

Let us consider two parallel-connected DC generators to a common load resistance R_L. The open circuits EMF of these generators are E_1 and E_2 and their internal resistances are, respectively, R_1 and R_2. We can write

$$\begin{aligned} E_1 &= I_1R_1 + (I_1 + I_2)R_L \\ E_2 &= I_2R_2 + (I_1 + I_2)R_L \end{aligned} \tag{4.36}$$

Solving we get

$$\begin{aligned} I_1 &= \frac{E_1R_2 + (E_1 - E_2)R_L}{R_1R_2 + (R_1 + R_2)R_L} \\ I_2 &= \frac{E_2R_1 - (E_1 - E_2)R_L}{R_1R_2 + (R_1 + R_2)R_L} \\ I = I_1 + I_2 &= \frac{E_1R_2 + E_2R_1}{R_1R_2 + (R_1 + R_2)R_L} \end{aligned} \tag{4.37}$$

The terminal voltage is

$$V = IR_L = \frac{E_1R_2 + E_2R_1}{\frac{R_1R_2}{R_L} + (R_1 + R_2)} \tag{4.38}$$

When load is not connected, the terminal voltage is

$$V_0 = \frac{E_1R_2 + E_2R_1}{(R_1 + R_2)} \tag{4.39}$$

When generator terminals are short circuited, the steady state current will be

$$I_{SC} = \frac{E_1R_2 + E_2R_1}{R_1R_2} = \frac{E_1}{R_1} + \frac{E_2}{R_2} \tag{4.40}$$

If the open-circuits EMF are different, the circulating current is given as

$$I_{cir} = \frac{E_1 - E_2}{R_1 + R_2} \tag{4.41}$$

Example 4.13 Two DC generators, Generator-I and Generator-II, each having a capacity of 500 kVA are connected in parallel. How will they share a load of 800 kVA when

(a) They have identical open circuit EMF of 500 V but their voltage falls to 2% and 3% when fully loaded;

(b) Their open circuit EMF are 500 V and 505 Volt, but their governors have identical speed regulation of 2%.

Solution

Case–(a): Open circuit EMF = 500 Volt.

Full load current = 1000 A.

The voltage in Generator-I falls to 2% when fully loaded = 500 − 0.02 × 500 = 490 V.

The voltage Generator-II falls to 3% when fully loaded = 500 − 0.03 × 500 = 485 V.

The actual load is VI = 800 kVA.

Their characteristics can be assumed linear,

For Generator –I:

$$V = 500 + \frac{500 - 490}{0 - 1000} I_1$$
$$V = 500 - 0.01 I_1$$

For Generator –II:

$$V = 500 + \frac{500 - 485}{0 - 1000} I_2$$
$$V = 500 - 0.015 I_2$$

When sharing a load of 800kVA at voltage V, the load current is

$$I = I_1 + I_2 = \frac{800 \times 1000}{V}$$

From above equations we have, $I_1 = 1.5\ I_2$.

Thus,

$$2.5 I_2 = \frac{800 \times 1000}{V}$$

Again we get

$$V = 500 - 0.015 I_2$$

$$V = 500 - 0.015 \frac{800 \times 1000}{2.5V}$$

Solving we get

$$25V^2 - 12,500V + 120,000 = 0$$

Thus, we get

$$V = \frac{12,500 \pm \sqrt{12,500 \times 12500 - 4 \times 25 \times 120,000}}{50}$$

Thus, the load voltage is 490.2 Volt, discarding lower voltage.
The currents are $I_1 = 978.6$ A and $I_2 = 652.39$ A.

Case–(b):

Open circuits EMFs are 500 and 505 V.

Let full load current be 1000 A.

The voltage in Generator-I falls to 2% when fully loaded = 500 − 0.02 × 500 = 490 V.

The voltage in Generator-II falls to 2% when fully loaded = 505 − 0.02 × 505 = 494.9 V.

Their characteristics can be assumed linear,

For Generator-I:

$$V = 500 + \frac{500 - 490}{0 - 1000} I_1$$
$$V = 500 - 0.01 I_1$$

For Generator-II:

$$V = 505 + \frac{505 - 494.9}{0 - 1000} I_2$$
$$V = 505 - 0.0101 I_2$$

When sharing a load of 800 kVA at voltage of V, the load current is

$$I = I_1 + I_2 = \frac{800 \times 1000}{V}$$

From above equations we have

$$V = 500 - 0.01 I_1$$

$$I_1 = \frac{500}{0.01} - \frac{V}{0.01} = 50,000 - 100\,\text{V}$$

$$V = 505 - 0.0101 I_2$$

$$I_2 = \frac{505}{0.0101} - \frac{V}{0.0101} = 50,000 - 99.0099\,\text{V}$$

Putting values of currents in above equation yields

$$V = \frac{100,000 \pm \sqrt{100,000 \times 100,000 - 4 \times 199.0099 \times 800,000}}{2 \times 199.0099}$$

Solving we get $V = 494.35V$ discard lower values.

Thus total current $I = 1618$ A and the individual generator currents are $I_1 = 565$ A and $I_2 = 1055$ A.

4.8 Characteristics of DC Motors

DC motor classifications are similar to that of generator. The shunt motor is normally referred to as *constant torque motor*, and series motor is referred to as *constant power motor*. In electrical machines terminology, their characteristics are sometimes called shunt *characteristics* and *series characteristics*, respectively, for other motors [1–4]. We know that back EMF is given as

$$E = k_e \phi \omega$$

where EMF constant k_e is

$$k_e = \frac{pZ_a}{2\pi a}$$

The torque can also be expressed as

$$T = k_t \phi I_a$$

The torque constant k_t is

$$k_t = \frac{pZ_a}{2\pi a}$$

Using these relationships, the torque-speed characteristics of different types of motors can be derived.

4.8.1 Shunt Motor

In case of shunt motor, the back EMF may be written as,

$$E = kI_f \omega = V - I_a(R_a + R_x) \tag{4.42}$$

It is assumed that the field flux is proportional to the field current, and the magnetic circuit is not saturated. R_x is the extra resistance in series with the armature circuit. The torque is given as,

$$T = kI_f I_a \tag{4.43}$$

Eliminating armature current I_a from the above two expressions, we get

$$\omega = \frac{V}{kI_f} - T\frac{(R_a + R_x)}{(kI_f)^2} \tag{4.44}$$

Equation (4.44) represents a drooping characteristic. When motor is loaded, its speed drops. Figure 4.34a shows the variation of speed and torque with respect to armature current. The torque-speed characteristic is shown in Fig. 4.34b. Dotted lines show the effects of demagnetization. The characteristics of shunt motor for different armature resistances are shown in Fig. 4.35. Shunt motors are used in applications, where load torque is constant with respect to speed. A slight variation in speed is observed on loading.

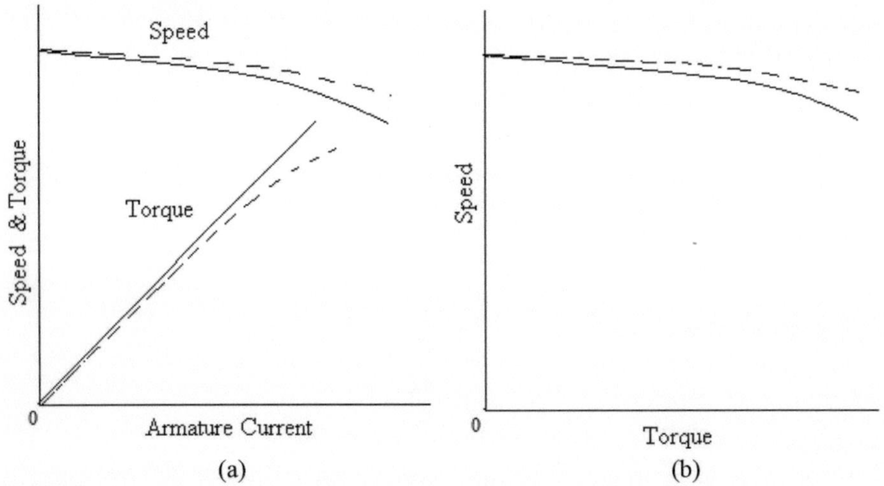

Fig. 4.34 Characteristics of shunt and separately excited DC motors **a** variation of speed and torque with respect to armature current **b** speed torque characteristics (*dotted lines show the effect of demagnetization*)

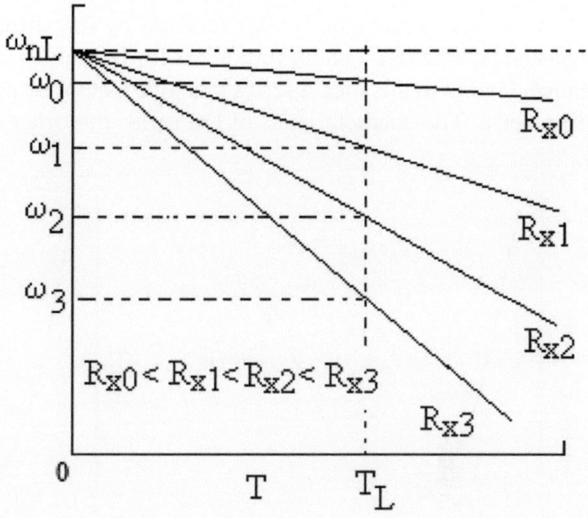

Fig. 4.35 Torque-speed characteristics of shunt motor with different armature resistances

4.8.2 Series Motor

In the case of DC series motor, the field current is same as the armature current. Thus, back EMF is

$$E = kI_a\omega = V - I_a\left(R_a + R_f + R_x\right) \tag{4.45}$$

Rearranging Eq. (4.45) we get

$$V = I_a\left\{k\omega + (R_a + R_f + R_x)\right\} \tag{4.46}$$

where R_f is the field resistance and R_x is the extra resistance in series with the circuit. The torque is given as

$$T = kI_a^2 \tag{4.47}$$

where k is a constant. Eliminating the armature current I_a, we get

$$T = k\frac{V^2}{\left\{k\omega + (R_a + R_f + R_x)\right\}^2} \tag{4.48}$$

Assuming $(R_a + R_f + R_x)$ to be negligible, we get

$$T \;\; \alpha \;\; \frac{V^2}{\omega^2} \tag{4.49}$$

A DC series machine is characterized by low-armature resistance and low-field resistance, whereas a shunt motor has low armature resistance and high field resistance. Owing to this fact, a series machine cannot be made to run as shunt motor and vice versa. The characteristics of DC series motor are shown in Fig. 4.36.

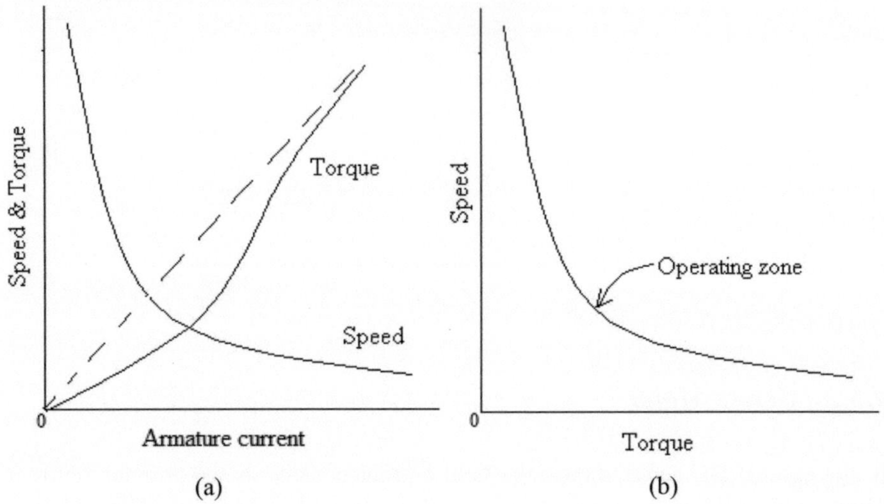

Fig. 4.36 Characteristics of DC series motors **a** variation of speed and torque with respect to armature current **b** speed-torque characteristics

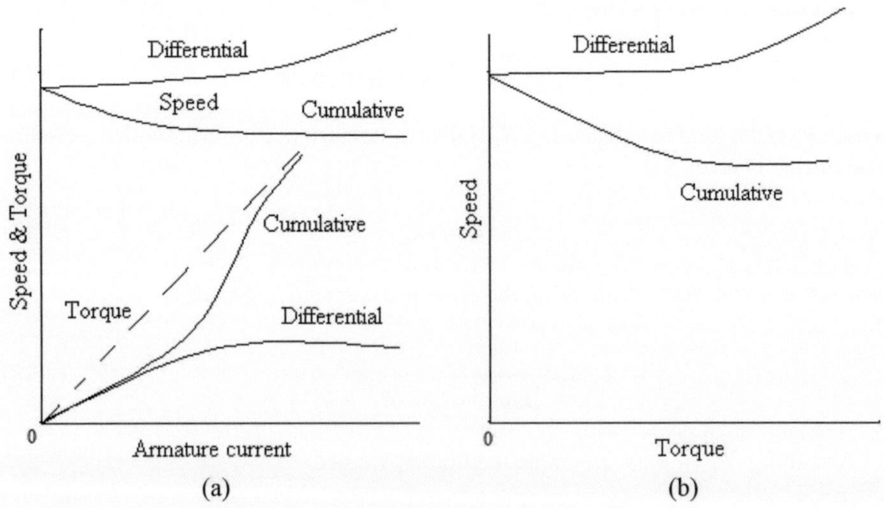

Fig. 4.37 Characteristics of DC compound motors-cumulative and differential. **a** Variation of speed and torque with respect to armature current. **b** Speed-torque characteristics

The speed and torque of DC series motors versus armature current is shown in Fig. 4.36a. At higher armature current the flux is almost constant thus torque is proportional to current. The Figure shows an asymptotic behavior.

The speed-torque characteristics of a DC series motor are shown in Fig. 4.36b, which depicts an inverse relationship. In the operating zone, the product of speed and torque is almost constant. As such, the DC motor is referred to as constant power motor. These motors are used in applications where the starting torque is very large and running torque is low, such as traction locomotives, crane application, EV and EHV.

In the event of free running or no-load, the machine attains very high-speed resulting in damage of mechanical coupling. To avoid this, the DC series motors should always be started and run on load. The series motors, which can be operated on AC / DC, are known as universal motor.

4.8.3 Compound Motor

In cumulatively compound motor/generator, the series field is in addition to the main shunt field. The series field carries larger current as that of shunt field and is designed to carry full load current. The number of turns is much less than that of shunt field. According to the type of connection, i.e., long shunt or short shunt, this field is connected in series either with the armature circuit or with the supply. Current flowing in the series field causes the flux in the same direction as that of shunt field. Such motors when loaded, makes the weak series flux to aid the main flux. As a result, there is a tendency to slow down the speed of the motor. The speed is inversely proportion to the net field flux.

In differential compound motors, the series field causes the flux in the opposition to the main shunt field flux. Current flowing in the series field causes the flux in the opposite direction to that of shunt field flux. Such motors, when loaded, make the weak series flux to oppose the main flux. As a result, there is tendency to increase the speed of the motor. When motor is heavily loaded, the opposing series flux will be more pronounced (torque being proportional to the armature current) so as to weaken the net flux. The machine will tend to race when overloaded. If machine is operated with overloaded condition, there are chances of starting the motor in the opposite direction. Due to these reasons, differential compound motors are rarely used. The characteristics of Compound motors are shown in Fig. 4.37a, b.

4.9 Starting of DC Motors

The starting current in a DC motor is given as

$$I_a = \frac{V - E}{R_a + R_x} \tag{4.50}$$

We know the back EMF is

$$E = k\phi n$$

where 'k' is the EMF constant, 'ϕ' is the flux per pole and 'n' is the rotation per seconds.

When machine is at rest, rotational speed 'n' is zero. Back EMF is also zero. Since, there is no opposing EMF, the current drawn by the armature in absence of an external resistance R_x is given as

$$I_a = \frac{V}{R_a} \tag{4.51}$$

In a typical 10 Horse Power (HP), 220 Volt DC shunt motor, the order of armature resistance R_a is (say) about 1.0 Ω. If this motor is started without any starter or resistance connected in series with the armature, it will cause roughly about 220 A armature current from the supply at the instant of starting. This huge current will result in the following:

(a) Blowing of the fuse
(b) Damage of rectifiers supplying DC power
(c) Damage of the commutator and brushes
(d) A hot-spot may be created in the winding, which results in the weakening of insulation and, hence, reduction in life of machine.
(e) Winding fault due to over-heating.
(f) Sudden dip in the line voltage.

If a machine is attempted to start frequently without limiting the starting current, the ohmic losses within the machine will be enormous and it may not be properly cooled down. This will ultimately result in the creation of hot spots within the winding due to over-heating. The life of insulation of the windings and the insulation of winding and core are temperature sensitive. A frequent starting in a similar way may cause insulation breakdown or melting of current carrying conductors of the windings. This will result in complete breakdown of the motor.

This large amount of current for a short duration (starting) needs to be supplied by a nearby substation. The conductors connecting the substation and the location of the motor have a finite resistance. It causes a finite voltage drop for shorter duration (starting) of time. If we consider a resistance of conductor as 0.1 Ohm of the transmission line, a voltage drop of 22 V is expected. This voltage drop slowly vanishes, once the machine is fully started. Consider a case of an ideal process industry, like printing paper or textile making where several motors are used. When one of the large motor is suddenly started using Direct-On-Line (DOL), the drop in the voltage will be experienced by each one of the motors connected to the same line, the speed of all the motors will be momentarily dipped. As soon as the voltage regains, the speed will also regain its original value. If proper speed control techniques are not arranged for controlling the speed of all the other motors, the quality of production during this time duration will be affected.

The Direct-On-Line (DOL) method of starting is normally preferred for smaller capacity motor below 3 HP. A DOL starter has an inbuilt bimetallic relay, which trips the supply in the event of excessively larger starting current.

4.9.1 Grading of Starting Resistances

The starters are designed with a condition that during starting, the current can be limited within a known limit. A higher current will cause a large power loss and will also result in higher starting torque; whereas a very low current will produce insufficient starting torque. Thus, graded resistances are needed while starting of DC motor so as to limit the starting current as well as to produce sufficient starting torque. If we assume that field winding is carrying a current of I_f when armature is connected to the supply, the governing equation is given as,

$$V = i_a(t) \times (r_a + r_1 + r_2 + r_3 + r_4 + r_5 + r_6 + \ldots) + L_a \frac{di_a(t)}{dt} + k_e I_f \omega(t) \tag{4.52}$$

where V is a step voltage. If we assume that the machine is connected to load, the torque equation is

$$T - T_L = J \frac{\mathrm{d}\omega(t)}{\mathrm{d}t} + B\omega(t) \tag{4.53}$$

Here T_L = Load torque (Nm)

$T(t)$ = Transient electromagnetic torque = $k_t\ I_f\ i_a(t)$.

$i_a(t)$ = Armature current as a function of time, (A).

$\omega(t)$ = Rotational speed during starting (radians per second).

L_a = Inductance of armature circuit (Henry).

r_a = Resistance of armature (Ω).

r_1, r_2, r_3, r_4, r_5 etc. = Graded resistances in armature circuit which are removed from the circuit after a time delay of Δt, $2\Delta t$, $3\Delta t$, $4\Delta t$, $5\Delta t$ etc., respectively.

J = Moment of inertia of rotating parts (MKS unit).

B = Damping coefficient (MKS unit).

The above set of differential equations can be easily solved using either Laplace's transform or MATLAB SIMULINK software. Interested readers may develop their own SIMULINK model. A typical variation of starting current and speed of the DC motor with graded resistances is shown in Fig. 4.38. The graded resistance of starter with DC shunt motor are shown in Fig. 4.39.

It is assumed that during starting, the upper current limit is I_1 and lower current limit is I_2. At the time of starting, all the resistances are in series. We define total

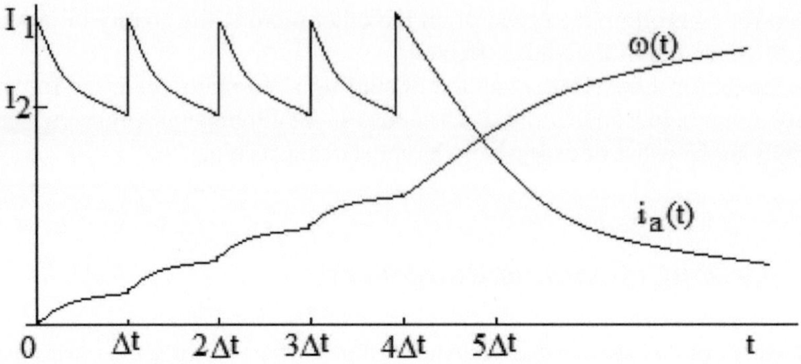

Fig. 4.38 Variation of starting current and speed of a DC motor

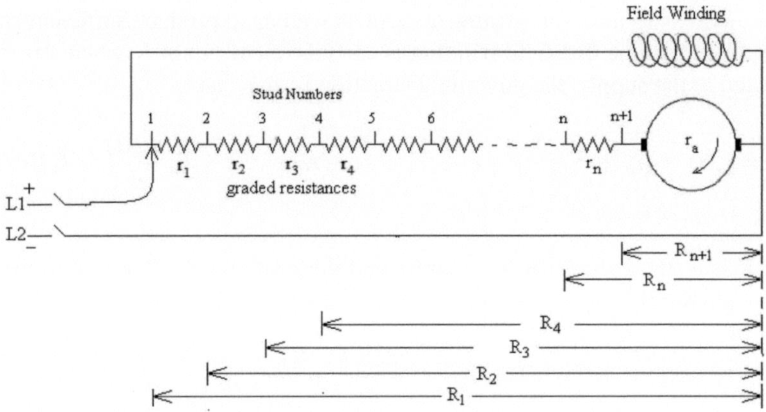

Fig. 4.39 Graded resistances in DC motor starter

resistances in armature circuit as

$$\begin{aligned} R_1 &= r_1 + r_2 + r_3 + r_4 + r_5 + \ldots\ldots. + r_n + r_a \\ R_2 &= r_2 + r_3 + r_4 + r_5 + \ldots\ldots. + r_n + r_a \\ R_3 &= r_3 + r_4 + r_5 + \ldots\ldots. + r_n + r_a \\ &\ldots\ldots\ldots\ldots\ldots \\ &\ldots\ldots\ldots\ldots\ldots \\ R_{n-1} &= r_a \end{aligned} \tag{4.54}$$

At the instant of starting, total resistance is

$$R_1 = \frac{\text{Terminal Voltage}}{\text{Maximum allowed armature current during starting}} = \frac{V}{I_1} \tag{4.55}$$

If the total resistance is too large, the starting current will be too low to produce sufficient torque for starting the motor. The purpose of starter is to limit the starting current as well as give sufficient electromagnetic torque to overcome the resisting torque so that motor can start.

After a known time delay of Δt seconds, the resistance r_1 is removed from the armature circuit. It is assumed that at that instant of time, current falls to I_2, the lowest permissible current during starting. Since the motor speed has increased, there is an appreciable EMF E_1 in the motor armature. Now, the total resistance is given as

$$R_2 = r_2 + r_3 + r_4 + r_5 + \cdots + r_n + r_a \tag{4.56}$$

The EMF at this instant of time is,

$$E_1 = V - I_2 R_1 \tag{4.57}$$

$$R_1 = \frac{V - E_1}{I_2} \tag{4.58}$$

As the resistance r_1 is removed from the circuit, the current again rises to I_1, thereby satisfying the equation

$$E_1 = V - I_1 R_2 \tag{4.59}$$

$$R_2 = \frac{V - E_1}{I_1} \tag{4.60}$$

Thus, we have

$$\frac{R_2}{R_1} = \frac{I_2}{I_1} \tag{4.61}$$

Similarly, when after time delay of $2\Delta t$ seconds, the resistance r_2 is also removed from armature circuit, the instantaneous current falls to I_2. The EMF E_2 in the motor armature corresponds to the instantaneous speed.

$$E_2 = V - I_2 R_2 \tag{4.62}$$

$$R_2 = \frac{V - E_2}{I_2} \tag{4.63}$$

The total resistance is given as

$$R_3 = r_3 + r_4 + r_5 + \cdots + r_n + r_a \tag{4.64}$$

As the resistance r_2 is removed from the circuit the current rises again to I_1, thereby satisfying the equation,

$$E_2 = V - I_1 R_3 \tag{4.65}$$

$$R_3 = \frac{V - E_2}{I_1} \tag{4.66}$$

Thus, we get

$$\frac{R_3}{R_2} = \frac{I_2}{I_1} \tag{4.67}$$

We can show that

$$\frac{R_2}{R_1} = \frac{R_3}{R_2} = \frac{R_4}{R_3} = \frac{R_5}{R_4} = \cdots = \frac{R_n}{R_{n-1}} = \frac{r_a}{R_n} = \frac{I_2}{I_1} = \gamma \tag{4.68}$$

Thus,

$$\frac{R_2}{R_1} \times \frac{R_3}{R_2} \times \frac{R_4}{R_3} \times \frac{R_5}{R_4} \times \cdots \times \frac{R_n}{R_{n-1}} \times \frac{r_a}{R_n} = \left(\frac{I_2}{I_1}\right)^n = \gamma^n \tag{4.69}$$

$$\text{Or} \quad \gamma^n = \frac{r_a}{R_1} \quad \text{or} \quad \gamma = \left(\frac{r_a}{R_1}\right)^{1/n} = \frac{I_2}{I_1} \tag{4.70}$$

From the above expressions, we get

$$R_2 = \gamma R_1$$
$$r_1 = R_1 - R_2 = (1 - \gamma) R_1$$

Similarly, we have

$$R_3 = \gamma R_2$$
$$r_2 = R_2 - R_3 = \gamma R_1 - \gamma R_2 = \gamma R_1 - \gamma^2 R_1 = \gamma R_1 (1 - \gamma) = \gamma r_1$$
$$r_3 = \gamma^2 r_1$$
$$r_4 = \gamma^3 r_1$$
$$\ldots\ldots\ldots\ldots$$
$$r_n = \gamma^{n-1} r_1 \tag{4.71}$$

The graded resistances are essential in starters. In the manual starters (as will be discussed in the subsequent section), if starter handle is moved faster, there is a possibility that the starting current rises beyond upper current limit, resulting in blowing of fuses. If the resistances are not properly graded, then also the starting current are likely to deviate from the current limit of 'I_1 & I_2'. These can be explained by solving differential Eqs. (4.52) and (4.53).

The above expressions are useful in designing starter of DC motors. The starter of DC series motor is a two-point starter normally referred as drum controller. A drum controller is useful in starting, speed control, speed reversal, and braking of DC series motor.

Example 4.14 Calculate the graded resistance of a 20 HP, 220 V shunt motor with five step starter. The full load efficiency is 90%. The lowest current ratings is 1.8 times the full load current. Total copper loss is 5% of input and field resistance is 220 Ω.

Solution: Given Output of motor $= 20 \times 746$ W

Input to the motor $= \frac{20 \times 746}{0.9} = 16.578$ KW

Full load current $= \frac{16,578}{220} = 75.35$ A

Copper loss $= 16,578 \times 0.05 = 829$ W

Ohmic loss in field $= \frac{220 \times 220}{220} = 220$ W

Armature copper loss $= 829 - 220 = 609$ W

Armature resistance $= \frac{609}{75.35 \times 75.35} = 0.1073\Omega$

Lower current, $I_2 = 1.8 \times 75.35 = 135.63$ A

Number of steps = 5.

Number of resistances, $n = 5 \ 1 = 4$.

Now, we have

$$\gamma = \left(\frac{I_2}{I_1}\right) = \left(\frac{R_a}{R_1}\right)^{1/n} = \left(\frac{R_a I_1}{R_1 I_1}\right)^{1/n}$$

$$\gamma \gamma^{1/n} = \left(\frac{R_a I_2}{V\gamma}\right)^{1/n} \gamma^{1/n}$$

$$\gamma^{\frac{n+1}{n}} = \left(\frac{R_a I_2}{V}\right)^{1/n}$$

$$\gamma = \left(\frac{R_a I_2}{V}\right)^{\frac{1}{n+1}}$$

$$\gamma = \left(\frac{0.1073 \times 135.63}{220}\right)^{\frac{1}{5}} = 0.5809$$

Initial current,

$$I_1 = \frac{I_2}{\gamma} = \frac{135.63}{0.5809} = 233.48 \text{ A}$$

Initial resistance,

$$R_1 = \frac{V}{I_1} = \frac{220}{233.48} = 0.9423\Omega$$
$$R_2 = \gamma R_1 = 0.5809 \times 0.9423 = 0.5474\,\Omega$$
$$r_1 = R_1 - R_2 = (1-\gamma)R_1 = 0.9423(1-0.5809) = 0.3949\,\Omega$$

Similarly, we have

$$R_3 = \gamma R_2$$

$$r_2 = \gamma r_1 = 0.5809 \times 0.3949 = 0.2294\,\Omega$$

$$r_3 = \gamma^2 r_1 = 0.5809^2 \times 0.3949 = 0.1333\,\Omega$$

$$r_4 = \gamma^3 r_1 = 0.5809^3 \times 0.3949 = 0.0774\,\Omega$$

Thus, the graded resistances are 0.3949 Ω, 0.2294 Ω, 0.1333 Ω and 0.0774 Ω.

Example 4.15 In a five-step automatic starter for a 500 V DC shunt motor, the resistances are cut from the circuit on the basis of current limit. If the initial starting current is 30A and peak current on each of the subsequent current peak is 50A, find the lower current limit and the resistance of each section. The armature circuit resistance is 1.0Ω.

Solution:

Given Initial current = 30 A.
Thus initial resistance $R_1 = 500/30 = 16.67\ \Omega$.
The current ratio,

$$\gamma = \left(\frac{r_a}{R_1}\right)^{1/4} = \left(\frac{1}{16.67}\right)^{1/4} = 0.4949$$

Upper current limit is $I_1 = 50$ A.
Lower current limit is.

$$\mathrm{I}_2 = \gamma \mathrm{I}_1 = 0.4949 \times 50 = 24.75\mathrm{A}$$

The graded resistances are

$$r_1 = R_1(1-\gamma) = 16.67\,(1-0.4949) = 8.24\,\Omega$$
$$r_2 = r_1\gamma = 0.4949 \times 8.24 = 4.1670\,\Omega$$
$$r_3 = r_2\gamma = 0.4949 \times 4.1670 = 2.0622\,\Omega$$
$$r_4 = r_3\gamma = 0.4949 \times 2.0622 = 1.0206\,\Omega$$

Thus, the graded resistances are 8.24 Ω, 4.1670 Ω, 2.0622 Ω and 1.0206 Ω.

Example 4.16 A 220 V, 10 HP DC series motor operates on the lower current and higher current limits of 40A and 60A, respectively. Total number of steps is five. The efficiency is 90%. The motor flux rises by 0.5% per ampere as the current rises from 40 to 60A. The motor terminal resistance is 0.8Ω.

Solution

Given: The total increase in the flux when the current rises from 40 to 60A = 0.5 × 20 = 10%

Let the induced EMF be E_1 and E_2, respectively, at 40 A and 60 A.

Considering the EMFs to be unchanged, the instant at which the current is changed from 40 to 60A in the last contact resistance, we get

$$E_1 = 500 - 40 \times 0.8$$
$$E_2 = 500 - 60 \times (0.8 + r_4) = 500 - 60 \times R_4$$

where r_4 is the fourth step resistance, and $R_4 = 0.8 + \mathrm{r}_4$.

$$E_1 = 1.1\,E_2$$
$$[500 - 40 \times 0.8] = 1.1 \times [500 - 60 \times R_4]$$
$$500 - 32 = 550 - 66 \times R_4$$

Thus, we get

$$R_4 = \frac{550 - 500 + 32}{66} = 1.24\,\Omega$$

Thus, fourth step resistance is

$$r_4 = 1.24 - 0.8 = 0.44\,\Omega$$

Similarly, we have

$$= 1.1 \times [500 - 60 \times R_3]$$
$$500 - 49.6 = 550 - 66 \times R_3$$
$$R_3 = \frac{550 - 500 + 49.6}{66} = 1.509\,\Omega$$

Thus, third step resistance,

$$r_3 = 1.509 - 1.24 = 0.269\,\Omega$$

Similarly, we have

$$= 1.1 \times [500 - 60 \times R_2]$$
$$500 - 60.36 = 550 - 66 \times R_2$$

Thus,

$$R_2 = \frac{550 - 500 + 60.36}{66} = 1.672\,\Omega$$

The second step resistance,

$$r_2 = 1.672 - 1.509 = 0.1631\Omega$$

Similarly, we get

$$= 1.1 \times [500 - 60] \times R_1$$
$$500 - 66.88 = 550 - 66 \times R_1$$

Thus,

$$R_1 = \frac{550 - 500 + 66.88}{66} = 1.771\,\Omega$$

Thus, first step resistance, $r_1 = 1.771 - 1.672 = 0.099\,\Omega$.

Therefore, the resistance steps in series motor starter are 0.099 Ω, 0.1631 Ω, 0.269 Ω and 0.44 Ω.

4.9.2 DC Shunt Motor Starters

A DC motor can be started using different methods:

(a) Manual starters
(b) Automatic starter
(c) Solid-state starting

The following are the manual starters for DC shunt motor:

(a) Three-point starter
(b) Four-point starter

(a) **Three Point Starter**: A typical three-point starter is shown in Fig. 4.40. The common terminals of armature and field winding are connected to 'L2' terminal of DC supply. The graded resistances are connected behind the studs. Initially, all the resistances are in series with the armature. These resistances are cut-off from the armature circuit and effectively connected in series with the field circuit. Since the graded value of resistances are too low, it does not cause any appreciable decrease in the field current. A mechanical spring is attached to the starter's handle, which keeps it in OFF position and provides sufficient restrain. The starter handle has an iron piece attached to it, which is attracted by Hold-On-Magnet also known as No Volt Release. As a result, the starter handle remains connected in ON position as long as power supply is available and machine is operating in healthy condition. The hold-on-coil or no Volt-release is a strong electromagnet connected in series with the field winding. In case of power failure, the hold-on-magnet is demagnetized causing the starter handle to move in OFF position due to spring action. In case of opening of the field winding by chance or by fault or when field winding current become very small, it causes the starter- handle to activate the switch 'OFF'. The terminal L1 is connected to Over-Load- Relay (OLR). In the case of over load conditions, the supply of the hold-on-magnet is short-circuited, thereby causing the magnet to demagnetize, which again initiates the switching OFF.

When a DC shunt motor is running its field should never be opened, otherwise the machine will start racing and speed of machine will be excessively large causing damage to the coupling and other mechanical parts. The field failure protection is automatically provided in three-point starter.

(a) **Four Point Starter**: When machine is used for speed control using field flux control, the field weakening is required. This causes the hold-on-coil of a three-point starter to produce very feeble attraction force, which may not overcome the spring force. As such, faulty disconnection of the supply is initiated,

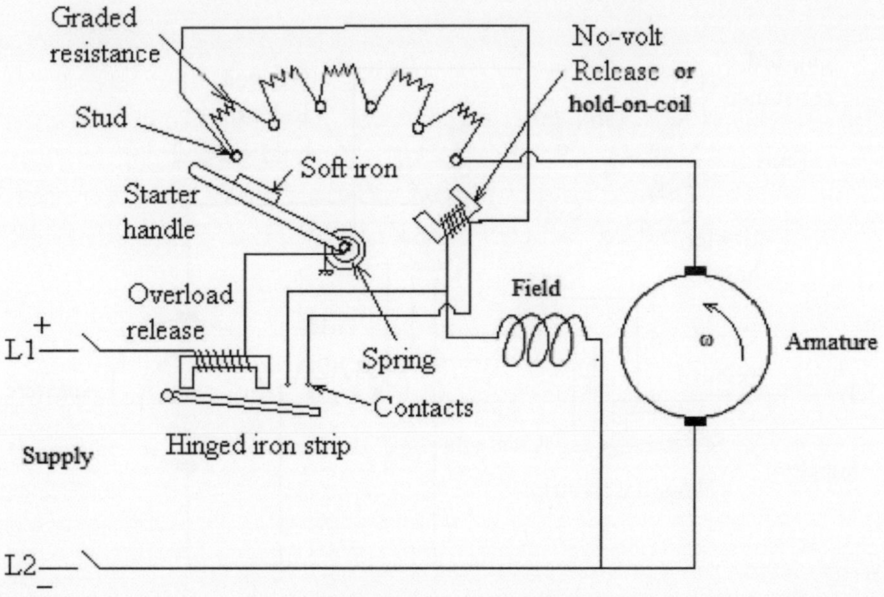

Fig. 4.40 Schematic diagram of 3-point DC shunt motor starter

due to demagnetization of the hold-on electromagnet. This can be avoided by connecting the hold-on-coil in parallel with the supply as shown in Fig. 4.41. The starter will fail to provide the necessary protection in the case of (accidental) opening of the field winding during running condition. Four-point starters are useful when it is desired to increase the speed of DC shunt motor using field flux weakening.

(b) **Automatic starters**: The three-point and four-point starters are conventional starters. The industrial automatic starters are based on industrial contactors and commercially available Over Load Relay (OLR). These programmable contactors effectively short circuit or bypass the graded resistances of the starter of a DC motors. These industrial contactors are based on the following:

(a) Electromagnetic contactor
(b) Electro-pneumatic contactors
(c) Mechanical contactor

Figure 4.42 shows a typical industrial automatic starter in which the starting resistances r_1, r_2, r_3, r_4 etc. are connected in series with the armature circuit. These resistances are taken out from the circuit by short-circuiting them or bypassing them by using industrial contactors. At the time of starting, all the contactors C1, C2, C3, C4 etc. are not connected (NO- normally open) or open, causing all the resistances to appear in series with the armature circuit. After a time delay of δt seconds, the contactor C1 is ON, thereby it shorts the resistance r_1. Similarly, after time delay of $2\delta t$ seconds C2 also shorts r_2 and so on. These contactors can be programmed depending on the requirement. The functioning of automatic starter is based on the following principle :-

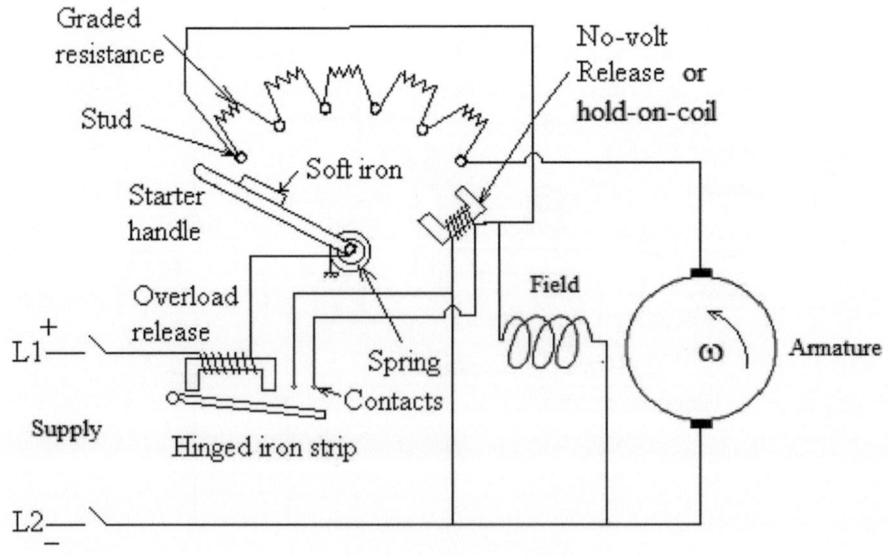

Fig. 4.41 Schematic diagram of 4-point DC shunt motor starter

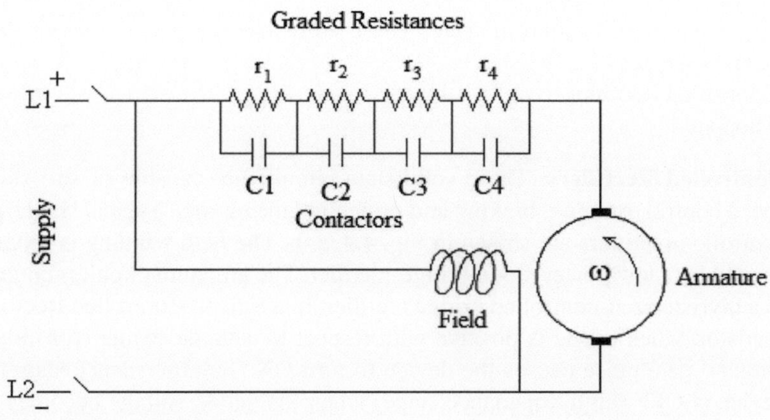

Fig. 4.42 Automatic DC shunt motor starter based on industrial contactors

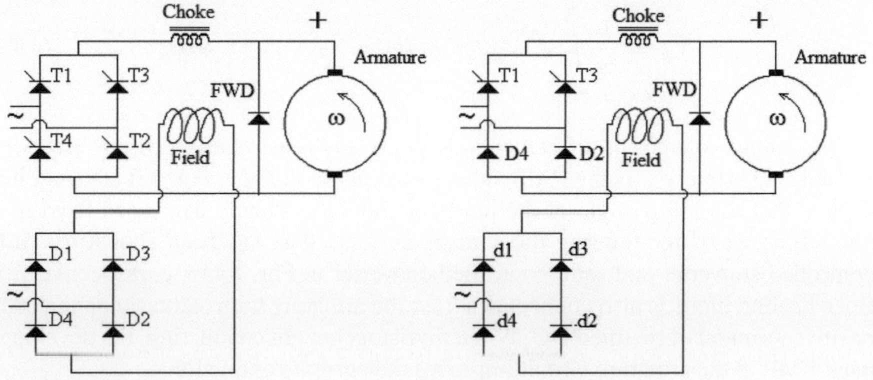

(a) Fully Controlled Converter (b) Semi-Controlled Converter

Fig. 4.43 DC motor using solid-state AC / DC converter (*firing circuit not shown*)

(i) Time delay
(ii) Current limit
(iii) Counter EMF

As the name suggests, in case of time delay starter, the contactors are operated after a predefined time step in open loop manner. These contactors can also be programmed according to the current limit or development of counter EMF. The operation of electromagnetic contactors requires reduced voltage either DC or AC supply for their operation. Large industries have availability of compressed air, which are normally used for the operation of electro-pneumatic contactors.

(a) **Solid-state starting**: Currently, large DC motors working in industries are operating on AC to DC converters. These solid-state converter converts 3-phase or single-phase AC to controlled DC. These solid-state starters-cum-controller can be of two types

(i) Controlled Rectifier
(ii) Chopper

(i) **Controlled Rectifiers**: These solid-state circuits are capable of soft-starting, speed control, running, braking and protecting the motors. Typical single-phase controlled rectifiers are shown in Fig. 4.43a, b. The field winding is separately excited by a low power diode-bridge-rectifier. The armature circuit is connected to a thyristorized-controlled bridge rectifier. In a Silicon Controlled Rectifier or thyristor, when anode is positive with respect to cathode- wider (for inductive circuit) gate pulse causes the device to turn ON (low impedance stage). The device is OFF (high impedance stage) when the anode voltage becomes sufficiently negative with respect to the cathode. When the thyristor is triggered at a firing angle, 'α', the average voltage across the load is given as

$$V_{dc} = \frac{1}{\pi}\int_{\alpha}^{\pi} \sqrt{2}V_{ac} \sin \omega t d\omega t = \frac{2\sqrt{2}}{\pi}V_{ac}(1 + \cos \alpha) \quad (4.72)$$

The output voltages of different single-phase and three-phase rectifiers are given in Table 4.3. Here V_{ac} is the RMS value of the phase voltage. For soft starting, it is desired that the firing angle of the thyristor should be slowly decreased from near 180° to about 0° for limiting the starting current. This has been shown for fully controlled converter and semi-controlled converter in Fig. 4.44b, c, respectively. A freewheeling diode is also connected across the armature with cathode connected to positive terminal of rectified DC. When thyristors are not conducting, the developed back EMF in the armature circuit improves the current conduction.

Since motor circuit is inductive, a train of wider gate pulse is required for successful operation. Gate signals are connected to thyristor/SCR (Silicon Controlled Rectifier) gate and cathode via opto-couplers. Figure 4.45 shows supply voltage and armature voltage when half controlled single-phase bridge converter is triggered at α firing angle ranging from 0 to 180°. The converter operating as controlled rectifier when firing angle α is changed. Free Wheeling Diode (FWD) circulates the inductive current of armature when Thyristors are OFF. Thyristors are line commutated at zero crossing voltage. It is assumed that DC machine is working as a motor driving certain hoist load. In full controlled converter for motoring operation α is changed,

Table 4.3 Output voltages of single-phase and three-phase controlled rectifier

Single-phase controlled converter		Three –phase controlled converter	
Full	Semi	Full	Semi
$\frac{2\sqrt{2}}{\pi}V_{ac}\cos\alpha$	$\frac{2\sqrt{2}}{\pi}V_{ac}(1+\cos\alpha)$	$\frac{3\sqrt{6}}{\pi}V_{ac}\cos\alpha$	$\frac{3\sqrt{6}}{2\times\pi}V_{ac}(1+\cos\alpha)$

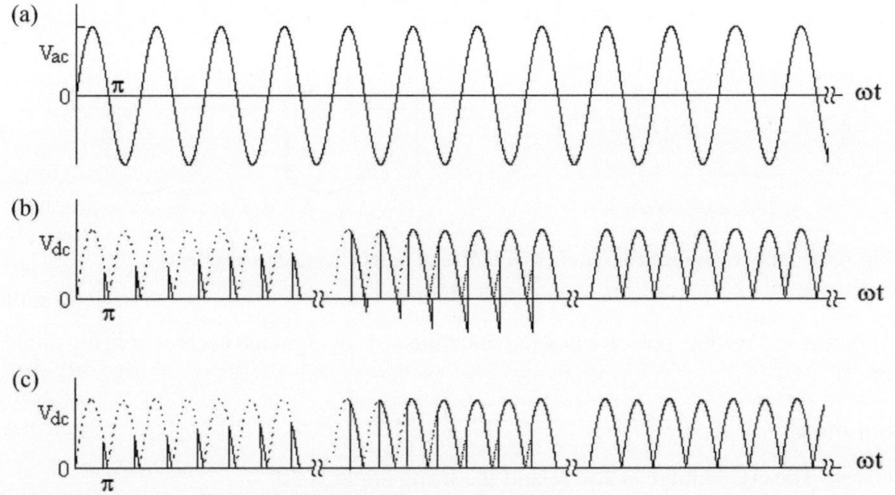

Fig. 4.44 Soft-start of DC motor using solid-state AC DC converter **a** AC voltage. **b** controlled DC voltage (fully controlled rectifier). **c** Half controlled rectifier

the thyristor is ON beyond π and commutated at a commutation angle β. The inductive current continues to flow even after π till angle β, when anode voltage becomes positive, as shown in Fig. 4.46. The DC machine works as a motor. When in DC machinc hoist load drives shaft (heavy mass being lowered due to gravity) such that motor over speeds beyond nominal speed and back EMF is greater than applied voltage, if α is now changed from 90° to 180°, the converter acts as an inverter, a sizable amount of energy is fed back to AC lines from π to angle β as shown in Fig. 4.46, this avoids free falling of heavy mass because of opposing torque. Both continuous and discontinuous operations are possible.

Example 4.17 The speed of a 20 Hp, 220 V 1000 RPM separately excited DC motor is controlled by single-phase fully controlled converter. The machine is operating at an efficiency of 90%. The armature resistance is 0.15 Ω and armature inductance is 20mH. The AC supply voltage is 250 V and the EMF constant is 0.2 V/RPM. Assuming the motor current to be continuous and ripple free. Determine the speed

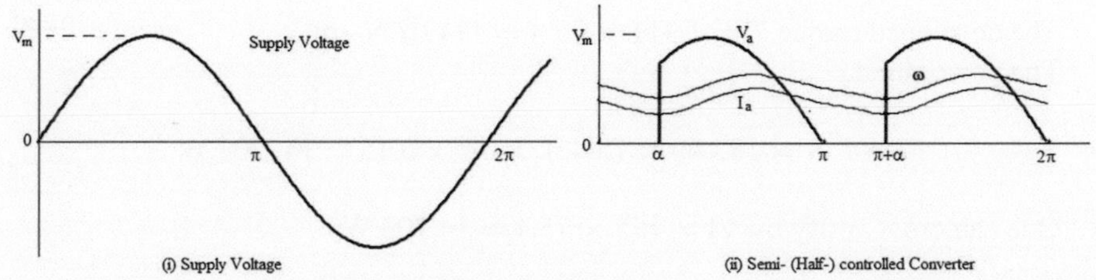

Fig. 4.45 Supply voltage waveform and semi controlled converter—DC motor load

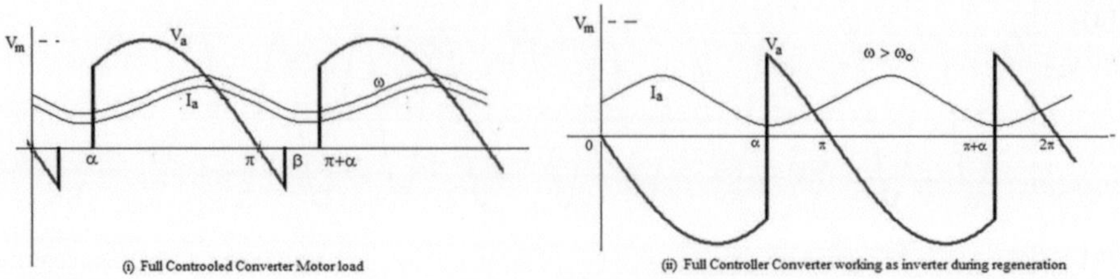

Fig. 4.46 Fully controlled converter DC motor load as rectifier and as inverter

of motor and torque, power when the machine is drawing rated current at firing angle of 30°.

Solution

Given: The AC voltage is 250 V and the firing angle is 30°

The average terminal voltage

$$V_t = \frac{250 \times 2\sqrt{2}}{\pi} \cos 30^\circ = 195 \text{ V}$$

Rated current is

$$I_a = \frac{746 \times 20}{220 \times 0.9} = 75.4 \text{ A}$$

The back EMF is.

$$E = 195 - 75.4 \times 0.15 = 195 - 11.31 = 183.69 \text{ V}$$

Hence the speed can be given as

$$N = \frac{184.69}{0.2} = 918.45 \, \text{RPM}$$

$$\text{The EMF constant } 0.2 \text{ V/RPM} = \frac{0.2 \times 60}{2\pi} = 1.911 \text{ Volt s/radians}$$

The developed torque $T = 1.911 \times 75.4 = 114.07$ N–m.

The power intake

$$EI + I^2R = 183.69 \times 75.4 + 75.4^2 \times 0.15 = 14,703 \text{ W}$$

One can cross verify by VI = 195 × 75.4 = 14,703 W.

Example 4.18 The speed of 100 HP, 500 V, 1500 RPM separately excited DC motor is controlled by three-phase fully controlled converter. The supply is three-phase

440 V, 50 Hz. The machine is operating at rated condition at 89% efficiency. The parameters of the motor are $R_a = 0.22$ Ohm, $L_a = 5.7$ mH and EMF constant is 0.35 Volt/RPM. Determine (a) the firing angle to obtain the rated speed at rated motor current. (b) The no-load speed at firing angle 45° and 55°. Assume that no-load current is about 12% of full rated current.

Solution

Rated current is

$I_a = 746 \times 100/500 \times 0.89 = 167.64$ A

Phase voltage is

$$V_{ph} = \frac{440}{\sqrt{3}} \text{ V}$$

The supply voltage is given by

$$V_t = \frac{3\sqrt{6}V_{ph}}{\pi} \cos\alpha = 594.5 \cos\alpha$$

(a) At rated current, the back EMF is

$$E = 1500 \times 0.35 = 525 \text{ V}$$

The terminal voltage of motor is

$V = 525 + 167.64 \times 0.22 = 488$ V

$$\cos\alpha = \frac{488}{594.50} = 0.82$$

Thus firing angle $\alpha = 34.83°$

(b) The no-load current = 0.12 × 167.64 = 20.11 A

Voltage at 45° firing angle is.

$V_t = 594.5 \cos 45° = 420.6$ V.

EMF is

$E_0 = 420.6 - 20.11 \times 0.22 = 420.6$ V.

No-load speed is

$$N = \frac{420.6}{0.35} = 1189 \text{ RPM}$$

At firing angle 55°, the terminal voltage is = 594.5 cos 55° = 341 V.

EMF is

$E_{30} = 341 - 20.1168 \times 0.22 = 336.6$ V.

The no-load speed is $N = \frac{336.6}{0.35} = 961.6$ RPM.

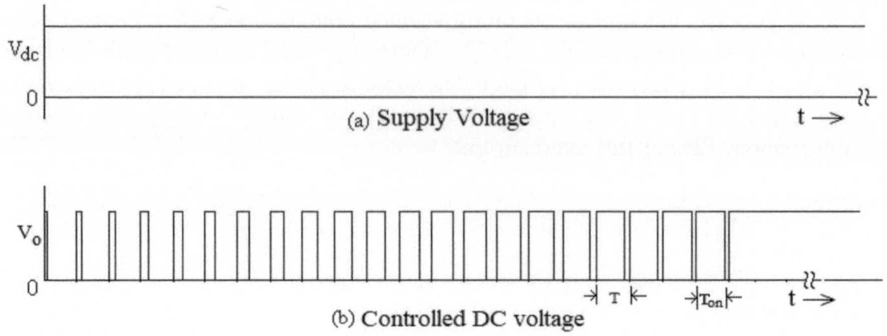

Fig. 4.47 Soft-starting of DC motor using solid-state DC DC converter. **a** Supply voltage. **b** Controlled DC voltage

(i) **(ii) Chopper**: A Chopper is a DC-DC solid-state switch. If the input voltage is V_{DC}, the output voltage is given as

$$V_o = \frac{T_{ON}}{T_{ON} + T_{OFF}} V_{DC} \tag{4.73}$$

The average DC voltage can be smoothly varied from zero voltage to the full voltage by changing the duty cycle as follows:

(a) Changing the ON time T_{ON} and keeping the time period '$T = T_{ON} + T_{OFF}$' constant
(b) Changing the time period and keeping 'T_{ON}' time constant

A typical controlled voltage waveform using chopper is shown in Fig. 4.47, in which the time period is constant and ON time is varied for limiting the starting current of a DC motor. DC–DC chopper circuit is mainly used when the supply source is a battery. Solid-state circuits for starting and speed control of DC motor are having better efficiency as compared to resistance control.

Example 4.19 The speed of a separately excited DC motor is controlled by a chopper circuit. The DC supply voltage is 250 V, armature circuit resistance is 1.0 Ohm, the armature circuit inductance is 30 mH and the motor constant is 0.19 V/RPM. The motor drives a constant torque load requiring an average armature current of 25A. Assuming, the motor current to be continuous, determine speed control range of the duty cycle.

Solution

Given that at standstill, back EMF E is zero. Thus,

$$V_1 = I_a R_a = 25 \times 1.0 = 25\,\text{V}$$
$$250\alpha = 25$$
$$\alpha = \frac{25}{250} = 0.1$$

For maximum speed $\alpha = 1.0$ thus $V_2 = 250$ Volt

$$E_2 = 250 - 25 \times 1.0 = 225\ \text{V}$$

Speed is given as

$$N = \frac{E_2}{\text{EMF constant}} = \frac{225}{0.19} = 1184.2\,\text{RPM}$$

The range of speed is 0 to 1184.2 RPM and the range of duty cycle α is 0.1–1.0.

4.10 Speed Control of Shunt Motor

All electrical machines have specified rated speeds. However, in many applications, it becomes necessary to change the speed of the driven unit depending on the requirements. The speed of a DC motor can be controlled by the following:

(a) Controlling the armature circuit
   - (i) Changing the armature circuit resistance and
   - (ii) Changing the armature voltage. This requires separate DC source of either battery or generator or solid-state controlled rectifier circuit.

(b) Controlling the field-flux by changing the field current

At design level, the rated speed can be changed by changing the EMF constant k_e. This includes the changing of total armature conductors, number of poles and number of parallel paths. But once a DC machine is fabricated, the EMF constant is difficult to alter.

Speed control below and above the rated speed is needed in many applications. It is required that the drive must be able to produce sufficient torque below rated speed. In this zone of speed control the machine is supplied by a constant voltage source and power intake may vary up to its rated power. When speed control is done above rated speed, it is desired that the power intake should remain constant. Thus the torque follows the inverse speed relation. This has been depicted in Fig. 4.48.

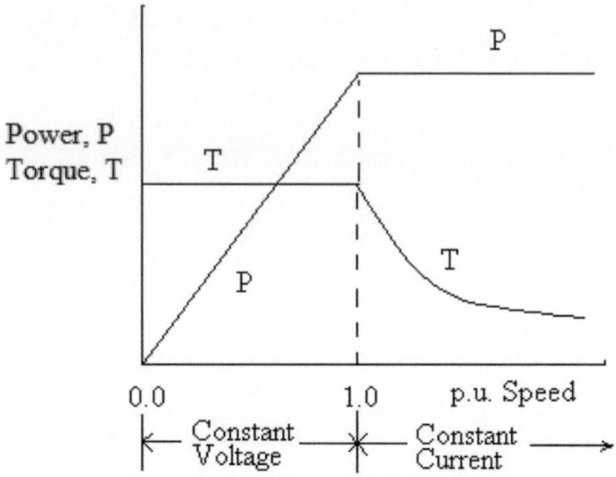

Fig. 4.48 Speed control of electrical motors

4.10.1 Armature Circuit Control

(a) Resistance control: By changing the resistance of armature circuit, the speed of a DC shunt motor can be changed according to the Eq. (4.74).

$$N = \frac{V - I_a(R_a + R_x)}{k_e^{/}\phi} \tag{4.74}$$

The characteristics of speed versus armature voltage have been shown in Fig. 4.49a. The speed of a DC shunt motor is proportional to the armature voltage, when its field current is constant. Here, speed and armature voltage are linearly related. Any change in field current results in a variation of speed as shown in Fig. 4.49a. The armature voltage control is preferred for lowering speed below rated speed. Using this method, it is possible to achieve near zero speed. The torque-speed characteristics of shunt motor for the different supply voltages are shown in Fig. 4.49b. In the method of armature resistance control of speed, the overall efficiency is very poor due to wastage of power in the resistance. One can show that for a DC shunt motor,

$$\frac{\omega_2}{\omega_1} = \frac{V - I_{a2}(R_a + R_{x2})}{V - I_{a1}(R_a + R_{x1})} = \frac{N_2}{N_1} \tag{4.75}$$

(b) Armature voltage control: The speed control using armature voltage control is quite attractive because of its increased efficiency, if solid-state circuits are used. The schemes suitable for starting, as discussed in Sect. 4.9.2, are also useful for speed control. The supply voltage can be changed using any of the following:

(a) Silicon Controlled Rectifier or Single-phase / three-phase controlled converters
(b) Chopper circuit DC to controlled DC

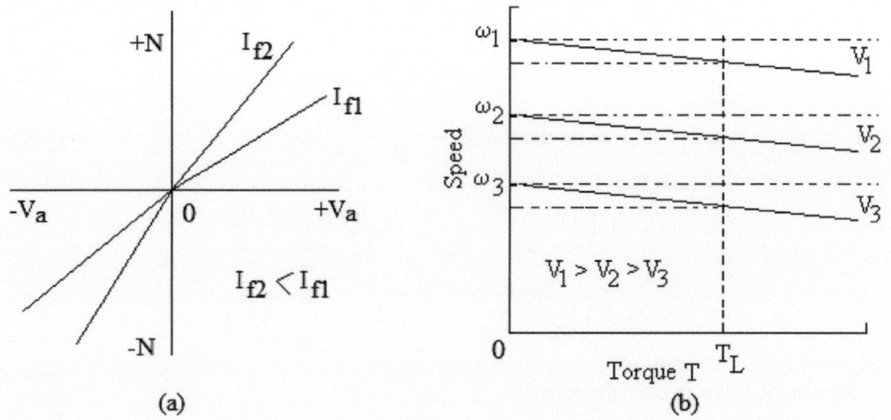

Fig. 4.49 **a** Armature voltage control of shunt motor. **b** torque-speed characteristics by varying voltage

(c) Motor Generator set
(d) Auto transformer and rectifier
(e) Series parallel control

The torque-speed characteristics of shunt motor for different supply voltage are shown in Fig. 4.49b.

(c) Speed control by shunting the armature: The speed control by series connection of resistance is helpful in reducing the speed below the base speed. However, the response is not very good. It exhibits more drooping characteristics. In the event of slight variation of torque, the changes in speed are large. A better alternative for running the DC shunt motor at low speeds is to shunt the armature by a suitable resistance. In such a connection as shown in Fig. 4.50a, the armature gets a reduced voltage. The resistances R_1 and R_2 act as a voltage divider for the armature. The overall efficiency of such type of speed control is expected to be poor. Here, we can write

$$V = E + R_a I_a + R_1 I_1 \tag{4.76}$$

$$V = R_2 I_2 + R_1 I_1 \tag{4.77}$$

$$I_2 = \frac{V - R_1 I_1}{R_2} \tag{4.78}$$

$$I_1 = I_a + I_2 \tag{4.79}$$

$$I_1 = I_a + \frac{V - R_1 I_1}{R_2}$$

$$I_1 = \frac{V + I_a R_2}{R_1 + R_2} \tag{4.80}$$

$$V = E + I_a R_a + \frac{V + I_a R_2}{R_1 + R_2} R_1 \tag{4.81}$$

$$E = V\left(1 - \frac{R_1}{R_1 + R_2}\right) - I_a R_a - \frac{I_a R_1 R_2}{R_1 + R_2}$$

$$E = V\left(\frac{R_2}{R_1 + R_2}\right) - I_a\left(R_a + \frac{R_1 R_2}{R_1 + R_2}\right)$$

The back EMF is expressed as

$$E = CV - I_a(R_a + CR_1) \tag{4.82}$$

where

$$C = \frac{R_2}{R_1 + R_2}$$

We already know that the back EMF is

$$E = k_e \phi \omega$$

Choosing torque constant $k_t = k_e$, the electromagnetic torque is

$$T = k_e \phi I_a$$

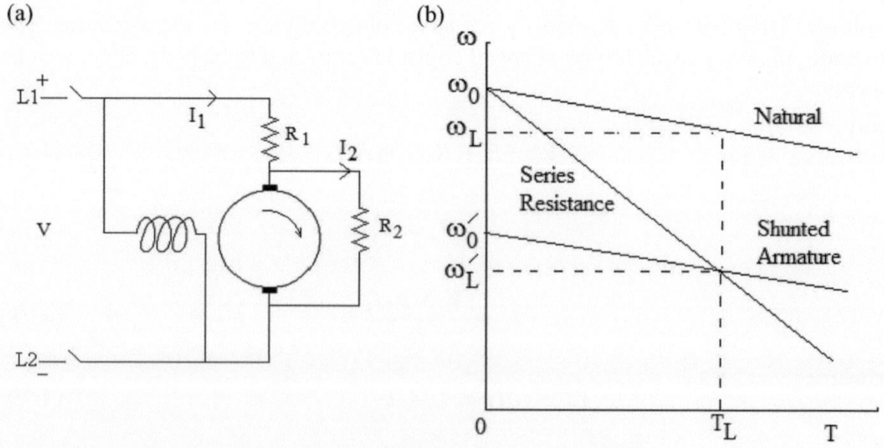

Fig. 4.50 **a** Shunted armature control. **b** Torque-speed characteristics

The torque-speed characteristics are given by,

$$\omega = \frac{1}{k_e\phi}\left(CV - \frac{T}{k_e\phi}(R_a + CR_1)\right) \tag{4.83}$$

Thus, the no-load speed is

$$\omega_0 = \frac{CV}{k_e\phi}$$

The speed torque characteristics are given by,

$$\omega = \left(\omega_0 - \frac{T}{(k_e\phi)^2}(R_a + CR_1)\right) \tag{4.84}$$

Figure 4.50b shows the natural, series resistance and shunted armature characteristics of shunt motor. The speed torque characteristic is harder (drop in speed is less when machine is loaded) with shunted armature. This is a conventional method of speed control and is useful in running the shunt motor at a very low speed. The efficiency is poor owing to the resistance loss. A better alternative is to use solid-state circuit for providing low voltage to the armature circuit, instead of a voltage division method. The use of solid-state controlled rectifier improves the overall efficiency.

(d) Ward Leonard Control: Armature voltage control is sometimes referred to as *continuous speed control* method as it is possible to quickly stop the motor at N_r=0 and restart it in reverse direction. Since the inductance of armature circuit is lower than that of field circuit, a quick reversal does not pose any problem. This is used in *Ward Leonard* method of speed control.

It is sometime required to run the motor is forward and backward direction in a continuous manner. When the length of travel is less, to-and-fro motion in a continuous manner is achieved using mechanical arrangements, either gear or rotary to linear converter (Slider Crank Mechanism). It is quite cumbersome to provide mechanical arrangements, when the length of travel is very large. In case of rolling mill application, a slab of molten metal is pressed between two rollers and rolled in forward direction. A re-rolling in backward direction further reduces the thickness of the metal thus formed. In hoist control also, it is required to run the motor frequently in forward and reverse direction. In milling/shaping of a large slab of molten metal, the tool drive also requires to-and-fro speed control.

Speed of a DC motor may be controlled by controlling the field current or the voltage applied to the armature of a DC motor. In Ward Leonard method, a variable DC voltage is applied to the armature of separately excited DC motor for starting, speed control and reversal. A typical set-up for voltage control is shown in Fig. 4.51. The variable DC voltage is obtained by separately excited DC generator, which is driven by a three-phase induction motor. The field winding of the generator is separately excited using a potentiometer type of regulator, as shown in the figure. By connecting one end of field winding to the midpoint of the regulator and the other to the moving contact, the magnitude of field current of generator can be varied in either direction from its null position. The field current is null, when the moving terminal '*K*' is at the midpoint '*C*' of the potentiometer. The movement of terminal '*K*' in either direction causes the field current to rise either in positive direction or in negative direction. Since the generator is driven by induction motor, any change in the field current of generator induces a corresponding DC voltage of either polarity. A controlled variable voltage can be obtained from the generator by controlling the potentiometer. A typical variation of field current during operation of a Ward Leonard control is shown in Fig. 4.52.

The voltage generated is applied to the armature of a separately excited DC motor, driving certain mechanical load. The field winding of this motor is connected to a separate DC source. Normally, its excitation is kept constant. When it is desired to change the forward and backward speeds of the driven unit, the field current of the motor is also changed accordingly.

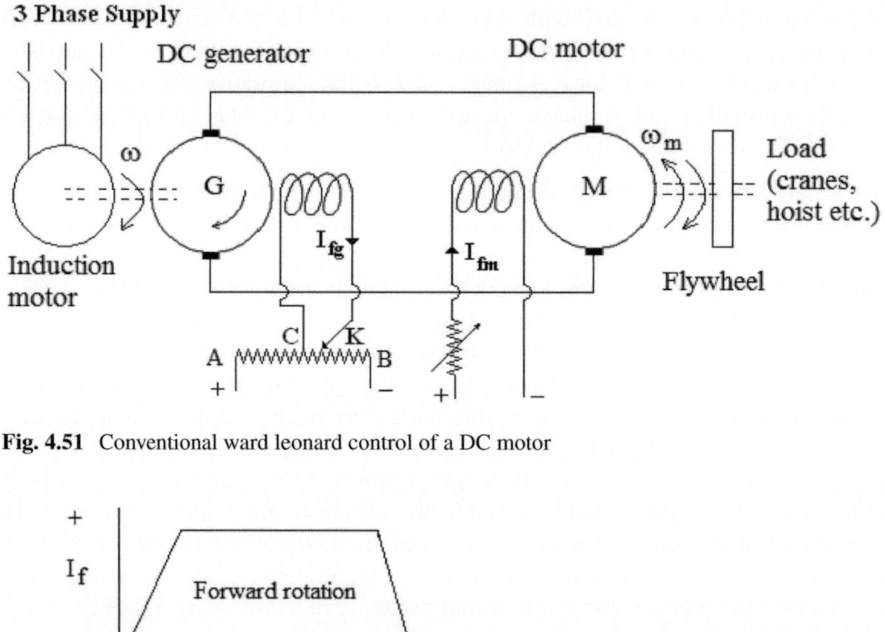

Fig. 4.51 Conventional ward leonard control of a DC motor

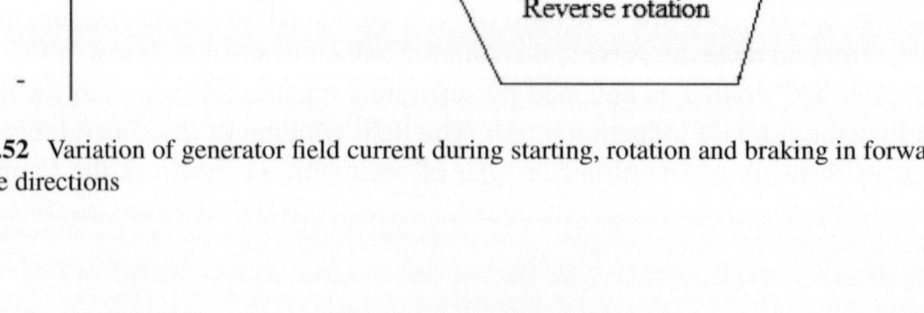

Fig. 4.52 Variation of generator field current during starting, rotation and braking in forward and reverse directions

As explained earlier, the field of the DC generator is controlled using center-tapped rheostat or potentiometer. When the moving contact of rheostat '*K*' is at center or neutral point, both the generator field current and the developed voltage are zero and the driven motor is at rest. When the moving contact of the rheostat '*K*' is moved toward point '*A*', the field current of the generator rises slowly, and so does the developed voltage. The motor starts in one direction. When the moving contact is quickly brought to neutral point, there is a possibility that motor's back EMF becomes more than generator's back EMF. Momentarily, the current reverses its sign. This causes the generator to receive current and act as a motor for a while above its nominal speed. Certain amount of power is fed back to the three-phase lines. Line excited induction generator operates in ($\omega_r > \omega_s$) super synchronous speed. During this process, the induction motor speed momentarily becomes more than synchronous speed and induction machine regenerates.

Any attempt to move the moving contact toward point '*B*', raises the field current of generator in reverse direction. The polarity of developed voltage reverses and the motor starts in the opposite direction. The stored mechanical energy of the motor is fed back to the lines, if the condition $E_m\ (t) > E_g\ (t)$ is satisfied. Under transient condition, we have

$$E_g(t) - E_m(t) = \left(R_{\text{ag}} + R_{\text{am}}\right)i_a(t) + \left(L_{\text{ag}} + L_{\text{am}}\right)\frac{\mathrm{d}i_a(t)}{\mathrm{d}t} \tag{4.85}$$

where the generated EMF is

$$E_g(t) = \pm k_{eg} i_{fg}(t)\omega(t)$$

and motor EMF is

$$E_m(t) = \pm k_{em} i_{fm}(t)\omega(t) \tag{4.86}$$

Here k_{eg} and k_{em} are the EMF constants of generator and motor, respectively. The instantaneous field current of generator is $\pm\ i_{\text{fg}}\ (t)$ which is being changed at regular interval of time and $\pm\ i_{\text{fm}}\ (t)$ is the motor field current, which can be constant. The generator is running at a speed of 'ω(*t*)' and the motor speed is '$\pm\ \omega_m\ (t)$'. Quantities *E* and *i* are generated EMF and current respectively and subscript *m* and *g* are for motor and generator respectively. The motor can be made to operate under steady state condition. The variations of speed-torque characteristics for different armature voltages are shown in Fig. 4.53a. The speed versus armature voltage for a typical to-and-fro drive having motor field control is shown in Fig. 4.53b.

This method of speed control is mostly used in large reversing rolling mill motors and for colliery winding. Here, much of the energy of motion of the motor and its load can be recuperated while bringing the speed to zero by causing the motor to set as a generator, and feed the energy back to the induction-motor-generator set.

The following are the merits of the Ward Leonard Control:

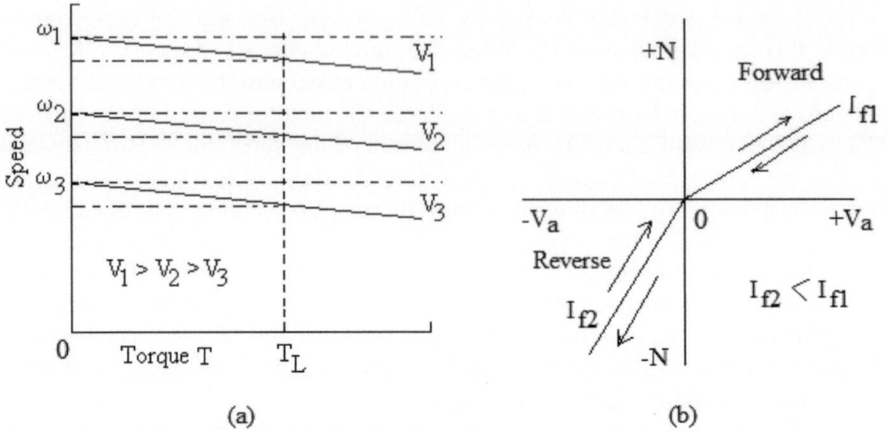

Fig. 4.53 **a** Speed torque characteristics of separately excited motor with variable supply. **b** To-and-fro control of DC motor

(a) Motor can be made to run in forward and reverse directions. This is referred to as *to- and –fro control of DC motor*.
(b) There is no need of starter.
(c) The stored energy can be recuperated when the motor is stopped by reducing generator field current (EMF) to zero. The stored mechanical energy of motor is converted into electrical power and fed back to lines.

However, Ward Leonard control has the following demerits as well:

(a) For controlling one DC motor, we require two additional machines.
(b) The total cost and maintenance cost are high in the conventional Ward Leonard control.
(c) Large floor area is required for installing three machines.

Similar type of control can be achieved, using solid-state circuit. Figure 4.54 shows a thyristorized four quadrant controlled converter of DC motor. Soft start technique can be adopted for starting and braking the motor in either direction. When, the controlled converter-I is operating, the motor runs in forward direction. Similarly, when the other converter-II is operating the motor gets reverse voltage and starts in opposite direction. Different types of control of four quadrant controlled converters are possible.

The solid-state control has the following merits:

(a) Large floor areas of installing two additional machines are not required.
(b) Compact size
(c) Solid-state circuitry is more reliable.
(d) Frequent maintenance is not required.
(e) A fully automated control is possible, which can be reprogrammed depending on requirements.

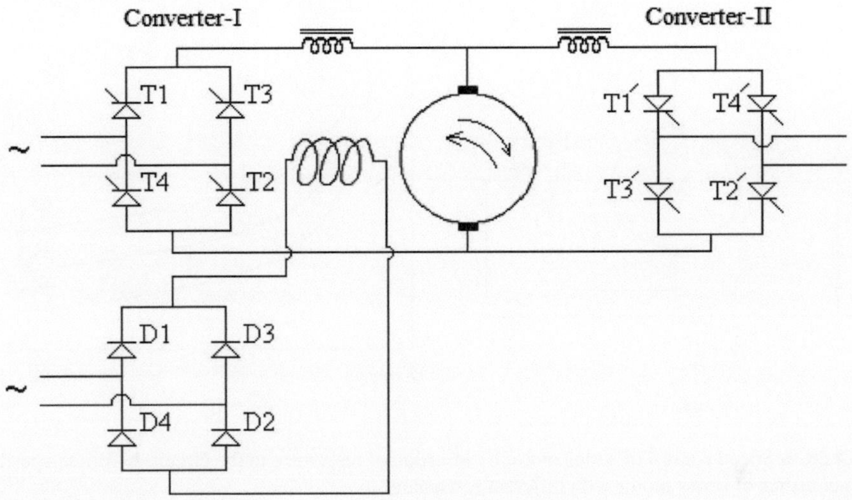

Fig. 4.54 Thyristor reversible DC drive (*Converter-II is OFF when Converter-I is operating and* vice Versa)

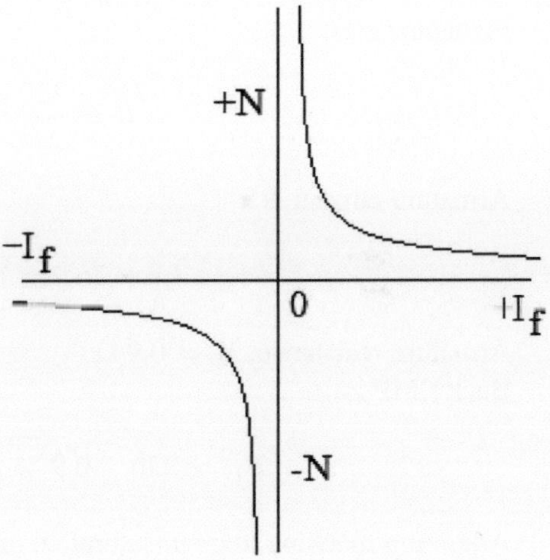

Fig. 4.55 Field flux control of speed of shunt motor-*discontinuous speed control method*

Example 4.20 A 20 HP, 220 V, 1000 RPM DC shunt motor is operating at an efficiency of 89% delivering the rated torque. The armature resistance is 0.9 Ω and field resistance is 200 Ω. A resistance of 0.2 Ω is inserted in the armature circuit. If the load torque does not change, calculate the new speed of motor.

Solution

Given: 20 HP, 220 V 1000 RPM.

Rated current is

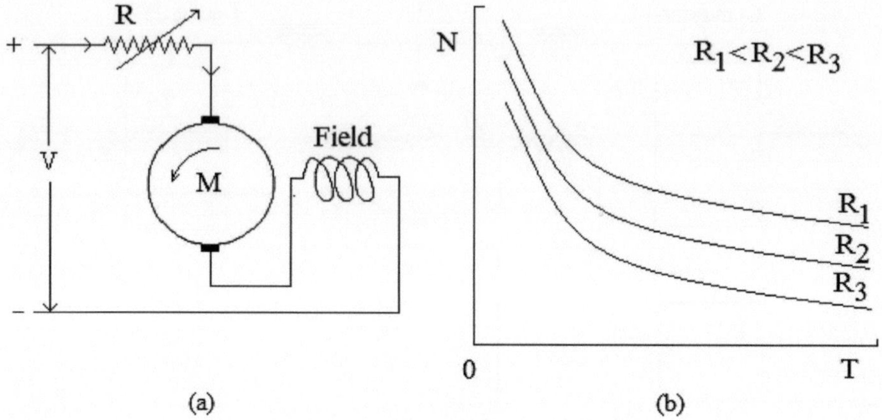

Fig. 4.56 **a** Speed control of series motor by insertion of resistance in the circuit. **b** Torque-speed characteristics of series motor with different resistances

$$I = \frac{20 \times 746}{220 \times 0.89} = 76.2\ \text{A}$$

Field current is

$$I_f = \frac{220}{200} = 1.1\ \text{A}$$

Armature current is

$$I_{a1} = 76.2 - 1.1 = 75.1A$$

Armature resistance, $R_a = 0.9\ \Omega$.
Back EMF is

$$E_1 = 220 - 0.9 \times 75.1 = 152.41\ \text{V}$$

Assuming machine to be unsaturated, the torque is given as

$$T = k_e I_f I_{a1}$$

Since torque is constant, as such armature current will also remain constant. The new EMF can be given as

$$E_2 = 220 - (0.9 + 0.2) \times 75.1 = 137.4\ \text{V}$$

Thus, we get

$$\frac{N_2}{N_1} = \frac{E_2}{E_1}$$

$N_2 = 1000\frac{137.4}{152.41} = 901.5\,\text{RPM}$ is the new speed.

4.10.2 Field Flux Control

In field flux control, the magnetic flux is reduced by reducing the field current. This requires either resistance control or change of supply voltage in the field circuit using solid-state controlled rectifier. The speed of DC motor is inversely proportional to the field flux (Eq. 4.87). If magnetic saturation is not reached, then the flux can be taken as proportional to the field current. Thus, speed of a shunt motor is given as

$$\omega = \frac{2\pi N}{60} \quad \alpha \quad \frac{1}{\phi} \quad \alpha \quad \frac{1}{I_f} \tag{4.87}$$

The speed versus field current of a shunt motor is shown in Fig. 4.55. It is sometimes known as *discontinuous control of speed*. The inductance of the field winding of a shunt motor is higher than that of armature. Any attempt to decrease the speed requires subsequent increase of field current. The sudden reversal of speed by field reversal demands that the direction of field current should be changed. Any attempt to quickly change the direction of field current, needs the discharging of stored energy in its inductance due to positive current, before a new field current of reverse polarity can flow. Any attempt of sudden reversal of the inductive current results in huge spark on the contactor. In reverse rotation, the armature current during starting is limited by resistance in the armature circuit. As such, the field flux control is not preferred for quick speed reversal.

The field flux speed control is only useful in increasing the speed beyond the rated speed. The method is referred as *discontinuous speed control method*, because smooth speed reversal is not possible.

The field winding of DC shunt motor should not be opened when the motor is running. A field current, '$I_f = 0$', causes the machine to race, since speed is inversely proportional to the field current. The solid-state controller and automatic starter are designed to switch OFF the supply in the event of field failure.

4.11 Speed Control of DC Series Motor

The following speed control methods are used for controlling speed of DC series motors:

(a) Variation of resistance in armature circuit

(b) Variation of applied voltage –Series/parallel control
(c) Variation of field excitation
(d) Shunting of field winding
(e) Use of tapping in the field winding

(a) **Variation of resistance in armature circuit:** Using this method, the speed can be reduced below base speed. A schematic diagram and speed torque characteristics with different rotor resistance have been shown in Fig. 4.56a, b. The method is used for constant torque load. The efficiency is poor because of high ohmic losses in the controlling resistance, which is carrying the motor armature current. This is one of the simplest methods and was commonly used in intermittent duty drives such as cranes. These resistances carry full load current on continuous basis.

(b) **Shunting of armature winding:** A DC series motor can be made to run at low speed by shunting the armature winding. The current through the armature is bypassed through a shunt resistance R_d, as shown in Fig. 4.57a. The effect of variation of shunt resistance on speed-torque characteristics is shown in Fig. 4.57b. When R_d tends to infinite, machine exhibits natural characteristics.

(c) **Shunting of Field Winding:** In certain application, it required to increase the speed of a DC series motor. We already know that speed in inversely proportional to field current (Eq. (4.87)). In the case of DC series motor, field flux can be reduced by shunting the field winding. The current through the field winding is bypassed through a shunt resistance R_d, as shown in Fig. 4.58a. The effect of variation of shunt resistance on speed torque characteristics is shown in

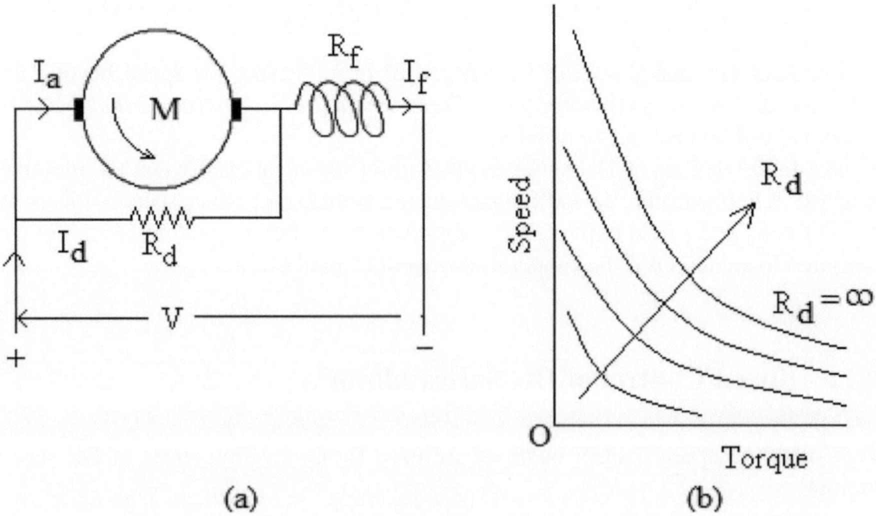

Fig. 4.57 **a** Shunting of armature winding of DC series motor. **b** Torque-speed characteristics

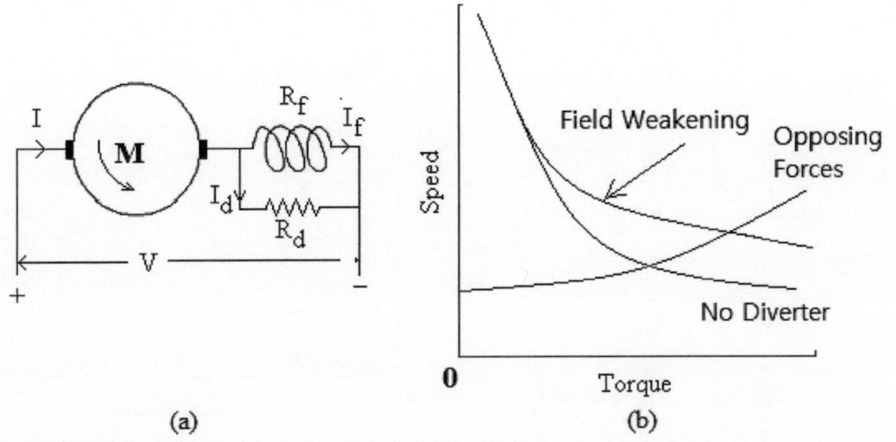

Fig. 4.58 **a** Shunting of Field winding of DC series motor. **b** Torque-speed characteristics with diverter resistance

Fig. 4.58b. When R_d tends to infinite, machine exhibits natural characteristics. The method is used in traction applications for running the train at higher speed.

The shunting resistance of the field is called as diverter resistance; it must be very low because the resistance of field winding of a series motor is also very low. The diverter resistance should also have high inductance for limiting high inrush of current after re-switching.

(d) **Series–parallel control**: In traction or tram application, large numbers of DC series motors are used. These motors are connected in pair to several wheels and provide torque in the same direction. Since the supply voltage is constant, a reduction in voltage shared per motor can be obtained by connecting these series motors in series and parallel combination. This is a superior method as compared to series resistance control. Consider two identical DC series motors, as shown in Fig. 4.59a. When these motors are connected in series, the voltage across each of the motor is half the supply voltage. Let the applied voltage be 'V' and terminal resistance of each motor 'R'. For a constant current operation, we can write

$$\frac{N_p}{N_s} = \frac{V - IR}{0.5V - IR} \tag{4.88}$$

where N_p and N_s are the speeds of motor, in parallel and series connections, respectively. The method is useful if at least two similar motors are connected to the same drive load. The characteristics of series parallel combination are shown in Fig. 4.59b. In traction locomotive, a connection of several motors in series helps low speed maneuvering.

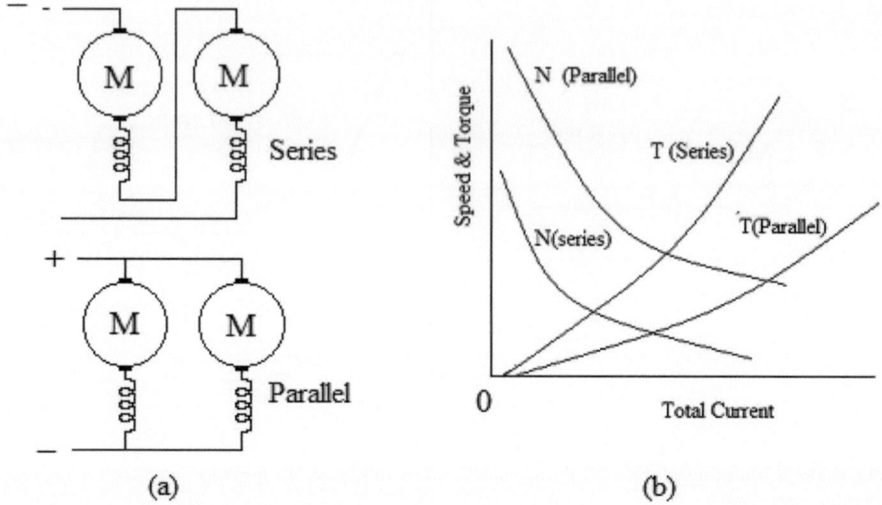

Fig. 4.59 **a** Series and parallel control and **b** Characteristics of Series parallel control of DC series motor

(e) **Tapping in Field winding**: The method is used in smaller capacity DC or AC/DC universal motor used in domestic appliances. The field windings have several tapping as shown in Fig. 4.60. The field weakening required for high speed operation is done by choosing lesser number of turns.

Example 4.21 A 10 HP, 220 V, 600 RPM DC shunt motor is operating at an efficiency of 88% delivering the rated torque. The armature resistance R_a is 1.5 Ω and field resistance R_f is 250 Ω. A resistance of 50 Ω is inserted in the field circuit. If the load is inversely proportional to speed, calculate the new speed of motor.

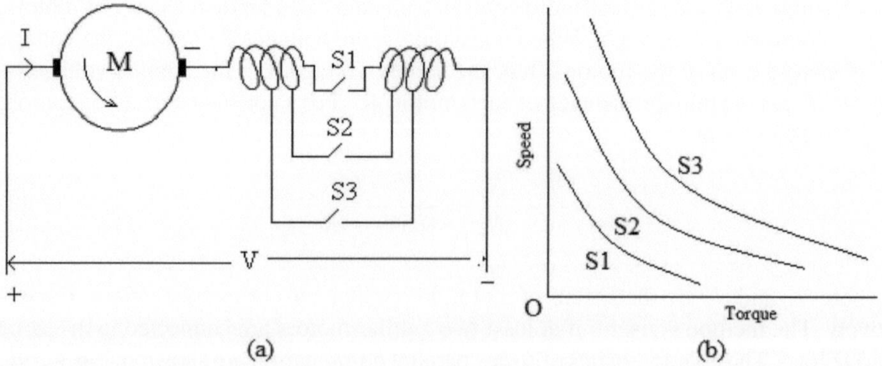

Fig. 4.60 Speed control of DC series motor using tapping in field windings. **a** Schematic diagram. **b** Torque-speed characteristics

Solution

Given 10 HP, 220 V,600 RPM, 88%, $R_a = 1.5$ Ohm, $R_f = 250$ Ohm.

Rated current is

$$I = \frac{10 \times 746}{220 \times 0.88} = 38.53 \text{ A}$$

Field current, $I_{fl} = \frac{220}{250} = 0.88\ \text{A}$.

Armature current 'I_{a1}' is given as

$$I_{al} = 38.53 - 0.88 = 37.65 \text{ A}$$

Back EMF is

$$E_1 = 220 - 37.65 \times 1.5 = 163.525 \text{ V}$$

New field current after increasing field resistance by 50 Ω.

$$I_{f2} = \frac{220}{250 + 50} = 0.733\ \text{A}$$

$$T_1 = kI_{f1}I_{a1}\ \ \alpha\ \ \frac{1}{\omega_1}$$

$$T_2 = kI_{f2}I_{a2}\ \ \alpha\ \ \frac{1}{\omega_2}$$

$$\frac{T_1}{T_2} = \frac{I_{f1}I_{a1}}{I_{f2}I_{a2}} = \frac{\omega_2}{\omega_1} = \frac{N_2}{N_1}$$

$$I_{a2} = \frac{I_{f1}I_{a1}}{I_{f2}} \times \frac{\omega_1}{\omega_2} = \frac{0.88 \times 37.65 \times 600}{0.733 \times N_2}$$

EMF

$$E_2 = V - I_{a2}R_a$$

$$\frac{E_1}{E_2} = \frac{kI_{f1}N_1}{kI_{f2}N_2}$$

$$E_2 = \frac{0.733 \times N_2 \times 163.53}{0.88 \times 600} = 220 - \frac{0.88 \times 37.65 \times 1.5 \times 600}{0.733 \times N_2}$$

$$0.22701N_2^2 - 220N_2 + 40680.5 = 0$$

$$N_2 = \frac{220 \pm \sqrt{220 \times 220 - 4 \times 0.22701 \times 40680.5}}{2 \times 0.22701}$$

$$N_2 = 720.35 \text{ RPM or } 248 \text{ RPM}$$

The new speed is 720 RPM or 248 RPM.

4.12 Braking of DC Motors

In several drive applications, it is sometimes required to quickly stop and restart the driven unit, either in the same direction or in the opposite direction. For example, stopping or braking of the electrical motor-operated vehicle, traction locomotive, drones, crane, printers, processing machines etc. If supply of a motor is switched OFF, the mechanical energy stored in the moving or rotating parts is dissipated in the opposing windage and frictional torque. As such an electrical motor will naturally come to rest. The mechanical braking methods are based on the principle of increasing frictional torque (rubbing action) and are efficient at comparatively lower speed of operation. The electrical braking methods are non-contact type and more efficient when motor is running at high speed. The electromagnetic braking torque is imparted to the rotating mass in a non-contact way. Apart from this, the control of electrical braking torque is an easier task. The term 'braking of electrical motor' implies the development of reverse torque, when motor is still rotating in the forward direction. The use of electrical braking reduces the time taken by the machine in coming to rest, by enhancing the opposing torque. Following are three methods of braking DC motor:

(a) Regenerating braking
(b) Rheostatic braking or dynamic braking
(c) Plugging or counter current braking.

4.12.1 Regenerative Braking

In certain applications, like a DC motor-operated vehicle moving down gradient due to gravity and lowering of load in hoisting mechanism, the connected mechanical load sometimes tends to drive the operating motor. The speed of the drive motor may become more than its no-load speed due to gravitational force. The DC machine will start generating when speed becomes greater than its no-load speed, and the back EMF becomes more than supply voltage. As a result, the directions of current and torque both reverse and the machine returns the electrical energy to the supply mains. During such operations, the direction of electromagnetic torque also reverses, which restricts the motor speed around its no-load speed. This is an economical method because the power is being returned to the supply. In absence of supply voltage, when the load drives the motor, the shaft may attain a dangerously high speed, because of absence of braking torque. The regenerative action restricts the speed from rising to such high level of runaway speed.

When the back EMF become more than the supply voltage, the stored mechanical energy is converted into electrical energy and fed back to the supply lines. The conditions in which these situations arise, however, are limited. The torque-speed characteristics of the machine restricts the over speeding of the motor. The motoring current, when $V > E$ is,

$$I_m = \frac{V - E}{R_a}$$

Consider a case, in which a vehicle using DC motor is going in down (slope) gradient. The vehicle will be running due to gravitational forces, causing the speed to become more than its no-load speed. Such that,

$$N > N_o \quad \text{or} \quad \omega > \omega_o$$

Under such a condition, the back EMF, E > V. The braking current is

$$I_{rb} = -\frac{E - V}{R_a} \tag{4.89}$$

The armature current in the machine changes its direction and so does the torque. The torque-speed characteristic is given in Fig. 4.61. Similar situations arise, when a crane is lowering its load. The regenerative braking avoids a free fall of the object due to gravity. As has been explained, in the case of Ward Leonard control, similar situation arises when the test motor is brought to zero speed. The regenerated power is sometimes used for charging the battery bank in traction applications. Braking torque in an electric vehicle can be increased by using the regenerated DC power for supplying to eddy current brakes.

4.12.2 Rheostatic Braking

In rheostatic braking or dynamic braking, the stored mechanical energy is dissipated in resistive or Ohmic losses. Here, armature circuit is disconnected from the supply and reconnected across a resistance R_D, as shown in Fig. 4.62. The motoring armature current with resistance R_D in series with the armature is given as

$$I_m = \frac{V - E}{R_a + R_D} \tag{4.90}$$

When supply voltage to armature is zero, the back EMF reverses the direction of current, as shown in Fig. 4.63. Since the speed falls with time, the back EMF also

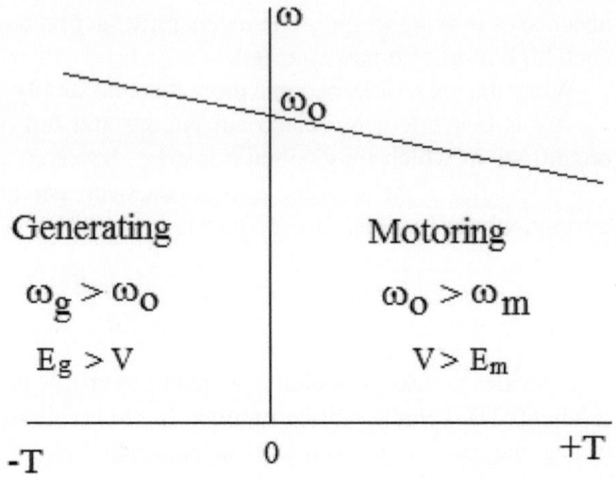

Fig. 4.61 Torque-speed characteristics during regenerating braking

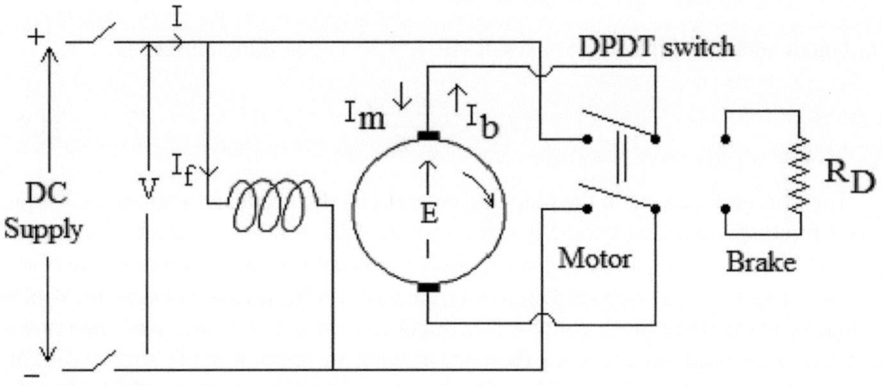

Fig. 4.62 Schematic diagram for dynamic braking of a DC shunt motor *(Starter not shown for clarity)*

varies with time. The time varying braking current is,

$$i_{db}(t) = \frac{E(t)}{R_a + R_D} \tag{4.91}$$

The braking torque is given by

$$T_{db}(t) = -k_t \times \phi \times \frac{E(t)}{R_a + R_D} \tag{4.92}$$

The motor acts as a separately excited generator (driven by its own moment of inertia). Both the armature current and torque reverse the sign of their directions. The braking current and speed are transient in nature, which fall from their respective

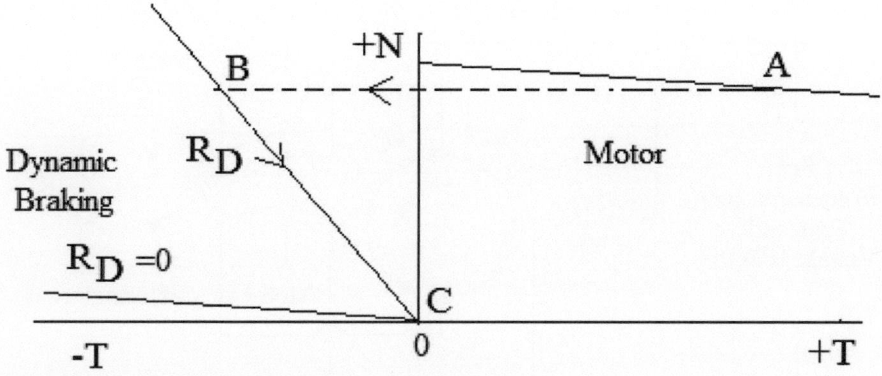

Fig. 4.63 Speed transition during dynamic braking

initial values to zero. Once the machine has stopped rotating**, the motor field supply should be switched OFF.** It can only then be started using a starter. Can you give suitable reasons?

The time taken by the machine to halt is directly proportional to the peak braking current and is inversely proportional to the braking resistance 'RD'. How does the braking time depend on the moment of inertia and load torque? The scheme is useful in braking machine-tool mechanism. In electric vehicle, the braking torque can be augmented by connecting eddy current brakes to the armature terminals of DC traction motors working as generator.

4.12.3 Plugging

In plugging or counter-current braking, the armature of the shunt motor is disconnected from the supply and reconnected in a reverse fashion to the supply terminals, as shown in Fig. 4.64. The motoring current is given by

$$I_m = \frac{V - E}{R_a}$$

The counter current is given as

$$I_{cc} = \frac{V + E}{R_a + R_C} \tag{4.93}$$

During plugging, the back EMF comes in series addition to the supply voltage. During this time, the motor is still rotating in the forward direction but motor armature connections are changed to run in the opposite direction. As a result, the effective voltage acting in the armature circuit is about twice the supply voltage. If

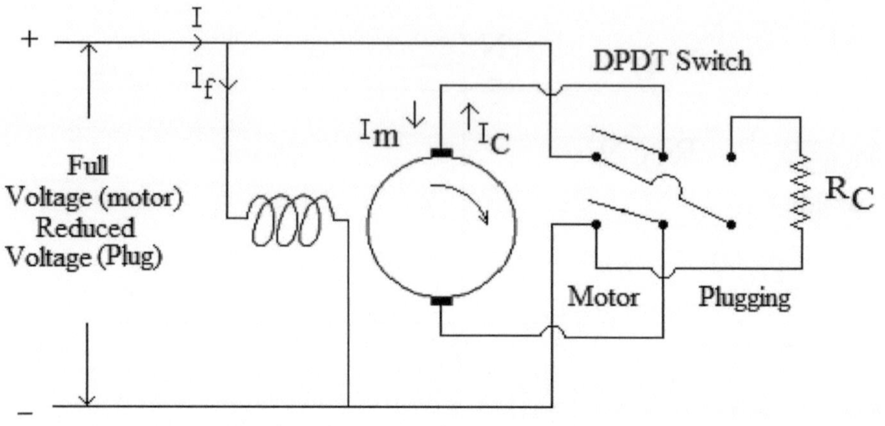

Fig. 4.64 Schematic diagram for plugging operation of DC shunt motor *(Starter not shown for clarity)*

extra resistance R_C is not added in series with the armature circuit, it causes large counter (braking) current. As a consequence, huge sparking results on the commutator segments, commutator and brush, thereby causing extensive damage to the machine. The braking time is directly proportional to the peak counter current. Since both the armature and the field winding are connected to the supply, the braking torque is higher in this case as that of dynamic braking. The reversal of field winding instead of armature is avoided, because it causes sparking owing to large inductance of the field circuit and also very large starting current.

Figure 4.65 shows the speed transition of a shunt motor during plugging operation. The DC motor is operating as motor at point 'A'. On counter current braking, the operation point shifts in the second quadrant at point 'B'. The current and torque change their signs. Due to negative torque, motor comes to rest at point 'C'. If motor is not switched OFF at point C, it restarts in the reverse direction. The machine must be switched OFF and restarted using its starter, to avoid any damage to the machine.

Example 4.22 A 220 V, 10 HP, 1000 RPM DC shunt motor is operating at 0.90 efficiency. Armature resistance is 0.5 Ω and field resistance is 100 Ω. If the peak braking current should not cross the rated current, obtain the suitable value of resistances for

(a) Dynamic braking and
(b) Counter current braking

Solution

Rated current is

$$I = \frac{10 \times 746}{220 \times 0.9} = 37.7\,\text{A}$$

The field current is

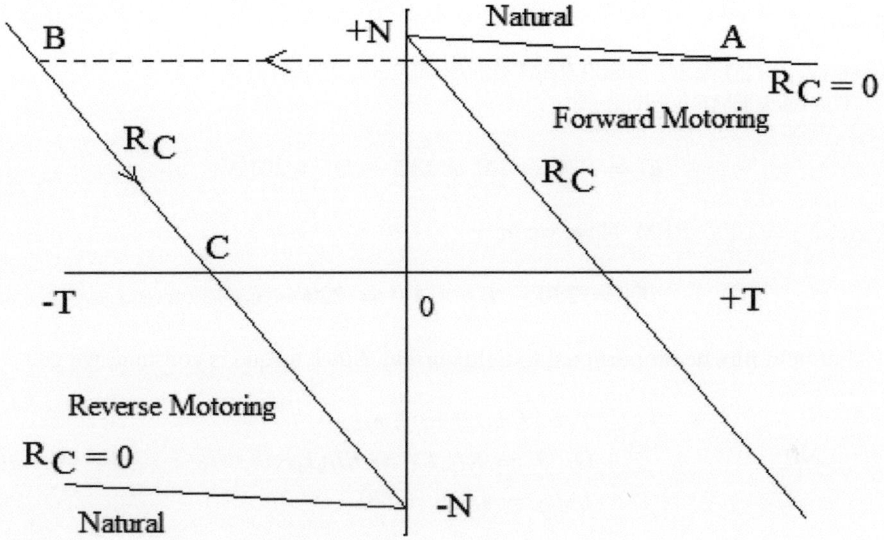

Fig. 4.65 Transition during plugging operation

$$I_f = \frac{220}{100} = 2.2\,\text{A}$$

Armature current is

$$I_a = 37.7 - 2.2 = 35.5\,\text{A}$$

The back EMF is given as

$$E = 220 - 35.5 \times 0.5 = 220 - 17.75 = 202.25\,\text{V}$$

(a) In the case of dynamic braking, the resistance is

$$R_d = \frac{202.25}{37.7} = 5.36\,\Omega$$

(b) The resistance in case of counter current braking is,

$$R_c = \frac{220 + 202.25}{37.7} = 11.2\,\Omega$$

Example 4.23 A 220 V DC shunt motor has an armature resistance of 1.0 Ω, and a field resistance of 220 Ω. It drives a constant torque load at 800 RPM, taking 20 A armature current. What resistance must be added in the field circuit for raising the speed to 1000 RPM. Assume linear magnetization curve.

Solution

Given $I_{a1} = 20$ A, $N_1 = 800$ RPM and $I_{f1} = 220/220 = 1.0$ A.

The back EMF is given as.

$$E_1 = 220 - 1.0 \times (20 - 1) = 201 \text{ V}$$

Here, $N_2 = 1000$ RPM. Now, we have

$$E_2 = 220 - I_{a2} \times 1.0 = 220 - I_{a2}$$

Let field flux be proportional to field current. Since torque is constant, we get

$$T = K\phi_1 I_{a1} = K\phi_2 I_{a2}$$
$$\text{Or, } T = KI_{f1}I_{a1} = KI_{f2}I_{a2}$$
$$I_{f2}I_{a2} = I_{f1}I_{a1} = 20$$

The new EMF is.

$$E_2 = 220 - 1.0 \times I_{a2}$$

Thus,

$$\frac{220 - I_{a2}}{201} = \frac{I_{f2}N_2}{I_{f1}N_1} = I_{f2}\frac{1000}{800 \times 1.0}$$
$$220 - I_{a2} = 201 \times \left(\frac{10}{8}\right) \times \left(\frac{20}{I_{a2}}\right) = \frac{5000}{I_{a2}}$$
$$I_{a2}^2 - 220I_a + 5000 = 0$$

$I_{a2} = 110 \pm 84.26 = 25.7$ A and 194.26 A, discarding higher value.

Thus, field current is given as

$$I_{f2} = \frac{20}{I_{a2}} = 0.7782 \text{ A}$$

Field resistance is

$$R_f = \frac{220}{0.7782} = 282.7\,\Omega$$

Extra resistance required is $(282.7 - 220) = 62.7\,\Omega$.

Example 4.24 A DC series motor having a resistance of 2.0 Ω between the terminals is used for driving a fan. At 220 V the fan runs at 500 RPM and takes 22 A. The speed is to be raised to 600 RPM by increasing the voltage. Find the voltage and current for the case when magnetic circuit is.

(a) Unsaturated
(b) Saturated.

Solution

Given that $V_1 = 220$ V, $I_1 = 22$ A, $N_1 = 500$ RPM and $R = 2.0$ Ohm.
The EMF is

$$E_1 = 220 - 22 \times 2.0 = 220 - 44 = 176\,\text{V}$$

Load toque is

$$T_L = K_L N^2$$

where K_L = constant
(a) When magnetic circuit is unsaturated, flux ϕ α I.
The torque developed is

$$T_1 = k\phi_1 I_1 = K I_1^2$$

Thus, we get

$$K I_1^2 = K_L N_1^2$$

Similarly, we have.

$$K I_2^2 = K_L N_2^2$$

$$\frac{I_2}{I_1} = \frac{N_2}{N_1}$$

$$I_2 = \frac{600}{500} \times 22 = 26.4A$$

$$I_2 = \frac{600}{500} \times 22 = 26.4A$$

Back EMF is
$E_2 = V_2 - 2 \times 26.4 = (V_2 - 52.8)$ Volt
Since back EMF, $E = K\,\phi\,N$, thus, we get

$$\frac{E_2}{E_1} = \frac{\phi_2 N_2}{\phi_1 N_1} = \frac{I_2 N_2}{I_1 N_1}$$

$$E_2 = \frac{176 \times 26.4 \times 600}{22 \times 500} = 253.44\,\text{V}$$

Thus, the applied voltage is
$V_2 = 253.44 + 52.8 = 306.24\,\text{V}$

(b) When magnetic circuit is saturated, flux is considered as constant and almost independent of field current. Thus

$$KI_1 = K_L N_1^2$$
$$KI_2 = K_L N_2^2$$

$$I_2 = \left(\frac{600}{500}\right)^2 \times 22 = 31.68\,\text{A}$$

$$E_2 = \left(\frac{600}{500}\right) \times 176 = 211.2\,\text{V}$$

The applied voltage is

$$V_2 = 211.2 + 31.68 \times 2 = 274.56\,\text{V}$$

Example 4.25 A 220 V DC series motor develops full load torque at 1800 RPM taking 20 A from source. A diverter resistance is connected across the field for increasing the speed. The armature and field resistances are 1.0 Ω each. The diverter resistance is 0.5 Ω. Obtain new speed and current if torque remains constant. Assume no saturation.

Solution

Given $V = 220$ V, $I_1 = 20$ A, $R_a = R_f = 1.0\,\Omega$ and $N_1 = 1800$ RPM.
Armature current is $I_1 = 20\text{ A} = I_f$ and flux per pole is ϕ,

$$E_1 = 220 - 2 \times 20 = 180\,\text{V}$$

When diverter resistance is connected, the current I is bypassed through R_d, the new field current is

$$R_f I_f = R_d I_d$$
$$\text{And} \quad I_f + I_d = I$$
$$R_f . I_f = R_d I - R_d I_f$$

Thus, we get

$$I_{f2} = \frac{R_d I}{(R_f + R_d)} = \frac{0.5 I_2}{1.5} = \frac{I_2}{3}$$

The torque remains constant, i.e.,

$$T = K\phi_1 I_1 = K\phi_2 I_2$$
$$or, \quad I_{f2} I_{a2} = I_{f1} I_{a1}.$$
$$I_2^2 = 3 I_1^2$$
$$I_2 = 1.732 \times 20 = 34.64$$
$$E_2 = 220 - (1 + (1/3)) \times 20 = 193.33\,\text{V}$$
$$E_1 = K I_{f1} N_1$$
$$E_2 = K I_{f2} N_2$$
$$I_{f2} = \frac{34.64}{3} = 11.55\,\text{A}$$

Thus, new speed is

$$N_2 = \frac{N_1 E_2 I_{f1}}{E_1 I_{f2}} = \frac{1800 \times 193.33 \times 22}{180 \times 11.55} = 3348\,\text{RPM}$$

4.13 Losses in DC Machine and Condition for Maximum Efficiency of DC Machines

Different losses present in DC machines are as follows:

1. Losses in Ferromagnetic material or core due to
   - (a) Hysteresis loss in armature core and slot teeth due to rotating magnetizing of core and
   - (b) Eddy current losses in armature core, slot teeth, and pole face due to rotating magnetization
2. Resistive losses in conducting circuits include
   - (a) Ohmic loss in the armature circuit or $I_a^2 R_a$ loss
   - (b) Ohmic loss in different field windings or $I_f^2 R_f = V_f I_f$ losses
   - (c) Ohmic loss in compensating windings,
   - (d) Ohmic loss in brushes and brush contact loss
3. The losses of mechanical origin include
   - (a) Friction loss in bearing. All types of bearing such as bush bearing, ball bearing or roller bearing, thrust bearing will have friction losses. These losses can be minimized using proper lubrication.
   - (b) Mechanical friction due to rubbing action of brush on the commutator segments.
   - (d) The rotating parts cut the air, and the power required to overcome this is known as windage loss

4. Stray load loss

The losses under no-load running conditions are

1. Core loss in armature because of main field flux
2. Pole face losses due to slot opening
3. Windage loss
4. Bearing friction loss
5. Brush friction
6. Copper losses in the conductors
7. Eddy current loss in steel wire bands of armature due to slight variation in main flux

The losses under load conditions are as follows:

(a) Copper losses in conductors carrying load current
(b) Stray Load loss: When machine is loaded the current drawn by motor or current supplied by generator increases. Due to this, the fringe flux and leakage flux also increase. A part of these flux links with metallic parts of the machine, either stationary (poles, frame, brush holder metallic bands etc.) or rotating (armature, commutator, shaft, bearings etc.) and cause magnetic losses. These losses are referred as stray load loss. Under no-load conditions, they are lesser as compared to loaded conditions. In DC machine, the actual flux density distribution in air-gap is having ripples due to slot opening on the armature, and it is never a smooth flux density distribution as shown while discussing armature reaction.

4.13.1 Efficiency of DC Machine

The efficiency of DC machine is given as

$$\eta = \frac{\text{Output Power}}{\text{Input Power}}$$

The output power in a motor is the mechanical power given to the shaft. The output power of a generator is the electrical power delivered to the electrical load. The efficiency is given as,

$$\text{Output Power} = \text{Input Power} - \text{Losses}$$

$$\text{Output Power} = \text{Input Power} - (\text{Fixed Losses} + \text{Variable Losses})$$

The fixed losses are the losses occurring at no-load and these losses are considered as constant. The variable losses are the losses which depend on the variation of load-current. Since these losses are the active losses or ohmic losses, these are considered to vary with respect to the square of armature current. Suppose the armature current is kth fraction of rated armature current. If at rated armature current, the constant loss and variable loss are P_C and $P_{V,}$ respectively. If the armature current is kth fraction of the rated armature current, the variable loss is given as '$k^2 P_V$'.

Let input power be S. The output power is,

$$P_o = \text{Input Power} - (\text{Fixed Losses} + \text{Variable Losses})$$
$$P_o = S - \left(P_C + k^2 P_V\right)$$

Thus the % efficiency is given as

$$\eta = \frac{P_o}{S} = \frac{S - \left(P_C + k^2 P_V\right)}{S} = 1 - \frac{\left(P_C + k^2 P_V\right)}{S}$$

The machine will give maximum efficiency at certain fraction of load. Differentiating the above equation with respect to 'k' and setting it to zero or,

$$\frac{d\eta}{dk} = 0$$

We get the condition for maximum efficiency,

$$k = \sqrt{\frac{P_C}{P_V}}$$

The condition of maximum efficiency is similar as that of transformer.

4.14 Testing of Machines

The efficiency of DC motor is determined by using different methods as mentioned below:

(a) Direct loading of DC motors
(b) Swinburne's test of DC shunt motor
(c) Hopkinson's test on two identical DC shunt motors
(d) Retardation test
(e) Field's Test on two identical DC series motors.

4.14.1 Direct Load Test

Direct load test on all types of motors is performed using any of the following.

(a) Pony brake system

(b) Separately excited DC generator connected to the motor shaft
(c) Eddy current brake

The schematic diagram for direct load test using a pony brake system is shown in Fig. 4.66. Here motor shaft is connected to a hollow drum. On the outer periphery of the drum, a leather belt does the rubbing (friction) action when shaft is rotating. The two ends of the belt are connected to spring balances. The belt tension can be controlled. The friction between belt and drum opposes rotation and produces large heat. Water is poured inside the drum for cooling purposes.

For testing of DC shunt motor or induction motor or engine, the belt may be loose at the instant of starting. When DC series motor is being tested, the belts must be tight enough to provide large opposing torque even at the instant of starting.

A slippage between drum and belt is observed when the belt becomes smooth due to overheating. This demand pouring of small quantity of sand in between rotating drum and belt.

Let the tension on the belt S_1 & S_2 in kg, outer diameter of drum be 'd' and thickness of belt be 't'.

Since the force exerted on 1.0 kg mass is 9.81 N, the Torque is given by,

$$T = (S_1 - S_2) \times \left(\frac{d+t}{2}\right) \times 9.81\,\mathrm{Nm} \tag{4.94}$$

The output power is

$$\mathrm{BHP} = \frac{2\pi NT}{60 \times 746} \tag{4.95}$$

Input power is

$$I\mathrm{HP} = \frac{\text{Voltage} \times \text{Current}}{746} \tag{4.96}$$

Percent efficiency is

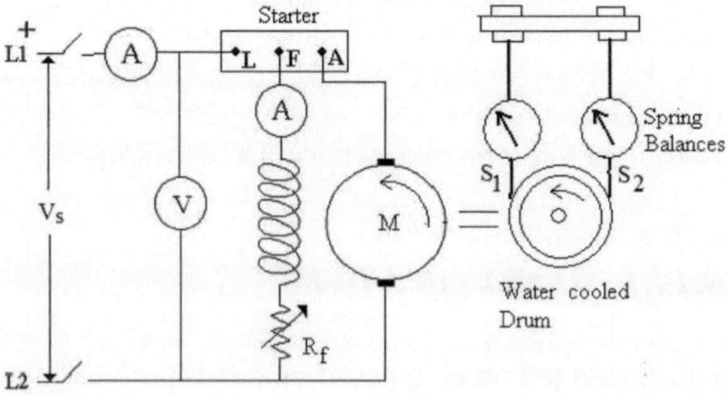

Fig. 4.66 Direct loading of DC shunt motor

$$\eta = \frac{\text{Output}}{\text{Input}} \times 100 \tag{4.97}$$

This method is useful in direct testing of small and medium capacity motor. It takes care of all the load losses. When test is performed using separately excited DC generator connected to shaft, the armature terminals are connected to resistive load box. When a DC series motor is tested, initially all the resistive loads must be connected, in order to provide sufficient counter torque to the motor. The output power is taken as product of generator (armature) voltage and current.

The output power is

$$P_o = V_{\text{g}} \cdot I_{\text{ag}} = \frac{2 \cdot \pi \cdot N}{60} T \tag{4.98}$$

Using above expression, torque can be calculated,

$$T = \frac{60 \times V_g \times I_{\text{ag}}}{2.\pi.N} \tag{4.99}$$

The testing with eddy current brake needs its calibration. The functioning of eddy current brake is explained in Chap. 7.

4.14.2 Swinburne's Test

Predetermination of efficiency of DC machine is done by conducting Swinburne's Test. Figure 4.67 shows the schematic diagram of the Swinburne's test. The motor is not mechanically loaded.

The % efficiency is calculated as,

$$\eta = \frac{\text{Output}}{\text{Input}} \times 100 = \frac{\text{Input} - (\text{constant} + \text{variable})\text{Losses}}{\text{Input}} \times 100$$

Let I_{LO} be the line current on no-load and V be the applied voltage. The total input on no-load will be VI_{LO}.

If I_f is the field current and I_{ao} is the armature current under no-load, then we get

$$I_{\text{LO}} = I_{\text{ao}} + I_f.$$

If R_a is the armature resistance, the sum of total iron loss and friction loss is given as

$$P_o = VI_{ao} - I_{ao}^2 R_a$$

Now we have

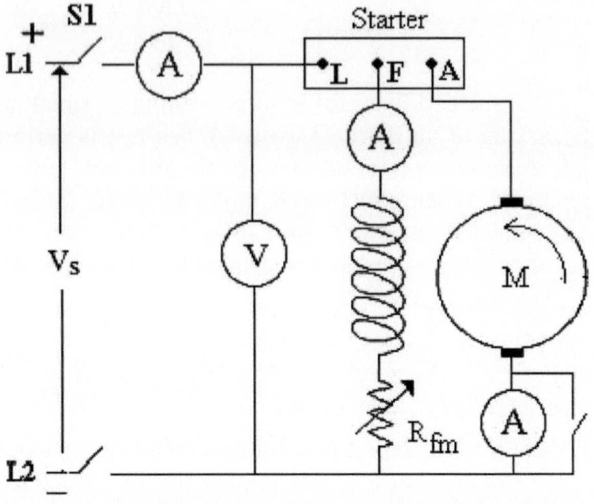

Fig. 4.67 Schematic diagram for Swinburne's Test

$$\text{Total constant losses} = \text{Losses in field} + \text{ Fixed losses} = P_T = VI_f + P_o$$

Now, for an input current 'I_L', the armature current is

$$I_a = I_L - I_f = I_L - \frac{V}{R_f}$$

Here resistance of field circuit is 'R_f'.

The input power is VI_L.

The copper loss in the armature circuit is

$$P_C = I_a^2 R_a$$

$$\%\eta = \frac{\text{Input} - (P_T + P_C)}{\text{Input}} \times 100 = \left[1 - \frac{(P_T + P_C)}{\text{Input}}\right] \times 100 \tag{4.100}$$

The efficiency is determined by using its no-load test data. The efficiency thus obtained does not account for stray load losses. It gives only a rough estimate of the efficiency.

4.14.3 Hopkinson's Test

Hopkinson's test requires two identical DC shunt machines. The method is useful for testing large capacity DC machine under loading conditions. The determined efficiency is more realistic in nature as it takes into account the stray load losses.

This test uses two DC machines working at motor-generator (M–G) set as shown in Fig. 4.68. When DC motor is loaded, the load current is partly supplied by supply terminals and partly by the DC generator. The M-G set is started using DC motor starter, when switch ‘S’ is in open condition. The motor and generator fields are separately connected to the supply. The voltmeter across the motor and generator may either read zero voltage or it may read twice the supply voltage. In the latter case the polarity of terminals of motor and generator connected across the switch are dissimilar. The switch should **never** be connected if the voltage across the switch ‘S’ is twice the supply voltage. The polarity of generated voltage should be changed by reversing either armature or field terminals. When the voltage across switch ‘S’ is zero, the terminals connected across the switch are at same polarity. If we close the switch, the generator ‘floats’ on the power supply, i.e., neither taking nor supplying electrical power to the system, though it is being rotated by the motor shaft. The motor receives a no-load current from the power supply.

When the generator ‘floats’ on the supply system, the motor is running at no-load condition. The generator can be made to deliver current to the armature of the motor by controlling different field currents. The motor receives current from the supply as well as from the generator. A counter torque appears on the motor due the generator’s loading. If we change either of the following, it will cause the generator to deliver the current to the motor:

(a) Increase the field current of generator
(b) Decrease the field current of motor

An increase in the generator’s field current causes the generator’s EMF, E_g, to rise above the supply voltage. This satisfies the condition, $E_g > V$ for power to flow from generator to motor. Similarly, a decrease in the motor’s field current causes the speed of motor to rise. As a consequence of this the generator’s EMF rises beyond the supply voltage. This again satisfies the condition, $E_g > V$, for power to flow from generator to motor.

Any decrease in the field current of generator or any increase of the field current of motor from the ‘floating’ condition causes the motor to act as generator and vice versa. When a generator delivers the current to the motor, the counter torque developed by the generator acts in opposition to motor-developed electromagnetic torque, thereby

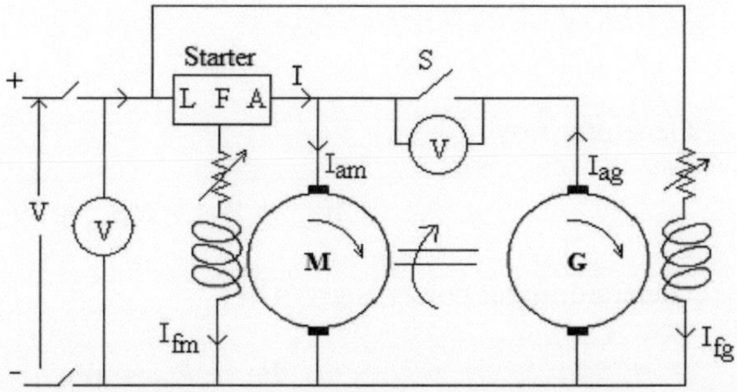

Fig. 4.68 Hopkinton’s test on two identical DC shunt machines

loading the motor. As a consequence of it motor demands more current from the electrical system.

Let the applied voltage be V, the motor and generator armature currents be I_{am} and I_{ag}, and their field currents be I_{fm} and I_{fg}, respectively. The resistances of armature and field winding for both the machines will be R_a and $R_{f,}$ respectively.

Motor armature currents is

$$I_{\text{am}} = I + I_{\text{ag}}$$

Motor armature circuit copper loss $= R_a I_{\text{am}}^2$
Generator armature copper loss $= R_a I_{\text{ag}}^2$
Total Iron and rotational losses $= VI - (R_a I_{\text{am}}^2 + R_a I_{\text{ag}}^2)$
Total iron and rotational losses for each machine,

$$W_C = 0.5 \times \left\{ VI - \left(R_a I_{\text{am}}^2 + R_a I_{\text{ag}}^2 \right) \right\}$$

Motor input is given as

$$P_{\text{im}} = VI_{\text{am}} + VI_{\text{fm}}$$

Motor loss is

$$W_{\text{mL}} = W_C + R_a I_{\text{am}}^2 + VI_{\text{fm}}$$

Motor output power is given as

$$P_{\text{om}} = P_{\text{im}} - W_{\text{mL}}$$

Motor efficiency is

$$\eta_m = 1 - \frac{W_{\text{mL}}}{P_{\text{im}}} \tag{4.101}$$

Generator output is

$$P_{\text{og}} = VI_{\text{ag}}$$

Generator loss

$$W_{gL} = W_C + R_a I_{\text{ag}}^2 + VI_{\text{fg}}$$

Generator input power is given as

$$P_{\text{ig}} = P_{\text{og}} + W_{gL}$$

Generator efficiency is

$$\eta_g = 1 - \frac{W_{gL}}{P_{\mathrm{ig}}} \tag{4.102}$$

In this test, both the machines are loaded up to their full capacities. The efficiencies can be determined at different load conditions. When an electrical machine is loaded, the fringe flux also increases. This interacts with the surrounding metallic parts and gives rise to additional load losses, which are not accounted in Swinburne's test. When the efficiency is determined using Hopkinson's test, both the machines are loaded. The efficiency determined using this method includes these stray load losses.

4.14.4 Retardation Test

Magnetic losses and rotational losses in DC machine can be separated using retardation test. The DC machine is run as motor slightly at higher than its rated speed using field flux weakening. Its speed is recorded after regular interval of time by switching OFF the supply. The kinetic energy of rotating parts is dissipated in terms of rotational losses.

Let P be the rotational loss, J the moment of inertia of rotating parts, and ω the angular velocity at any instant of time.

Then we can write,

$$P = \frac{\mathrm{d}}{\mathrm{d}t}\{\text{Kinetic Energy}\} = \frac{\mathrm{d}}{\mathrm{d}t}\left\{\frac{1}{2}J\omega^2\right\} = J\omega\frac{\mathrm{d}\omega}{\mathrm{d}t}$$

If N is the speed in RPM, the rotational speed is

$$\omega = \frac{2\pi N}{60}$$

Thus

$$\frac{\mathrm{d}\omega}{\mathrm{d}t} = \frac{2\pi}{60}\frac{\mathrm{d}N}{\mathrm{d}t}$$

Or

$$P = \left(\frac{2\pi}{60}\right)^2 JN\frac{\mathrm{d}N}{\mathrm{d}t}$$

If we know the moment of inertia 'J', the rotational losses may be determined. The machine may be allowed to come to halt by either of the following two ways:

(a) Both armature and field supply are disconnected. If machine is allowed to halt (come to standstill) when both armature and field supply is disconnected (without any excitation), power P will have rotational losses only.

Or

(b) Only armature is disconnected, but excitation or field terminals are still connected to supply. When field is remained excited, then P will consists of both the rotational losses plus ferromagnetic losses due to rotational magnetization. Here the field flux cuts the rotating armature. The time taken for the machine to come to halt will be less than that in case (a), because kinetic energy is dissipated in rotational loss as well as ferromagnetic loss.

This test requires the measurement of speed versus time curve during retardation. In case (a), a tachometer or tacho-generator is used to determine speed at different instants of time. However, in case (b) since field excitation is present, the induced EMF on the armature is calibrated with speed. The plot of speed versus time curve can be easily made. Depending on availability, one can use instruments such as data logger, storage oscilloscope, X–Y plotter, strip chart recorder, or even using dSpace kit. In the absence of these instruments, the speed is recorded manually using tachometer at certain intervals of time.

The moment of inertia can be calculated using physical dimensions of rotating part with some approximation. The moment of inertia can be experimentally determined by connecting a flywheel of known inertia '$J^{/}$' to the shaft and retarding the machine after switching OFF both armature and field excitation. For the same speed, losses will be same with or without flywheel.

In other words,

$$P = \left(\frac{2\pi}{60}\right)^2 JN\frac{\mathrm{d}N}{\mathrm{d}t} = \left(\frac{2\pi}{60}\right)^2 (J + J')N\frac{\mathrm{d}N}{\mathrm{d}t'}$$

Here we perform two retardation tests: one without fly wheel and the other with flywheel. When the supply is switched OFF, the kinetic stored energy of rotating parts is more when flywheel is present. As such, the time taken in the machine to halt is more in this case. The moment of inertia of the motor is given in terms of known moment of inertia,

$$J = J'\frac{\mathrm{d}N/\mathrm{d}t'}{\mathrm{d}N/\mathrm{d}t - \mathrm{d}N/\mathrm{d}t'}$$

For separation of iron losses, the retardation test is performed by switching OFF the armature only, but without additional flywheel. When armature is disconnected, motoring torque reduces to zero and a part of kinetic energy stored is wasted in iron losses. The latter is due to rotational magnetization.

The speed time curves of retardation tests have been shown in Fig. 4.69. The slope of the individual curves can be conveniently determined as shown in Fig. 4.69.

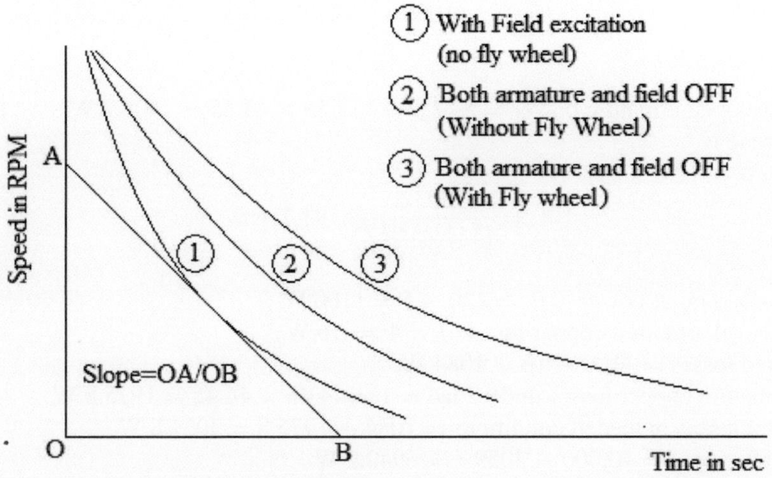

Fig. 4.69 Speed time curves of retardation tests

Example 4.26 A 220 V DC shunt motor takes 5.0 A at no-load and runs at 1000 RPM. The armature resistance is 1.0 Ω and field resistance is 220 Ω. Find the speed, the torque developed and efficiency when the motor input is 10 kW.

Solution

In No-load condition, we have $I_{\text{nl}} = 5$ A and $I_f = \frac{220}{220} = 1.0$ A.

The no-load armature current, $I_{\text{anl}} = 5 - 1 = 4$ A.

Back EMF is

$$E = 220 - 4 \times 1.0 = 216\,V$$

Under loaded conditions, we have

Motor input = 10 kW = 10,000 W

$$\text{Motor input current} = \frac{10,000}{220} = 45.45\text{ A}$$

$$I_a = 45.45 - 1 = 44.45\text{ A}$$

Back EMF is

$$E_2 = 220 - 44.45 \times 1.0 = 220 - 44.45 = 175.55\text{ V}$$

Since I_f is assumed to be constant, back EMF will be proportional to speed. Therefore, we get

$$N_2 = \frac{N_1 E_2}{E_1} = \frac{1000 \times 175.55}{216} = 813\,\text{RPM}$$

Developed armature power $= E_2\, I_{a2} = 175.55 \times 44.45 = 7803.2$ W.
Torque is

$$T = \frac{7803.2}{\left(\frac{2\times 3.14\times 813}{60}\right)} = 91.7\ \text{Nm}$$

No-load input power is $P_i = 220 \times 5 = 1100$ W.
No-load armature copper loss $= 1 \times 4^2 = 16$ W.
Fixed losses $= 1100 - 16 = 1084$ W.
Armature copper losses under load $= 1 \times 44.45 \times 44.45 = 1975.8$ W.
Total losses in loaded conditions $= 1084 + 1975.8 = 3059.8$ W.
Shaft output $= 10{,}000 - 3059.8 = 6940.2$ W.
Shaft torque,

$$T = \frac{6940.2}{\left(\frac{2\times 3.14\times 813}{60}\right)} = 81.56\,\text{Nm}$$

Percentage efficiency, $\eta = \frac{\text{Output}}{\text{Input}} \times 100 = \frac{6940.2}{10{,}000} \times 100 = 69.40\%$

Example 4.27 The Hopkinson's test on two shunt machines for full load gave the following results: Line voltage = 220 V; Line current excluding fields currents = 12A; motor armature current = 72A; field currents = 1.5A & 1.0A for generator and motor, respectively; the armature resistance of each machine is 0.2Ω. Calculate the efficiency of each machine.

Given $V = 220$ Volt, Line current $I = 12$A, motor armature current $I_{\text{am}} = 72$A, motor field currents $I_{\text{fm}} = 1$A, generator field current $I_{fg} = 1.5$A and resistance of each armature $R_a = 0.2\Omega$.
Generator armature current, $I_{\text{ag}} = I_{\text{am}} - I = 72 - 12 = 60$ A.
Loss in motor field winding, $P_{\text{fm}} = V\, I_{\text{fm}} = 220 \times 1.0 = 220$ W.
Loss in generator field winding, $\text{P}_{\text{fg}} = \text{V}\ \text{I}_{\text{fg}} = 220 \times 1.5 = 330$ W.
Loss in motor armature circuit $= I^2{}_{\text{am}}\, R_a = 72^2 \times 0.2 = 1036.8$ W.
Loss in generator armature $= I^2{}_{\text{ag}}\, R_a = 60^2 \times 0.2 = 720$ W.
Motor EMF $= V - I_{\text{am}}\, R_a = 220 - 72 \times 0.2 = 205.6$ V.
Generator EMF $= V + I_{\text{ag}}\, R_a = 220 + 60 \times 0.2 = 220 + 12 = 232$ V

$$\text{Total Iron and rotational losses} = \text{VI} - \left(R_a I_{\text{am}}^2 + R_a I_{ag}^2\right) = 220 \times 12 - 1036.8 - 720$$
$$= 883.2\ \text{W}$$

Core loss per machine, $W_c = 0.5 \times 883.2 = 441.6\,\text{W}$.
Motor Input,

$$P_{\text{im}} = VI_{\text{am}} + VI_{\text{fm}} = 220 \times 72 + 220 \times 1 = 16,060\,\text{W}$$

Motor losses,

$$W_{\text{m}} = W_c + R_a I_{\text{am}}^2 + VI_{\text{fm}} = 441.6 + 0.2 \times 72^2 + 220 \times 1 = 1698.4\,\text{W}$$

Motor output power, $P_{\text{om}} = P_{\text{im}} - W_m = 16,060 - 1698.4 = 14,361.6\,\text{W}$
Motor efficiency, $\eta_m = 1 - \frac{W_m}{P_{\text{im}}} = 1 - \frac{1698.4}{16060} = 89.4\%$
Generator output, $P_{\text{og}} = VI_{\text{ag}} = 220 \times 60\,\text{W}$.
Generator losses, $W_g = W_c + R_a I_{\text{ag}}^2 + VI_{\text{fg}} = 441.6 + 0.2 \times 60^2 + 220 \times 1.5 = 1491.6\,\text{W}$.
Generator input power, $P_{\text{ig}} = P_{\text{og}} + W_g = 220 \times 60 + 1491.6 = 14691.6\,\text{W}$.
Generator efficiency, $\eta_m = 1 - \frac{W_g}{P_{\text{ig}}} = 1 - \frac{1491.6}{14691.6} = 89.84\%$.

4.14.5 Field's Test on Two Identical DC Series Motors

In this method, two identical DC series motors are required, which are mechanically coupled. Such a combination of DC series motors is also used in traction, in which both the machines act as motors. In traction / electric vehicle application, they can be connected either in series or in parallel for starting and as well as for speed control purposes.

In Field's test one of the machine acts as a motor, whereas the other machine acts as a loading generator. Since the field winding resistance of both the series machines is low, the separate excitation of the generator from available voltage source becomes difficult. If the field winding resistance of certain DC series motor is (say) 1.0 Ω and its rated current is 30A, then for separate excitation, we require a DC source of 30 V capable of supplying 30A DC current. However, in Fields test, the generator is separately excited by connecting its field winding in series with the motor circuit, as shown in Fig. 4.70.

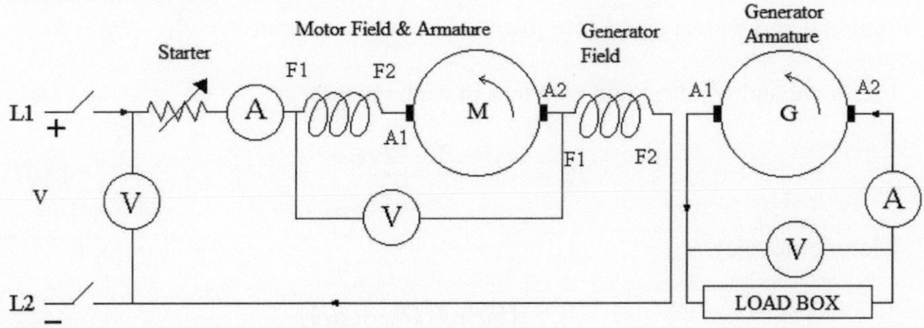

Fig. 4.70 Field's test on two identical DC series motors

Since two identical machines have similar field and armature windings, it can be assumed that their iron losses will be approximately identical. The motor is started using 2-point resistance type DC series motor starter. This starter is normally referred as *drum controller* (because of its shape and rotary mechanical switches). Unlike to the shunt motor starter, it is used for starting, speed control, braking as well as reverse rotation of the series motor. The resistance in series starter of the DC series motor carry full load current on continuous basis.

A DC series motor should not be started and run on no-load, otherwise it will attain dangerously high speed, thereby damaging commutator, carbon brushes and coupling. Therefore, it must be ensured that full load is connected across the generator terminals at the instant of starting.

Now, let us assume that V is supply voltage V_1 is the motor terminal voltage, I_1 is motor input current, V_2 is generator output voltage, I_2 is generator current. Let the resistances of the armature and field, respectively, be R_a & R_f.

Power Input to the entire set,

$$P_1 = VI_1 \tag{4.103}$$

Power Input to the motor,

$$P_{\mathrm{im}} = V_1 I_1 \tag{4.104}$$

Power Output of the generator,

$$P_g = V_2 I_2 \tag{4.105}$$

$$\text{Losses in the entire set} = P_1 - P_g$$

Total copper losses for both the machines,

$$P_{\mathrm{CU}} = I_1^2 (R_a + 2R_f) + I_2^2 R_a \tag{4.106}$$

$$\text{Frictional and windage and core losses of both the machines } = P_1 - P_g - P_{\mathrm{CU}}$$

Frictional and windage and core loss of each machine,

$$P_o = \frac{P_1 - P_g - P_{\mathrm{CU}}}{2} \tag{4.107}$$

Motor efficiency,

$$\eta_m = 1 - \frac{(P_o + I_1^2 (R_a + R_f))}{P_m} \tag{4.108}$$

$$\text{Generator input} = P_g + P_o + I_2^2 R_a + I_1^2 R_f$$

Therefore, generator efficiency is

$$\eta_g = \frac{P_g}{P_g + P_o + I_2^2 R_a + I_1^2 R_f} \tag{4.109}$$

Torque is

$$T = \frac{V_2 \times I_2 \times 60}{2 \times \pi \times N} \text{ Nm} \tag{4.110}$$

The method is useful for determining efficiency of larger capacity series motors. Using Field's test, the torque-speed characteristics of DC series motor can also be determined under different operating conditions.

Example 4.28 Two identical DC series motors are connected for Fields Test. Following is the test data: Armature resistance $R_a = 1.0\Omega$ and field resistance $R_f = 0.82\Omega$ for both the machines; $V_s = 274$ V; Motor voltage 102 V; generator voltage 110 V; motor current 9.5 A and generator current 17.5 A. Set is running at a speed of 1440RPM.

Calculate the efficiencies of the motor and the generator. Also, calculate the torque.

Solution

Given $V_s = 274$ V; $V_m = 102$ V; $V_g = 110$ V; $I_m = 9.5$A; $I_g = 17.5$ A; $R_a = 1.0\ \Omega$; $R_f = 0.82\ \Omega$.

Input Power is

$$P_I = V_s I_m = 274 \times 9.5 = 2603 \text{ VA}$$

Output power is

$$P_g = V_g I_g = 110 \times 17.5 = 1925 \text{ VA}$$

Power input to motor

$$P_{\text{im}} = V_m I_m = 102 \times 9.5 = 969 \text{ VA}$$

Losses in entire set are given as

$$P_L = P_I - P_g = 2603 - 1925 = 678 \text{ VA}$$

Total copper losses for both the machines are

$$P_{\text{CU}} = I_m^2 (R_a + 2R_f) + I_g^2 R_a = 9.5^2 \times (1.0 + 2 \times 0.82) + 17.5^2 \times 1.0 = 544.51 \text{ W}$$

Frictional, windage and core losses of both the machines are given as

$$P_1 - P_g - P_{cu} = 678 - 544.51 = 133.5\,\text{W}$$

Frictional, windage and core loss of each machine is given as

$$P_o = \frac{P_1 - P_g - P_{CU}}{2} = \frac{133.5}{2} = 66.75\,\text{W}$$

Motor efficiency,

$$\eta_m = 1 - \frac{\left(P_o + I_1^2(R_a + R_f)\right)}{P_m} = 1 - \frac{66.75 + 9.5^2(1.0 + 0.82)}{969} = 76.16\%$$

Generator input $= P_g + P_o + I_g{}^2\,R_a + I_m{}^2 R_f$
$= 1925 + 66.75 + 17.5^2 \times 1.0 + 9.5^2 \times 0.82 = 2433.75$ VA
Therefore, generator efficiency is

$$\eta_g = \frac{P_g}{P_g + P_o + I_2^2 R_a + I_1^2 R_f} = \frac{1925}{2433.75} = 79.1\%$$

Torque is

$$T = \frac{V_g \times I_g \times 60}{2 \times \pi \times N} = \frac{1925 \times 60}{2 \times 3.14 \times 1440} = 12.77\,\text{Nm}$$

4.15 Cross-Field Machines

There are mainly two types of rotating amplifiers or cross-field machines:

(a) Amplidyne
(b) Metadyne

These machines are being rotated using prime mover and act as controlled generators. A schematic diagram of an amplidyne is shown in Fig. 4.71. The main field flux ϕ_f is in the polar axis or d-axis, which induces a rotational voltage across q-axis brushes. The q-axis winding is made with thicker conductors and has very low resistance and is short circuited. A very low induced EMF in q-axis winding E_q results in a very large current. This sets up an additional flux ϕ_q in q-axis, quadrature to d-axis. The d-axis winding interacts with the very large flux ϕ_q and induces a large rotational EMF E_d across the d-axis brushes. If a load is connected across the d-axis

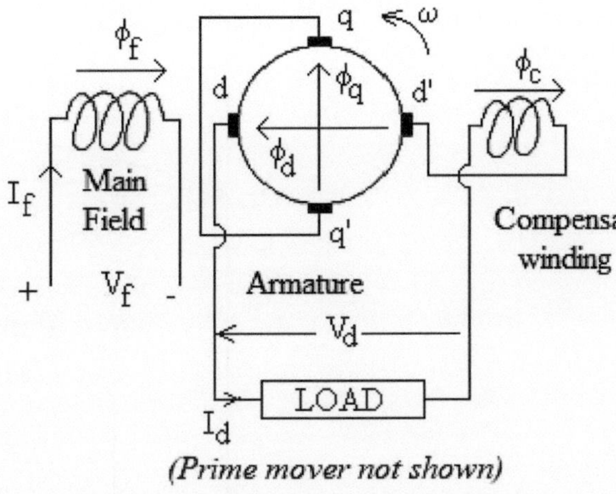

Fig. 4.71 Cross-field machine-amplidyne

brushes a current I_d flows. This current sets up additional flux ϕ_d in the armature opposite to the main field flux ϕ_f. Since two fluxes ϕ_d and ϕ_q are in quadrature or cross each other, the machine is referred to as *cross-field machine*.

When the d-axis flux ϕ_d opposes the main field flux ϕ_f, the voltage V_d falls. The machine is known as **Metadyne** having 0% compensation. The terminal characteristics of cross-field machines are shown in Fig. 4.72. A Metadyne acts as a constant current generator. A class of such machine has major applications in battery charging in train lighting. These generators were also used to supply power to the rotating machines, which undergo into stalling mode of operation during its duty cycle like dredging machines used to clean river bed. Currently being replaced by Current Source Inverter fed induction motor.

If an additional winding is put in the d-axis carrying the load current I_d, the effect of armature reaction can be nullified. This produces sufficient flux ϕ_c in the polar axis, which attempts to counter the effect of load current or ϕ_d. With compensating winding, the machine is known as **Amplidyne**. The Amplidyne may have several control- windings, which may carry the control-signals for controlling certain processes.

Let the main field is energized by a voltage source V_f, the current be i_f. We can write

$$V_f = R_f.i_f + \frac{d\phi_f}{dt} \tag{4.111}$$

The flux linkage ψ_f gives rise to rotational voltage in q-axis winding.

$$E_q = k_q \phi_f \omega \tag{4.112}$$

The q-axis winding is short circuited, which gives rise to q-axis flux linkage Φ_q.

Fig. 4.72 Terminal characteristics of cross-field machines

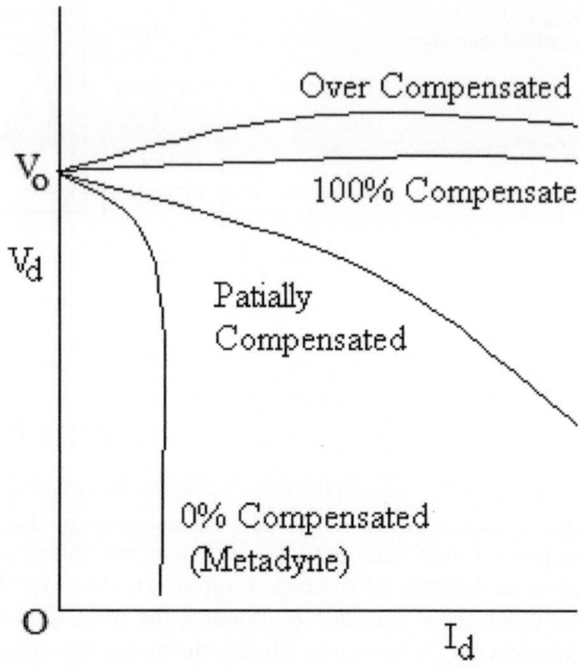

$$E_q = R_q.i_q + \frac{d\phi_q}{dt} \tag{4.113}$$

Since the resistance of q-axis winding is very low the short circuit current i_q results in a very strong flux linkage ϕ_q in q-axis. This results in d-axis rotational induced EMF, given by

$$E_d = k_d \phi_q \omega \tag{4.114}$$

The voltage E_d is connected across load. The load current is I_d. This sets up an additional flux in d-axis and is opposition to the main field flux linkage ϕ_f. As already mentioned, the voltage drastically falls supplying a constant load current. The machine is known as **Metadyne** generator. In the presence of compensating windings carrying load current, a flux ϕ_c opposing the d-axis flux is created. This overcomes the effect of armature reaction. If the compensating winding is having N_c number of turns carrying current I_d, then (4.111–4.114) will accordingly modify. The machine is referred to as **Amplidyne**. Depending on the number of turns in the compensating windings, the terminal characteristics can be changed accordingly.

A classic example of closed loop control using Amplidyne is shown in Fig. 4.73, where the voltage of an alternator is being maintained constant. Here an Amplidyne is used to control **the alternator field current according to the generated voltage. The voltage generated is stepped down using a transformer** and converted into

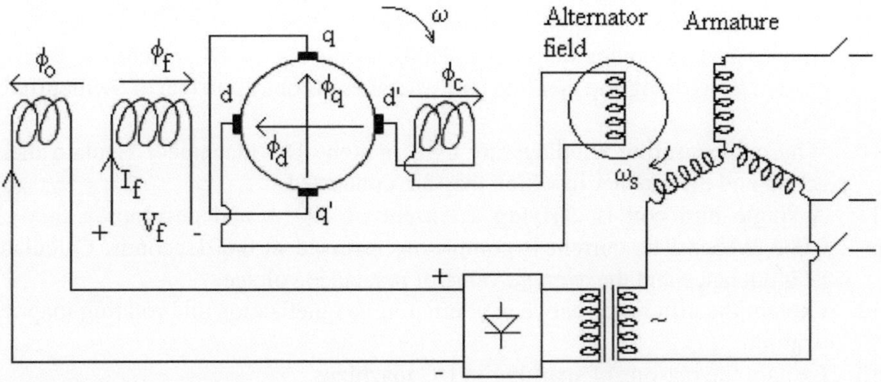

Fig. 4.73 Application of cross-field machines in classical control problem- Closed loop control of an alternator terminal voltage

DC using a rectifier. This voltage is proportional to the generated voltage, which is fed into a control winding in the polar axis. The flux produces by the control winding ϕ_o opposes the main field flux ϕ_f. When alternator is loaded, the developed voltage falls. As a result the flux ϕ_o also falls, but the net flux ($\phi_f - \phi_o$) rises, thereby causing the field current of the alternator to rise for maintaining the original voltage. When the load of alternator is withdrawn, the voltage tends to rise. The flux of control winding ϕ_o tends to rise, causing the net flux ($\phi_f - \phi_o$) to fall. As a result, the Amplidyne output, i.e., the field current of the alternator falls for keeping alternator output at rated voltage. The system stability depends on many factors, like turn ratio of step-down transformer, number of turns in control field winding and moment of inertia.

Cross-field machines have limited applications owing to the development of solid-state controllers. These machines are useful in battery charging in locomotives and in applications where large DC power is to be controlled.

Problems

1. Discuss the conditions of self-excitation of DC shunt generator.
2. Why a starter is required for starting a DC motor? Discuss the problems associated with operating a DC shunt motor, when the grading of starting resistance is incorrect.
3. Explain the need for a commutator in a DC machine.
4. Comment on the statement:

   'The field of a DC shunt motor operating at a rated speed should not be opened'. Also, discuss the field failure protection provided in a three-point starter.
5. Can regulation of a DC generator be negative? Discuss in brief.
6. Discuss the salient features and applications of cumulative and differential compound DC motors and generators.
7. Discuss, in brief, the constructional details of DC machine.

8. What is the object of equalizer connection?
9. Draw a neat schematic diagram of Hopkinson's test on DC machines. Briefly explain its working and mention its advantages in contrast to that of Swinburne's test.
10. Why compensating windings are used in some DC machines? Explain their action and the manner in which they are connected.
11. A single turn coil is carrying a current of 100 A and develops a flux of 0.003 Weber. The current is completely reversed in 0.001seconds. Calculate its inductance and the average value of reactance voltage.
12. Explain the difference between alternating magnetization and rotating magnetization.
13. Explain the reasons of sparking in DC machines.
14. What is the function of interpoles?
15. What is the purpose of providing 'brush-lead' and 'brush-lag' in a DC machine? Explain briefly.
16. Discuss the various methods of speed control of DC shunt motor, giving its torque-speed characteristics.
17. In conventional DC motor having p number of poles running at N rotations per minute, what is the frequency of armature rotating magnetization?
18. Compare three point and four point starters of DC shunt motor.
19. Explain the counter current braking of a DC shunt motor using its torque-speed characteristics. What are essential precautions?
20 Explain the terms 'critical resistance' and 'critical speed'.
21. A shunt motor operates on AC supply 220 V, 50 Hz. The motor parameters are $R_a = 0.9\ \Omega$; $R_f = 180\ \Omega$; $L_{\text{ff}} = 20$ H; $L_{\text{aa}} = 0.05$ H; $L_{\text{fa}} = 0.9$ H. Discuss its operation as a motor

    Note: The armature torque is insignificant in comparison with the effective armature current. The inductance of field winding is much higher than that of armature. Thus when field flux is at its peak value the armature current is feeble. The field current lags the armature current. At $N = 0$, θ_a=1.51° lagging and θ_f = 88.36° lagging.
22. A series motor operates on AC supply 220 V, 50 Hz. The motor parameters are $R_a = 0.9\ \Omega$; $L_{\text{aa}} = 0.05$ H; $R_f = 0.09\ \Omega$; $L_{\text{ff}} = 0.09$ H and $L_{\text{fa}} = 1.5$ H calculate the average torque and speed and power developed when input current is 10 A.

    Note: Motor operates on DC as well as on AC with almost identical performance. This type of motor is referred universal motor. The phase angle of field current is identical to armature current. The field flux and armature current are in the same phase.
23. Calculate the maximum value of the induced EMF in the conductors of an armature having 60 cm diameter and 30cm axial length, rotating at 1000 RPM. The maximum value of flux density in air-gap is 0.9 T.
24. A ten-pole armature is having 1500 conductors and 8 number of parallel path. The flux per pole is 0.8 Wb. The armature is rotating at 500 RPM. Calculate induced EMF.

25. Determine the value of torque exerted by an armature of six-pole motor having 800 conductors and two parallel paths. The flux per pole is 0.02 Wb and total current is 20 A.
26. A 500 kW, 500 V, six-pole DC generator has a lap connected armature with 90 slots and six conductors per slot. Estimate the number of turns required on each interpole if the effective interpole air-gap is 1.2 cm and the flux density in air-gap under the inter pole is 0.3 Wb/m^2.
27. The no-load EMF of a separately excited DC generator is 440 V, the voltage drop across the brushes is 2 V, the armature resistance R_a is 0.2 Ω, mechanical losses P_m is 200 W and iron loss is P_{fe} is 100 W. Calculate

    (a) The voltage across the terminals for a rated current of 30A
    (b) the load resistance
    (c) the machine efficiency, assuming that the excitation power represents 3% of the useful output power
    (d) the mechanical power required to drive the generator.

28. Two identical separately excited DC generators has four poles, wave wounded total conductors 1000 and R_a of 1.0 Ω. Both are connected in parallel and supply a common load. The excitation of first machine is 10 mWb while that of the second machine is 10.5 mWb. Both are running at 1000 RPM. How much of the total load current of 40 A is supplied by each of the two generators? The voltage drop across the brushes and armature reaction may be ignored.
29. A two-pole DC shunt motor operates from a constant supply voltage of 220 V. The total current fed is 25 A and the excitation current is 1.25 A. The simple lap winding has 500 conductors and flux is 20 mWeber. If the machine efficiency is 0.87 and the armature resistance is 0.30 Ω. Calculate the motor speed and the useful torque.
30. Two 220 V, 10HP separately excited DC motors are operating on a constant voltage supply of 220 V. The motors are coupled to the same shaft and their combined electromagnetic torque is T. How do these two motors share the load torque, when all their parameters are identical but for the following:

    (a) Armature resistances are different, having a ratio of 1.1
    (b) Field resistances have a ratio of 1.5.

    Ignore the effect of armature reaction.
31. Two separately excited DC motors operate at a load torque T_L, which is directly proportional to speed ω. The armature of first motor is connected to constant voltage source, while second armature is connected to a constant current source. The field windings of both motors are supplied by constant voltage source. How should the field current be varied for increasing the speed of two motors?
32. A separately excited DC generator runs at a constant sped of 1000RPM. The ratings are 20 kW, 250 V, R_a is 0.9Ω and L_{aa} is 5.0mH. The generator was supplying rated power when the terminals were short-circuited. Calculate.

    (a) Variation with time of the armature current after short circuit has occurred.

(b) For the protection of the generator, a switch opens the armature circuit 0.09 s after the current has reached two times the rated current. What current will be flowing when circuit is switched OFF?

33. A DC series motor has rating of 500 V, 100 A and 800 RPM. The resistance of field winding and armature are 0.045 Ω and 0.20 Ω, respectively. Determine the value of armature resistance, which is connected for developing one-half of full-load torque at 1000 RPM. The machine is operating at an efficiency of 89%.
34. Four identical DC series motors are connected in series and in parallel connections in a traction application. Obtain the expressions for speed and torque when

    (a) All are in series
    (b) All are in parallel.

    Choose your own parameters.
35. Four identical DC generators, each of rating 220 V, 100 kW, are connected in parallel. They are running at 1000RPM. They exhibit speed drop from no-load to full-load as 1, 1.25, 1.75 and 2%. If they are supplying a common load of 400 kW.

    (a) determine the load supplied by each of the generators
    (b) which of the generators will supply maximum and minimum shares of load.
    (c) if one of the generator exhibits a rise in speed instead of fall in speed when loaded. How will these generators behave?

36. Two DC generators each having a capacity of 500 kVA, 500 V, are connected in parallel to supply a power of 600kVA. Their open circuits EMFs are 510 V and 508 V, respectively. Their internal resistances are 0.010Ω and 0.011Ω, respectively. Determine.

    (a) the common terminal voltage
    (b) current shared by individual generators
    (c) steady state short circuit current
    (d) circulating current

37. The speed of a separately excited DC motor is controlled by a chopper circuit. The DC supply voltage is 220 V, armature circuit resistance is 0.9 Ω, the armature circuit inductance is 33 mH and the motor EMF constant is 0.23 Volt /RPM. The motor drives a constant torque load, requiring an average armature current of 22 A. Assuming, the motor current to be continuous, determine the speed control range of the duty cycle.
38. The speed of a 30 HP, 440 V, 600 RPM separately excited DC motor is controlled by single-phase fully controlled converter. The machine is operating at an efficiency of 0.87%. The armature resistance is 0.15 Ω and armature inductance is 10 mH. The AC supply voltage is 440 V and the EMF constant is 0.197 V/

RPM. Assuming the motor current to be continuous and ripple free, determine the speed of motor, torque, power when the machine is drawing rated current at firing angle of 40°.

39. Hopkinson's test on two DC machines gave the following test results for full load: Line voltage 250 V; line current excluding field current 50 A; motor armature current 380 A; field currents are 5 A each for both the machines. The armature resistance of the motor is 0.02 Ω and that of generator is 0.025 Ω. Draw schematic diagram and calculate the efficiency of each machine. State the assumptions made.

40. The speed of 75 HP, 400 V separately excited DC motor driven by three-phase full converter is 1000RPM. The supply being three-phase 440 V, 50 Hz. The machine is operating at rated condition at 87.5% efficiency. The parameters of the motor are $R_a = 0.0635\ \Omega$, $L_a = 7.7$ mH and EMF constant is 0.335 V/RPM. Determine.

    (a) the firing angle to obtain the rated speed at rated motor current
    (b) the no-load speed at firing angle 0° & 30°

    Assume that no-load current is about 12 % of full rated current.

41. A DC shunt generator has the following data of its OCC at 1000RPM:

I_f (A)	0.2	0.4	0.6	0.8	1.0	1.2	1.4
V_{oc} (V)	140	275	375	440	500	540	570

Obtain (a) the no-load induced EMF vs speed characteristics with field circuit resistance (b) Critical field resistance.

42. Design the graded resistances for a six-contact starter for starting a 50 HP, 500 V shunt motor. The lower current limit is to be 120% of the full load current. The full load efficiency is 91% and total copper losses are 6% of the input. The field resistance is 200 Ω.

43. A five-step automatic starter for a 20 HP, 220 V shunt motor operates on the current limit principle. The armature resistance is 1.0 Ω. If the initial starting current is 2.5 times its full load rated current, find the lower current limit and resistances of the starter.

44. A 250 V DC shunt motor takes a line current of 60 A at 1000 RPM. Its armature and field resistances are 0.2 and 125 Ω, respectively, and brush contact drop is 2.0 V. Calculate.

    (a) No-load speed if the no-load current is 6A
    (b) Percentage reduction in the flux per pole so that the speed may be 1000RPM when armature takes 40A current.

45. Two DC generators, each having linear characteristics, are operating in parallel and supply a total load of 160 A. The terminal potential difference of one machine falls from 240 V on no-load to 220 V when current output is 100A.

The terminal potential difference of second machine falls from 240 V on no-load to 216 V when current output is 80A. Determine.

(a) Current supplied by each machine
(b) Terminal voltage

46. Two DC generators, each having a capacity of 100 kVA are connected in parallel. How will they share a load of 200 kVA when.

    (a) they have identical open circuit EMF of 500 V but their voltage falls to 2.5% and 3.0% when fully loaded
    (b) Their open circuit EMF are 500 and 495 V, but their governors have identical speed regulation of 1.8%.

47. A 250 V shunt motor has a five-step starter. The maximum current during starting should not exceed 50 A. Armature resistance is 0.54 Ω. Find the resistance of different steps.
48. The armature of a six-pole DC machine has 125 turns and runs at 1000 RPM. The EMF generated in open circuit is 500 V. Find the useful flux per pole when the armature is Lap wound. For the same value of useful flux per pole, determine the speed of the machine to generate 600 V during open circuit.
49. A DC shunt motor is being operated from 250 V mains. Its no-load speed is 1200 RPM. When fully loaded, it delivers a torque of 400 Nm and its speed drops to 1100 RPM. Find the speed and power output when delivering the same torque, if operated with an armature voltage of 500 V. Excitation is assumed to remain unchanged, i.e., motor field is still excited at 250 V. State the assumption you are required to make.
50. A 500 V DC shunt motor takes 9.0 A at no load and runs at 800 RPM. The armature resistance is 0.10 Ω and field resistance is 125 Ω. Find the speed and the torque developed when the motor input is 15 kW.

Objective-type Questions:

1. Conventional DC machines are

    (a) Homo polar machines
    (b) Hetro-polar machines
    (c) Uni-polar machines
    (d) Claw pole machines

2. In DC motors, the commutation of coil is done using

    (a) Electronic circuits
    (b) Commutator
    (c) Position based switching
    (d) Mechanically

3. Following statement is correct

(a) In Lap winding, 'finish' of one coil is connected to 'start' of another coil one pole pitch apart.
(b) In Lap winding, 'finish' of one coil is connected to 'start' of another coil two-pole pitch apart.
(c) In Wave winding, 'finish' of one coil is connected to 'start' of another coil one pole pitch apart.
(d) In Wave winding, 'finish' of one coil is connected to 'start' of another coil two-pole pitch apart.

4. DC shunt motor and DC series motor are, respectively, referred as

(a) Constant Speed and constant power
(b) Constant Power and Constant Torque
(c) Constant Torque and constant power
(d) Constant power and constant speed

5. In a shunt motor, if the armature and field windings are connected in series, the machine will satisfactorily work as a Series motor.

(a) Yes
(b) No
(c) Doubtful
(d) Sometimes

6. A DC series motor can work satisfactorily on

(a) Pure DC supply
(b) DC supply having ripples
(c) AC supply
(d) All the above

7. A three-point starter is designed to operate in the current limit of I_1 and I_2 during the starting period. If the starter resistances are cut off in reverse order,

(a) The starting current is limited to $I_1 - I_2$.
(b) The starting current will shoot the highest current limit
(c) The starting current will always be less than lowest current limit
(d) None

8. In three-point starter of dc shunt motor, which of the following statement is correct

(a) Can be used when wide range of speed control is desired.
(b) There is an inherent field failure protection.
(c) Can be used to start DC series motor also.
(d) None

9. In Field's Test on two identical series machines, the machine acting as loading generator is

(a) Self-excited

   (b) Its field winding is connected in series with the motor circuit
   (c) Separately excited and connected to a DC source
   (d) Its Field winding is connected in series with the generator circuit.

10. In Hopkinson's Test, the machines can be loaded by

   (a) Increasing the field current of generator or decreasing the field current of motor
   (b) Making the two field current equal
   (c) Increasing the field current of motor or decreasing the field current of generator
   (d) Opening the field terminal of generator

11. Electrical motor used in traction locomotive may be

   (a) DC Shunt motor
   (b) DC series motor
   (c) PMDC motor
   (d) DC arc motor

12. Following is incorrect statement

   (a) A DC Series motor should not be started without load
   (b) The disconnection of field winding from circuit should be avoided when DC motor is running
   (c) A DC shunt motor should not be started without load.
   (d) Three-point and four-point starter are used for starting DC shunt and compound motors.

References

1. E. Openshaw Taylor: The Performance and design of A.C. commutator motors, Wheeler Publications, India, 11th Print, 1990.
2. A. Fransua, R. Magureanu, Electrical Machines and Drive Systems, TP Technical Press, Oxford, 1984
3. Clayton & Hancock, Performance & design of direct current machines, ELBS
4. M.G. Say & E.O. Taylor, Direct current machines, ELBS, 1980
5. V Del Toro, Electric machines and power systems, PHI, 1980
6. S.A.Nasar, Electromagnetic energy conversion & Systems, Prentice-Hall, Inc., N.J., 1970
7. A. Draper, Electrical Machines, ELBS, 1st Ed, 1956
8. J. R. Hendershot Jr. & THE Miller, Design of Permanent-Magnet Motors, Clarendon Press, Oxford, 1994
9. Stephen J Chapman, Electric Machinery Fundamentals, Indian Edition, 4e, Mcgraw hill education (Indi) Pvt Ltd, 2005
10. Siskind, Electrical Machines, 2nd edition International student edition, McGraw hill international book company, 12th printing, 1983

Chapter 5
Induction Machines

Abstract Constructional details-Three-phase armature winding:-integral- and fractional-; MMF distribution; distribution factor; EMF generation; chording; EMF induced and elimination of harmonics; pitch factor; Rotating magnetic field; Concept of time and space harmonics; Concept of forward and reverse rotating magnetic field; Effect of slotting-elimination of slot harmonics-skewing; Cogging and crawling; Principle of operation of induction machines; Equivalent circuit; phasor diagram; power intake, air-gap power and torque; Maximum torque and the slip at maximum torque; Ratio of V/f and maximum torque; Complete torque slip characteristics; Free running or no-load test; Separation of friction and windage loss; Locked rotor test-skin effect; Circle diagram; Use of circle diagram; Starting of three-phase induction motor; Speed control and braking; Cascade connection; injection of EMF in rotor circuit; Kramer's and Scherbius cascade; Induction generator Self- and grid- excited; Georges's phenomenon; Deep bar and double cage rotor; Single-phase induction motor-equivalent circuit and starting.

Keywords Induction · Squirrel cage · Slip ring · NEMA · Armature · Windings MMF distribution · Pitch factor · Short pitch · Rotating magnetic field · Space and time · Slot harmonics · Cogging and crawling · Equivalent circuit · Determination of equivalent circuit parameters · Torque-speed · Load test · Circle diagram · Starting · Speed control and braking · Scherbius and Kramer cascade · Induction generator · George's phenomenon · Deep bar rotor · Single phase induction motor

5.1 Induction Machine

Induction machines work on the principle of electromagnetic induction. An elementary form of induction motor was invented by Nicolas Tesla in 1891. A cage induction motor was developed by Dobrowolsky. The slip-ring induction motors came in the beginning of twentieth century. Induction motors are supplied by three-, two- or

R. K. Srivastava, *Electrical Machines*, https://doi.org/10.1007/978-981-96-5039-2_5

single-phase power. Three-phase squirrel cage induction motors have many advantages, like rugged construction, lesser maintenance, and absence of brush and associated sparking and hence, they are widely used. Squirrel cage induction motors are preferred over DC motors. About 80% motors in the whole world are squirrel cage induction motors. DC motors give superior control in wider range of speed. On the other hand, the speed of induction motor is restricted by the supply frequency. Nowadays, due to research and development in solid-state power devices and computer control of drives, induction motors are being applied even to areas where wide range of speed control is desired. This has been achieved by solid-state inverter, VVVF (Variable Voltage Variable Frequency) control, vector control and direct torque control etc. The power to the rotor is transferred by means of induction, without any electrical power connection to the rotor. Induction motors are increasingly being considered in the application areas, where earlier DC motors were used.

5.2 Constructional Details

Three-phase armature winding is normally on the outer side, on the *stator*. Schrage-motors and certain single-phase induction motors have an inverted structure. In the case of the Schrage-motors, the rotating armature winding is on the inner rotating side of the machine, whereas in the case of single-phase induction motors used in ceiling fans, the stationary armature is on the inner stationary side. Most of the induction motors have stationary armature winding on outer side [1–5 and 7–9].

The magnetic core of armature is made of laminated electrical grade hot rolled non-grain oriented sheet steel stampings having regularly distributed semi-closed slots on the inner periphery of the annular stampings. These slots carry three-phase distributed windings. The stamping of a typical three-phase armature is shown in Fig. 5.1. The windings may be connected in star or delta configuration.

The rotor of induction motor is of following two types:

(a) Squirrel cage
(b) Slip ring

Figure 5.2 shows the stamping of squirrel cage rotor. The outer periphery carries regularly distributed closed-slots (or semi closed-slots). On each of these slots, a rotor-bar of bare conductor of copper alloys is inserted/hammered from one end to the other end. The cross-sectional area of rotor slot and conductor are almost identical. Both the ends of these conductors of each of the slots are short-circuited by means of two end-rings made of conducting material like copper. The squirrel cage winding is short-circuited from both the ends. A typical squirrel cage which is embedded in rotor slots is shown in Fig. 5.3. The rotor currents develop rotor MMF, corresponding to the number of stator poles. Since the power to the rotating winding is transferred by means of induction, there is a complete absence of sparking in its operation. The cost of cage type induction motor is lesser than that of any other motor. No maintenance required and faults are extremely rare to occur in cage windings. The

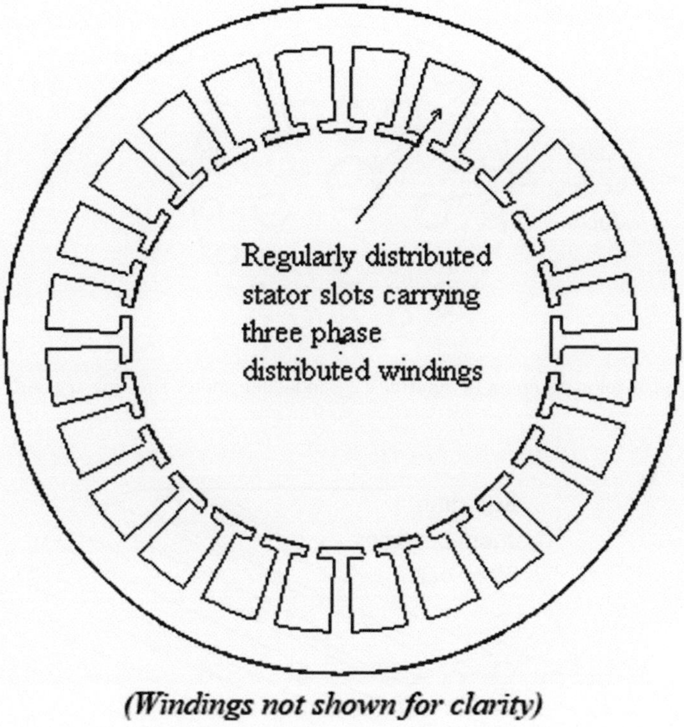

Fig. 5.1 Stator stamping showing regularly distributed slots

only difficulty with cage rotor is that the rotor electrical parameters and thus overall characteristics of the machine cannot be changed, while running on power frequency source. However, due to advent of solid-state Variable Voltage Variable Frequency (VVVF) and vector control drives, these motors are finding greater attention in newer areas of applications. The control of torque and speed are achieved by changing both the supply voltage and frequency.

Figure 5.4 shows the rotor stamping of slip –ring induction motor. The regularly distributed semi-closed slots on the outer periphery carry three-phase star-connected distributed winding, as that of stator. In case of slip-ring induction motors, the rotor parameters and the characteristics of the induction motor can be changed by varying rotor resistances. The star-connected three-phase slip-ring winding is connected to the three slip rings provided on the shaft. These conducting slip rings are insulated from the shaft. A set of three stationary carbon brushes is electrically connected to rotating rotor through slip rings. When these slip rings are not short-circuited and three-phase power is supplied on the stator winding, it results in the induction of three-phase voltage across slip rings at standstill condition. The slip-ring induction motor with its open rotor windings acts as a transformer. The induced EMF across slip rings depends on the number of turns on stator and rotor as well as on the layout of their windings. For the operation as induction motors, the three-phase slip rings are short-circuited, which allows the rotor currents to flow and interact with the

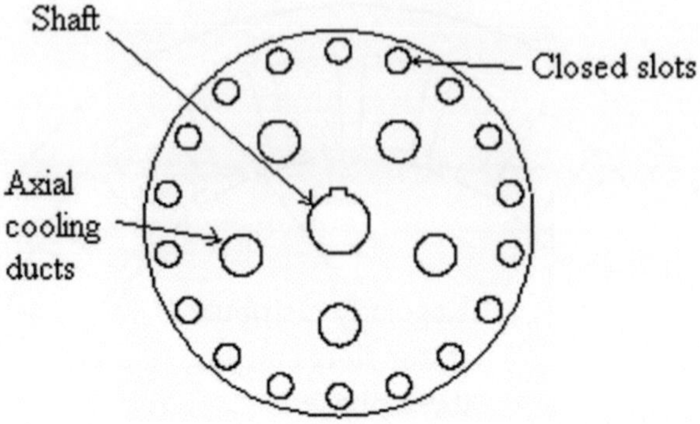

Fig. 5.2 Typical rotor stamping of squirrel cage induction motor showing regularly distributed closed slots

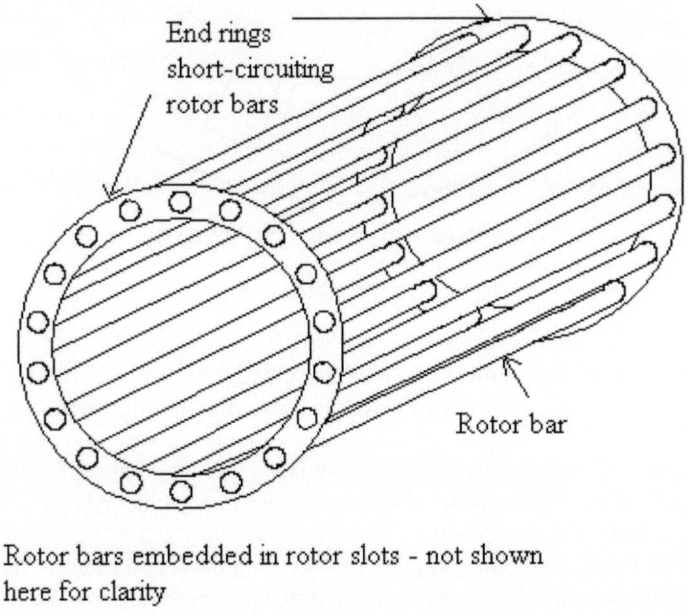

Fig. 5.3 A typical squirrel cage rotor embedded in rotor stampings (*stampings not shown for clarity*)

stator flux. For changing the rotor parameters, three-phase star-connected balanced resistances can be connected to the three rotor terminals. The number of slots in the stator and rotor differs from each other to avoid locking of rotor and stator teeth, a phenomenon known as *cogging*. Certain space harmonics results in *crawling*, where the induction motor continues to run at about on one-seventh of its synchronous speed. The rotor shaft is supported by two bearings provided on the end covers, which are fixed to stator frame.

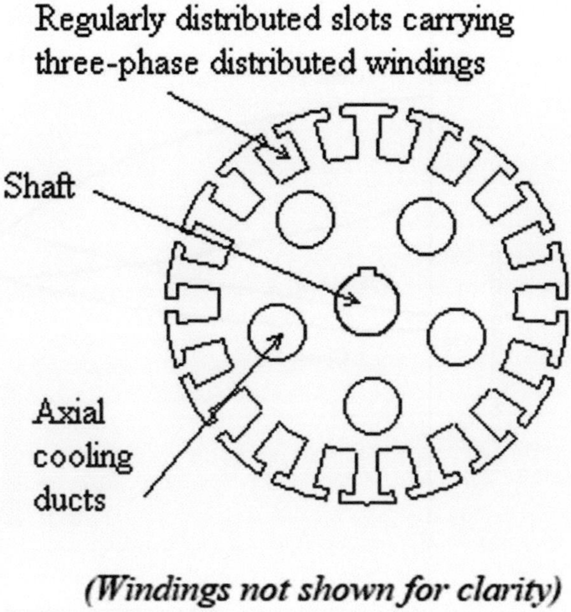

Fig. 5.4 Rotor stamping of slip-ring induction motor showing regularly distributed semi-closed slots

5.2.1 NEMA Specifications

National Electrical Manufacturers Association (NEMA) is the trade association of choice for the electrical manufacturing industry. It is headquartered near Washington, D.C. It provides a forum for the development of the technical standards for products used in the generation, transmission and distribution, control and end-use of electricity. These products are used in utility, medical imaging, industrial, commercial, institutional, and residential applications. These technical standards are in the best interests of the industry and users.

Some of the NEMA designs of induction motors are as follows:

NEMA design A: High torque, low slip, high locked amperes (Low r_2/x_2)

NEMA design B: Normal torque, normal slip, normal locked amperes

NEMA design C: High torque, normal slip, normal locked amperes

NEMA design D: High locked rotor torque, high slip (High r_2/x_2)

The above designs have different shapes for the rotor slots. NEMA design A is a general-purpose induction motor having rectangular rotor slots. Similarly, design B is deep bar rotor or double cage. Design C is a double cage rotor and Design D is semi open circular slots. There are NEMA standards for Wound rotor induction motor, multi-speed induction motor, and single-phase induction motor. Figure 5.5 shows typical torque-speed characteristics of NEMA three-phase squirrel cage induction motor.

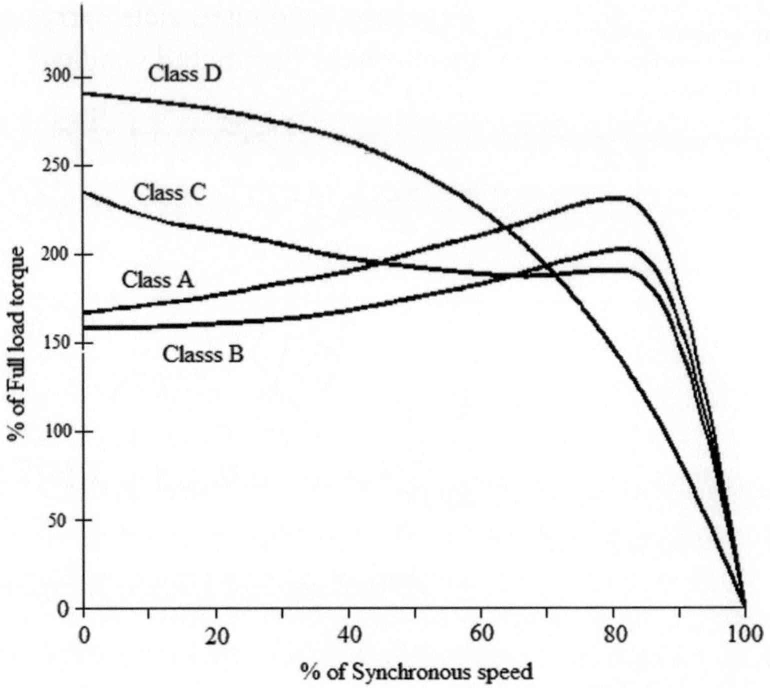

Fig. 5.5 Typical torque-speed characteristics of NEMA Class A, B, C, and D squirrel cage induction motors

5.3 Armature Winding

The armature of polyphase induction and synchronous machine windings are similar in many respects. The conducting material used in winding has been discussed in Sect. 1.4.1. The designed polyphase winding must be balanced in all respect, magnetic as well as electric. Three-phase windings are shown in Fig. 5.6a–c. The darker lines show the upper coil sides and lighter lines show the lower coil sides in a slot. All the phases are shown separately for clear understanding. Although here open slots are shown, in actual machine the slots are semi-closed type for reducing the leakage reactance. Each coil has multiple turns. The conductors are placed in the slot one by one through slot opening. In a full pitch winding the winding factor is same as distribution factor [2].

A multiturn coil is made by wrapping S.W.G. wire of appropriate number on a former. A typical 60° phase spread, 12-slot, three phase, two-pole windings are shown separately for different phases in Fig. 5.6a–b. The patterns of winding in different phases are identical, except the marking of terminals like R1-R2, Y1-Y2, B1-B2. The terminals R1- R2 are marked as that of B1- B2. Since the winding are 60° phase spread, the reverse connection of one of the phase group Y1-Y2, makes

Fig. 5.6 **a** R-phase winding 12 slot, two pole three phase, **b** Y-phase windings 12 slot, two pole three phase, **c** B-phase windings 12 slot, two pole three phase (all the phases are shown separately for clarity and clear understanding)

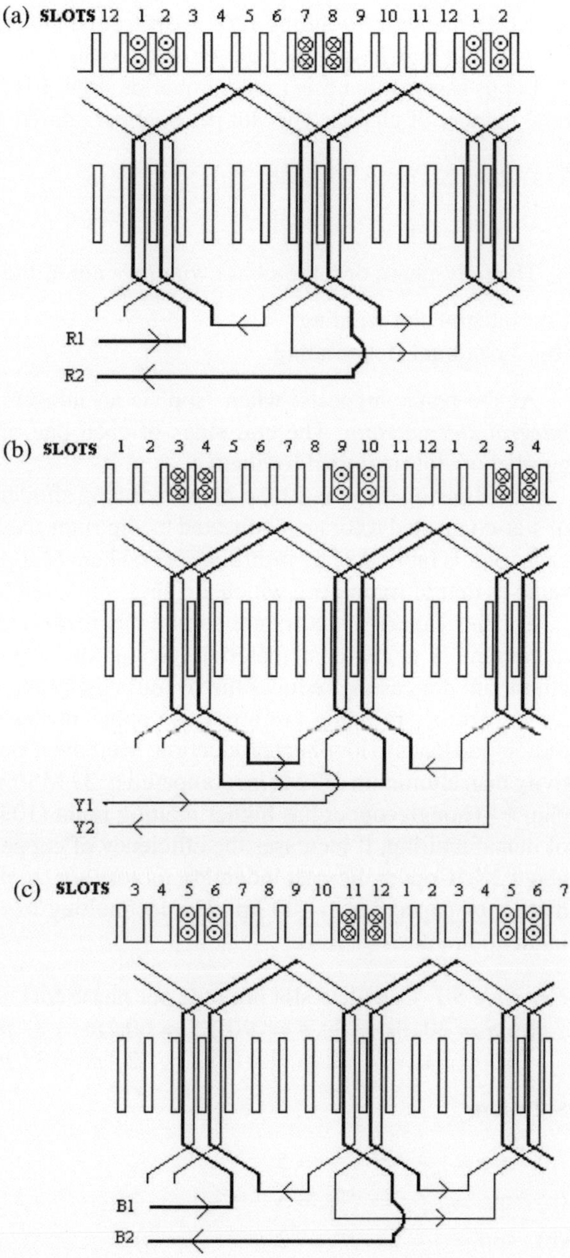

the MMF due to the winding 120° apart from other two phases. The three phases of the winding may have either star or delta connection.

Let us assume that S is total number of slots, p is total number of poles and m is total number of phases. The slot per pole per phase (spp) is

$$\text{spp} = \frac{S}{p \times m}$$

The poly-phase or three-phase windings are of the following two types:

(a) Integral slot winding
(b) Fractional slot winding.

As the name suggests, when 'spp' is an integer, the winding is referred to as *integral slot winding*. The coil sides of each one of the phases are having equal number of slots assigned for them.

The slip-ring rotor has three-phase winding similar to that of stator, the terminals of star-connected rotor are connected to slip rings. As mentioned earlier, the squirrel cage rotor is fabricated by inserting several bars of copper alloy (having cross section same as that of rotor slot), which are short-circuited by two end rings.

Smaller capacity motors are sometimes made with aluminum die-casting. The aluminum is a low-cost material having lower (660 °C) melting point. Using aluminum die-casting reduces the overall cost of the motor. In recent years, copper die-cast rotor are being preferred as copper increases the efficiency of induction motors and leads to low-heat production. Moreover, since copper has a higher conductivity than aluminum (57 MS/m compared to 37 MS/m), the use of copper is a natural choice. Though copper has higher melting point (1085 °C), resulting in higher cost of manufacturing. It increases the efficiency of copper die-cast induction motors by about 3%. Copper die-cast induction motors are chiefly used in industrial induction motors of capacities 1–100 hp. The technology has a potential for fractional HP induction motor.

Example 5.1 Calculate slot per pole per phase for
(a) $S = 30,\ m = 3,\ p = 2$ (b) $S = 60,\ m = 3,\ p = 4$
(c) $S = 24,\ m = 3,\ p = 4$ (d) $S = 12,\ m = 3,\ p = 2$

Solution:

(a) $\text{spp} = \frac{S}{p \times m} = \frac{30}{2 \times 3} = 5$
(b) $\text{spp} = \frac{S}{p \times m} = \frac{60}{4 \times 3} = 5$
(c) $\text{spp} = \frac{S}{p \times m} = \frac{24}{4 \times 3} = 2$
(d) $\text{spp} = \frac{S}{p \times m} = \frac{12}{2 \times 3} = 2$

All the above are integral slot windings.

Let us design a 12-slots, three phase, two-pole winding. The three-phase windings are designed in such a way so that each of the phase groups have equal number of conductors. The balanced winding means that all the phase groups must be identical for interchangeability. Since 'spp' is 2, the slots allotted for each phase in single-pole are two. Thus, for 12 slot windings the phase band is given as,

Slot No.	1	2	3	4	5	6	7	8	9	10	11	12
UC	**R**	**R**	**−Y**	**−Y**	**B**	**B**	**−R**	**−R**	**Y**	**Y**	**−B**	**−B**
LC	R	R	−Y	−Y	B	B	−R	−R	Y	Y	−B	−B

Since each slot has two coil sides, the winding is a double layer winding. Here the bold-face (**R-Y-B**) shows the upper coil side (UC) and regular-face (R-Y-B) show the lower coil side (LC) in a particular slot. As pointed out in Fig. 5.6, the coil having upper coil side **−B** in slot No.11 has other side of the coil +B in slot No. 5. Similarly, the coil having lower coil side −B in slot No. 11 has its other side +**B** again in slot No.5. In case of large number of slots and poles, the coil sides are expected to be in different slots.

If the winding is short pitched by one slot, the phase band is given as

Slot No.	1	2	3	4	5	6	7	8	9	10	11	12
UC	**R**	**R**	**−Y**	**−Y**	**B**	**B**	**−R**	**−R**	**Y**	**Y**	**−B**	**−B**
LC	R	−Y	−Y	B	B	−R	−R	Y	Y	−B	−B	R

Here in short-pitched winding, instead of placing coil sides in slots 1−7 slots, coil sides are placed in 1−6 slots for R-phase. Similarly for other phases. In over-pitched winding, coil sides are placed in 1−8 slots.

It is costly and uneconomical to design and manufacture stampings according to once own choice. There are already standardizations in commercially available stampings like, the stampings size, slot size, diameter etc. For unconventional machines like, Linear Induction Motor (LIM), Linear Synchronous Motor (LSM), Axial Flux Induction/Synchronous Motors the magnetic circuit of stator core require customized design. Slot removal from stampings can be done using shear cutting machine, CNC machines or Wire Electrical Discharge Machining (WEDM). The final assembly of such customized core requires varnishing of each stamping, before reassembling. Care should be taken to avoid the short-circuiting of stampings on the machined surface, to avoid eddy current losses. There are also standardizations in the commercially available materials, like diameter or thickness of (SWG) winding wire and cross section of bare copper strips, thickness of insulation, size of frame etc. To design and manufacture of stampings of our own choice is quite cumbersome. It is an easier task to design a conventional machine, which can be easily fabricated with the commercially available materials. Due to commercially available stampings, there is a possibility of fractional slot per pole per phase (spp), when number of phases and/or number of poles are changed. As such, if 'spp' is a fraction, the winding is known as *fractional slot windings*. The designed windings must be a balanced, symmetrical three-phase winding.

Example 5.2 Calculate slot per pole per phase for
(a) $S = 54,\ m = 3,\ p = 8$(b) $S = 32,\ m = 3,\ p = 4$

Obtain phase allocation series. Also, state whether balanced winding is possible.

Solution:

(a) $\text{spp} = \frac{54}{3\times8} = \frac{9}{4} = 2\frac{1}{4}$

The winding will be fractional slot winding. The phase allocation series is calculated as below:

$0\times2\frac{1}{4}$	$1\times2\frac{1}{4}$	$2\times2\frac{1}{4}$	$3\times2\frac{1}{4}$	$4\times2\frac{1}{4}$ 4
0	2	4	6	9

(2–0)	(4–2)	(6–4)	(9–6)
2	2	2	3

The phase allocation series is 2, 2, 2, 3. This says that 2, 2, 2, 3, 2, 2, 2, 3, 2, … slots are allocated, respectively, to R,Y, B, R,Y,B,R,Y,B,…. phases in sequence. It can be shown that the slot per phase is an integer as such balanced winding may be possible.

(b) $\text{spp} = \frac{32}{3\times4} = \frac{8}{3} = 2\frac{2}{3}$

The winding will be fractional slot winding. The phase allocation series is determined as below:

$0\times2\frac{2}{3}$	$1\times2\frac{2}{3}$	$2\times2\frac{2}{3}$	$3\times2\frac{2}{3}$
0	2	5	8

(2–0)	(5–2)	(8–5)
2	3	3

The phase allocation series is 2, 3, 3. Since slots per phase are not integer, *R* phase will have 8 slots whereas Y and B phases will have 12 slots as such, balanced winding may not be possible.

5.3.1 MMF Distribution

Figure 5.7 shows the MMF distribution for a single coil. For a coil system, the Fourier analysis is the standard technique for determining slot harmonies [2, 4].

The MMF due to single coil may be approximated as a rectangular distribution, trapezoidal distribution or triangular distribution. The analysis of MMF distribution is helpful in obtaining the harmonic contents in MMF or air-gap flux density distribution in space. When the machine has a large number of slots, it is convenient to assume

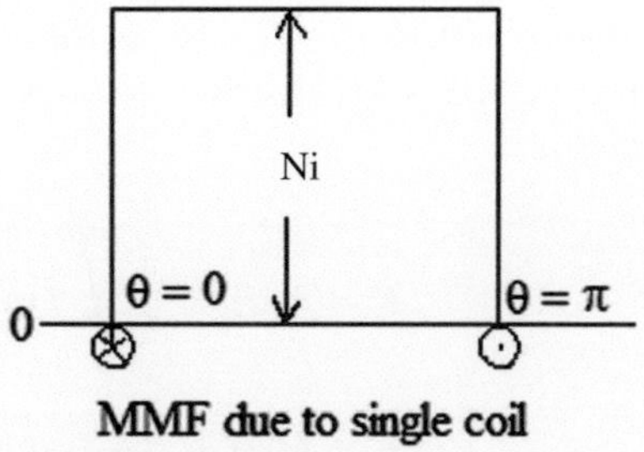

Fig. 5.7 MMF due to single coil

a rectangular distribution. The full pitch rectangular distribution of MMF is shown in Fig. 5.8a, in which the coil sides are 180° electrical apart. Figure 5.8b shows the MMF distribution due to short-pitch coils. Here the coil sides are less than 180° electrical apart. The short-pitch coil is also known as *chorded coil*. An over-pitch coil will have coil sides which are more than 180° electrical apart.

These MMF distributions have

(a) Half wave symmetry or $F(\alpha) = -F(180 + \alpha)$. Thus, all even harmonics are absent.
(b) Quarter wave symmetry or $F(\alpha) = F(180 - \alpha)$. Thus, all cosine terms are zero.
(c) The distribution is symmetric in positive and negative halves.

It is assumed that total number of conductors in a coil is 'N' which is carrying current 'i' and order of harmonics is 'n'.

Full-pitch rectangular MMF distribution (F) can be analyzed using Fourier series

$$F = f(\alpha) = \frac{4}{\pi} \int_0^{\pi/2} f(\alpha) \sin n\alpha \quad d\alpha$$
$$F = f(\alpha) = \frac{4Ni}{\pi n}[-\cos n\alpha]_0^{\pi/2} = \frac{4Ni}{\pi n} \tag{5.1}$$

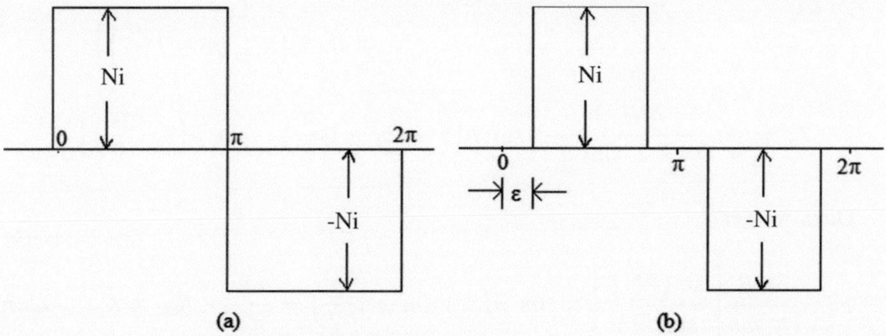

Fig. 5.8 **a** Rectangular distribution of MMF due to Full pitch coil. **b** Rectangular distribution of MMF due to short-pitch or chorded coil

Here $n = 1, 3, 5, 7, \ldots\ldots.$ Similarly, the Fourier analysis of the fractional pitch rectangular MMF distribution shown in Fig. 5.8b, due to short-pitch coil or chorded coil is

$$F = f(\alpha) = \frac{4}{\pi}\int_{\varepsilon}^{\pi/2} f(\alpha)\sin n\alpha \; d\alpha$$
$$F = f(\alpha) = \frac{4Ni}{\pi n}[-\cos n\alpha]_{\varepsilon}^{\pi/2} = \frac{4Ni}{\pi n}\cos n\varepsilon = \frac{4Ni}{\pi n}K_p \tag{5.2}$$

Here 'K_p' is the pitch factor and 'n' is the order of harmonics.

Consider a case of actual winding, where m number of slots have coils carrying unidirectional current, as shown in Fig. 5.9. If the number of turn per pole is N and current is i, the peak MMF is F = Ni. The peak value of nth space harmonic MMF is the sum of the series of MMFs, where each of ampere turn has rectangular distributed (Ni/m). The Fourier series analysis yields,

$$F_n = \frac{4Ni}{\pi \text{nm}}[\cos n\varepsilon + \cos n(\varepsilon+\alpha) + \cos n(\varepsilon+2\alpha) + \cos n(\varepsilon+3\alpha) + \cos n(\varepsilon+4\alpha) + \cdots\cdots\cdots + \cos n(\varepsilon+(m-1)\alpha)]$$

$$F_n = \frac{4Ni}{\pi\, nm \times 2\sin\left(\frac{n\alpha}{2}\right)}\left[2\sin\left(\frac{n\alpha}{2}\right)\{\cos n\varepsilon + \cos n(\varepsilon+\alpha) + \cos n(\varepsilon+2\alpha) + \cos n(\varepsilon+3\alpha) + \cos n(\varepsilon+4\alpha) + \cdots\cdots\cdots + \cos n(\varepsilon+(m-1)\alpha)\}\right]$$

$$F_n = \frac{4Ni}{\pi\, nm \times 2\, \sin\left(\frac{n\alpha}{2}\right)}\left[\left\{\sin n\left(\varepsilon+\frac{\alpha}{2}\right) - \sin n\left(\varepsilon-\frac{\alpha}{2}\right) + \sin n\left(\varepsilon+\frac{3\alpha}{2}\right) - \sin n\left(\varepsilon+\frac{\alpha}{2}\right) + \cdots\cdots\cdots\cdots\cdots\cdots\cdots\cdots\cdots\cdots + \sin n\left(\varepsilon+(m-1)\frac{\alpha}{2}\right)\right]\right.$$

$$F_n = \frac{4Ni}{\pi\, nm \times 2\, \sin\left(\frac{n\alpha}{2}\right)}\left[\sin n\left(\varepsilon+(m-1)\frac{\alpha}{2}\right) - \sin n\left(\varepsilon-\frac{\alpha}{2}\right)\right]$$

Thus, we get

$$F_n = \frac{4Ni}{\pi\, nm}\left[\frac{\sin\left(\frac{nm\alpha}{2}\right)}{\sin\left(\frac{n\alpha}{2}\right)}\right] \times \cos n\left(\varepsilon+(m-1)\frac{\alpha}{2}\right) = \frac{4Ni}{\pi n} \times K_{dn} \times K_{pn} \tag{5.3}$$

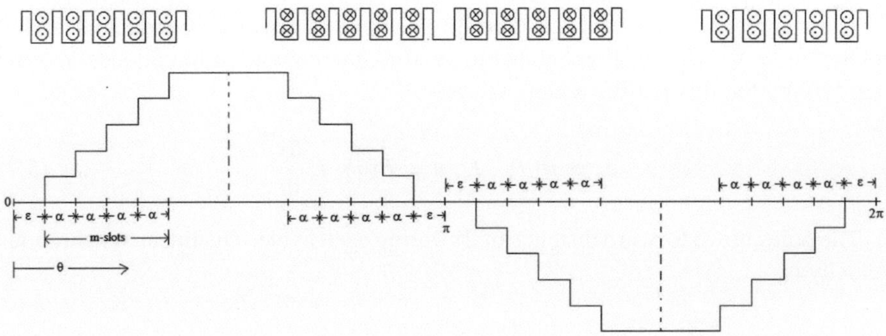

Fig. 5.9 MMF distribution of a fractional pitch winding of a phase of poly-phase winding

Here K_{dn} is the distribution factor for *n*th-order space harmonics and K_{pn} is the pitch factor for *n*th-order space harmonics. Similar expressions can be derived with assumptions of triangular or trapezoidal distribution of MMF due to single coil. The study of MMF distribution is helpful in studying the effects of slotting on the air-gap flux density distribution. The behavior of machine during fault condition can also be studied using *winding function theory*, which is based on MMF distribution in air-gap due to faulted windings. The discussion is not within the scope of the present course contents.

5.4 EMF Generation

As can be seen from Fig. 5.9, the fundamental and harmonic MMF has a cosine distribution of space angle, θ. The flux density distribution is a sine function of space angle, θ. The flux density distribution in air-gap can be given as

$$B(\theta) = \sum_{n=1}^{n=\infty} B_n \sin(n\theta + \phi_n) \tag{5.4}$$

Here, subscript n is the order of harmonics and angle ϕ_n is some space angle. Since North and South poles of the machine are essentially similar, we can write

$$B(\theta) = -B(\theta + \pi) \tag{5.5}$$

Thus, only odd harmonics are present. The flux density distribution is symmetric about pole axis. Therefore, we get

$$B(\theta) = B(\pi - \theta) \tag{5.6}$$

Thus, angle $\phi_n = 0$ for all n. The flux passes the armature conductors with a velocity $u = \frac{dx}{dt}$. The EMF generated in an armature conductor in coil side-a in the Fig. 5.10 by the flux cutting action will be,

$$e_a = B(\theta)\ .\ L\ .\ u = B(\theta).\ L.\frac{dx}{dt} \tag{5.7}$$

The poles move forward distance of 2τ during each cycle. The linear synchronous velocity is,

$$u = \frac{dx}{dt} = \frac{2\tau}{(1/f)} = 2\tau f \tag{5.8}$$

If the machine is having p number of poles, the pole pitch is

$$\tau = \frac{\pi D}{p},$$

where D is the diameter of the active surface of the machine. The linear synchronous velocity is,

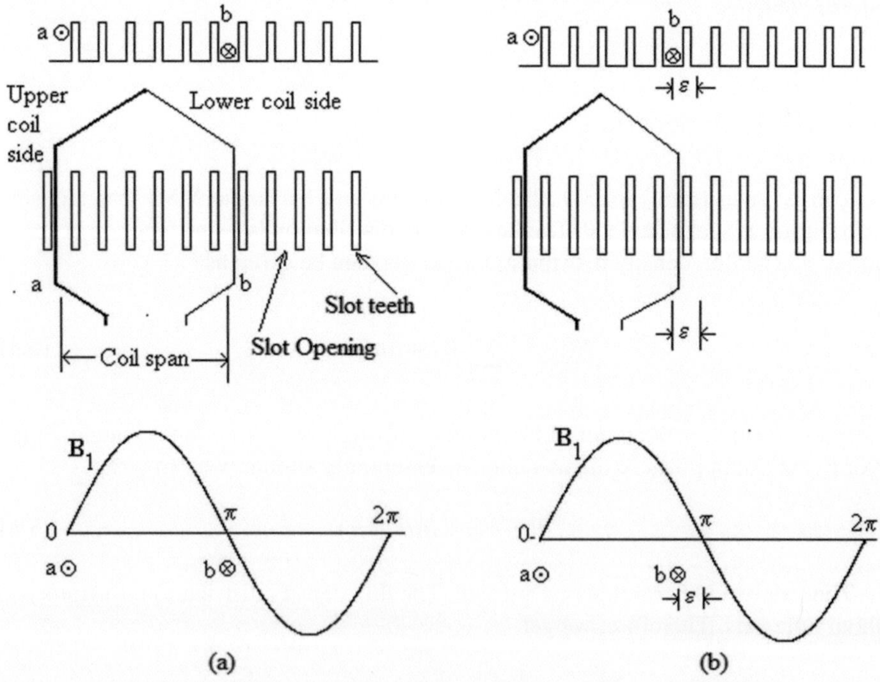

Fig. 5.10 **a** Full-pitch coil and induced EMF. **b** Short-pitch coil and induced EMF

$$u = \frac{2\pi Df}{p}$$

A single complete revolution of πD distance takes a time $\frac{p}{2f}$. In other words time $\frac{p}{2f}$ is taken to complete one revolution. Thus in one second, we have $n_s = \frac{2f}{p}$ revolution per second, and this speed, i.e., n_s is known as *synchronous speed per second*. The synchronous speed per minute in RPM (Rotation Per Minute) is

$$N_s = \frac{2 \times 60 \times f}{p} = \frac{120f}{p} \text{ RPM}$$

The rotational speed for two-pole is $\omega = \frac{d\theta}{dt} = 2\pi f$, where $\theta = \omega t$. If speed is maintained constant then linear speed u, or rotational speed ω or frequency are all constant. Thus, the induced EMF will vary according to the $B(\theta)$. Figure 5.10a, b show full-pitched and short-pitched coils, respectively, passing over a sinusoidal flux density B_1 distribution in space. The waveform of the induced EMF in conductor 'a', to a time base will be identical with the waveform of the flux distribution.

$$e_a = L\,.\,u \sum B_n \sin\ n\omega t \tag{5.9}$$

The EMF induced in the other side-b of the coil, which is displaced by coil-span angle β from coil side-a is,

$$e_b = L\,.\,u \sum B_n \sin\ n(\omega t + \beta) \tag{5.10}$$

The EMF induced in a coil a–b is given by, $e_{ab} = e_a - e_b$.
The flux per pole,

$$\begin{aligned}\phi_n &= (B_n \sin\ n\theta)_{\text{average}}\,.L\tau \\ &= \frac{L\tau}{\pi}\int_0^{\pi} B_n \sin\ n\theta.d\theta \\ &= \frac{2B_n}{\pi\,.\,n} L\,.\,\tau \quad \alpha \quad \frac{1}{n}\end{aligned} \tag{5.11}$$

$$B_n = \frac{\pi}{2} \times \frac{n\,.\,\phi_n}{L\,.\,\tau} \tag{5.12}$$

The value of nth-order harmonics voltage induced in a conductor is given as

$$\begin{aligned}e_n &= L.u.B_n \sin\ n\omega t \\ &= L \times 2.\tau.f \times \frac{\pi}{2} \times \frac{n.\phi_n}{L.\tau} \times \sin\ n\omega t\end{aligned}$$

$$= \pi.n.f.\phi_n \sin\ n\,\omega t \tag{5.13}$$

The coil is having T_c number of turns. The EMF induced in a coil is

$$e_{ab} = T_c(e_a - e_b) \tag{5.14}$$

For full-pitch coil angle span, $\beta = \pi$, the fundamental EMF is

$$\begin{aligned} e_a &= \pi.f.\phi_1 \sin\,\omega t \\ e_b &= \pi.f.\phi_1\ \sin(\omega t + \pi) \\ e_{ab} &= T_c(e_a - e_b) = 2\pi.f.\phi_1.T_c\ \sin\ \omega t \end{aligned} \tag{5.15}$$

The RMS value of the fundamental component of the coil voltage is

$$E_1 = \sqrt{2}.\pi.f.\phi_1.T_c \tag{5.16}$$

The above expression is similar to transformer induced voltage.

5.4.1 Pitch Factor

In a full-pitch coil, the coil span angle β is equal to pole-pitch or π electrical angle. If the coil-span is more than full-pitch, it is referred to as *over-pitched coil.* If the coil-span is less than π or pole-pitch, the coil is said to be *short-pitched* or *chorded.* In a short-pitched coil, the EMF induced on one side of the coil is not in exact phase opposition of the EMF induced on the other side, it will be short of angle ε by which the coil is short-pitched. Figure 5.11a and b show, respectively, the full-pitch winding and short-pitch winding EMF phasor in these coils. Thus the fundamental EMF induced in the short-pitched coil will be

$$|E_{ab}| = |E_a - E_b| = |E|e^{j0} - |E|e^{j\varepsilon} = 2|E|\cos\frac{\varepsilon}{2}$$

The effect of short-pitching is that the coil induced EMF is reduced by a factor of

$$K_p = \cos\frac{\varepsilon}{2} \tag{5.17}$$

Here, K_p is referred to as *pitch factor* or *coil span factor*. For nth-order harmonic the pitch factor is,

$$K_{pn} = \cos\frac{n.\varepsilon}{2} \tag{5.18}$$

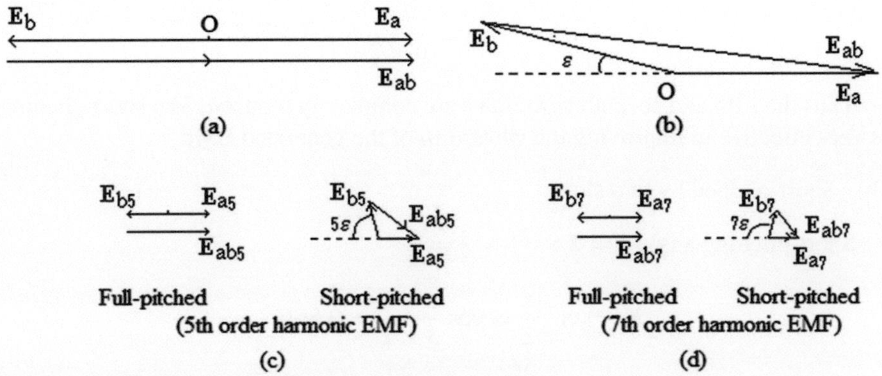

Fig. 5.11 **a** Voltage across a full-pitched coil. **b** Voltage across a short-pitched coil. **c** Voltage due to 5th-order harmonic. **d** Voltage due to 7th-order harmonic

Short-pitching of coil has the following effects:

(a) The fundamental induced EMF gets reduced by a factor of $\cos\frac{\varepsilon}{2}$.
(b) The nth harmonic induced EMF gets reduced by a factor of $\cos\frac{n.\varepsilon}{2}$. The harmonic voltage has a greater reduction than that of fundamental EMF. If the winding is designed such that $\cos\frac{n.\varepsilon}{2} = 0$, then a particular harmonic will be completely eliminated from the waveform. This has been explained in Fig. 5.11c, d.
(c) The winding requires slightly lesser copper wires due to shorter coil span.

Example 5.3 In a 24slot, four pole alternator, calculate the pitch factor if the winding is short pitched by (a) one slot (b) two slots.

Solution:

(a) Short-pitching by one slot

$$\text{Slot per pole} = \frac{24}{4} = 6$$

The coil span is 1–7 slots, or if one coil side is placed in 1st slot the other coil side will be placed in 7th slot. Between a full pitched coil sides, six slot teeth are present.

$$\text{Slot angle } \varepsilon = \frac{2\pi}{24} \times \frac{4}{2} = \frac{\pi}{6}\text{electrical}$$

Pitch factors

$$K_p = \cos\frac{\varepsilon}{2} = \cos\frac{\pi}{6 \times 2} = 0.9659$$

$$K_{p5} = \cos\frac{5\varepsilon}{2} = \cos\frac{5\pi}{6 \times 2} = 0.2588$$

$$K_{p7} = \cos\frac{7\varepsilon}{2} = \cos\frac{7\pi}{6\times 2} = -0.2588$$

Thus the fifth and seventh harmonics are completely reduced. The short pitching is very effective in improving the waveform of the generated EMF.

(b) Short-pitched by two slots

Short-pitching angle $\varepsilon = 2 \times \frac{2\pi}{24} \times \frac{4}{2} = \frac{\pi}{3}$

$$K_p = \cos\frac{\varepsilon}{2} = \cos\frac{\pi}{3\times 2} = 0.866$$
$$K_{p5} = \cos\frac{5\varepsilon}{2} = \cos\frac{5\pi}{3\times 2} = -0.866$$
$$K_{p7} = \cos\frac{7\varepsilon}{2} = \cos\frac{7\pi}{3\times 2} = -0.8660$$

The short pitching by two slots reduces the pitch factor. But the fifth and seventh harmonics are not canceled. Thus, short pitching by one slot is more effective.

5.4.2 Spread Factor

Let there are m number of coils connected in series to form a phase band. The voltage of these coils is equal in magnitude but they differ in phase by one slot pitch, as shown in Fig. 5.12. The resultant voltage across the series-connected phase band can be obtained using summation of the voltage phasor.

Let m be the number of slots per pole per phase. Now, we have

$$\sigma = \text{phase spread} = \frac{\pi}{\text{number of phases}} = \frac{\pi}{3}$$

The spread factor or breadth factor is defined as

$$K_d = \frac{\text{Voltage of a phase band}}{\text{Voltage induced per coil} \times \text{Number of coils}} \tag{5.19}$$

In a lap winding, the finish of the first coil is connected to the beginning of the second coil and so on to form the phase band. The phasor addition is shown in Fig. 5.12.

Refer Fig. 5.12, the voltage phasor are displaced by slot angle 'ε'. Let m number of such coils of a phase band are connected in series. These coils are having their individual voltages $E_1 = Ee^{j0}$,$E_2 = Ee^{j\varepsilon}$,$E_3 = Ee^{j2\varepsilon}$,.., $E_m = Ee^{j(m-1)\varepsilon}$.

The voltage across a phase band is

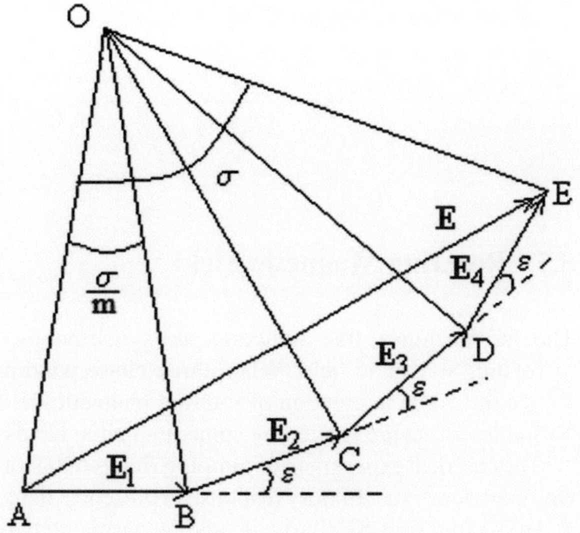

Fig. 5.12 Voltage across a phase group

$$= Ee^{j0} + Ee^{j\varepsilon} + Ee^{j2\varepsilon} + \ldots.. + Ee^{j(m-1)\varepsilon} = \frac{E \sin \frac{m\varepsilon}{2}}{\sin \frac{\varepsilon}{2}} \quad (5.20)$$

Thus, the distribution factor is

$$K_d = \frac{\sin \frac{m\varepsilon}{2}}{\sin \frac{\varepsilon}{2}} \quad (5.21)$$

Or,

$$K_d = \frac{\sin \frac{m\varepsilon}{2}}{\sin \frac{\varepsilon}{2}} \approx \frac{\sin \frac{m\varepsilon}{2}}{\frac{\varepsilon}{2}} \quad (5.22)$$

Due to the presence of slot teeth, the conductors are not perfectly spread over the periphery of the armature. Thus, the voltage induced in the phase band is reduced by a factor of distribution factor, K_d. The winding factor K_w is described as product of the distribution factor and pitch factor.

$$K_w = K_p \times K_d \quad (5.23)$$

Example 5.4 Calculate distribution factor of a 60 slot, three-phase, four-pole winding.

Solution:

The slot per pole per phase, spp, $m = \frac{60}{3\times4} = 5$

Slot angle, $\varepsilon = \frac{2\pi}{\text{slot}} \times \frac{\text{Poles}}{2} = \frac{2\times180^\circ\times4}{60\times2} = 12^\circ$ electrical

The distribution factor,

$$K_d = \frac{\sin \frac{m\varepsilon}{2}}{m \sin\frac{\varepsilon}{2}} = \frac{\sin 30^\circ}{5 \times \sin 6^\circ} = 0.9566$$

5.5 Rotating Magnetic Field

The AC machines, like induction and synchronous motors work on the principle of rotating magnetic field. When three-phase windings are excited by three-phase source, it results in creation of rotating magnetic field. It appears as if the alternate N-S poles are being rotated by some invincible hands.

The classical experiment of jumping ring is of historical importance. A conducting ring experiences oscillatory response in presence of alternating current magnetic field and a shading ring. Similarly, in other experiment, two solenoid coils mounted at the ends of a ferromagnetic bar are excited by AC sources having slightly different frequencies. When a conducting ring is inserted in the iron bar, it oscillates at slip frequency.

A magnetic field whose magnitude alternates with respect to time is known as the *alternating magnetic field*. For example, the magnetic field of a solenoid excited by an AC source. On the other hand, a *traveling magnetic field* or a *rotating magnetic field* is defined as the magnetic field which is traveling or rotating in space as the time is elapsed. The concept of rotating magnetic field can be explained with the help of the following: *

(a) Ferraris principle
(b) Analytical method

5.5.1 *Ferraris Principle*

Ferrari's principle states that an alternating magnetic field may be represented as two counter rotating magnetic fields, one in the anti-clockwise direction and the other in clockwise direction. A single coil carrying AC current is assumed to have two rotating magnetic fields in forward and in reverse directions.

Consider a two-phase coil system, as shown in Fig. 5.13a, excited by two-phase source. At any time $t = 0$, it is assumed that the magnetic field is maximum in coil A. The pulsating magnetic field is maximum at time $t = 0$ and is aligned in the axis of coil A. The pulsating magnetic field is split into two components 1 and 2, which are rotating in opposite direction to each other with synchronous speed. At this instant of time, the field due to coil-B current is zero in the axis of coil-B. The pulsating magnetic field in coil-B is assumed to be 90° behind their maximum. This is considered to have components 3 and 4, which are to travel 90° before they meet at the axis of coil-B for maximum field. They are rotating in opposite direction at synchronous speed. Thus, the rotating magnetic field produced on the coil-B will be 90° behind its maximum. At any instant of time, two components 1 and 4 are canceling each other. They are rotating in same direction and will cancel each other at any instant of time. The other two components 2 and 3 are rotating in same direction. Their alignment is coincident. This gives a magnetic field rotating in clockwise direction. For changing the direction of rotation, the phase sequence of applied voltage is changed. Fig. 5.13b shows three identical coils in space at 120° apart, and which are excited by RYB phases.

Consider the instant, when current in R-phase is maximum positive. The magnetic field will be aligned in the axis of the coil. The components 1 and 2 due to R-phase are rotating in opposite direction. At time $t = 0$, they are assumed to be aligned in the axis of R-phase coil.

The current in B-phase coil will be 120° lagging. The rotating magnetic field components 3 and 4 of the B-phase will have to travel 120° before they are coincident along the axis of coil B.

Similarly, the current in Y-phase coil will be 240° lagging. The rotating magnetic field components 5 and 6 of the Y-phase will have to travel 240° before they are coincident along the axis of coil-Y.

Components 1, 4, 5 are rotating in the counter clockwise direction and they are displaced by space angle 120°. At any instant, they cancel each other. Components 2, 3, 6 are rotating in clockwise direction, and they are always aligned in the same direction. This results in a rotating magnetic field. For reversing the direction of rotation of rotating magnetic field, the phase sequence has to be reversed. Thus, a single-phase winding does not have a rotating magnetic field. This is the reason that single-phase induction motors do not have a natural starting torque. But once single-phase induction motors are started by any means, they continue to run in the same

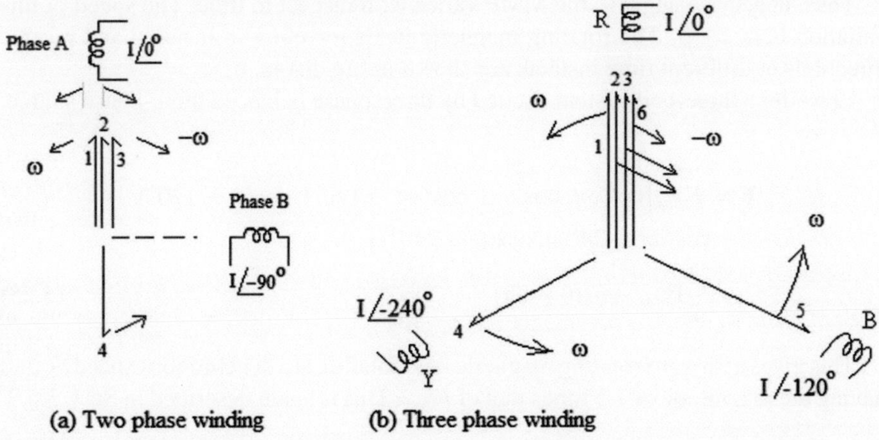

Fig. 5.13 Rotating magnetic field due to **a** two-phase winding and **b** three-phase winding

direction. On the other hand, two-phase, three-phase and poly phase windings have rotating magnetic field, with inherent starting torque.

5.5.2 Analytical Treatment

It is assumed that the MMF of a single-phase coil energized by a single-phase source is given by

$$\mathbf{F} = F_{\max} \cos\ \omega t\ .\ \cos\ \theta = F_{\max}\left(\frac{e^{+\omega t} + e^{-\omega t}}{2}\right)\left(\frac{e^{+\theta} + e^{-\theta}}{2}\right) = \frac{F_{\max}}{2}\left[e^{\omega t+\theta} + e^{\omega t-\theta}\right]$$

The terms containing $\frac{F_{\max}}{2}e^{\omega t-\theta}$ is forward rotating MMF, while terms $\frac{F_{\max}}{2}e^{\omega t+\theta}$ is a reverse or backward rotating MMF. This is in line with the Ferraris principle which states that an alternating magnetic field can be expressed as two rotating magnetic field both rotating in opposite direction with a synchronous speed ω.

Consider two coils in quadrature (space angle 90°) excited by a two-phase AC current. The resultant MMF is given as,

$$\begin{aligned}\mathbf{F} &= F_{\max}\left[\cos\ \omega t.\cos\ \theta + \cos(\omega t - 90°).\cos(\theta - 90°)\right]\\ &= F_{\max}[\cos\ \omega t.\cos\ \theta + \{\sin\ \omega t\}.\{\sin\ \theta\}]\\ &= F_{\max}.\cos(\omega t - \theta)\end{aligned} \tag{5.24}$$

---

* Both are identical since $cos\,\theta = \frac{e^{+\theta}+e^{-\theta}}{2}$ and $\sin\ \theta = \frac{e^{+\theta}-e^{-\theta}}{2}$

Thus, at a space angle θ, the MMF varies with respect to time. The speed of this variation is $\omega = \frac{d\theta}{dt}$. The rotating magnetic fields traveling in forward and reverse directions at different time instants are shown in Fig. 5.14a, b.

Consider a three-coil system excited by three-phase balanced three-phase source. We have

$$\begin{aligned}\mathbf{F} &= F_{\max}\left[\cos\ \omega t.\cos\ \theta + \cos(\omega t - 120°).\cos(\theta - 120°)\right.\\ &\quad \left.+\cos(\omega t - 240°).\cos(\theta - 240°)\right]\\ &= \frac{3}{2}\mathrm{F}_{max}.\ cos(\theta - \omega t)\end{aligned} \tag{5.25}$$

This gives a forward rotating magnetic field rotating at a synchronous speed ω_s and having the magnitude of 1.5 times that of $F_{\max}$. This is again described in Sect. 5.5.3.

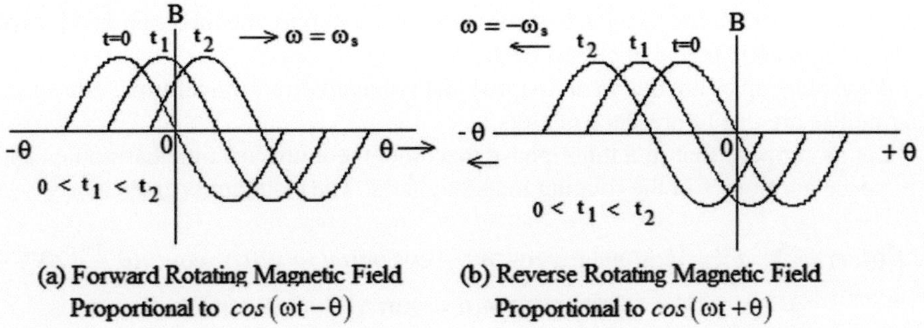

Fig. 5.14 **a** Forward rotating magnetic field- proportional to cos $(\omega t - \theta)$. **b** Reverse rotating magnetic field-proportional to cos $(\omega t + \theta)$

5.5.3 Presence of Time and Space Harmonics

Consider a three-phase balanced winding having three-phase balanced currents, i_1, i_2 and i_3.

$$\begin{aligned} i_1 &= I_m \cos \omega t \\ i_2 &= I_m \cos(\omega t - 120^\circ) \\ i_3 &= I_m \cos(\omega t - 240^\circ) \end{aligned} \tag{5.26}$$

The three-phase windings are uniformly spread in the armature. Thus, the MMF distribution in a three-phase winding can be given as

$$\begin{aligned} F_1(\theta, t) &= N.\ I_m \cos\ \omega t\ .\ \cos \theta = F_{\max} \cos \omega t\ .\ \cos \theta \\ F_2(\theta, t) &= F_{\max} \cos(\omega t - 120^\circ).\ \cos(\theta - 120^\circ) \\ F_3(\theta, t) &= F_{\max} \cos(\omega t - 240^\circ) \cdot \cos(\theta - 240^\circ) \end{aligned} \tag{5.27}$$

The resultant MMF in space due to three-phase currents in three-phase distributed windings is.

$$\begin{aligned} F(\theta, t) &= F_1(\theta, t) + F_2(\theta, t) + F_3(\theta, t) \\ &= \frac{3}{2} F_{\max}.\ \cos(\theta - \omega t) \end{aligned} \tag{5.28}$$

Thus, a three-phase balanced winding excited by a balanced three-phase source results in rotating magnetic field in forward direction, and the magnitude of resultant MMF is 1.5 time the individual maximum MMF. A forward rotating MMF $1.5F_{\max}.\ \cos(\theta - \omega t)$ has been plotted in Fig. 5.14a. The rotating magnetic field travels in forward direction as time t is increased with a rotational synchronous speed $\omega_s = d\theta/dt$. A reverse rotating magnetic field will be given by $1.5F_{\max}.\ \cos(\omega t + \theta)$.

By changing the phase sequence of the three-phase supply the direction of forward rotating magnetic field can be reversed.

Time harmonies are due to non-sinusoidal voltage/current waveforms. The space harmonics are due to presence of slots.

Let us suppose that in a three-phase machine, there are 'm' time harmonics and 'n' space harmonics in the rotating magnetic field. The resultant is given as,

$$F(\theta, t) = \sum F_{\max}.\big[\cos\ m\omega t\ .\ \cos\ n\theta + \cos\ m(\omega t - 120^\circ).\ \cos\ n(\theta - 120^\circ) + \cos\ m(\omega t - 240^\circ).\ \cos\ n(\theta - 240^\circ)\big]$$

By selectively putting the values of $m = 3, 5, 7 \ldots$ (for $n = 1$) and $n = 3, 5, 7 \ldots.$ (for $m = 1$) in the above expression, we can prove that in a three-phase machine:

(a) The third-order space and time harmonics are absent.
(b) Seventh-order or '$6q + 1$' space and time harmonics are running in the forward direction. ($q = 1, 3, 5, , \ldots.$)
(c) Fifth and '$6q - 1$' space and time harmonics are running in the reverse direction.
(d) All even harmonics are absent.
(e) The synchronous speed of nth-order space harmonic corresponds to $n \times p$ number of poles.
(f) The synchronous speed of mth-order time harmonics corresponds to supply frequency of $m \times f$.

Example 5.5 Show that in a three-phase machine excited by three-phase balanced source, third-order space and time harmonics are absent.

Solution:

For third harmonic time harmonics,

$$\mathbf{F} = F_{\max}\big[\cos\ 3\omega t \cdot \cos\ \theta + \cos\ 3(\omega t - 120^\circ) \cdot \cos(\theta - 120^\circ) + \cos\ 3(\omega t - 240^\circ) \cdot \cos(\theta - 240^\circ)\big] = 0$$

Similarly, for third harmonic space harmonics,

$$\mathbf{F} = F_{\max}\big[\cos\ \omega t \cdot \cos\ 3\theta + \cos(\omega t - 120^\circ) \cdot \cos\ 3(\theta - 120^\circ) + \cos(\omega t - 240^\circ) \cdot \cos\ 3(\theta - 240^\circ)\big] = 0$$

Example 5.6 Show the expression of rotating MMF of a three-phase winding excited by a three-phase source when one phase is disconnected.

Solution:

$$\mathbf{F} = F_{\max}\big[\cos\ \omega t \cdot \cos\ \theta + \cos(\omega t - 240^\circ) \cdot \cos(\theta - 240^\circ)\big]$$

$$= F_{\max}\left[\cos\,\omega t\cdot\cos\,\theta+\left\{\frac{1}{2}\cos\,\omega t+\frac{\sqrt{3}}{2}\sin\,\omega t\right\}\left\{\frac{1}{2}\cos\,\theta+\frac{\sqrt{3}}{2}\sin\,\theta\right\}\right]$$

$$= F_{\max}\left[\frac{3}{4}\cos(\omega t-\theta)+\frac{1}{2}\cos\,\omega t\;\cos\,\theta+\frac{\sqrt{3}}{4}\sin(\omega t+\theta)\right]$$

$$= F_{\max}\left[\cos(\omega t-\theta)+\frac{1}{4}\cos(\omega t+\theta)+\frac{\sqrt{3}}{4}\sin(\omega t+\theta)\right]$$

The above expression shows the presence of both the forward $(\omega t-\theta)$ and reverse $(\omega t+\theta)$ rotating magnetic fields. Thus, when a fault is present in a three-phase electrical machine-either on the windings or on the supply, both forward and reverse rotating magnetic fields are present in the machine. A forward field gives rise to positive torque and backward rotating magnetic field gives rise to negative torque in a motor.

5.6 Effect of Slotting

Slot opening is desired so that thin multiple conductors can be embedded in different slots. It also ensures that the magnetic flux created by the current carrying coil crosses the air-gap and goes to the secondary member of the machine, as permeability of ferromagnetic material is higher than that of air. Local air-gap flux density in air-gap decreases over the slot opening. Similarly, the presence of slot teeth (iron) has the effect of slight increase in the local flux density is air-gap. Figure 5.15 shows ideal and actual flux density distributions in air-gap over slotted AC windings. The rotating magnetic field, which is assumed to be a space-sinusoid, has similar ripples due to slotting. These ripples in the rotating magnetic field do not travel in either direction; rather they glide over the rotating magnetic field. Its net effect is to induce high frequency EMF in the rotor conductors, which ultimately results in high frequency noise and vibrations in the machine [1, 2].

While manufacturing the rotor, the high frequency induced EMF in it is eliminated by skewing the rotor stampings by certain angle. Due to skewing, half the length of rotor conductor faces a rising flux density whereas the other half length of rotor conductor faces a decreasing flux density. Due to this, the generated high frequency induced EMF cancel each other as shown in Fig. 5.15. This eliminates net high frequency induced EMF in the rotor circuit. The slot is twisted in the axial direction by an electrical angle γ. A typical range of γ is 0.1–0.5 radian (5°−30°).

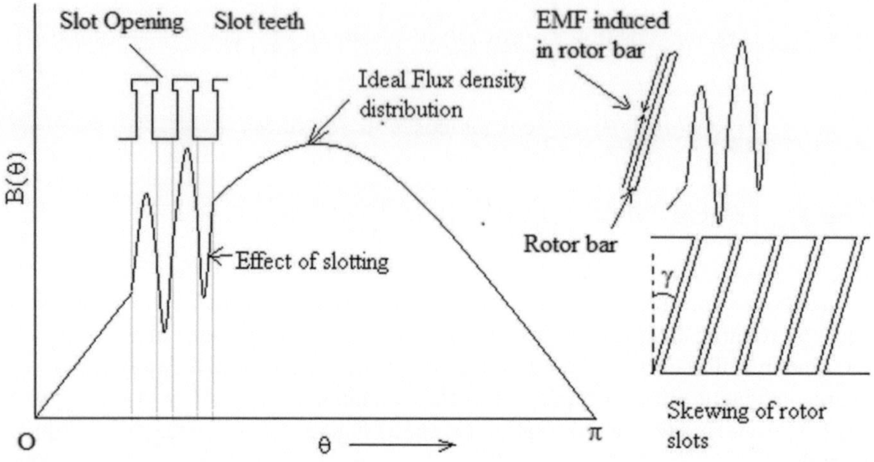

Fig. 5.15 Effect of slotting on the flux density distribution in air-gap and skewing of rotor slot-cancelation of slot harmonic EMF in rotor conductor

5.6.1 Cogging and Crawling

If the number of slots in the rotor (S_2) is wrongly selected, strong magnetic forces between the stator teeth and rotor teeth develop. Due to this, the stator teeth are magnetically locked with the rotor teeth and the induction motor cannot be started. This phenomenon is referred to as cogging. To avoid the cogging in induction motor, the following rules are used for selecting the rotor slots:

(a) $(S_1 - S_2)/2p$ an integer
(b) $S_1 \ / \ S_2$ fractional
(c) S_2 not widely different from S_1

Here S_1 and S_2 are, respectively, total number of slots on stator and rotor.

The seventh-order time and space harmonics travels in the forward direction. The space harmonics are due to slotting in the machine. The torque produced due to seventh -order space harmonics is superimposed with the fundamental torque slip characteristics as shown in Fig. 5.16. The seventh-order space harmonics appears as if there are seven time the actual number of poles in the machine. In other words, it appears as if in a two pole induction motor, another fourteen-pole induction motor runs in the same direction connected to the same shaft. Thus the synchronous speed of seventh-order space harmonics is

$$N_{s7} = \frac{120f}{7p} = \frac{1}{7}N_s$$

The torque-speed characteristic shows a saddle or a dip in the torque at speed near N_{s7}. The machine can be started, but it not be able to achieve its speed closer

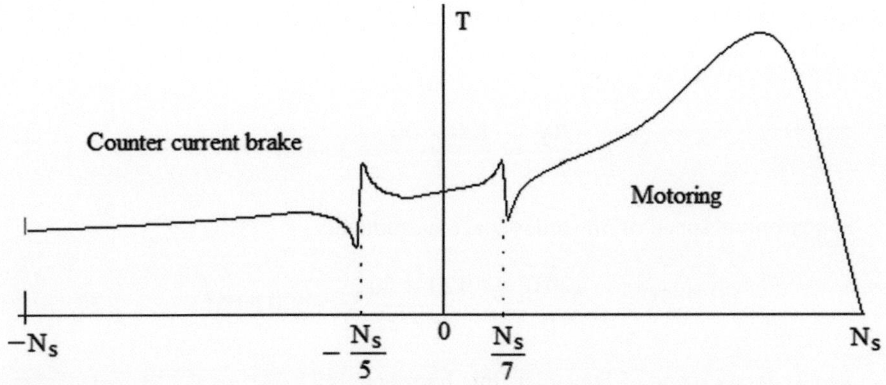

Fig. 5.16 Effect of 7th- and 5th-order space harmonics on torque-speed characteristics

to synchronous speed. It will continue to run near speed N_{s7}, which is about one-seventh of synchronous speed N_s. The phenomenon is referred to as crawling in induction motor. To avoid crawling, the rotor slots should be selected on the basis of the following rule:

(a) S_2 not to exceed 1.25 S_1
(b) $S_2 \neq 6\ px\ or\ 6\ px \pm 2p,\ \ x = 1, 2, 3, 4\ldots.$
(c) $S_2 \neq S_1 \pm 2p$
(d) $S_2 \neq S_1 \pm p$
(e) $S_2 \neq 0.5\ S_1 \pm p$

The fifth-order space harmonic is rotating in reverse direction. As such its effect is present in the counter-current braking zone of the torque-speed characteristics.

The synchronous speed of seventh-order time harmonics corresponds to the synchronous speed of seven times the supply frequency. The synchronous speed of seventh-order time harmonic is

$$N_7 = \frac{120 \times 7 \times f}{p} = 7\ N_s$$

The torque contributed by the seventh-order time harmonic in the active region aids the fundamental torque characteristics. In a two-pole 50 Hz induction motor, the synchronous speed is 3000 RPM, whereas for seventh-order time harmonics, it is 21000 RPM. The seventh-order time harmonics will be present due to distortion in the AC supply. Its net effect is to marginally increase the overall torque. This is helpful in increasing net motoring torque when induction motor is being run with the help of solid-state inverter supply.

Example 5.7 Calculate the synchronous speed of 50 Hz, 2-pole induction motor. What are the synchronous speeds of 5th- and 7th-order space and time harmonics?

Solution:

Synchronous speed is

$$N_s = \frac{120f}{p} = \frac{120 \times 50}{2} = 3000 \text{ RPM}$$

Synchronous speed of 5th-order space harmonic is

$$N_{s5} = -\frac{120\,\mathrm{f}}{5\,\mathrm{p}} = -\frac{120 \times 50}{5 \times 2} = -600 \text{ RPM}$$

Synchronous speed of 5th-order time harmonic is

$$N_5 = -\frac{120 \times 5 \times f}{p} = -\frac{120 \times 5 \times 50}{2} = -15,000 \text{ RPM}$$

Synchronous speed of 7th-order space harmonic is

$$\mathrm{N_{s7}} = \frac{120\mathrm{f}}{7\mathrm{p}} = \frac{120 \times 50}{7 \times 2} = 428.57 \text{ RPM}$$

Synchronous speed of 7th-order time harmonic is

$$N_7 = \frac{120 \times 7 \times f}{p} = \frac{120 \times 7 \times 50}{2} = 21,000 \text{ RPM}$$

5.7 Principles of Operation of Induction Motor

The rotating or traveling magnetic field set-up by the three-phase currents is similar to that of Halbach array (already discussed in Chap. 1), in which the motion is due to certain invincible hands. It appears as if the alternate N-S poles are being rotated in forward direction. Consider a slip-ring induction motor in which rotor windings are open. When the stator winding is excited by a three-phase source, the rotating magnetic field of stator cuts the standstill conductors of the rotor and induces EMF in the rotor winding. The frequency of this induced EMF is same as that of supply frequency.

The magnitude of the per phase induced voltage at standstill condition, in the rotor winding is,

$$\mathbf{E}_2 = \frac{N_2 K_{w2}}{N_1 K_{w1}} \mathbf{E}_1 \tag{5.29}$$

where N_1 and N_2 are the per phase number of turns on stator and rotor windings, respectively. Similarly, K_{W1} and K_{W2} are the respective winding factors. The phase angle of rotor induced EMF slightly differs from the stator-induced EMF due to rotor leakage inductance. This is considered to be negligible in the analysis. When rotor is open circuited, there is no rotor currents and machine is at rest. The slip-ring induction motor at standstill condition behaves similar to the open-circuit three-phase transformer.

When the rotor terminals are short-circuited, the induced EMF on the rotor causes the rotor current to flow. These currents set up a rotating magnetic field of the rotor, which tries to align itself with the stator rotating magnetic field. But due to Hysteresis effect, the rotating magnetic field of rotor always lags behind the rotating magnetic field of stator. The rotor is free to rotate. The rotor current develops torque due to presence of stator magnetic field. According to Lenz's law the rotor starts rotating in the direction, trying to minimize the cause. When the rotor is running at speed N_r in the direction of stator rotating magnetic field, the *relative speed* or *slip speed* between the rotating magnetic field and rotor is $(N_s - N_r)$. The stator rotating magnetic field now cuts the rotor conductors at slip speed. The frequency of rotor induced EMF is *slip frequency*, i.e.,

$$\frac{120 f_r}{p} = N_s - N_r$$

$$\frac{120 f_r}{p} \times \frac{p}{120\, f} = \frac{N_s - N_r}{N_s}$$

$$f_r = \frac{N_s - N_r}{N_s} f = sf \tag{5.30}$$

Here slip is defined as

$$s = \frac{N_s - N_r}{N_s}$$

When rotor speed N_r is equal to synchronous speed, there is no flux cutting action. Therefore, the slip is zero or in other words, rotor-induced frequency, rotor-induced EMF, rotor current and torque are all zero. As such induction motor speed decreases from synchronous speed (slip is positive), the torque developed again rises to a finite value.

When the speed of induction motor rises above the synchronous speed due to any external reason, the slip and torque both change their signs. A sizable amount of electrical energy is fed back to the three-phase lines due to regenerating action.

When the machine is allowed to run in opposite direction by a prime mover, the slip (2−s) exhibits the *counter current zone,* an open circuit EMF as well as frequency are accordingly modified as f_r=(2-s)f. The induction machine has the following three regions of operation:

(a) **Motoring** speed range: 0 to N_s (or slip $s = 1.0$ to 0.0), torque is positive
(b) **Generating** speed range $N_r > N_s$ (or slip $s = 0.0$ to -1.0), torque is negative
(c) **Counter current braking- reverse rotation speed or plugging** range $N_r < 0$ (or slip $s = 1.0$ to 2.0), torque is counter-current braking torque or plugging torque opposite to motoring torque.

In motoring region, the machine runs in the direction of rotating magnetic field, when the armature winding is on the stator. When armature winding is on the rotor side, which is also free to rotate-the machine runs in the direction opposite to the rotating magnetic field. This is common with the machines having armature winding on the rotor, as in case of Schrage motor.

Example 5.8 A six-pole alternator running at 900 RPM is supplying power to a three-phase 4 pole induction motor. If induction motor is running at 1250 RPM, determine the frequency of rotor current.

Solution

The supply frequency of six-pole alternator running at 900 RPM is

$$f = \frac{N_s p}{120} = \frac{900 \times 6}{120} = 45 \text{ Hz}$$

The synchronous speed of 4-pole induction motor at 45 Hz is

$$N_s = \frac{120f}{p} = \frac{120 \times 45}{4} = 1350 \quad \text{RPM}$$

Slip is given as

$$s = \frac{1350 - 1250}{1350} = 0.074$$

The frequency of rotor current is

$$f_r = sf = 0.074 \times 45 = 3.33 \text{ Hz}$$

Example 5.9 A three-phase 50 Hz 400 V, 2 pole induction motor is running at 2800 RPM. Determine the slip and frequency of rotor currents.

Solution

Synchronous speed $N_s = \frac{120f}{p} = \frac{120 \times 50}{2} = 3000 \quad \text{RPM}$
Slip, $s = \frac{N_s - N_r}{N_s} = \frac{3000-2800}{3000} = 0.0667 = 6.67\%$
Rotor frequency $f_r = sf = 0.0667 \times 50 = 3.33$ Hz

Example 5.10 A three-phase induction motor is running at 3% of slip. If the number of poles is 4 and supply frequency is 50 Hz. Find the speed of rotor.

Solution

Synchronous speed $N_s = \frac{120f}{p} = \frac{120 \times 50}{4} = 1500 \quad \text{RPM}$

$$N_r = (1 - s)N_s = (1 - 0.03) \times 1500 = 1455 \text{ RPM}$$

Example 5.11 A-6 pole, three-phase 50 Hz induction motor develops a fault in stator winding, giving rise to forward and reverse rotating magnetic field of stator. Determine the speed of forward and reverse rotating magnetic field with respect to stator and rotor. The rotor is assumed to run in forward direction at slip 3%.

Solution

Synchronous speed $N_s = \frac{120f}{p} = \frac{120\times 50}{6} = 1000 \quad \text{RPM}$

The speed of forward rotating magnetic field with respect to stator is $+1000$ RPM

The speed of backward rotating magnetic field with respect to stator is -1000 RPM

Rotor speed, $N_r = (1 - s)N_s = (1 - 0.03) \times 1000 = +970$ RPM

The speed of forward rotating magnetic field with respect to rotor surface is the relative speed between forward rotating magnetic field and rotor surface $= 1000 - 970 = +30$ RPM. Therefore, the rotating magnetic field cuts the rotor surface at speed of $+30$ RPM and inducing the current of slip frequency.

The speed of backward rotating magnetic field with respect to rotor surface is the relative speed between backward rotating magnetic field and rotor surface, and is 1970 RPM (1000+970) in the backward direction. The backward rotating magnetic field cuts the rotor surface at speed of 1970 RPM and induces the EMF of frequency of about twice ($2.0 - 0.03 = 1.97$ times) the supply frequency.

Example 5.12 A 2 pole, three-phase 50 Hz induction motor develops a fault in stator winding, giving rise to forward and reverse rotating magnetic field of stator. Determine the speed of forward and reverse rotating magnetic field with respect to stator and with respect to rotor. The rotor is assumed to run in forward direction at slip 5%.

Solution

Synchronous speed, $N_s = \frac{120f}{p} = \frac{120\times 50}{2} = 3000 \quad \text{RPM}$

The speed of forward rotating magnetic field with respect to stator is $+3000$ RPM

The speed of backward rotating magnetic field with respect to stator is -3000 RPM

Rotor speed, $\mathrm{N_r = (1 - s)N_s} = (1 - 0.05) \times 3000 = +2850$ RPM

The speed of forward rotating magnetic field with respect to rotor surface is the relative speed between forward rotating magnetic field and rotor surface $= sN_s = 0.05 \times 3000 = 150$ RPM

The rotating magnetic field is cutting the rotor surface at speed of $+150$ RPM and inducing the current of frequency of slip frequency $= 0.05 \times 50 = 2.5$ Hz

The speed of backward rotating magnetic field with respect to rotor surface is the relative speed between backward rotating magnetic field and rotor surface

$$= (2 - s)N_s = (2 - 0.05) \times 3000 = 5850 \text{ RPM}$$

The backward rotating magnetic field induces a current of frequency $= (2-s)f = (2 - 0.05) \times 50 = 97.5Hz$

This technique of generating high frequency (100 Hz) is sometimes used by running the slip-ring induction motor at synchronous speed in reverse direction using a prime mover and exciting stator windings with three-phase 50 Hz supply.

Example 5.13 A 4-pole, three-phase 50 Hz induction motor develops a fault in rotor winding, giving rise to forward and reverse rotating magnetic field of rotor. Determine the speed of forward and reverse rotating magnetic field with respect to rotor and with respect to stator. The rotor is assumed to run in forward direction at slip 5%.

Solution

Synchronous speed $N_s = \frac{120f}{p} = \frac{120\times 50}{4} = 1500 \quad \text{RPM}$

The rotor induced frequency of forward field is $f_r = \text{sf} = 0.05 \times 50 = 2.5Hz$

Rotor forward rotating magnetic field with respect to rotor surface is rotating at a speed of $sN_s = 0.05 \times 1500 = 75$ RPM

The rotor induced frequency of backward field is $f_{2r} = (2 - s)f = (2 - 0.05) \times 50 = 97.5Hz$

Rotor backward rotating magnetic field with respect to rotor surface is rotating at a speed of $-sN_s = -0.05 \times 1500 = -75$ RPM

But the rotor is running in forward direction at

$$N_r = (1 - s)N_s = (1 - 0.05) \times 1500 = 1425 \text{ RPM}$$

The speed of forwards rotating magnetic field of rotor with respect to stator is

$$N_r + sN_s = (1 - s)N_s + sN_s = N_s = 1425 + 75 = 1500 \text{ RPM}$$

The speed of backward rotating magnetic field of rotor with respect to stator is $-sN_s + N_r = -75 + 1425 = 1350$ RPM

One can see that, if slip is 50%, then the backward rotating magnetic field of rotor appears to be stationary with respect to stator. Or in other words, the speed of backward rotating magnetic field of rotor with respect to stator for 50% slip is $-sN_s + N_r = -0.5 \times 1500 + (1 - 0.5)1500 = -750 + 750 = 0$ RPM

Example 5.14 A 2 pole slip-ring induction motor has its stator excited by 50 Hz source the rotor is connected to a three-phase source of 10 Hz, determine the speed of rotor.

Solution

The synchronous speed is $N_s = \frac{120f}{p} = \frac{120\times 50}{2} = 3000 \quad \text{RPM}$

Slip frequency is 10 Hz, the slip $s = \frac{f_r}{f} = \frac{10}{50} = 0.2$

Rotor speed, $N_r = (1 - s)N_s = (1 - 0.2)3000 = 2400 \quad \text{RPM}$

The motor will develop torque only at this speed. If the speed drops from 2400-RPM, the torque will become zero.

5.8 Equivalent Circuit of an Induction Motor

The operation of three-phase induction motors has been explained on the basis of rotating magnetic fields. The equivalent circuit of induction motor is derived similar to that of a transformer. The machine has core losses, i.e., Hysteresis and eddy current losses, copper losses in stator and rotor windings, friction and windage losses and load losses. The rotor core losses assumed to be neglected, since the frequency of rotor induced EMF under running condition is quite low. The power flow in an induction motor has been shown in Fig. 5.17. An induction motor has two separate windings namely— stator (primary) and rotor (secondary), their per phase resistances, respectively, are R_1 and R_2. The flux produced by one winding links the other winding and vice versa. Small amount of flux produced by these windings do not link the other winding. This leakage is reflected in equivalent circuit as leakage reactance of these two windings, i.e., X_1 and X_2. An induction motor can be easily analyzed using simple circuit known as per phase equivalent circuit. If the equivalent circuit parameter is known then the input-output relation can be easily established [5].

The equivalent circuit of induction motor is shown in Fig. 5.18. Here stator equivalent circuit is at supply frequency f_1 and rotor equivalent circuit is at a slip frequency $f_r = sf_1$. The rotor induced EMF is sE_2. Where E_2 is the rotor induced EMF at standstill or slip $s = 1.0$. The per phase rotor circuit current is now given by,

$$\mathbf{I}_2 = \frac{s\mathbf{E}_2}{R_2 + jsX_2} \quad \text{at frequency } f_r = sf_1$$

$$\text{Or,} \quad \mathbf{I}_2 = \frac{\mathbf{E}_2}{\left(\frac{R_2}{s}\right) + jX_2} \quad \text{at frequency } f_1 \tag{5.31}$$

The corresponding equivalent circuit is given in Figs. 5.19 and 5.20. The circuit parameters are

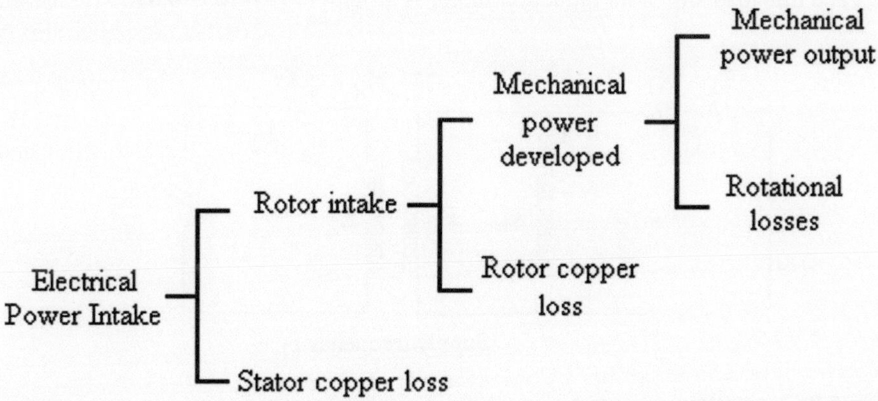

Fig. 5.17 Losses in induction motor

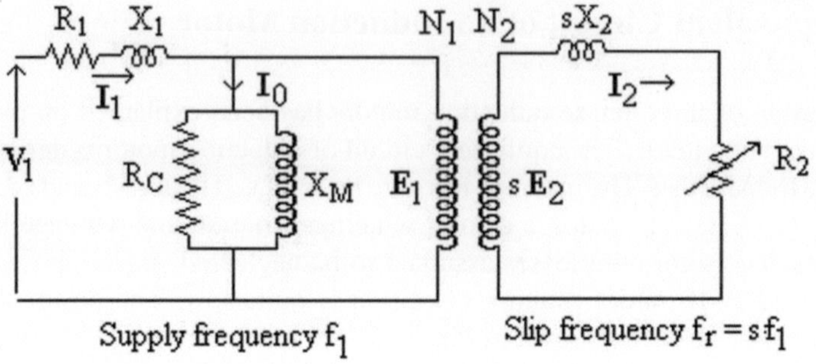

Fig. 5.18 Equivalent circuits of Stator and rotor of induction motor

R_1	the per phase resistance of the stator,Ω
X_1	ωl_1 per phase leakage reactance of stator, Ω
R_2	the per phase rotor resistance, Ω
X_2	ωl_2 per phase rotor leakage reactance, Ω
R_c	per phase iron loss component of resistance referred to stator, Ω
X_M	per phase magnetizing reactance referred to stator, Ω
V_1	per phase stator voltage, V
N_1	per phase stator number of turns
N_2	per phase rotor number of turns
I_1	per phase stator current, A
I_0	per phase no-load current, A
I_2	per phase rotor current, A.

$$\%\ \text{slip}\ s = \frac{N_s - N_r}{N_s} \times 100\% = \frac{\text{Synchrounous speed} - \text{Rotor speed}}{\text{Synchrounous speed}} \times 100\%$$

The full-load slip of an induction motor is around 0.0 to 0.03–0.07.

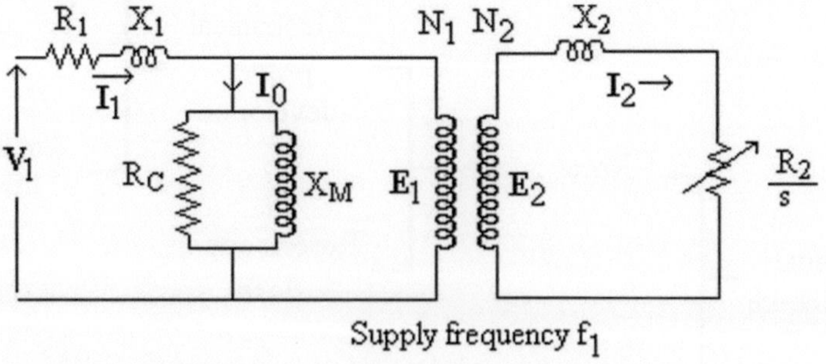

Fig. 5.19 Equivalent circuits of stator and rotor at supply frequency f_1

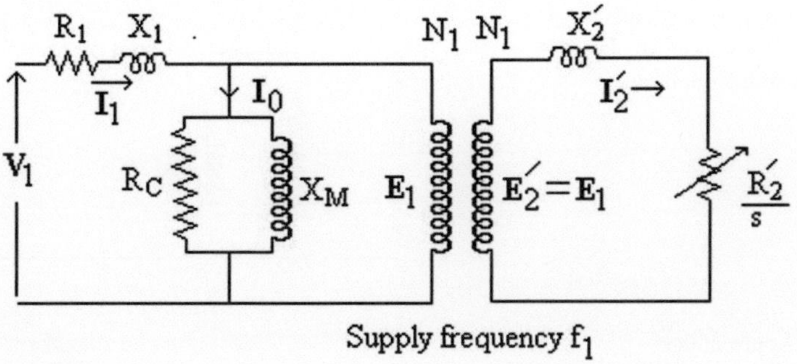

Fig. 5.20 Equivalent circuits of stator and rotor at supply frequency with equal number of turns in both the members.

If we assume that both the stator and rotor have same number of turn, the equivalent circuit will modify to Fig. 5.20. The rotor circuit parameters referred to stator are calculated using

$$\text{Transformation ratio a} = \frac{\text{stator number of turns}}{\text{rotor number of turns}} = \left(\frac{N_1}{N_2}\right)$$

R_2'	per phase resistance of the rotor referred to stator $R_2' = a^2 R_2$
$X_2 = \omega l_2$	per phase leakage reactance of rotor
$X_2' = a^2 X_2$	per phase leakage reactance of rotor referred to stator
$\mathbf{I}_2' = (1/a)\mathbf{I}_2$	per phase rotor current referred to stator
$\mathbf{E}_2' = a\mathbf{E}_2 = \mathbf{E}_1$	rotor per phase EMF referred to stator

Since $E_1 = E_2'$, the ideal coupling is removed in equivalent circuit of induction motor. The exact per phase equivalent circuit of induction motor is shown in Fig. 5.21. Similar to transformer, the parallel branches may be shifted to supply side, thus the per phase approximate equivalent circuit of induction motor is as shown in Fig. 5.22. This simplifies the tedious circuit calculations. It is assumed that

$$\mathbf{I}_1 = \mathbf{I}_0 + \mathbf{I}_2' \tag{5.32}$$

Fig. 5.21 Per phase equivalent circuit of three-phase induction motor

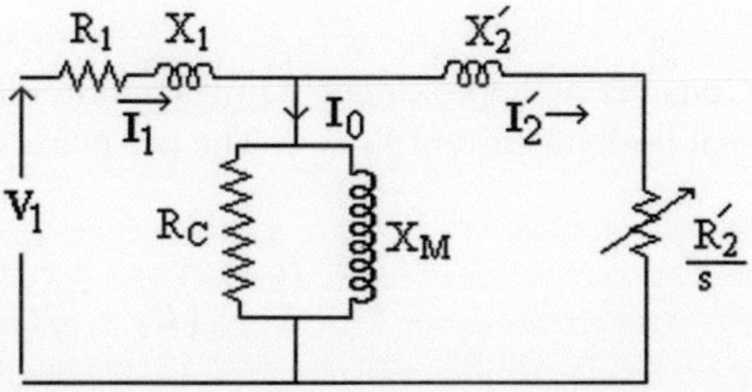

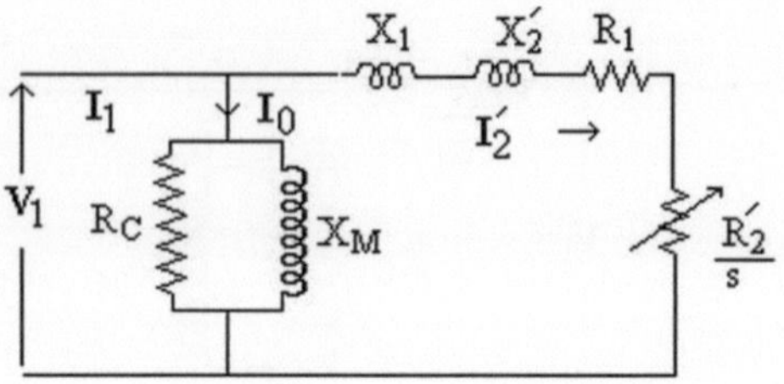

Fig. 5.22 Approximate per phase equivalent circuit of three-phase induction motor

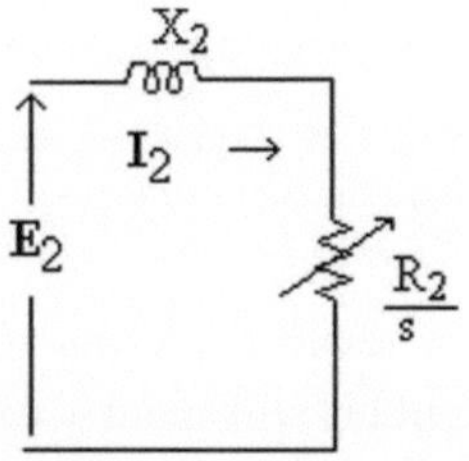

Fig. 5.23 Approximate rotor equivalent circuit

The rotor equivalent circuit shown in Fig. 5.23 is quite helpful, which further simplifies the calculation.

5.8.1 Phasor Diagram

The phasor diagram of three-phase induction motor representing equations (5.31) have been shown in Fig. 5.24. Here for stator

$$\mathbf{V}_1 = \mathbf{E}_1 + \mathbf{I}_1(R_1 + jX_1)$$

For rotor EMF at slip frequency $s\mathbf{E}_2^{/} = \mathbf{I}_2^{/}\left(R_2^{/} + jsX_2^{/}\right)$

Rotor EMF at supply frequency $\mathbf{E}_2^{/} = \mathbf{I}_2^{/}\left(\frac{R_2^{/}}{s} + jX_2^{/}\right)$

5.8.2 Power Intake, Air-Gap Power, and Torque

Consider an approximate equivalent circuit as shown in Fig. 5.21. Here we assume that no-load current $\mathbf{I}_0 = 0$. The per phase current is given by,

$$I_2^{/} = \frac{V_1}{\left(R_1 + \frac{R_2^{/}}{s}\right) + j\left(X_1 + X_2^{/}\right)} \tag{5.33}$$

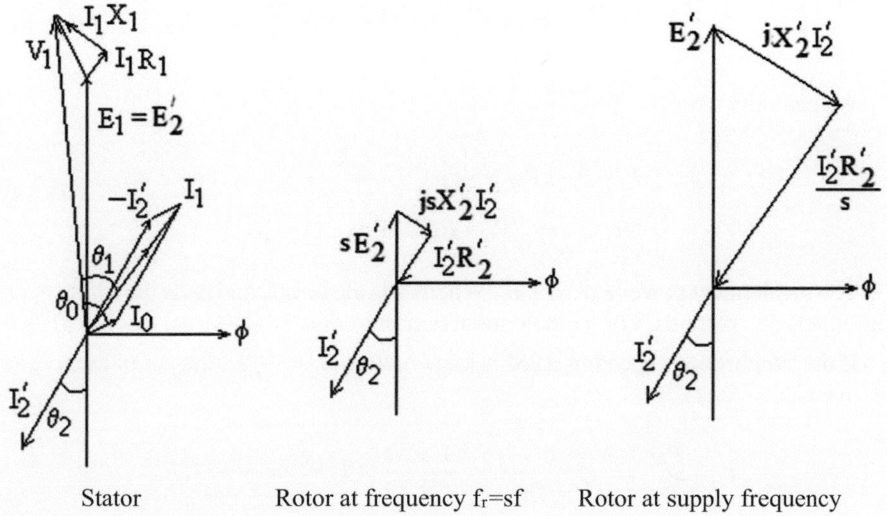

Fig. 5.24 Phasor diagrams of three-phase induction motor

$$I_2^/ = \frac{V_1}{\sqrt{\left(R_1 + R_2^/ / s\right)^2 + \left(X_1 + X_2^/\right)^2}}$$

The per phase rotor intake is given by

$$P_I = \frac{\left(I_2^/\right)^2 R_2^/}{s} \tag{5.34}$$

This is the per phase maximum air-gap power transferred from stator to rotor winding at slip s. The rotor resistance results in certain copper loss associated with them, because of rotor currents.

The per phase rotor copper loss,

$$P_C = \left(I_2^/\right)^2 R_2^/ \tag{5.35}$$

Thus,

$$\frac{P_I}{P_C} = \frac{1}{s} \tag{5.36}$$

The rotor intake is accounted for mechanical power or torque and rotor loss. Thus, the per phase rotor power resulting in mechanical power is,

$$P_M = P_I - P_C = \frac{(1-s)}{s}\left(I_2^{/}\right)^2 R_2^{/} = \frac{(1-s)}{s} P_C = (1-s)P_I \tag{5.37}$$

We can thus write,

$$\begin{aligned} P_I : P_C : P_M &:: \frac{1}{s} : 1 : \frac{1-s}{s} \\ \text{Or,} \quad P_I : P_C : P_M &:: 1 : s : (1-s) \end{aligned} \tag{5.38}$$

The mechanical power $P_M = T\omega_r$. Where T is the torque and ω_r is the rotor speed in radians per seconds. For a p pole induction motor $\omega_r = (1-s)\omega_s = (1-s)\frac{4\pi f}{p}$

If the synchronous speed in RPM is known, then $\omega_s = \frac{2\pi N_s}{60}$, the per phase torque is given as,

$$\begin{aligned} T &= \frac{P_M}{\omega_r} = \frac{1}{(1-s)\omega_s} \times \frac{(1-s)}{s} \times I_1^2 R_2^{/} = \frac{1}{s \cdot \omega_s} I_1^2 R_2^{/} \\ &= \frac{V_1^2}{s \cdot \omega_s \left[\left(R_1 + \frac{R_2^{/}}{s}\right)^2 + \left(X_1 + X_2^{/}\right)^2\right]} R_2^{/} \quad \text{(Nm)} \end{aligned} \tag{5.39}$$

The complete torque-speed characteristics of a three-phase induction motor is shown in Fig. 5.25. The torque is some time expressed in terms of Synchronous Watts, where the expression of torque is multiplied by synchronous rotational speed in radians per second. In other words,

$$T = \frac{V_1^2}{s\left[\left(R_1 + \frac{R_2^{/}}{s}\right)^2 + \left(X_1 + X_2^{/}\right)^2\right]} R_2^{/} \quad \text{Synchronous Watts}$$

Differentiating the equation (5.39) with respect to slip s we get,

$$\frac{\mathrm{d}}{\mathrm{d}s} T = 0$$

The value of slip at which maximum torque occurs is obtained as,

$$s_m = \pm \frac{R_2^{/}}{\sqrt{R_1^2 + \left(X_1 + X_2^{/}\right)^2}} \tag{5.40}$$

Thus, the maximum torque during motoring is given as,

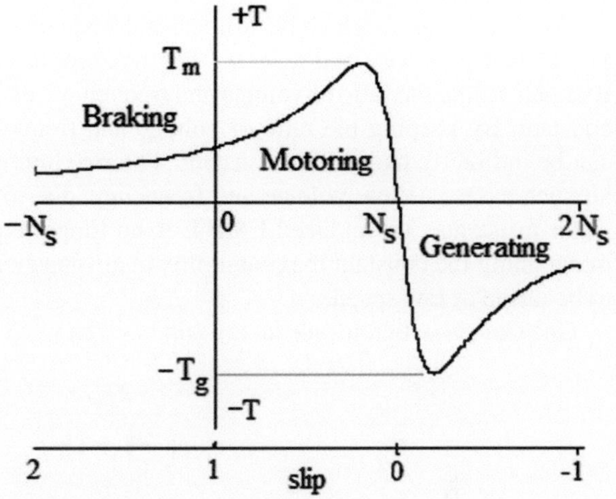

Fig. 5.25 Complete torque-speed characteristics of three-phase induction motor

$$T_m = \frac{V_1^2}{2\omega_s\left[R_1 + \sqrt{R_1^2 + \left(X_1 + X_2^{/}\right)^2}\right]} \tag{5.41}$$

The maximum torque during generating as,

$$T_g = \frac{V_1^2}{2\omega_s\left[-R_1 + \sqrt{R_1^2 + \left(X_1 + X_2^{/}\right)^2}\right]} \tag{5.42}$$

The ratio of maximum torque during generating and motoring is,

$$\frac{T_g}{T_m} = \frac{R_1 + \sqrt{R_1^2 + \left(X_1 + X_2^{/}\right)^2}}{-R_1 + \sqrt{R_1^2 + \left(X_1 + X_2^{/}\right)^2}} \tag{5.43}$$

The maximum torque in generating region is more than that of maximum torque in motoring region. It can be shown that maximum torque is proportional to the square of ratio of Voltage and frequency. Here we know that leakage reactance and synchronous speed are proportional to the supply frequency. One can assume the stator per phase resistance to be negligible,

$$T_m \;\alpha\; \left(\frac{V_1}{f}\right)^2 \tag{5.44}$$

Consider a case in which the Voltage and frequency both are reduced by a factor of x, such that $V_x = xV_1$ and $f_x = xf$. The synchronous speed is reduced to $\omega_x = x\omega_s$. By maintaining the ratio of voltage and frequency one can keep the maximum torque constant. By keeping the ratio of voltage and frequency constant, the air-gap flux can be indirectly maintained constant. The maximum torque will remain same for all such cases. At low voltage and frequency, due to considerable voltage drop in stator resistance, the induced EMF E is no longer equal to applied voltage V. For maintaining the constant maximum flux in air-gap and peak torque the voltage need to be raised at low speed.

One can show that torque at any slip is given by,

$$T = \frac{2T_m}{\left(\frac{s}{s_m}\right) + \left(\frac{s_m}{s}\right)} \tag{5.45}$$

A complete torque slip characteristics is shown in Fig. 5.25. The torque slip characteristic of an induction motor has two portions—unstable and stable. The torque slip characteristic has following regions of operation:

(a) When slip is near 0.0, torque $T \quad \alpha \quad s$, which is a stable portion of the characteristics. This can be verified using Eq. (5.45). Upon loading the speed drops and slip increases. From slip s_m to 0.0, it is the stable portion of the characteristics.
(b) When slip is near 1.0, torque $T \quad \alpha \quad \frac{1}{s}$, which is unstable portion of the characteristics. Thus from slip 1.0 to s_m, it is the unstable portion of the characteristics.

Starting torque: The starting torque is the torque which a machine produces at standstill or zero speed condition. A machine is started in the direction of starting torque. If a machine is having zero starting torque it cannot be started naturally. The load torque at standstill condition should be less than the developed starting torque.

Maximum torque or pull-out torque is the torque which induction motor produces at certain critical slip. If the load torque at critical slip is more than maximum torque, the machine fails to start.

5.9 Determination of Equivalent Circuit Parameters

Following tests are performed for determination of equivalent circuit parameters:

(a) Free running or No-load Test
(b) Locked rotor test.

These are described in the following subsections.

5.9.1 Free Running Test or No-Load Test

In free running test, machine is allowed to run without any mechanical load connected to the shaft. During this test, the speed of induction motor is nearly synchronous speed and slip tends to 0.0. In the rotor circuit, the effective rotor resistance $R_2^{/}/s$ is very high and the rotor currents are negligibly small. The equivalent circuit during free running test is given in Fig. 5.26. The stator current is a feeble current sufficient to provide the necessary losses and torque for overcoming frictional and windage.

The circuit diagram for free running test is given in Fig. 5.27. Since the power factor is low, a Low-Power-Factor (LPF) Wattmeter is used for measurement of power. The ammeter range is selected so as to read about 35−40% of rated current. In comparison with the no-load current of transformer, the no-load current of induction motor is comparatively large. This is because of air-gap or *entrefer* (French word for iron-to-iron distance), which raises the leakage reactance. The no-load current supplies necessary losses and torque for running. The induction motor is started using three-phase auto transformer, by initially giving lower voltage. The starting current may be quite high. As such for protecting the ammeter and current coil of wattmeter, these are short circuited using switch S1. When motor picks up speed, switch S1 is made open.

The single wattmeter method for measuring three-phase power is applicable for balanced conditions. Switch S2 is connected accordingly for recording W_1 and W_2. The purpose of cross-connected Double-Pole-Double-Throw (DPDT) switch is to read the negative deflection in the wattmeter. The stator is connected in star. Here no-load line /phase current is, I_0 and per phase voltage is $V_0 = \frac{1}{\sqrt{3}} V_{LL}$. The net power intake is

$$W_0 = W_1 \pm W_2 \tag{5.46}$$

The Wattmeter reading may be taken as rated core loss. The minus sign '-' is taken when after changing S2, DPDT switch is also operated to read positive reading of wattmeter. The per phase power intake,

$$P_0 = \frac{1}{3} W_0$$

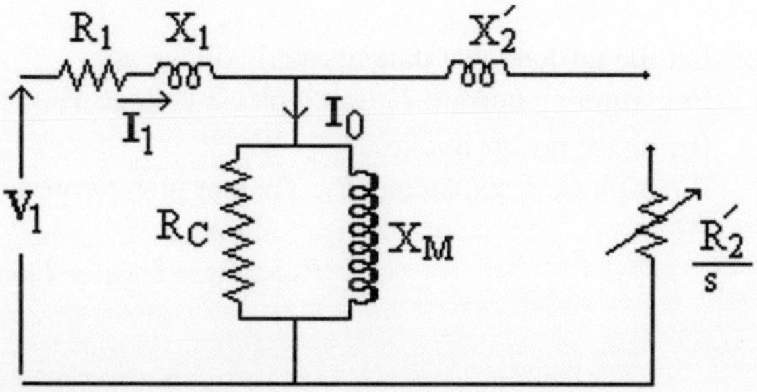

Fig. 5.26 Equivalent circuit during free running test

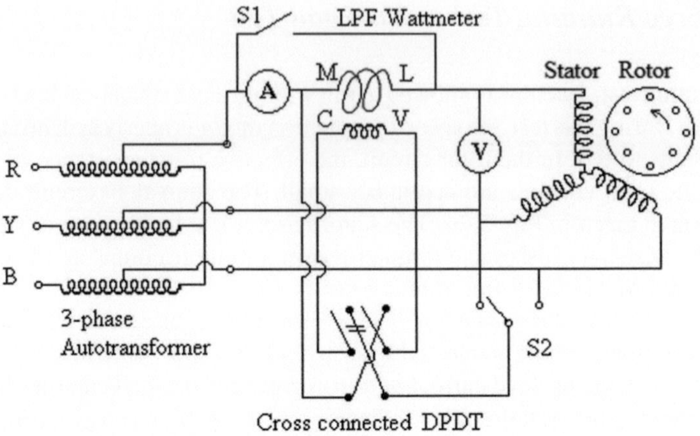

Fig. 5.27 Circuit diagram for Free running test

If the per phase winding resistance is R_1, then the per phase fixed loss (core loss + friction and windage loss) is

$$P_c + P_{fw} = P_0 - I_0^2 R_1 \tag{5.47}$$

The per phase friction and windage loss P_{fw} is separated using variable voltage test as discussed earlier.

Separation of friction and windage loss When the rotor is running, it cuts the surrounding air. Apart from this, well lubricated bearing also has certain friction. Certain power is wasted in overcoming the friction and windage; this is referred as friction and windage loss. For separation of P_{fw}, the free running test is performed at different voltages. When machine is started using autotransformer, the voltage is lowered in steps till the no-load current starts rising. Figure 5.28 depicts the per phase $P_0 - I_1^2 R_1$ versus applied voltage. The extrapolated loss corresponding to $V = 0$ is taken as friction and windage loss. The curve $P_0 - I_0^2 R_1$ versus V^2 is a straight line. The per phase friction and windage loss is assumed to remain constant at all speed. Thus, the core loss can be separated

$$P_c = P_0 - I_0^2 R_1 - P_{fw}$$

Let the no-load test data at rated voltage are
Free running current I_0; rated phase voltage $V_0 = \frac{1}{\sqrt{3}} V_{LL}$;
Per phase power intake, $P_0 = W_0/3$;
If per phase resistance is R_1. The per phase core loss

$$P_c = P_0 - I_0^2 R_1 - P_{fw} \tag{5.48}$$

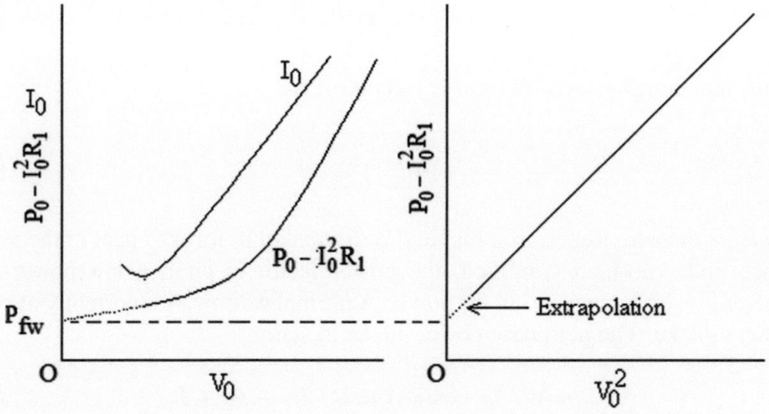

Fig. 5.28 Separation of frictional and windage loss per phase by extrapolation

The core loss resistance and magnetizing reactance can now be determined similar to that of transformer.

At free running, the current $\mathbf{I}_0$ is assumed to flow in the parallel combination of R_c and X_m. The free running current,

$$\mathbf{I}_0 = \mathbf{I}_c + \mathbf{I}_m \tag{5.49}$$

The real component of no-load current is

$$I_c = \frac{V_0}{R_c} = I_0 \cdot \cos(\phi_0) \tag{5.50}$$

where ϕ_0 is no-load power factor angle. The imaginary component of open-circuit current is,

$$I_m = \frac{V}{j.X_m} = I_0.\ \sin(\phi_0) \tag{5.51}$$

The per phase induced EMF in primary winding is,

$$\mathbf{E}_0 = \mathbf{V}_0 - \mathbf{I}_0(R_1 + jX_1) \tag{5.52}$$

If the winding resistance R_1 is known, then the EMF $\mathbf{E}_0$

$$\mathbf{E}_0 \approx \mathbf{V}_0 - \mathbf{I}_0\mathrm{R}_1$$

Normally $\mathbf{I}_0 R_1$ drop is quite less and is ignored. However, if winding resistance R_1 is known, the voltage drop due to it can be accounted for. Though, it is assumed that $\mathbf{E}_0 \approx \mathbf{V}_0$. Thus, the real power loss,

$$P_c = E_0 I_c \tag{5.53}$$

Thus, real component of current is calculated as,

$$I_c = \frac{P_c}{E_0} \tag{5.54}$$

where P_c is the core loss or iron loss and is dissipated in form of heat in the system.

Since under no-load condition, the power factor is poor, a low-power-factor wattmeter is used for measuring the power. A normal wattmeter incorrectly measures the power intake. The per phase power intake in stator,

$$P_{in} = V_1 \cdot I_0 \ \cos(\phi_0) = I_0^2 \cdot R_1 + P_c + P_{fw} \tag{5.55}$$

If the drop in primary winding is assumed to be negligible then,

$$E_0 \approx V_0 \tag{5.56}$$

Thus using open-circuit test, the two equivalent circuit parameters can be calculated as,

$$R_c = \frac{E_0}{I_c} = \frac{E_0}{I_0 \cdot \cos(\phi_0)} = \frac{P_c}{(I_0 \cdot \cos(\phi_0))^2} \approx \frac{V_0}{I_0 \cdot \cos(\phi_0)} \tag{5.57}$$

$$X_m = \frac{E_0}{I_m} = \frac{E_0}{I_0 \cdot \sin(\phi_0)} \approx \frac{V_0}{I_0 \cdot \sin(\phi_0)} \tag{5.58}$$

5.9.2 Blocked Rotor Test

For obtaining the series parameters of equivalent circuit, the motor is tested at either standstill condition, $N_r = 0$ or slip $=1.0$ or nearly standstill condition. The equivalent circuit at blocked (locked) rotor condition is given in Fig. 5.29. The rotor is not allowed to rotate. A small voltage is impressed so that the machine takes the rated current. The test is analogous to short-circuit test of a transformer. In case of large induction motor, the equivalent circuit parameters can be calculated using design data or can also be numerically obtained using FEM or BEM software.

The circuit diagram for locked rotor test is shown in Fig. 5.30. In locked rotor test, the per phase locked rotor copper loss is

$W_{sc} = (W_{sc1} \pm W_{sc2})/3$	Here Wattmeter reading in locked rotor test at rated current or rated copper loss is $(W_{sc1} \pm W_{sc2})$ Watt
V_{sc}	applied per phase voltage, Volt
I_{sc}	locked rotor (rated) per phase current, A

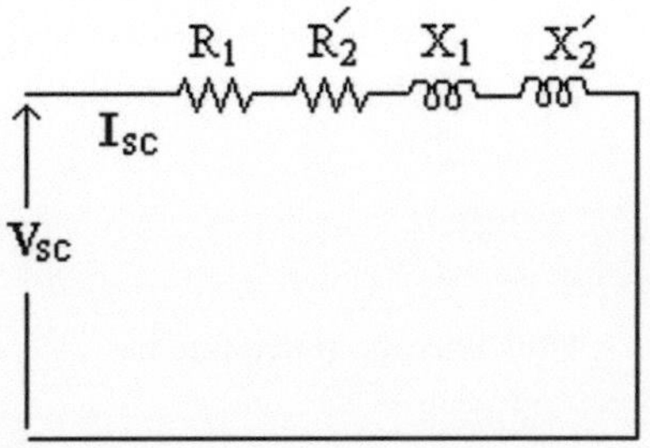

Fig. 5.29 Locked rotor per phase equivalent circuit

Considering approximate equivalent circuit, we have the per phase resistances and leakage reactance referred to stator are,

$$R_s = R_1 + R_2^{/} = R_1 + a^2.R_2 \tag{5.59}$$

$$X_s = X_1 + X_2^{/} = X_1 + a^2.X_2 \tag{5.60}$$

According to simple circuit analysis, we have

$$R_s = \frac{W_{\text{sc}}}{I_{\text{sc}}^2} \tag{5.61}$$

If per phase stator resistance R_1 is known then the per phase rotor resistance referred to stator side is $R_2^{/} = R_s - R_1$

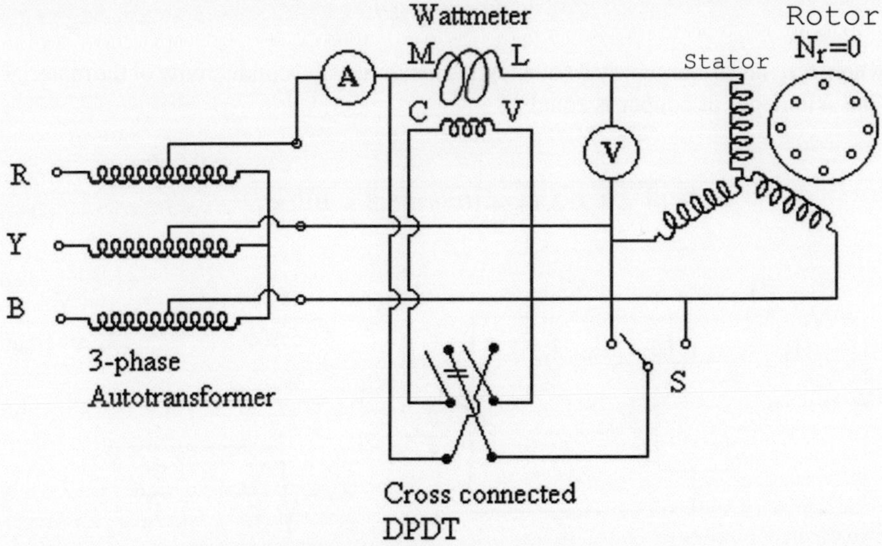

Fig. 5.30 Circuit diagram for locked rotor test

If the per phase stator resistance is not known, one can choose, $R_1 = R_2^{/} = \frac{1}{2}R_s$ The impedance seen from the terminal,

$$Z_s = \frac{V_{\text{sc}}}{I_{\text{sc}}} \tag{5.62}$$

Thus leakage reactance is,

$$X_s = \sqrt{(Z_s^2 - R_s^2)}$$

One can choose,

$$X_1 = X_2^{/} = \frac{1}{2}X_s$$

Figure 5.31 shows per phase equivalent circuit of a three-phase induction motor. The phenomenon known as *skin effect* causes the AC current to flow on outer surface of the conductor. The effective cross section carrying the AC current is only the area of an outer skin of the conductor. The resistance of the conductor is increased accordingly. The high frequency resistance of a conductor is higher than its DC resistance or low frequency resistance. When a 50 Hz machine is operated on 400 Hz supply, the winding resistance is expected to be very large owing to skin effect. Though, the core losses– Hysteresis and eddy current are also increased. The skin depth is defined as

$$\delta = \frac{1}{\sqrt{\pi f \mu \sigma}} \tag{5.63}$$

where f, μ, and σ are supply frequency, permeability, and conductivity of the material. The skin depth of copper is equal to

$$\delta = \frac{1}{\sqrt{3.14 \times 4 \times 3.14 \times 10^{-7} \times 5.8 \times 10^7 \times f}} = \frac{0.066}{\sqrt{f}}\ \text{m}$$

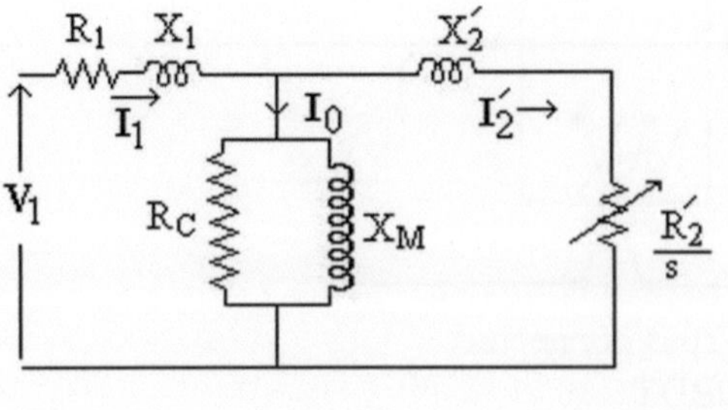

$$R_c = \frac{V_0}{I_0 \cdot \cos(\phi_0)} \qquad X_m = \frac{V_0}{I_0 \cdot \sin(\phi_0)}$$

$$R_2^{/} = \frac{W_{sc}}{I_{sc}^2} - R_1$$

$$X_s = \sqrt{\left(\frac{V_{sc}}{I_{sc}}\right)^2 - \left(\frac{W_{sc}}{I_{sc}^2}\right)^2}$$

Fig. 5.31 Per phase equivalent circuit referred to stator side

The skin depth of copper at 50 Hz and 400 Hz for copper are, respectively, 9.33 mm and 3.33 mm. For accounting the skin effect in induction motor two methods are adopted. If three phase variable frequency source is available, the locked rotor test is also conducted at low frequency. The per phase stator winding resistance is obtained using DC power supply. Thus,

DC winding resistance per phase $= R_{DC} = \frac{V_{DC}}{I_{DC}}$

The winding made with thin SWG wire having small skin effect. This can be determined as follows:

The AC resistance from the locked rotor test at supply frequency

$$\left(R_1 + R_2^{/}\right)\Big|_{50\text{ Hz}} = \frac{W_{sc}}{I_{SC}^2}\Big|_{50\text{ Hz}}$$

Thus, per phase rotor resistance referred to stator

$$R_2^{/} = \left(R_1 + R_2^{/}\right)\Big|_{50\text{ Hz}} - R_{DC}$$

The per phase impedance from locked rotor test at reduced frequency say 15 Hz, The resistance at low frequency locked rotor test at reduced supply frequency

$$\left(R_1 + R_2^{/}\right)\Big|_{15\text{ Hz}} = \frac{W_{sc}}{I_{sc}^2}\Big|_{15\text{ Hz}}$$

Here W_{sc} is per phase copper loss during blocked rotor test. The corrected value of AC stator winding has the value,

$$R_{1ac} = \frac{\left(R_1 + R_2^{/}\right)\Big|_{50\text{ Hz}}}{\left(R_1 + R_2^{/}\right)\Big|_{15\text{ Hz}}} \times R_{DC}$$

The second rotor parameter rotor resistance referred to stator is at low frequency, Thus

$$R_{2ac}^{/} = \left(R_1 + R_2^{/}\right)\Big|_{15\text{Hz}} - R_{DC} \tag{5.64}$$

The per phase DC resistance rotor can also be experimentally obtained by variable frequency locked rotor test on induction motor. The DC value is obtained by extrapolation at $f = 0$ Hz. The latter method is more realistic in nature. When the test data at low frequency is not available it is convenient to use correction factor. A correction factor accounting for power frequency resistance is used. The stator winding resistance is increased by a correction factor of 1.1 for accounting the eddy current within the conductors.

Using the above per phase equivalent circuit, the characteristics of induction motor can be easily computed. Figure 5.32 shows the variation of current at different

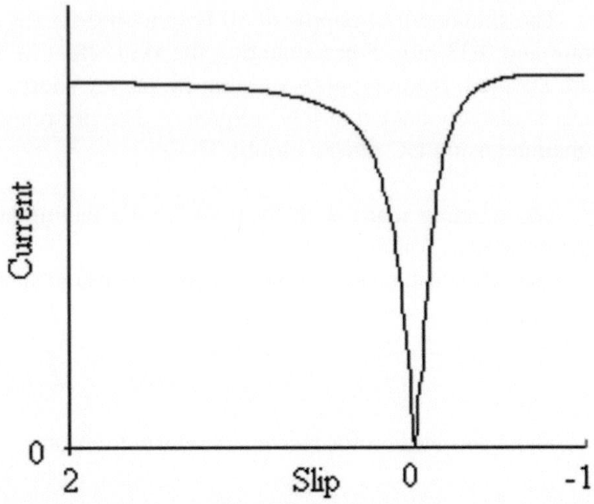

Fig. 5.32 Variation of current with respect to slip for constant voltage in braking motoring and generating region

speed of operation. When slip is zero, the current drawn is lowest. Fig. 5.24 shows the complete torque-speed/slip characteristics. The torque is positive in motoring region, whereas it is negative in regenerating zone.

5.9.3 Loading of Induction Motor

The load test on induction motor is performed for determining its performance characteristics in stable portion of its torque slip characteristics. Normally, the loading of induction motor can be conducted using following methods:

(a) Direct load test using belt and pulley
(b) Separately excited DC generator
(c) Eddy current brake.

Direct load test using belt and pulley or using eddy current brake is useful in testing smaller capacity induction motors.

If load cell or torque transducer is installed in swinging frame machine, a signal proportional to force or torque can be obtained. It is also possible to obtain current, voltage and speed signals. Data acquisition system (DAS) can be used for recording torque-speed characteristics and variation of current with speed.

The eddy current brake (ECB) need to be calibrated using tested model. For a known air-gap and ECB current the torque-speed characteristics of eddy current brake is useful in getting the torque exerted by the induction motor. When eddy current brake is used for loading induction motor, the speed of induction motor can be precisely controlled by controlling the ECB current. Such a technique is helpful in determining the harmonic torque of induction motor.

In absence of sophisticated instruments, the spring balance gives acceptable results at steady state conditions. When pulley and belt system is used, the calculation of torque and output power can be done as suggested in Sect. 4.14.

The direct load test gives the performance of induction motor in the stable portion of its torque-speed characteristics. The induction motor halts when one attempts to run it in unstable portion of its torque-speed characteristics.

Under loading conditions, the current supplied to the stator of induction motor is determined by three factors. Firstly, it has to supply almost constant magnetizing current required to maintain the rotating magnetic field. The second component of current is supplying the core losses–Hysteresis and Eddy current losses. The third component of current produces a field and this current is equal and opposite to that formed by the rotor current. The rotor current increases with loading. As a consequence of this, the stator current therefore also rises with load.

At no load, the power factor of induction motor is very low because of comparatively high value of lagging magnetizing current. When machine is loaded, the power factor improves, because of the fact that active power component of current is increased. Machine absorbs both active and reactive (lagging) power. One of the disadvantages of an induction motor is its low power factor operation.

As mentioned earlier, an induction moor draws larger magnetizing current as compared to transformer due to presence of air-gap between stator and rotor. The magnetizing current in case of induction motor can be as high as 35% of the full load current compared to about 5% of full load current in case of a transformer. One can recall that there is no air-gap in the magnetic circuit of a transformer. For reducing the magnetizing current the air-gap in rotary induction motor is kept as small as possible. In induction motor the air-gap is smaller than any other type of electrical machines.

The losses occurring in an induction motor are:

(a) Ohmic losses in stator and rotor windings
(b) Friction and windage losses
(c) Iron losses in stator and rotor core.

The output power P is equal to $P = T\omega_r$. The speed of the motor does not vary appreciably with load, the torque increases with increase in the load and output power.

Consider the induction machine is connected to a separately excited DC generator. The generator is connected to a resistive load; the current in the DC generator produces a counter torque which is opposite to the motoring torque developed by induction motor. Let the output power of generator be

$$P_O = V_{\mathrm{dc}} I_{\mathrm{dc}}$$

The power drawn by the induction motor from three phase supply is

$$P_{\mathrm{in}} = 3 V_{\mathrm{ph}} I_{\mathrm{ph}} \cos\,\theta$$

The input power can be measured using wattmeter. The input power factor can be calculated using above expression. The efficiency can be calculated as

$$\eta = \frac{\text{Output power of DC generator}}{\text{Input power of induction motor}} = \frac{P_O}{P_{\text{in}}} \tag{5.65}$$

When an induction motor is at no load, its speed is slightly below the synchronous speed. The current due to induced EMF in the rotor winding is responsible for production of torque required to overcome the frictional and windage loss. When running at exact synchronous speed, the rotor induced EMF is zero and the developed torque is also zero. The motor speed falls below the synchronous speed.

The no-load test at exact synchronous speed gives correct value of magnetizing reactance. For this the speed of induction motor is maintained at synchronous speed using another motor / DC motor connected to the shaft of induction motor and running in the same direction. This test is sometimes useful when induction machine is used as Self Excited Induction Generator (SEIG).

The speed of rotating shaft of any rotating machine can be measured or sensed using any one method mentioned below:

(a) Mechanical tachometer
(b) Tacho generator connected to the shaft
(c) Optical tachometer - contact type and reflection type
(d) Magnetic pick up which requires a spoked wheel to be connected to the shaft
(e) Hall Effect based speed measurement. The Hall Effect sensor (uni polar or bipolar) senses tiny magnetic pole / poles (single or alternate N−S) rotating with the shaft.
(f) Stroboscopic RPM meter working on the principle of changing (ON/OFF) frequency of light. The shaft appears to be stationary when frequency of light and shaft RPM is matched.
(g) Encoder
(h) Resolver.

Mechanical tachometer, tacho-generator and tachometers based on cutting of light or optical shaft encoding principle are contact type. The optical tachometer based on sensing reflected light, magnetic pick RPM meter, Hall effect RPM meter and stroboscopic RPM meter are all non-contact type of instruments. Encoder and resolver are connected to one end of shaft.

If we know the DC generator output and speed of induction motor N_r, the torque developed by induction motor can be calculated,

$$T = \frac{P_O}{\omega_r} = \frac{P_O}{\left\{\frac{2\pi N_r}{60}\right\}} \text{Nm}$$

Using above method torque−speed/slip characteristics of induction motor in stable portion can be experimentally determined. This requires changing excitation (field current) of DC generator and changing of resistive load connected to DC generator or both.

For experimental determination of complete torque-speed characteristics of induction motor, a separate arrangement is needed. It requires two identical induction motor-separately excited DC motor sets. These two sets are electrically coupled. Both the induction motors are connected to a common three phase supply. The test motor is supplied by the voltage sufficient to flow rated current under locked rotor condition. The armatures of DC generators are initially tested for floating condition and are joined together. (Refer Sec 4.14.3 Hopkinson test). By changing the magnitude and direction of field current of auxiliary DC generator, the test induction motor can be made to run as motor generator and brake. The induction machine can be tested for determining torque- speed/slip, current-speed/slip characteristics in operating slip of 2.0 to 1.0 to 0.0 to −1.0 at low voltage corresponding to rated current under locked rotor condition. The results obtained are according modified at rated voltage. The circuit diagram is shown in Fig. 5.33. Test induction machines and auxiliary induction machine may be either squirrel cage induction machine or slip-ring induction machine.

Initially both the induction motors are started using autotransformer. The Field currents of DC generators are adjusted such that the voltmeter reads zero. If it reads twice the rated DC voltage necessary changes need to be done in the field circuit. The switch 'S' is closed, when the series voltmeter reads zero. Both the generators are in floating conditions and developed voltages of DC generators are equal. The input voltage of test induction machine is adjusted to the voltage required to flow rated current under short circuit test. A reduced voltage is required such that maximum current during test can be limited to rated current.

When the two generators are in floating conditions, if we reduce the field current of second (auxiliary) generator, it causes the developed voltage of second machine to reduce and current flows from First (main) DC generator to the second DC (auxiliary) generator. The power flowing in the DC generator is recorded and so are the input current, input voltage and input power and speed of main induction machine. What will happen when the field current of second generator is increased?

If we change the direction of field current of second machine using cross connected DPDT, it will cause the machine to run from motoring (critical slip) to counter current braking zone.

The ammeter used in armature circuit is of center-zero type, which can measure the DC current in both the directions. Using the direction of armature current, one can easily judge whether machine is working as motor or generator.

By changing the field current of second generator in both the direction a complete operation of induction motor in slip range 2.0−1.0−0.0− 1.0 can be obtained.

The setup is also particularly helpful in experimentally judging the machine behavior in the event of unequal rotor resistance of slip-ring induction motor.

As a precautionary measure the test is performed at reduced voltage and the obtained results are multiplied by a factor corresponding to the rated voltage and the test voltage.

Example 5.15 A three phase 6 pole, 100 hp, 440 V star-connected squirrel cage induction motor has following test data: No-load test: 440 V, 10 A, 4.5 kW

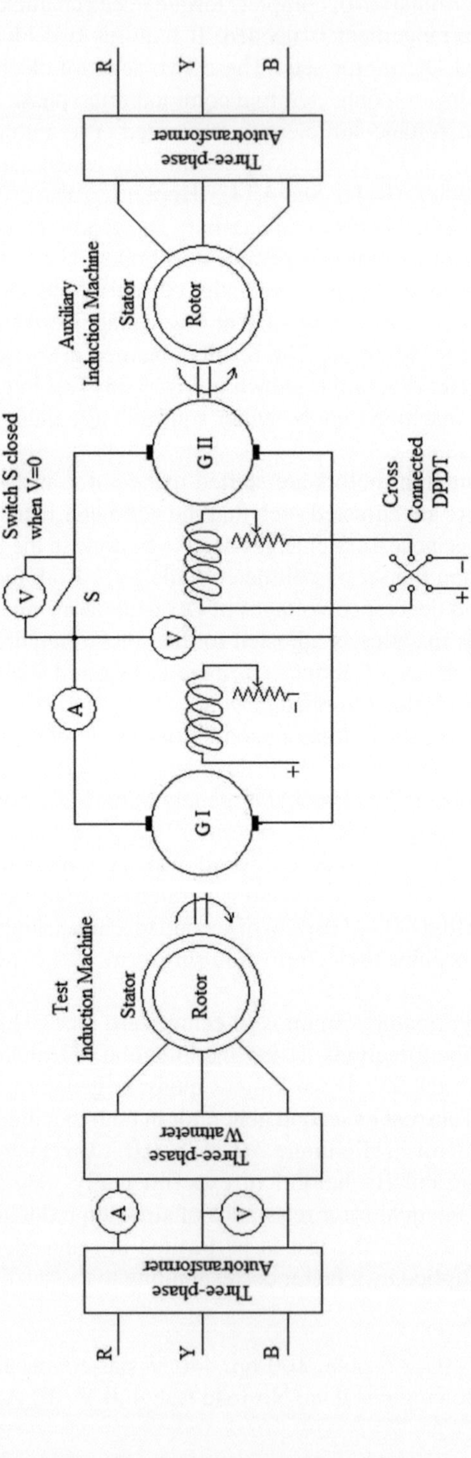

Fig. 5.33 Circuit diagram for experimental determination of complete torque slip characteristics of three phase induction motor

Blocked rotor test: 90 V, 50 Hz, 120 A, 16.0 kW. Average DC resistance between the terminals is 0.15 Ω. Determine the equivalent circuit parameters. Estimate the current, power factor, torque, output power when motor is running at 5 % slip.

Solution

The DC winding resistance per phase $= R_1 = \frac{0.15}{2} = 0.075\,\Omega$

Per phase blocked rotor test loss, Wsc = (16000/3) W. The AC resistance referred to stator from the locked rotor test at supply frequency

$$\left(R_1 + R_2^/\right) = \frac{W_{sc}}{I_{sc}^2} = \frac{16000}{3(120)^2} = 0.3703\,\Omega$$

Thus per phase rotor resistance referred to stator

$$R_2^/ = 0.3703 - 0.075 = 0.2954\ \Omega$$

The per phase impedance from locked rotor test,

$$Z_{sc} = \frac{V_{sc}}{I_{sc}\sqrt{3}} = \frac{90}{120\sqrt{3}} = 0.433\,\Omega$$

Per phase leakage reactance referred to stator,

$$X_s = \sqrt{Z_{sc}^2 - \left(R_1 + R_2^/\right)^2} = \sqrt{0.433^2 - 0.3703^2} = 0.2245\,\Omega$$

From no-load test the no-load power factor angle,

$$\theta_0 = \cos^{-1}\left(\frac{W_0}{V_0 I_0\sqrt{3}}\right) = \cos^{-1}\left(\frac{4500}{440 \times 10\sqrt{3}}\right) = 53.8°$$

The reactive component of no-load current $I_m = I_0 \sin\theta_0 = 10 \sin 53.8° = 8.1\ A$
Thus the magnetizing reactance $X_m = \frac{V_{ph}}{I_m} = \frac{440}{8.1\sqrt{3}} = 31.36\ \Omega$
The total core loss $P_{cT} = P_0 - P_{fw} - 3I_0^2 R_1$
Assuming the frictional and windage loss to be zero, the per phase core loss is,

$$P_c = [4500 - 3 \times 10 \times 10 \times 0.075]/3 = 1492.5\ \text{W}$$

Thus, per phase core loss resistance $R_c = \left(\frac{440}{\sqrt{3}}\right)^2\left(\frac{1}{1492.5}\right) = 43.24\ \Omega$
The per phase voltage, $V_{ph} = \frac{440}{\sqrt{3}} = 254\text{V}$
The core loss current, $I_c = \frac{V_{ph}}{R_c} = \frac{254}{43.24} = 5.87\angle 0°\ \ A$
Thus magnetizing current, $\mathbf{I}_m = \frac{V_{ph}}{jX_m} = \frac{254}{31.36j} = -8.1j = 8.1\angle -90°\ \ A$

Thus, no-load current,

$$\mathbf{I}_0 = 5.87 - 8.1j = 10.19\angle -54^\circ \text{ A}$$

The current

$$\mathbf{I}_2^/ = \frac{V_{ph}}{\left(R_{lac} + \frac{R_{2dc}^/}{s}\right) + jX_s} = \frac{254}{(0.075 + \frac{0.2954}{0.05}) + j0.2245} = 42.42\angle -2.15^\circ \quad \text{A}$$

Thus, the input current

$$\mathbf{I}_1 = \mathbf{I}_0 + \mathbf{I}_2^/ = 5.87 - 8.1\,j + 42.43 - 1.59\,j$$
$$= 48.3 - 9.69\,j = 49.26\angle -11.3^\circ \text{ A}$$

Power factor is cos 11.3° = 0.98 lagging
Three-phase air-gap power or rotor intake is

$$\mathrm{P_I} = 3\frac{\left(\mathrm{I}_2^/\right)^2 \mathrm{R}_2^/}{\text{slip}} = 3\frac{(42.42)^2 0.2954}{0.05} = 31.89 \text{ kW}$$

Output power $P_o = P_I(1 - s) = 31.89(1 - 0.05) = 30.3\,\text{kW}$
Synchronous speed $\omega\mathrm{s} = 2\pi\left(\frac{120\mathrm{f}}{\mathrm{p}} \times \frac{1}{60}\right) = \frac{4\pi\mathrm{f}}{\mathrm{p}} = \frac{4\times3.14\times50}{6} = 104.66\ \text{Rad/sec}$
Torque. $T = \frac{P_I}{\omega_s} = \frac{P_o}{\omega_r} = \frac{31890}{104.66} = 289.6$ Nm

Example 5.16 Develop MATLAB computer program for verifying the results of previous Example.
% MATLAB program for Example 5.14

```
RT = 0.15                 % resistance across terminal
R1 = RT/2                 % per phase stator resistance
WSC = 16000               % Blocked rotor test data
ISC = 120
VSC = 90
W0 = 4500                 % Free running test data
V0 = 440
I0 = 30
WFW = 0                   % friction and windage loss
SLIP = 0.05               % slip
RSC = WSC/(3*ISC*ISC)     % per phase stator + rotor resistance
                          referred to stator
```

```
RT = 0.15                        % resistance across terminal
R2D = RSC − R1                   % per phase rotor resistance referred
                                 to stator
ZSC = VSC/(1.732 ∗ ISC)
XSC =                            % per phase sum of leakage
sqrt(ZSC∗ZSC−RSC∗RSC)            reactance
PF0 = W0/(1.7321∗I0∗V0)          % free running power factor
SINPF0 =                         % sin of no load p.f. angle
sqrt(1.0 − PF0 ∗ PF0)
IM = I0 ∗ SINPF0                 % magnetizing current
XM = V0/(1.732 ∗ IM)             % per phase magnetizing reactance
PC =                             % total core loss
W0 − 3 ∗ I0 ∗ I0 ∗ R1 − WFW
RC = V0 ∗ V0/PC                  % per phase core loss resistance
IC = V0/(1.7321 ∗ RC)            % current in core loss resistance
IO = IC − j. ∗ IM                % no-load current in complex form
I2D = V0/(1.7321 ∗ ((R1 +        % rotor current referred to stator
(R2D/SLIP)) + j. ∗ XSC))
IR = real(I2D)                   % REAL of I2D
II = imag(I2D)                   % Imaginary of I2D
I2 = abs(I2D)                    % Absolute value
POIN =                           % total air-gap power or rotor intake
3 ∗ I2 ∗ I2 ∗ R2D/SLIP
POUT = POIN ∗ (1 − SLIP)         % total mechanical power
WS = 4 ∗ 3.14 ∗ 50/6             % synchronous rotational speed
                                 radians/s
T = POIN/WS                      % total torque (Nm)
```

Example 5.17 The following no-load test is performed on a three-phase four pole, 50 Hz, 20 hp, 440 V induction motor. The test data are

No-load test:	440 V	4.0 A	500 W	50 Hz
Blocked rotor test at rated frequency	110 V	18 A	2500 W	50 Hz
	55 V	18 A	2050 W	15 Hz
DC test on stator per phase	10 V	15 A		

The friction and windage loss is known to be 200 W. Determine the per phase core loss resistance and magnetizing reactance of equivalent circuit.

Solution: The DC winding resistance per phase $= \frac{V_{dc}}{I_{dc}} = \frac{10}{15} = 0.6667\Omega$

The winding made with thin SWG wire having small skin effect. This can be determined as follows:

The AC resistance from the locked rotor test at supply frequency is

$$\left(R_1 + R_2^{/}\right)\Big|_{50\text{ Hz}} = \frac{W_{sc}}{3I_{sc}^2}\Big|_{50\text{ Hz}} = \frac{2500}{3(18)^2} = 2.572\ \Omega = R_{sc}$$

The resistance at low frequency locked rotor test at reduced supply frequency
The AC resistance from the locked rotor test at reduced frequency.

$$\left(R_1 + R_2^{/}\right)\Big|_{15\,\text{Hz}} = \frac{W_{sc}}{3I_{sc}^2}\Big|_{15\,\text{Hz}} = \frac{2050}{3(18)^2} = 2.11\,\Omega$$

The corrected value of AC stator winding has the value,

$$R_{1ac} = \frac{\left(R_1 + R_2^{/}\right)\Big|_{50\text{Hz}}}{\left(R_1 + R_2^{/}\right)\Big|_{15\,\text{Hz}}} \times R_{1dc} = \frac{2.572}{2.11} \times 0.6667 = 0.8127\,\Omega$$

The second rotor parameter- rotor resistance referred to stator is at low frequency,
Thus $R_{2\ dc}^{/} = \left(R_1 + R_2^{/}\right)\Big|_{15\,\text{Hz}} - R_{dc} = 2.11 - 0.6667 = 1.4433\,\Omega$
The per phase impedance at power frequency locked rotor test,

$$Z_{sc} = \frac{V_{sc}}{I_{sc}\sqrt{3}} = \frac{110}{18\sqrt{3}} = 3.53\,\Omega$$

Per phase leakage reactance referred to stator,

$$X_s = \sqrt{Z_{sc}^2 - \left(R_1 + R_2^{/}\right)^2} = \sqrt{3.53^2 - 2.57^2} = 2.42\,\Omega$$

From no-load test the no-load power factor angle,

$$\theta_0 = \cos^{-1}\left(\frac{W_0}{V_0 I_0\sqrt{3}}\right) = \cos^{-1}\left(\frac{500}{440 \times 4\sqrt{3}}\right) = 80.6°$$

The reactive component of no-load current $I_m = I_0 \sin\theta_0 = 4 \sin 80.6° = 3.946$ A

Thus the magnetizing reactance $X_m = \frac{V_{ph}}{I_m} = \frac{440}{3.946\sqrt{3}} = 64.4\ \Omega$

The total core loss $P_c = P_0 - P_{fw} - 3I_0^2 R_1$

$$P_c = 500 - 200 - 3 \times 4 \times 4 \times 0.8127 = 261\,\text{W}$$

Thus per phase core loss resistance $R_c = \left(\frac{440}{\sqrt{3}}\right)^2 \left(\frac{3}{261}\right) = 742\,\Omega$

Example 5.18 A three-phase, four pole, 50 hp, 440 V, 50 Hz, star connected induction motor has the following parameters per phase:

$$R_1 = 0.10\ \Omega\ ;\ X_1 = 0.35\ \Omega\ ;\ R_2^/ = 0.11\ \Omega\ ;\ X_2^/ = 0.40\ \Omega$$

The stator core losses amount to 1000 W and the rotational losses are equal to 900 W. The motor draws a line current of 15 A at a power factor of 0.2 lagging. Calculate the (a) input line current and power factor (b) the developed electromagnetic torque (c) the horse power output (d) the efficiency at 5% slip.

Solution

The phase voltage is, $V_{\text{ph}} = \frac{440}{\sqrt{3}} = 254\ \text{V}$

The rotor per phase current referred to stator side is,

$$I_2^/ = \frac{V_{\text{ph}}}{R_1 + \left(\frac{R_2^/}{s}\right) + j\left(X_1 + X_2^/\right)}$$

$$I_2^/ = \frac{254}{0.10 + \left(\frac{0.11}{0.05}\right) + j(0.35 + 0.40)} = \frac{254}{2.3 + j0.75} = 104.99\angle - 18.1^\circ\ \text{A}$$

$$I_2^/ = 104.99\angle - 18.1^\circ = (99.79 - j32.62)\ \text{A}$$

The no-load current is

$$I_0 = 15\angle - cos\ (0.2) = 15\angle - 78.46^\circ\ \text{A} = 15(0.2 - 0.9797\,\text{j})\ \text{A}$$

The input line current is,

$$\begin{aligned} I_1 = I_0 + I_2^/ &= 15(0.2 - 0.9797j) + (99.79 - j32.62) \\ &= 102.79 - j47.32 = 113.2\angle - 24.71^\circ\ \text{A} \end{aligned}$$

The power factor = cos 24.71° = 0.9084 lagging

The rotational speed $\omega_s = \frac{2\pi N_s}{60} = \frac{2\pi \times 1500}{60} = 157\ \ \text{Rad/s}$

The developed torque

$$T = \frac{P_g}{\omega_s} = \frac{3I_2^{/2}R_2^/}{s\omega_s} = \frac{3 \times (104.99)^2 \times 0.11}{0.05 \times 157} = 450\ \ \text{Nm}$$

The power output

$$\begin{aligned} P_o &= P_m - \text{Rotational loss} \\ &= (1 - s)\frac{3I_2^{/2}R_2^/}{s} - \text{Rotational loss} \end{aligned}$$

$$= (1 - 0.05)\frac{3 \times (104.99)^2 \times 0.11}{0.05} - 900$$
$$= 68213.6 \text{ W}$$

The horse power is therefore Horsepower $= \frac{68213.6}{746} = 91.43$ hp

The efficiency $= \frac{\text{Output}}{\text{Input}} \times 100 = \frac{\text{Output}}{\text{Output} + \sum \text{Losses}} \times 100 = \frac{68213.6}{68213.6+1000+900+3\times 113.2^2\times 0.10} = 92.23\%$

Example 5.19 A three-phase, six-pole, 50 hp, 440 V, 50 Hz induction motor has the following parameters per phase:

$$R_1 = 0.25\ \Omega\ ;\ X_1 = 0.75\ \Omega\ ;\ R_2^{/} = 0.25\ \Omega\ ;\ X_2^{/} = 0.75\ \Omega$$
$$R_c = 100\ \Omega\ ;\ X_m = 1000\ \Omega$$

The stator core loss is 800-W and rotational losses are 450-W. Calculate the (a) input line current and power factor (b) the developed electromagnetic torque (c) the horse power output (d) the efficiency at 2.5% slip. Assume approximate equivalent circuit.

Solution

The phase voltage is, $V_{\text{ph}} = \frac{440}{\sqrt{3}} = 254$ V

The rotor per phase current referred to stator side is,

$$I_2^{/} = \frac{V_{\text{ph}}}{R_1 + \left(\frac{R_2^{/}}{s}\right) + j\left(X_1 + X_2^{/}\right)}$$
$$I_2^{/} = \frac{254}{0.25 + \left(\frac{0.25}{0.025}\right) + j(0.75 + 0.75)} = \frac{254}{10.25 + j1.50} = 24.52\angle - 8.33^\circ \text{ A}$$
$$I_2^{/} = 24.52\angle - 8.33^\circ = (24.26 - j3.55) \text{ A}$$

The no-load current is

$$I_c = \frac{254}{100} = 2.54\ A \text{ and } I_m = \frac{254}{j1000} = -0.254j\ \text{ A}$$
$$I_0 = 2.54 - 0.254j\ \text{ A}$$

The input line current is,

$$I_1 = I_0 + I_2^{/}$$
$$= (2.54 - 0.254j) + (24.26 - 3.33j) = 26.8 - 3.584j = 27\angle - 7.62^\circ \text{ A}$$

The power factor or $P.F = cos\,7.62 = 0.9911$

Synchronous speed $N_s = \frac{120f}{P} = \frac{120\times 50}{6} = 1000$ RPM

The rotational speed $\omega_s = \frac{2\pi N_s}{60} = \frac{2\pi \times 1000}{60} = 104.67 \quad \text{Rad/s}$
The developed torque

$$T = \frac{P_g}{\omega_s} = \frac{3I_2^{/2}R_2^{/}}{s\omega_s} = \frac{3 \times (24.52)^2 \times 0.25}{0.025 \times 104.67} = 172.3 \quad \text{Nm}$$

The power output

$$\begin{aligned} P_o &= P_m - \text{ Rotational loss} \\ &= (1-s)\frac{3I_2^{/2}R_2^{/}}{s} - \text{ Rotational loss} \\ &= (1-0.05)\frac{3 \times (24.52)^2 \times 0.25}{0.025} - 450 \\ &= 17136 \text{ W} \end{aligned}$$

The horse power is therefore

$$\text{Horse power} = \frac{17136}{746} = 22.97 \text{ hp}$$

$$\text{The efficiency} = \frac{\text{Output}}{\text{Input}} \times 100 = \frac{\text{Output}}{\text{Output } + \sum \text{Losses}}$$

$$\times 100 = \frac{17136}{17136 + 800 + 450 + 3 \times 0.25 \times 27^2} = 90.5\%$$

The above can be solved using exact equivalent circuit and approximate equivalent circuit (ignoring R_c and X_m) and also rotor equivalent circuit. Students are expected to solve using these different equivalent circuits and compare the calculated results.

Example 5.20 A four-pole, 50 Hz, three-phase induction motor has maximum torque 500 N-m and the starting torque is 200 N-m. Estimate the torque and output power at slip 0.05.

Solution

We know that torque of an induction motor is given by

$$T = \frac{2T_m}{\left(\frac{s}{s_m}\right) + \left(\frac{s_m}{s}\right)}$$

Here $T_m = 500Nm$ and $T_{st} = 200\text{Nm}$ at slip 1.0. Therefore

$$200\left(\frac{1}{s_m} + s_m\right) = 2 \times 500$$

Solving this we get $s_m = 0.21$

Thus torque at 0.05 slip is

$$T = \frac{2 \times 500}{\left(\frac{0.05}{0.21}\right) + \left(\frac{0.21}{0.05}\right)} = \frac{1000}{0.238 + 4.2} = 225.53 \text{ Nm}$$

Synchronous speed $N_s = \frac{120 \times 50}{4} = 1500 \, \text{RPM}$
Power output $P = T\omega_r = 225.53 \times 2\pi(1 - 0.05)\frac{1500}{60} = 33,607.5 \text{ W} = 45.05 \text{ hp}$

5.10 Circle Diagram

We know that performance of induction motor can be determined using equivalent circuit parameters. Another method to quickly determine the performance is by using graphical method known as *circle diagram.* The method is based on the locus of a current in a reactive circuit. In a reactive circuit, current is given as

$$I = \frac{V}{R + jX}$$

If R is varied, the trace of the current is a circle. In case of three-phase induction motor, the rotor current referred to stator is,

$$I_2^{/} = \frac{V_1}{\sqrt{\left(R_1 + R_2^{/}/s\right)^2 + \left(X_1 + X_2^{/}\right)^2}}$$

In an equivalent circuit, we know that the rotor resistance $R_2^{/}/s$, which undergoes a change from standstill ($s = 1.0$) to running conditions ($s \rightarrow 0.0$). The slip varies from 1.0 to 0.0 for motoring operation. The variation of current is depicted in Fig. 5.34. Here segment OP is the current flowing through the stator, which is the phasor sum of no-load current OA and rotor current referred to stator AP. The angle θ($\angle$VOP) is the power factor angle. Here y Axis is taken as voltage axis. Thus, we have

$$\sin\,\theta = \frac{\left(X_1 + X_2^{/}\right)}{\sqrt{\left(R_1 + R_2^{/}/s\right)^2 + \left(X_1 + X_2^{/}\right)^2}} \tag{5.66}$$

This gives us the current

$$I_2^{/} = \frac{V_1}{\left(X_1 + X_2^{/}\right)} \sin\,\theta \tag{5.67}$$

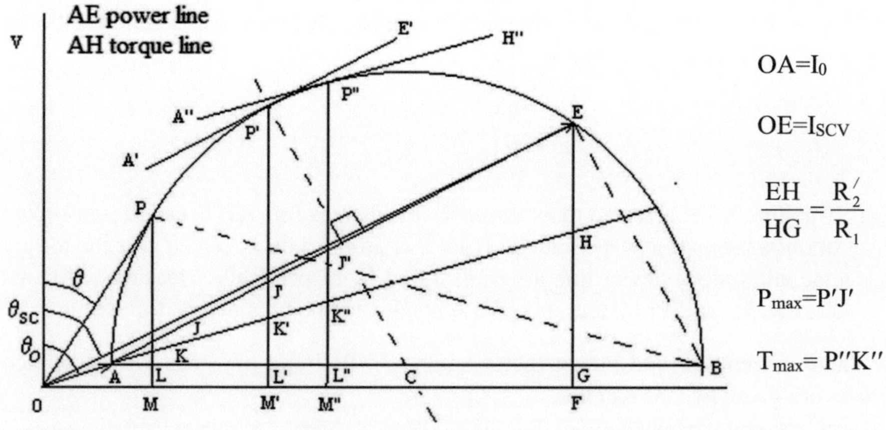

Fig. 5.34 Circle diagram of three-phase induction motor

In the circle diagram (Fig. 5.34), point P is the operating point. From no-load or free running test and locked rotor test we have

$$I_0 \cos \theta_0 = \frac{V_0}{R_c} \tag{5.68}$$

$$I_{sc} = \frac{W_{sc}}{V_{sc}} \tag{5.69}$$

Here W_{sc} is the per phase copper loss and V_{sc} is the per phase voltage during locked rotor test. V_1 is per phase rated voltage.

Using test data, the circle diagram can be constructed. Following are the steps for drawing circle diagram:

(a) Define current scale, say, $a = 1.0$ A/cm Draw no-load current OA at no-load power factor angle from $\theta(\angle VOP)$ from the voltage or y –axis. Define power scale as aV_{ph} W/cm.
(b) Draw the short circuit current OE, required for full voltage at blocked (locked) rotor power factor. The short circuit current corresponding to full voltage condition is OE $= I_{sc}\frac{V_1}{V_{sc}}$ at short circuit / locked rotor power factor angle $\theta_{SC}(\angle VOE)$.
(c) Now, join AE and have its perpendicular bisector.
(d) Draw line AB parallel to x Axis, which cuts the perpendicular bisector at point C.
(e) Taking C as center and AC as the radius and draw semicircle. Point P is the operating point, which moves to no-load condition at point A and short circuit condition at E.
(f) Draw perpendicular from E on x Axis. The line EG is divided so that

$$\frac{\text{EH}}{\text{HG}} = \frac{R_2'}{R_1}$$

If R_x is extra resistance in each rotor phase, its referred value is added with the rotor resistance referred to stator. Thus

$$\frac{\text{EH}}{\text{HG}} = \frac{R_2^{/} + R_x^{/}}{R_1}$$

(g) The line AH is referred to as *torque line*, whereas line AE is called *power line* or *output line*. If a perpendicular from P is dropped on x Axis, it cuts the torque line and output power line at points J and K, respectively. Measure segments PJ, PK, PL and PM. Here, per phase results can be obtained as follows:

Input current OP and Power factor angle ∠VOP. Input power is PM and Fixed losses are given by LM and GF.

Applied voltage per phase = $V_1 \quad \alpha \quad \text{OV} \quad \alpha \quad \frac{V_{LL}}{\sqrt{3}}$

Per phase Rotor intake = PK $\alpha \quad \frac{I_2^{/} \times I_2^{/} \times R_2^{/}}{s}$

Per phase Mechanical power output = PJ $\alpha \quad (1-s)\frac{I_2^{/} \times I_2^{/} \times R_2^{/}}{s}$

Per phase Rotor loss = JK $\alpha \quad \mathrm{I}_2^{/} \times \mathrm{I}_2^{/} \times \mathrm{R}_2^{/}$

From this slip and efficiency are obtained

$$\text{slip, s} = \frac{\text{JK}}{\text{PK}} \quad \text{and} \quad \text{Efficiency} = \frac{\text{PJ}}{\text{PM}} \quad \text{and} \quad (1-\text{slip}) = \frac{\text{PJ}}{\text{PK}}$$

Per phase Torque = OV × PK in Synchronous Watts

If the machine is having p poles, the rotational synchronous speed, $\omega_s = \frac{4\pi f}{p} = \frac{2\pi N_s}{60}$

Thus, per phase torque in Nm is

$$T = \frac{\text{OV} \times \text{PK}}{\omega_s} \text{ Nm}$$

$$\text{Total rotor intake} = 3 \times \text{OV} \times \text{PK} \times \text{Scale(A/cm) (W)}$$

$$\text{Total rotor loss} = 3 \times \text{OV} \times \text{JK} \times \text{Scale(A/cm) (W)}$$

$$\text{Total mechanical power output} = 3 \times \text{OV} \times \text{PJ} \times \text{Scale(A/cm) (W)}$$

$$\text{Total torque, } T = 3 \times \frac{\text{OV} \times \text{PK} \times \text{Scale(A/length)}}{\omega_s} \text{ Nm} \tag{5.70}$$

5.10.1 *Use of Circle Diagram*

Circle diagram is very useful in qualitative assessment of the induction motor performance. The diameter of the circle is given by $\frac{V_1}{\left(X_1 + X_2^{/}\right)}$.

Thus new circle diagrams can be easily plotted for different cases such as

(i) Different supply voltage
(ii) Different frequency
(iii) Different leakage impedances
(iv) Different rotor resistance.

The diameter of the circle increases when voltage increases and frequency and/or leakage impedance decrease. The results obtained from equivalent circuit approach is some how comparable with those obtained using circle diagram. The equivalent circuit parameters of induction motor can be calculated using test results. Let us assume that machine is operating at slip s. The input per phase current I_1 can be computed and torque can be determined as discussed in Sect. 5.8.2.

The maximum power and maximum torque are determined as follows (see Fig. 5.34):

(a) Draw tangents, $A'E'$ and $A''H''$, on the circle diagram, which are, respectively, parallel to output power line AE and torque line AH.
(b) Tangent $A'E'$ touches circle at point P' and tangent $A'H'$ touches circle at point P''.
(c) Drop perpendicular from points P' and P'' on x Axis.
(d) Maximum output power corresponds to segment P′J′
(e) Maximum torque corresponds to segment P″K″.

The values of power and torque can be easily determined using expressions (5.70) mentioned above. Develop MATLAB computer program for obtaining performance of induction motor using circle diagram and equivalent circuit. Cross verify results obtained from computer, equivalent circuit, test result and hand drawn circle diagram.

Example 5.21 The result of the no-load and blocked rotor test on a three phase, 50 Hz, four pole, 220 V, star-connected induction motor are as follows:

No-load test: $V_{oc} = 220$ V; $W_{oc} = 52$ W; $I_{oc} = 2.6$ A; Rotational loss is negligible

Blocked rotor test: $V_{sc} = 56$ V; $W_{sc} = 1000$ W; $I_{sc} = 18\ A$

The per-phase stator resistance is 0.65 Ω. If machine is running at 5% slip, determine current, power factor and torque from approximate equivalent circuit. Draw circle diagram and validate the results for the same current.

Solution

The per phase voltage is

$$V_{ph} = \frac{220}{1.7321} = 127 \text{ V}$$

No-load current is

$$I_{oc} = 2.6 \text{ A}$$

Per phase no-load loss is given as

$$W_0 = \frac{52}{3} = 17.33 \text{ W}$$

No-load pf angle $\theta_0 = \cos^{-1} \frac{17.33}{2.6 \times 127} = 87° \theta_0 = \cos^{-1} \frac{17.33}{2.6 \times 127} = 87°$
Per phase locked rotor voltage, $V_{sc} = \frac{56}{1.7321} = 32.33$ V
Per phase locked rotor losses, $W_{sc} = \frac{1000}{3} = 333.33$ W
Locked rotor p.f. angle is given as

$$\theta_{sc} = \cos^{-1} \frac{333.33}{18 \times 32.33} = 55°$$

The locked rotor current at full voltage is $\frac{18 \times 220}{56} = 70.7$ A
The resistance can be given as $r_1 + R_2$

$$R_e = \frac{333.33}{18^2} = 1.029\ \Omega$$

The per phase AC stator resistance is

$$R_1 = 0.65 \times 1.1 = 0.715\ \Omega$$

The per phase rotor resistance $R_2^/ = 1.029 - 0.715 = 0.314\,\Omega$
The per phase impedance $Z_{sc} = \frac{V_{sc}}{I_{sc}} = \frac{32.33}{18} = 1.796\ \Omega$
The leakage reactance is

$$X_1 + X_2^/ = \sqrt{(1.796)^2 - (1.029)^2} = 1.472\ \Omega$$

Using approximate equivalent circuit ignoring core loss resistance R_c and magnetizing reactance X_m, the current is

$$I_2^/ = \frac{V_1}{\sqrt{\left(R_1 + R_2^/ / s\right)^2 + \left(X_1 + X_2^/\right)^2}}$$

$$= \frac{127}{\sqrt{(0.715 + 0.314/0.05)^2 + 1.472^2}} = 17.77 \text{ A}$$

$$\text{Power factor} = \cos\left[\tan^{-1}\left\{\frac{1.472}{(0.715 + 0.314/0.05)}\right\}\right] = \cos\ 11.88 = 0.978$$

The rotational speed is given as

$$\omega_s = \frac{2\pi N_s}{60} = \frac{2\pi \times 1500}{60} = 157 \quad \text{Rad/s}$$

Torque at 0.05 slip is

$$T = \frac{P_g}{\omega_s} = \frac{3I_2^{/2}R_2^{/}}{s\omega_s} = \frac{3 \times (17.77)^2 \times 0.314}{0.05 \times 157} = 37.89 \text{ Nm}$$

For drawing circle diagram, we have, OA=2.6 A at 87° from y axis, OE=70.7 A at 55° from y Axis and current OP=17.77 A. The circle diagram is shown in Fig. 5.35. From circle diagram, following can be calculated:

Total rotor intake = 3×OV×PK×Scale(A/length) = 3×127×11.6 = 4419.6 W
Total rotor loss = 3 × OV × JK × Scale(A/length) = 3 × 127 × 0.8 = 304.8 W
Total mechanical power output

$$= 3 \times \text{OV} \times \text{PJ} \times \text{Scale(A/length)} = 3 \times 127 \times 10.8 = 4114.8 \text{ W}$$

$$\text{Total torque is } T = 3 \times \frac{\text{OV} \times \text{PK} \times \text{Scale}(A/\text{length})}{\omega_s} = 3\frac{127 \times 11.6}{157} = 28.15 \text{ Nm}$$

$$\text{slip, } s = \frac{\text{JK}}{\text{PK}} = 0.069 = 7\% \quad \text{and} \quad \text{Efficiency} = \frac{\text{PJ}}{\text{PM}} = 93.1\%$$

Thus, we can see that the circle diagram yields lesser torque as calculated from equivalent circuit. But, if circle diagram is made with larger dimensions, the error can be minimized. For identical currents, the calculated slip is 7% from circle diagram and 5% for equivalent circuit. The circle diagram is meant for quick prediction of performance of induction motor. If we actually load the induction motor, the experimentally obtained torque is expected to be in between the two computed values. Both the equivalent circuit method and circle diagram are approximate methods. They do not account for load losses. An accurate method is to analyze the three-dimensional induction motor using FEM or BEM software. These are quite cumbersome, costly and time taking approaches and are beyond the scope of the book.

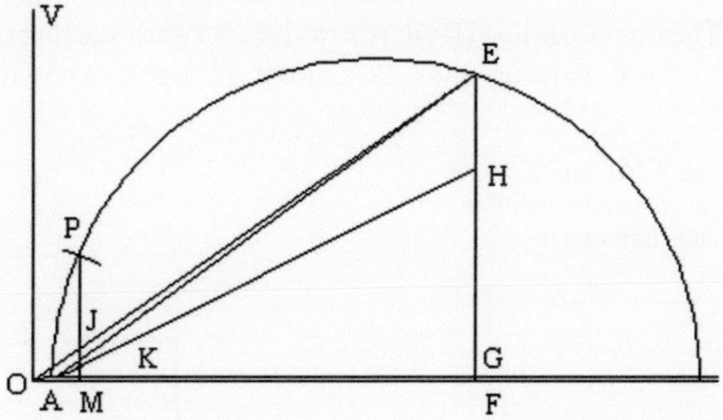

Fig. 5.35 Circle diagram for Example 5.21

5.11 Starting of Three-Phase Induction Motor

Figure 5.36 shows the per phase equivalent circuit of an induction motor. At slip s=1, the impedance seen from the terminals is given by [4–9]

$$Z|_{s=1} = \left(R_1 + R_2^/\right) + j\left(X_1 + X_2^/\right)$$

When machine is running as a motor near synchronous speed, slip tend to be zero and the impedance is

$$Z|_{s\to 0} = R_1 + \frac{R_2^/}{s} + j\left(X_1 + X_2^/\right)$$

During starting when machine is at rest ($s = 1$), the impedance of induction machine, as seen from the terminal is low. This initially increases the starting current. The machine offers large impedance when it runs closer to synchronous speed ($s \to 0.0$). At running condition, the current drawn from supply is reduced [Refer Fig. 4.36]. As machine picks up the speed, slip reduces causing the effective impedance to rise and, hence the current drops. It is clear from the Fig. 5.32 that when an induction motor is running in counter current region or near standstill conditions, the braking and starting currents, both are extremely large. Any deviation from slip 0.0 causes the current taken by induction motor to rise.

Since the initial starting current is higher than rated current in medium and large capacity induction motors, suitable methods are adopted to limit the starting current as well as to provide sufficient starting torque so that machine can overcome opposing static friction and load torque.

5.11.1 Direct-on-Line Starting

The direct-on-line (DOL) or straight forward starting of induction motor is preferred for lower capacity induction motor. If large capacity (say 100 HP) induction motor

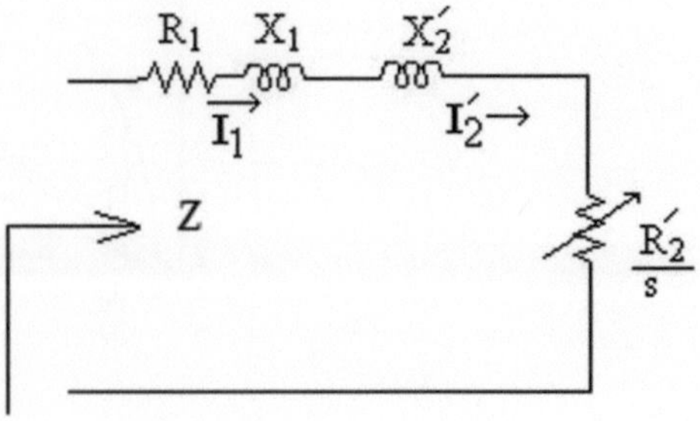

Fig. 5.36 Approximate equivalent circuit of an induction motor

is subjected to DOL starting, the starting current becomes extremely high, thereby causing large ohmic losses during starting period. As a consequence, the winding may get damaged due to overheating. The torque developed may become too large and may even damage the mechanical coupling. A large starting current drawn from power supply can cause significant voltage drop in the distribution network, disturbing the consumer, and causing malfunctioning of sensitive equipment. In large capacity motors, strong electromagnetic forces are produced between the overhang portions of the stator coils.

DOL starting is a simple and cheapest method in which three-phase windings are directly connected to the rated voltage. The starting current is equal to the locked rotor current at rated voltage. This current slowly comes to around rated current. Squirrel cage induction motor of capacity 3 KW, 440 V and 5.5 KW 380 V can be started using DOL starter. In induction motor, the starting current with DOL starting is about 5-8 times the rated current. This type of starting is used in starting machine tools, ventilators, polishing machine, etc. The DOL starter has in-built protection for over-current, thermal protection and over-load protection.

If Z refers to per phase equivalent impedance of the motor at slip $s = 1.0$, then the starting current during straight forward (DOL) starting when windings are connected in star is

$$I_{s(Y)} = \frac{V_{\text{ph}}}{Z} = \frac{V_{L-L}}{Z\sqrt{3}}$$

The torque, when machine is connected in star, is

$$T_{s(Y)} = \frac{3R_2^{/}}{s\omega_s} I_{s(y)}^2$$

When windings are connected in delta, the line current at the instant of starting is,

$$I_{s(\Delta)} = \frac{V_{L-L}}{Z}\sqrt{3}$$

The torque in delta connected machine is

$$T_{s(\Delta)} = \frac{3R_2^{/}}{s\omega_s}\left(\frac{I_{s(\Delta)}}{\sqrt{3}}\right)^2$$

Thus, the ratio of starting line current in delta and star connections can be given as

$$\frac{I_{s(\Delta)}}{I_{s(Y)}} = 3 \tag{5.71}$$

The ratio of starting torque in delta and star connections,

$$\frac{T_{s(\Delta)}}{T_{s(Y)}} = 3 \tag{5.72}$$

5.11.2 Star/Delta Starting

This method consists of supplying the rated voltage to the three-phase star-connected winding, initially at the time of starting. When speed reaches 90-95% of the synchronous speed in forward direction, the three-phase stator windings are reconnected in delta for forward running. The changeover from star to delta may be either manual or automatic. A schematic star / delta starter have been shown in Figs. 5.37 and 5.38.

The ratio of starting line-current when windings are in star and delta are

$$\frac{I_{s(Y)}}{I_{s(\Delta)}} = \frac{\left(\frac{V_{L-L}}{Z\sqrt{3}}\right)}{\left(\frac{V_{L-L}}{Z}\sqrt{3}\right)} = \frac{1}{3}$$

Thus, when windings are connected in star, the starting current will be one third of delta connection. This is because the per phase voltage is reduced by a factor of $1/\sqrt{3}$. Since the phase voltage in star connection is $V_{L-L}/\sqrt{3}$, the electromagnetic torque is 3 times smaller than that of delta connection. The starting electromagnetic torque being proportional to V^2, we get

$$\frac{T_{\mathrm{st}(Y)}}{T_{\mathrm{st}(\Delta)}} = \frac{1}{3}$$

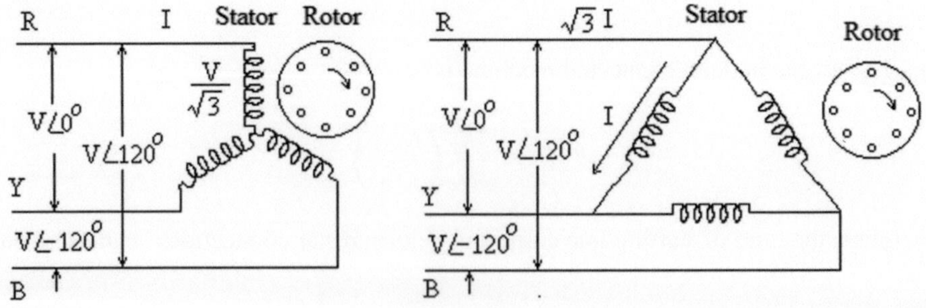

Fig. 5.37 Starting of induction motor using star-delta starter

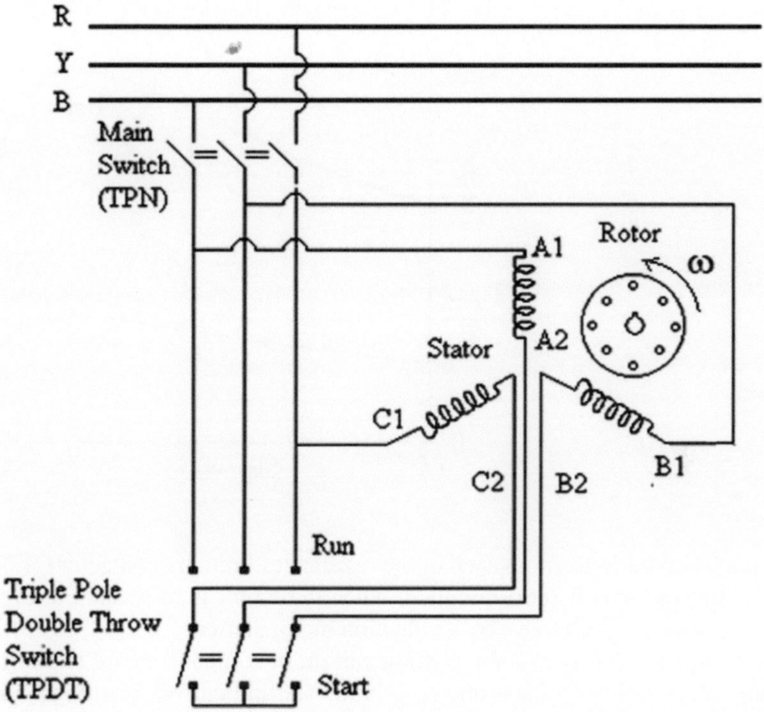

Fig. 5.38 Schematic diagram of star-delta starter

The transient peak current as well as transient peak torque occur when the windings are reconnected to delta. The transient peak current is much higher if windings are reconnected into delta before 90–95% of speed is achieved.

If the load torque is less than 30% of the rated motor torque, the star-connected motor may be used even in running condition. As a result, power factor and efficiency are improved, and losses are also considerably reduced.

For star /delta starting, it is required that all the six terminals of the stator windings are accessible and the rated voltage of the stator phase winding is equal to the line voltage of the supply. Figure 5.39 shows the torque-speed characteristics during star / delta starting. It is generally used for medium capacity motors.

5.11.3 Stator Resistance Starting

Like DC motor starting, the terminal voltage on the three-phase induction motor may be reduced during starting by means of extra resistances in three phase line. Using inductances for limiting the starting current lowers the power factor during starting. Can you tell, what will happen when series capacitors are added? The machine draws excessive current when capacitors are added in series. Identical resistances R are connected in series with each of the phase. These resistances are short-circuited by

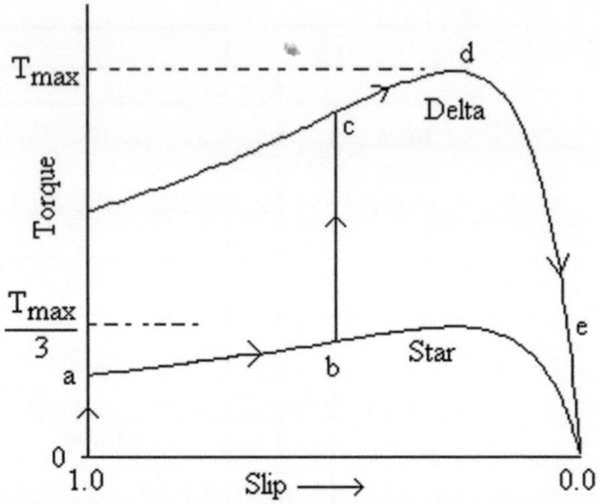

Fig. 5.39 Torque-slip characteristics during star/delta starting

contactor C1 connected across each of the resistances. Initially contactor C1 is OFF. These contactors are all short-circuited, when the motor is fully started. Fig. 5.40 shows a resistance type three-phase induction motor starter.

Let us suppose that $I_{s(Y)}$ is the starting current in straight forward starting when windings are connected in star and $I_{s(\Delta)}$ is the starting current in straight forward starting when windings are connected in delta. A reduction in the supply voltage can be used for reducing the current. Let $I_{sR(Y)}$ and $I_{sR(\Delta)}$ be the starting current when the windings are connected in star and delta, respectively, with resistance type starter. Let these currents are reduced by a factor $K_R > 1.0$. Thus, we get

$$I_{sR(Y)} = \frac{I_{s(Y)}}{K_R} \quad \text{and} \quad I_{sR(\Delta)} = \frac{I_{s(\Delta)}}{K_R} \tag{5.73}$$

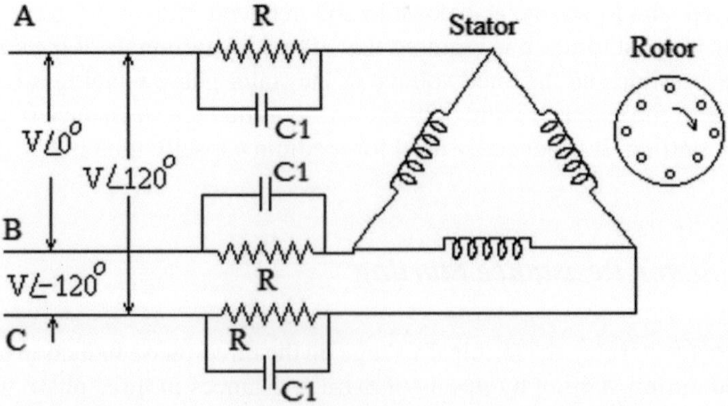

Fig. 5.40 Resistance type starter of induction motor

The respective starting torques are,

$$\frac{T_{stR(Y)}}{T_{st(Y)}} = \frac{1}{K_R^2} \text{ and } \frac{T_{stR(\Delta)}}{T_{st(\Delta)}} = \frac{1}{K_R^2} \tag{5.74}$$

where $T_{stR(Y)}$ and $T_{stR(\Delta)}$ are the starting torques when full voltage is applied in DOL starting with resistance starting. Thus, a reduction of starting current by a factor of $K_R > 1.0$, reduces the starting torque by a factor of K_R^2. The three resistances must be identical in order to avoid unbalancing. In case of unbalanced resistances, the starting torque will be further reduced. This method is preferred for starting lower capacity motors where load torque is very small. Excessive amount of energy is wasted as ohmic loss during starting. The method has merely of historical importance.

5.11.4 Autotransformer Starting

The applied voltage in case of AC can be easily changed using autotransformer. The autotransformer starting is more costly as compared to resistance type starter, but losses during starting can be effectively controlled. Thus, autotransformer starting is a superior method as far as power loss during starting is concerned and can be used for large horse power motors. Let us assume that the phase voltage is changed to

$$V_{\text{au}} = \frac{V_{\text{ph}}}{K_A}, \tag{5.75}$$

where K_A is the transformation ratio of the auto transformer. The phase starting current of the motor is given as

$$I_{\text{st}(A)} = \frac{V_{\text{au}}}{Z} = \frac{V_{\text{ph}}}{K_A} \cdot \frac{1}{Z} = \frac{I_{\text{st}}}{K_A} \tag{5.76}$$

Thus, the phase current is reduced by a factor of K_A. The starting torque is reduced by a factor of K_A^2.

$$T_{\text{st}(A)} = \frac{T_{\text{st}}}{K_A^2} \tag{5.77}$$

Figures 5.41 and 5.42 show the two different schemes using autotransformer for starting Induction motors. In Fig. 5.41, the motor is supplied a reduced voltage during starting and then terminals of the motor are reconnected to full voltage of same phase sequence.

In Fig. 5.42, a reduced voltage is applied during starting by connecting the switch C1 and C2. When motor is started, switch C2 is opened. There is no autotransformer action when the neutral point of the autotransformer is opened by switch C2. As

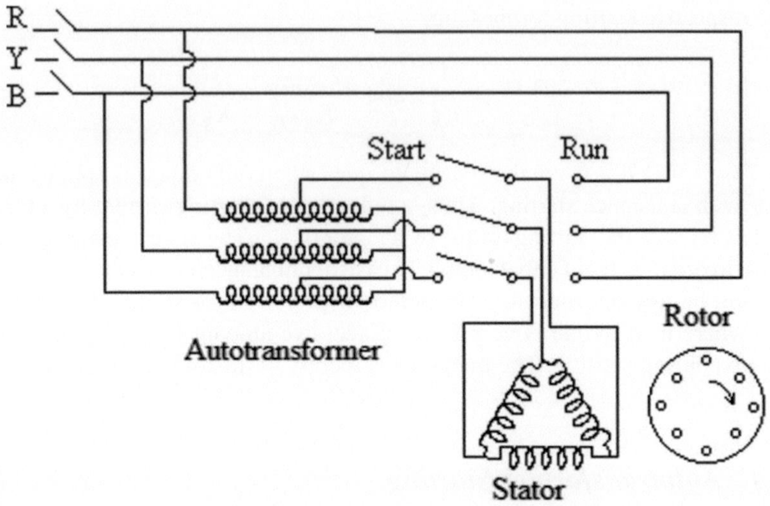

Fig. 5.41 Autotransformer starting of induction motor using Triple pole double throw switch (TPDT)

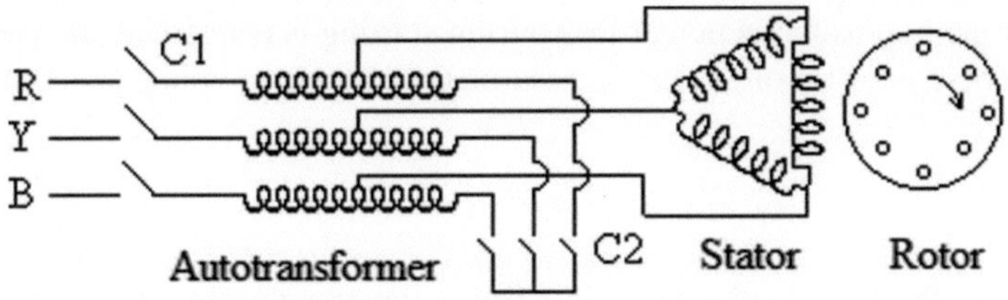

Fig. 5.42 Auto transformer starting of induction motor without disconnecting supply from start to run

a result, the motor gets full voltage less a small voltage drop in the three common windings of the autotransformer. Thus, the demerit of autotransformer is used as an advantage here for starting the induction motor.

In Fig. 5.41, it takes a fraction of second to reconnect the windings manually to the full voltage of the same phase sequence. During this time the machine does not lose its magnetization and since machine is still running it works as an induction generator for a fraction of second. When the terminals are reconnected to full voltage, the polarity of voltage on the terminals of the machine and those of the supply slightly differ. This causes a transient current to flow in the machine. The transient current is not pronounced in smaller capacity motors. But in case of large capacity (horse power) induction motors, the transient current may be very large-which tries to overload the transformer supplying power to it. Therefore, the scheme of Fig. 5.42 is normally preferred for starting large capacity induction motor. The only drawback is that the autotransformer has to be remained connected to the supply when the induction motor is running.

5.11.5 VVVF Starting

Due to the advent of solid-state devices and their control, it is convenient to run the induction motor on Variable Voltage Variable Frequency (VVVF) inverter. It has already been shown that the maximum torque remains constant if during starting the ratio of voltage and frequency is maintained constant.

VVVF method is used for starting as well as for speed and torque control during induction motor operation. Here, both the voltage and the frequency are slowly increased, keeping their ratios a constant. This limits the starting current. Maximum torque is also maintained constant. Since, V/f or E/f is maintained constant, the peak torque remains constant. The change in the frequency changes the synchronous speed. Below base speed, the speed is controlled by changing both voltage and frequency. It gives a constant torque operation of induction motor, similar to that of DC shunt motor characteristics. The voltage cannot be increased beyond base (rated) voltage for raising the speed. Thus, above base speed the machine can be operated by keeping the voltage constant and increasing the frequency. It gives a constant power operation of three-phase induction motor, similar to DC series motor characteristics. Fig. 5.43 shows a schematic diagram for DC link inverter controlled induction motor drives. The torque-speed characteristics with VVVF control is shown in Fig. 5.50. The control of voltage and frequency is normally achieved with the help of inbuilt microcontroller.

Considering the exact equivalent circuit of induction motor, the maximum torque is given as

$$T_m \;\alpha\; \left(\frac{V_1}{f}\right)^2 \;\alpha\; \left(\frac{E_1}{f}\right)^2 \;\alpha\; \phi_{\max}^2$$

By keeping the V/f constant, the peak value of air-gap flux is indirectly maintained constant. From the EMF equation, we know that $\phi_{\max} \;\alpha\; \frac{E_1}{f}$. At higher frequencies, when applied voltage and frequency are close to the base values, one can assume that $\phi_{\max} \;\alpha\; \frac{V_1}{f}$

However, when the applied voltage and frequency both are low,

$$E_1 = V_1 - I_1(R_1 + jX_1) \quad \text{and} \quad E_1 \neq V_1 \tag{5.78}$$

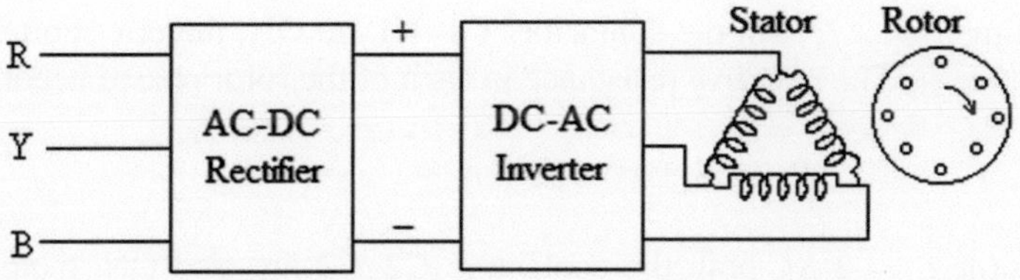

Fig. 5.43 Schematic diagram for DC link inverter- induction motor drive

Since the voltage drop (resistance and leakage reactance) in stator parameters is significant with low applied voltage, the maximum torque cannot be maintained constant if the ratio of voltage and frequency is maintained constant during starting. It is essential to maintain the ratio of E_1/f rather than V_1/f a constant during initial starting condition. During starting period, the airgap flux is maintained constant by increasing the ratio of voltage and frequency.

5.11.6 Rotor Resistance

Using per phase approximate equivalent circuit of rotor, one can show that the starting torque of a three-phase induction motor is

$$T_{\text{st}} = \frac{3V_1^2}{\omega_s\left[\left(R_2' + R_x\right)^2 + \left(X_2'\right)^2\right]} R_2' \tag{5.79}$$

Here, R_x is the extra resistance per phase starting resistance in the rotor circuit referred to stator side. It can be shown that the starting torque is highest when $R_2' + R_x = X_2'$.

This method is applicable to slip-ring induction motor only for starting as well as speed controlling. The graded rotor resistances are removed from the rotor circuit for controlling the starting current. It is essential that all the three resistances which are simultaneously removed from the circuit must be identical in their Ohmic values. If the rotor resistances in different phases are uneven, it will result in Georges's phenomenon, which we will discuss latter. The resistances are removed from the circuit using either (mechanical switch) drum controller or electronics contactors. These contactors are programmed similar to that of DC motor's automatic starter.

Figure 5.44 shows the graded resistances r_1, r_2, r_3, r_4, r_5,...connected in each of the rotor phases. These resistances are successively short-circuited using industrial contactors—$C1, C2, C3, C4, C5$, etc. Initially, when motor is at rest all the contactors are open. At this instant, when stator supply is switched ON, the effective resistance in each of the rotor phase is given as

$$R_1 = r_1 + r_2 + r_3 + r_4 + r_5 + \cdots\cdots + R_{2r}$$

After time $t = \Delta t$, both the contactors $C1$–$C1$ are ON, thereby short-circuiting the resistance r_1. The effective resistance in each of the rotor phases becomes

$$R_2 = r_2 + r_3 + r_4 + r_5 + \cdots\cdots + R_{2r}$$

After time $t = 2\,\Delta t$, both the contactors $C2$–$C2$ are also ON, thereby short-circuiting the resistance r_2, and the effective resistance in each of the rotor phase becomes

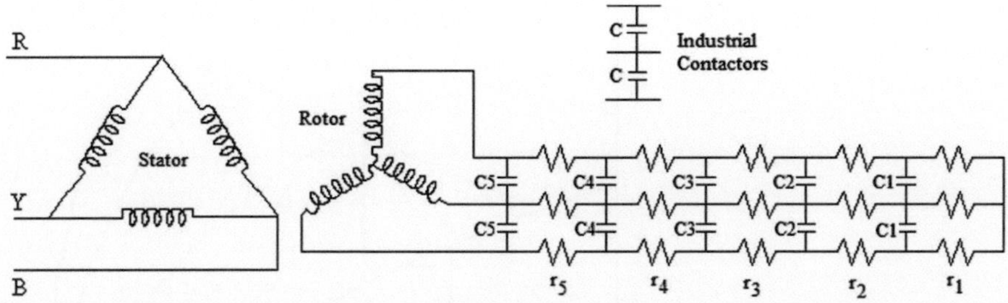

Fig. 5.44 Starting and speed control of slip-ring induction motor using rotor resistance control

$$R_3 = r_3 + r_4 + r_5 + \cdots\cdots + R_{2r}$$

After time t = 3 Δt, both the contactors $C3$–$C3$ are also ON, thereby short-circuiting the resistance r_3. The effective resistance in each of the rotor phase is given as

$$R_4 = r_4 + r_5 + \cdots\cdots + R_{2_r}$$

After time $t = 4\,\Delta t$, both the contactors $C4$–$C4$ are also ON, thereby short-circuiting the resistance r_4. The effective resistance in each of the rotor phase is

$$R_5 = r_5 + \cdots\cdots + R_{2_r}$$

As per Fig. 5.46, the rotor resistance starter has extra resistances r_1, r_2, r_3, r_4, r_5 and contactors $C1$, C2, C3, $C4$ and $C5$. After time $t = 5\,\Delta t$, both the contactors $C5$–$C5$ are also ON, thereby short-circuiting the resistance r_5. The effective resistance in each of the rotor phases is

$$R_r = R_{2_r}$$

Here R_r is the resistance of each of the rotor phase. Large number of resistances may be included in the rotor resistance starters.

The grading of resistances may be done like that of DC machine starter by choosing the current limits during starting. This requires the lower and upper starting currents. The contactors are controlled using different methods like drum controller, electronic circuit, programmed using state of art micro controllers or Programmable Logic Controller (PLC). Fig. 5.45 shows the torque-speed characteristics using rotor resistance starters.

Design of Rotor resistance starter: The rotor of slip-ring induction motor may be connected in star or delta. The end terminals of three phase rotor windings are connected to three slip rings. Through these slip rings stationary three phase resistances are connected to individual rotor terminals by means of three carbon brushes in contact with the rotating slip rings.

Fig. 5.45 Torque-speed characteristics using rotor resistance starters

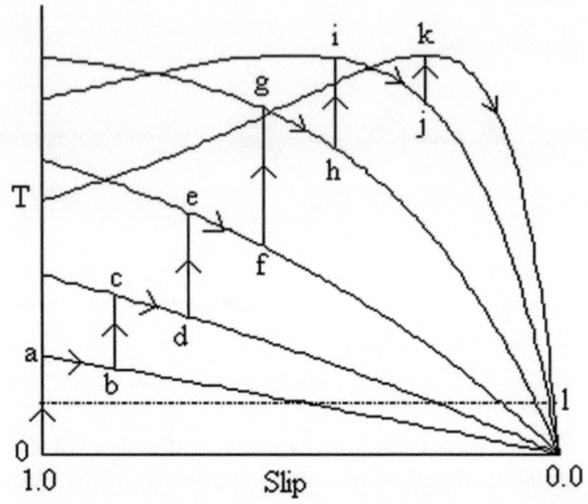

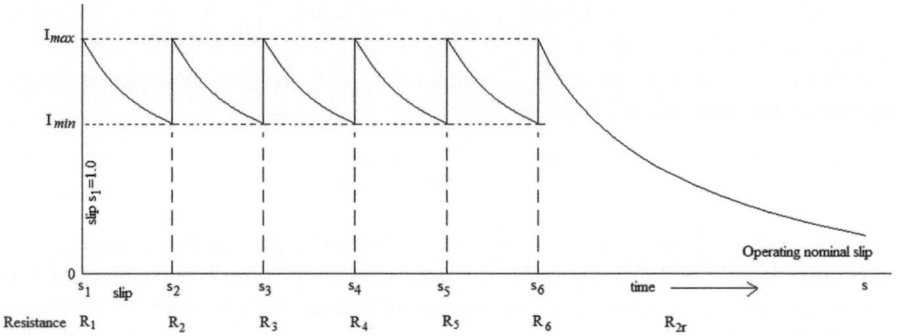

Fig. 5.46 Variation of stator current with rotor resistance starting

Let one of the rotor phase windings of slip-ring induction motor is having resistance R_{2r} and leakage reactance is x_2. Identical resistances in all the three phases are to be connected, otherwise negative sequence torque will cause Georges's phenomenon in the machine. Let $r_1, r_2, r_3 \ldots\ldots r_n$ be the step resistances of n-step starter. As usual the number of studs will be $n + 1$.

Let the total resistance of rotor circuit on first, second, third, ... $(n + 1)$th studs, respectively, are:

$$R_1 = r_1 + r_2 + r_3 + r_4 + r_5 + \cdots\cdots r_n + R_{2r}$$
$$R_2 = r_2 + r_3 + r_4 + r_5 + \cdots\cdots r_n + R_{2r}$$
$$R_3 = r_3 + r_4 + r_5 + \cdots\cdots r_n + R_{2r}$$
$$\cdots\cdots\cdots\cdots\cdots\cdots\cdots\cdots\cdots\cdots$$
$$\cdots\cdots\cdots\cdots\cdots\cdots\cdots\cdots\cdots\cdots$$

$$R_n = r_n + R_{2r}$$
$$R_{n+1} = R_{2r}$$

It is assumed that the input current fluctuates between maximum and minimum limits of I_{max} and I_{min} and torque slowly rises during the starting period. The variation of starting current versus slip is shown in Fig. 5.46.

Initially when supply is switched ON, the slip, $s_1 = 1.0$ and the machine draws maximum current with total resistance R_1 in the rotor circuit. The rotor resistance referred to stator is $R_1^/$. Here r_s and x_s are stator resistance and leakage reactance per phase. Graded resistances are shown in Fig. 5.47. Considering approximate equivalent circuit of induction motor, the input current is (Fig. 5.47)

$$I_{max} = \frac{V_{ph}}{\sqrt{\left[\left\{r_s + \frac{R_1^/}{s_1}\right\}^2 + \left(x_s + x_2^/\right)^2\right]}}$$

When the motor is started, its speed rises (or slip falls) and current also falls. When the circuit resistance unchanged (R_1) new slip s_2, current falls to lower limit, or we can write

$$I_{min} = \frac{V_{ph}}{\sqrt{\left[\left\{r_s + \frac{R_1^/}{s_2}\right\}^2 + \left(x_s + x_2^/\right)^2\right]}} \quad (5.80)$$

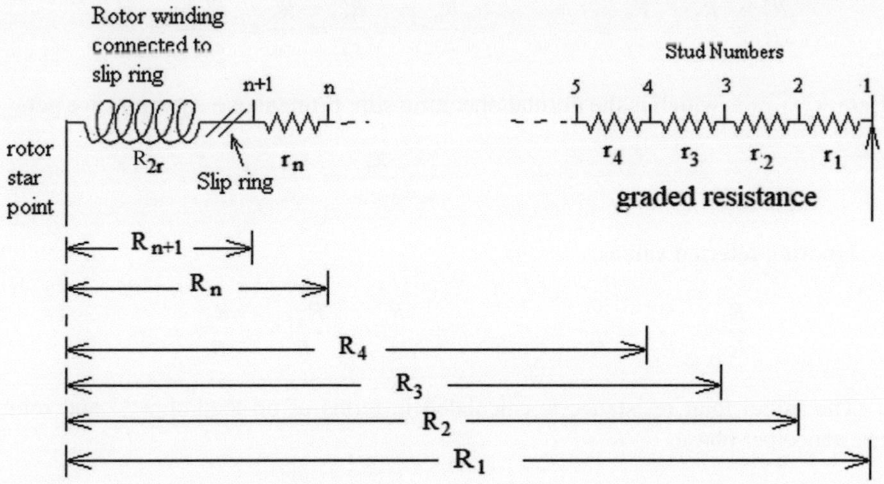

Fig. 5.47 Graded resistance in rotor of slip-ring induction motor (only one phase is shown other phases have identical rotor resistances)

when the resistance r_1 is removed (or cut out from circuit or short-circuited using contactors) from individual rotor phases at slip s_2, such that current again rises to $I_{\max}$. The rotor phase resistance is R_2. Thus, we can write that

$$I_{\max} = \frac{V_{\mathrm{ph}}}{\sqrt{\left[\left\{r_s + \frac{R_2^/}{s_2}\right\}^2 + \left(x_s + x_2^/\right)^2\right]}} \tag{5.81}$$

Again at slip s_3, with R_2, the new current is

$$I_{\min} = \frac{V_{\mathrm{ph}}}{\sqrt{\left[\left\{r_1 + \frac{R_2^/}{s_3}\right\}^2 + \left(x_1 + x_2^/\right)^2\right]}}$$

When the circuit resistance changed to R_3 for new slip is s_3, the current rises, we can write

$$I_{\max} = \frac{V_{\mathrm{ph}}}{\sqrt{\left[\left\{r_s + \frac{R_3^/}{s_3}\right\}^2 + \left(x_s + x_2^/\right)^2\right]}}$$

Similarly, these expressions can be written, when other resistances are removed from rotor circuit at different slips. Using above expressions, we can write that

$$\frac{R_1^/}{s_1} = \frac{R_2^/}{s_2} = \frac{R_3^/}{s_3} = \ldots\ldots.. = \frac{R_{n-1}^/}{s_{n-1}} = \frac{R_n^/}{s_n} = \frac{R_n^/}{s_n} = \frac{R_{n+1}^/}{s_{n+1}} = \frac{r_{2R}^/}{s}$$

Here, $s_{n+1} = s$, which is the normal operating slip. From above equations we get,

$$\frac{s_2}{s_1} = \frac{s_3}{s_2} = \frac{s_4}{s_3} = \ldots\ldots.. = \frac{s_n}{s_{n-1}} = \frac{s_{n+1}}{s_n} = \frac{s}{s_{n+1}}$$

Ignoring referred values,

$$\frac{R_2}{R_1} = \frac{R_3}{R_2} = \frac{R_4}{R_3} = \ldots\ldots.. = \frac{R_n}{R_{n-1}} = \frac{R_{n+1}}{R_n} = \frac{R_{2r}}{R_n} = \gamma$$

The initial total resistance is calculated in terms of no-load slip ‘s’ and rotor resistance per phase,

$$R_1 = \frac{s_1}{s} \times R_{2r} = \frac{1.00}{s} \times R_{2r}$$

Further, Fig

$$R_2 = \gamma \times R_1$$
$$R_3 = \gamma \times R_2 = \gamma^2 \times R_1$$
$$R_{n+1} = \gamma \times R_n = \gamma^n \times R_1$$
$$R_{2r} = \gamma^n \times R_1$$

One can show that no-load slip is equal to $\gamma = \sqrt[n]{s}$.

Using these calculations, the resistances between the studs can be calculated.

$$r_1 = R_1 - R_2 = (1-\gamma)R_1$$
$$r_2 = R_2 - R_3 = (1-\gamma)R_2 = (1-\gamma)\gamma R_1 = \gamma r_1$$

Similarly, other resistances between different studs are

$$r_3 = R_3 - R_4 = (1-\gamma)R_3 = (1-\gamma) \times \gamma \times R_2 = \gamma^2(1-\gamma) \times R_1 = \gamma^2 r_1$$

Similarly,

$$\begin{aligned} r_4 &= \gamma^3 r_1 \\ r_n &= \gamma^{n-1} r_1 \end{aligned} \tag{5.82}$$

The graded values of resistance need to be calculated for rotor resistance starter

Example 5.22 A three-phase, two pole, 440 V, 50 Hz induction motor has following parameters

$$\mathrm{R}_1 = 0.2\ \Omega \;\; \mathrm{R}_2^{/} = 0.4\ \Omega$$
$$\mathrm{X}_1 = 1.0\ \Omega \;\; \mathrm{X}_2^{/} = 1.5\ \Omega$$

Using approximate equivalent circuit, determine per phase impedance seen from the terminals at standstill. Estimate the initial starting current and starting torque when

(a) induction motor is connected in star or delta
(b) started using autotransformer with initial voltage reduced to 50%
(c) Both voltage and frequency are reduced to 10% of their rated values. Compare starting current and torque with rated frequency and rated voltage.

Solution

Synchronous angular speed $\omega_s = 2\pi\mathrm{f} = 314$ Rad/s

At slip s=1, the impedance seen from the terminals is given by

$$Z = \left(R_1 + R_2^{/}\right) + j\left(X_1 + X_2^{/}\right)$$
$$Z = (0.2 + 0.4) + j(1.0 + 1.5) = 0.6 + 2.5j = 2.571\angle + 76.5^\circ\Omega$$

(a) When winding is connected in star, we have

$$I_{s(Y)} = \frac{V_{\text{ph}}}{Z} = \frac{V_{L-L}}{Z\sqrt{3}} = \frac{440}{2.571\sqrt{3}} = 98.81 \text{ A}$$

$$T_{s(Y)} = \frac{3I_{s(y)}^2 R_2^{/}}{s\omega_s} = \frac{3 \times 98.81^2 \times 0.4}{1.0 \times 314} = 37.3 \text{ Nm}$$

When windings are connected in delta, the line current at the instant of starting is, given by

$$I_{s(\Delta)} = \frac{V_{L-L}}{Z}\sqrt{3} = \frac{440}{2.571}\sqrt{3} = 296.42 \text{ A}$$

$$T_{s(\Delta)} = \frac{3I_{s(\Delta)}^2 R_2^{/}}{s\omega_s} = \frac{3 \times (296.43/\sqrt{3})^2 \times 0.4}{1.0 \times 314} = 111.93 \text{ Nm}$$

Ratio of line currents is $\frac{I_{s(\Delta)}}{I_{s(Y)}} = \frac{296.4}{98.81} = 3$

Ratio of starting torque is $\frac{T_{s(\Delta)}}{T_{s(Y)}} = \frac{111.9}{37.31} = 3$

(b) Assuming that winding is connected in star and voltage is 50%.

$$I_{s(Y)} = \frac{0.5\ V_{ph}}{Z} = \frac{0.5 \times 440}{Z\sqrt{3}} = \frac{0.5 \times 440}{2.571\sqrt{3}} = 49.4 \text{ A}$$

$$T_{s(Y)} = \frac{3I_{s(y)}^2 R_2^{/}}{s\omega_s} = \frac{3 \times 49.4^2 \times 0.4}{1.0 \times 314} = 9.32 \text{ Nm}$$

When windings are connected in delta and 50% voltage is applied, the line current at the instant of starting is,

$$I_{s(\Delta)} = \frac{0.5V_{L-L}}{Z}\sqrt{3} = \frac{0.5440}{2.571}\sqrt{3} = 148.22 \text{ A}$$

$$T_{s(\Delta)} = \frac{3I_{s(\Delta)}^2 R_2^{/}}{s\omega_s} = \frac{3 \times (148.22/\sqrt{3})^2 \times 0.4}{1.0 \times 314} = 28 \text{ Nm}$$

(c) Voltage and frequency both are reduced to 10% of their rated value.

Synchronous angular speed at 5 Hz, two pole motor, $\omega_s = 2\pi f = 31.4$ Rad/s

At slip $s = 1$, the impedance seen from the terminals is given by $Z = \left(R_1 + R_2^{/}\right) + j\left(X_1 + X_2^{/}\right)\left(\frac{5}{50}\right)$

$$Z = (0.2 + 0.4) + j(1.0 + 1.5) \times 0.1 = 0.6 + 0.25j = 0.65\angle + 22.62^\circ\Omega$$

Let us suppose windings are in star. Then we have

$$I_{s(Y)Vf} = \frac{0.1 \times V_{\text{ph}}}{Z} = \frac{0.1 \times 440}{Z\sqrt{3}} = \frac{0.1 \times 440}{0.65\sqrt{3}} = 39.01 \text{ A}$$

$$T_{s(Y)Vf} = \frac{3I_{s(y)}^2 R_2^/}{s\omega_s} = \frac{3 \times 39.01^2 \times 0.4}{1.0 \times 31.4} = 58.13\,\text{Nm}$$

When windings are connected in delta, the line current at the instant of starting is,

$$I_{s(\Delta)Vf} = \frac{0.1\ V_{L-L}}{Z}\sqrt{3} = \frac{0.1 \times 440}{0.65}\sqrt{3} = 117.25 \text{ A}$$

$$T_{s(\Delta)Vf} = 175.13 \text{ Nm}$$

Comparing the calculate values for starting current and torque with rated voltage and rated frequency, we get

	Star		Delta	
	440 V, 50Hz	44 V, 5Hz	440 V, 50Hz	44 V, 5Hz
Starting current	98.91 A	39 A	254 A	117.25 A
Starting torque	37.3 Nm	58.13 Nm	112 Nm	175.13 Nm

Thus for the present case, when machine is excited by 10% of rated Voltage and frequency the starting current drops to 39 A and starting torque jumps to 58.13 Nm. These results are based on approximate equivalent circuit. By starting an induction motor with reduced voltage and reduced frequency, the starting current drops whereas the starting torque rises. The power factor during starting also improves.

Example 5.23 A three-phase, star connected, 440 V, 50 Hz induction motor has the following per phase equivalent circuit parameters:

$$R_1 = 0.2\ \Omega\ R_2^/ = 0.4\ \Omega\ R_C = 150\ \Omega$$
$$X_1 = 1.0\ \Omega\ X_2^/ = 1.5\ \Omega\ X_m = 30\ \Omega$$

Compare the initial starting current during DOL starting with the following cases

(a) A per phase resistance of 0.5 Ω is added in series with the stator circuit
(b) A per phase rotor resistance of 0.5 Ω is added in series with the rotor circuit
(c) Applied voltage is reduced by 50%
(d) At standstill, the induction motor is supplied by reduced voltage 44 V and reduced frequency of 5 Hz. Choose ratio of stator to rotor turns as 1.0.

Solution

Taking exact equivalent circuit of induction motor at standstill condition, as shown in Fig. 5.48. the rated per phase voltage can be given as

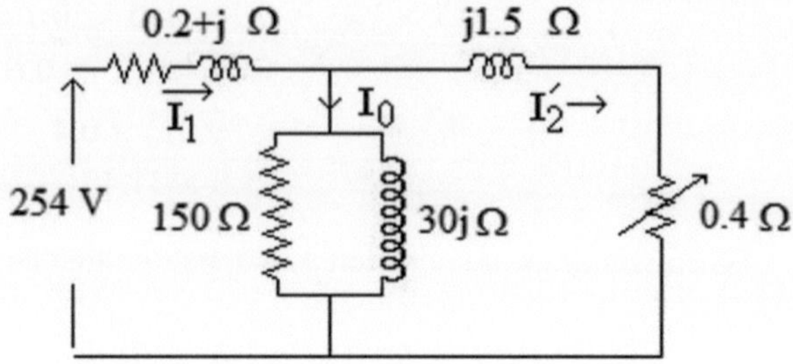

Fig. 5.48 Per phase equivalent circuit of induction motor at standstill $s = 1.0$

$$V_1 = \frac{440}{\sqrt{3}} = 254\ \text{V}$$

The equivalent impedance per phase in DOL starting is

$$Z_{\text{DOL}} = (R_1 + jX_1) + R_c| \ |jX_m| \ |\left(R_2^/ + jX_2^/\right)$$
$$Z_{\text{DOL}} = (0.2 + j) + 150| \ |j30| \ |(0.4 + j1.5)$$
$$Z_{\text{DOL}} = (0.2 + j) + \frac{150 \times j30 \times (0.4 + j1.5)}{150 \times j30 + 150 \times (0.4 + j1.5) + j30 \times (0.4 + j1.5)}$$
$$= 2.59\angle 76.6^\circ \Omega$$

Starting current in DOL starting is

$$I = \frac{254\angle 0^\circ}{2.59\angle 76.7^\circ} = 98.06\angle - 76.6^\circ \quad \text{A}$$

(a) When 0.5-Ω is added in series with the stator terminals, the total impedance seen from the terminals can be given as

$$I_2^/ = \frac{V_1}{\sqrt{\left(R_1 + R_2^/ / s\right)^2 + \left(X_1 + X_2^/\right)^2}}$$
$$= \frac{127}{\sqrt{(0.715 + 0.314/0.05)^2 + 1.472^2}} = 17.77\ \text{A}$$
$$\text{Power factor} = \cos\left[\tan^{-1}\left\{\frac{1.472}{(0.715 + 0.314/0.05)}\right\}\right]$$
$$= \cos\ 11.88 = 0.978$$

Starting current in DOL starting is

$$I = \frac{254\angle 0^\circ}{2.6538\angle 66.13^\circ} = 95.7\angle - 66.13^\circ \quad \text{A}$$

(b) When 0.5Ω is added in series with the rotor terminals the total impedance seen from the ter05minals. Here it is assumed that stator to rotor turn ration is 1.0.

$$Z_{rr} = (R_1 + jX_1) + R_c| \ |jX_m| \ |\left(R_r^{/} + R_2^{/} + jX_2^{/}\right)$$

$$Z_{rr} = (0.2 + j) + \frac{150 \times j30 \times (0.5 + 0.4 + j1.5)}{150 \times j30 + 150 \times (0.9 + j1.5) + j30 \times (0.9 + j1.5)}$$

$$= 1.87\angle 54.12^\circ \Omega$$

Starting current in DOL starting

$$I = \frac{231\angle 0^\circ}{1.87\angle 66^\circ} = 96.11\angle - 54.12^\circ \quad \text{A}$$

(c) Applied voltage is 50%

Starting current in DOL starting, $I = \frac{0.5 \times 254\angle 0^\circ}{2.4935\angle 76.7^\circ} = 50.95\angle - 76.7^\circ \quad \text{A}$

(d) When voltage is 40 V and frequency is 5 Hz, the parameters of equivalent circuit are modified accordingly. Thus

$$X_1 = \left(\frac{5}{50}\right)1.0 = 0.1 \ \Omega \qquad X_2^{/} = \left(\frac{5}{50}\right)1.5 = 0.15 \ \Omega \qquad X_m = \left(\frac{5}{50}\right)30 = 3 \ \Omega$$

The resistance will not vary. Thus, the impedance seen from the terminals on per phase basis is given as

$$Z_{\text{DOL}} = \left(R_1 + j\left(\frac{5}{50}\right)X_1\right) + R_c| \ |j\left(\frac{5}{50}\right)X_m| \ |\left(R_2^{/} + j\left(\frac{5}{50}\right)X_2^{/}\right)$$

$$Z_{\text{DOL}} = (0.2 + j0.1) + 150| \ |j3| \ |(0.4 + j0.15)$$

$$Z_{\text{DOL}} = (0.2 + j\ 0.1) + \frac{150 \times j3 \times (0.4 + j0.15)}{150 \times j3 \times (0.4 + j0.15)} = 0.9518 + 0.4549j$$

$$= 1.0549\angle 25.6^\circ \Omega$$

Starting current with reduced voltage and reduced frequency is $I = \frac{25.4\angle 0^\circ}{1.0549\angle 25.6^\circ} = 24.1\angle - 25.6^\circ A$

Thus, by reducing voltage as well as frequency, the peak current at the instant of starting is reduced by a factor of $\frac{24.1}{96.11} = 0.25$ of starting current with rated voltage and frequency.

5.12 Speed Control

Some applications require either marginal speed changes or stepped speed changes. The speed of induction motor is given as

$$N_r = (1 - s)\frac{120f}{p} \tag{5.83}$$

By changing the frequency and number of poles, the speed of induction motor can be changed by changing the synchronous speed. There are many other methods for controlling the speed of induction motor. We will now discuss some of these methods.

5.12.1 Change of Stator Voltage

The maximum torque of an induction motor is proportional to square of applied voltage, i.e.,

$$T_{\max} \alpha V^2 \tag{5.84}$$

If the supply voltage drops to 80% of rated voltage, the maximum torque drops to 64% of its rated maximum value. If voltage control of speed is applied, the induction motor having hard torque-speed characteristics yields little change in its speed. The method may be useful for changing speed of induction motor having soft torque-speed characteristics, like the induction motor having higher rotor resistance.

The change in the supply voltage can be obtained using either autotransformer or solid-state three-phase AC voltage controller.

A typical torque-speed characteristics for varying voltage is shown in Fig. 5.49.

5.12.2 Changing the Frequency

The synchronous speed of an induction motor can be changed by changing frequency and number of poles. The change in supply frequency requires a motor Alternator set for supplying variable frequency power to an induction motor. But such an arrangement is quite cumbersome and is rarely used-except in experiments. Due to advent of solid-state devices like power transistors, MOSFET, IGBT etc. and also computer control of solid-state inverters, the frequency can now be easily and smoothly controlled over a wide range. The synchronous speed is

$$N_s = \frac{120f}{p}$$

Fig. 5.49 Effect of change of supply voltage on Torque slip characteristics

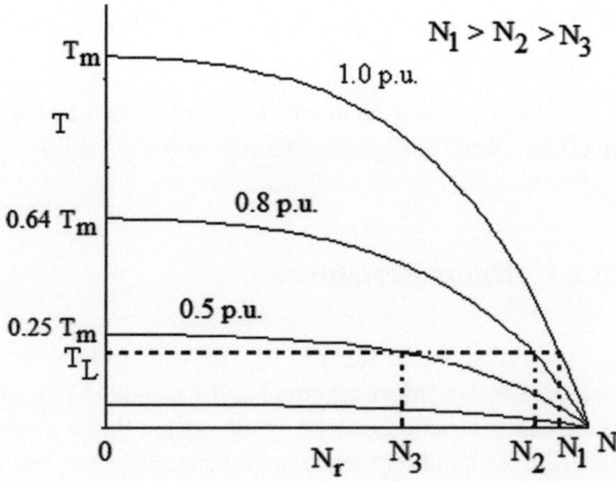

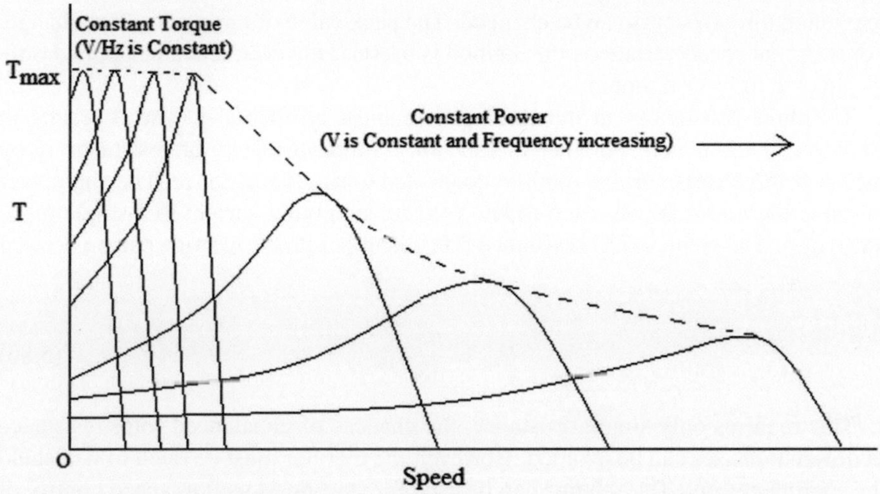

Fig. 5.50 Torque-speed characteristics of three-phase induction motor with VVVF control

and the maximum torque of induction motor is

$$T_m \ \alpha \ \left(\frac{V_1}{f}\right)^2 \tag{5.85}$$

Thus, the control of frequency brings a simultaneous change in the maximum value of air-gap flux and hence, maximum torque. As already explained, there are two methods of speed control of induction motor with changing the supply frequency. These methods are as follows:

(a) Constant torque control by keeping ratio of voltage and frequency constant

(b) Constant power operation by keeping voltage constant but changing the frequency.

The torque-speed characteristics are shown in Fig. 5.50. The VVVF starting of induction motor has been explained in Sect. 5.11.5.

5.12.3 Rotor Resistance

The rotor resistance starting of slip-ring induction motor has been discussed in Sect. 5.11.6. By inserting equal rotor resistance in all the rotor phases, the torque-speed characteristics can be made softer (large speed variation when machine is loaded) from a harder torque-speed characteristic. We know that maximum torque at starting occurs when per phase rotor resistance is equal to per phase rotor reactance. By inserting the extra resistance in each of the rotor phases, the critical slip at which maximum torque occurs can be changed. The peak value of torque does not change. For marginal speed variations the method is useful. However, it can be applied only in slip-ring induction motor.

The three-resistances in the rotor phases must be identical; else it results in George's phenomenon. Another method of controlling the rotor resistance is by means of three-phase bridge rectifier connected to single resistance. The slip power of induction motor is converted to DC voltage, supplying current to a single resistance R_{DC}. The value to DC resistance referred to per phase AC side can be derived as,

$$R_{ac} = \frac{\pi^2}{18} R_{DC} \tag{5.86}$$

This requires only single resistance, the chances of unbalanced rotor resistance in different phases can be avoided. However, the rectifier must be rated to withstand the starting current. The scheme can be used for starting as well as speed control of induction motor.

As already explained, identical resistances are required in all the three rotor phases. Fig. 5.51 shows the effect of rotor resistance on the speed variation for a constant torque load. The following section discusses the rotor chopper control of three-phase slip-ring induction motor.

5.12.4 Rotor Chopper Control

In rotor resistance control, it is desired to vary the rotor resistance in symmetrical manner; otherwise, there is a possibility of Georges's phenomenon in induction motor, making the slip-ring induction motor to run at about 50% of its synchronous

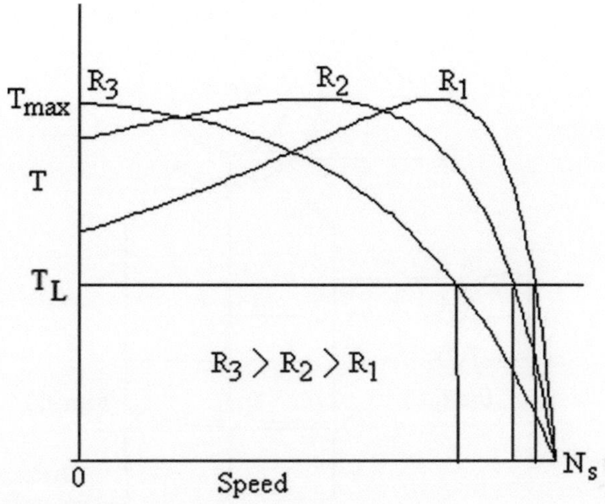

Fig. 5.51 Effect of changing rotor resistance on Torque-speed characteristics

speed. An alternative method is to vary a single resistance across the rectifier connected to the slip rings.

Consider the schematic circuit diagram of rotor chopper control in Fig. 5.52. The Machine is started using rotor resistance control by joining switch C1. When the motor is started, switch C1 is made OFF and switch C2 is ON. Now, the three-phase bridge rectifier converts the slip frequency voltage into DC voltage. The purpose of choke coil is to smoothen the ripples contents. An electronic ON/OFF switch short-circuits resistance R at certain duty cycle. When short (or T_{ON}) period is increased, the effective resistance decreases causing the rotor current to increase. However, if ON period is decreased, the effective resistance increases and as a consequence, the rotor current decreases. By changing the ON/OFF period or duty cycle, the effective resistance on the each of the rotor phases can be indirectly varied. The torque-speed characteristics of induction motor are similar to that of rotor resistance control. Depending on load torque, the speed of induction motor can be precisely controlled.

Figure 5.53 shows the effect of changing the duty cycle on the variation of rotor current. Here T_{ON} is the duration for which the electronic ON/OFF switch short-circuits the resistance R and T_{OFF} is the duration for which this switch is open, i.e., the current also flows through resistance R. The ON/OFF control of the switch or duty cycle is done using electronic circuit.

When $T_{ON} = 0$, effective DC resistance is given as

$$R_{eff} = R + r$$

And when $T_{OFF} = 0$, effective DC resistance becomes

$$R_{eff} = r$$

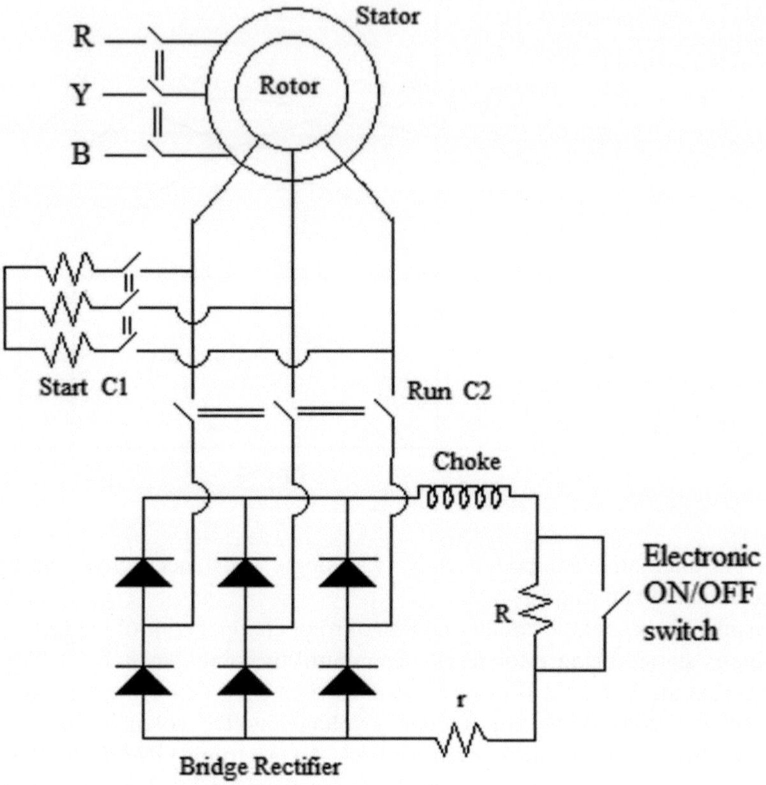

Fig. 5.52 Schematic diagram for rotor chopper control

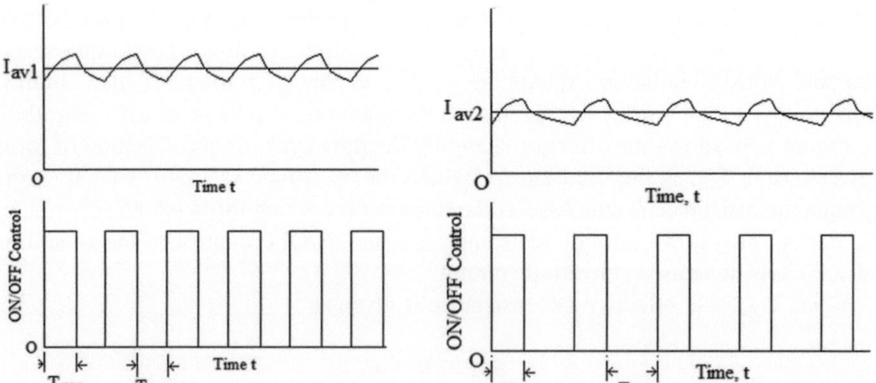

Fig. 5.53 Variation of rotor current with duty cycle in rotor chopper control

The effective DC resistance is

$$R_{\text{eff}} = r + R(1 - D) \tag{5.87}$$

Thus, by controlling the duty cycle,

$$D = \frac{T_{\text{ON}}}{T_{\text{ON}} + T_{\text{OFF}}}$$

the effective resistance on DC side can be varied from 'r' to '$R + r$'. In Fig. 5.53a, when T_{OFF} is less, the average current is comparatively high, thereby giving lower effective resistance. In Fig. 5.53b, when T_{OFF} is high, the average current is less, giving higher effective resistance. The effective value of the per phase resistance also changes accordingly, resulting in speed variation of induction motor.

The method is useful in controlling speed below base speed of smaller and medium capacity three-phase slip-ring induction motor. Though, the method offers smooth speed control with a provision of closed loop control, the overall efficiency is poor owing to loss of energy in the resistances.

5.12.5 Pole Amplitude Modulation

An induction motor may have two separate windings having different number of poles. However, only one set of winding may be used at a time. An induction motor can be made to run at different speed by changing the winding connections resulting in the change in effective number of poles. This requires larger slot area for accommodating two separate windings. Since at a time only one of the winding will be energized, the slot area is not completely utilized.

Instead of two separate windings of different number of poles (in the stator slots), a single winding can give 2:1 pole-ratio. Another approach is to rearrange the same windings for different number of poles. This approach is referred to as *pole amplitude modulation.* The windings are rearranged in following fashion.

(a) If the windings are having two distinct parts, the second part is reversed in whole with respect to the first part as shown in Fig. 5.54a
(b) A section of winding is omitted from each half and the remaining portion is then reversed with respect the other half as shown in Fig. 5.54b

Let us suppose that the flux density distribution $B_\theta = A \sin\left(\frac{p}{2}\theta\right)$ of original machine having p-number of poles is modulated. Then, we get

$$B_\theta = C \sin(k\theta)\sin\left(\frac{p}{2}\theta\right) = \frac{C}{2}\left[\cos\left(\frac{p}{2} - k\right)\theta - \cos\left(\frac{p}{2} + k\right)\theta\right] \tag{5.88}$$

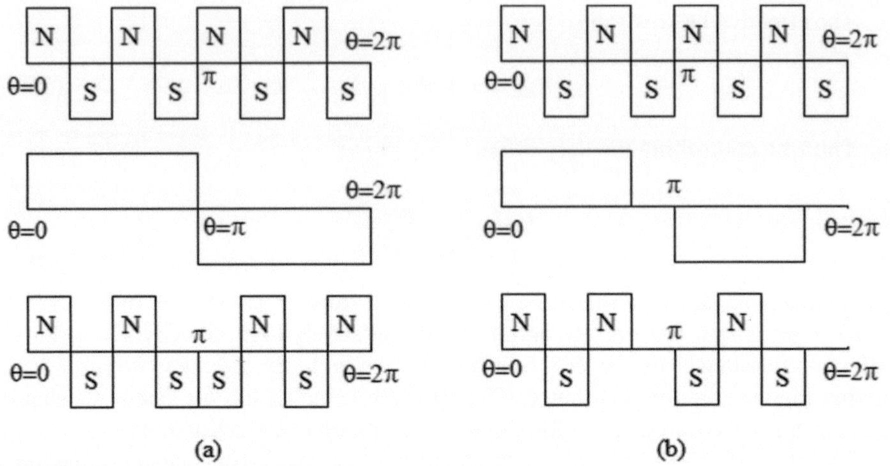

Fig. 5.54 Pole amplitude modulation-MMF distribution

Total numbers of poles are '$p - 2k$' or '$p + 2k$'. If an eight pole machine is modulated with two poles, the pole numbers will be either six or ten.

Consider, a three-phase, two-pole 24-slot winding. The winding can be connected into 4 poles. On the two-pole connection, each phase belt of a normal three-phase winding spans 60° electrical, which on 4-pole connection, the same phase belts are 120° electrical.

If a phase winding is having two distinct parts, they can be joined together as shown in Fig. 5.55a. In Fig. 5.55b–d, respectively, show the connections for constant torque, constant horse power and variable torque have been, respectively, shown. In these connections, power can be given by either T1–T2–T3 or T4–T5–T6 for low speed and high speed, respectively.

In pole Amplitude modulation, the MMF is non-symmetrically distributed in space. This results in large number of odd harmonics giving forward and reverse torque.

5.12.6 Cascade Connection

This is a classical method of speed control of induction motor. It requires an auxiliary induction motor for its operation apart from main slip-ring induction motor. Auxiliary motor may be either slip-ring or squirrel cage induction motor. The slip power of main motor is fed to the auxiliary induction motor. The auxiliary induction motor also needs a starter. The main motor is started using any method for starting induction motor—DOL starter or star/delta starter or rotor resistance starter or an autotransformer. A schematic diagram of cascade connected three-phase induction motor is shown in Fig. 5.56.

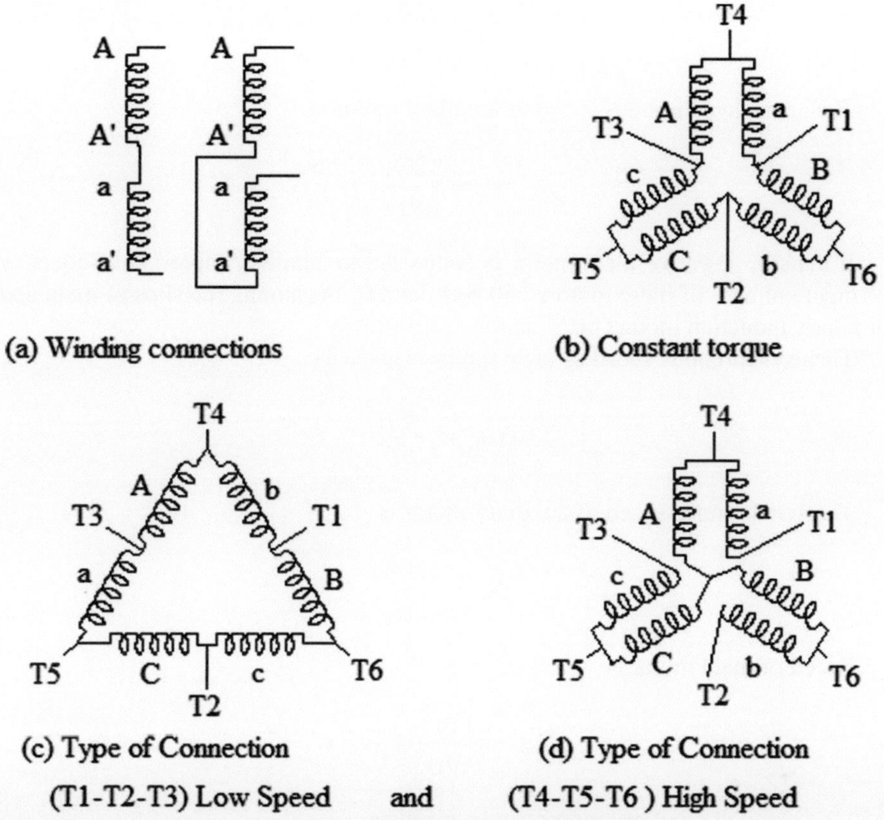

Fig. 5.55 Types of winding connections in pole amplitude modulation

Fig. 5.56 Schematic diagram of cascade connected three-phase induction motor

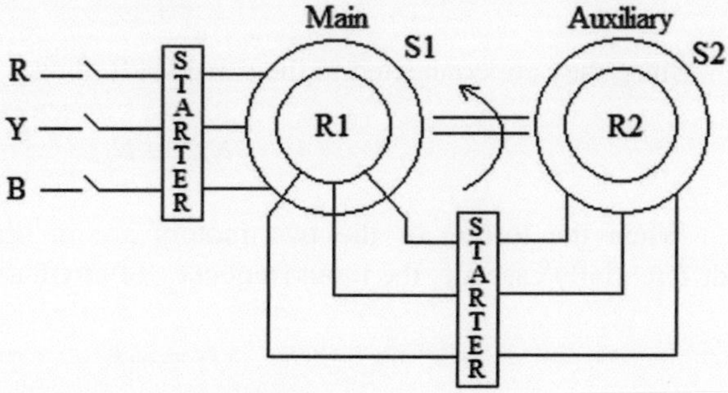

Let us assume that p_1 and p_2 are the number of poles of main and auxiliary induction motors, respectively. If these motors are operated alone at a supply frequency f_1, then one can have two synchronous speeds according to their number of poles. Thus, actual synchronous speed of main motor is

$$N_m = \frac{120f_1}{p_1}$$

The actual synchronous speed of auxiliary motor is

$$N_a = \frac{120f_1}{p_2}$$

If the slip power of main motor is connected to auxiliary motor, the frequency of input voltages of these motors will be f_1 and f_2. Assuming the slips of main and auxiliary induction motors are s_1 and s_2, respectively.

Then synchronous speed of main motor is given as

$$N_{m1} = \frac{120f_1}{p_1}$$

The synchronous speed of auxiliary motor is

$$N_a = \frac{120f_2}{p_2}$$

Speed of main motor is

$$N_{mr} = \frac{120f_1}{p_1}(1 - s_1)$$

and the speed of auxiliary motor can be given as

$$N_{ar} = \frac{120f_2}{p_2}(1 - s_2)$$

Since they are connected to the same shaft, thus

$$N_{mr} = N_{ar}$$

When the torque of the two motors are in opposite directions referred as differentially cascade, the input frequency of auxiliary motor is given as

$$f_2 = s_1 f_1$$

When the torque of the two motors are in the same direction, referred as cumulatively cascade we get

$$\frac{120f_1}{p_1}(1 - s_1) = \frac{120f_2}{p_2}(1 - s_2)$$

$$\frac{(1-s_1)}{p_1} = s_1\frac{(1-s_2)}{p_2}$$

$$s_1 = \frac{p_2}{p_1+p_2-p_1s_2}$$

The slip s_2 of auxiliary motor is to be very small as such $p_1\ s_2$ can be neglected. Thus motor speed

$$N = \frac{120f_1}{p_1}(1-s_1) = \frac{120f_1}{p_1+p_2} \tag{5.89}$$

Similarly, when the torque of both the motors are in the opposite direction, referred to as differential cascade, then one of the motor is said to operate in the counter current braking mode and is slip s_2 will be replaced by $(2-s_2)$. Thus

$$\frac{(1-s_1)}{p_1} = s_1\frac{(1-(2-s_2))}{p_2}$$

$$s_1 = \frac{p_2}{p_1-p_2+p_1s_2}$$

Considering p_1s_2 to be negligible, we get

$$s_1 = \frac{p_2}{p_2-p_1}$$

Therefore, speed can be given as

$$N = \frac{120f_1}{p_1}(1-s_1) = \frac{120f_1}{p_1-p_2} \tag{5.90}$$

If both the machines are having equal number of poles, i.e., $p_1 = p_2$, then in case of differential cascade (opposing torque), the speed of the set up will fail to rise above the synchronous speed of main motor. As soon as the speed of main induction motor increases beyond its synchronous speed, the torque reverses and motor decelerate. The net speed becomes slightly less than synchronous speed. Thus, it is evident that if supply frequency is 50 Hz and $p_1 = 4$ and $p_2 = 4$, the induction motors cascade set can be run on speeds according to synchronous speed of four pole and eight pole induction motors only. But if number of poles are $p_1 = 4$ and $p_2 = 6$, then the set can be made to run according to synchronous speeds of two, four, six, eight poles.

Such connections inherently have low power factor, efficiency and breakdown torque. These connections were applied for AC ship propulsion systems requiring low maneuvering speed. The cascade connection can be used in motoring and generating regions. It offers great possibility in low-speed wind-power or variable speed wind generation.

Example 5.24 Two 50 Hz three-phase induction motors having six and ten poles are connected in cascade. Calculate the range of speed which can be obtained from the set.

Solution: Here frequency $f = 50$ Hz and $p_1 = 6$ and $p_2 = 10$

When single motors are running, then synchronous speed are

$$N_{s1} = \frac{120\,\mathrm{f_1}}{p_1} = 120\frac{50}{6} = 1000 \text{ RPM}$$
$$N_{s2} = \frac{120 f_1}{p_2} = 120\frac{50}{10} = 600 \text{ RPM}$$

These two motors can be connected in cascade the speed during cumulative and differential cascade are

$$N_{\mathrm{scu}} = \frac{120\mathrm{f_1}}{\mathrm{p_1 + p_2}} = \frac{120 \times 50}{6 + 10} = 375 \text{ RPM}$$
$$N_{\mathrm{diff}} = \frac{120\mathrm{f_1}}{\mathrm{p_2 - p_1}} = \frac{120 \times 50}{10 - 6} = 1500 \text{ RPM}$$

Thus, the range of speed is 375–600–1000–1500 RPM

5.12.7 *Injection of EMF in the Rotor Circuit*

Consider a three-phase induction motor in which the power is supplied through both stator and rotor. It is assumed that the machine is running at constant torque load. The speed control of three-phase slip-ring induction motor can be achieved by injected EMF in the rotor circuit. In the Schrage motor, slip frequency EMF is produced and injected into secondary winding on the stator by means of brushes. The Schrage motor has been discussed in Sect. 7.3.4.

Under steady state condition,

$$\mathbf{V}_1 = \mathrm{R_1}\mathbf{I}_1 + \mathrm{jX_1}\mathbf{I}_1 - \mathbf{E}_1$$

$$\mathbf{V}_2^{/} = \mathrm{R}_2^{/}\mathbf{I}_2^{/} + \mathrm{j\,sX}_2^{/}\mathbf{I}_2^{/} - \mathrm{s}\mathbf{E}_1 \tag{5.91}$$

Here, $\mathbf{V}_2^{/}$ is the injected EMF referred to stator. The rotor quantities are referred to stator side. The rotor voltage referred to stator is

$$V_2^{/} = \frac{N_1 K_{w1}}{N_2 K_{w2}} V_2$$

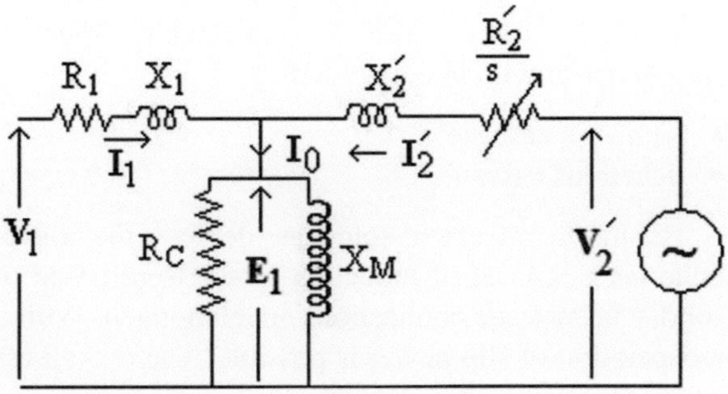

Fig. 5.57 Equivalent circuit of induction motor with injected EMF in rotor circuit

Here N_1 and N_2 are the number of turns per phase in stator and rotor, respectively. K_{W1} and K_{W2} are the respective winding factors. The equivalent circuit of induction motor with injected EMF in rotor circuit is shown in Fig. 5.57.

Ignoring the voltage drop in the stator, we can write

$$\mathbf{V}_1 = -\mathbf{E}_1 = \text{ Constant}$$

The expression for torque is given as

$$T = \frac{3sE_1I_2^{/}\cos\varphi_{12}}{\omega_s - \omega_r} = \frac{3E_1I_2^{/}\cos\varphi_{12}}{\omega_s} = \text{constant} \tag{5.92}$$

Here φ_{12} is angle between $\mathbf{E}_1$ and $\mathbf{I}_2^{/}$. The synchronous speed ω_s is constant. For constant torque, $\mathbf{E}_1$ should be constant. Thus $I_2^{/}\cos\varphi_{12} \approx$ constant. If rotor voltage $\mathbf{V}_2^{/}$ has same phase shift as the EMF $\mathbf{E}_1$, then if slip is very small

$$\cos\varphi_{12} = \frac{R_2^{/}}{\sqrt{\left(R_2^{/}\right)^2 + \left(\mathrm{s}X_2^{/}\right)^2}} \approx 1.0$$

Thus, current $\mathrm{I}_2^{/} \approx$ constant . We can write that

$$\mathrm{R}_2^{/}\mathbf{I}_2^{/} \approx \mathbf{V}_2^{/} + \mathrm{s}\mathbf{E}_1 = \mathbf{V}_2\frac{N_1K_{w1}}{N_2K_{w2}} + \mathrm{s}\mathbf{E}_1 \approx \text{constant} \tag{5.93}$$

Though the injected EMF is having slip frequency, it may have slightly different phase difference with the rotor voltage. The machine can be made to run in either sub-synchronous or super-synchronous speed apart from normal induction motor operation.

Certain industrial application requires a lower speed of operation. The use of rotor resistance control decreases the overall efficiency. The energy lost in slip can be recuperated electrically or mechanically. There are two solid-state methods of speed control with injected EMF:

- Kramer's cascade
- Scherbius cascade.

Before the advent of sold state devices, the conventional Kramer's cascade and Scherius's cascade of induction motor were based on rotary machines. Here only solid-state cascade connections are elaborated. With these cascade connections, the recuperation of slip power is possible. The overall efficiency is higher because slip power is not wasted as resistive loss.

(a) **Kramer's cascade:** In Kramer's cascade, the slip-ring induction motor is started using rotor resistance starter. Fig. 5.58 shows a schematic diagram of Kramer's cascade. When machine is running, the rotor circuit EMF is rectified and connected to armature of a separately excited DC motor. The DC motor is connected to the main shaft of induction motor by means of gears. By varying the field current of DC motor, the speed of shaft can be varied in sub-synchronous region. The starting of similar induction motor=Schrage motor has already been explained in Sect. 5.12.4.

Let the effective phase voltage to the rectifier is V_2 and the average output voltage is V_R. For a three-phase bridge rectifier the voltage relation is

$$V_2 = \frac{\pi}{3\sqrt{6}} V_R = k \cdot V_R$$

The referred value of V_2 on stator side is,

$$V_2^{/} = \frac{N_1 K_{w1}}{N_2 K_{w2}} V_2$$

In no-load operation, we have

$$V_2^{/} = s_0 E_1 \approx -s_0 V_1 = \text{constant}$$

The voltage V_R supplies power to DC motor. Thus, we get

$$V_R = k_e \phi_e N_{dc} = k_e \phi_e N_r \cdot i$$

Here transmission coefficient of the reduction gear is

$$i = \frac{N_{dc}}{N_r}$$

Let the synchronous speed of induction motor be N_s. then

$$N_r = N_s(1 - s_0)$$

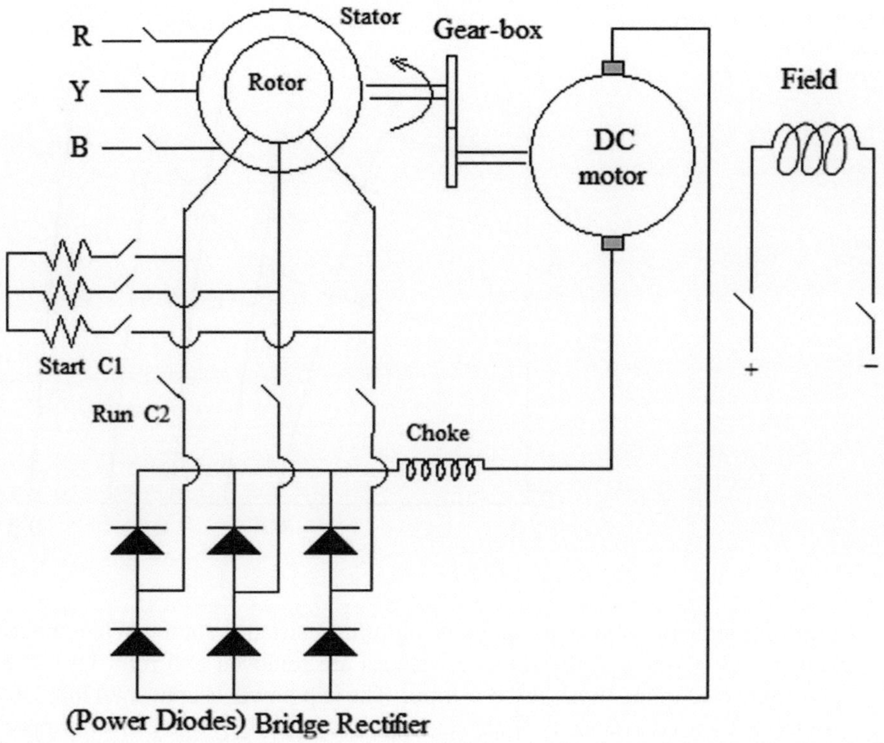

Fig. 5.58 Schematic diagram for Kramer's cascade

Thus

$$s_0 = \frac{kk_eN_si\frac{N_1K_{w1}}{N_2K_{w2}}\phi_e}{E_1 + kk_eN_si\frac{N_1K_{w1}}{N_2K_{w2}}\phi_e} = \frac{\phi_e}{\phi_e + \phi_{en}} \quad (5.94)$$

When $\phi_e = 0$, or motor field is not excited, then the no-load slip is $s_0 = 0$. The induction motor runs on its natural torque –speed characteristics. However, when $\phi_e = \phi_{en}$, then the no-load slip is $s_0 = 0.5$. In the Kramer's cascade, the no-load speed of induction motor can be reduced up to 50% of the synchronous speed. The torque slip characteristics have been shown in Fig. 5.59. The Kramer's cascade is used in drives requiring large torque at low speeds, such as cement mills and rolling mills. This scheme has also gained importance in the areas of variable speed wind power generations.

(b) **Scherbius cascade**: In Scherbius cascade, the slip power is converted into DC and then into three-phase AC, which is fed back to three-phase lines. The schematic diagram for solid-state Scherbius cascade is given in Fig. 5.60. The

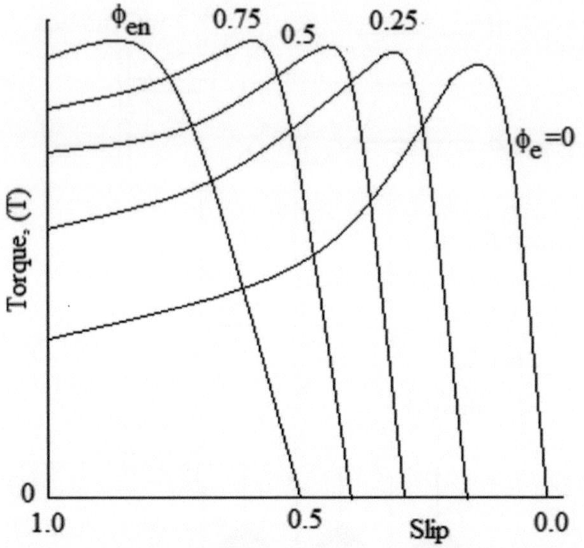

Fig. 5.59 Torque-speed characteristic- effect of varying field current

slip-ring induction motor is started using rotor resistance starter. When induction motor is running, the rotor resistances are removed and rotor terminals are connected to the three-phase rectifier. The slip power is converted into DC, which is boosted and again connected to a three-phase bridge converter operating as an inverter, in which the firing angle is more than 90°. The logic for gate pulses for different thryristors is obtained from three-phase lines. The (SCR) converter converts the DC power into three-phase AC power having frequency same as line frequency. The slip power is fed back to the lines using Y-Y transformer having a definite turn ratio. The SCR converter can be replaced by three phase inverters having power MOSFET or IGBT or GaN or SiC power devices

The speed of induction motor is controlled by changing the firing angle of the converter. For a three-phase bridge rectifier the per phase voltage relation is

$$V_2 = \frac{\pi}{3\sqrt{6}} V_R = k \cdot V_R$$

The per phase injected EMF is

$$E_1 = \frac{V_2^{/}}{s_0} = \frac{1}{s_0} \times \frac{N_1 K_{w1}}{N_2 K_{w2}} V_2 \tag{5.95}$$

The inverter input and output voltages satisfy the relationship,

$$k \cdot V_I = V_T |\cos\ \alpha|$$

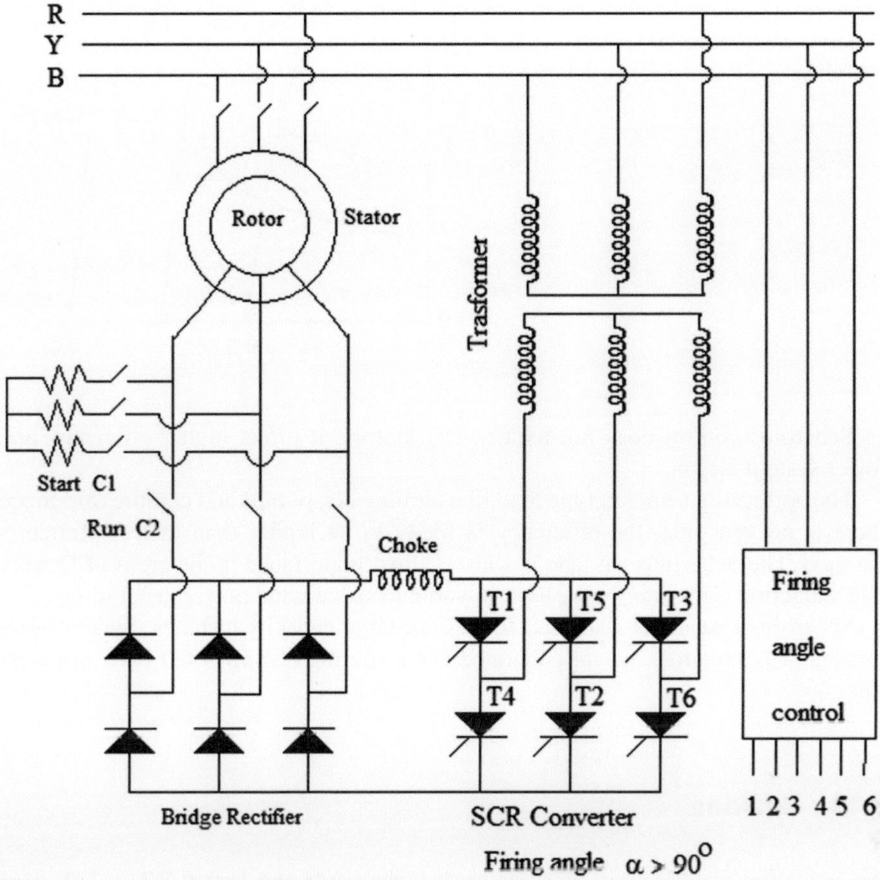

Fig. 5.60 Schematic diagram of solid-state Scherbius control of induction motor

Here V_I and V_T are the inverter input and output per phase voltage

The transformer voltage is $V_I = k_T\, V_T$,

Here, k_T is the coefficient of transformation. Since V_R=V_I thus we get

$$s_0 = \frac{V_2}{E_1} \times \frac{N_1 K_{w1}}{N_2 K_{w2}} V_2 = \frac{V_I}{k_T E_1} \times \frac{N_1 K_{w1}}{N_2 K_{w2}} |\cos \alpha| \tag{5.96}$$

This shows that no-load slip is proportional to $|\cos\ \alpha|$. When firing angle $\alpha = 90°$, the no-load slip $s_0 = 0.0$. This corresponds to the natural torque slip characteristics. The maximum recommended slip is 0.5. This uses lower capacity power rectifier, inverter and transformer. Fig. 5.61 shows the torque-slip characteristics of Scherbius control. Here, slip power less rotor loss is rectified, then converted back into the power of supply frequency and returned to the three-phase lines through transformer.

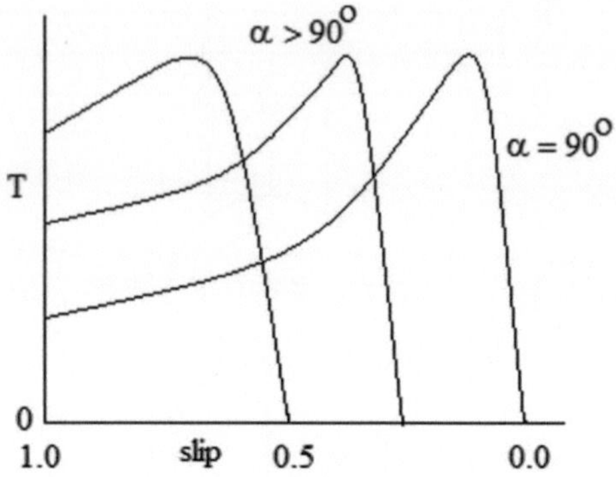

Fig. 5.61 Torque slip characteristics of Scherbius cascade

Scherbius control does not require DC motors. It offers higher efficiency and lower system inertia.

The applications are fan type load like air blowers, pumps and compressor. Since there is no gear box, the efficiency is likely to be higher than that of Kramer's cascade. The Scherbius cascade has also gained importance in the areas of Doubly Fed Induction Generator being used in variable speed wind power generations.

Scherbius cascade is also used for testing large capacity turbines, Synchronous Generator (alternator), where generated power during testing is fed back to power line.

5.13 Braking

An induction machine can work as motor, generator and brake. When induction motor is disconnected from supply, stored mechanical energy in rotating parts is dissipated in overcoming the friction and 'windage losses and opposing load torque. The use of electrical braking minimizes the time between initiation of the brake and the halt. Electrical methods of braking are superior to mechanical braking methods. Mechanical brakes are based on friction (physical contact – rubbing action by moving parts) and are applicable for low speed of operation. On the other hand, the electrical braking methods are non-contact braking methods having superior control of braking torque and are applicable at high speed of operation. A closed loop control of braking torque is also possible. In some cases, there is a possibility of recuperation of electrical energy. There are different methods of braking an induction motor.

5.13.1 Regenerative Braking

When load drives the motor such that rotor speed N_r becomes more than synchronous speed N_s, the induction machine works as an induction generator ($N_r > N_s$). The rotor rotates faster than rotating magnetic field. As a result, the slip becomes negative.

A sizable amount of rotor power is converted into electrical energy and fed back to three-phase lines. The torque becomes negative which then opposes or resists the torque causing the rotor to over speed beyond synchronous speed.

The condition in which the load attempts to drive the motor rarely occurs. Though it is an economical form of braking, it is effective only in the speed range above synchronous speed. The regenerative braking is effective in the stable portion of torque slip characteristics of induction machine.

Consider a pole changing induction motor (50 Hz) running in two-pole configuration near its synchronous speed of 3000 RPM, when the number of poles are changed to six-pole. Refer Fig. 5.62. The synchronous speed corresponding to six-pole is 1000 RPM. It works as a regenerative brake causing the shaft speed to quickly slow down according to new synchronous speed of 1000-RPM. When speed is higher than 1000 RPM, an opposing torque appears on the shaft. The braking torque quickly brings the speed from 3000 RPM to around 1000 RPM. Apart from that a sizeable amount of electrical energy is fed back to the three-phase lines.

Similar situations may occur in crane applications when load is being lowered. The shaft over speeds due to gravitational forces, whereas the induction motor works as a regenerative brake, thereby avoiding a free fall of the load due to gravitational forces.

If we consider a case of induction motor driven electric vehicle (EV) going in down gradient, the gravitational forces attempt to over speed the motor such that the speed crosses the synchronous speed. The braking torque opposes the over speeding

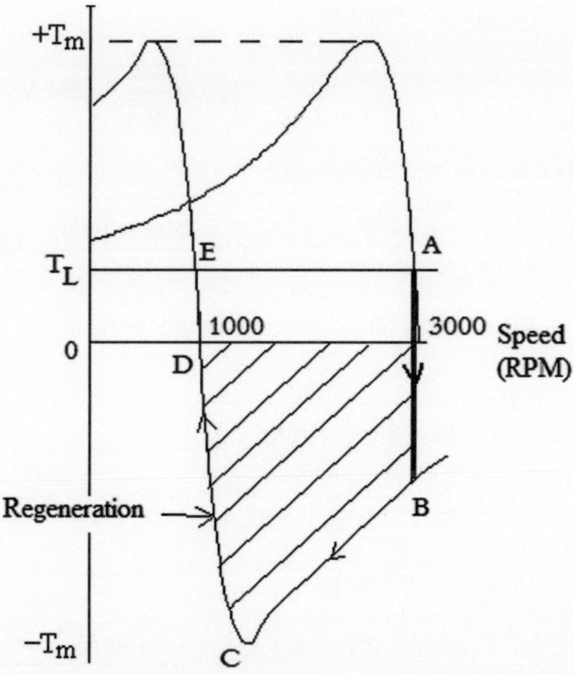

Fig. 5.62 Regenerative braking of induction motor

due to gravitational forces and restricts the speed near about synchronous speed. The stored energy is converted into electrical energy and fed back power line, which can ultimately be used for charging the battery or storing energy in flywheel system.

Example 5.25 A three-phase four pole, 440 V, 50 Hz, star-connected induction motor has following parameters

$$R_1 = 0.25\,\Omega \;\; R_2^/ = 1.5\,\Omega$$
$$X_1 = 1.5\,\Omega \;\; X_2^/ = 1.5\,\Omega$$

Using approximate equivalent circuit, determine the current and torque when machine is running at (a) 1400 RPM (b) 1600 RPM (c) -100 RPM

Solution

Synchronous speed is given as

$$N_s = \frac{120\text{f}}{\text{p}} = \frac{120 \times 50}{4} = 1500 \text{ RPM}$$

The rotational speed

$$\omega_s = \frac{2\pi N_s}{60} = \frac{2\pi \times 1500}{60} = 157 \quad \text{Rad/s}$$

The phase voltage is

$$V_{\text{ph}} = \frac{440}{\sqrt{3}} = 254 \text{ V}$$

(a) Slip is

$$s = \frac{1500 - 1400}{1500} = 0.067$$

The rotor per phase current referred to stator side is,

$$I_2^/ = \frac{V_{\text{ph}}}{R_1 + \left(\frac{R_2^/}{s}\right) + j\left(X_1 + X_2^/\right)}$$

$$I_2^/ = \frac{254}{0.25 + \left(\frac{0.5}{0.066}\right) + j(1.5 + 1.5)} = \frac{254}{7.75 + j3.0} = 30.56\angle - 21.2^\circ \text{ A}$$

The developed torque

$$T = \frac{P_g}{\omega_s} = \frac{3I_2^{/2}R_2^/}{s\omega_s} = \frac{3 \times (30.56)^2 \times 0.5}{0.0666 \times 157} = 133.84 \text{ Nm}$$

(b) $s = \frac{1500-1600}{1500} = -0.067$

The rotor per phase current referred to stator side is,

$$I_2^{/} = \frac{V_{\text{ph}}}{R_1 + \left(\frac{R_2^{/}}{s}\right) + j\left(X_1 + X_2^{/}\right)}$$

$$I_2^{/} = \frac{254}{0.25 + \left(\frac{0.5}{-0.066}\right) + j(1.5 + 1.5)} = \frac{254}{-7.25 + j3.0} = 32.37\angle 22.48^\circ \text{ A}$$

The developed torque is generating torque,

$$T = \frac{P_g}{\omega_s} = \frac{3I_2^{/2}R_2^{/}}{s\omega_s} = \frac{3 \times (32.37)^2 \times 0.5}{-0.0666 \times 157} = -150.16 \text{ Nm}$$

(c) $s = \frac{1500-(-100)}{1500} = 1.0667$

$$I_2^{/} = \frac{254}{0.25 + \left(\frac{0.5}{1.066}\right) + j(1.5 + 1.5)} = \frac{254}{0.72 + j3.0} = 82.32\angle -76.5^\circ \quad \text{A}$$

The developed torque is counter current braking torque,

$$T = \frac{P_g}{\omega_s} = \frac{3I_2^{/2}R_2^{/}}{s\omega_s} = \frac{3 \times (82.32)^2 \times 0.5}{1.0666 \times 157} = 60.7 \text{ Nm}$$

It is counter current opposing torque as such $T = -60.7\,\text{Nm}$.

5.13.2 DC Dynamic Braking

Figure 5.63 shows a schematic diagram of DC dynamic braking of induction motor. The induction motor is connected to three-phase supply using triple-pole switch C1. The machine is assumed to run at steady state speed. When DC dynamic braking is initiated, supply to the motor is cut OFF by switching OFF switch C1. Simultaneously, the double pole switches C2 and C3 are made ON. A single-phase two-winding transformer is used to reduce the single-phase voltage, which is rectified and connected to the terminals of the induction motor. The turn ratio of transformer is suitably chosen so as to limit the braking current. The magnitude of braking current decides the braking torque and the time taken for the machine to halt. The windings may be connected in various ways. Fig. 5.64 shows the some of the winding connections, which can be connected to DC supply. The braking torque also depends on the type of winding connections. The torque slip characteristics of three-phase induction motor is given as

$$T = \frac{2T_m}{\left(\frac{s}{s_m}\right) + \left(\frac{s_m}{s}\right)} \tag{5.97}$$

When the three-phase supply of an induction motor is disconnected and the winding terminals are connected across a low voltage DC source, the DC magnetic poles are formed on the stator active surface. Since the rotor is already rotating due to its moment of inertia, this static magnetic fields cuts the rotating conductors. This is similar to induction motor at standstill conditions, when the rotating magnetic fields are cutting the stationary rotor conductors. When the induction motor is running at synchronous speed, there is no flux cutting action. In case of DC dynamic braking, there is no flux cutting action, when machine is at rest. The torque-speed characteristics exhibited in this type of braking is shown in Fig. 5.65. The shape of characteristics is similar to that of motor, but is inverted and skewed on both the axes. An induction motor has zero torque at synchronous speed, but DC dynamic braking has zero torque at zero speed. A motor has maximum motoring torque at slip s_m, whereas DC dynamic braking has maximum braking torque at S_m. If we write braking critical slip

$$S_m = 1 - s_m$$

Then we can write expression for braking Torque slip characteristics as

$$T = -\frac{2T_{mdc}}{\left(\frac{S}{S_m}\right) + \left(\frac{S_m}{S}\right)} \tag{5.98}$$

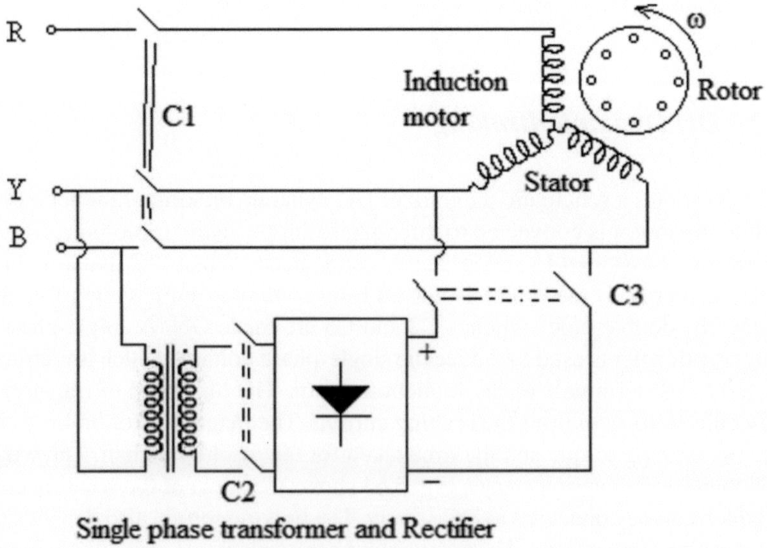

Fig. 5.63 DC dynamic braking of induction motor

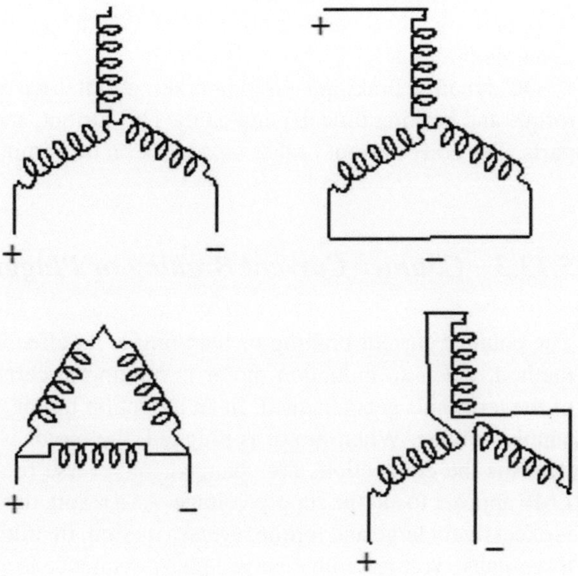

Fig. 5.64 Winding connections in DC dynamic braking

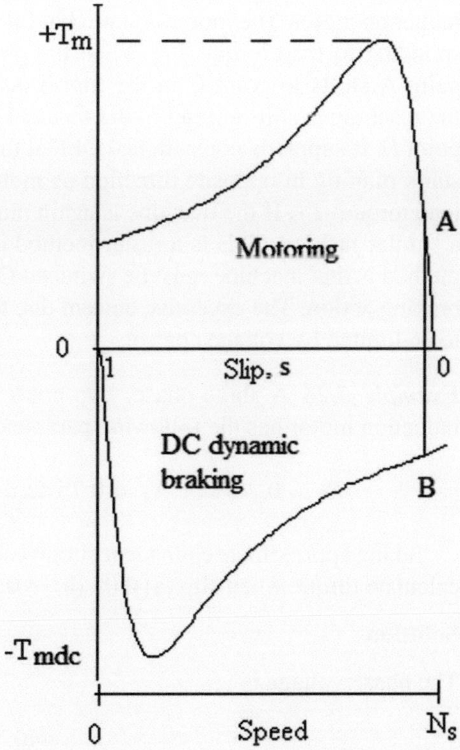

Fig. 5.65 DC dynamic braking of three-phase induction motor

Here braking slip S varies from 1.0 to 0.0 and T_{mDC} is the maximum torque in DC dynamic braking.

DC dynamic braking is used in braking tool drive applications. Both, the braking torque and braking time depend on the DC current, and the stored energy of rotating parts. The eddy current brakes also work on the similar principle.

5.13.3 Counter Current Braking or Plugging

The counter current braking or plugging is an effective method of braking. In this method, when an induction motor is running at certain slip s, the phase sequence of the terminals gets changed. In an induction motor, the induced EMF opposes the supply voltage. When motor is plugged, the motor is running in forward direction whereas the connections are changed for reverse rotation or slip 2-s. The induced EMF appears to aid the supply voltage. As a result, the current drawn from the supply is excessively large and torque reverses its sign. In order to limit the current, a reduced three-phase voltage with changed phase sequence is applied.

Figure 5.66 shows the four quadrant torque-speed characteristics of a three-phase induction motor. The motor is started in forward direction and runs at point A corresponding to load torque T_L. When the phase sequence is changed, the operating point A shifts to point C in the fourth quadrant. The torque reverses its sign but the machine is still running in the forward direction. The machine comes to halt at point D. If supply is not switched OFF at this instant of time, the induction machine starts running in opposite direction as motor and runs at point E corresponding to load torque $-T_L$. If the machine is again plugged, operating point shifts to F-G A in a similar fashion. This is a faster method of braking. The only drawback with the method is that machine must be switched OFF as soon as it is brought to rest due to braking action. The excessive current due to counter current braking or plugging is to be limited by voltage control.

Example 5.26 A three-phase, two-poles, 50 hp, 440 V, 50 Hz star-connected induction motor has the following parameters per phase:

$$R_1 = 0.25\ \Omega\ ;\ X_1 = 0.75\ \Omega\ ;\ R_2^{/} = 0.25\ \Omega\ ; X_2^{/} = 0.75\ \Omega$$

Taking approximate equivalent circuit calculate the slip at maximum torque, hence calculate torque when slip (a) 0.05 (b) − 0.05

Solution

The phase voltage is

$$V_1 = \frac{440}{\sqrt{3}} = 254\,\text{V}$$

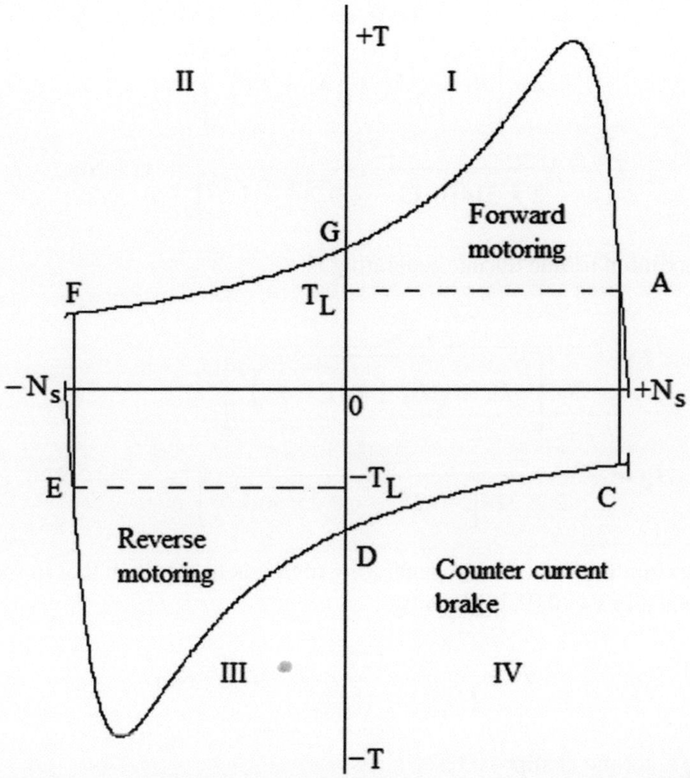

Fig. 5.66 Counter current braking of induction motor

Rated current is

$$I = \frac{50 \times 746}{440\sqrt{3}} = 48.94 \text{ A}$$

Slip at maximum torque is given by

$$s_m = \pm \frac{R_2^{/}}{\sqrt{R_1^2 + \left(X_1 + X_2^{/}\right)^2}}$$

$$s_m = \pm \frac{0.25}{\sqrt{0.25^2 + (0.75 + 0.75)^2}} = \pm 0.164$$

Thus the maximum torque during motoring is given as,

$$T_m = 3\frac{V_1^2}{2\omega_s\left[R_1 + \sqrt{R_1^2 + \left(X_1 + X_2^/\right)^2}\right]}$$

$$T_m = 3\frac{254^2}{2 \times 314\left[0.25 + \sqrt{0.25^2 + 1.5^2}\right]} = 174\ \text{Nm}$$

The maximum torque during generating is

$$T_g = 3\frac{V_1^2}{2\omega_s\left[-R_1 + \sqrt{R_1^2 + \left(X_1 + X_2^/\right)^2}\right]}$$

$$T_g = -3\frac{254^2}{2 \times 314\left[-0.25 + \sqrt{0.25^2 + 1.5^2}\right]} = -242.5\ \text{Nm}$$

The maximum torque during generating region is higher than that in motoring. Torque at slip $s = 0.05$ is given by,

$$T = \frac{2 \times 175}{\left(\frac{0.05}{0.164}\right) + \left(\frac{0.164}{0.05}\right)} = 97.07\ \text{Nm}$$

Similarly torque at slip -0.05

$$T = -\frac{2 \times 242.5}{\left(\frac{-0.05}{-0.164}\right) + \left(\frac{-0.164}{-0.05}\right)} = -135.29\ \text{Nm}$$

5.14 Induction Generator

Induction machine operating as generator is increasingly being preferred over conventional alternators for power generation using renewable energy sources due to its overall maintenance, operational simplicity, lower cost and availability. There are basically two types of induction generators [1, 2, 7]:

(a) Self-Excited Induction Generator (SEIG)
(b) Line or grid Excited Induction Generator

5.14.1 Self-excited Induction Generator (SEIG)

As the name suggests, SEIG is standalone induction generators, as it does not require extra three-phase power supply for excitation. The induction generator is driven by a prime mover. The magnetizing reactive power is provided by a capacitor bank, as shown in Fig. 5.67.

The per phase equivalent circuit of SEIG connected to load is given in Fig. 5.68. The operation of SEIG demands that there should be a remanent (residual) magnetic flux density in the rotor. Rotor frequency of an induction motor is low. When an induction motor is switched OFF, the rotor retains residual magnetism. If an induction machine is being rotated using a prime mover, a small voltage of the order of 5 to 10 V is generated on stator terminals. This is a check that the rotor is having residual magnetism. On no-load, the per phase capacitor current $I_c = V\omega C = \frac{V}{X_c}$ must be equal to the per phase magnetizing current, $I_m = \frac{V}{X_m}$.

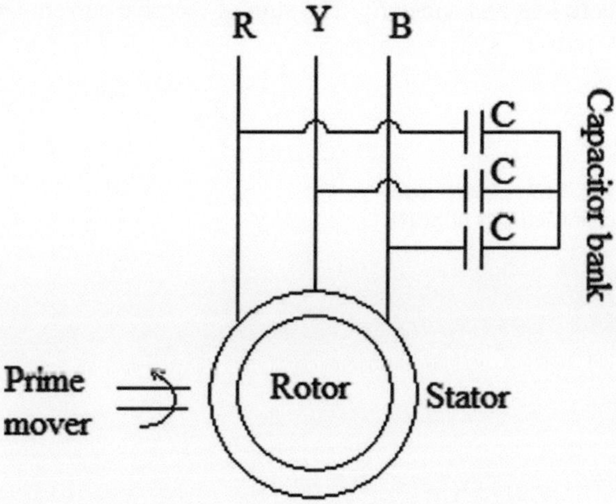

Fig. 5.67 Schematic representation of SEIG

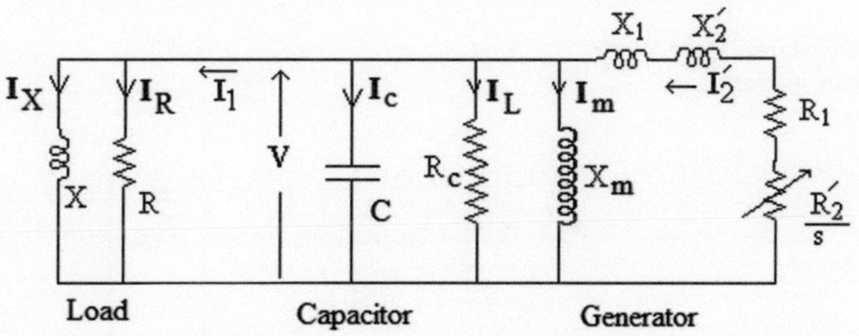

Fig. 5.68 Per phase equivalent circuit of SEIG connected to load

The operating frequency and the generated voltage depend on speed, load and capacitor ratings. For any lagging load, the capacitor must also supply the reactive power of the connected load.

However, the developed voltage V is a function of magnetizing current I_m. For self- excited induction generation, it is desired that the straight line X_c versus I_m should cut the V versus I_m characteristics as shown in Fig. 5.69. The voltage versus magnetizing susceptance $(1/X_m)$ is shown in Fig. 5.70.

The capacitor bank may be connected in star or delta. The variation of no-load voltage versus capacitance values for star and delta connected capacitor is shown in Fig. 5.71. The relation between delta and star-connected capacitance is

$$C_Y = 3C_\Delta \tag{5.99}$$

The operating point is fixed by the corresponding capacitive susceptance, $\omega C = \frac{1}{X_c}$.

Under load condition, the generated power VI $cos\ \varphi$ gives load power in R_L and core loss resistance R_c. The sum of reactive currents must be zero, or,

$$\frac{V}{X} + \frac{V}{X_m} + I\ \sin\ \varphi = \frac{V}{X_c} \tag{5.100}$$

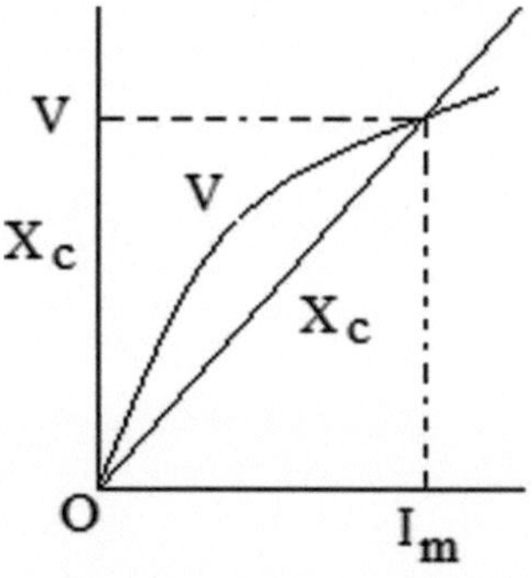

Fig. 5.69 Open circuit characteristics of SEIG

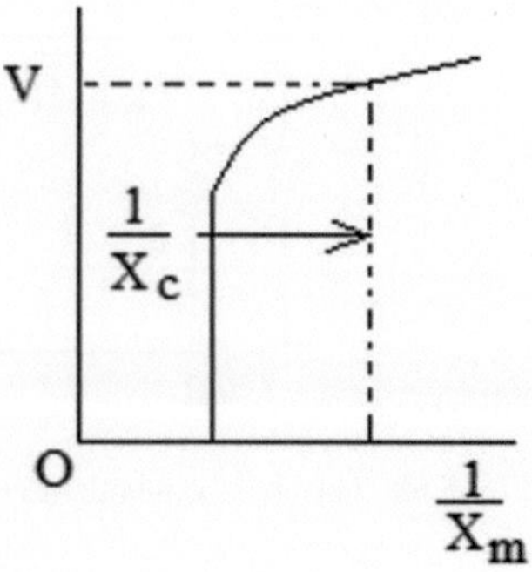

Fig. 5.70 Developed Voltage versus magnetizing susceptance

Fig. 5.71 Effect of capacitor connection in Δ and Y on no-load voltage

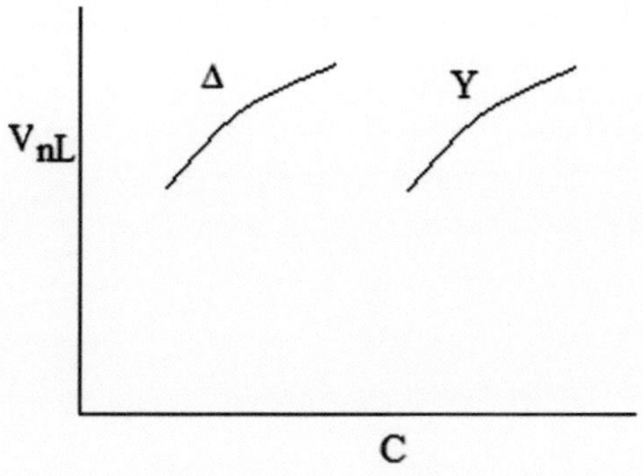

The voltage regulation may be large in SEIG. The voltage waveform is sometimes distorted. Self-excited induction generators connected to a three-phase resistance can also be used as brake. Figure 5.72 shows the variation of no-load voltage versus speed for different capacitances. When the value of capacitance is large, the developed voltage is better. The load voltage versus developed power for different speed and fixed capacitance is shown in Fig. 5.73. The developed voltage and developed power both are high at higher speeds.

Determination of the magnetizing reactance X_m, the induction motor is run at exact synchronous speed by another motor connected to the shaft. Normally, the DC shunt/ separately excited motor is used to run the induction motor at exact synchronous speed for no-load or free running test of induction motor. These machines are used for standalone induction generator serving load in remote locations, where normal AC supply is not available. The three phase induction generator serving single-phase load has been quite common. These SEIG machines have found applications in starter-generator technology. The machine works as a motor for starting the engine, and once the engine is started, it acts as a generator.

Fig. 5.72 Variation of no-load voltage with speed for different capacitors

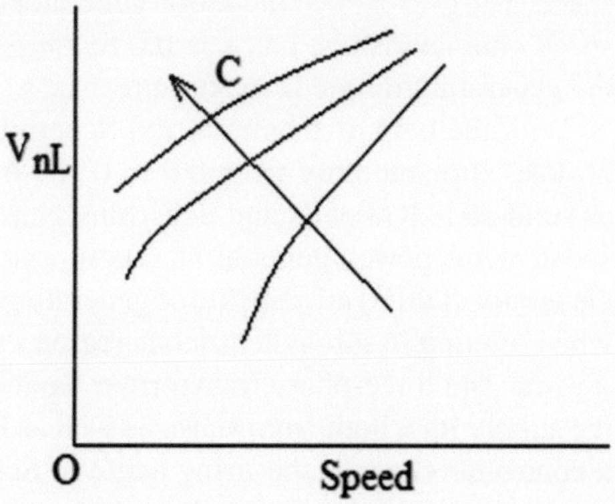

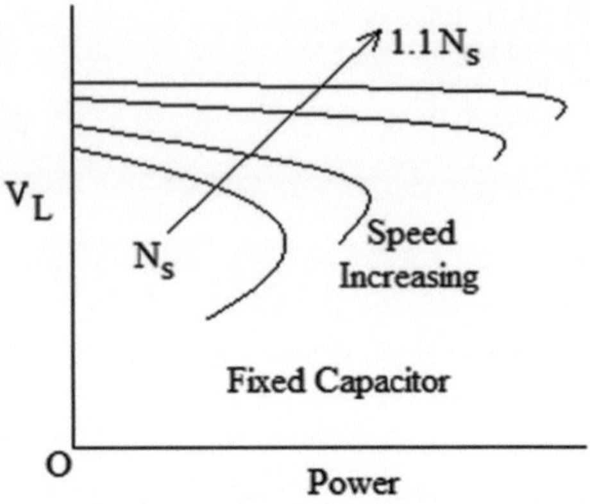

Fig. 5.73 Variation of load voltage and power for different speed with fixed capacitor

5.14.2 Line or Grid-Excited Induction Generator

The grid excited induction generators are of different types:

(a) Squirrel cage type
(b) Slip-ring induction machine-Solid-state Scherbius and Kramer's cascade.

The squirrel cage induction motor running in super-synchronous speed operates as grid excited induction generator. We know that an induction machine works as a generator when $N_r > N_s$ or slip is negative. The frequency of generation is the grid frequency. Depending on the prime mover power, the electrical power is fed back to the lines. The counter torque developed in induction generator opposes the prime mover torque. If the speed falls below the synchronous speed the induction machine works as a motor, wherein it receives electrical power from the grid instead of delivering the power to the grid. Its operation essentially requires the availability of three-phase power source and the no-load speed must always be more than synchronous speed. The grid excited induction generator operates in the linear portion of its torque-speed characteristics, i.e., $s = 0.0$ to $s = -s_{\text{cr}}$. Here s_{cr} is the critical slip at which the generating torque is maximum.

With the help of Kramer's and Scherbius's cascade, it is possible to change the no-load slip smoothly from 0.0 to 0.5. Nowadays, the cascaded connections, such as solid-state Kramer's and Scherbius's cascade have found applications in variable speed wind power generation. These machines known as Doubly Fed Induction Generator (DFIG) are capable of generating power frequency three phase power even when running in sub-synchronous region. A Scherbius's cascade requires solid-state inverter and three-phase transformer having variable turn ratio. The power is fed to the supply line, both from stator as well as from rotor. As the speed of the rotor falls, a controller changes the firing angle in Scherbius's cascade or field current of DC motor in case of the Kramer's cascade, so as to enable the machine to generate. A typical characteristic is shown in Fig. 5.74.

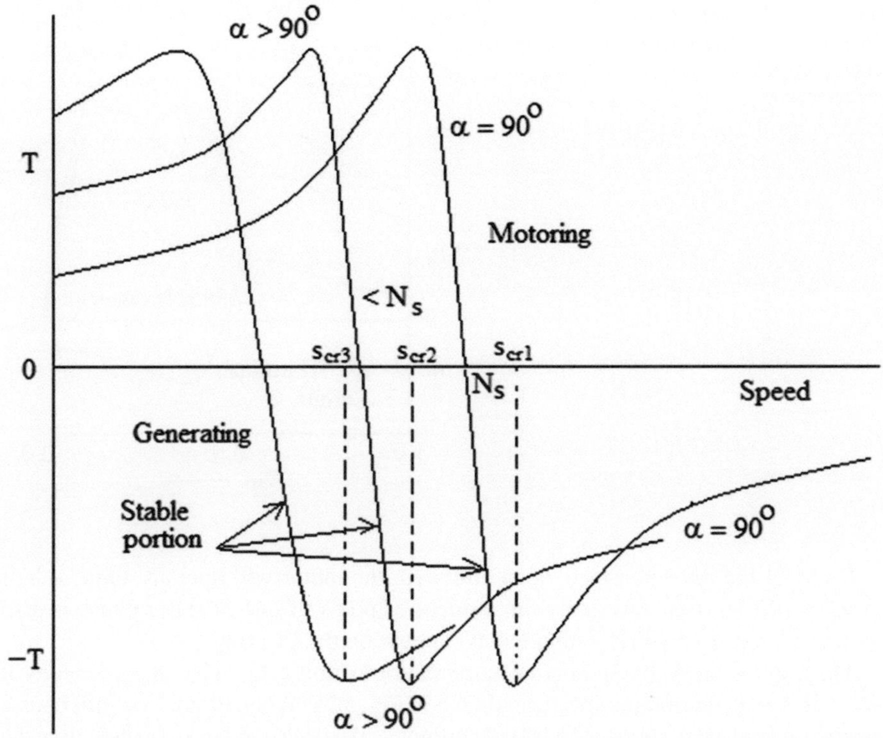

Fig. 5.74 Grid connected variable speed generation using Scherbius's control

5.15 GEORGES'S Phenomenon

When a slip-ring induction motor is started using rotor resistance starter with fault in the rotor circuit or with unequal rotor resistance. The machine exhibits a peculiar characteristics. The single phasing in rotor circuit results in the creation of forward and reverse rotating magnetic fields in the rotor. These rotating magnetic fields run at slip speed in forward and reverse direction with respect to the rotor surface. At about 50% of the synchronous speed, the backwards-rotating magnetic field appears to be standstill with respect to the stator surface.

Any unbalance in the rotor circuit produces backward rotating magnetic field component of the fundamental rotating magnetic field. This backward component of rotating magnetic field of rotor generates a stator induced EMF of frequency $(1-s)f-sf$ or $(1-2s)f$. Therefore, at half speed, when $1–2s = 0$, this backward rotating magnetic field of the rotor appears standstill with respect to the stator. Above half speed, the $1–2s$ frequency induced in the stator will cause a rapid dip in the torque-speed characteristics. It also causes flicker in the impressed stator voltage. The latter is very objectionable.

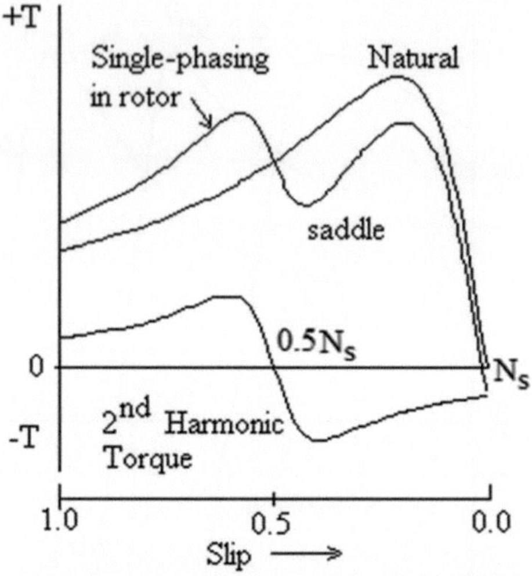

Fig. 5.75 Effect of fault in rotor of slip-ring induction motor-Georges' phenomenon

If one of the rotor phases is open circuited, the motor will operate stable at half speed as if it has twice as many the number of poles. If one of stator phase is also open-circuited, a very high voltage may appear on the slip ring.

The torque-speed characteristics of the motor exhibit a dip in its characteristics at about 50% of synchronous speed, as shown in Fig. 5.75. When motor has fault before starting, it runs near about 50% of synchronous speed. When fault appears after the motor has started, it runs near about synchronous speed. The noise is produced due to second harmonic torque in the machine.

5.16 Deep Bar and Double Cage Induction Motor

In many applications, it is desired that the induction motor should have large starting torque. The slip-ring induction motor may have large starting torque, and when machine runs near synchronous speed, the rotor resistance should be made low to have a hard torque-speed characteristic. Squirrel cage induction motors with high starting torque characteristics have a soft torque-speed characteristic. This means when motor is mechanically loaded, a large variation in speed takes place. The double-cage or deep-bar induction motors have a high starting torque and low rated slip.

In deep bar squirrel cage induction motor, at standstill conditions, the rotor induced EMF is of supply frequency and skin depth is less. As a result, effective resistance is large. This results in large starting torque. At any slip, the skin depth increases, resulting in lesser rotor resistance, and the motor exhibits large starting torque and hard torque-slip characteristics. The double cage induction motor may have two types of slotting arrangements as shown in Fig. 5.76:

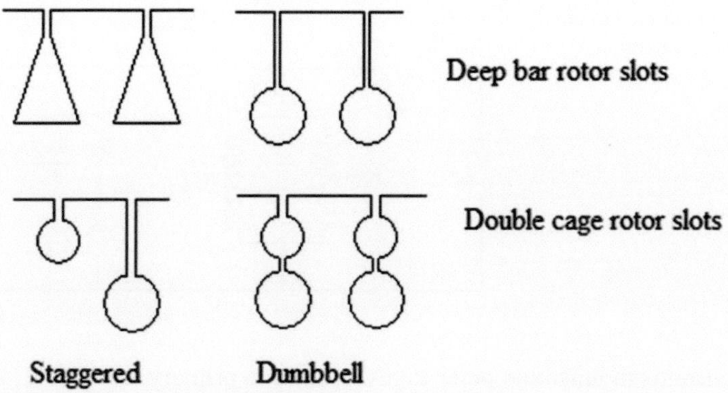

Fig. 5.76 Deep bar and double cage rotor slots

(a) Dumbbell
(b) Staggered.

Double cage induction motor has two separate cage windings. The rotor bars in these cage windings are of different conducting materials. The outer cage has a higher resistance and lower leakage reactance, whereas the inner cage has a low resistance and high leakage reactance. The slit is necessary so that the flux is able to link the inner cage. The pull-out torque (maximum torque) is smaller because both the cages produce their pull-out torques at different slips. The high-resistance outer-cage produces large starting torque. The inner-cage with lesser resistance makes the characteristics harder. The torque-speed characteristic are shown in Fig. 5.77. The double cage induction motor may be considered to have the equivalent circuit as shown in Fig. 5.78. The reactance X_{23} is the leakage reactance between the two cages. Here subscripts 2 and 3 are for inner- and outer-cages, respectively.

Fig. 5.77 Torque-speed characteristics of double cage induction motor

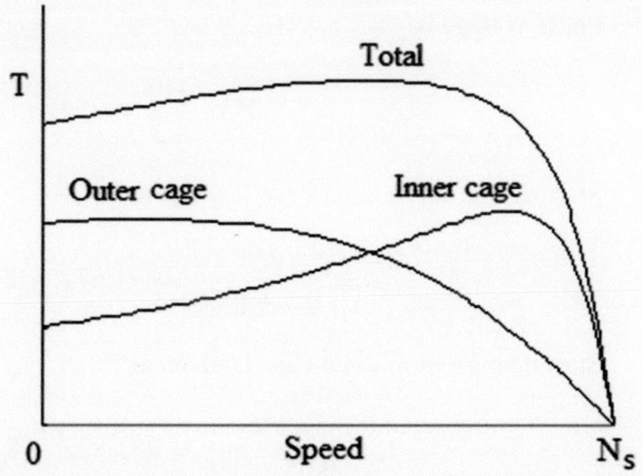

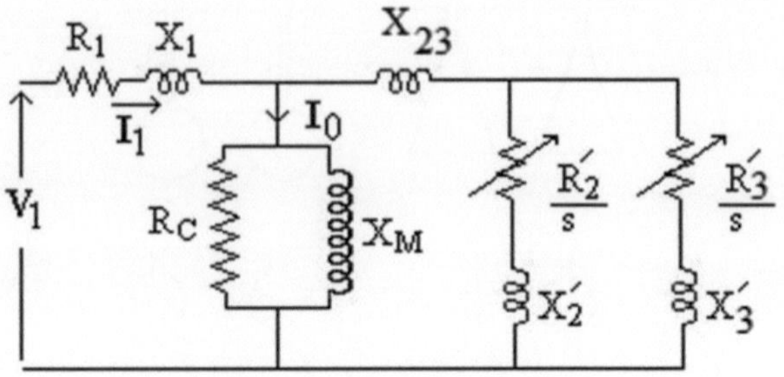

Fig. 5.78 Equivalent circuit of double cage induction motor

If the currents in inner and outer cages referred to primary are $I_2^{/}$ & $I_3^{/}$, the torque in a three-phase double cage induction motor can be given as

$$T = \frac{3\left(I_2^{/}\right)^2 R_2^{/}}{s\omega_s} + \frac{3\left(I_3^{/}\right)^2 R_3^{/}}{s\omega_s}$$

Example 5.27 A 440 V, 50 Hz, 2-pole star-connected double cage three-phase induction motor has following parameters of rotor equivalent circuit:

$$R_o = 0.5\ \Omega \quad X_o = 0.2\ \Omega \quad R_i = 0.05\ \Omega \quad X_i = 0.9\ \Omega$$

Calculate the starting torque and running torque, when machine is running at 0.07 slip.

Solution

Synchronous speed

$$\omega_s = 2\pi f = 314\ \text{Rad/s}$$

Phase voltage is

$$V = \frac{440}{\sqrt{3}} = 254V$$

Starting current in outer cage is given as

$$I_{\circ} = \frac{254}{0.5 + 0.2j} = 471.67\angle - 21.8^{\circ}\ \ \text{A}$$

Starting current in inner cage is given as

$$\text{I}_\text{i} = \frac{254}{0.05 + 0.9\text{j}} = 281.82\angle - 86.82^{\circ}\ \ \text{A}$$

The torque is $T = \frac{3I_o^2 R_o}{s\omega_s} + \frac{3I_i^2 R_i}{s\omega_s}$
At standstill s=1. Thus, we get

$$T_{st} = \frac{3 \times 471.67 \times 471.67 \times 0.5}{314} + \frac{3 \times 281.82 \times 281.82 \times 0.05}{314}$$
$$T_{st} = 1062.76 + 37.94 = 1100.7 \text{ Nm}$$

At slip 0.07 the current is

$$I_{os} = \frac{254}{\frac{0.5}{0.07} + 0.2j} = 35.55\angle - 1.6° \text{ A}$$
$$I_{is} = \frac{254}{\frac{0.05}{0.07} + 0.9j} = 221.06\angle - 51.6° \text{A}$$
$$T = \frac{3 \times 35.55 \times 35.55 \times 0.5}{0.07 \times 314} + \frac{3 \times 221 \times 221 \times 0.05}{0.07 \times 314}$$
$$T = 86.25 + 333.31 = 419.56 \text{ Nm}$$

Example 5.28 A 25 KW, 440 V, 50 Hz, 4-poles delta connected double cage induction motor has the following equivalent circuit parameters:

$$R_c = 500 \ \ \Omega, \ \ X_m = 50 \ \ \Omega, \ \ R_1 = 0.2 \ \ \Omega, \ \ X_1 + X_{23} = 2.0 \ \ \Omega,$$
$$R_o = 2.5 \ \Omega, \ \ X_o = 1.0 \ \Omega, \ \ R_i = 0.5 \ \Omega, \ \ X_i = 5.0 \ \Omega$$

Calculate torque when machine is running at slip 0.05.

Solution

The synchronous speed is

$$\omega_s = 2\pi \frac{120f}{p} = 157 \text{ Rad/s}$$

The equivalent circuit of double cage induction motor is shown in Fig. 5.79.

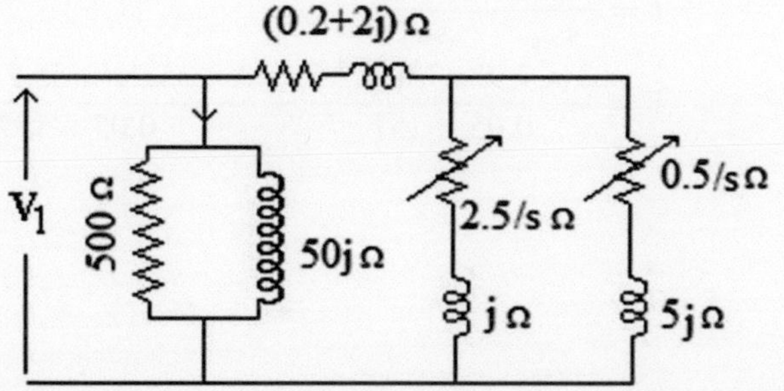

Fig. 5.79 Equivalent circuit of double cage induction motor for Example 5.28

Ignoring the no-load current, outer cage impedance at 0.05 slip can be given as

$$Z_o = \frac{2.5}{0.05} + j = 50 + j = 50.01\angle 3.5 \times 10^{-4}\ \Omega$$

Inner cage impedance at .05 slip is given as

$$Z_i = \frac{0.5}{0.05} + 5j = 10 + 5j = 11.18\angle 26.56^\circ\ \Omega$$

The total impedance,

$$\begin{aligned} Z &= (0.2 + 2j) + \frac{50.01 \times 11.18\angle 26.56^\circ}{10 + 5j + 50 + j} \\ &= (0.2 + 2j) + \frac{559.11\angle 26.56^\circ}{60.3\angle 5.71^\circ} = (0.2 + 2j) + 9.27\angle 20.85^\circ \\ &= 8.86 + 5.3j = 10.32\angle 30.9^\circ\ \Omega \end{aligned}$$

Thus current in the cage windings is

$$\mathbf{I} = \frac{440}{10.32\angle 30.9^\circ} = 42.2\angle -30.9^\circ\ \text{A}$$

The current in the outer cage is

$$\mathbf{I}_o = \frac{\mathbf{I}(Z_o||Z_i)}{Z_o} = \frac{(42.2\angle -30.9^\circ)(9.27\angle 20.85^\circ)}{50.01\angle 0^\circ} = 7.9\angle -10.05^\circ\ \text{A}$$

The current in the inner cage is

$$\mathbf{I}_i = \frac{\mathbf{I}(Z_o||Z_i)}{Z_i} = \frac{(42.2\angle -30.9^\circ)(9.27\angle 20.85^\circ)}{11.18\angle 26.56^\circ} = 35.21\angle -36.61^\circ\ \text{A}$$

The torque at any slip is given as

$$\mathrm{T} = \frac{3\mathrm{I}_o^2\mathrm{R}_o}{s\omega_s} + \frac{3\mathrm{I}_i^2\mathrm{R}_i}{s\omega_s}$$

$$\mathrm{T} = \frac{3 \times 7.9 \times 7.9 \times 2.5}{0.05 \times 157} + \frac{3 \times 35.21 \times 35.21 \times 0.5}{0.05 \times 157} = 296.52\ \text{Nm}$$

5.17 Single-Phase Induction Motor

Single-phase induction motors are used in many applications in household, shops, offices and farms, where three phase supply is not available. These motors have built in fractional horse power range or a few horse power. Single-phase induction motor consists of single-phase stator winding and a squirrel cage rotor. Sometimes, the rotor is a die-cast aluminum rotor. In recent days, the use of copper die-cast rotor is also being used in energy efficient motors. The single-phase AC voltage applied to the stator windings creates a field distribution. The single phase air-gap flux is characterized by being fixed in space and alternating in magnitude. If hysteresis is neglected, the flux is maximum when the instantaneous current is maximum and zero when the instantaneous current is zero. Such an arrangement gives a zero starting torque.

Consider a three-phase induction motor running at light load. If one of the supply line is disconnected, the motor will continue to run, although at slightly different speed. The noise produced is because of large number of space harmonics introduced in flux density distribution in space due to single phasing. Since one of the line is disconnected, the three-phase winding on a single-phase supply gives improper conductor distributions. Such an operation of a three-phase induction motor may be considered as the operation of a single-phase induction motor. A single-phase induction motor has a proper distribution of conductors in the slots.

Split phase starting of a three-phase induction motor is possible. The torque-speed characteristics of such motors are similar to a single-phase induction motor. This has only historical importance.

The alternating magnetic field in single-phase induction motor is of the form

$$B(\theta, t) = B_m \cos\ \theta\ \ \sin\ \omega t$$

It can be written as

$$B(\theta, t) = \frac{1}{2} B_m \sin(\omega t - \theta) + \frac{1}{2} B_m \sin(\omega t + \theta) \tag{5.101}$$

Here, the first term is forward rotating magnetic field, whereas the second term corresponds to backward rotating magnetic field. The theory based on such resolution of an alternating magnetic field into two counter-rotating magnetic field is known as the double-revolving field theory. The direction of rotation of forward field is considered to be the same as that of rotor. Thus, if the rotor runs at N-RPM and N_s is the synchronous speed in RPM, the slip s_f of the rotor with respect to the forward rotating magnetic field is same as s, thus

$$s_f = s = \frac{N_s - N}{N_s} = 1 - \frac{N}{N_s} \tag{5.102}$$

The back ward slip s_b, of the rotor with respect to the backward rotating magnetic field is given by

$$s_b = \frac{N_s - (-N)}{N_s} = 1 + \frac{N}{N_s} = 2 - s_f = 2 - s \tag{5.103}$$

We know from the operation of poly-phase motors that for $N < N_s$, the slip corresponds to the motoring, operation and N < 0 corresponds to braking operation. The two resulting torques have an opposite influence on the rotor. Thus, the resultant torque of a single-phase induction motor can be written as

$$T = \frac{I_f^2 R_2^/}{s\omega_s} - \frac{I_b^2 R_2^/}{(2-s)\omega_s} \tag{5.104}$$

where I_f is the forward current and I_b is the backward current as shown in the equivalent circuit in Fig. 5.80.

A single-phase induction motor does not have starting torque. An induction motor having single-phase winding is treated as two identical 3-phase induction motors connected to the same shaft, but running in opposite direction. The characteristic of such a machine is shown in Fig. 5.81. The induction motor does not have any starting torque, therefore, these motors are not self-starting. If a mechanical jerk is given in one of the direction, the motor starts running in the same direction. We already know that a single-phase winding will give rise to two components of rotating magnetic field rotating in opposite directions at synchronous speed.

A three-phase induction motor gives rise to steady torque, but a single-phase induction motor produces a pulsating torque which gives rise to noise and vibration.

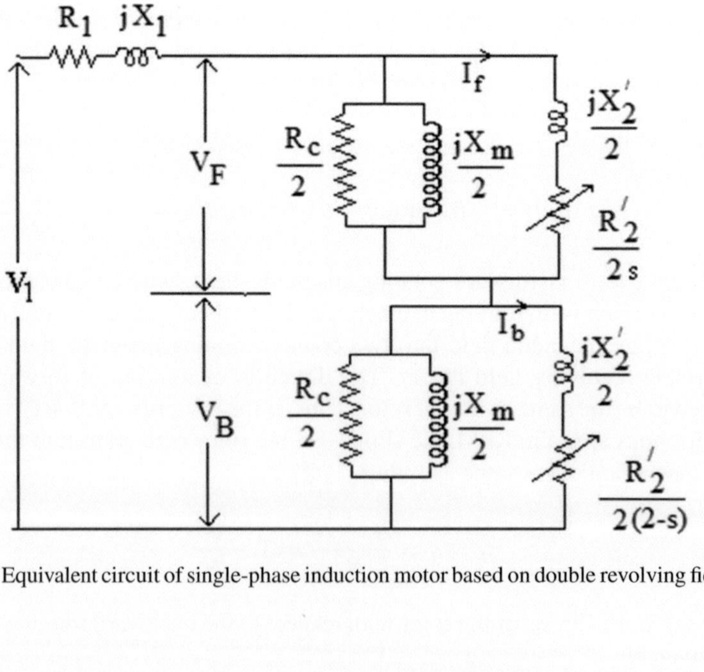

Fig. 5.80 Equivalent circuit of single-phase induction motor based on double revolving field theory

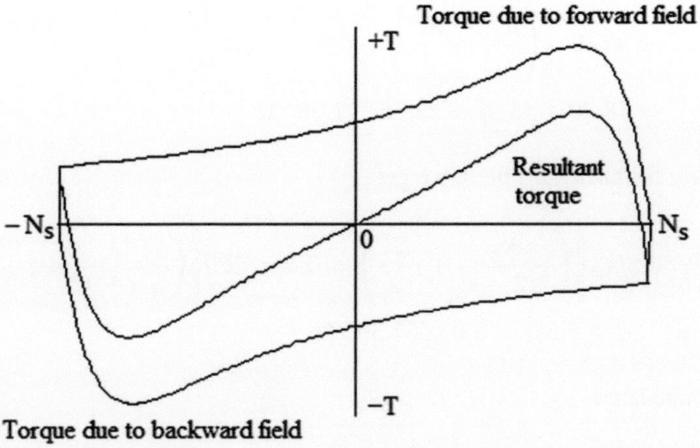

Fig. 5.81 Torque-speed characteristics of single-phase induction motor-torque due to forward and backward rotating magnetic fields

A portion of torque is produced due to double frequency rotor currents with the backward field components. Thus this part of torque pulsates at double frequency.

5.17.1 Double Revolving Field Theory

The single-phase induction motors do not have natural starting torque. A single-phase induction motor may be regarded as two three-phase induction motors connected to the shaft but running in opposite directions. The torque-speed characteristic of such motor is similar to that of an induction motor with single-phase windings. It exhibits zero starting torque. The equivalent circuit based on double revolving theory of single-phase induction motor has been given in Fig. 5.80. The characteristics of a single-phase induction motor may be obtained using this equivalent circuit.

Example 5.29 A 220 V, 50 Hz, 2-pole single-phase induction motor has following equivalent circuit parameters:

$$R_1 = 10\ \Omega \quad X_1 = 12\ \ \Omega \quad X_m = 125\ \Omega \quad R_2^/ = 11\ \Omega \quad X_2^/ = 12\ \Omega$$

Calculate input current and power factor and torque when machine is running at slip 0.03.

Solution

From the given data, the impedance offered by forward field,

$$Z_f = 0.5jX_m \left\| \left(\frac{R_2^/}{2s} + j\frac{X_2^/}{2} \right) = 0.5 \times 125j \right\| \left(\frac{11}{2 \times 0.03} + j\frac{12}{2} \right)$$
$$= 18.70 + 55.5j = 58.56\angle 71.38^\circ \ \Omega$$

Similarly for backward field, we get

$$Z_b = 0.5jX_m \left\| \left(\frac{R_2^/}{2(2-s)} + j\frac{X_2^/}{2} \right) = 0.5 \times 125j \right\| \left(\frac{11}{2 \times 1.97} + j\frac{12}{2} \right)$$
$$= 2.32 + 5.57j = 6.03\angle 67.39^\circ \ \Omega$$

Total impedance

$$Z = 10 + 12.0j + 18.70 + 55.5j + 2.32 + 5.57j = 31.02 + 73.08j = 79.4\angle 67^\circ \ \Omega$$

$$\text{Input current } I = \frac{V}{Z} = \frac{220}{79.4\angle 67^\circ} = 1.08 - 2.55j = 2.76\angle -67^\circ \quad \text{A}$$

Power factor $\cos 67^\circ = 0.39$ lagging
Since slip is small (0.03), the forward voltage is given as

$$V_f = IZ_f = 2.76 \times 58.56 = 161.62 \text{ V}$$

The backward voltage $V_b = IZ_b = 2.76 \times 6.03 = 16.64\,\text{V}$
In order to avoid long calculations, it is considered that forward and backward current are $\mathrm{I_f} = \frac{\mathrm{V_f}}{0.5\mathrm{R}_2^/ / \mathrm{s}} = \frac{161.62}{(0.5 \times 11/0.03)} = 0.88 \text{ A}$

$$\mathrm{I_b} = \frac{\mathrm{V_b}}{0.5\mathrm{R}_2^/ /(2-\mathrm{s})} = \frac{16.64}{(0.5 \times 11/(2-0.03))} = 5.96 \text{ A}$$

The developed torque

$$T = \frac{1}{\omega_s}\left(\frac{0.5R_2^/}{s} I_f^2 - \frac{0.5R_2^/}{2-s} I_b^2 \right) \approx \frac{1}{2\pi \times 50}\left(\frac{0.5 \times 11}{0.03} 0.88^2 - \frac{0.5 \times 11}{2-0.03} 5.96^2 \right)$$
$$\simeq (141.97 - 99.17)\frac{1}{100\pi} = 0.130 \text{ Nm}$$

Example 5.30 The open-circuit and short-circuit tests are conducted on a 220 V, 50 Hz, four-pole single-phase induction motor. The test data are:
No-load test: 220 V, 1.5 A, 60-W
Locked rotor test: 200 V, 5.0 A, 900-W

Resistance of main primary winding is 12.0-Ω. Determine the equivalent circuit parameters of induction motor. When motor is running at 7% slip, determine current and power factor.

Solution:

From the locked rotor test, we have

$$Z = \frac{200}{5} = 40\ \Omega$$
$$R = \frac{900}{5^2} = 36\ \Omega$$
$$X = \sqrt{(40^2 - 36^2)} = 17.44\ \Omega$$

Choosing the leakage reactance of stator and rotor equal, we get

$$X_{L1} = X_{L2}^{/} = \frac{17.44}{2} = 8.72\ \Omega$$

The equivalent resistance of rotor referred to stator

$$R_2^{/} = R - R_1 = 36 - 12 = 24\ \Omega$$

Thus, we have

$$\frac{R_2^{/}}{2} = 12\ \Omega$$

From no load test, we have

$$Z = \frac{220}{1.5} = 146.67\ \Omega$$
$$R_c = \frac{60}{1.5^2} = 26.67\ \Omega$$
$$X_0 = \sqrt{(146.67^2 - 26.67^2)} = 144.22\ \Omega$$
$$W_{Loss} = W_0 - I_0^2\left(R_1 + R_2^{/}\right)$$
$$W_{Loss} = 60 - 1.5^2(12 + 12) = 6\ W$$

Ignoring R_m in Fig. 5.80, we get

$$\frac{X_m}{2} = X_0 - X_{L1} - \frac{X_{L2}^{/}}{2}$$
$$\frac{X_m}{2} = 144.22 - 8.72 - \frac{8.72}{2} = 131.14\ \Omega$$

Ignoring R_c in the equivalent circuit in Fig. 5.80, for a slip of 0.07, we get

$$R2F = \frac{R_2^/}{2s} + j\frac{X_{L2}^/}{2} = \frac{24}{2 \times 0.07} + j\frac{8.72}{2} = 171.43 + j4.36 = 171.49\angle 1.45^\circ\ \Omega$$

The total forward impedance is

$$Z_f = j\frac{X_m}{2} \| R2F = \frac{(131.14\angle 90^\circ)(171.45\angle 1.45^\circ)}{171.43 + j\ 4.36 + 131.14j}$$
$$= \frac{(131.14\angle 90^\circ)(171.45\angle 1.45^\circ)}{218.5\angle 38.32^\circ} = 102.9\angle 53.13^\circ = 61.74 + j\ 82.32\ \Omega$$

For the back ward flux, slip s is replaced by $2 - s = 2 - 0.07 = 1.93$. Thus, we get

$$R2B = \frac{R_2^/}{2(2-s)} + j\frac{X_{L2}^/}{2} = \frac{24}{2 \times 1.93} + j\frac{8.72}{2} = 6.22 + j4.36 = 7.59\angle 35.02^\circ\ \Omega$$

The total backward impedance

$$Z_b = j\frac{X_m}{2} \| R2B = 5.81 + 4.49j = 7.34\angle 37.7^\circ\ \Omega$$

Total impedance of motor is

$$Z = R_1 + jX_1 + Z_f + Z_b$$
$$= 12 + 8.72j + 61.74 + 82.32j + 5.81 + 4.49j$$
$$= 79.56 + 95.54j = 124.33\angle 50.21^\circ\ \Omega$$

The motor current is

$$I = \frac{V}{Z} = \frac{220}{124.33\angle 50.21^\circ} = 1.77\ A \quad at \quad 0.64 \quad \text{lag}$$

Example 5.31 The open-circuit and short-circuit test were conducted on a 220 V, 50 Hz, four-pole single-phase induction motor. The test data are:

No-load test: 220 V, 6.5 A, 360-W

Locked rotor test: 125 V, 15.0 A, 600-W

The resistance of primary main winding is 1.2-Ω. Determine the equivalent circuit parameters of induction motor. Estimate the current, power factor, forward and backward torque and net torque at slip 0.05.

Solution

From the locked-rotor test, we have

$$Z_{LR} = \frac{125}{15} = 8.33\ \Omega$$
$$R_{LR} = \frac{600}{15^2} = 2.67\ \Omega$$
$$X_{LR} = \sqrt{\left(8.33^2 - 2.67^2\right)} = 7.89\ \Omega$$

Choosing the leakage reactance of stator and rotor equal, we get

$$X_{L1} = X_{L2}^{/} = \frac{7.89}{2} = 3.95\ \Omega$$

The equivalent resistance of rotor referred to stator.

$$R_2^{/} = R_{LR} - R_1 = 2.67 - 1.2 = 1.47\ \Omega$$

Thus $\frac{R_2^{/}}{2} = 0.735\ \Omega$
From no-load test, we have

$$Z_0 = \frac{220}{6.5} = 33.85\ \Omega$$
$$R_0 = \frac{360}{6.5^2} = 8.52\ \Omega$$
$$X_0 = \sqrt{\left(33.85^2 - 8.52^2\right)} = 32.76\ \Omega$$
$$W_{Loss} = W_0 - I_0^2\left(R_1 + R_2^{/}\right)$$
$$W_{Loss} = 360 - 6.5^2 \times (2.67) = 247.19\ \text{W}$$

Ignoring $\frac{R_m}{2}$ in the equivalent circuit, shown in Fig. 5.80 we get

$$\frac{X_m}{2} = X_0 - X_{L1} - \frac{X_{L2}^{/}}{2}$$
$$\frac{X_m}{2} = 32.76 - 3.95 - \frac{3.95}{2} = 26.835\ \Omega$$

Ignoring R_m in the equivalent circuit in Fig. 5.80, we get for a slip of 0.05

$$\frac{R_2^{/}}{2s} + j\frac{X_{L2}^{/}}{2} = \frac{1.47}{2 \times 0.05} + j\frac{3.95}{2} = 14.7 + j1.975 = 14.83\angle 7.65^\circ\ \Omega$$

The total forward impedance is

$$Z_f = j\frac{X_m}{2} \left\| \left(\frac{R_2^{/}}{2s)} + j\frac{X_{L2}^{/}}{2}\right) \right. = \frac{(26.835\angle 90^\circ)(14.83\angle 7.65^\circ)}{14.7 + j\ 1.975 + 26.835\ j}$$

$$= \frac{(26.835\angle 90^\circ)(14.83\angle 7.65^\circ)}{32.34\angle 62.97^\circ} = 12.31\angle 34.68^\circ = 10.103 + 6.98j\ \Omega$$

For the back ward flux slip s is replaced by 2−s = 2−0.05 = 1.95

$$\frac{R_2^/}{2(2-s)} + j\frac{X_{L2}^/}{2} = \frac{1.47}{2 \times 1.95} + j\frac{3.95}{2} = 0.377 + j1.975 = 2.01\angle 79.19^\circ\ \Omega$$

The total backward-impedance

$$Z_b = j\frac{X_m}{2}\left\|\left(\frac{R_2^/}{2(2-s)} + j\frac{X_{L2}^/}{2}\right) = \frac{(26.835\angle 90^\circ)(2.01\angle 79.19^\circ)}{0.377 + j\ 1.975 + 26.835\ j}\right.$$

$$= \frac{(26.835\angle 90^\circ)(2.01\angle 79.19^\circ)}{26.81\angle 89.25^\circ} = 2.011\angle 79.94^\circ = 0.35 + 1.98j\ \Omega$$

Total impedance of motor is

$$\begin{aligned} Z &= R_1 + jX_1 + Z_f + Z_b \\ &= 1.2 + 3.95\ j + 10.103 + 6.98\ j + 0.35 + 1.98\ j \\ &= 11.673 + 12.77j = 17.27\angle 47.60^\circ\ \Omega \end{aligned}$$

The motor current

$$I = \frac{V}{Z} = \frac{220}{17.27} = 12.74\ \ A\ \ at\ \ 0.6743\ \ p.f.\ \ \text{lag}$$

Voltage across forward impedance,

$$\mathbf{V}_f = \mathbf{I}Z_f = 12.74\angle -47.60 \times 12.31\angle 34.68^\circ = 156.83\angle -12.92^\circ\ \ \text{V}$$

Forward current producing torque,

$$\mathbf{I}_f = \frac{\mathbf{I}Z_f}{\left[\frac{R_2^/}{2s} + j\frac{X_{L2}^/}{2}\right]} = \frac{156.83\angle -12.92^\circ}{14.83\angle 7.65^\circ} = 10.58\angle -20.57^\circ\ \text{A}$$

Forward Torque $T_f = \frac{I_f^2 R_2^/}{2s} = \frac{10.58^2 \times 1.47}{2 \times 0.05} = 1643.9$ Synchronous Watts

Voltage across backward impedance,

$$\mathbf{V}_b = \mathbf{I}Z_b = 2.011\angle 79.94^\circ \times 12.74\angle -47.6^\circ = 23.83\angle 32.34^\circ\ \ \text{V}$$

Backward Current producing backward torque,

$$\mathbf{I}_b = \frac{\mathbf{I}Z_b}{\left[\frac{R_2^{/}}{2(2-s)} + j\frac{X_{L2}^{/}}{2}\right]} = \frac{23.83\angle 32.34^\circ}{2.01\angle 79.94^\circ} = 11.855\angle -47.6^\circ \quad \text{A}$$

Backward Torque is

$T_b = \frac{I_b^2 R_2^{/}}{2(2-s)} = \frac{11.855^2 \times 1.47}{2\times 1.95} = 52.97$ Synchronous Watts

Net torque $T = T_f - T_b = 1643.9 - 52.97 = 1590.93$ Synchronous Watts

5.17.2 *Starting of Single-Phase Induction Motor*

We have seen that a single-phase induction motor is not self-starting. These motors are started using certain techniques. The following are the methods of starting single-phase induction motor:

(a) Split phase motors- creation of rotating magnetic field
(b) Shaded pole induction motor
(c) Commutator and brushes—Universal motor and Repulsion motor.

These are described below:

(a) **Split-phase induction motor:** In split-phase motor, there are two windings which are displaced in space by 90° electrical. One of the winding is a main-winding and other winding is referred to as auxiliary-winding. The first MMF arises from the main-winding. The auxiliary winding (MMF) is having a current, which is not in phase with the main winding current. Due to this, the rotating magnetic field produced is somewhat elliptical in nature. Fig. 5.82 shows the fundamental MMF or fundamental flux density distribution in space due to two windings. Let the two flux densities be:

$$B_m(\theta, t) = B_m \cos\theta \, sin\, \omega t$$
$$B_a(\theta, t) = B_m \cos(90 - \theta)\sin(\omega t \pm \varphi)$$

Thus the net magnetic field is

$$B(\theta,\ t) = B_m \cos\theta \, sin\, \omega t + B_m \cos(90 - \theta)\sin(\omega t \pm \varphi) \tag{5.105}$$

Fig. 5.83 shows a typical stamping in single-phase induction motor (inner stator) of ceiling fan. The outer slots carry main winding and inner slots carry auxiliary windings.

The split phase induction motor has one additional winding which is displaced by 90° in space. The auxiliary winding carries a series resistance or capacitance so as to cause a phase difference between main winding and auxiliary winding currents as shown in Figs. 5.84 and 5.85, respectively. This results in elliptical rotating magnetic

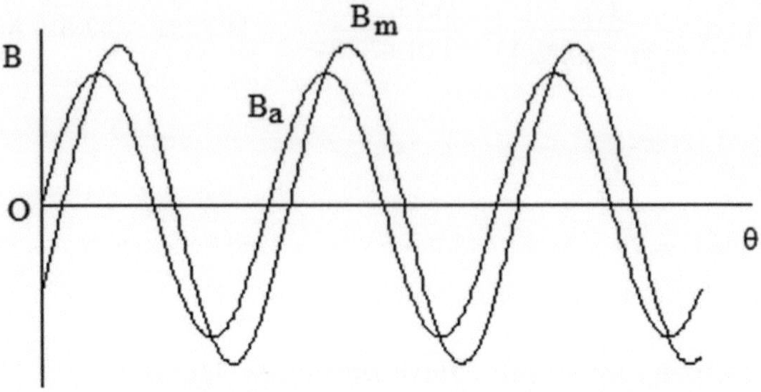

Fig. 5.82 Flux density distribution of main and auxiliary windings of single-phase induction motor at any time instant

Fig. 5.83 Slots in a single-phase induction motor-inner stator

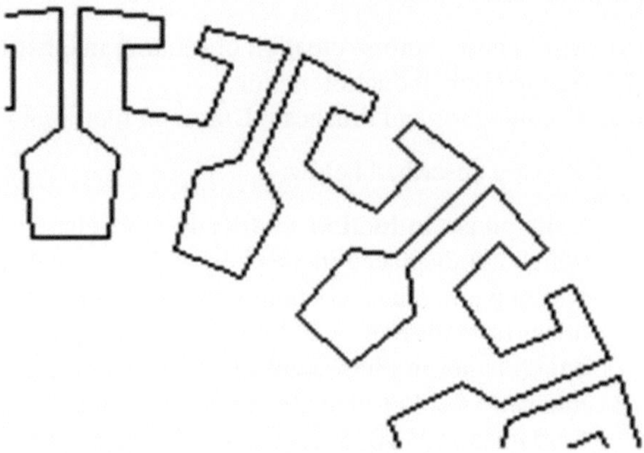

field resulting in net starting torque. Once the machine is started as motor, the auxiliary winding is either removed from the circuit by means of centrifugal switch mounted on the shaft or remains connected. The following are the types of split phase induction motors:

(a) Resistance split phase motor
(b) Capacitor start capacitor runs induction motor
(c) Capacitor start motor.

The use of inductor is avoided as it deteriorates the power factor. The split phase induction motors normally carry a centrifugal switch 'S' which disconnects the auxiliary winding when speed reaches near synchronous speed. The switch is automatically ON when machine is switch OFF.

The fractional horse power single-phase induction motors are mostly of capacitor start induction motor. The auxiliary winding is disconnected by the centrifugal switch, when the machine is fully started. The auxiliary winding is designed to

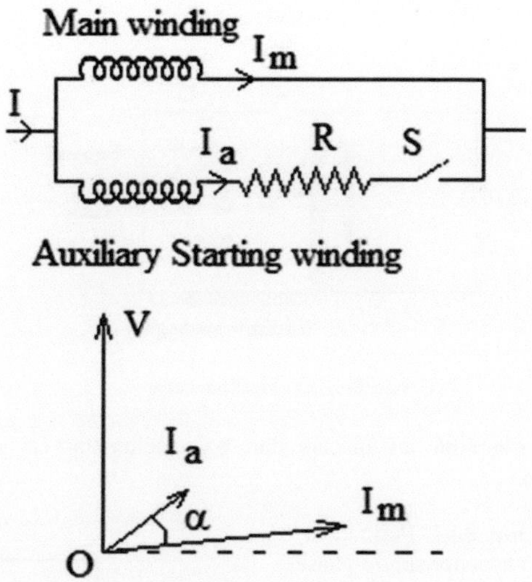

Fig. 5.84 Resistance split phase motor

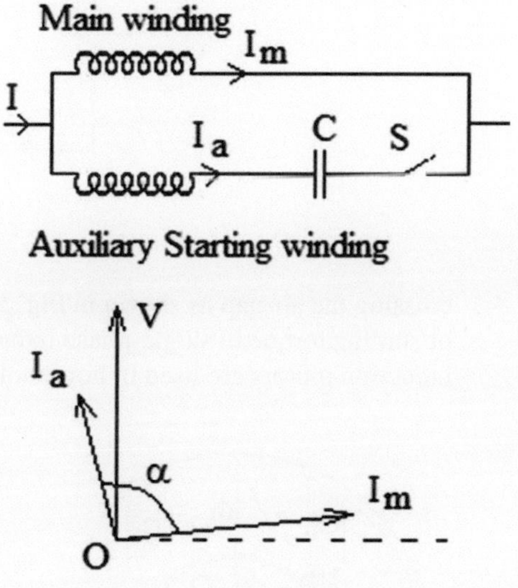

Fig. 5.85 Capacitor split phase motor

carry current during starting only. The capacitors are useful in improving the overall power factor of a machine. In capacitor start capacitor run induction motor, there is no centrifugal switch. The auxiliary winding remains connected to the supply and is designed accordingly. Capacitor start, capacitor run, and single-phase induction motors are shown in Figs. 5.86a, b and 5.87.

(b) **Shaded pole induction motor**: In case of shaded pole induction motor, a shading ring is provided in the stamping of magnetic core as shown in Fig. 5.88. The short-circuited copper ring creates a phase difference in the main stator flux

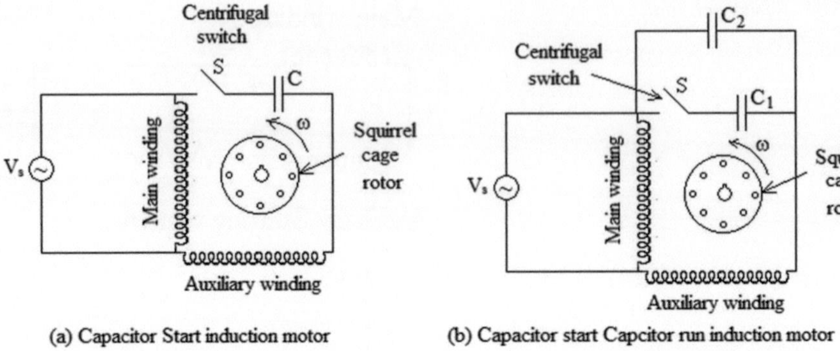

Fig. 5.86 **a** Capacitor start. **b** Capacitor start capacitor run single-phase induction motor

Fig. 5.87 Permanent capacitor single-phase induction motors

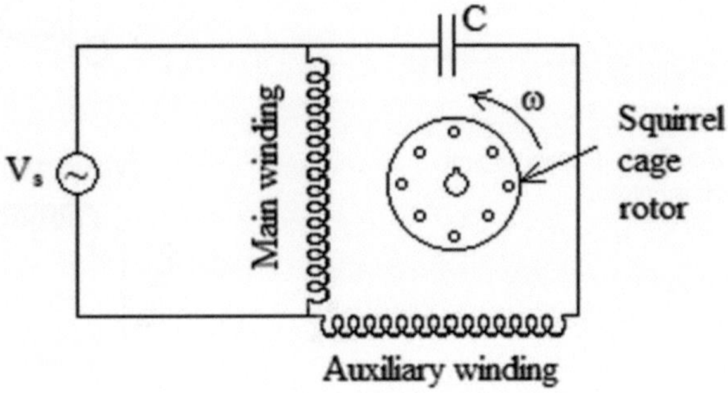

crossing the air-gap as shown in Fig. 5.88. This is responsible for development of starting torque in single-phase induction motor. Such fractional horse power induction motors are used in household appliances, such as fans.

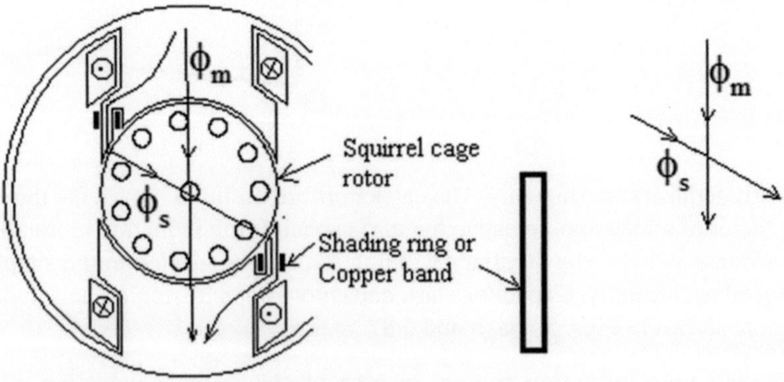

Fig. 5.88 Shaded pole single-phase induction motor

The Hysteresis motors and reluctance synchronous motors are constant speed motors are explained in Chap. 6. The repulsion motor and universal motor are commutator motors operating on single-phase or three-phase supply and have been explained in Chap. 7.

5.17.3 Single-Phase Induction Motor—Free Running (No-Load) and Blocked Rotor Test—Performance Analysis

The equivalent circuit parameters of single-phase induction motor may be approximately determined from no-load and blocked rotor tests, similar to the tests performed on polyphase induction motor. In the no-load or free-running test, the auxiliary winding is open due to operation of centrifugal switch. In capacitor run motor the auxiliary winding will remain connected. The frictional and windage losses can be determined using variable voltage no-load test or free running test. The power, speed, current and voltage are recorded. The extrapolation of curve between Power versus voltage for zero voltage yields the frictional and windage loss, similar to that of three-phase induction motor (refer Sect. 5.9). The circuit diagram for free running and locked rotor tests on a typical single-phase induction motor is shown in Fig. 5.89. Since the no-load slip is small, the equivalent circuit will take a form as shown in Fig. 5.90a The equivalent circuit under locked rotor test is shown in Fig. 5.90b.

Let the instruments reading during no-load test at rated voltage and blocked rotor test at rated current are as follows:

No-load or free-running test at rated voltage:

$$\text{Rated Voltage } V_0, \quad \text{No load current } I_0 \text{ and no load power } P_0$$

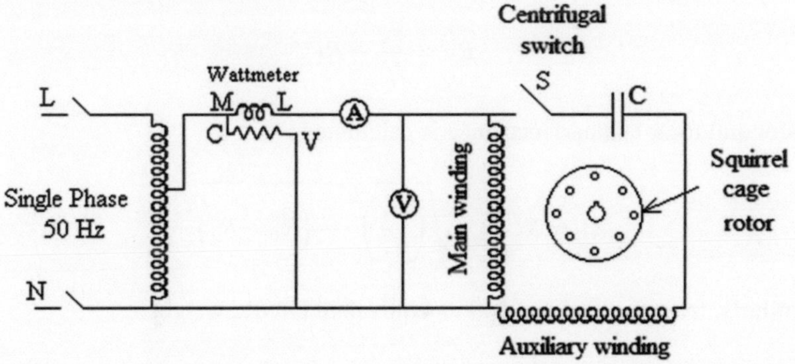

Fig. 5.89 Circuit diagram for Free running and locked rotor tests on a typical single-phase induction motor

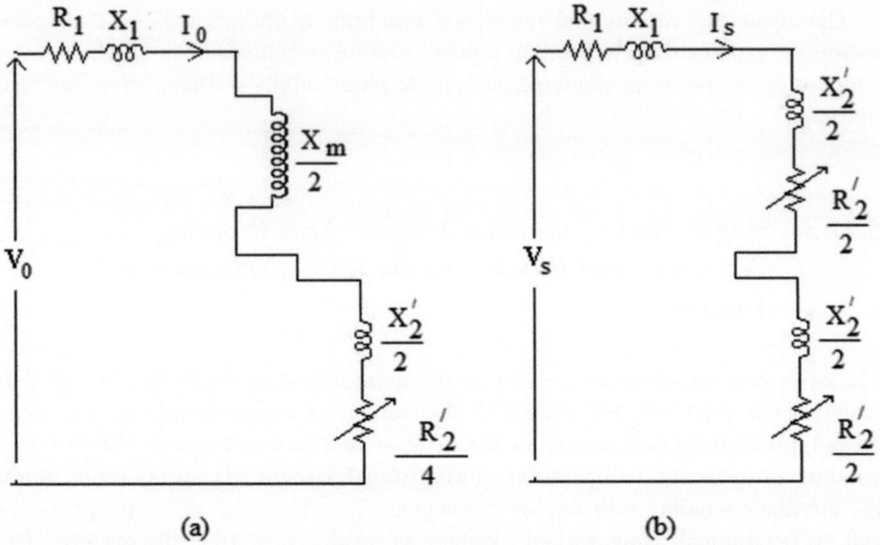

Fig. 5.90 Equivalent circuit of single-phase induction motor under **a** No-load or free running test. **b** Blocked rotor test

Blocked rotor test at rated stator current

Applied volatge V_s, Rated current I_s and input power P_s

From blocked rotor test, we get

$$R_1 + R_2^{/} = \frac{P_s}{I_s^2}$$

The stator resistance is known as such rotor resistance referred to stator is

$$R_2^{/} = \frac{P_s}{I_s^2} - R_1$$

Stator and rotor leakage reactance is calculated as

$$X_1 = X_2^{/} = \frac{1}{2}\sqrt{\left(\frac{V_s}{I_s}\right)^2 - \left(R_1 + R_2^{/}\right)^2}$$

Similarly, from no-load test and its equivalent circuit, we get

$$\overline{V}_0 = \overline{I}_0\left\{R_1 + \frac{R_2^{/}}{4} + j\left(X_1 + \frac{X_2^{/}}{2}\right) + j\frac{X_m}{2}\right\}$$

Taking voltage $\overline{V}_0$ as reference, the current $\overline{I}_0$ is written as

$$\overline{I}_0 = I_0 \angle -\theta_0$$

where the no-load power factor angle or phase angle is

$$\theta_0 = \cos^{-1}\left(\frac{P_0}{V_0 I_0}\right)$$

Solving we get expression for magnetizing reactance,

$$X_m = 2\left[\text{Imag}\left(\frac{\overline{V}_0}{\overline{I}_0}\right) - \left(X_1 + \frac{X_2^{/}}{2}\right)\right]$$

Here *Imag* denotes the imaginary part of quantity in bracket. Having determined the equivalent circuit parameters, one can calculate – forward voltage and current i.e., V_f and I_f, respectively, backward voltage and current i.e., V_b and I_b, respectively, at different slips 1.0 to 0.0. The forward and backward currents are assumed to flow through the rotor resistances $R_2^{/}/2s$ and $R_2^{/}/2(2-s)$, respectively. The torque can be calculated using expression (5.104) at different slips. Following curves- torque versus speed, current versus speed, power and power factor versus speed curves etc. can be plotted and compared with the experimentally obtained curves.

5.18 Synchronous Induction Motor

A three-phase slip-ring induction motor can be made to run as synchronous motor. Such motors are known as induction start synchronous motor. We know that a synchronous motor does not possess natural starting torque, but once started it is capable of developing torque at synchronous speed.

A slip-ring induction motor may be started as using star/delta starter on stator and also rotor resistance starter. This helps in limiting starting current of large capacity motors. Fig. 5.91 shows starting of slip-ring induction motor using rotor resistance starter. When speed is close to synchronous speed, the rotor terminals first short circuited and then connected to DC supply produced by the excitor or pilot generator. The machine runs as a synchronous motor and exhibits the characteristics as that of synchronous motor. In the event of failure of the excitor or pilot generator either due to its field failure or armature circuit failure or due to loss of residual magnetism of the field poles of excitor, the rely in the panel runs the machine as slip-ring induction motor. For example, in the event of excitor failure the three step rotor resistance starter automatically switches to step 2, and the machine runs as slip-ring induction motor. The panel of synchronous motor normally has starting control and protection of motor, which are based on either relay logic or PLC (Programmable logic controllers).

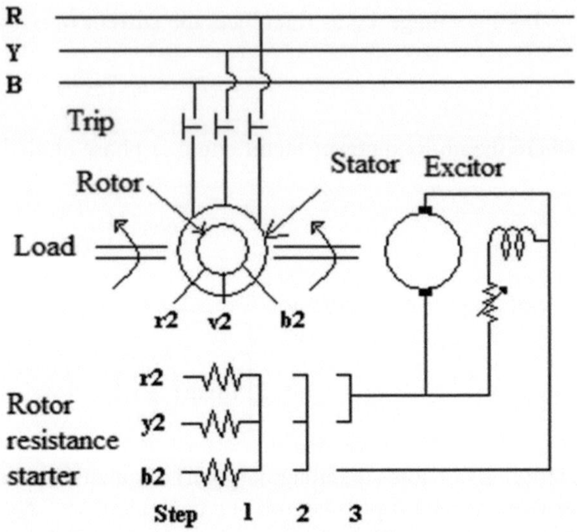

Fig. 5.91 Induction start synchronous motor

Problems

1. Calculate the number of slots in a three-phase induction motor for
   (a) 2-pole spp =2, 3, 4, 5
   (b) 16-poles spp=2.5, 3.5, 4.5, 5.5
2. Obtain phase allocation series for two pole three-phase induction motor having spp=4.5.
3. Develop MMF diagram of individual phases of a three-phase induction motor. Divide the MMF diagram when the winding is divided into two halves.
4. In a 36 slot, 4 pole induction machine, calculate the pitch factor if the winding is short pitched by one slot.
5. Calculate the distribution factor of a 72 slot, three-phase, eight pole winding.
6. What are the differences between time and space phasors?
7. Show that in a three-phase induction machine excited by three-phase balanced source, seventh-order space and time harmonics are running in forward direction.
8. Show that in a three-phase induction machine excited by three-phase balanced source, fifth-order space and time harmonics are running in backward direction.
9. Show that the following expression will yield reverse rotating MMF.

$$\mathbf{F} = F_{\max}[\cos\omega t \cdot \cos\theta + \cos(\omega t - 240^\circ)\cdot\cos(\theta - 120^\circ) + \cos(\omega t - 120^\circ)\cdot\cos(\theta - 240^\circ)]$$

10. Show the expression of rotating MMF of a three-phase winding excited by a two-phase source will result in forward and backward fields.

$$\mathbf{F} = F_{\max}\left[\cos\ \omega t \cdot \cos\theta + \cos(\omega t - 90^\circ)\cdot\cos(\theta - 120^\circ)\right]$$

11. Calculate the synchronous speed of 50 Hz, four pole induction motor. What is the synchronous speed of 5th- 7th-,11th- and 13th-order space and time harmonics?
12. A three-phase, 50 Hz, 400 V, four-pole induction motor is running at 1450 RPM. Determine the slip and frequency of rotor current.
13. A three-phase induction motor is running at 7% of slip. If the number of poles is two and supply frequency is 50 Hz. Find rotor's speed.
14. A two-pole, three-phase, 50 Hz induction motor develops a fault in stator winding, giving rise to forward and reverse rotating magnetic field of stator. Determine the speed of forward and reverse rotating magnetic fields with respect to stator and with respect to rotor. The rotor is assumed to run in forward direction at slip 5%.
15. A 4-pole, three-phase 50 Hz induction motor develops a fault in rotor winding, giving rise to forward and reverse rotating magnetic field of rotor. Determine the speed of forward and reverse rotating magnetic fields with respect to stator and with respect to rotor. The rotor is assumed to run in forward direction at slip 7%.
16. An eight pole, slip-ring induction motor has its stator excited by 50 Hz source. The rotor is connected to a three-phase source of 25 Hz. Determine the speed of rotor.
17. A four-pole, three-phase, 50 Hz induction motor develops a fault in rotor winding, giving rise to forward and reverse rotating magnetic field of rotor. Determine the speed of forward and reverse rotating magnetic fields with respect to rotor and with respect to stator. The rotor is assumed to run in forward direction at slip 50%.
18. Develop Thevenin's equivalent circuit of a three-phase induction motor.
19. A three-phase, 400 V, 50 Hz, star-connected induction motor has the following per phase equivalent circuit parameters:

$$R_1 = 0.25\ \Omega \quad R_2^{/} = 0.45\ \Omega \quad R_C = 2000\ \Omega$$
$$X_1 = 1.5\ \Omega \quad X_2^{/} = 1.7\ \ \Omega \quad X_m = 500\ \Omega$$

The transformation ratio is $a = 1$. Compare the initial starting current with the initial starting current during direct online starting. When (a) a per phase resistance of 1.5 Ω is added in series with the stator circuit, (b) a per phase rotor resistance of 1.5 Ω is added in series with the rotor circuit,

(c) applied voltage is reduced by 70%,
(d) rotor windings are connected in delta, and
(e) At standstill the motor is supplied by reduced voltage 150 V and reduced frequency of 15 Hz

20. A three-phase, six pole 100 hp, 440 V, star-connected, squirrel cage induction motor has following test data:
No-load test 440 V, 30 A, 4.5 KW
Blocked rotor test 90 V, 50 Hz, 120 A, 16.0 KW

Average DC resistance between the terminals is 0.125 Ω. Determine the equivalent circuit parameters. Also, estimate the current, power factor, torque, output power when motor is running at 3.5 % slip.

21. The following no-load and blocked-rotor tests are performed on a three-phase four-pole, 50 Hz, 20 hp, 440 V induction motor. The test data are

No-load test:	440 V	4.0 A	609 W	50Hz
Blocked rotor test at rated frequency	110 V	18 A	2750 W	50 Hz
	55 V	18 A	1550 W	15 Hz
DC test on stator per phase	10 V	15 A		

The friction and windage loss is 250 W. Determine the per phase equivalent circuit parameters.

22. Using the equivalent circuit parameters of Problem 21, determine the current, power factor, torque, output power when motor is running at 3.25 % slip. Use approximate Equivalent circuit.
23. A 220 V, 3-phase, 50 Hz, star-connected induction motor has a per phase stator resistance of 0.24 Ω. The no-load and blocked rotor test results are:
    Free running test: 220 V, 3.0 A, 650 W
    Blocked rotor test: 45 V, 18 A, 1200 W
    The frictional and windage losses are 350 W

    (a) Determine equivalent circuit parameters,
    (b) Calculate the current, power factor, rotor losses, developed power, torque and efficiency, when motor is running at 12.5% slip.
    (c) Calculate maximum torque and slip at maximum torque.

24. Draw circle diagram using data given in Problem 5.23. Determine current drawn by motor at 4.5 % slip using approximate equivalent circuit. From circle diagram, determine rotor losses, developed power, torque and efficiency for the calculated current.
25. A three-phase, twelve-poles, 1000 hp, 2200 V, 50 Hz induction motor has the following parameters per phase:

$$R_1 = 0.10\ \Omega\ ;\ X_1 = 0.55\ \Omega\ ; R_2 = 0.08\ \Omega\ ; X_2 = 0.30\ \Omega$$

The stator to rotor turns ratio is 1.45. The motor is operating at a slip of 2%. The machine develops a rated torque at rated stator current. The friction and windage loss are 5.5 KW. Determine the magnitude and phase of minimum value of injected rotor EMF per phase for changing the slip at 0.50 at constant torque.

Hint: For constant torque E_1 will be constant.

26. A three phase, 50 Hz four pole induction motor has rated speed of 1445 RPM. The rotor resistance per phase is 0.65 Ω. What rotor resistance must be added in series with the rotor circuit so that rated torque is developed at starting?

27. A three-phase slip-ring induction motor delivers rated torque and draws a line current of 50 A at a power factor of 0.8 lagging. The effective turn ratio is 1.40. The rotor resistance per phase is 0.09 Ω. Calculate the value of rotor injected EMF needed to increase the line power factor to 0.9 lagging. Find the power supplied by the source of injected voltage.
28. The rotor resistance per phase of a 50 Hz, 100 hp, 20 pole induction motor is 0.018 Ω. The full load slip of motor is 2.8%. It is assumed that the torque-slip characteristic is a straight line from no-load to full loads. The motor drives a fan load requiring 100 hp at full load speed of the motor. The load torque of fan is proportion to the square of speed. The slip power is converted into DC using three-phase bridge rectifiers. What value of rotor resistance must be connected across the output of rectifier so that the fan runs at 250 RPM?
29. Determine the ratio between the starting and rated torque of a six-pole, three-phase 50 Hz three-phase induction motor. The full-load slip is 3.5% and ratio of maximum to full load torque is 4.0.
30. A three-phase, six-pole, star-connected, 380 V, 50 Hz induction motor has following equivalent circuit parameters:

$$R_1 = 0.75\ \Omega\ ; X_1 = 2.4\ \Omega\ ; R_2' = 1.0\ \Omega\ ; X_2' = 2.8\ \Omega$$
$$R_c = 80\ \Omega\ ; X_m = 30\ \Omega\ ; s_{fl} = 0.05$$

Calculate:

(a) stator current and power factor
(b) active absorbed power
(c) mechanical power absorbed and electromagnetic torque
(d) the slip at maximum torque
(e) the starting torque and maximum torque.

31. A three-phase, 400 V, 50 Hz, four pole slip-ring induction motor has following parameters:

$$R_1 = 0.25\ \Omega\ ; X_1 = 0.66\ \Omega\ ; R_2' = 0.22\ \Omega\ ; X_2' = 0.66\ \Omega$$

The no-load current is 7.0 A and iron losses at rated voltage is 300 W. The total load torque is 70 N m. Calculate:

(a) the starting torque
(b) maximum torque at slip at which it occurs.
(c) the resistance must be added in each of the rotor phase for obtaining maximum starting torque

32 In Problem 31, three capacitors are connected in series with the rotor phases. Their values are chosen such that the rotor currents have no phase-shift with the induced EMF per phase at slip s=1.0. Determine the values of the capacitors and the starting torque.

33. A six-pole, 50 Hz three-phase induction motor has maximum torque 1000 Nm and the starting torque is 500 Nm. Estimate the torque and output power at slip 0.03.
34. A three phase, 440 V, 50 Hz slip-ring induction motor with four poles is running at 1445 RPM. The useful EMF developed in the rotor is 50 V and phase current is 50A. An EMF of 15 V is injected in each of the rotor phase opposite to the induced EMF. If the rotor leakage reactance is ignored and the load torque is kept constant, what is the new speed?
35. A three phase, four poles, star connected, 3 hp, 440 V, 50 Hz squirrel cage induction motor has the following parameters per phase:

$$R_1 = 2.5\ \Omega\ ;\ X_1 = 5.1\ \Omega\ ;\ R_2^{/} = 2.25\ \Omega\ ;\ X_2^{/} = 4.8\ \Omega$$
$$G_0 = 3.3 \times 10^{-3}\ \text{mho}\ ;\ B_0 = 0.0015\ \text{mho}\ ;\ s_{fl} = 3\%$$

Calculate:

(a) current and power factor
(b) slip at maximum torque
(c) the maximum developed electromagnetic torque
Assume exact equivalent circuit. How these results will change when stator connected in delta.

36. A 15 HP, 440 V, 50 Hz three-phase star-connected squirrel cage induction motor has an efficiency of 88% and a power factor is 0.88 lagging when the motor is running at full load. The starting current at full voltage is 5 times the full load current. The starting current is to be reduced to two times the full load current. Calculate

(a) value of resistance to be inserted in the phases
(b) % tapping of three-phase auto transformer.

37. A 50 hp, 400 Volt, 50 Hz, 10 pole squirrel cage motor has the following design parameters on per phase basis:

$$R_1 = 0.25\ \Omega\ ;\ X_1 = 1.12\ \Omega\ ;\ R_2^{/} = 0.29\ \Omega\ ;\ X_2^{/} = 1.12\ \Omega$$
$$R_c = 100\ \Omega\ ;\ X_m = 100\ \Omega\ ;$$

The motor is tested with free running and locked rotor conditions. Calculate the expected reading of the instruments in these two tests, the friction and windage loss is 1.0 KW.
38. Determine the starting current for motor in Problem 37, when it is started using star / delta starter. At what speed the machine should be connected to delta such that the current peak in the supply line is the initial starting current with the star connection.

39. Write a MATLAB program for plotting the torque-speed characteristics of the induction machine given in Problem 37 using (a) equivalent circuit (b) Circle diagram (use coordinate geometry) for slip 2.0 to 1.0.
40. The induction machine of Problem 37 is as induction generator and is connected to 400 V, 50 Hz. The machine is running at a speed of 690 RPM. Determine (a) the power in kW flowing into the supply system. (b) the reactive kVA supplied to the induction generator by the supply system.
41. The motor is supplied by a reduced voltage for starting, maintaining the ratio of voltage and frequency constant. Show that the current drawn by the motor remains almost constant.
42. The current in the rotor circuit on the basis of rotor equivalent circuit, is

$$I_2 = E_2/(R_2/s + jX_2)$$

Obtain the expression for

(a) critical slip
(b) Maximum torque
(c) condition for maximum starting torque
(d) maximum torque, $T_m \alpha \left(\frac{E_2}{f}\right)^2$, and
(e) torque, $T(s)$ as a function of critical slip.

43. Two 50 Hz, three-phase induction motors having poles four and six, are connected in cascade. Calculate the range of speed which can be obtained from the set.
44. A 400 V, 50 Hz, two-pole, star-connected double cage three-phase induction motor has following parameters:

$$R_o = 0.55\ \Omega \quad X_o = 0.25\ \Omega \quad R_i = 0.07\ \Omega \quad X_i = 1.0\ \Omega$$

Calculate the starting torque and running torque at 0.03 slip.
45. A 25 KW, 440 V, 50 Hz, four-pole, delta connected double cage induction motor has the following equivalent circuit parameters:

$$R_c = 550\ \Omega \quad X_m = 55\ \Omega \quad R_1 = 0.5\ \Omega \quad X_1 + X_{23} = 2.0\ \Omega$$
$$R_o = 2.5\ \Omega \quad X_o = 1.0\ \Omega \quad R_i = 0.5\ \Omega \quad X_i = 5.0\ \Omega$$

Calculate

(a) starting current, power factor and torque when machine is started using star / delta starter
(b) current and power factor, the torque and output power at slip 0.05.

46. A 440 V, 50 Hz, two-pole, star-connected double-cage three-phase induction motor has following parameters:

$$R_o = 1.0\ \Omega\quad X_o = 0.4\ \Omega\quad R_i = 0.07\ \Omega\quad X_i = 1.5\ \Omega$$

Calculate

(a) starting current, power factor and torque, when machine is started with 50% voltage
(b) current and power factor, the torque and output power, when induction motor is running at slip 0.035.

47. Write a MATLAB program for plotting the torque versus slip characteristics of double cage induction motor. Choose data of Problem 46.
48. A three-phase, four-pole, 440 V, 50 Hz induction motor has following parameters:

$$R_1 = 0.55\ \Omega\ R_2^/ = 0.45\ \Omega$$
$$X_1 = 2.5\ \Omega\quad X_2^/ = 2.9\ \Omega$$

Using approximate equivalent circuit, determine per phase impedance seen from the terminals at standstill. Estimate the starting current and starting torque when machine is connected in

(a) star or delta
(b) started using autotransformer with initial voltage reduced to 40%
(c) Voltage and frequency both are at 20% of their rated values.

49. The result of the no-load and blocked-rotor test on a three-phase, 50 Hz, six pole, 10HP, 440 V, star- connected induction motor are as follows:
No-load test: $V_{oc} = 440$ V, $W_{oc} = 900\ W$, $I_{oc} = 7.0$ A, Friction and windage loss 200 W
Blocked rotor test: $V_{sc} = 100\ V$, $W_{sc} = 1000$ W $I_{sc} = 15$ A
The per-phase DC stator resistance is 0.50 Ω. The per phase AC resistance is 30% greater. Using approximate equivalent circuit determine current and torque at 5 % slip. Draw circle diagram and determine performance of motor for the same current. Why there is difference in the two results? Justify
50. Two 50 Hz, three-phase induction motors having poles two and six are connected in cascade. Calculate the range of speed which can be obtained from the set.
51. A 440 V, 50 Hz, two-pole, star connected, double cage, three-phase induction motor has following parameters: $R_o = 1.0\ \Omega$ $X_o = 0.4\ \Omega$ $R_i = 0.07\ \Omega$ $X_i = 1.5\ \Omega$
What is the starting and running currents and power factors? Calculate the starting torque and running torque when machine is running at 0.05 slip.
52. A four pole, 30 hp, three phase, 400 V, 50 Hz, star-connected induction motor has the following test results:
Free running test: 415 V, 8.9 A, 1300 W
Blocked rotor test: 52 V, 21.5 A 760 W

Frictional and windage loss is 400 W. Using exact equivalent circuit, determine the current and torque and power when machine is running at

(a) 1400 RPM
(b) 1600 RPM
(c) 100 RPM

53. The following no-load and blocked-rotor test data are available for a 220V, 50 Hz single-phase induction motor:
Blocked rotor test: 55 V, 6.7 A, 212 W
No-load test: 220 V, 4 A, 180 W
The stator resistance is 1.50 Ω and the rotational losses are 50 W. Determine
(a) the parameters of the equivalent circuit
(b) Core loss and core loss resistance
(c) If the machine is running at 5% slip, calculate current, power factor and torque.

54. A 0.5 HP, 220 V, 50 Hz, four pole single-phase induction motor has the following equivalent circuit parameters:

$$R_1 = 2.75\,\Omega \quad X_1 = 5.2\,\Omega \quad R_2^{/} = 2.2\,\Omega \quad X_2^{/} = 2.5\,\Omega \quad X_m = 100\,\Omega$$

The core loss is 25 W and the rotational losses are 45 W. Determine for slip 0.045

(a) line current and power factor
(b) torque
(c) output power
(d) efficiency

55. Write a MATLAB program for plotting the torque versus slip characteristics of single-phase induction motor. Choose data of Problem 54.
56. Show that in a resistance split phase induction motor,
(a) the condition for maximum starting torque is

$$r_a = (r_m + x_m)\frac{x_a}{x_m}$$

(b) Maximum torque per Ampere is

$$r_a = \frac{r_m x_a + z_m\sqrt{x_a^2 + x_a x_m}}{x_m}$$

where subscript 'a' is for auxiliary winding and subscript 'm' is for main winding.
57. Show that in a capacitor split phase induction motor,

(a) the condition for maximum starting torque is

$$x_c = \frac{r_a . r_m}{x_m + z_m} + x_a^{\prime}$$

(b) Maximum torque per Ampere is

$$x_c = \frac{-x_m r_a + z_m \sqrt{r_a^2 + r_a r_m}}{r_m} + x_a^{\prime}$$

where $x_a^{/}$ subscript is the inductive reactance of starting winding.

58. In a 50 Hz split phase induction motor following impedances at standstill have been obtained

    Main winding: $r_m = 2.2\,\Omega$, $x_m = 5.5\,\Omega$

    Auxiliary winding: $r_a = 3.5\,\Omega$, $x_a = 7.5\,\Omega$

    Determine the

    (a) resistance
    (b) capacitance in series with the auxiliary winding for maximum starting torque
    (c) Determine the ratio of the maximum starting torques of capacitor split phase and resistor split phase induction motor.

59. A 220 V, four pole, 50 Hz induction motor has following parameters at standstill conditions:

    Main winding: $r_m = 1.6\,\Omega$, $x_m = 4.7\,\Omega$

    Auxiliary winding: $r_a = 2.5\,\Omega$, $x_a = 6.25\,\Omega$

    Determine the value of capacitance required in series with the auxiliary winding to get

    (a) Maximum starting torque
    (b) Maximum starting torque per ampere of current.

60. A 220 V, four pole, 60 Hz induction motor has following parameters at standstill conditions:

    Main winding: $r_m = 1.1\,\Omega$, $x_m = 5.7\,\Omega$

    Auxiliary winding: $r_a = 2.7\,\Omega$, $x_a = 7.5\,\Omega$

    Determine the value of resistance required in series with the auxiliary winding to get

    (a) Maximum starting torque
    (b) Maximum starting torque per ampere of current.

Objective Type Questions:

1. The torque in induction motor at synchronous speed is
   (a) Breakdown torque
   (b) Maximum torque
   (c) Counter current torque
   (d) Zero
2. The power factor correction of small and medium capacity induction motor can be done using
   (a) Inductive bank and capacitive bank
   (b) Inductive bank
   (c) Capacitive bank
   (d) None
3. In a two pole, 50 Hz slip-ring induction motor, when one of the slip ring is disconnected, the motor will run at
   (a) Nearly 3000 RPM when the slip ring is disconnected when motor is already running
   (b) Nearly 1500 RPM when one of the slip ring is already disconnected before starting
   (c) Both (a) and (b)
   (d) Always less than 3000 RPM
4. Using slip power energy recovery scheme, it is possible to use induction machine as
   (a) Induction motor and line excited Induction generator
   (b) Induction motor and self-excited Induction generator
   (c) Induction motor and standalone Induction generator
   (d) Induction motor and both Line excited and self-excited Induction generator
5. The essential condition for the operation of self-excited induction generator is
   (a) Residual magnetism
   (b) Availability of three phase voltage
   (c) Large rotor resistance
   (d) Less rotor resistance
6. A four pole, 50 Hz three-phase slip-ring induction motor exhibits a no-load speed of around 750 RPM. Which of the following is the correct choice?
   (a) There is a single phasing on stator side
   (b) There is single phasing on rotor side
   (c) This is a normal operation
   (d) None

7. A three-phase star-connected induction motor is started on a single-phase supply using split resistance technique.
   (a) All the three terminals will exhibit balanced three phase voltage.
   (b) All the three terminals will exhibit unbalanced three phase voltage.
   (c) The open terminal not connected to supply will have zero induced voltage.
   (d) The terminals connected to single-phase supply will give single-phase voltage, between phase and neutral.
8. Single-phase to three phase conversion is done using
   (a) Induction generator
   (b) Arno Converter
   (c) Scott connected transformer
   (d) T-T connected transformer
9. The hysteresis and eddy current losses in rotor of induction motor can be considered to be negligible, because
   (a) Rotor frequency is zero
   (b) Rotor frequency is low as such core losses can be ignored for all cases.
   (c) Rotor frequency is high during starting period as such core losses are present during starting period. The core losses can be ignored only when motor is running near about synchronous speed.
   (d) None
10. An induction motor has its marked stator terminals as A1 A2, $B1-B2$, $C1-C2$. For running the motor in star connection. Terminals $A2-B1-C2$ are made neutral or star point and supply is given from A1, B2, C1. The induction motor will
    (a) Not run
    (b) It will run but this is an unbalanced operation.
    (c) It will run, this is a perfectly balanced operation.
    (d) Runs at very high speed because number of poles will change.
11. In an application of induction motor in which fly wheel is also provided on the shaft. It is essential that
    (a) The speed should essentially increase when motor is loaded and speed should essentially decrease when motor is running on light load.
    (b) Constant speed irrespective of loading
    (c) The speed should essentially decrease when motor is loaded and speed should essentially increase when motor is running on light load.
    (d) None
12. Two three-phase induction motors having p_1 and p_2 poles are connected in cascade, the set can be used as
    (a) Induction motor

(b) Induction generator
(c) Induction motor and generator
(d) One machine as induction motor, the other as induction generator

References

1. Philip L. Alger, Induction Machines, Gordon & Breach Science Pub. N.Y. 2nd Ed. 1965
2. M.G.Say, Alternating current Machines, 4th ELBS edition, 1976
3. E. Openshaw Taylor: The Performance and design of A.C. commutator motors, Wheeler Publications, India, 11th Print, 1990.
4. A. Fransua, R. Magureanu, Electrical Machines and Drive Systems, TP Technical Press, Oxford, 1984
5. V Del Toro, Electric machines and power systems, PHI, 1980
6. S.A.Nasar, Electromagnetic energy conversion & Systems, Prentice-Hall, Inc., N.J., 1970
7. A.S.Langsdorf, Theory of alternating current machinery, TMH,2001
8. Stephen J Chapman, Electric Machinery Fundamentals, Indian Edition, 4e, Mcgraw hill education (Indi) Pvt Ltd, 2005
9. Siskind, Electrical Machines, 2nd edition International student edition, McGraw hill international book company, 12th printing, 1983

Chapter 6
Synchronous Machines

Abstract Constructional details of synchronous machines, generation of three-phase voltages; determination of induced EMF of alternator using general (ZPF), EMF, MMF, and ASA methods; determination of machines parameters and characteristics-OCC, SCC, ZPF, X_d & X_q; power angle diagram and torque angle diagram; synchronizing power and torque; V-curves; swing equation; equal area criterion; synchronous motor; synchronous condenser; hunting; sudden short circuit of alternator—sub-transient and transient currents; reluctance synchronous motor and hysteresis motor have been dealt.

Keywords Synchronous machines · Alternator · Regulation · Synchronous reactance · Leakage reactance · Salient pole · Cylindrical pole · OCC · SCC · ZPF · Power angle · Torque angle · V-curve · Power · Torque · Exciter

6.1 Introduction

Synchronous machines are doubly fed machines which have both AC and DC windings. These machines are classified as:

(a) Synchronous motor.
(b) Synchronous condenser.
(c) Synchronous generator (alternator).

Synchronous motors have large built-in capacities. They also have the capability of power factor correction and can change power factor from lagging to leading and vice versa. A synchronous motor is a constant speed drive, in which torque develops only at synchronous speed. As long as the supply frequency is constant, a synchronous motor continues to run at a constant steady-state speed. In the absence of damper bars, the natural starting torque is zero. As such, cylindrical rotor synchronous motor is not self-starting motor. Though, a salient pole synchronous motor develops starting torque due to saliency. Such a motor without any field winding is known as reluctance synchronous motor (RSM), which is also a constant speed motor. Earlier, for starting a cylindrical rotor synchronous motor, it was brought near to the synchronous

R. K. Srivastava, *Electrical Machines*, https://doi.org/10.1007/978-981-96-5039-2_6

speed with the help of some pony-motors before rotor being *pulled into synchronism.* Nowadays, part squirrel cage windings or damper bar provided on the rotor pole face produces induction torque for starting purpose. The synchronous motors are started using induction principle. At around 90–95% synchronous speed, the DC field winding is excited for synchronizing the rotor with the rotating magnetic field, developed by the armature windings. The DC field poles are locked with the rotating magnetic field. This procedure is also referred as rotor *pulled into synchronism.* If we compare it with induction motor, a three-phase induction motor has a finite starting torque but torque is zero at synchronous speed. An induction motor always exhibits a lagging power factor, which limits its use in high-power drive applications. Synchronous motors have the provision of controlling the power factor by overexcitation or under-excitation. Majority of motors used in large industries are induction motors. An overexcited synchronous motor at no load is run for overall power factor correction of a large industry, where several other lagging power factor loads such as induction motors and arc furnaces are also present. The machine is referred to as *synchronous condenser* [1, 2, 4–6].

The main bulk power generation is done through an alternator (synchronous generator), which converts mechanical power of prime mover into electrical power. The prime mover can be hydro-turbine, steam turbine, wind turbine, petrol/diesel/gas engine or electrical motor, etc. The alternator has a DC field winding and a three-phase (AC) distributed armature winding. One of the winding is rotated using a prime mover. This results in continuous variation of flux linkages in the three-phase distributed winding and, hence, results in three-phase-induced EMF in spatially distributed armature winding. The operation is based on the Faraday's law of induction. In an alternator, the armature current results in a counter torque, which opposes the torque provided by the prime mover. The capacity of these alternators varies from 1.0 kVA to 2000 MVA.[1]

In permanent magnet synchronous motor, (PMSM) and permanent magnet synchronous generator (PMSG), the DC flux is provided either by rare earth magnets, SmCo, or NdFeB magnets. The main application of PMSG is for power generation in renewable energy resources and main generator of aircraft. PMSG is also used in brushless excitation of main alternator and wind power generation. PMSM operating on Variable Frequency Drive (VFD) are used in variable speed drive applications. These PMSM and PMSG do not have carbon brushes. We will now discuss the salient features of alternators, which is also common for other synchronous machines.

[1] Nowadays, photo voltaic (PV) generation using solar cell is fully accepted technology for solar power generation. The initial cost is very high. The bulk power developed in solar farm is connected to the national grid using state-of-the-art inverter circuits. PV generation and fuel-cell-based power generation do not require rotating machines for electricity generation.

6.2 Alternator

Synchronous generators are used for generating AC power in hydropower stations as well as in thermal power stations based on coal, lignite, nuclear energy, wind power and geothermal energy etc. These machines are commercially known as *alternators*. The motive power to the alternators is provided from steam turbines, water turbine and wind turbine etc.

Thermal power stations use treated water and either coal or lignite for steam generation. Water is treated to prevent the formation of scales inside the metallic tubes carrying water for steam formation. Steam is also produced for running steam turbine in a nuclear power station, where nuclear reaction results in heat generation.

Hydro-turbines are categorized as follows:

(a) High head, Pelton turbine.
(b) Medium head, Francis turbine.
(c) Low head, Kaplan turbine.

Hydro-turbines require either the storage of water in large dams or a continuous stream of water for their operation. The cost of power generation is low, but the cost of installation and time required from its inception to actual operation is quite large. Though there is no pollution involved, but social factors and environmental factors are equally important.

The wind power generation also utilizes these alternators for power generations. Non conventional power generation, like wind-mills, micro-hydel and the gas, petrol or diesel-based power generation either in large scale or in portable form utilizes these machines for power generation. Apart from power generation, similar machines in modified form are used for high voltage pulse power generation in nuclear and defense applications. These machines are known as compulsator.

6.3 Constructional Details

A synchronous machine has AC armature windings similar to that of induction motor. Figure 6.1 shows a typical stator of a synchronous machine. The stator is made with electrical grade silicon steel laminations for limiting the core losses. The size/diameter of the stator of a salient pole machine used in hydropower generation is made large with several sectors of stampings. These sectors are again interleaved for minimizing the leakage flux.

The regularly distributed slots on the inner periphery of the stator carry balanced three-phase distributed windings. The large capacity synchronous machine windings are made with the help of hollow conductors. In each slot, several conductors are laid which are electrically connected in series or in parallel. Treated water is pumped through the hollow conductors for cooling purposes. In certain large capacity alternators, hydrogen cooling is also used. The balanced three-phase windings are connected

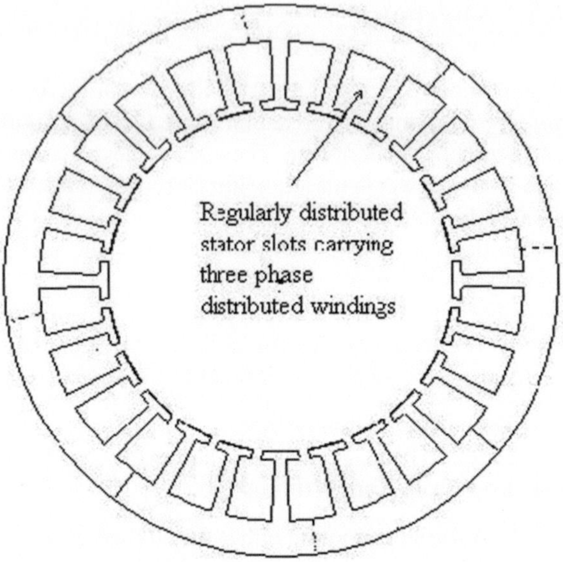

Fig. 6.1 Stamping of stator showing regularly distributed slots carrying spatially distributed three-phase winding (*windings not shown for clarity*)

either in lap or in wave winding. Each phase has multiple winding segments. They are joined together either in series or parallel. There can be many parallel paths in each of the phases. The design of windings has already been explained in Chap. 5.

The purpose of the field winding is to create even number of alternate North and South poles. In case of alternator, these alternate N–S poles are being rotated by prime mover. This changes the flux linkages of the armature conductors. As a result, three-phase voltages are induced in the stationary armature conductors. The field poles (rotor) of synchronous machines are mainly of the following two types:

(a) Salient poles.
(b) Cylindrical rotor.

Figure 6.2 shows a cylindrical rotor, made with a single piece of iron, in which slots are created by milling process for accommodating the DC field winding. The cylindrical rotor can be rotated at high speed without causing higher centrifugal forces. Therefore, cylindrical rotor construction is preferred for steam power generation and high-head hydrogenation. They are designed for rotational speed of 3000 or 1500 RPM for generating 50 Hz supply and are made with either two or four poles, respectively.

A salient pole rotor has been shown in Fig. 6.3. When a salient pole rotor, made with several pieces, is rotated at high speed, the high-centrifugal forces on the projected rotating mass attempts to tear it apart from the hub. The mechanical stresses are very high when a salient pole structure is rotated at high speed. These machines are used in low and medium head hydropower generation and are designed to operate at rotational speed of 40–1500 RPM. A salient pole alternator has large number of poles. Figure 6.4 shows a schematic representation of an alternator. The

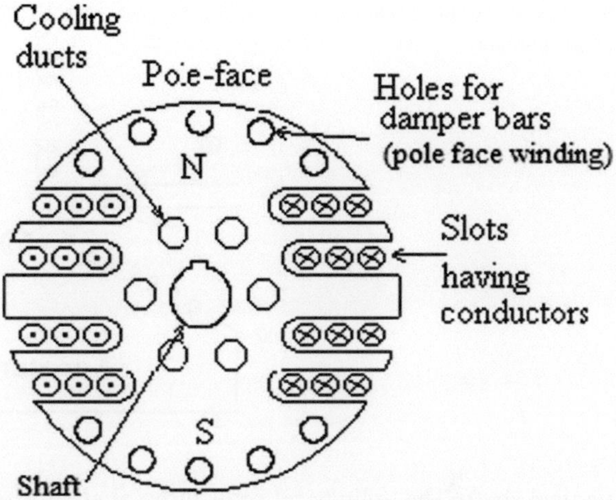

Fig. 6.2 Cylindrical rotor

preparation of stator with the help of several sectors of stampings causes the uneven leakage or fringe magnetic flux, which interacts with the shaft, leading to an induced EMF in the shaft. This gives rise to shaft currents. The shaft current flowing through the bearing may cause extensive damage to the bearings. These currents are limited by insulating their paths.

As mentioned earlier, the salient pole construction is preferred for hydro-generation. The machine runs at low speed; it requires larger number of poles on the rotor for 50 Hz power generation. For accommodating large number of poles and field windings, diameter of these machines is comparatively larger than that of cylindrical rotor machines. If D is the diameter and L is the length of machine, the output power is proportional to D^2L. A salient pole alternator has large diameter and lesser axial length. The larger diameter machines are normally constructed with a

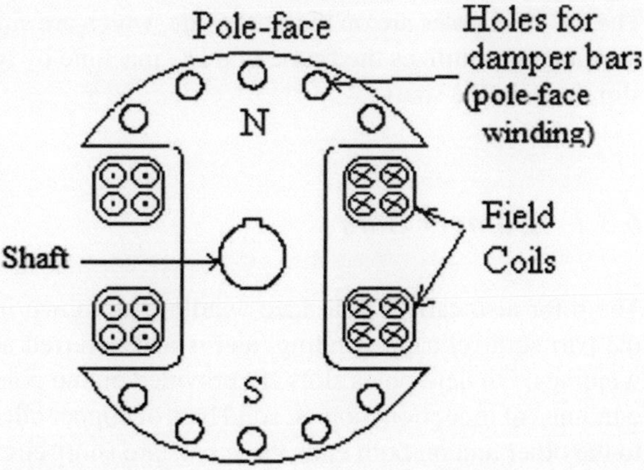

Fig. 6.3 Salient pole rotor

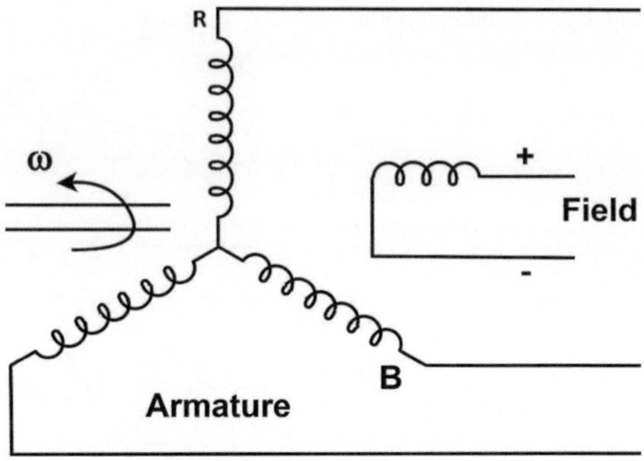

Fig. 6.4 Schematic diagram of an alternator

vertical shaft. The cylindrical rotor turbo alternator using steam turbines has large length. They require additional care to check the bending of the shaft.

In large-sized alternators, the armature winding is on stator and DC field winding is on the rotor side. The DC power is supplied to the rotating field windings with the help of following:

(a) Slip rings and brush arrangements.
(b) Brushless excitation.

In alternators and synchronous motors, when the AC winding is on the stator, the field winding is provided on the rotor. This requires an arrangement of two slip ring and two carbon brushes for feeding DC current to the rotating field winding. The DC field current in large-sized alternators and synchronous motors is provided using brushless excitation or contactless system.

In smaller capacity synchronous machines and laboratories, the three-phase AC armature winding is some time on the rotor side. For collection of AC power, three slip rings attached with the shaft and three stationary carbon brushes are provided. The DC field poles are on the stator side, which are similar to that of DC machine. A manufacturer utilizes the frame of a DC machine by removing commutator by three slip rings on the shaft.

6.3.1 Damper Bars

The rotor also carries pole face winding, as shown in Fig. 6.5, which is similar to the part squirrel cage winding, and is also referred as damper bars or *amortisseur* windings. On field poles slots are provided on the pole face. Similar to squirrel cage windings of induction motors, solid bars of copper alloy are hammered from one side to the other and on both ends these bars are short-circuited using either complete or part end rings. The winding is similar to part squirrel cage winding. These damper windings have many beneficial effects on the machine, as discussed below:

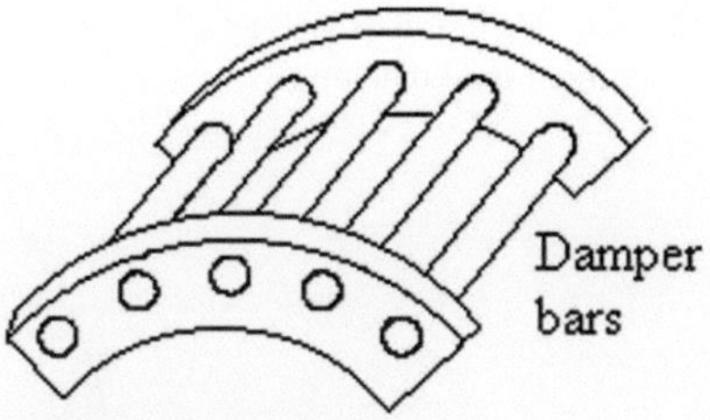

Fig. 6.5 Damper bar or part squirrel cage winding on pole face embedded with the rotor (rotor stampings are shown in Figs. 6.2 and 6.3)

(a) In synchronous motor, the damper bar or part squirrel cage winding provides starting torque. This avoids the use of auxiliary motor as pony drive for starting a synchronous motor.
(b) The synchronous motor and alternator are subjected to hunting phenomenon. In the absence of damper bars, the machine exhibits sustained oscillations in speed and current, which may pull the machine *out of synchronism.* Due to sudden variation of the load, a synchronous machine has a tendency to oscillate or hunt around the synchronous speed. The machine may lose synchronism, in case of large variation of load, and may ultimately stop operating. In the presence of damper bars, these oscillations in an alternator or in a motor are damped out. As the machine tries to accelerate beyond the synchronous speed, the induced currents in the damper bars produce a retarding torque. Similarly, as the machine decelerates below the synchronous speed, the induced currents in the damper bars produce a counter torque, which tries to oppose the deceleration. The induced current in damper bars flows only when certain disturbances in the supply or in the load are present. These currents try to damp out any hunting in the machine. The presence of damper bar helps restore the stability. The damper bar currents are also present in case of winding faults.
(c) Damper bars help in smoothing out the torque when the alternator is driven by reciprocating engines such as diesel or petrol engine.
(d) In cylindrical rotor alternators connected to long transmission lines, sometimes a characteristics phenomenon known as *sub-synchronous resonance* occurs. It results in low frequency mechanical oscillation of the shaft, which may cause extensive mechanical damage to a long shaft of turbo-alternator. The provision of damper bars in cylindrical rotor alternators prevents the damage due to *sub-synchronous resonance.*

6.3.2 Air-Gap Length

The air-gap length in large capacity alternators is kept large, for increasing the leakage reactance. This limits the maximum steady-state short-circuit current to about 2.0 per unit. The flux density distribution in air-gap due to the field poles must be as close

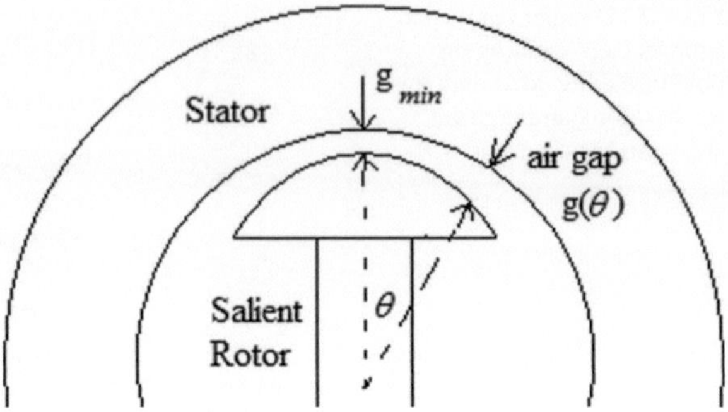

Fig. 6.6 Variation of air-gap in a salient pole machine

to a space sinusoid as possible. A sinusoidal-distributed magnetic flux in air-gap induces sinusoidal EMF in the armature winding. As shown in Fig. 6.6, the length of air-gap between stator and rotor of a salient pole machine is controlled during manufacturing process such that the gap length at any angle θ is given by

$$g(\theta) = \frac{g_{\min}}{\cos\theta} \tag{6.1}$$

where $g_{\min}$ is the minimum air-gap length under the pole face. The angle θ is the angle from the polar-axis in either direction. The scheme is used in salient pole machine.

In cylindrical rotor machine, the resultant air-gap flux contains several harmonics. Some of these harmonics in the induced voltages are minimized using short-pitched armature windings. For example, in a four-pole 24 slots machine, if the full-pitch coil is made by placing the coil sides in first to seventh slots and short pitching is done by placing the coil sides in first to sixth slots, the seventh-order harmonics would disappear from coil and phase voltage. This has been explained in Chap. 5.

6.3.3 Synchronous Speed

In a p-pole machine, the synchronous speed (in RPM) by an electrical source having frequency f is defined as

$$N_s = \frac{120f}{p} \tag{6.2}$$

The synchronous speed per second is

$$n_s = \frac{2f}{p}$$

In radians per second,

$$\omega_s = 2\pi\frac{2f}{p} = \frac{4\pi f}{p}$$

For a two-pole machine, the synchronous speed is $\omega_s = 2\pi f$ radians per second. The frequency of power generation of an alternator is dependent on the speed and number of poles. A six-pole alternator will generate 40–60 Hz frequency, if the speed is varied from 800 to 1200 RPM. If a rotor is rotating at speed of ω_r, the slip is given as

$$s = \frac{\omega_s - \omega_r}{\omega_s}$$

A synchronous motor is running at a constant speed of N_s, as long as the supply frequency is constant. The motor can be made to run at different synchronous speed by changing the supply frequency using solid state three-phase inverter. Under certain disturbances, the speed of synchronous machine oscillates or hunts near synchronous speed. The variation in the speed causes an induction of slip frequency EMF in the windings and damper bars.

6.3.4 Pilot Generator

In case of large capacity alternator and synchronous motor, the field winding requires large DC power. A pilot generation is required for supplying DC power to the field winding of alternator or synchronous machine. The pilot generator or exciter is connected to the same shaft. It is being done in following ways:

(a) Self-excited DC generator.
(b) Alternator.

The self-excited DC generator is normally lap wound for supplying DC power at low voltage high current. In many cases cascaded connected pilot generators are also used, in which a DC generator supplies power to the field winding of another DC generator connected to the same shaft.

In certain cases, a pilot alternator having stationary field winding and rotating three-phase armature is connected to the same shaft. The developed three-phase AC power is connected to three-phase bridge rectifier also rotating along with the shaft. The rectified DC voltage output is then connected to the rotating field winding of main machine.

Theoretically, it is possible to generate a small three-phase voltage due to residual magnetism in an alternator. The voltage may be rectified and however regulated using control circuit before connected to the field winding.

6.3.5 Brushless Excitation

For brushless excitation, the primary of exciter is stationary and is excited by high frequency source having frequency of 400 Hz. A distributed rotating winding is provided on the same shaft of alternator, which is similar to that of slip-ring induction motor. The main shaft is rotated by prime mover at a speed corresponding to number of poles and supply frequency of 50 Hz. The maximum shaft speed is 3000 RPM. The high frequency supply on the stator or exciter creates a rotating magnetic field, which cuts the secondary three-phase winding being rotated by the prime mover. It induces a slip frequency voltage on the secondary winding. For two-pole configuration, the slip is either 0.875 or (2.0 − 0.875). Thus the rotor-induced voltage and induced frequency will be 0.875 or 1.125 times their standstill values for a transformation ratio of 1:1. On the shaft, the high frequency-induced voltage in the rotating secondary of exciter is rectified using power rectifiers, which are also rotating along with the shaft. The DC output of rectifier is connected to the rotating field winding of the main synchronous machine. The DC field current of the main synchronous machine is controlled by controlling the 400 Hz excitation of the pilot generator. A schematic diagram has been shown in Fig. 6.7.

The DC generator as exciter is preferred for small and medium size synchronous machines, where a pilot generator (DC generator) is used to energize the field winding of the main exciter using automatic voltage regulator (AVR). The output of main exciter (separately excited DC generator) is connected to the field winding of main synchronous machine. All the rotating machines are connected to the main shaft. In

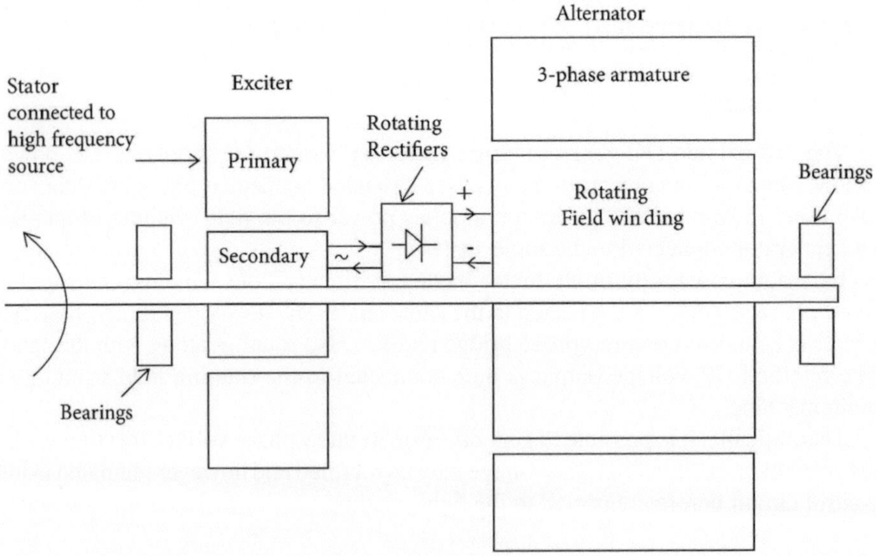

Fig. 6.7 Schematic representation of brushless excitation

certain cases, even a single exciter (self-excited DC generator) is used to provide DC current to the field winding of main synchronous machines. The main drawback of DC machine is that there are chances of sparking on the trailing edge of brushes and their use require frequent maintenance and replacement of carbon brushes. For large capacity synchronous machines brushless excitation is necessary.

In a typical brushless excitation of large capacity two-pole alternator, there is a pilot generator of 16 pole permanent magnet rotor three-phase alternator (PMG), an exciter of six-pole three-phase alternator with rotating armature and three-phase bridge rectifier mounted on the rotating shaft. All the components are being rotated by the same prime mover. The PMG serves the purpose of generation of three-phase 400 Hz voltage. The output of the PMG is connected to automatic voltage regulator (AVR) which checks the output of main alternator and suitably adjusts the magnitude of the output voltage of AVR. The controlled voltage of AVR is then rectified by another set of (stationary) three-phase bridge rectifier. The DC output of bridge rectifier is then connected to the field winding (on stator) of exciter alternator. The armature of exciter alternator develops three-phase voltage, which is rectified using rotating rectifiers. The output of rotating rectifiers is connected to the field winding of main alternator. The scheme does not requires any slip rings and brushes.

In another scheme, the exciter alternator being similar to slip-ring induction motor. The output of pilot generator (PMG) at 400 Hz is adjusted using AVR before being connected to three-phase stator windings of exciter. The three-phase rotor-induced voltage is rectified using rotating rectifier on shaft. The output of rotating rectifier is then connected to the field winding of main alternator. The scheme also does not require any sip ring and brushes.

Example 6.1 Calculate the number of poles required for generating 50 Hz using a turbine running at (a) 3000RPM (b) 1000RPM (c) 300RPM (d) 40RPM.

Solution

Given $f = 50$ Hz.

We know that $N_s = \frac{120f}{p}$

Thus numbers of poles for different speed are $p = \frac{120f}{N_s}$

(a) If $N_s = 3000$ RPM $p = 2$
(b) If $N_s = 1000$ RPM $p = 6$
(c) If $N_s = 300$ RPM $p = 20$
(d) If $N_s = 40$ RPM $p = 150$

Example 6.2 A 60 Hz, 1200 RPM alternator is running at 1000 RPM. Calculate the frequency of generated EMF.

Solution

The number of poles are $p = \frac{120f}{N_s} = \frac{120\times60}{1200} = 6$.

When it is running at 1000 RPM the frequency is given by

$$f = \frac{pN_s}{120} = \frac{6 \times 1000}{120} = 50\,\text{Hz}$$

Example 6.3 A 2 pole synchronous motor is supplied by frequency of (a) 10 Hz (b) 50 Hz (c) 60 Hz (d) 100 Hz and (e) 400 Hz. Calculate the synchronous speeds.

Solution

The synchronous speed is given by, $N_s = \frac{120f}{p}$.

(a) $f = 10$ Hz $N_s = \frac{120 \times 10}{2} = 600$ RPM
(b) $f = 50$ Hz $N_s = \frac{120 \times 50}{2} = 3000$ RPM
(c) $f = 60$ Hz $N_s = \frac{120 \times 60}{2} = 3600$ RPM
(d) $f = 100$ Hz $N_s = \frac{120 \times 100}{2} = 6000$ RPM
(e) $f = 400$ Hz $N_s = \frac{120 \times 400}{2} = 24000$ RPM.

6.4 Generation of Three-Phase Voltage

Let us assume two-pole structure of an alternator. The field winding is energized by DC source for creating the pole flux. The profiles of pole faces are shaped to produce a sinusoidal flux density distribution in air-gap. The stator carries a three-phase balanced winding, with axis of each phase winding displaced by 120° electrical. Let us further assume that rotor is rotating in counter clockwise direction with a constant speed by a prime mover. See Fig. 6.8. At any instant, when coil sides **aa′** are located closely to the maximum flux density, the induced EMF in phase **a** is maximum. At that instant of time, coil sides **c′ & b′** are facing equal magnetic flux. As a result, the EMF induced in the two phases is identical, as shown in Fig. 6.8 at time $t = 0$. The instantaneous values of EMFs in **b-b′** and **c-c′** phases are half of their maximum value or cos 60° times its maximum values [4, 5].

The rotor is rotating at synchronous speed in anti-clockwise direction. Once the rotor is moved by 60°, the **N** pole is closer to coil side **b′**, the phase **b-b′** has its

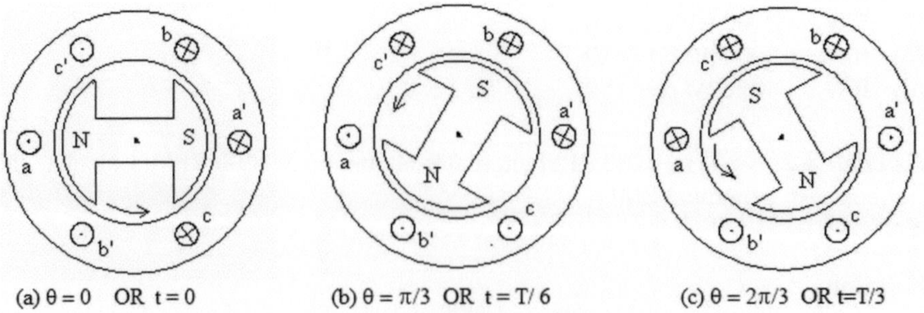

Fig. 6.8 Generation of induced voltage in a three-phase two-pole alternator

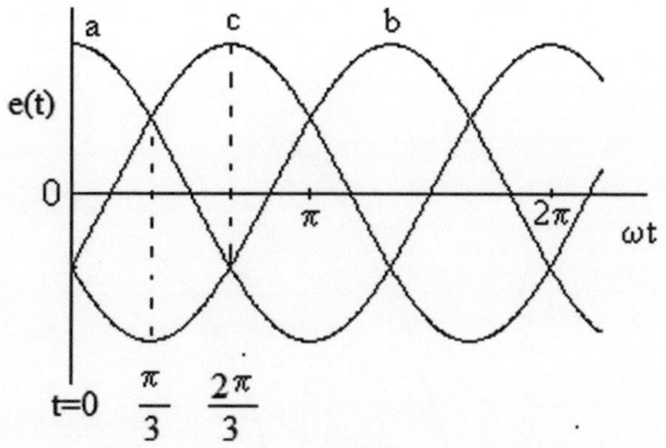

Fig. 6.9 Waveforms of generated three-phase-induced voltage of an alternator

maximum negative value of induced EMF. At this instant of time, phase **c-c′** finds the flux of reverse polarity; hence the EMF of phase **c** changes its sign at time $t = T/6$ or $\theta = \pi/3$ (Fig. 6.9).

Once the rotor is moved by another 60°, the **N** pole is beneath coil side **c** and the phase **c-c′** has its maximum value of induced EMF. At this instant of time, phase **aa′** finds the flux of reverse polarity, hence the EMF of phase **aa′** changes its sign.

The induced EMF variation for each phase is identical with the other two phases, i.e., **b** & **c**, except for the time displacements of 120° and 240°, respectively. Although salient pole rotor has been assumed similar explanation can be given for cylindrical rotor machine as well.

6.4.1 *Characteristics of Alternator*

The resultant air-gap flux in synchronous machine depends on the magnitude of the exciting DC current in the field windings (located on the rotor structure) as well as on the value of AC armature current. It is essential to know the variation of open-circuit EMF or magnetization characteristics with respect to the field current. The curve can be experimentally obtained. The rotor is driven at synchronous speed with the help of a prime mover. The DC field current is varied and the open-circuit EMF is recorded. Figure 6.10 shows the OCC or magnetization characteristics of alternator. The straight line portion of the curve corresponds to unsaturated magnetic circuit and is known as *air-gap line*.

The short-circuit characteristics (SCC) shown in Fig. 6.10 are obtained by measuring the DC field current required for rated AC armature current during short-circuit at rated speed. The SCC is linear, because the air-gap flux required to produce it is so small as to cause little or no saturation of the magnetic circuit. Flux is needed to produce EMF, which in turn is required to overcome the leakage impedance drop. We will later discuss the details of experimental testing.

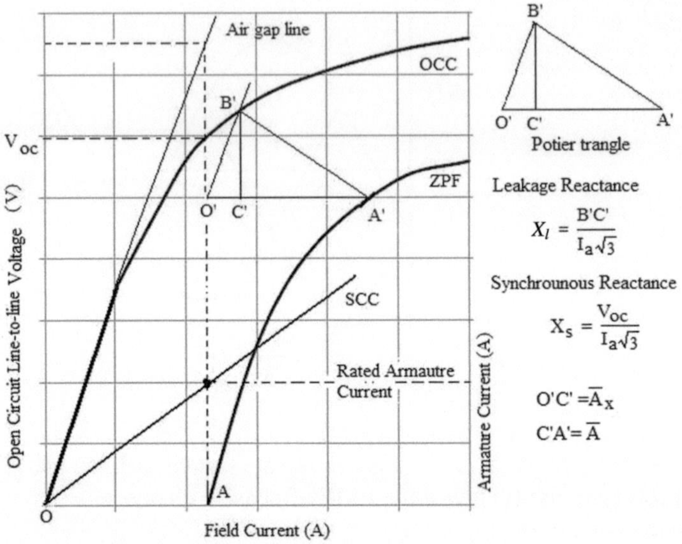

Fig. 6.10 Open-circuit, short-circuit, and zero power factor characteristics of alternator and calculation of leakage reactance and synchronous reactance

Figure 6.10 briefly describes the determination of the per phase leakage reactance and per phase synchronous reactance from OCC, SCC, and ZPF characteristics. Following points should be noted in this Fig:

(a) OA is the field current required to flow rated armature current during short circuit test at rated speed.
(b) Point A′ corresponds to the rated terminal voltage at certain field current, when alternator is supplying rated armature current at zero power factor (ZPF) lagging (inductive) load.
(c) Segment OA = O′A′.
(d) O′B′ is parallel to air-gap line.
(e) In triangle O′A′B′, the perpendicular B′C′ shows the leakage reactance drop, or

$$B'C' = I_a X_l \sqrt{3}$$

(f) C′A′ is the field current due to armature reaction MMF $\mathbf{F}_A$ or corresponding field current due to armature reaction **A.**
(g) O′C′ is the field current required to overcome the leakage reactance drop or $\mathbf{A}_x$. This gives us OA = O′A′ = $\mathbf{A} + \mathbf{A}_x$.
(h) Sometimes OA is taken as completely due to armature reaction MMF or OA = **A.**
(i) ZPF characteristics is the locus of point A′ and is obtained by sliding the triangle O′A′B′, such that vertex B′ always rests the OCC.

6.4.2 Concept of MMF

The field winding is on the rotor, and the DC current through the field winding creates a sinusoidal air-gap flux density distribution in space, which is rotating in counter clock wise direction at synchronous speed. At any instant, the North Pole is aligned as shown in Fig. 6.11, and it produces maximum EMF induced on the coil sides on the stator. The axis of the pole is marked as vertical. The coil $\mathbf{e_f}$-$\mathbf{e_f}'$ can be considered to represent the EMF induced in the entire winding. As the field rotates, a sinusoidal EMF will result. The entire three-phase winding is represented by coil $\mathbf{e_f}$-$\mathbf{e_f}'$. The axis of the coil $\mathbf{e_f}$-$\mathbf{e_f}'$ is 90° displaced by the field MMF $\mathbf{F}_F$. Thus, the field MMF $\mathbf{F}_F$ leads the direction of the excitation EMF phasor. When the induced EMF in $\mathbf{e_f}$-$\mathbf{e_f}'$ is at maximum, the flux linkage is zero.

The coil $\mathbf{i_a}$-$\mathbf{i_a}'$ is having maximum current at that instant of time. It is assumed that alternator is supplying a lagging current. The MMF $\mathbf{F}_F$ is space distributed MMF due to field poles being rotated by prime mover at speed ω_S. The armature conductors are distributed in space. The peak current lags the EMF phasor by and angle $\psi°$. The axis of this coil coincides with the location of corresponding armature MMF $\mathbf{F}_A$ and is space displaced from the field flux axis $\mathbf{F}_F$ by $(90° + \psi°)$. If the phase angle between armature current and excitation EMF changes, a corresponding change occurs in the axis $\mathbf{F}_A$. Armature MMF is placed in coincidence with the armature current phasor. The MMF $\mathbf{F}_A$ is due to the three-phase currents in the armature. The armature reaction MMF $\mathbf{F}_A$ is rotating in space at same speed ω_S.

The space sinusoids may be summed up in a similar manner as phasor summation. Two space phasor $\mathbf{F}_F$ and $\mathbf{F}_A$ are having same frequency of rotation and same number of poles. These two can be added to get resultant space sinusoid denoted by $\mathbf{F}_R$, which is responsible for generating the resultant induced EMF in the armature winding. It

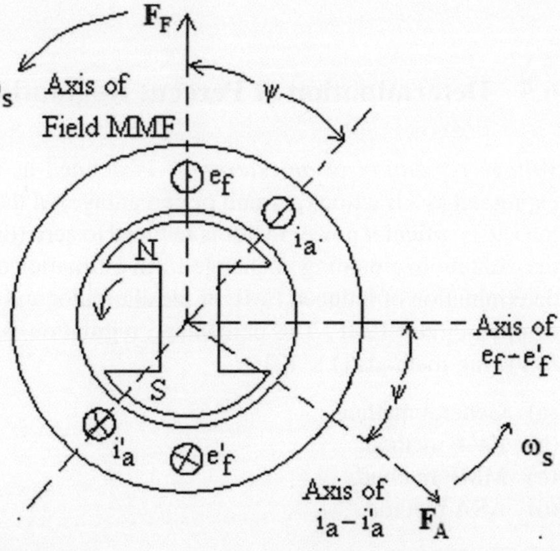

Fig. 6.11 Representation of MMF in an alternator connected to lagging load

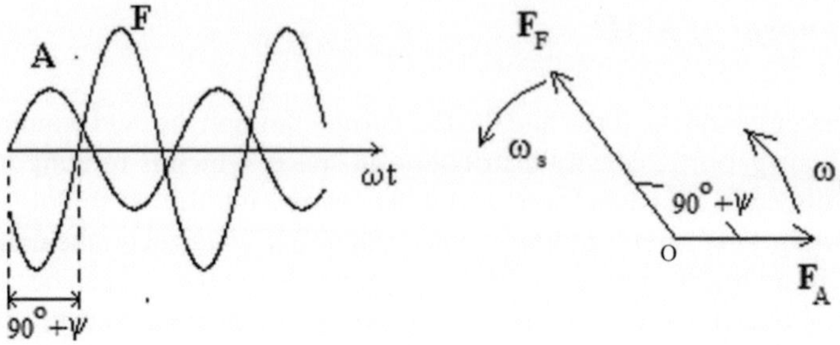

Fig. 6.12 Rotating MMF—armature and field MMF of alternator and their phasor for lagging power factor

is assumed that they are space sinusoids and not time varying sinusoids. It is further assumed that the alternator has per phase resistance, R_a and leakage reactance on per phase basis, X_l. The armature is assumed to deliver a balanced current of **I** which lags the applied voltage **V** by an angle $\psi°$. The three-phase winding current leads to rotating magnetic field MMF $\mathbf{F}_A$, which is rotating in space at constant synchronous speed (Fig. 6.12). The resultant MMF is given as

$$\mathbf{F}_R = \mathbf{F}_F + \mathbf{F}_A \tag{6.3}$$

The resultant MMF $\mathbf{F}_R$ is responsible for resultant air-gap flux, which, in turn, produces actual terminal voltage at certain load. The induced voltage in armature per phase is given as

$$\mathbf{E}_r = \mathbf{V} + \mathbf{I}(R_a + jX_l) \tag{6.4}$$

6.5 Determination of Percent Regulation

Voltage regulation of an alternator is defined as the change in terminal voltage expressed as a fraction (per unit or percentage) of the rated voltage, when a specified load at a particular power factor is reduced to zero (disconnected), with the speed and the excitation remaining unchanged. An estimation of percentage regulation requires determination of induced EMF of the alternator and also referred to as no-load EMF or open-circuit EMF. The percentage regulation is estimated with the help of the following methods [11, 12]:

(a) General method.
(b) EMF method.
(c) MMF method.
(d) ASA method.

These methods have different treatment of saturation, leakage reactance, and armature reaction. These are explained in the following subsections.

6.5.1 General Method

This method is also referred as *ZPF method or Potier method*. The leakage reactance X_l is obtained from ZPF characteristic, as depicted in Fig. 6.10. The details of finding are given in the later section. Figures 6.13 and 6.14 show phasor diagrams for two different power factors. According to the phasor diagram, the field current **R** (or MMF $\mathbf{F}_R$) required for obtaining voltage $\mathbf{E}_r$ is determined from OCC. The field current **F** (or MMF $\mathbf{F}_F$) required for induced EMF is obtained using MMF phasor diagram.

Corresponding to field current **F**, the no-load EMF is obtained from OCC. The armature reaction field current **A** (or MMF $\mathbf{F}_A$) is obtained from the SCC. The field currents **F**, **R,** and **A** are proportional to their respective MMF. They are regarded as space phasors. The resultant MMF is obtained from their phasor sum, i.e.,

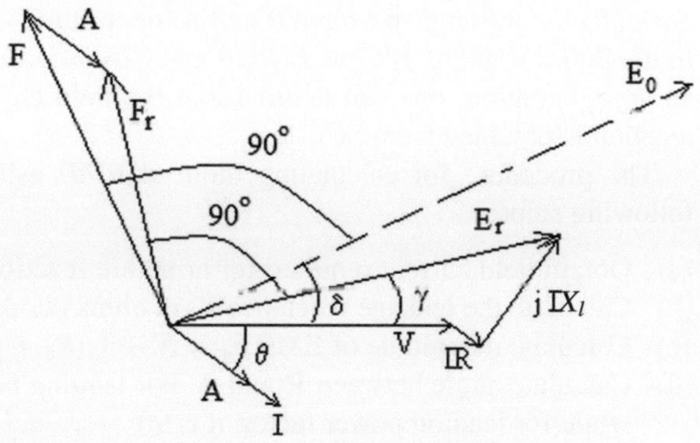

Fig. 6.13 Phasor diagram of alternator at lagging power factor—general method

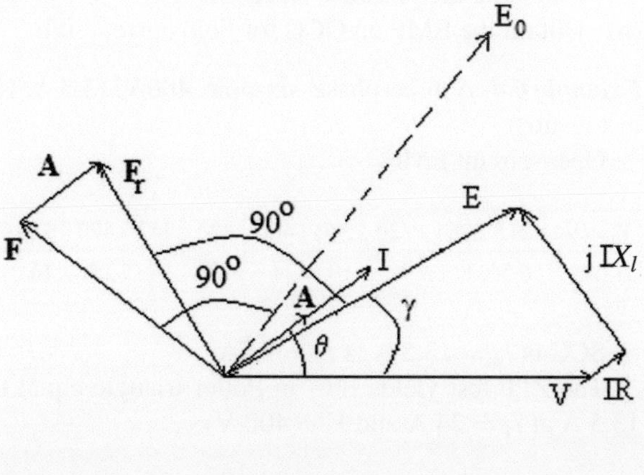

Fig. 6.14 Phasor diagram of alternator at leading power factor—general method

$$\mathbf{F}_{\mathrm{R}} = \mathbf{F}_{\mathrm{F}} + \mathbf{F}_{\mathrm{A}}$$

or,

$$\mathbf{R} = \mathbf{F} + \mathbf{A} \tag{6.5}$$

Here, $\mathbf{F}_{\mathrm{R}} \quad \alpha \quad R = R\angle 90^\circ + \gamma$ and $\mathbf{F}_{\mathrm{A}} \quad \alpha \quad \mathbf{A} = A\angle - \theta$ for lagging load.

The field current required for generating open-circuit EMF is given by

$$\mathbf{F} = \mathbf{R} - \mathbf{A} = R\angle\left(90^o + \gamma\right) - A\angle - \theta \tag{6.6}$$

The voltage is

$$\mathbf{E}_r = \mathbf{V} + \mathbf{I}_{\mathrm{a}}(R_{\mathrm{a}} + jX_l) \tag{6.7}$$

$$E_{\mathrm{r}}\angle\gamma = V\angle 0 + I_{\mathrm{a}}\angle - \theta(R_{\mathrm{a}} + jX_{\mathrm{l}}) \tag{6.8}$$

This is shown in Figs. 6.13 and 6.14 for lagging and leading power factors, respectively. The angle between **R** and **A** for lagging power factor is $90^\circ + \gamma + \theta$. For leading power factor, the angle between **R** and **A** (or current **I**) is $90 - (\theta - \gamma) = 90 - \theta + \gamma$. In the Potier triangle, $\mathrm{B'C'} = I_{\mathrm{a}}X_{\mathrm{l}}\sqrt{3}$ and $\mathrm{C'A'}$ = armature reaction MMF or **A**. As an approximation, one can assume that the field current **A** required for armature reaction is obtained from SCC test.

The procedure for calculating induced EMF using ZPF method involves the following steps:

(a) Obtain field current required for armature reaction $\mathbf{A} = A\angle - \theta$ from SCC.
(b) Calculate the leakage reactance X_{l} in ohms per phase.
(c) Calculate magnitude of EMF $\mathbf{E}_{\mathrm{r}} = \mathbf{V} + \mathbf{I}_{\mathrm{a}}(R_{\mathrm{a}} + jX_l)$ or $\mathrm{E}_{\mathrm{r}}\angle\gamma$.
(d) Calculate angle between **R** and **A**. For lagging power factor, it is $90^\circ + \gamma + \theta$, while for leading power factor, it is $90^\circ + \gamma - \theta$. Here θ is power factor angle.
(e) From OCC, get the field current **R** required for E_{r}.
(f) Calculate the phasor **F** = **R - A.**
(g) Obtain the EMF on OCC for field current, |**F**|.

Example 6.4 A three-phase, six-pole, 400 V, 13.5 A, 1000 RPM, yields the following test results:

Open-circuit EMF

V_{oc} (V)	215	284	320	380	400	422	452	472	488	508	520	532	540	552	560
I_{f} (A)	6.5	8	9	10	11	12	14	15	16	17	18	19	20	22	24

SCC is $I_{\mathrm{a}} = 13.5$ A at $I_{\mathrm{f}} = 9.5$A.

The ZPF test yields $\mathrm{B'C'}$ in Potier triangle equal to 120 Volt for test data ($I_{\mathrm{a}} =$ 13.5 A at $I_{\mathrm{f}} = 24$ A and $V = 400$ V).

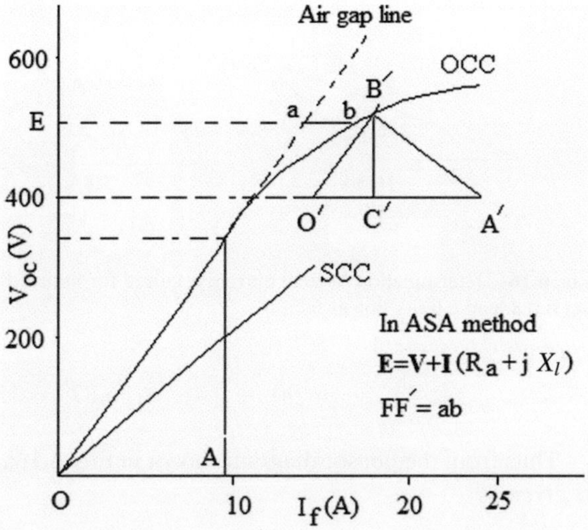

Fig. 6.15 OCC & SCC of the Example 6.4

Determine

(a) Leakage reactance, when armature resistance is $R_a = 1.0\ \Omega$.

(b) Open-circuit EMF using general or ZPF method, when the alternator is delivering rated current at

(i) 0.8 P.F. lagging (ii) 0.8 leading (iii) U.P.F. (iv) Z.P.F. lagging (v) Z.P.F. leading.

Solution

The OCC, SCC, and Potier triangle are plotted in Fig. 6.15.

(a) From the Potier triangle B′C′ = 120 V. Thus per phase leakage reactance is $X_l = \frac{120}{13.5\sqrt{3}} = 5.13\Omega$

(b) General method:

(i) 0.8 P.F. lagging

The induced EMF E is

$$\begin{aligned}
\mathbf{E}_r &= \mathbf{V} + \mathbf{I}(R_a + jX_1) \\
&= 231 + 13.5(0.8 - j0.6)(1.0 + j5.13) \\
&= 283.353 + j47.304 \\
&= 287.2\angle 9.48^\circ \text{Volt}
\end{aligned}$$

From OCC the field current required for $287.27\sqrt{3} = 497.56$ V (line to line) is **R** = 16.5 A. The armature reaction **A** = 9.5A from SCC.

The angle between the **R** and **A** is $90^\circ + 9.48^\circ + \cos^{-1}0.8$

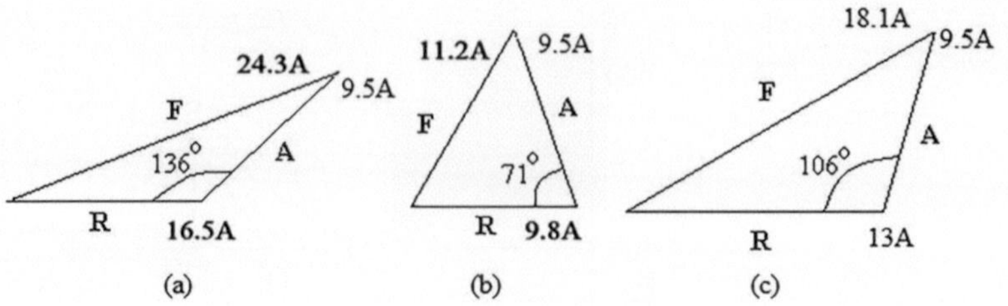

Fig. 6.16 Determination of field current required for induced EMF-general or ZPF method **a** 0.8 lag **b** 0.8 lead **c** unity power factor

$$= 90^\circ + 9.48^\circ + 36.87^\circ = 136.35^\circ$$

Thus from the phasor diagram shown in Fig. 6.16a, the field excitation necessary is given as

$$\begin{aligned} F &= \sqrt{\left[(R)^2 + (A)^2 - 2RA\cos 136.35^\circ\right]} \\ &= \sqrt{\left[(16.5)^2 + (9.5)^2 - 2 \times 16.5 \times 9.5\cos 136.35^\circ\right]} \\ &= 24.27\ \text{A} \end{aligned}$$

Corresponding to field current 24.27 A, the open-circuit EMF from OCC is 565 V. The percentage regulation can now be given as

$$= \frac{565 - 400}{400} \times 100 = 41.25\%.$$

(ii) 0.8 P.F. leading.

The induced EMF E,

$$\begin{aligned} \mathbf{E}_r &= \mathbf{V} + \mathbf{I}(R_a + jX_l) \\ &= 231 + 13.5(0.8 + j0.6)(1.0 + j5.13) \\ &= 231 + 10.8 + 55.404j + j8.1 - 41.553 \\ &= 200.247 + j63.504 \\ &= 210.1\angle 17.48^\circ \text{Volt} \end{aligned}$$

From OCC, the field current required for $210.1\sqrt{3}$ V (line to line) = 363.71 V is $\mathbf{R} = 9.8$ A and $\mathbf{A} = 9.5$ A. The angle between the **R** and **A** is $90^\circ + 17.48^\circ - \cos^{-1}0.8 = 90^\circ + 17.48^\circ - 36.87^\circ = 70.61^\circ$ Thus, from the phasor diagram shown in Fig. 6.16b, the field excitation necessary is

$$\text{F} = \sqrt{\left[(\text{R})^2 + (\text{A})^2 - 2\text{RA}\cos 70.61^\circ\right]}$$

$$= \sqrt{\left[(9.8)^2 + (9.5)^2 - 2 \times 9.8 \times 9.5 \cos 70.61°\right]}$$
$$= 11.15\,\text{A}$$

Corresponding to field current 11.15 A, the line-to-line open-circuit EMF from OCC is 402.5 V.

The % regulation $= \frac{402.5-400}{400} \times 100 = 0.625\%$

(iii) Unity P.F.

The induced EMF **E**,

$$\begin{aligned}\mathbf{E}_\mathrm{r} &= \mathbf{V} + \mathbf{I}(R_\mathrm{a} + jX_l) \\ &= 244.5 + j69.255 \\ &= 254.1\angle 15.81°\text{Volt}\end{aligned}$$

From OCC, the field current required for $254.1\sqrt{3}\,\text{V} = 440.3\,\text{V}$ (line to line) is.

R = 13 A and armature reaction **A** = 9.5 A. The angle between the **R** and **A** is 90° + 15.81° = 90° + 15.81° = 105.81°. Thus, from the phasor diagram shown in Fig. 6.16c, the field excitation necessary is

$$\mathrm{F} = \sqrt{\left[\mathrm{R}^2 + \mathrm{A}^2 - 2\mathrm{RA}\cos 105.81°\right]}$$
$$= \sqrt{\left[(13)^2 + (9.5)^2 - 2 \times 13 \times 9.5 \cos 105.81°\right]}$$
$$= 18.1\,\text{A}$$

Corresponding to field current 18.0 A, the line-to-line open-circuit EMF from OCC is 520 V.

The % regulation $= \frac{520-400}{400} \times 100 = 30\%$.

(iv) Z.P.F. lagging

The induced EMF E is given as

$$\begin{aligned}\mathbf{E}_\mathrm{r} &= \mathbf{V} + \mathbf{I}(R_\mathrm{a} + jX_l) \\ &= 300.255 - j13.5 \\ &= 300.3\angle - 2.57°\text{Volt}\end{aligned}$$

From OCC, the field current required for $300.3\sqrt{3}$ V (line to line) = 521 V is.

R = 18 A and armature reaction **A** = 9.5 A. The angle between the **R** and **A** is 90°-2. 57 ° + $\cos^{-1}0.0$ = 90° − 2.57° + 90° = 177.43°. Thus, the field excitation necessary is

$$F = \sqrt{\left[R^2 + A^2 - 2RA\cos 177.57°\right]}$$

$$= \sqrt{\left[(18)^2 + (9.5)^2 - 2 \times 18 \times 9.5 \cos 177.43^\circ\right]}$$
$$= 27.5\,\text{A}$$

Corresponding to field current 27.5 A, the open-circuit EMF from OCC is 570 V. The percentage regulation is given as

$$\frac{570 - 400}{400} \times 100 = 42.5\%$$

(v) Z P.F. leading

The induced EMF E is

$$E_{\text{f}} = V + I(R_{\text{a}} + jX_l)$$
$$= 162.3\angle 4.77^\circ \text{ Volt}$$

From OCC, the field current required for $162.3\sqrt{3} = 281.11$ V (line to line) is **R** = 7.6 A and armature reaction is **A** = 9.5 A. The angle between the **R** and **A** is $90^\circ + 4.77^\circ\text{-}\cos^{-1}0.0 = 90^\circ + 4.77^\circ - 90^\circ = 4.77^\circ$. Thus, the field excitation necessary is

$$F = \sqrt{R^2 + A^2 - 2RA\cos 4.77^\circ}$$
$$= \sqrt{\left[(7.6)^2 + (9.5)^2 - 2 \times 7.6 \times 9.5 \cos 4.77^\circ\right]}$$
$$= 2.02\text{A}$$

Corresponding to field current 2.02A the line-to-line open-circuit EMF is 75 V. The regulation $= \frac{75-400}{400} \times 100 = -81.25\%$.

Thus % regulation using general method or ZPF method or Pottier method are:

P.F	0.8 lag	0.8 lead	1.0	Z.P.F. lag	Z.P.F. lead
% regulation	41.25	0.625	30	42.5	−81.25

6.5.2 *EMF Method*

MMF in an alternator produces their corresponding magnetic fluxes in the machines. These MMFs are rotating in space at synchronous speed and their resultant MMF can be obtained by using their phasor additions. If we assume magnetic circuit is to be unsaturated, then we can further assume that the individual MMFs produce their proportional fluxes. These fluxes, in turn, are assumed to develop their proportional EMFs. These EMF can be combined to yield the resultant EMF. The EMF method

Fig. 6.17 Phasor diagram at lagging power factor for EMF method

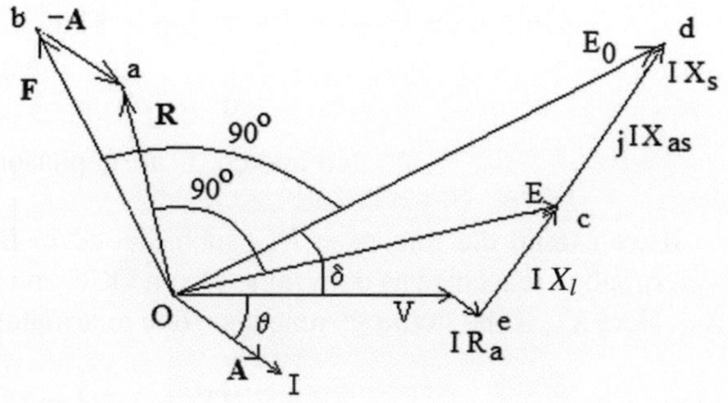

is regarded as *optimistic method*, as the percentage regulation calculated is higher (i.e., poorer) than expected.

Consider the phasor diagram of Fig. 6.17. The field current **R** is proportional to the MMF required to produce the terminal voltage and can be obtained from the OCC. It is desired to know the open-circuit EMF, when the load is withdrawn from the alternator. This requires field current **F**, which is proportional to MMF producing open-circuit EMF. The armature reaction field current **A** corresponds to the MMF required to flow rated short-circuit current. Thus, the field current required to obtain the open-circuit voltage is given by,

$$\mathbf{F} = \mathbf{R} - \mathbf{A} \quad \text{or} \quad \mathbf{R} = \mathbf{F} + \mathbf{A} \tag{6.9}$$

If saturation is neglected, then no-load EMF is proportional to **F,** i.e., ed and cd ⊥ ab ⊥ phasor **I.**

$$\mathbf{E}_0 \;\alpha\; \mathbf{F} \quad \text{and} \quad \mathbf{E} \;\alpha\; \mathbf{R} \tag{6.10}$$

where

$$\mathbf{E} = \mathbf{V} + \mathbf{I}(R_\mathrm{a} + jX_l) \tag{6.11}$$

where R_a is per phase armature resistance in ohms and X_l is the per phase leakage reactance in ohms obtained from ZPF test (already discussed).

In triangles ΔOab and ΔOcd (see Fig. 6.17), we assume that the MMFs are proportional to the respective EMFs they produce. Let Oa = k · Oc and Ob = k · Od, where k is the proportionality constant. In the phasor diagram (Fig. 6.17), we have angles.

$$\angle \mathrm{aOc} = 90^\circ$$

$$\angle \mathrm{bOd} = 90^\circ$$

$$\angle \mathrm{aOb} = \angle \mathrm{cOd}$$

Here, Δ Oab and Δ Ocd are similar triangles. Thus, we get

$$\text{cd} \ \alpha \ \text{ab} \ \alpha \ \mathbf{I}$$

$$\text{ed and cd} \perp \text{ab} \perp \text{phasor } I$$

If we extend the line ec $= IX_l$, such that ed $= IX_s$, where X_s is the per phase synchronous reactance as determined from OCC and SCC, then the segment cd $= I X_{as}$. Here X_{as} is the fictitious reactance due to armature reaction. Thus, we get

$$\text{ed} = j \cdot (X_l + X_{ag})I = jX_s I$$

The synchronous impedance per phase is given as

$$Z_s = R_a + jX_s$$

From EMF method, we get

$$\mathbf{E}_0 = \mathbf{V} + (R_a + jX_s)\mathbf{I} \tag{6.12}$$

Under steady-state short-circuit conditions $\mathbf{V} = 0$. Thus, we get

$$\mathbf{E}_{SC} = (R_a + j \cdot X_s)\mathbf{I}_{SC} \tag{6.13}$$

The synchronous impedance per phase is

$$Z_s \text{ in p.u.} = \frac{Z_s}{\left(V_r/I_r\right)} = \frac{1}{\text{SCR}} \tag{6.14}$$

Here, V_r and I_r are per phase rated voltage and current, respectively. When machine is under steady-state short-circuit condition, a very small MMF is required to produce sufficient flux for developing necessary EMF, $\mathbf{E}_{sc}$ for circulating rated short-circuit current $\mathbf{I}_{SC}$. The synchronous impedance calculated on the basis of OCC and SCC relates to the unsaturated conditions and gives a very high value of the regulation. The regulation thus computed is higher (poorer) than that in practice. Due to these reasons, the EMF method is referred as *an optimistic method*.

(a) **Saturated synchronous reactance**: The value of synchronous impedance varies with the field current. The unsaturated value is higher, whereas the saturated value is lower. The regulation calculated on the basis of either of the values may yield erroneous results. It is, therefore, better to calculate the regulation on the basis of corrected synchronous reactance taking saturation into account.

If X_s' = unsaturated per phase synchronous reactance calculated from the air-gap line, the corrected synchronous reactance computed with correction factor is given as

$$X_s = \frac{X_s' - X_l}{k} + X_l$$

where k is saturation factor, which is determined from the OCC. For this EMF $E_f = V + I(R_a + jX_l)$ is calculated. The value of saturation factor is given as

$$k = \frac{\text{Field current on OCC for } |E_f|}{\text{Field current on air-gap line for } |E_f|} = \frac{|R|}{|R_{ag}|} \tag{6.15}$$

Here |**R**| is the magnitude of field current for EMF |$\mathbf{E}_r$| on OCC, while |$\mathbf{R}_{ag}$| is the magnitude of field current for EMF |$\mathbf{E}_r$| on air-gap line.

(b) **Short-circuit ratio:** The saturated short-circuit ratio (SCR) (see Fig. 6.18) is defined as

$$\begin{aligned} SCR &= \frac{\text{Field current required for rated voltage on OCC}}{\text{Field current required for rated current on SCC}} \\ &= \frac{OF_V}{OF_I} \end{aligned} \tag{6.16}$$

Unsaturated SCR corresponds to air-gap line. Thus, we have

$$SCR = \frac{\text{Field current required for rated voltage from airgap line}}{\text{Field current required for rated current on SCC}}$$

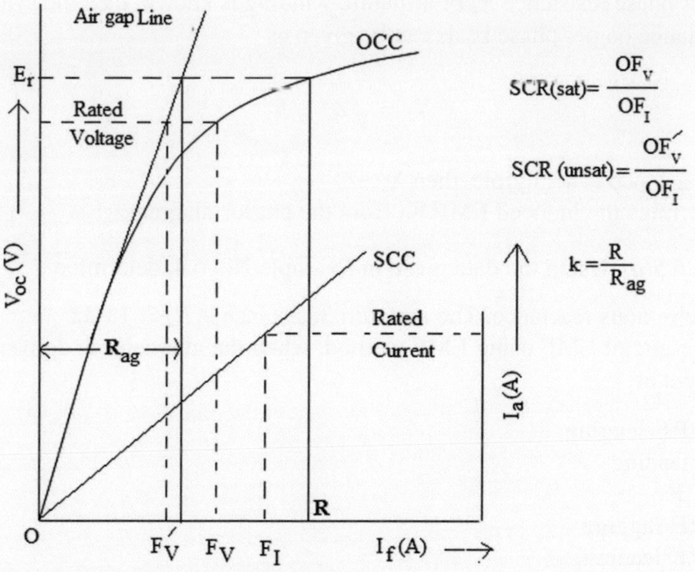

Fig. 6.18 Short-circuit ratio and saturation factor

$$= \frac{\mathrm{OF}_{\mathrm{V}}^{/}}{\mathrm{OF}_{\mathrm{I}}} \tag{6.17}$$

Machine is more sensitive, when SCR is small. This results in large armature reaction. The SCR for high-speed turbo alternator is 0.6 to 0.9 and that for hydro-alternators running at low speed is 0.9 to 1.5.

(c) **ZPF characteristics:** For purely inductive load, we can say that

$$|\mathbf{V}| \approx |\mathbf{E}_0| - |\mathbf{X}_s\mathbf{I}|$$

or,

$$V = E_0 - IX_s \tag{6.18}$$

The ZPF characteristics can be obtained from EMF method by subtracting IX_s from the OCC. For obtaining the induced EMF using EMF method, following procedure is adopted.

(a) The synchronous impedance Z_s is computed from OCC and SCC, keeping field excitation constant. The per phase synchronous impedance is calculated as

$$Z_s = \left.\frac{\text{Open circuit EMF per phase}}{\text{Rated current per phase}}\right|_{I_f=\text{Constant}} \tag{6.19}$$

(b) If per phase resistance R_a of armature winding is known, then the synchronous reactance on per phase basis can be given as

$$X_s = \sqrt{Z_s^2 - R_a^2} \tag{6.20}$$

(c) If resistance is negligible, then $X_s \simeq Z_s$
(d) Determine the induced EMF $\mathbf{E}_0$ from the phasor diagram.

Example 6.5(a) Using the data given in Example No. 6.4, determine

(a) Synchronous reactance. The armature resistance is $R_a = 1.0\ \Omega$.
(b) Open-circuit EMF using EMF method, when the alternator is delivering rated current at

(i) 0.8 P.F. lagging
(ii) 0.8 leading
(iii) U.P.F.
(iv) Z.P.F. lagging
(v) Z.P.F. leading.

Tabulate induced EMF per phase, percentage regulation, and power angle.

Solution
Rated voltage is given as

$$V = \frac{400}{\sqrt{3}} = 231\ \text{V}$$

(a) From graph shown in Fig. 6.15 open-circuit line-to-line EMF is 345 V for $I_f =$ 9.5 A. The unsaturated synchronous impedance is

$$Z_s = \frac{\text{Open circuit EMF per phase}}{\text{Rated current per phase}}\bigg|_{I_f=9.5A} = \frac{345}{13.5\sqrt{3}} = 14.75\ \Omega$$

$$X_s = \sqrt{Z_s^2 - R_a^2} = \sqrt{14.75^2 - 1.0^2} = 14.72\ \Omega$$

The synchronous impedance corresponding to rated voltage is given as

$$Z_s = \frac{\text{Rated Open Circuit EMF per phase}}{\text{Current from SCC}}\bigg|_{V_{oc}=400\ \text{V}} = \frac{400}{15.75\sqrt{3}} = 14.66\ \Omega$$

The synchronous reactance is now given as

$$X_s = \sqrt{Z_s^2 - R_a^2} = \sqrt{14.66^2 - 1.0^2} = 14.63\ \Omega$$

Let us compute open-circuit EMF using unsaturated synchronous reactance of 14.72 Ω.

(b) Percentage regulation using EMF method.

Fig. 6.19a shows the phasor diagram for lagging power factor. The induced EMF per phase for 0.8 power factor lagging is

$$E = \sqrt{(V + I_a R_a \cos\theta + I_a X_s \sin\theta)^2 + (I_a X_s \cos\theta - I_a R_a \sin\theta)^2}$$

$$E = \sqrt{(231 + 13.5 \times 1.0 \times 0.8 + 13.5 \times 14.72 \times 0.6)^2 + (13.5 \times 14.72 \times 0.8 - 13.5 \times 1.0 \times 0.6)^2}$$

$$= \sqrt{(231 + 10.8 + 119.23)^2 + (158.976 - 8.1)^2}$$

$$= \sqrt{(361.03)^2 + (150.876)^2} = 391.26\ \text{V}$$

Now, we have

$$\delta = \tan^{-1}\left(\frac{150.876}{361.03}\right) = +22.68^\circ.$$

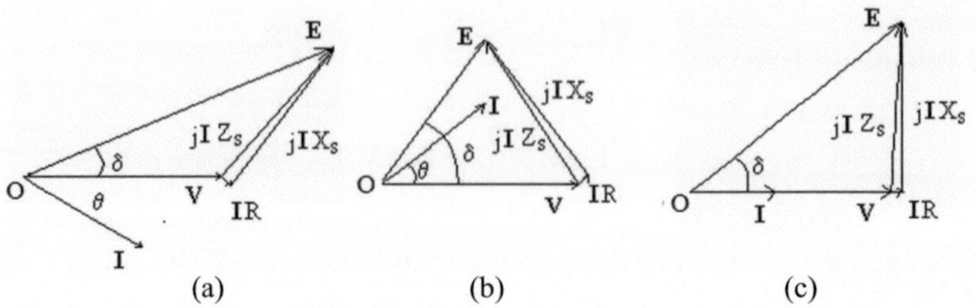

Fig. 6.19 Phasor diagrams for different power factors—EMF method **a** 0.8 lag power factor **b** 0.8 lead power factor **c** unity power factor

Percentage regulation $= \frac{391.26-231}{231} \times 100 = 69.38\%$.

Fig. 6.19b shows the phasor diagram for leading power factor. The induced EMF per phase for 0.8 power factor leading is

$$E = \sqrt{(V + I_a R_a \cos\theta - I_a X_s \sin\theta)^2 + (I_a X_s \cos\theta + I_a R_a \sin\theta)^2}$$

$$E = \sqrt{(231 + 13.5 \times 1.0 \times 0.8 - 13.5 \times 14.72 \times 0.6)^2 + (13.5 \times 14.72 \times 0.8 + 13.5 \times 1.0 \times 0.6)^2}$$

$$= \sqrt{(231 + 10.8 - 119.23)^2 + (158.976 + 8.1)^2}$$

$$= \sqrt{(122.57)^2 + (167.078)^2} = 207.21\ \text{V}$$

Now,

$$\delta = \tan^{-1}\left(\frac{167.078}{122.57}\right) = +53.74^\circ$$

Percentage regulation

$$= \frac{207.21 - 231}{231} \times 100 = -10.30\%$$

Fig. 6.19c shows the phasor diagram for UPF. The induced EMF per phase for unity power factor is

$$E = \sqrt{(V + I_a R_a)^2 + (I_a X_s)^2}$$

$$E = \sqrt{(231 + 13.5 \times 1.0)^2 + (13.5 \times 14.72)^2}$$

$$= \sqrt{(231 + 13.5)^2 + (198.72)^2}$$

$$= \sqrt{(244.5)^2 + (198.72)^2} = 315.1\ \text{V}$$

Now we have

$$\delta = \tan^{-1}\left(\frac{198.72}{244.5}\right) = +39.10^\circ$$

$$\text{Percentage regulation} = \frac{315.1 - 231}{231} \times 100 = 36.41\%$$

(i) In Z.P.F. lagging, the induced EMF is given as

$$E = \sqrt{(V + I_a X_s)^2 + (-I_a R_a)^2}$$

$$E = \sqrt{(231 + 13.5 \times 1.0 \times 0.0 + 13.5 \times 14.72 \times 1.0)^2 + (13.5 \times 14.72 \times 0.0 - 13.5 \times 1.0 \times 1.0)^2}$$

$$= \sqrt{(231 + 198.72)^2 + (-13.5)^2}$$

$$= \sqrt{(429.72)^2 + (13.5)^2} = 429.93\ \text{V}$$

Now

$$\delta = \tan^{-1}\left(\frac{-13.5}{429.72}\right) = -1.8^\circ$$

$$\text{Percentage regulation} = \frac{429.72 - 231}{231} \times 100 = 86.12\%$$

(ii) In Z.P.F. leading, the induced EMF is given as

$$E = \sqrt{(V - I_a X_s)^2 + (I_a R_a)^2}$$

$$E = \sqrt{(231 - 13.5 \times 14.72 \times 1.0)^2 + (13.5 \times 1.0 \times 1.0)^2}$$

$$= \sqrt{(231 - 198.72)^2 + (13.5)^2}$$

$$= \sqrt{(32.28)^2 + (13.5)^2} \quad = 34.98\ \text{V}$$

Now we can write

$$\delta = \tan^{-1}\left(\frac{13.5}{32.28}\right) = +22.70^\circ$$

$$\text{Percentage regulation} = \frac{34.98 - 231}{231} \times 100 = -86.84\%$$

Thus, the induced EMF per phase, % regulation, and power angle using EMF method are:

P.F	0.8 lag	0.8 lead	1.0	Z.P.F. lag	Z.P.F. lead
EMF	391.26	207.28	315.1	429.93	34.98
% regulation	69.38	−10.3	36.41	86.12	−86.84

(continued)

(continued)

P.F	0.8 lag	0.8 lead	1.0	Z.P.F. lag	Z.P.F. lead
Power angle δ	22.68°	53.74°	39.10°	−1.8	22.70°

Here the power angle δ is negative for ZPF lagging load or purely inductive load. The percentage regulation is negative for leading power factors.

Example 6.5(b) Write a computer program using MATLAB software for finding the induced EMF, power angle, and percentage regulation using EMF method for Example 6.5(a).

Solution

```
V = 400/1.7321
  I = 13.5
  XS = 14.72
  R = 1.0
  PF = 0.8
  PFS = sqrt(1-PF*PF)
  % Lagging PF 0.8
  ER = V + I*R*PF + I*XS*PFS
  EI = I*XS*PF-I*R*PFS
  E = sqrt(ER*ER + EI*EI)
  DEL = (180/3.14)*atan(EI/ER)
  REG = (E - V)*100/V
  PF = 0.8
  PFS = sqrt(1-PF*PF)
  % Leading PF 0.8
  ER = V + I*R*PF-I*XS*PFS
  EI = I*XS*PF + I*R*PFS
  E = sqrt(ER*ER + EI*EI)
  DEL = (180/3.14)*atan(EI/ER)
  REG = (E - V)*100/V
  % Upf
  PF = 1
  PFS = 0
  ER = V + I*R*PF-I*XS*PFS
  EI = I*XS*PF + I*R*PFS
  E = sqrt(ER*ER + EI*EI)
  DEL = (180/3.14)*atan (EI/ER)
  REG = (E - V)*100/V
  % ZPF lagging
  PF = 0
  PFS = 1
  ER = V + I*XS*PFS
```

```
EI = I*XS*PF-I*R*PFS
E = sqrt(ER*ER + EI*EI)
DEL = (180/3.14)*atan(EI/ER)
REG = (E - V)*100/V
% ZPF leading
PF = 1
PFS = 0
ER = V-I*XS*PFS
EI = I*XS*PF + I*R*PFS
E = sqrt(ER*ER + EI*EI)
DEL = (180/3.14)*atan(EI/ER)
REG = (E - V)*100/V
```

6.5.3 MMF Method

We have seen that EMF method is based on the fact that each MMF in the unsaturated magnetic circuit can be replaced by the corresponding EMF it produces, if it existed alone (Sect. 6.5.2). The resultant EMF is then obtained using superposition. The converse of this assumption is used in the MMF method. Here it is assumed that in an unsaturated magnetic circuit, the EMF produced in the electrical circuit is replaced by MMF which produces it. The voltage is given as

$$\mathbf{E} = \mathbf{V} + \mathbf{I}(R_a + jX_l) \tag{6.21}$$

This voltage (Eq. 6.21) is produced by the resultant field current **R**, or resultant MMF $\mathbf{F}_R$.

where

$$\mathbf{F} = \mathbf{R} - \mathbf{A} \quad \text{or} \quad \mathbf{R} = \mathbf{F} + \mathbf{A} \tag{6.22}$$

With the assumption of replacing the EMF by MMF, taking k as constant of proportionality we can write,

$$\mathbf{E} = -j \cdot k \cdot \mathbf{R} \tag{6.23}$$

Let us assume that

$$\mathbf{E}_1 = \mathbf{V} + \mathbf{I} \cdot R_a \tag{6.24}$$

We can write,

$$\mathbf{E}_1 = -j \cdot k \cdot \mathbf{R}_1 \tag{6.25}$$

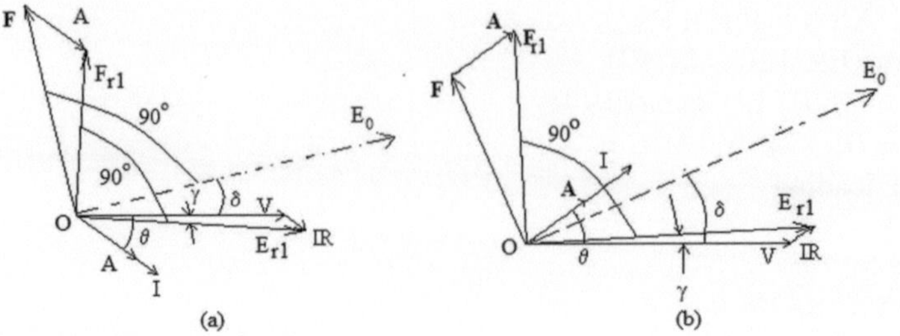

Fig. 6.20 Phasor diagram of alternator at **a** lagging **b** leading power factor for MMF method

Here, **R** and $\mathbf{R_1}$ are the field currents required to produce **E** and $\mathbf{E}_l$ voltages, respectively. The per phase armature resistance is given by R_a. Here triangles OAA1 and OCC1 are similar triangles in Fig. 6.20. Let AA1 be equal to field current or MMF $\mathbf{A}_x$ so that

$$j\mathbf{I}\,\mathbf{X}_l = jk\mathbf{A}_x$$

$$\mathbf{E} = \mathbf{E}_l + j \cdot IX_l$$

$$-jk\mathbf{R} = -jk\mathbf{R}_l + jk\mathbf{A}_x$$

$$\mathbf{R} = \mathbf{R}_l - \mathbf{A}_x \tag{6.26}$$

$$\mathbf{R} = \mathbf{F} + \mathbf{A} \tag{6.27}$$

$$\mathbf{R} = \mathbf{R}_l - (\mathbf{A} + \mathbf{A_x}) \tag{6.28}$$

The effective armature reaction has been increased by $\mathbf{A}_x$, to take care of leakage reactance drop. It can be stated that in EMF method, the armature reaction is replaced by an equivalent leakage reactance. On the other hand, in MMF method the armature leakage is replaced by an equivalent MMF. From OCC (Fig. 6.18), the field current $\mathbf{R}_l$ corresponds to field current due to $\mathbf{E}_l$. Under SCC, we have

$$\mathbf{E}_l = 0;\ \ \mathbf{R} = 0;\ \ \mathbf{R}_l = 0$$

$$\mathrm{OF_I} = \mathbf{A} + \mathbf{A}_x$$

The field current, (Figs. 6.10 and 6.18) O′A′ = $\mathrm{OF_I}$ = $\mathbf{A} + \mathbf{A}_x$, corresponds to the rated short-circuit current in SCC. The ZPF characteristics using MMF method are obtained by adding ($\mathbf{A} + \mathbf{A}_x$) into the OCC as shown in Fig. 6.23.

The EMF and MMF methods are based on the assumption that the magnetic circuit is unsaturated. However, in alternators, saturation cannot be avoided. Therefore, these methods are not expected to give reliable results. If the OCC was a straight line, all the three methods—general, EMF and MMF methods would give similar results. For obtaining the induced EMF using MMF method, following procedure is adopted.

(a) The EMF $\mathbf{E}_1 = \mathbf{V} + \mathbf{I}\mathrm{R}_a$ is computed. The field current $\mathbf{R}_1$ (proportional to MMF $\mathbf{F}_{r1}$) is obtained on OCC corresponding to $|\mathbf{E}_1|\angle\gamma$.
(b) The field current **A** corresponding to armature reaction $\mathbf{F}_A$ is obtained from SCC for rated armature current.
(c) For lagging power factor if $\mathbf{E}_1 = E_1 \angle\text{-}\gamma$ and $\theta = \cos^{-1}$(P.F.), the angle between $\mathbf{F}_{r1}$ and $\mathbf{F}_A$ is $90\text{-}\gamma + \theta$.

The induced EMF is obtained on the OCC corresponding to the field current **F**, which is proportional to MMF $\mathbf{F_F}$ of MMF diagram. The MMF diagram is shown in Fig. 6.16 for lagging and leading power factors. The regulation predicted by MMF is considered lesser than the actual values. The method is, therefore, known as pessimistic method.

The method for finding out the regulation by synchronous impedance (EMF) method is based on the assumption that the alternator is cylindrical rotor machine. In synchronous impedance method, the value of Z_s used for computation is more or less unsaturated, because a rather small value of field current can cause full load current to flow at short circuit. The value of field current under actual working conditions is usually much higher so that current value of synchronous impedance used should has been at higher saturation. Since the unsaturated value of X_s is more than the saturated value, the regulation predicted by synchronous impedance method is considerably higher than the actual values.

Example 6.6 In Example 6.4, determine open-circuit EMF using MMF method, when the alternator is delivering rated current at

(i) 0.8 P.F. lagging
(ii) 0.8 leading
(iii) U.P.F.
(iv) Z.P.F. lagging
(v) Z.P.F. leading.

Solution

(i) 0.8 power factor lagging

The induced EMF E is

$$\begin{aligned}\mathbf{E}_1 &= \mathbf{V} + \mathbf{I}R_a \\ &= 231 + 13.5(0.8 - \mathrm{j}0.6)1.0 \\ &= 241.8 - \mathrm{j}8.1\end{aligned}$$

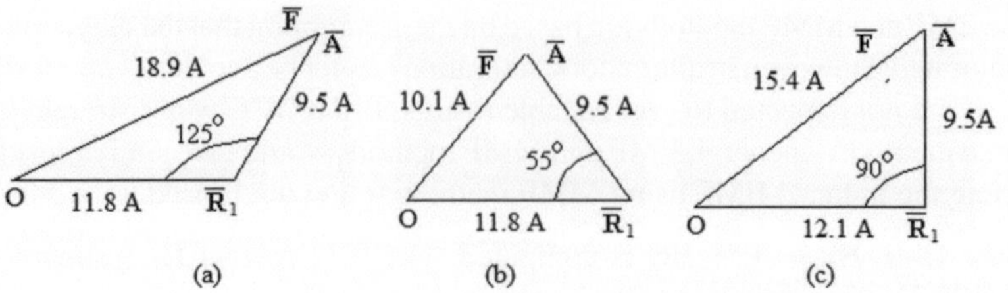

Fig. 6.21 Determination of field current required for induced EMF–MMF method at **a** 0.8 lag **b** 0.8 lead **c** unity power factor

$$= 241.94\angle - 1.92° \text{ Volt}$$

From OCC the field current required for $241.94\sqrt{3} = 419.05$ V (line to line) is $\mathbf{R}_1 = 11.8$ A and armature reaction is $\mathbf{A} = 9.5$ A. The angle between the $\mathbf{R}_1$ and $\mathbf{A}$ is $90° - 1.92° + \cos^{-1}0.8 = 90° - 1.92° + 36.87° = 124.95°$. Thus from the phasor diagram shown in Fig. 6.21a, the necessary field excitation is

$$\begin{aligned} F &= \sqrt{\left[R_1^2 + A^2 - 2R_1 A \cos 124.95°\right]} \\ &= \sqrt{\left[(11.8)^2 + (9.5)^2 - 2 \times 11.8 \times 9.5 \cos 124.95°\right]} \\ &= 18.94 \text{ A} \end{aligned}$$

Corresponding to field current 18.94 A, the line-to-line open-circuit EMF from OCC is 538 Volt.

The percentage regulation is

$$= \frac{538 - 400}{400} \times 100 = 34.5\%$$

(ii) 0.8 Power factor leading

The induced EMF E is given as

$$\begin{aligned} \mathbf{E}_1 &= \mathbf{V} + \mathbf{I}R_a \\ &= 231 + 13.5(0.8 + j0.6)1.0 \\ &= 241.94\angle 1.92° \text{Volt} \end{aligned}$$

From OCC the field current required for $241.94\sqrt{3} = 419.1$ V (line to line) is $\mathbf{R}_1 = 11.8$ A and armature reaction $\mathbf{A} = 9.5$A. The angle between $\mathbf{R}_1$ and $\mathbf{A}$ is $90° + 1.92° - \cos^{-1}0.8 = 90° + 1.92° - 36.87° = 55.07°$

Thus, from the phasor diagram shown in Fig. 6.21b, the field excitation necessary is

$$
\begin{aligned}
F &= \sqrt{\left[R_1^2 + A^2 - 2R_1A\ \cos\ 55.07^\circ\right]} \\
&= \sqrt{\left[(11.8)^2 + (9.5)^2 - 2 \times 11.8 \times 9.5\ \cos\ 55.07^\circ\right]} \\
&= 10.1\,\text{A}
\end{aligned}
$$

Corresponding to field current 10.1 A the line-to-line open-circuit EMF from OCC is 382 Volt.

The percentage regulation is given as

$$
= \frac{382 - 400}{400} \times 100 = -4.5\%
$$

(iii) Unity power factor.

The induced EMF E is

$$
\begin{aligned}
\mathbf{E}_1 &= \mathbf{V} + \mathbf{I}R_\text{a} \\
&= 231 + 13.5 \times 1.0 \times 1.0 \\
&= 244.5\angle 0^\circ \text{Volt}
\end{aligned}
$$

From OCC, the field current required for $244.5\sqrt{3} = 423.5$ V (line to line) is $\mathbf{R}_1 = 12.1$ A and armature reaction $\mathbf{A} = 9.5$ A. The angle between $\mathbf{R}_1$ and $\mathbf{A}$ is 90°. Thus, from the phasor diagram shown in Fig. 6.21c, the necessary field excitation is

$$
\begin{aligned}
F &= \sqrt{\left[R_1^2 + A^2 - 2R_1A\ \cos\ 90^\circ\right]} \\
&= \sqrt{\left[(12.1)^2 + (9.5)^2 - 2 \times \times 9.5\ \cos\ 90^\circ\right]} \\
&= 15.38\,\text{A}
\end{aligned}
$$

Corresponding to field current 15.38 A, the open-circuit EMF from OCC is 480 Volt.

The percentage regulation is given as

$$
= \frac{480 - 400}{400} \times 100 = 20\%
$$

(iv) Z.P.F. lagging

The induced EMF E is

$$
\begin{aligned}
\mathbf{E}_1 &= \mathbf{V} + \mathbf{I}R_\text{a} \\
&= 231 + 13.5(0.0 - j1.0)1.0 \\
&= 231 - j13.5
\end{aligned}
$$

$$= 231.39\angle - 3.34° \text{ Volt}$$

From OCC, the field current required for $231.39\sqrt{3} = 400.8$ V (line to line) is $\mathbf{R}_1 = 11.2$ A and armature reaction $\mathbf{A} = 9.5$ A.

The angle between the $\mathbf{R}_1$ and $\mathbf{A}$ is $90° - 0.6° + \cos^{-1}0.0 = 90° - 3.34° + 90° = 177.66°$. Thus, from the phasor diagram, the necessary field excitation is

$$\begin{aligned} F &= \sqrt{\left[R_1^2 + A^2 - 2R_1A \cos 177.66°\right]} \\ &= \sqrt{\left[(11.2)^2 + (9.5)^2 - 2 \times 11.2 \times 9.5 \cos 177.66°\right]} \\ &= 20.8 \text{ A} \end{aligned}$$

Corresponding to field current 20.8 A the open-circuit EMF from OCC is 545 Volt.

The percentage regulation

$$= \frac{545 - 400}{400} \times 100 = 36.25\%$$

(v) Z P.F. leading

The induced EMF E is

$$\begin{aligned} \mathbf{E}_1 &= \mathbf{V} + \mathbf{I}R_a \\ &= 231 + 13.5(0.0 + j1.0)1.0 \\ &= 231 + j13.5 \\ &= 231.39\angle 3.34° \text{ Volt} \end{aligned}$$

From OCC the field current required for $231.39\sqrt{3} = 400.8$ V (line to line) is $\mathbf{R}_1 = 11.2$ A and armature reaction $\mathbf{A} = 9.5$ A.

The angle between the $\mathbf{R}_1$ and $\mathbf{A}$ is $90° + 3.34° - \cos^{-1}0.0 = 90° + 3.34° - 90° = 3.34°$.

Thus from the phasor diagram, the field excitation necessary is

$$\begin{aligned} F &= \sqrt{\left[R_1^2 + A^2 - 2R_1A \cos 3.34°\right]} \\ &= \sqrt{\left[(11.2)^2 + (9.5)^2 - 2 \times 11.2 \times 9.5 \cos 3.34°\right]} \\ &= 1.81 \text{ A} \end{aligned}$$

Corresponding to field current 1.81 A, the line-to-line open-circuit EMF from OCC is 70 Volt.

$$\% \text{ regulation} = \frac{70 - 400}{400} \times 100 = -82.5\%$$

Thus the % regulations computed from MMF method for different power factor are:

P.F.	0.8 lag	0.8 lead	1.0	Z.P.F. lag	Z.P.F. lead
% regulation	34.5	– 4.5	20	36.25	– 82.5

6.5.4 ASA Method

American Students Association (ASA) method is a modification of MMF method. The method is also referred to as American Standards Association (ASA) method. It yields good regulation for both cylindrical rotor and salient pole alternators. This method adopts the following procedure:

(a) The field current **R**, corresponding to MMF $\mathbf{F}_{r2}$ is obtained on air-gap line corresponding to rated voltage.
(b) The armature reaction field current **A** corresponding to armature reaction MMF $\mathbf{F}_A$ is obtained using SCC for rated armature current.
(c) The difference in field current is obtained on air-gap line and OCC for EMF $\mathbf{E} = \mathbf{V} + \mathbf{I}(R_a + jX_l)$ as shown by a-b in the Fig. 6.15.
(d) MMF diagram for different power factors are shown in Fig. 6.22. The angle between **R** and **A** (or MMF $\mathbf{F}_{r2}$ and $\mathbf{F}_A$) is taken as $90° \pm \theta$, where θ is power factor angle. Here '+' sign is taken for lagging power factor and '–' sign is taken for leading power factor. After obtaining the phasor OF, it is extended by field current FF′ = ab for obtaining OF′. The no- load EMF is obtained corresponding to OF′ on OCC.

Example 6.7 In Example 6.4, determine open-circuit EMF using ASA method, when the alternator is delivering rated current at

(i) 0.8 P.F. lagging

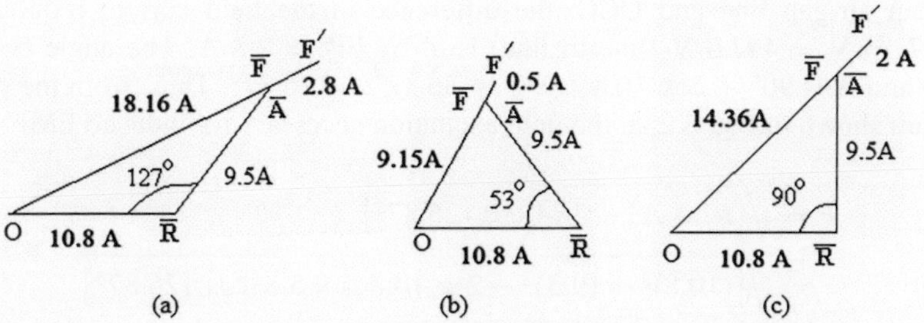

Fig. 6.22 Determination of field current for induced EMF–ASA method at **a** 0.8 lag **b** 0.8 lead and **c** unity power factor

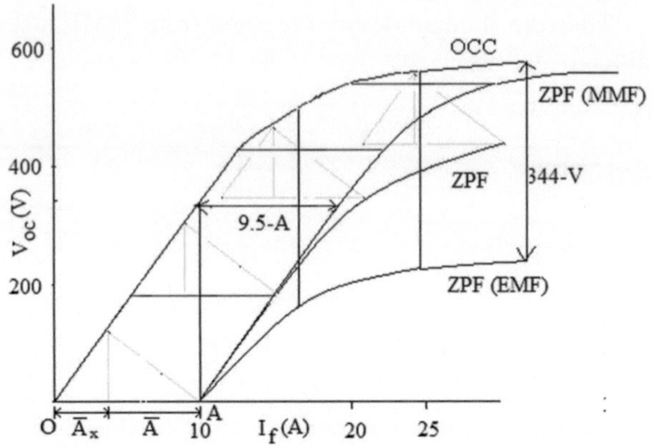

Fig. 6.23 ZPF characteristics obtained from different methods

(ii) 0.8 leading
(iii) U.P.F.
(iv) Z.P.F. lagging
(v) Z.P.F. leading.

Solution
From the OCC shown in Fig. 6.15, the field current required for rated voltage from air-gap line is **R** = 10.8 A and, the armature reaction field current required for rated short-circuit current from SCC is **A** = 9.5 A. Here armature resistance per phase R_a = 1 Ω and leakage reactance X_l= 5.13 Ω per phase (calculated in Example 6.4).

(i) 0.8 P.F. lagging

The induced EMF is **E**,

$$\begin{aligned}\mathbf{E} &= \mathbf{V} + \mathbf{I}(R_a + jX_l)\\ &= 283.353 + j47.304\\ &= 287.27\angle 9.48^\circ \text{ Volt}\end{aligned}$$

From air-gap line and OCC, the difference in the field current required for $287.27\sqrt{3}$ V = 497.6 V (line to line) is ab = FF′ = 2.8 A. The angle between the **R** and **A** is 90° + $\cos^{-1}0.8$ = 90° + 36.87° = 126.87°. Thus, from the phasor diagram shown in Fig. 6.22a, the field excitation necessary for induced EMF is

$$\begin{aligned}F &= \sqrt{\left[R^2 + A^2 - 2RA\cos 126.87^\circ\right]}\\ &= \sqrt{\left[(10.8)^2 + (9.5)^2 - 2\times 10.8\times 9.5\times \cos\ 126.87^\circ\right]}\\ &= 18.16\text{A} = \text{OF}\end{aligned}$$

Corresponding to field current OF′ = OF + FF′ = 18.16 + 2.8 = 20.96 A, the open-circuit EMF from OCC is 545 Volt. The percentage regulation is

$$\% \text{ regulation } = \frac{545 - 400}{400} \times 100 = 36.25\%$$

(ii) 0.8 P.F. leading

The induced EMF E is

$$\begin{aligned} \mathbf{E} &= \mathbf{V} + \mathbf{I}(R_a + jX_l) \\ &= 231 + 13.5(0.8 + j0.6)(1.0 + j5.13) \\ &= 200.247 + j63.504 \\ &= 210.1\angle 17.48^\circ \text{ Volt} \end{aligned}$$

From air-gap line and OCC, the difference in the field current required for $210.1\sqrt{3}$ V = 363.9 V (line to line) is ab = FF′ = 0.5 A.

The angle between the **R** and **A** is $90^\circ - \cos^{-1}0.8 = 90^\circ - 36.87^\circ = 53.13^\circ$.

Thus, from the phasor diagram shown in Fig. 6.22b, the necessary field excitation is

$$\begin{aligned} F &= \sqrt{\left[R^2 + A^2 - 2RA \cos 53.13^\circ\right]} \\ &= \sqrt{\left[(10.8)^2 + (9.5)^2 - 2 \times 10.8 \times 9.5 \cos 53.13^\circ\right]} \\ &= 9.15 \text{ A} = \text{OF} \end{aligned}$$

Corresponding to field current OF′ = OF + FF′ = 9.15+0.5 =9,65 A the open-circuit EMF from OCC is 345 Volt. The parentage regulation is

$$\% \text{ regulation} = \frac{345 - 400}{400} \times 100 = -13.75\%$$

(iii) Unity P.F.

The induced EMF E is

$$\begin{aligned} \mathbf{E} &= \mathbf{V} + \mathbf{I}(R_a + jX_l) \\ &= 244.5 + j69.255 \\ &= 254.1\angle 15.81^\circ \text{ Volt} \end{aligned}$$

From air-gap line and OCC, the difference in the field current required for $254.1\sqrt{3}$ V = 440.11 V (line to line) is ab = FF′ = 2.0 A. The angle between the **R** and **A** is 90°. Thus from the phasor diagram shown in Fig. 6.22c, the necessary field excitation is

$$F = \sqrt{\left[R^2 + A^2 - 2RA \cos 90°\right]}$$
$$= \sqrt{\left[(10.8)^2 + (9.5)^2 - 2 \times 10.8 \times 9.5 \; cos \; 90°\right]}$$
$$= 14.38 \text{ A} = \text{OF}$$

Corresponding to field current OF′ = OF + FF′ = 14.38 + 2 = 16.28 A, the open-circuit EMF from OCC is 495 Volt. The percentage regulation is

$$\% \text{ regulation} = \frac{495 - 400}{400} \times 100 = 23.75\%$$

(iv) Z.P.F. lagging

The induced EMF E is

$$\mathbf{E} = \mathbf{V} + \mathbf{I}(R_a + jX_l)$$
$$= 231 + 13.5(0.0 - j1.0)(1.0 + j5.13)$$
$$= 231 - j1.35 + 69.255$$
$$= 300.6\angle - 2.43° \text{Volt}$$

From air-gap line and OCC, the difference in the field current required for $300.6\sqrt{3} = 520.6$ V (line to line) is ab = FF′ = 4.0 A.

The angle between the **R** and **A** is $90° + \cos^{-1}0.0 = 90° + 90° = 180°$. Thus, from the phasor diagram the necessary field excitation is

$$F = \sqrt{\left[R^2 + A^2 - 2RA \cos 180°\right]}$$
$$= \sqrt{\left[(10.8)^2 + (9.5)^2 - 2 \times 10.8 \times 9.5 \cos 180°\right]}$$
$$= 20.3 \text{ A} = \text{OF}$$

Corresponding to field current OF′ = OF + FF′ = 20.3 + 4 = 24.3 A, the open-circuit EMF from OCC is around 565 Volt. The percentage regulation is

$$\frac{565 - 400}{400} \times 100 = 41.25\%$$

(v) Z P.F. leading

The induced EMF E is

$$\begin{aligned}\mathbf{E} &= \mathbf{V} + \mathbf{I}(R_a + jX_l) \\ &= 231 + 13.5(0.0 + j1.0)(1.0 + j5.13) \\ &= 161.745 + j13.5 \\ &= 162.3\angle 4.77^\circ \text{ Volt}\end{aligned}$$

From air-gap line and OCC, the difference in the field current required for $162.3\sqrt{3} = 281.1$ V (line to line) is ab = FF′ ⟶ 0.0A.

The angle between the **R** and **A** is $90^\circ - \cos^{-1}0.0 = 90^\circ - 90^\circ = 0^\circ$.

Thus from the phasor diagram, the necessary field excitation is

$$\begin{aligned}F &= \sqrt{\left[R^2 + A^2 - 2RA\ \cos\ 0^\circ\right]} \\ &= \sqrt{\left[(10.8)^2 + (9.5)^2 - 2 \times 10.8 \times 9.5\ \cos\ 0^\circ\right]} \\ &= 1.3\text{ A} = \text{OF}\end{aligned}$$

Corresponding to field current OF′ = OF + FF′ = 1.3 A, the open-circuit EMF is 13 V.

The percentage regulation is

$$= \frac{13 - 400}{400} \times 100 = -96.75\%$$

Thus, percentage (%) regulation using ASA method are as follows:

P.F	0.8 lag	0.8 lead	1.0	Z.P.F. lag	Z.P.F. lead
Percentage regulation	36.25	−13.75	27.75	41.25	−96.75

Example 6.8 From the data of Example 6.4, draw ZPF characteristics using EMF and MMF method and test results.

Solution

In EMF method, we can write

$$|\mathbf{V}| \approx |\mathbf{E}_0| - |X_s \cdot \mathbf{I}|$$

or,

$$V = E_0 - IX_s = E_0 - 13.5 \times 14.72 = E_0 - 198.72$$

The ZPF characteristics according to EMF method can be obtained by subtracting $IX_s\sqrt{3} = 13.5 \times 14.72\sqrt{3} = 198.72\sqrt{3} = 344.2$ V line to line on y-axis from the OCC.

The ZPF characteristics using MMF method is obtained by adding the $(A + A_x) = 9.5$ A in x-Axis of the OCC. The actual characteristics are plotted by sliding the Potier triangle on the OCC and plotting the locus of point A′. The ZPF characteristics obtained as above are shown in Fig. 6.23.

Example 6.9 A 1500 RPM, 50 Hz, 40 MVA, 11 kV alternator has the following test data available:

V_{oc} (kV)	10	10.8	11.5	12.1	12.6	13	14	14.5	14.8
I_f (A)	175	200	225	250	275	300	400	450	500

$$\text{Short - circuit test : } I_f = 200\text{ A} \quad I_a = 2099\text{ A}$$

$$\text{Potier test: } I_f = 450\text{ A} \quad I_a = 2099\text{ A} \quad V = 11\text{ kV}$$

Compute the induced EMF, power angle, % regulation at 0.8 power factor lagging using general method and ASA method.

Solution

Figure 6.24 shows the OCC, SCC, and Potier triangle.

The rated terminal voltage per phase is given as

$$V = 11000/1.7321 = 6351\text{ V}$$

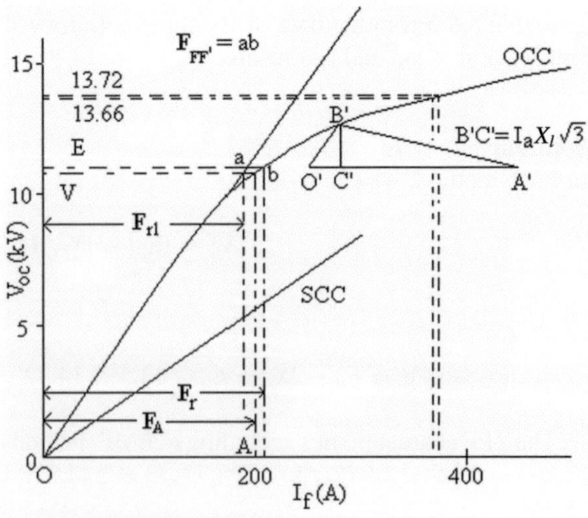

Fig. 6.24 OCC, SCC. and Pottier triangle—determination of F_r, $F_{FF'}$, F_{r1} and open-circuit EMF corresponding to (OF) in MMF method and (OF′) in ASA method

Full load per phase current is

$$I = 40 \times 1{,}000{,}000/\sqrt{3} \times 11{,}000 = 2099.35\ \text{A} = 2099\ \text{A}$$

From the Potier triangle, we get

$$I_a X_l = 1.6292/\sqrt{3}$$

Thus,

$$X_1 = 0.4481\ \Omega \text{ per phase}$$

General method:

The induced EMF due to air-gap flux, when power factor is lagging and **V** is taken as the reference phasor, is

$$E = V + I(\cos\theta - j\sin\theta)(0.0 + jX_1)$$

$$\begin{aligned} E &= 6351 + 2099(0.8 - j0.6)(0.0 + j0.4481) \\ &= 6956 + j752.5 \\ &= 6956\angle 6.2^\circ \text{ Volt} \end{aligned}$$

The line-to-line EMF is $\sqrt{3} \times 6956 = 12048$ V. The field excitation corresponding to 12.048 kV on airgap line gives **R** = 208.4 A. The field excitation corresponding to the armature reaction is **A** = 200 A (field current required for rated armature current in SCC). The angle between the **R** and **A** is $90^\circ + \cos^{-1}0.8 + 6.2 = 133^\circ$. Thus, from the phasor diagram shown in Fig. 6.25a, the necessary field excitation is

$$F = \sqrt{\left[R^2 + A^2 - 2RA\ \cos\ 133^\circ\right]}$$

$$\begin{aligned} \text{F} &= \sqrt{\left[(208.4)^2 + (200)^2 - 2 \times 208.4 \times 200 \times \cos\ 133^\circ\right]} \\ &= \sqrt{\left[(208.4)^2 + (200)^2 - 2 \times 208.4 \times 200 \times (-0.6683)\right]} \\ &= 374.54\ \text{A} \end{aligned}$$

The value of open-circuit line-to-line EMF from OCC corresponding to field current of 374 A is 13.72 kV. Therefore percentage regulation is (13.72–11)/11 = 24.72%.

ASA method gives fairly accurate regulation for both the salient pole and cylindrical rotor machines. Corresponding to this line EMF is $\sqrt{3} \times 6956 = 12.048$ kV, the difference in excitation between air-gap line and OCC is $F_{FF'} = 33$ A = ab.

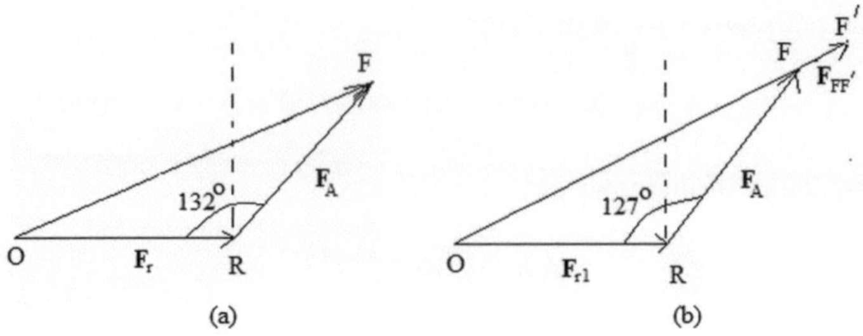

Fig. 6.25 MMF diagram for **a** general method, **b** ASA method

The field excitation F_{R1} is obtained from the air-gap line corresponding to terminal voltage. The field current required for rated voltage from air-gap line is $\mathbf{F}_{r1} = 193$ A and the field current required for rated short-circuit current from SCC is $\mathbf{F}_A = 200$A.

Form the graph, $F_{r1} = 193$ A. the angle between F_{R1} and $A + A_x$ is roughly 90° + p.f. angle. Here this angle is $90° + 36.87° = 126.87°$. Thus, we get

$$\mathrm{F} = \sqrt{\left[(193)^2 + (200)^2 - 2 \times 193 \times 200 \times \cos 126.87°\right]}$$

$$= \sqrt{\left[(193)^2 + (200)^2 - 2 \times 193 \times 200 \times (-0.60)\right]} = 351.5\,\mathrm{A}$$

Thus total excitation required is 351.5 +33 = 384.5 A. The EMF corresponding to this field current is 13.66 kV. The percentage regulation is (13.66 − 11)/11 = 24.18%.

6.6 Experimental Determination of Regulation—OCC, SCC and ZPF Characteristics

The *open-circuit characteristic* (OCC), also called open-circuit saturation curve or magnetizing curve, is a plot of the terminal voltage as a function of field excitation, with alternator is running at rated speed without any load. The experimental set-up is shown in Fig. 6.26; the alternator is being rotated at synchronous speed ω_s by a prime mover. Under these conditions, the alternator induced voltage is directly proportional to the flux per pole and the characteristic is effectively a plot of flux per pole as a function of field excitation (i.e., field current of the field winding or AT/pole). The shape of the characteristic is, therefore, a typical B-H curve or magnetization curve as shown in Figs. 6.10 and 6.27.

The short-circuit characteristic (SCC) is a plot of steady-state AC armature current as a function of DC field current, with symmetrical three-phase short-circuit applied across the alternator terminals, with alternator running at rated speed. The experimental set-up is shown in Fig. 6.28. Under this condition, the current in the armature winding totally depends on the internal impedance (resistance and leakage reactance

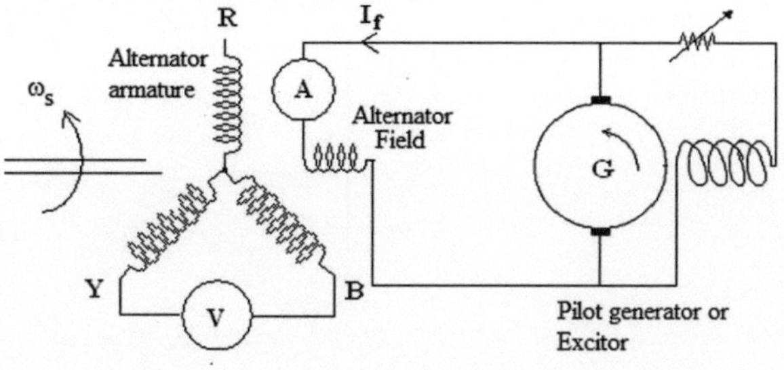

Fig. 6.26 Experimental set-up for open-circuit test of alternator

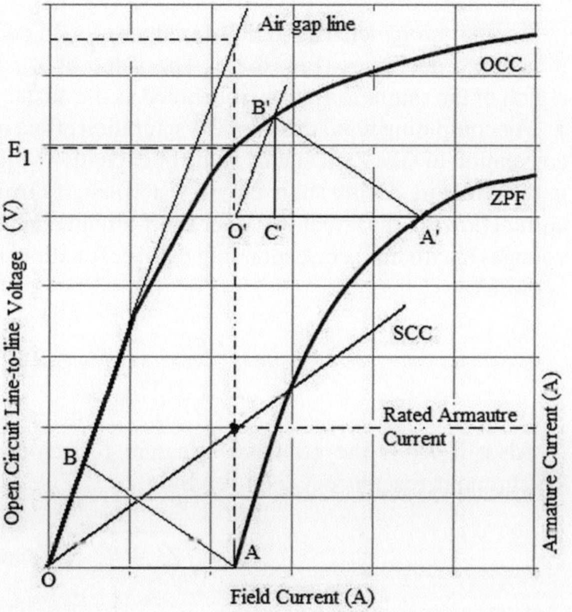

Fig. 6.27 Open-circuit, short-circuit and zero power factor characteristics

of armature) and the armature reaction. The phase angle between the short-circuit current and the induced voltage may be taken to be nearly 90° lagging, as armature resistance is negligible. Therefore, the armature reaction MMF is almost totally demagnetizing. When the steady-state full load rated current is flowing under short-circuit conditions, the resultant excitation acting on the magnetic circuit is low and the magnetic circuit is unsaturated. The short-circuit characteristic is therefore a straight line as shown in Figs. 6.10 and 6.27.

At any value of excitation, if E_0 is the open-circuit voltage, and I_{sc} is the steady-state short-circuit current, then for this value of field current ($A + A_x$), the unsaturated synchronous impedance is given as

$$Z_s = \left.\frac{E_0}{I_{sc}}\right|_{I_f=\text{Const}} \tag{6.29}$$

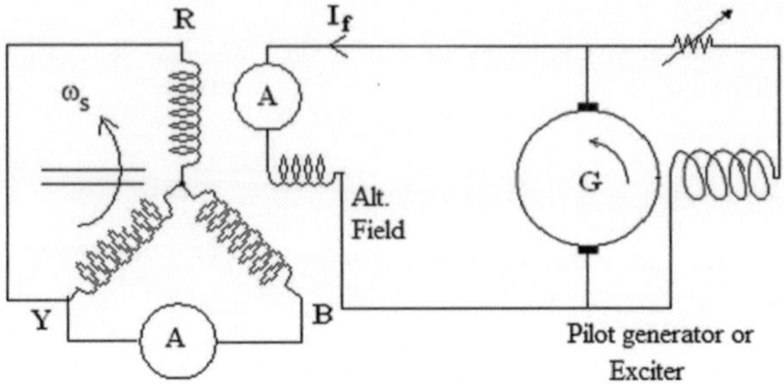

Fig. 6.28 Experimental set-up for short-circuit test of alternator

As saturation increases at higher values of field current, the value of synchronous impedance decreases (Fig. 6.27). The value of Z_s, calculated for the unsaturated region of the saturation curve, is denoted as the unsaturated synchronous impedance.

For computing regulation, unless saturation effect is otherwise corrected for, it is a convention to take Z_s at such a value of excitation which causes rated current to flow in the armature during short-circuit test. Then, if OA is the excitation at which rated current flows in the armature under short-circuit conditions and E_0 is the open-circuit voltage (line to line) corresponding to this excitation, the synchronous impedance Z_s is given by

$$Z_s = \left.\frac{E_1}{I_{\text{Rated}}\sqrt{3}}\right|_{I_f=\text{Const.}} = R_a + jX_s \tag{6.30}$$

Also, if R_a is the effective armature resistance of armature per phase, then synchronous reactance X_s is given by,

$$X_s = \sqrt{Z_s^2 - R_a^2} \ \ \Omega/\text{phase} \tag{6.31}$$

The calculation of open-circuit EMF and percentage (%) regulation has already been discussed. Figure 6.29 shows a test set-up for ZPF characteristics or Potier test. The load is a purely three-phase inductive load. The alternator is overexcited (increased field current more than normal excitation) for flowing rated current at rated voltage. In absence of suitable inductive load, a three-phase under-excited synchronous motor, which acts as an inductive load is used. In such a case, the alternator is overexcited but synchronous motor is under-excited so that alternator delivers rated current with lagging power factor at rated terminal voltage. These tests are performed on both cylindrical rotor and salient pole alternators/synchronous machines.

The phasor diagram for zero power factor lagging load is shown in Fig. 6.21.

It is clear that if we ignore the resistance drop, then the numerical value of induced EMF is equal to the applied voltage plus leakage reactance drop in an alternator. The procedure for drawing the Potier triangle is given below:

(a) OA is the field current required in short-circuit test for rated armature current.

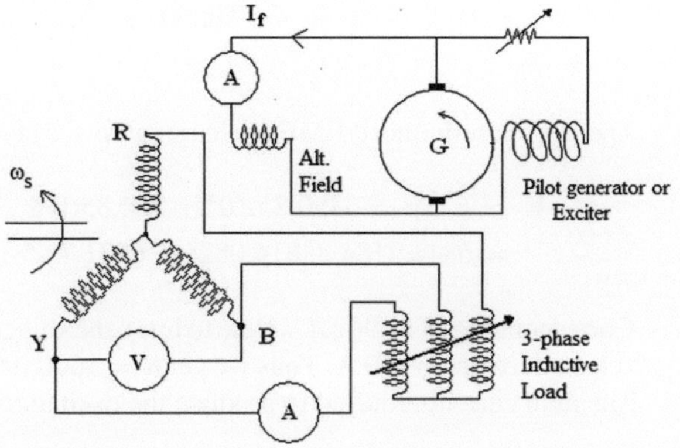

Fig. 6.29 Test set-up for zero power factor characteristics—Potier test

(b) A′ is the field current at rated voltage for flowing rated current in ZPF test.
(c) Shift OA at O′A′, such that OA = O′A′.
(d) From point O′, draw a line parallel to the air-gap line, which cuts open-circuit characteristics at B′.
(e) Join B′ and A′. Triangle O′A′B′ is known as Potier triangle.
(f) Drop perpendicular B′C′ on line O′A′ from point B′ or B′C′⊥ O′A′.
(g) Measure B′C′.
(h) Calculate per phase leakage reactance as

$$X_l = \frac{B'C'}{I_a\sqrt{3}} \tag{6.32}$$

(i) Obtain the locus of point A′ by sliding the Potier triangle such that B′ always rests on OCC. This is referred as *zero power factor or ZPF characteristics*.

Example 6.10 A 10-MVA three-phase star connected, two-pole, 50 Hz, 11 kV line-to-line turbo alternator has the air-gap characteristics as a straight line represented by 12 kV for field current of 160 A. The armature winding resistance is 0.05 Ω per phase and the winding leakage reactance per phase is 1.5 Ω. Find the field current required to produce rated terminal voltage when rated current is delivered at 0.8 power factor lagging balanced load. The armature MMF at rated current is equivalent to field current of 150 A.

Solution

The rated per phase voltage is

$$V_t = \frac{11000}{\sqrt{3}} = 6350.85\text{ V}$$

The rated armature current

$$I_a = \frac{10000}{11\sqrt{3}} = 524.86\ \text{A}$$

The per phase-induced EMF for lagging power factor rated current

$$\begin{aligned}\mathbf{E}_r &= E_r\angle\gamma = 6350.85\angle 0° + 524.85(0.8 - j0.6)(0.05 + 1.5j)\\ &= 6844.2184 + 614.082j = 6871.7\angle 5.12°\text{V}\end{aligned}$$

Corresponding to 11,902.4 V (line to line), the value of field current **R** from the air-gap characteristics is 160 A. Thus we get $R = 160\angle(90° + 5.12°) = 160\angle 95.12°$A.

The field current required to produce the excitation EMF

$$\begin{aligned}\text{F} = \text{R} - \text{A} &= 160\angle 95.12^{\circ} - 150\angle - 36.8^{\circ}\\ &= 283.21\angle 118.39^{\circ} A\end{aligned}$$

The required field current is 283.21 A.

Example 6.11 A 1500 RPM, 50 Hz 50,000 kVA, 11 kV synchronous generator has the following test data.

I_f (A)	175	200	225	250	275
V_{oc}	12	13	13.8	14.5	15.1

SC Test: $I_f = 200$ A at rated short-circuit current

ZPF Test: $I_f = 400$ A at rated voltage and rated current.

Determine leakage reactance, synchronous reactance and hence determine the field current required for maintaining the rated terminal voltage for rated kVA at 0.8 lagging power factor using different methods.

Solution

Line-to-line voltage 11 kV. The OC, SC. and ZPF are shown in Fig. 6.30.

The rated per phase voltage is

$$V_t = \frac{11000}{\sqrt{3}} = 6350.85\ \text{V}$$

The rated armature current

$$I_a = \frac{50000}{11\sqrt{3}} = 2624.3\ \text{A}$$

The Synchronous reactance

$$X_s = \frac{13000}{2624.3\sqrt{3}} = 2.86\ \Omega$$

Fig. 6.30 Characteristics OCC SCC of Example No. 6.11

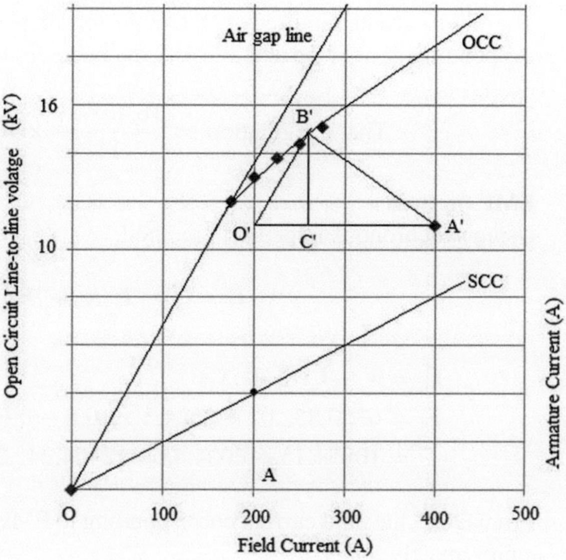

The leakage reactance,

$$X_l = \frac{B'C'}{2624.3\sqrt{3}} = \frac{4000}{2624.3\sqrt{3}} = 0.88\ \Omega$$

Using general method
The induced EMF E

$$\mathbf{E}_r = \mathbf{V} + \mathbf{I}_a(R_a + jX_l)$$

$$\begin{aligned}\mathbf{E}_r &= 6350.85\angle 0^\circ + 2624.3 \times (0.8 - 0.6j) \times (0.0 + j0.88) \\ &= 6350.85 + 1847.51j + 1385.6 \\ &= 7736.48 + 1847.51j = 7954\angle 13.43^\circ\ \text{V}\end{aligned}$$

From OCC, the field current corresponding to EMF $7954\sqrt{3}$ V = 13.776 kV on OCC is 220 A or **R** = 220 A. The armature reaction from SCC is **A** = 200 A. The angle between the **R** and **A** is $90^\circ + 13.43^\circ + \cos^{-1}0.8 = 90^\circ + 13.43^\circ + 36.87^\circ = 140.3^\circ$. Thus, the necessary field excitation is

$$\begin{aligned}F &= \sqrt{\left[R^2 + A^2 - 2R \cdot A\cos 140.3^\circ\right]} \\ &= \sqrt{\left[(220)^2 + (200)^2 - 2 \times 220 \times 200\cos 140.3^\circ\right]} \\ &= 395.1\ \text{A}\end{aligned}$$

The field current required to maintain terminal voltage at 11 kV is 395.1 A. Corresponding to field current 395.1 A the open-circuit EMF from OCC is 18.4 kV.

$$\text{The \% regulation} = \frac{18.4 - 11}{11} \times 100 = 67.27\%$$

EMF method

Using synchronous impedance method,

$$\mathbf{E}_\mathrm{r} = \mathbf{V} + \mathbf{I}_\mathrm{a}(R_\mathrm{a} + jX_\mathrm{s})$$

$$\begin{aligned}
\mathbf{E}_\mathrm{r} &= \mathbf{V} + \mathbf{I}_\mathrm{a}(R_\mathrm{a} + jX_\mathrm{s}) \\
\mathbf{E}_\mathrm{r} &= 6350.85\angle 0^\circ + 2624.3 \times (0.8 - 0.6j) \times (0.0 + j2.86) \\
&= 10854.15 + 6004.4j = 12404.24\angle 28.95^\circ \text{ V}
\end{aligned}$$

From OCC, the field current corresponding to EMF 12,404.24$\sqrt{3}$ V = 21.404-kV is approximately 500A.

MMF method

The induced EMF E,

$$\mathrm{E} = \mathrm{V} + \mathrm{I}R_a$$

$$E = 6350.85\angle 0^\circ + 2624.3 \times (0.8 - 0.6j) \times 0.0$$

$$E = 6350.85 \text{ V}\angle 0^\circ \text{ Volt}$$

From OCC the field current required for 6350.85-$\sqrt{3}$ = 11 kV (line to line) is $\mathbf{R} = 160$ A and $\mathbf{A} = 200$ A

The angle between the **R** and **A** is $90^\circ - 0^\circ + \cos^{-1}0.8 = 90^\circ - 0^\circ + 36.87^\circ = 126.87^\circ$.

Thus from the phasor diagram, the field excitation necessary is

$$\begin{aligned}
F &= \sqrt{\left[R^2 + A^2 - 2R \cdot A \cos 126.87^\circ\right]} \\
&= \sqrt{\left[(160)^2 + (200)^2 - 2 \times 160 \times 200 \cos 126.87^\circ\right]} \\
&= 322.5 \text{ A.}
\end{aligned}$$

The field current required to maintain the voltage is 322.5 A, from MMF method.

ASA Method

The induced EMF E, $\mathbf{E}_\mathrm{r} = \mathbf{V} + \mathbf{I}_\mathrm{a}(R_\mathrm{a} + jX_l)$

$$\mathbf{E}_r = 6350.85\angle 0^\circ + 2624.3 \times (0.8 - 0.6j) \times (0.0 + j0.88)$$
$$= 7736.48 + 1847.51j = 7954\angle 13.43^\circ \text{ V}$$

From air-gap line and OCC, the difference in the field current required for $7954\sqrt{3} = 13.8$ kV (line to line) is ab = FF′ = 30 A. The angle between the $\mathbf{R}_1$ and $\mathbf{A}$ is $90^\circ + \cos^{-1}0.8 = 90^\circ + 36.87^\circ = 126.87^\circ$. Thus, from the phasor diagram, the field excitation necessary is

$$F = \sqrt{\left[R_1^2 + A^2 - 2R_1 A\cos 126.87^\circ\right]}$$
$$= \sqrt{\left[(220)^2 + (200)^2 - 2 \times 220 \times 200 \times \cos 126.87^\circ\right]}$$
$$= 376 \text{ A}$$

Corresponding to field current required to keep the terminal voltage as rated voltage from ASA method. is 376 + 30 = 406 A.

6.7 Dynamic Equations of Synchronous Motor

For studying the dynamic operation of synchronous motor, Park-Blondel equations are used. For this also refer to generalized machine theory explained in Sect. 7.10 and also in the next section on two-axis theory.

Consider, a three-phase synchronous machine with salient poles having squirrel cage or damper bars located in the pole face with complete short-circuiting ring, so that the cage winding/damper bars acts along both the direct-axis (polar-axis) and the quadrature-axis (interpolar-axis) [3, 6, 10, 12].

It is assumed that

(a) Core permeability is infinite and magnetic circuit is not saturated.
(b) Iron losses are negligible.
(c) Machine is having perfect electrical symmetry
(d) Width of air-gap is uniform under the poles.
(e) Only fundamental harmonics of excitation and reaction fields are considered.

The schematic diagram for deriving Park-Blondel equations is shown in Fig. 6.31. The electrical layout of the machine is shown. It has three stator windings in space round interior periphery of stator. They are uniformly spaced. The rotor has the excitation windings. The axis of North Pole of the rotor is known as direct-axis and is denoted by d. The field or excitation winding is denoted by F or f. The leading axis with respect to the direction of speed of interpolar (large air-gap) zone is known as quadrature-axis and is denoted by q. One of two short-circuited windings on rotor corresponds to D-axis damper winding in the direction of direct-axis is an additional winding in d-axis along with field winding and is denoted by D. The Q-axis damper windings work in the q-axis and is denoted by Q. Space distribution of different magnetic axes has been shown in Fig. 6.32.

Let ω be the instantaneous angular speed of rotor in counterclockwise direction. The geometrical angle between the d-axis and the symmetry axis of '1' phase winding of the stator is considered as space reference and this angle is β. Thus,

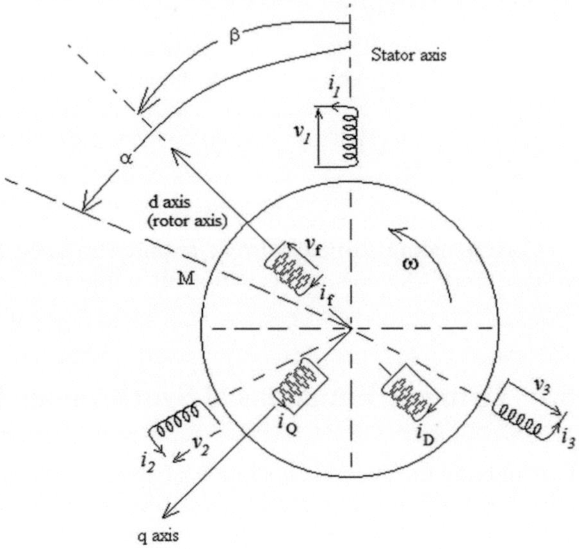

Fig. 6.31 Diagram for deriving the Park-Blondel equations

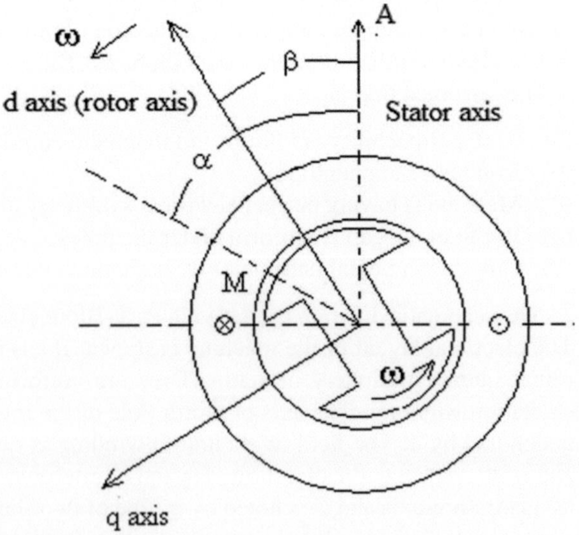

Fig. 6.32 Space distribution of magnetic axes of a salient pole synchronous machine

$$\omega = \frac{d\beta}{dt} \tag{6.33}$$

The electrical angle corresponding to rotor position is $\gamma = p\beta$ There are three stator windings 1, 2, 3 and three rotor windings—f (or F), D, and Q.

The self-inductances of these windings are L_{11}, L_{22}, L_{33}, and mutual inductances are L_{12}, L_{23}, L_{31}. These inductances vary with rotor position because of saliency. When applied to three-phase induction motor or cylindrical rotor synchronous machines, they are constant with rotor position. Let phase winding '1' carries the current i_1, if the air-gap is constant 'g'(for cylindrical rotor machine), it produces a space sinusoidal magnetic field as

$$B_0(p\alpha) = \frac{4}{\pi}\mu_0 \frac{\mathrm{Ni}}{\mathrm{pg}} k_w \cos p\alpha \tag{6.34}$$

where N is the number of turns of a single-phase winding, k_w is the winding factor, and α is the coordinate of a given point M on a stator phase winding with respect to the axis shown in the figure. Here p is the pole pairs and for two pole $p = 1$.

In the interpolar region, the magnetic flux density is almost zero because of very large air-gap here results in large reluctance which opposes the flux. As such the flux density distribution due to phase currents i_1, in air-gap is non-sinusoidal. For obtaining the inductances one has to find the fundamental component of flux density distribution. Therefore,

$$B_1(p\alpha) = C_1 \sin p\alpha + C_2 \cos p\alpha$$

Since flux density $B_1(p\alpha + \pi) = -B_1(p\alpha)$ and in interpolar region $B_1(p\alpha) = 0$. One can show that

$$C_1 = \frac{1}{\pi}\int_0^{2\pi} B(p\alpha) \sin p\alpha.d(p\alpha) = \frac{2}{\pi}\int_0^{\pi} B(p\alpha) \sin p\alpha.d(p\alpha)$$

$$C_2 = \frac{1}{\pi}\int_0^{2\pi} B(p\alpha)\cos p\alpha.d(p\alpha) = \frac{2}{\pi}\int_0^{\pi} B(p\alpha) \cos p\alpha.d(p\alpha)$$

Thus

$$C_1 = \frac{4\mu_0 N k_w i_1}{\pi^2 pg} \sin \pi\alpha_p \sin 2\gamma$$

And

$$C_2 = \frac{4\mu_0 N k_w i_1}{\pi^2 pg} [\pi\alpha_p + \sin \pi\alpha_p \cos 2\gamma]$$

Here the electrical angle corresponding to instantaneous rotor position is $\gamma = p\beta$. The fractional polar arc α_p is given with respect to pole pitch τ and pole b_p arc as $\alpha_p = b_p/\tau$. In salient pole machine, the air-gap 'g' is variable, i.e., constant under electrical angular aperture $\pi\alpha_p$. The flux linkage of phase '1' due to its own currents is given by

$$\lambda_{11} = Nk_w L\tau \frac{1}{\pi}\int_0^{\pi} B^1(p\alpha)d(p\alpha) = \frac{8\mu_0(Nk_w)^2 L\tau i_1}{\pi^3 pg}\left[\pi\alpha_p + \sin\pi\alpha_p\cos 2\gamma\right] \tag{6.35}$$

From the above expression the self-induction of phase winding is

$$L_{11} = \frac{\lambda_{11}}{i_1} = \left[L_1 + L_2\cos 2\gamma\right]$$

Here L_1 and L_2 are shown in Fig. 6.33. If leakage inductance L_l is also considered, then

$$L_{11} = \left[L_l + L_1 + L_2\cos 2\gamma\right] \tag{6.36}$$

Similarly, it can be shown that flux linkage produced by phase '1' linking the phase '2' winding is

$$\lambda_{12} = Nk_w L\tau \frac{1}{\pi}\int_{2\pi/3}^{5\pi/3} B^1(p\alpha)d(p\alpha) = \frac{Nk_w L\tau}{\pi}\left[-C_1\sqrt{3} - C_2\right]$$

Thus mutual inductance

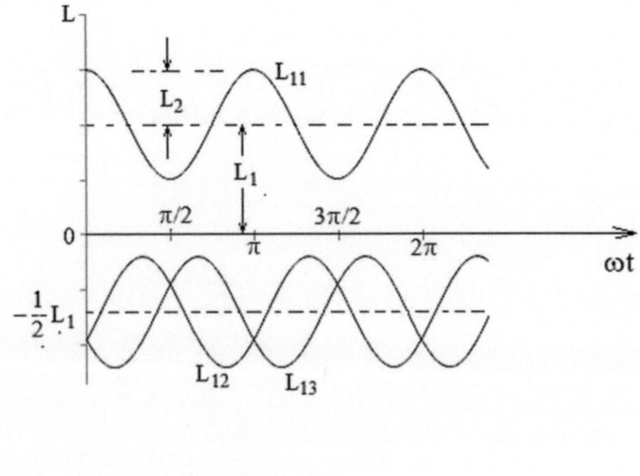

Fig. 6.33 Variation of self and mutual inductances of the stator with respect to time

$$L_{12} = \left[-\frac{1}{2}L_1 + L_2 \cos\left(2\gamma - \frac{2\pi}{3}\right)\right] \tag{6.37}$$

The variation of self and mutual inductances of the stator with respect to time is given in Fig. 6.33.

Based on above derivations one can write that the self-inductances are

$$L_{11} = [L_l + L_1 + L_2 \cos 2\gamma]$$

$$L_{22} = \left[L_l + L_1 + L_2 \cos\left(2\gamma - \frac{4\pi}{3}\right)\right]$$

$$L_{33} = \left[L_l + L_1 + L_2 \cos\left(2\gamma - \frac{2\pi}{3}\right)\right]$$

And mutual inductances are

$$\begin{aligned} L_{12} = L_{21} &= \left[-\frac{1}{2}L_1 + L_2 \cos\left(2\gamma - \frac{2\pi}{3}\right)\right] \\ L_{13} = L_{31} &= \left[-\frac{1}{2}L_1 + L_2 \cos\left(2\gamma - \frac{4\pi}{3}\right)\right] \\ L_{23} = L_{32} &= \left[-\frac{1}{2}L_1 + L_2 \cos 2\gamma\right] \end{aligned} \tag{6.38}$$

Similarly, one can write the mutual inductances between stator and field windings on rotor,

$$\begin{aligned} L_{1\mathrm{F}} &= L_{1\mathrm{FM}} \cos\gamma \\ L_{2\mathrm{F}} &= L_{1\mathrm{FM}} \cos\left(\gamma - \frac{2\pi}{3}\right) \\ L_{3\mathrm{F}} &= L_{1\mathrm{FM}} \cos\left(\gamma - \frac{4\pi}{3}\right) \end{aligned} \tag{6.39}$$

where the maximum mutual inductance is,

$$L_{1\mathrm{FM}} = \frac{2\mu_0 N_\mathrm{F} k_\mathrm{F} N k_\mathrm{w} L\tau}{\pi\ g}.$$

Here N_F is the number of turns in field winding. Similarly, mutual inductances between stator windings and damper windings D and Q are

$$L_{1\mathrm{D}} = L_{1\mathrm{DM}} \cos\gamma$$

$$L_{2\mathrm{D}} = L_{1\mathrm{DM}} \cos\left(\gamma - \frac{2\pi}{3}\right)$$

$$
\begin{aligned}
L_{3D} &= L_{1DM}\cos\left(\gamma - \frac{4\pi}{3}\right)\\
L_{1Q} &= -L_{1QM}\sin\gamma\\
L_{2Q} &= L_{1QM}\sin\left(\gamma - \frac{2\pi}{3}\right)\\
L_{3Q} &= -L_{1QM}\sin\left(\gamma - \frac{4\pi}{3}\right)
\end{aligned}
\tag{6.40}
$$

The inductances L_{FF}, L_{DD}, L_{QQ}, $L_{FD} = L_{DF}$ are constant irrespective of rotor position. The flux produced by Q damper winding does not link the field winding and D damper windings. As such no mutual inductance exists between Q and F, and Q and D,

or

$$L_{FQ} = 0 \text{ and } L_{QD} = 0$$

The differential circuit equations relating the instantaneous voltages and currents can be written as

$$
\begin{aligned}
v_1 = Ri_1 &+ \frac{d}{dt}(L_{11}i_1) + \frac{d}{dt}(L_{12}i_2) + \frac{d}{dt}(L_{13}i_3)\\
&+ \frac{d}{dt}(L_{1F}i_F) + \frac{d}{dt}(L_{1D}i_D) + \frac{d}{dt}(L_{1Q}i_Q)
\end{aligned}
$$

$$
\begin{aligned}
v_2 = Ri_2 &+ \frac{d}{dt}(L_{21}i_1) + \frac{d}{dt}(L_{22}i_2) + \frac{d}{dt}(L_{23}i_3)\\
&+ \frac{d}{dt}(L_{2F}i_F) + \frac{d}{dt}(L_{2D}i_D) + \frac{d}{dt}(L_{2Q}i_Q)
\end{aligned}
$$

$$
\begin{aligned}
v_3 = Ri_3 &+ \frac{d}{dt}(L_{31}i_1) + \frac{d}{dt}(L_{32}i_2) + \frac{d}{dt}(L_{33}i_3)\\
&+ \frac{d}{dt}(L_{3F}i_F) + \frac{d}{dt}(L_{3D}i_D) + \frac{d}{dt}(L_{3Q}i_Q)
\end{aligned}
$$

$$
\begin{aligned}
v_F = R_Fi_F &+ L_{FF}\frac{d}{dt}(i_F) + \frac{d}{dt}(L_{1F}i_1) + \frac{d}{dt}(L_{2F}i_2)\\
&+ \frac{d}{dt}(L_{3F}i_3) + L_{FD}\frac{d}{dt}(i_D)
\end{aligned}
$$

$$
\begin{aligned}
0 = R_Di_D &+ L_{DD}\frac{d}{dt}(i_D) + \frac{d}{dt}(L_{1D}i_1) + \frac{d}{dt}(L_{2D}i_2)\\
&+ \frac{d}{dt}(L_{3D}i_3) + L_{FD}\frac{d}{dt}(i_F)
\end{aligned}
$$

$$0 = R_Q i_Q + L_{QQ}\frac{d}{dt}(i_Q) + \frac{d}{dt}(L_{1Q} i_1) + \frac{d}{dt}(L_{2Q} i_2) + \frac{d}{dt}(L_{3Q} i_3)$$

The above set of equations can be shown in matrix form as

$$\begin{bmatrix} v_1 \\ v_2 \\ v_3 \\ v_F \\ v_D = 0 \\ v_Q = 0 \end{bmatrix} = \begin{bmatrix} R + \frac{d}{dt}L_{11} & \frac{d}{dt}L_{12} & \frac{d}{dt}L_{13} & \frac{d}{dt}L_{1F} & \frac{d}{dt}L_{1D} & \frac{d}{dt}L_{1Q} \\ \frac{d}{dt}L_{21} & R + \frac{d}{dt}L_{22} & \frac{d}{dt}L_{23} & \frac{d}{dt}L_{2F} & \frac{d}{dt}L_{2D} & \frac{d}{dt}L_{2Q} \\ \frac{d}{dt}L_{31} & \frac{d}{dt}L_{32} & R + \frac{d}{dt}L_{33} & \frac{d}{dt}L_{3F} & \frac{d}{dt}L_{3D} & \frac{d}{dt}L_{3Q} \\ \frac{d}{dt}L_{F1} & \frac{d}{dt}L_{F2} & \frac{d}{dt}L_{F3} & R_F + \frac{d}{dt}L_{FF} & \frac{d}{dt}L_{FD} & 0 \\ \frac{d}{dt}L_{D1} & \frac{d}{dt}L_{D2} & \frac{d}{dt}L_{D3} & \frac{d}{dt}L_{DF} & R_D + \frac{d}{dt}L_{DD} & 0 \\ \frac{d}{dt}L_{Q1} & \frac{d}{dt}L_{Q2} & \frac{d}{dt}L_{Q3} & 0 & 0 & R_Q + \frac{d}{dt}L_{QQ} \end{bmatrix} \begin{bmatrix} i_1 \\ i_2 \\ i_3 \\ i_F \\ i_D \\ i_Q \end{bmatrix} \tag{6.41}$$

In these equations shown in matrix form, the operator, $\frac{d}{dt}$ applies to both inductances and currents. The inductances depend on the rotation of the rotor and currents; both are dependent on time. This makes the system of equations highly nonlinear set of differential equations. It is quite cumbersome to obtain its direct solution. Due to availability of high-speed computers, the solution is possible with little difficulties. For simplifying the solution, a change in variables (mathematical transform) as suggested by Park is applied by Blondel for salient pole machine. According to this, any multiphase balanced symmetric system can be represented in new set of variables v_d, i_d and v_q, i_q. The details have been elaborated in Sect. 7.10. In case of balanced electrical system

$$\begin{aligned} v_1 + v_2 + v_3 &= 0 \\ i_1 + i_2 + i_3 &= 0 \end{aligned} \tag{6.42}$$

The above expressions are also valid for a balanced and symmetrical synchronous machine. The three-phase system of voltage equations is represented inverse Parks's transformation as

$$\begin{aligned} v_1 &= v_{\mathrm{d}} \cos\gamma - v_{\mathrm{q}} \sin\gamma \\ v_2 &= v_{\mathrm{d}} \cos\left(\gamma - \frac{2\pi}{3}\right) - v_{\mathrm{q}} \sin\left(\gamma - \frac{2\pi}{3}\right) \\ v_3 &= v_{\mathrm{d}} \cos\left(\gamma - \frac{4\pi}{3}\right) - v_{\mathrm{q}} \sin\left(\gamma - \frac{4\pi}{3}\right) \end{aligned} \tag{6.43}$$

Similarly, currents are also represented. The power balance equations show that the instantaneous power in the three-phase and transformed d-q quantities must be identical or,

$$v_1 i_1 + v_2 i_2 + v_3 i_3 = v_{\mathrm{d}} i_{\mathrm{d}} + v_{\mathrm{q}} i_{\mathrm{q}} \tag{6.44}$$

Using above equation, the system of transformed d-q set of voltages and currents can be obtained. Two equations given below are known as Park's transformation.

$$\begin{aligned} v_{\mathrm{d}} &= \frac{2}{3}\left[v_1 \cos\gamma + v_2 \cos\left(\gamma - \frac{2\pi}{3}\right) + v_3 \cos\left(\gamma - \frac{4\pi}{3}\right)\right] \\ v_{\mathrm{q}} &= -\frac{2}{3}\left[v_1 \sin\gamma + v_2 \sin\left(\gamma - \frac{2\pi}{3}\right) + v_3 \sin\left(\gamma - \frac{4\pi}{3}\right)\right] \end{aligned} \tag{6.45}$$

The three-phase system of equations is replaced by d-q set of equations. The new system of equations is as follows:

$$\begin{aligned} v_{\mathrm{d}} &= R i_{\mathrm{d}} + \frac{\mathrm{d}\lambda_{\mathrm{d}}}{\mathrm{d}t} - \lambda_{\mathrm{q}} \frac{\mathrm{d}\gamma}{\mathrm{d}t} \\ v_{\mathrm{q}} &= R i_{\mathrm{q}} + \frac{\mathrm{d}\lambda_{\mathrm{q}}}{\mathrm{d}t} + \lambda_{\mathrm{d}} \frac{\mathrm{d}\gamma}{\mathrm{d}t} \\ v_{\mathrm{F}} &= R_{\mathrm{F}} i_{\mathrm{F}} + \frac{\mathrm{d}\lambda_{\mathrm{F}}}{\mathrm{d}t} \\ 0 &= R_{\mathrm{D}} i_{\mathrm{D}} + \frac{\mathrm{d}\lambda_{\mathrm{D}}}{\mathrm{d}t} \\ 0 &= R_{\mathrm{Q}} i_{\mathrm{Q}} + \frac{\mathrm{d}\lambda_{\mathrm{Q}}}{\mathrm{d}t} \end{aligned} \tag{6.46}$$

In the above set of equations, voltage terms containing $\mathrm{d}\lambda/\mathrm{d}t$ are the transformer-induced voltages and voltage terms containing $\lambda(\mathrm{d}/\mathrm{d}t)$ are the speed-induced voltage. Further, in these equations,

$$
\begin{aligned}
\lambda_d &= L_d i_d + L_{dF} i_F + L_{dD} i_D \\
\lambda_q &= L_q i_q + L_{qQ} i_Q \\
\lambda_F &= \frac{3}{2} L_{dF} i_d + L_{FF} i_F + L_{FD} i_D \\
\lambda_D &= \frac{3}{2} L_{dD} i_D + L_{FD} i_F + L_{DD} i_D \\
\lambda_Q &= \frac{3}{2} L_{qQ} i_q + L_{QQ} i_Q
\end{aligned}
\tag{6.47}
$$

In the above equations, the term L_d is known as d- axis inductance and L_q is known as q-axis inductance. These parameters are experimentally determined using slip test. Figure 6.34 shows the equivalent scheme corresponding to system of equations. The electromagnetic torque in d-q frame is obtained from

$$T = \frac{3}{2} p \left(\lambda_d i_q - \lambda_q i_d \right) \tag{6.48}$$

The expression can also be derived using matrix expression

$$T = [i][G][i]' \tag{6.49}$$

where $[G]$ is the matrix containing speed-induced terms and $[i]'$ is the transpose of current matrix $[i]$. If the load torque is known,

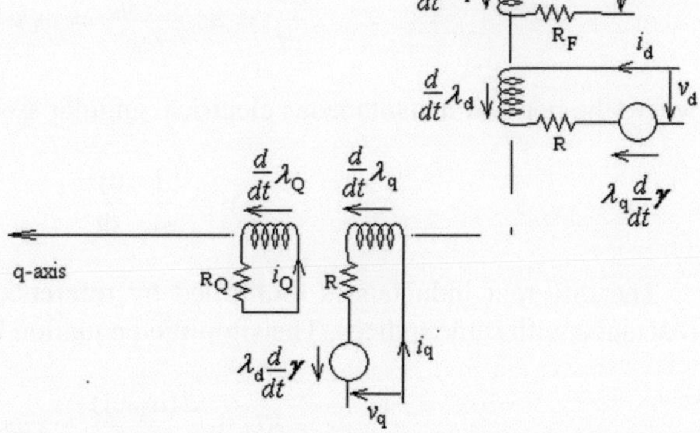

Fig. 6.34 Electrical layout of salient pole synchronous machine corresponding to the d-axis and q-axis

$$T - T_L = J\frac{d\omega}{dt} = \frac{J}{p}\frac{d^2}{dt^2}\gamma \tag{6.50}$$

The electromagnetic torque has three components.

(a) Due to excitation current even in steady-state conditions.
(b) Geometric asymmetry of machine due to saliency $L_d \neq L_q$. Also true in steady state. In cylindrical rotor machine this torque is absent.
(c) Damping torque exerted by damper bars. This is present in disturbances and is totally absent in steady-state conditions.

6.7.1 *Per Unit Quantities*

The set of equation can be simplified using per unit system. For this certain reference quantities must be chosen. Taking ω_b as reference for angular frequency and speed, the time derivative is

$$\frac{d}{dt} = \omega_b \frac{d}{d(\omega_b t)} = \omega_b \frac{d}{d\tau}$$

where the synchronous time is $\tau = \omega_b t$

6.7.2 *Steady-State Equation and Derivation of Phasor Diagram: The Equation for Steady State*

The derivative of angle γ with respect to time t is the instantaneous electrical angular speed,

$$\frac{d\gamma}{dt} = \omega_b \frac{d\gamma/dt}{\omega_b} = \omega_b \nu$$

where the per unit instantaneous electrical angular speed is

$$\nu = \frac{1}{\omega_b}.\frac{d\gamma}{dt} \tag{6.51}$$

The different inductances multiplied by reference angular frequency give the reactance with same indices. The simplified equation becomes

$$v_d = Ri_d + \frac{d(\omega_b \lambda_d)}{d\tau} - \nu\omega_b\lambda_q$$

$$v_q = Ri_q + \frac{d(\omega_b \lambda_q)}{d\tau} + \nu\omega_b\lambda_d$$

$$v_F = R_F i_F + \frac{d(\omega_b \lambda_F)}{d\tau}$$
$$0 = R_D i_D + \frac{d(\omega_b \lambda_D)}{d\tau}$$
$$0 = R_Q i_Q + \frac{d(\omega_b \lambda_Q)}{d\tau} \tag{6.52}$$

Further

$$\omega_b \lambda_d = X_d i_d + X_{dF} i_F + X_{dD} i_D$$
$$\omega_b \lambda_q = X_q i_q + X_{dQ} i_Q$$
$$\omega_b \lambda_F = \frac{3}{2} X_{dF} i_d + X_{FF} i_F + X_{FD} i_D$$
$$\omega_b \lambda_D = \frac{3}{2} X_{dD} i_D + X_{FD} i_F + X_{DD} i_D$$
$$\omega_b \lambda_Q = \frac{3}{2} X_{qQ} i_q + X_{QQ} i_Q \tag{6.53}$$

Here X_d is known as direct-axis synchronous reactance, and X_q is known as quadrature-axis synchronous reactance $(X_d > X_q)$. Here we assume that the stator reference quantities are V_b, I_b , Z_b such that

$$V_b = I_b Z_b$$

The reference quantities for excitation are assumed to be V_{Fb}, I_{Fb} , Z_{Fb} such that

$$V_{Fb} = I_{Fb} Z_{Fb}$$

Similarly we have

$$V_{Db} = I_{Db} Z_{Db}$$

$$V_{Qb} = I_{Qb} Z_{Qb}$$

The analysis can be done on per unit basis, thus the quantities in above set of equations will change as

$$v_d = \frac{v_d}{V_b};\ i_d = \frac{i_d}{I_b};\ i_q = \frac{i_q}{I_b};\ i_F = \frac{i_F}{I_{Fb}};\ i_D = \frac{i_D}{I_{Db}};\ i_Q = \frac{i_Q}{I_{Qb}}$$

Similarly the quantities like flux linkages, resistances, and reactance are represented as per unit basis. The equation for steady-state symmetrical and balanced operation of machine can be derived using above set of equation. Under steady-state operation when damper bar currents are zero, or $i_D = 0$; $i_Q = 0$, the field excitation is $i_F = I_F$; $\frac{d}{dt} = 0$; $\frac{d\gamma}{dt} = \omega = p\frac{d\beta}{dt} = \text{constant}$

Thus, in terms of RMS values,

$$V_d = RI_d - \omega\lambda_q = RI_d - X_q I_q$$

$$V_q = RI_q + \omega\lambda_d = RI_q + X_d I_d + \frac{\omega}{\sqrt{2}} L_{dF} I_F$$

$$\begin{aligned} V_F &= R_F I_F \\ \lambda_d &= L_d I_d + \frac{1}{\sqrt{2}} L_{dF} I_F \\ \lambda_q &= L_q I_q. \end{aligned} \tag{6.54}$$

In steady-state operation, the Park-Blondel equation becomes

$$\overline{V} = \sqrt{2}(V_d + jV_q) = v_d + jv_q$$

And

$$\overline{I} = \sqrt{2}(I_d + jI_q) = i_d + ji_q$$

This gives us

$$\overline{V} = RI + jX_d I_d - X_q I_q + j\frac{1}{\sqrt{2}} L_{dF} I_F \tag{6.55}$$

The above equation leads to phasor diagram of salient pole machine. The phasor diagram has also been derived in the next section using two -axis theory.

6.8 Two-Axis Theory

We have already studied that air-gap is uniform in cylindrical rotor machines. The saliency due to the slots is ignored. The generalized machine theory has been studied in Sects. 6.7 and 7.10. Figure 6.35a shows the model of three-phase salient pole machine. An equivalent two-pole salient pole machine is shown in Fig. 6.35b. The generalized machine model of salient pole synchronous machine is arbitrary d-q reference frame as show in Fig. 6.35c. The air-gap is non-uniform in case of salient pole machine, i.e., air-gap is minimum in polar or *d*-axis, while it is maximum in the interpolar or *q*-axis. The rotating magnetic field due to armature currents encounters a change in reluctance. A salient pole machine has two axes of symmetry. The MMF acting on the direct-axis produces a d-axis flux. Similarly, the MMF acting on q-axis produces q-axis flux. From two-axis theory, the equation of motion under steady state can be written as

$$\begin{bmatrix} v_d \\ v_q \\ v_f \end{bmatrix} = \begin{bmatrix} R_1 + L_d p & \dot{\theta} L_q & L_{ad} p \\ -\dot{\theta} L_d & R_1 + L_q p & -\dot{\theta} L_{ad} \\ L_{ad} p & 0 & R_f + L_f p \end{bmatrix} \begin{bmatrix} i_d \\ i_q \\ -i_f \end{bmatrix} \tag{6.56}$$

The inductances L_d and L_q are, respectively, known as direct-axis and quadrature-axis inductances. Here $\dot{\theta} = \omega_m$ and $\theta = \omega t + \delta$. Since the voltages are balanced, we can write the two phase voltage $v_a = V \sin \omega t$ and $v_b = V \cos \omega t$. The variation of self-inductances with respect to θ is shown in Fig. 6.36. Thus, we get

$$\begin{bmatrix} v_d \\ v_q \end{bmatrix} = V \begin{bmatrix} \cos\theta & -\sin\theta \\ \sin\theta & \cos\theta \end{bmatrix} \begin{bmatrix} \sin \omega t \\ \cos \omega t \end{bmatrix} \tag{6.57}$$

This gives us,

$$v_d = V \sin\delta \quad \text{and} \quad v_q = V \cos\delta \tag{6.58}$$

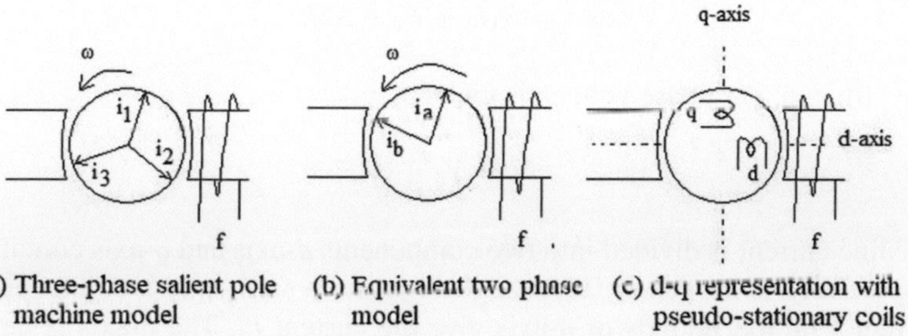

Fig. 6.35 d-q representation of a three and two phase salient pole alternator without damper bars

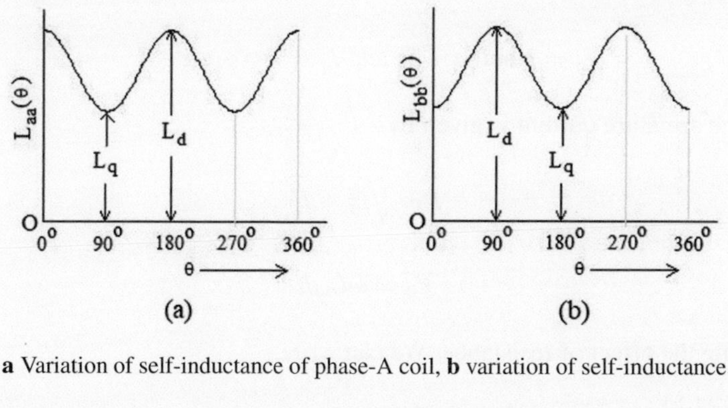

Fig. 6.36 **a** Variation of self-inductance of phase-A coil, **b** variation of self-inductance of phase-b coil

For a balanced two-phase, two-pole synchronous machine operating under steady-state condition, the operator $p = \mathrm{d}/\mathrm{d}t$ is ignored for armature. Thus, under steady-state conditions, we can write for armature $p = 0$ and $\dot{\theta} = \omega$ and $\theta = \omega t + \delta$. Here we define the following:

X_d $= \omega L_\mathrm{d}$ d-axis reactance.
X_q $= \omega L_\mathrm{q}$ q-axis reactance.
X_ad $= \omega L_\mathrm{ad}$ Mutual reactance.
I_d $=$ steady-state d-axis component of current $\boldsymbol{I}$.
I_q $=$ steady-state q-axis component of current $\boldsymbol{I}$.

The matrix (6.56) reduces to

$$\begin{bmatrix} v_\mathrm{d} \\ v_\mathrm{q} \\ v_\mathrm{f} \end{bmatrix} = \begin{bmatrix} R_1 & X_\mathrm{q} & 0 \\ -X_\mathrm{d} & R_1 & -X_\mathrm{ad} \\ L_\mathrm{ad}p & 0 & R_\mathrm{f} + L_\mathrm{f}p \end{bmatrix} \begin{bmatrix} i_\mathrm{d} \\ i_\mathrm{q} \\ -i_\mathrm{f} \end{bmatrix} \tag{6.59}$$

Here 'p' is an operator. The above matrix (6.56) yield,

$$\begin{aligned} v_\mathrm{d} &= V \sin\delta = R_1 I_\mathrm{d} + \omega L_\mathrm{q} I_\mathrm{q} = R_1 I_\mathrm{d} + X_\mathrm{q} I_\mathrm{q} \\ v_\mathrm{q} &= V \cos\delta = R_1 I_\mathrm{q} + E_0 - X_\mathrm{d} I_\mathrm{d} \end{aligned} \tag{6.60}$$

The armature per phase voltage is given by

$$\mathbf{V} = V_\mathrm{d} + jV_\mathrm{q} \tag{6.61}$$

The line current is divided into two components d-axis and q-axis curents I_d and I_q, respectively. If the power factor angle is φ, and power angle is δ, then projection of current $\mathbf{I}$ on the $\mathbf{E}_0$ axis or q-axis give the current I_q. The d-axis is defined as the -axis normal to q-axis. The projection of current $\mathbf{I}$ on the d-axis gives the I_d. Thus, when salient pole alternator is delivering lagging current, the d-axis and q-axis currents are given as

$$I_\mathrm{d} = I\ \sin(\varphi + \delta) \text{ and } I_\mathrm{q} = I\ \cos(\varphi + \delta)$$

And the armature current is given by

$$I = \sqrt{I_\mathrm{d}^2 + I_\mathrm{q}^2} \tag{6.62}$$

$$E_0 = \omega L_\mathrm{ad} I_\mathrm{f}$$

Ignoring the effect of resistance. We can write,

$$I_\mathrm{d} = \frac{1}{X_\mathrm{d}}(E_0 - V\cos\delta) \tag{6.63}$$

or,

$$E_0 - jX_d I_d = V \cos\delta$$

$$I_q = -\frac{V}{X_q} \sin\delta \tag{6.64}$$

or,

$$jX_q I_q = -V \sin\delta$$

Using the above two Eqs. (6.63) and (6.64), phasor diagram of a salient pole machine can be developed. The developed phasor diagrams have been shown in Figs. 6.40, 6.42, 6.43.

6.9 Slip Test

The slip test is used for experimental determination of d-axis and q-axis reactance. Figure 6.37 shows the set-up for slip test. Here, a salient pole alternator is connected to a three-phase balanced source and also to a prime mover. The direction of rotation of prime mover is noted. When set-up is in standstill condition a small three-phase AC voltage is injected into the armature windings. The salient pole machine starts in one direction as a reluctance synchronous motor (RSM). The direction of rotation is recorded. If the direction of rotation is same as that of prime mover, the prime mover is also started and the speed is brought near synchronous speed. A small AC voltage is injected into armature circuit, keeping field winding open. Now, a speed of prime mover is smoothly varied in sub-synchronous region. The oscillations in the current and voltage are recorded. These oscillations have been shown in Fig. 6.38. The d-axis and q-axis reactance are calculated as [11, 12]

$$X_d = \frac{1}{\sqrt{3}} \times \frac{\text{Maximum voltage}}{\text{Minimum current}}$$

$$X_q = \frac{1}{\sqrt{3}} \times \frac{\text{Minimum voltage}}{\text{Maximum current}} \tag{6.65}$$

The current and voltage oscillate only in sub-synchronous region. When armature windings carry three-phase balanced current, it sets up a rotating magnetic field, which is rotating with synchronous speed as decided by the source frequency. The prime mover is rotated at a speed less than synchronous speed in the direction of rotating magnetic field.

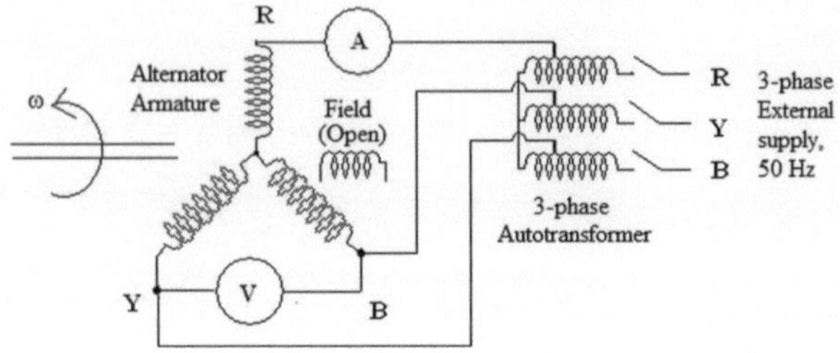

Fig. 6.37 Determination of direct-axis and quadrature-axis reactance—slip test

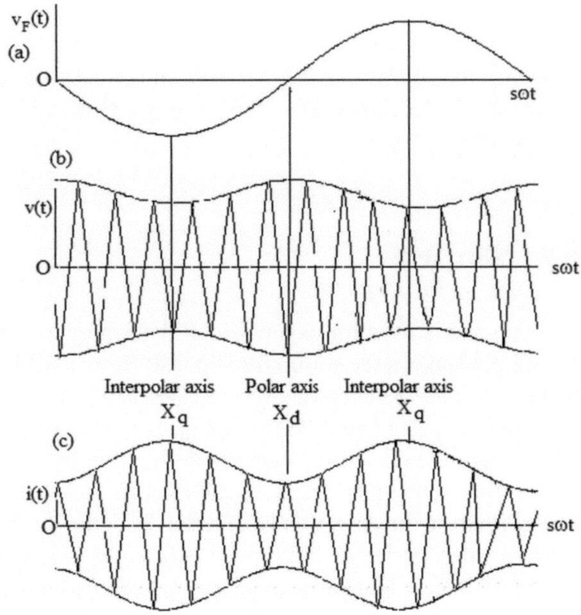

Fig. 6.38 Determination of X_d and X_q from slip test: **a** slip frequency voltage-induced in field winding, **b** armature voltage, **c** armature current

The air-gap flux is maximum when the crest of the rotating magnetic field is aligned with the polar-axis. As the rotor tries to move away from this position, the current jumps to a higher value for maintaining the flux constant (Lenz's law).

When the crest of the rotating magnetic field is aligned in the interpolar-axis, the air-gap is maximum, while the air-gap flux is minimum. According to Lenz's law, as the rotor tries to move away from this position, the current falls to maintain the flux constant. These oscillations are at slip frequency, which can also be observed by recording the voltage across the field windings.

Example 6.12 Cylindrical rotor synchronous alternator has the ratings of Cylindrical rotor synchronous alternator has the ratings of 20 KVA, 433 V, 3000 RPM, 50 Hz, three-phase star connected. The open-circuit, short-circuit, and ZPF test data are as follows:

I_f (A)	10	17	20	23	30	38	50	60
V_{oc} (V)	175	250	280	300	330	350	370	380
V_{ZPF} (V)	–	–	–	0	130	210	265	280

Determine the per phase leakage reactance, per phase synchronous reactance, and field current needed to provide rated voltage when rated current is delivered at power factors 0.8 lagging. Use EMF method.

Solution

The open-circuit, short-circuit characteristics, and ZPF characteristics have been plotted in Fig. 6.39. The rated line-to-line voltage is 433 Volt.

The rated per phase voltage is

$$V_t = \frac{433}{\sqrt{3}} \approx 250\ \text{V}$$

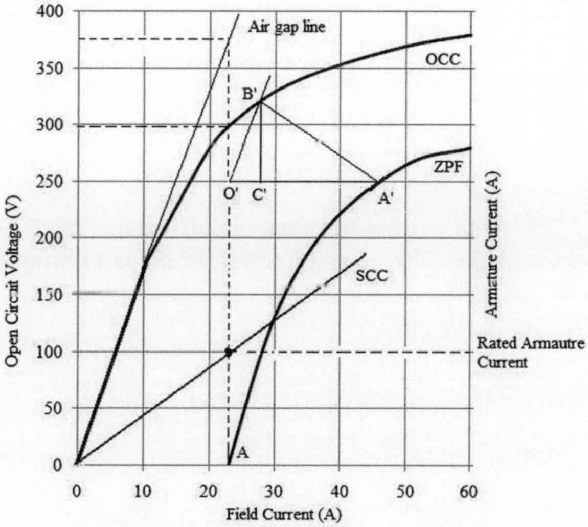

Fig. 6.39 OCC, SCC, and ZPF characteristics

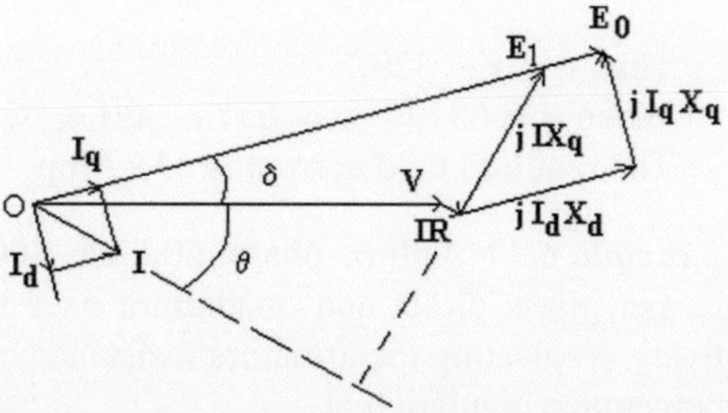

Fig. 6.40 Phasor diagram of salient pole alternator—lagging power factor

The rated armature current is

$$I_a = \frac{20000}{433\sqrt{3}} = 26.67\ \text{A}$$

The leakage reactance

$$X_l = \frac{B'C'}{2624.3\sqrt{3}} = \frac{70}{26.67} = 1.52\ \Omega$$

The synchronous reactance

$$X_s = \frac{298}{26.67\sqrt{3}} = 6.45\ \Omega$$

Using synchronous impedance method,

$$\text{E} = \text{V} + \text{I}_a(R_a + jX_s)$$

$$\begin{aligned} \text{E} &= \text{V} + \text{I}_a(R_a + jX_s) \\ \text{E} &= 250\angle 0^\circ + 26.67 \times (0.8 - 0.6j) \times (0.0 + j6.45) \\ &= 250 + 137.6j + 103.2 \\ &= 379\angle 20.24^\circ V \end{aligned}$$

The field current for open-circuit voltage of $379\sqrt{3} = 657$ V is obtained by extrapolation. The equation of line of the last portion of OCC is

$$y = \frac{380 - 370}{60 - 50} x + c$$

$$y = x + c$$

When

$$380 = 60 + c$$

$$c = 380 - 60 = 320$$

Thus $y = x + 320$.
When $y = 657\ V \quad x = 657 - 320 = 337$ A
The required field current is 337 Amp.

Example 6.13 A three-phase, 60 kVA, 400 V, 50 Hz star-connected alternator has its per phase direct and quadrature axes reactances as 1.5 Ω and 0.6 Ω, respectively. Neglecting the armature resistance calculate induced EMF, power angle, and percentage regulation at

(a) 0.8 lagging power factor (b) unity power factor (c) 0.8 leading power factor.

Solution

Given alternator ratings are 60 kVA, 400 V, 50 Hz. The reactances are $X_d = 1.5\Omega$ and $X_q = 0.6\Omega$.

Rated current is

$$I = 60,000/\left(\sqrt{3} \times 400\right) = 86.60\ \text{A}$$

Rated phase voltage is

$$V = 400/\sqrt{3} = 231\ \text{V}$$

(a) 0.8 Lagging power factor (Fig. 6.40).

Power factor angle, $\theta = \cos^{-1}(0.8) = 36.87°$.
Power angle,

$$\begin{aligned}\delta &= \tan^{-1}\left[IX_q \cos\theta/\left[V + IX_q \sin\theta\right]\right]\\ &= \tan^{-1}[86.60 \times 0.6 \times 0.8/[231 + 86.6 \times 0.6 \times 0.6]]]\\ &= \tan^{-1}[41.568/262.116] = 9.011°\end{aligned}$$

Thus, we get

$$\begin{aligned}I_q &= I\cos(\theta + \delta) = 86.6\cos(36.87 + 9.011) = 60.29\ \text{A}\\ I_d &= I\sin(\theta + \delta) = 86.6\sin(36.87 + 9.011) = 62.16\ \text{A}\\ I_dX_d &= 62.16 \times 1.5 = 93.39\\ I_qX_q &= 60.29 \times 0.6 = 36.174\end{aligned}$$

$$\begin{aligned}E_0 &= \sqrt{\left[\left(V + I_dX_d\cos\delta - I_qX_q\sin\delta\right)^2 + \left(I_dX_d\sin\delta + I_qX_q\cos\delta\right)^2\right]}\\ &= \sqrt{[(231 + 93.39 \times 0.9876 - 36.174 \times 0.15662)^2 + (93.39 \times 0.15662 + 36.174 \times 0.9876)^2]}\\ &= \sqrt{\left[317.51^2 + 50.388^2\right]}\\ &= 321.48\ \text{V}\end{aligned}$$

Thus,

$$\text{percentage regulation} = 100 \times (321.48 - 231)/231 = 39.16\%$$

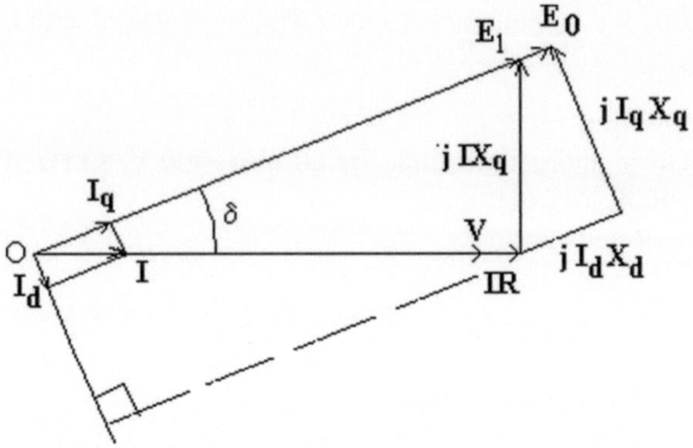

Fig. 6.41 Phasor diagram of salient pole alternator—unity power factor

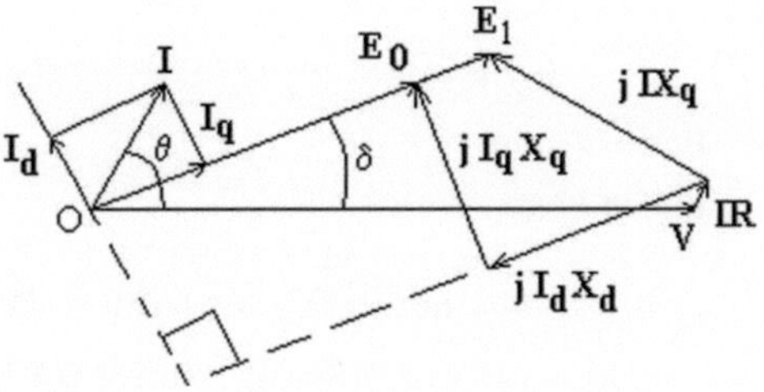

Fig. 6.42 Phasor diagram of salient pole alternator leading power factor

(b) Unity power factor (Fig. 6.41).

Power factor angle, $\theta = \cos^{-1}(1.00) = 0°$

$$\text{Power angle},\delta = \tan^{-1}\left[IX_q \cos\theta / \left[V + IX_q \sin\theta\right]\right]$$
$$= \tan^{-1}[86.60 \times 0.6 \times 1.0/[231 + 86.6 \times 0.6 \times 0.0]]$$
$$= \tan^{-1}[41.568/262.116] = 12.68°$$

$$I_q = I\cos(\theta + \delta) = 86.6\cos(0 + 12.68) = 84.49\,\text{A}$$

$$I_d = I\sin(\theta + \delta) = 86.6\sin(0 + 12.68) = 19.00\,\text{A}$$

$$I_dX_d = 19.00 \times 1.5 = 28.5$$

$$I_qX_q = 84.49 \times 0.6 = 50.694$$

$$E_0 = \sqrt{\left[\left(V + I_dX_d\cos\delta - I_qX_q\sin\delta\right)^2 + \left(I_dX_d\sin\delta + I_qX_q\cos\delta\right)^2\right]}$$
$$= \sqrt{\left[(231 + 28.5 \times 0.9756 - 50.694 \times 0.2195)^2 + (28.5 \times 0.2195 + 50.694 \times 0.9756)^2\right]}$$
$$= \sqrt{\left[247.62^2 + 55.71^2\right]} = 253.81\,\text{V}$$

Thus, percentage regulation is

$$\text{percentage regulation} = 100 \times (353.8 - 231)/231 = 9.46\%$$

(c) 0.8 leading power factor (Fig. 6.42).

Power factor angle, $\theta = \cos^{-1}(0.8) = 36.87°$

$$\begin{aligned}\text{Power angle}, \delta &= \tan^{-1}\left[IX_q \cos\theta / \left[V - IX_q \sin\theta\right]\right] \\ &= \tan^{-1}[86.60 \times 0.6 \times 0.8/[230.94 - 86.6 \times 0.6 \times 0.6]] \\ &= \tan^{-1}[41.568/199.764] = 11.75°\end{aligned}$$

Thus

$$I_q = I\cos(\theta - \delta) = 86.6\cos(36.87 - 11.75) = 78.41\,\text{A}$$

$$I_q = I\sin(\theta - \delta) = 86.6\sin(36.87 - 11.75) = 36.76\,\text{A}$$

$$I_d X_d = 36.76 \times 1.5 = 55.14$$

$$I_q X_q = 78.41 \times 0.6 = 47.046$$

$$\begin{aligned}E_0 &= \sqrt{\left[(V - I_d X_d \cos\delta - I_q X_q \sin\delta)^2 + (-I_d X_d \sin\delta + I_q X_q \cos\delta)^2\right]} \\ &= \sqrt{\left[(230.94 - 55.14 \times 0.9790 - 47.046 \times 0.2036)^2 + (-55.14 \times 0.2036 + 47.046 \times 0.9790)^2\right]} \\ &= 170.97\,\text{V}\end{aligned}$$

Thus % regulation, $= (170.97 - 230.94) \times 100/230.94 = -25.97\%$

Example 6.14 A three-phase salient pole star-connected alternator has its direct and quadrature axes reactance are 1.0 pu and 0.5 pu, respectively. Neglecting the armature resistance calculate induced EMF, power angle, and percentage regulation at (a) 0.8 lagging, (b) unity, and (c) leading power factor.

Solution

Let the phase voltage be $V = 1.0$ pu; Rated current $I = 1.0$ pu
$X_d = 1.0$ pu and $X_q = 0.50$ pu

(a) 0.8 Lagging power factor (Fig. 6.40).

Power factor angle, $\theta = \cos^{-1}(0.8) = 36.87°$

$$\text{Power angle}, \delta = \tan^{-1}\left[IX_q \cos\theta / \left[V - IX_q \sin\theta\right]\right]$$

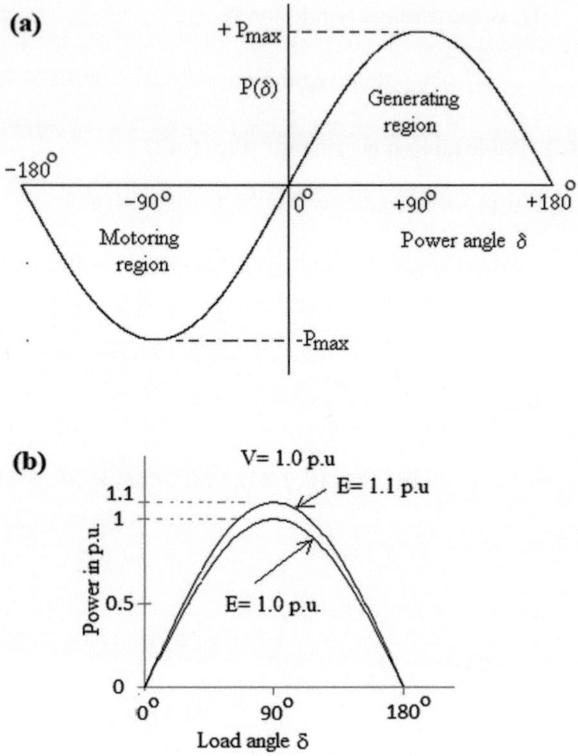

Fig. 6.43 **a** Power angle characteristics of cylindrical rotor machine **b** P−δ characteristics of alternator for two different values of excitation EMF

$$= \tan^{-1}[1.0 \times 0.5 \times 0.8/[1.0 + 1.0 \times 0.5 \times 0.6]]$$
$$= \tan^{-1}[0.4/1.3] = 17.1^\circ$$

Thus, we get

$$I_q = I\cos(\theta + \delta) = 1.0\cos(36.87 + 17.1) = 0.5881\,\text{pu}$$

$$I_d = I\sin(\theta + \delta) = 1.0\sin(36.87 + 17.1) = 0.8087\,\text{pu}$$

$$I_dX_d = 0.8087 \times 1.0 = 0.8087$$

$$I_qX_q = 0.5881 \times 0.5 = 0.29405$$

$$\cos 17.1 = 0.9558$$

$$\sin 17.1 = 0.2940$$

$$E_0 = \sqrt{\left[(V + I_d X_d \cos\delta - I_q X_q \sin\delta)^2 + (I_d X_d \sin\delta + I_q X_q \cos\delta)^2\right]}$$
$$= \sqrt{\left[(1.0 + 0.8087 \times 0.9558 - 0.29405 \times 0.2940)^2 + (0.8087 \times 0.2940 + 0.29405 \times 0.9558)^2\right]}$$
$$= \sqrt{\left[(1.6865)^2 + (0.5188)^2\right]} = 1.75\,\text{pu}$$

Thus percentage regulation is given as

$$(1.75 - 1.0) \times 100/1.0 = 0.75\,\text{pu} = 75\%$$

(b) Unity power factor (Fig. 6.41).

Power factor angle $\theta = \cos^{-1}(1.00) = 0°$

$$\begin{aligned}\text{Power angle, } \delta &= \tan^{-1}\left[IX_q \cos\theta / \left[V + IX_q \sin\theta\right]\right]\\ &= \tan^{-1}[1.0 \times 0.5 \times 1.0/[1.0 + 1.0 \times 0.5 \times 0.0]]\\ &= \tan^{-1}[0.5/1.0] = 26.57°\end{aligned}$$

Thus, we get

$$I_q = I\cos(\theta + \delta) = 1.0\cos(26.57) = 0.8944\,\text{pu}$$

$$I_d = I\sin(\theta + \delta) = 1.0\sin(26.57) = 0.4472\,\text{pu}$$

$$I_d X_d = 0.4472 \times 1.0 = 0.4472$$

$$I_q X_q = 0.8944 \times 0.5 = 0.4472$$

$$\cos 26.57 = 0.8944$$

$$\sin 26.57 = 0.4472$$

$$E_0 = \sqrt{\left[(V + I_d X_d \cos\delta - I_q X_q \sin\delta)^2 + (I_d X_d \sin\delta + I_q X_q \cos\delta)^2\right]}$$
$$= \sqrt{\left[(1.0 + 0.4472 \times 0.8944 - 0.4472 \times 0.4472)^2 + (0.4472 \times 0.4472 + 0.4472 \times 0.8944)^2\right]}$$
$$= \sqrt{\left[(1.19998)^2 + (0.59996)^2\right]} = 1.3416\,\text{pu}$$

Thus, percentage regulation, is $(1.3416 - 1.0)/1.0 = 0.3416\text{pu} = 34.16\%$

(c) 0.8 Leading power factor (Fig. 6.42).

Power factor angle, $\theta = \cos^{-1}(0.8) = 36.87°$

$$\begin{aligned}\text{Power angle, } \delta &= \tan^{-1}\left[I\,X_{\rm q}\cos\theta/\left[V - I\,X_{\rm q}\sin\theta\right]\right] \\ &= \tan^{-1}[1.0 \times 0.5 \times 0.8/[1.0 - 1.0 \times 0.5 \times 0.6]] \\ &= \tan^{-1}[0.4/0.7] = 29.74^\circ\end{aligned}$$

Thus, we get

$$I_{\rm q} = I\cos(\theta + \delta) = 1.0\cos(36.87 - 29.74) = 0.9923\,\text{pu}$$

$$I_{\rm d} = I\,\sin(\theta + \delta) = 1.0\sin(36.87 - 29.74) = 0.1240\,\text{pu}$$

$$I_{\rm d}X_{\rm d} = 0.1240 \times 1.0 = 0.1240$$

$$I_{\rm q}X_{\rm q} = 0.9923 \times 0.5 = 0.49615$$

$$\cos 29.74 = 0.8683$$

$$\sin 29.72 = 0.4961$$

$$\begin{aligned}E_0 &= \sqrt{\left[(V - I_{\rm d}X_{\rm d}\cos\delta - I_{\rm q}X_{\rm q}\sin\delta)^2 + (-I_{\rm d}X_{\rm d}\sin\delta + I_{\rm q}X_{\rm q}\cos\delta)^2\right]} \\ &= \sqrt{\left[(1.0 - 0.1240 \times 0.8683 - 0.49615 \times 0.4961)^2 + (-0.1240 \times 0.4961 + 0.49615 \times 0.8683)^2\right]} \\ &= \sqrt{\left[(0.64619)^2 + (0.36929)^2\right]} = 0.7442\,\text{pu}\end{aligned}$$

Thus % regulation, $= (0.7442 - 1.0)/1.0 = -0.2557\text{pu} = -25.57\%$

6.10 Power

The output power in an alternator is given as [12]

$$P_{\rm e} = 3 \cdot V \cdot I \cdot \cos\theta = 3\mathbf{V} \cdot \mathbf{I} \tag{6.66}$$

The power expression in a synchronous machine can be expressed in terms of induced EMF and power angle.

6.10.1 Cylindrical Rotor Machine

Consider a cylindrical rotor synchronous alternator. Figure 6.19a–c shows phasor diagrams at different power factors. Let the phase voltage be **V** and the induced EMF per phase is **E**, the power angle is δ. The electromagnetic power, P_{m}, or the mechanical power converted into electrical power, P_{e} of an alternator can be expressed in terms of the phase voltage, induced EMF per phase and power angle. Thus, the mechanical power can be given as

$$P_{\mathrm{m}} = 3\left(V \cdot I \cdot \cos\theta + I^2R\right) = 3[\mathbf{V} \cdot \mathbf{I} + (\mathbf{I} \cdot \mathrm{Z_s}) \cdot \mathbf{I}] = (\mathbf{V} + \mathbf{I} \cdot \mathrm{Z_s}) \cdot \mathbf{I}$$

Here, '.' shows the dot product of two phasor shown in bold face.

From the phasor diagram we get,

$$\mathbf{E}_0 = (\mathbf{V} + \mathbf{I} \cdot Z_s) \tag{6.67}$$

Thus, current phasor can be given as

$$\mathrm{I} = \frac{\mathrm{E}_0 - \mathrm{V}}{Z_s}$$

$$\mathrm{I} = \frac{E_0\angle\delta - V\angle 0}{Z_s\angle\gamma}$$

$$\mathrm{I} = \frac{E_0\angle\delta - V\angle 0}{Z_s\angle\gamma} = \frac{E_0\angle\delta - \gamma}{Z_s} - \frac{V\angle - \gamma}{Z_s} \tag{6.68}$$

Here, the internal angle is given as

$$\gamma = \tan^{-1}\frac{X_{\mathrm{s}}}{R}$$

Let the voltage is the reference phasor. The dot product of **V.I** is the product of |**V**| and real parts of current **I**. Thus, electrical power can be given as

$$P_{\mathrm{e}} = 3\mathbf{V} \cdot \mathbf{I} = \frac{3VE_0}{Z_{\mathrm{s}}}\cos(\gamma - \delta) - \frac{3V^2}{Z_{\mathrm{s}}}\cos\gamma \tag{6.69}$$

For obtaining the electromagnetic power P_{m}, $\mathbf{E}_0$ is taken as a reference. The current is

$$\mathrm{I} = \frac{E_0\angle 0 - V\angle - \delta}{Z_{\mathrm{s}}\angle - \gamma} = \frac{E_0\angle\gamma}{Z_{\mathrm{s}}} - \frac{V\angle\gamma - \delta}{Z_{\mathrm{s}}} \tag{6.70}$$

Since $\mathbf{E}_0$ is the reference phasor, $\mathbf{E}_0$.**I** dot product is the product of **E** and real part of **I.** Thus, we get

$$P_m = 3\mathbf{E}_0 \cdot \mathbf{I} = \frac{3E_0^2}{Z_s}\cos\gamma - \frac{3E_0^2 V}{Z_s}\cos(\gamma + \delta) \tag{6.71}$$

If the resistance is negligible, angle $\gamma = 90° \, and \quad Z_s = X_s$.
Thus, we get

$$P = P_e = P_m$$

$$P = P_e = P_m = \frac{3VE_0}{X_s}\sin\delta \tag{6.72}$$

Figure 6.43a shows the power angle diagram for cylindrical rotor machine

Figure 6.43b shows the variation of power P with load angle δ and **E.** here, **E** is the induced EMF measured in per unit. The value of **E** is obtained from the phasor diagram. The induced EMF **E** is proportional to excitation field current. It assumes that machine is unsaturated. Since the speed of synchronous motor or generator is constant at synchronous speed, the per unit expression of maximum power also represents the available torque in per unit.

For a given induced EMF **E**, if synchronous motor is loaded higher than the maximum torque available for that excitation, the motor tries to retard as such it may fall out of step and ultimately loose synchronism. Similarly, in case of a generator connected to grid, if driven with torque higher than maximum electrical torque it would accelerate and as such may fall out of step and loose synchronism.

In case of cylindrical rotor synchronous machine, the maximum real power is available at $\delta = 90°$ electrical. In generating mode, the operation is stable if with increase of load angle δ - both power and torque increase. After the stability limit has been reached, an increase in load angle δ decreases the power available from the machine, and operation is unstable.

In actual practice, it is desirable to remain quite far from the stability limit when generator is operating under full load condition. The machine is normally designed for operation at load angle varying from 30° to 45° electrical, so that enough reserve power is available. An under-excited alternator has a smaller value of maximum power P_{max}.

6.10.2 Salient Pole Alternator

The phasor diagram of the salient pole alternator is shown in Figs. 6.40, 6.41, and 6.42. Here X_d and X_q are d- and q-axis per phase reactance of the machine. In the phasor diagram, we can write

$$\mathbf{E}_1 = \mathbf{V} + (R + jX_q)\mathbf{I} = \mathbf{V} + Z_q\mathbf{I} \tag{6.73}$$

The power is obtained by replacing $\mathbf{E}_0$ and Z_s of Eq. (6.69) of cylindrical rotor machine by $\mathbf{E}_1$ and Z_q, respectively. Thus, the electrical and electromagnetic powers of a salient pole generator are

$$P_e = 3\mathbf{V}\cdot\mathbf{I} = \frac{3VE_1}{Z_q}\cos(\gamma - \delta) - \frac{3V^2}{Z_q}\cos\gamma = \frac{3VE_1}{X_q}\sin\delta \tag{6.74}$$

$$P_m = 3\mathbf{E}_1\cdot\mathbf{I} = \frac{3E_1^2}{Z_q}\cos\gamma - \frac{3E_1V}{Z_q}\cos(\gamma + \delta) = \frac{3E_1V}{X_q}\sin\delta \tag{6.75}$$

Here,

$$Z_q = \sqrt{R^2 + X_q^2}$$

and

$$\gamma = \tan^{-1}\left(\frac{X_q}{R}\right)$$

The power is also given by $P = 3VI\cos\theta$. From the phasor diagram of alternator at lagging power factor (Fig. 6.40), the current I_d and I_q are obtained as

$$I_q = I\cos(\theta + \delta) \quad \text{and} \quad I_d = I\sin(\theta + \delta)$$

Here $I = |\mathbf{I}|$, it can also be written that

$$I\cos\theta = I_d\sin\delta + I_q\cos\delta$$

or,

$$I_d = \frac{E_0 - V\cos\delta}{X_d} \text{ and } I_q = \frac{V\sin\delta}{X_q} \tag{6.76}$$

From the phasor diagram of Fig. 6.40, we get

$$E_1\cos\delta = V + IX_q\sin\theta$$

$$E_1\sin\delta = IX_q\cos\theta \tag{6.77}$$

The above expressions give us the value of power angle δ. Also, we have

$$\begin{aligned} E_0\cos\delta &= V + IR\cos\theta + I_dX_d\cos\delta - I_qX_q\sin\delta \\ E_0\sin\delta &= I_dX_d\sin\delta + I_qX_q\cos\delta - IR\sin\theta \end{aligned} \tag{6.78}$$

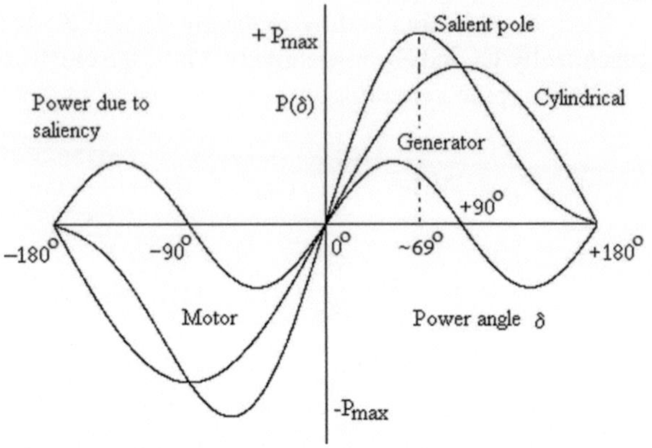

Fig. 6.44 Power angle characteristics of salient pole synchronous machine

These expressions are useful in finding the expressions for induced EMF E_0 and power angle δ. From the above expressions we get,

$$P = P_\mathrm{e} = P_\mathrm{m} = \frac{3VE_0}{X_\mathrm{d}}\sin\delta + \frac{3V^2}{2}\left(\frac{1}{X_\mathrm{q}} - \frac{1}{X_\mathrm{d}}\right)\sin 2\delta$$

$$P = P_\mathrm{e} = P_\mathrm{m} = P_{\max}\sin\delta + P_{\mathrm{sl}}\sin 2\delta$$

Here,

$$P_{\max} = \frac{3VE_0}{X_\mathrm{d}} \ \& \ P_{\mathrm{sl}} = \frac{3V^2}{2}\left(\frac{1}{X_\mathrm{q}} - \frac{1}{X_\mathrm{d}}\right) \tag{6.79}$$

Here $P_{\max}$ is the maximum power of a cylindrical rotor machine and P_{sl} is the power due to saliency. Figure 6.44 shows the plot of power angle characteristics of salient pole synchronous machines. The angle for maximum power is obtained by using the relation.

$$\frac{\mathrm{d}P}{\mathrm{d}\delta} = 0$$

$$\frac{VE_0}{X_\mathrm{d}}\cos\delta + V^2\left(\frac{1}{X_\mathrm{q}} - \frac{1}{X_\mathrm{d}}\right)\cos 2\delta = 0 \tag{6.80}$$

The above Eq. (6.80) can be solved to get the power angle corresponding to maximum power of salient pole machine. The angle for maximum power is less than 90°, and typically its values lies around 69° as compared to 90° for cylindrical rotor machines. When field current is zero, the induced EMF is also zero, i.e., $E_0 = 0$. The machine has a finite power due to saliency, which is proportional to sin 2δ. When a synchronous motor without field winding is connected to three-phase

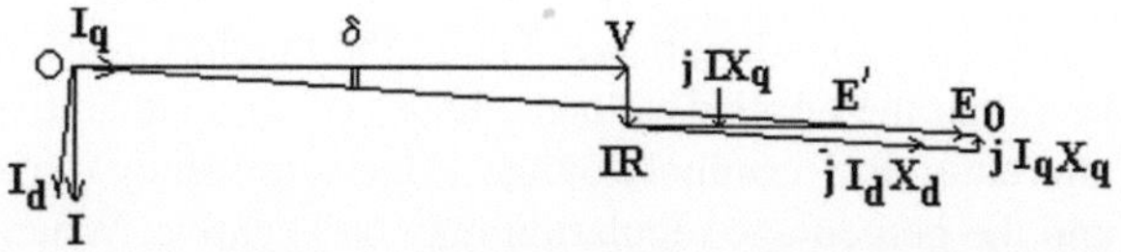

Fig. 6.45 Phasor diagram of salient pole alternator—zero lagging power factor

source, the machine develops a starting torque due to saliency. The motor is referred to as *reluctance synchronous motor* (RSM).

The electrical power is given as $P = 3VI\cos\theta$. When the phase angle between current **I** and voltage **V** changes to a more than 90°, the power P becomes negative. The machine takes power from the power system and acts as a motor. When the power angle δ is zero the power P is zero, an alternator floats on the system, neither taking nor receiving any power from the electrical system. In the event of grid failure, an alternator may behave as a motor.

The EMF $\mathbf{E}_1$ for lagging power factor (Fig. 6.45) is given by

$$E_1 = \sqrt{\left(V + IR\cos\theta + IX_{\mathrm{q}}\sin\theta\right)^2 + \left(IX_{\mathrm{q}}\cos\theta - IR\sin\theta\right)^2}$$

The induced EMF $\mathbf{E}_0$ is given as for a lagging power factor

$$E_0 = \sqrt{(V + IR\cos\theta + I_{\mathrm{d}}X_{\mathrm{d}}\cos\delta)^2 + \left(I_{\mathrm{d}}X_{\mathrm{d}}\sin\delta + I_{\mathrm{q}}X_{\mathrm{q}}\cos\delta - IR\sin\theta\right)^2} \quad (6.81)$$

In case of lagging and unity power factors the magnitude of induced EMF is more than terminal voltage or $E_0 > V$. Thus when a lagging or resistive load is connected across any alternator the terminals voltage drops from no-load EMF E_0 to V, and the voltage regulation is positive. When a lagging load or unity power factor load is drawing a current I and the terminal voltage of alternator is V, a disconnection of load makes the open-circuit terminal voltage of the alternator rises to E_0. This is true for both types of alternators.

Similarly for leading power factor, the EMF $\mathbf{E}_1$ is given by

$$E_1 = \sqrt{\left(V + IR_{\mathrm{a}}\cos\theta - IX_{\mathrm{q}}\sin\theta\right)^2 + \left(IX_{\mathrm{q}}\cos\theta + IR_{\mathrm{a}}\sin\theta\right)^2}$$

The currents I_{d} and I_{q} are

$$I_{\mathrm{q}} = I\cos(\theta - \delta) \text{ and } I_{\mathrm{d}} = I\sin(\theta - \delta)$$

The induced EMF $\mathbf{E}_0$ is given

$$E_0 = \sqrt{(V + IR_{\mathrm{a}}\cos\theta - I_{\mathrm{d}}X_{\mathrm{d}}\cos\delta)^2 + \left(I_{\mathrm{q}}X_{\mathrm{q}}\cos\delta - I_{\mathrm{d}}X_{\mathrm{d}}\sin\delta + IR_{\mathrm{a}}\sin\theta\right)^2} \quad (6.82)$$

In case of leading power factor, the magnitude of induced EMF E_0 is likely to be less than the applied voltage. Thus, when a capacitive load is connected across an alternator, the terminal voltage V has a tendency to rise above the no-load value E_0 and the percentage regulation may be negative. When a leading load (capacitive) is drawing a current I and the terminal voltage of alternator is V, a disconnection of load makes the open-circuit terminal voltage of the alternator to fall to E_0.

Consider the phasor diagram of a salient pole alternator at lagging power factor, as shown in Fig. 6.40. Here, it can be shown that,

$$\tan\delta = \frac{IX_q\cos\theta - IR_a\sin\theta}{V + IR\cos\theta + IX_q\sin\theta}$$

Similarly, no-load EMF is

$$E_0 = V\cos\delta + IR_a\cos(\theta+\delta) + I_dX_d \tag{6.83}$$

Similar expressions can be written for unity power factor. Figure 6.41 shows the phasor diagram for UPF. Figure 6.42 shows the phasor diagram for leading power factor for salient pole alternator. Similarly for leading power factor,

$$\tan\delta = \frac{IX_q\cos\theta + IR_a\sin\theta}{V + IR_a\cos\theta - IX_q\sin\theta}$$

$$E_0 = V\cos\delta + IR_a\cos(\theta-\delta) - I_dX_d \tag{6.84}$$

Similar expressions can be written for zero power factors. The phasor diagram for ZPF lagging load is given in Fig. 6.45. In Potier test, the effect of resistance is minimum, when an alternator is connected to a balanced zero power factor (ZPF) lagging (inductive) load.

Example 6.15 An alternator has short-circuit characteristics such that rated armature current is obtained at 0.75 per unit excitation. Rated voltage on the air-gap line is obtained at 1.25 per unit excitation. What per unit excitation is required to supply a load at rated voltage, rated kVA and 0.8 power factor lagging.

Solution

Given that rated voltage at air-gap line for $I_f = 1.25$ per unit

Rated current in short-circuit test for $I_f = 0.75$ per unit

The open-circuit voltage on air-gap line at 0.75 per unit

$$V_{oc} = 1.0 \times 0.75/1.25 = 0.6 \text{ per unit}$$

Now, per unit synchronous reactance is given as

$$X_s = V_{oc}/I_a = 0.6/1.0 = 0.6 \text{ per unit}$$

The rated load current, $I_a = 1.0$ per unit.
The open-circuit EMF $E = \mathbf{V} + j\mathbf{I}_a X_s = 1.0 + 1.0(0.8 - j0.6)(0.0 + 0.6j)$

$$= 1.36 + 0.48j$$
$$= 1.44\angle 19.44° \text{per unit}$$

The field current required for 1.44 per unit voltage on air-gap line is $= 1.44 \times 1.25 = 1.8$ per unit.

Example 6.16 An alternator has the following ratings of 20kVA, 440 V, 3000 RPM, 50 Hz three-phase star connected has $R_a = 0,\ X_l = 0.25\ \Omega$ per phase. The air-gap line is described by the equation $E = 16I_f$. The short circuit is described by $I_{sc} = I_f$. Determine induced EMF at 0.8 power factor lagging using EMF method, power angle, power, and counter torque.

Solution
Given 20 kVA, 440 V, 3000 RPM,
Number of poles,

$$p = \frac{120f}{N} = \frac{120 \times 50}{3000} = 2$$

Rotational speed is

$$\omega = 2\pi f = 2 \times 3.14 \times 50 = 314\,\text{Rad/s}$$

$R_a = 0$ and $X_l = 0.25\ \Omega$ per phase.
Phase voltage is given by

$$V = 440/\sqrt{3} = 254\,\text{V}$$

Rated current is

$$I = 20000/(440 \times \sqrt{3}) = 26.243\,\text{A}$$

In SCC, the field current for rated current is

$$I_f = 26.243\,\text{A}$$

The open-circuit EMF at this field current is

$$E = 16 \times 26.243 = 419.88\,\text{V}$$

Synchronous reactance

$$X_s = 419.88/\left(26.243 \times \sqrt{3}\right) = 9.24\,\Omega$$

For 0.8 power factor lagging, the induced EMF is

$$E = \sqrt{(V + I_a R_a \cos\,\theta + I_a X_s \sin\theta)^2 + (I_a X_s \cos\theta - I_a R_a \sin\,\theta)^2}$$
$$= \sqrt{\left[(254 + 26.242 \times 9.24 \times 0.6)^2 + (26.243 \times 9.24 \times 0.8)^2\right]}$$
$$= 443.7\,V$$

Power angle is given by

$$\delta = \tan^{-1}\{193/399.5\} = 25.8^\circ$$

Percentage regulation is

$$= (443.7 - 254) \times 100/254 = 74.68\%$$

Power

$$P = \frac{3VE_0}{Z_s}\sin\delta = \frac{3 \times 254 \times 443.7}{9.24}\sin 25.8^\circ = 15925\,\text{W} = 15.925\,\text{kW}$$

Torque is given by

$$T = \frac{P}{\omega_s} = \frac{15925}{314} = 50.71\,\text{Nm}$$

6.11 Torque in Synchronous Machines

In case of an alternator, the field or rotor MMF **F** leads its armature MMF **A** by an angle $90^\circ + \theta + \gamma$, which is always less than 180°. It is assumed that the poles due to these MMF exist separately and are rotating in the same direction. In alternators, the field poles are attracted by armature poles in the direction opposite to their motion. This requires mechanical power to be supplied to the generator's shaft so that machine can run against the attractive forces of the armature. A generator when loaded results in a counter torque [12].

In case of synchronous motors, the armature MMF **A** leads the field MMF **F** by an angle $90^\circ + \theta + \gamma$, which is less than 180°. The three-phase current in the armature winding results in the creation of rotating magnetic field, rotating in space at synchronous speed. The rotor poles are attracted by the stator poles, causing the rotor to move. The rotor poles are magnetically locked with those of stator fictitious poles of the rotating magnetic field. If the supply frequency is f Hz and the stator MMF **A** is rotating in space at synchronous speed N_s. This requires an electrical power for setting up rotating magnetic field in space. The interlocking of the stator and rotor poles is possible when the field system is already near synchronous speed.

6.11.1 Cylindrical Rotor Machine

The phasor diagram of overexcited cylindrical rotor synchronous motor drawing leading current is shown in Fig. 6.46. The electromechanical power supplied to the machine is given by

$$P_m = 3\mathbf{E}_0 \cdot \mathbf{I} = \frac{3E_0^2}{Z_s}\cos\gamma - \frac{3E_0V}{Z_s}\cos(\gamma+\delta) = \frac{3E_0V}{Z_s}\sin\delta$$

The machine is rotating at synchronous speed. Let the toque be T, then we can write

$$T \times \frac{2\pi N_s}{60} = P_m$$

$$T = \frac{60}{2\pi N_s}P_m = \frac{60}{2\pi N_s} \times 3E_0 \cdot I = \frac{60}{2\pi N_s} \times 3E_0 I\cos(\theta+\delta) = \frac{60}{2\pi N_s} \times \frac{3E_0V}{X_s}\sin\delta \tag{6.85}$$

Here, θ is the power factor angle (angle between **V** and **I**), and δ is the torque (power) angle (angle between **E** and **V**).

Another expression can be obtained using the basic equation of electromagnetics. The force on a conductor is $F = \text{BIL}$. Assuming that the flux is sinusoidally distributed in air-gap, the torque developed due to conductors of an armature phase can be given as

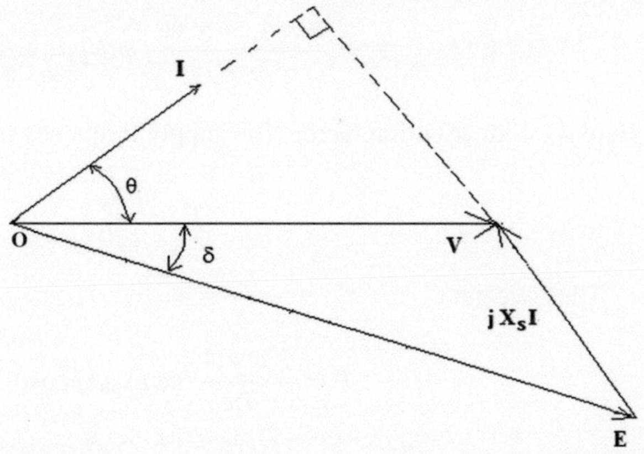

Fig. 6.46 Phasor diagram of cylindrical rotor synchronous motor drawing leading power factor current

$$T_{\mathrm{ph}} = \left(\sum \mathrm{B} \cdot \mathrm{I}\right) L \frac{D}{2}$$

Here, $\sum \mathrm{B}$ is the phasor sum of the flux density cutting the different conductors, L is the axial length of the conductor, and D is the diameter of the armature. Further, we know that $E = \sum \mathrm{B}Lv$, where E is the phase voltage developed due to the air-gap flux and v is the peripheral speed of armature conductors. Thus we get torque per phase

$$T_{\mathrm{ph}} = \mathrm{E} \cdot \mathrm{I} \frac{D}{2v}$$

where peripheral speed, v, is given as

$$v = \frac{2\pi N_{\mathrm{s}} D}{2 \times 60}$$

Torque developed

$$T = 3T_{\mathrm{ph}} = 3(\mathrm{E} \cdot \mathrm{I}) \frac{60}{2\pi N_{\mathrm{s}}} \tag{6.86}$$

Here,

$$\mathrm{E} \cdot \mathrm{I} = EI \cos(\theta + \delta)$$

The EMF per phase can be expressed as

$$E = 4.44 f \phi N_{\mathrm{ph}} k_{\mathrm{d}}$$

Torque is

$$T = \frac{3 \times 4.44 \times 60}{2\pi N_{\mathrm{s}}} f \phi I N_{\mathrm{ph}} k_{\mathrm{d}} \cos(\theta + \delta) \tag{6.87}$$

Here, k_{d} = distribution factor. The supply frequency is,

$$f = \frac{p N_{\mathrm{s}}}{120}$$

Thus, we get

$$T = \frac{3 \times 4.44}{4\pi} p \phi I N_{\mathrm{ph}} k_{\mathrm{d}} \cos(\theta + \delta) \tag{6.88}$$

In case of generator, θ is the angle by which the generating current lags the supply voltage. However in case of synchronous motor, θ is the angle by which the motoring current leads the supply voltage.

6.11.2 Salient Pole Machine

The phasor diagram of overexcited salient pole synchronous motor, drawing leading current shown in Fig. 6.47. In case of salient pole machine, the power is given as

$$P = P_{\mathrm{e}} = P_{\mathrm{m}} = \frac{3VE_0}{X_{\mathrm{d}}}\sin\delta + \frac{3V^2}{2}\left(\frac{1}{X_{\mathrm{q}}} - \frac{1}{X_{\mathrm{d}}}\right)\sin 2\delta$$

$$P = P_{\mathrm{e}} = P_{\mathrm{m}} = P_{\mathrm{max}}\sin\delta + P_{\mathrm{sl}}\sin 2\delta$$

Here,

$$P_{\mathrm{max}} = \frac{3VE_0}{X_{\mathrm{d}}} \;\&\; P_{\mathrm{sl}} = \frac{3V^2}{2}\left(\frac{1}{X_{\mathrm{q}}} - \frac{1}{X_{\mathrm{d}}}\right)$$

The expression for torque is

$$T = \frac{P_{\mathrm{max}}}{\omega_{\mathrm{s}}}\sin\delta + \frac{P_{\mathrm{sl}}}{\omega_{\mathrm{s}}}\sin 2\delta \tag{6.89}$$

Here the first term is due to the field current and the second term is due to the saliency in the rotor. In case of cylindrical rotor machine $X_{\mathrm{d}} = X_{\mathrm{q}} = X_{\mathrm{s}}$, the torque or power due to saliency is absent in cylindrical rotor machines.

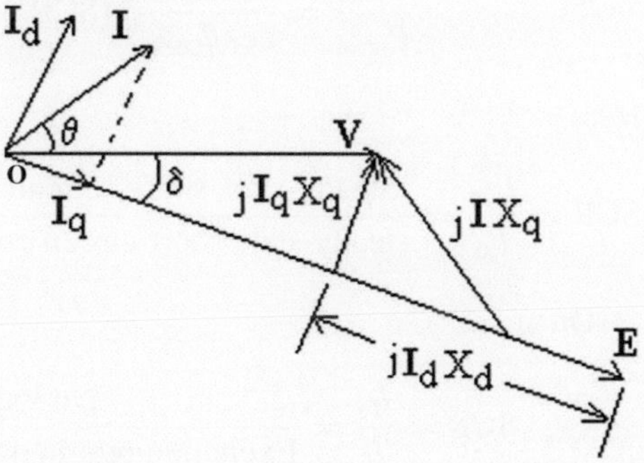

Fig. 6.47 Phasor diagram of salient pole synchronous motor drawing leading power factor current

6.11.3 Active/Reactive Powers and Overload Capacity of Salient Pole Machine

Here it is assumed that X_d is constant and all losses are neglected. The active current of synchronous machine is

$$I_P = I\cos\theta = \frac{E}{X_d}\sin\delta;\ \text{(in per unit)}$$

The reactive component of current

$$I_Q = I\sin\theta = \frac{E}{X_d}\cos\delta - \frac{V}{X_d}\ \text{(in per unit)}$$

As such for a given voltage, active power in per unit is

$$P_P = I_P \cdot V = \frac{EV}{X_d}\sin\delta$$

And reactive power in per unit is

$$P_Q = I_Q \cdot V = \frac{EV}{X_d}\cos\delta - \frac{V^2}{X_d} \tag{6.90}$$

If $V = 1.00$ per unit, the maximum available power

$$P_{\max} = \frac{E}{X_d} = \frac{P_{\text{rated}}}{\cos\theta_{\text{rated}}} \times \frac{E}{X_d} = \frac{P_{\text{rated}}}{\cos\theta_{\text{rated}}} \times \frac{I_{sc}}{I_{\text{rated}}}$$

The overload capacity of machine can be shown as

$$\frac{P_{\max}}{P_{\text{rated}}} = \frac{1}{\cos\theta_{\text{rated}}}\sqrt{\left(\text{SCR} + \sin^2\theta_{\text{rated}}\right)^2 + \cos^2\theta_{\text{rated}}}$$

Here

$$\text{SCR} = \frac{I_{sco}}{I_{sc}} = \frac{\text{Steady-state short - circuit current with no-load excitation current}}{\text{steady-state short-circuit current for an induced EMF } E \text{ under load}}$$

Or,

$$\text{SCR} = \frac{I_{fo}}{I_f} = \frac{\text{No load excitation}}{\text{Excitation required for rated full load current}} \tag{6.91}$$

The overload capacity of cylindrical rotor synchronous machine depends upon the power factor angle and SCR of synchronous machine. Hence, the overload capacity increases with an increase in the value of nominal field excitation I_{fo} which depends on the air-gap of machine. An increase in air-gap length increases the stability of the machine. The overload capacity depends upon $1/\cos\theta$ and as such a low power factor synchronous machine has higher overload capacity. Synchronous motors required to operate at unity power factor have to be designed with larger air-gaps length.

6.11.4 Overload Capacity of Cylindrical Rotor Synchronous Machines

The per power in per unit is given as

$$P = I_{ph} \cdot V_{ph} = \frac{EV}{X_d}$$

The rated terminal voltage is 1.00 per unit, then

$$P = \frac{E}{X_d} = \frac{P_{rated}}{\cos\theta} \times \frac{E}{X_d} = \frac{P_{rated}}{\cos\theta} \times \frac{I_{sc}}{I_{rated}}$$

The overload capacity of machine is

$$I_{sco} = \frac{V}{X_d} = \text{constant}$$

$$\frac{P}{P_{rated}} = \frac{1}{\cos\theta} \times \frac{I_{sc}}{I_{rated}} = \frac{1}{\cos\theta} \times I_{sco} \tag{6.92}$$

6.11.5 Maximum Power at Constant Voltage and Excitation in Case of the Salient Pole Synchronous Machines

The maximum power at constant voltage and excitation in the case of salient pole synchronous machine can be easily obtained from the expression of power. The maximum power would be available with E and V as constant quantity,

$$\frac{d}{d\delta}P = 0$$

On differentiation, we get

$$\frac{dP}{d\delta} = \frac{EV}{X_d} \times \cos\delta + V^2 \frac{(X_d - X_q)}{X_d X_q} \cos 2\delta = 0$$

$$\frac{EV}{X_d} \times \cos\delta = -V^2 \frac{(X_d - X_q)}{X_d X_q} \cos 2\delta$$

$$\frac{\cos 2\delta}{\cos\delta} = -\frac{V}{E} \frac{(X_d - X_q)}{X_q} \tag{6.93}$$

The angle delta corresponds to the maximum power with both V and E constant and hence maximum stable load angle δ_s. Solving the equation and substituting in power expression would result

$$P_{max} = \frac{EV}{X_d} \times \sin\delta_s + \frac{V^2}{2} \frac{(X_d - X_q)}{X_d X_q} \sin 2\delta_s$$

Another approach would be

$$\frac{dP}{d\delta} = 0 = \frac{EV}{X_d} \times \cos\delta + V^2 \frac{(X_d - X_q)}{X_d X_q} \left(\cos^2\delta - \sin^2\delta\right) = 0$$

Multiplying the equation by $\tan\delta$, we get

$$\frac{EV}{X_d} \times \sin\delta_s + \frac{V^2}{2} \frac{(X_d - X_q)}{X_d X_q} \sin 2\delta_s = V^2 \frac{(X_d - X_q)}{X_d X_q} \sin^2\delta_s \tan\delta_s$$

The theoretical stability curve

$$P_{max} = V^2 \frac{(X_d - X_q)}{X_d X_q} \frac{\sin^3\delta_s}{\cos\delta_s} \tag{6.94}$$

can be drawn for various values of delta. For delta 45°

$$P_{max} = 0.5V^2 \frac{(X_d - X_q)}{X_d X_q} \tag{6.95}$$

6.11.6 Capability Curve

Since the interconnected system of generator or power grid is liable to voltage fluctuations, it is desired that the machine should be in a position to supply the rated kVA at rated power factor irrespective of voltage fluctuations. Though the grid voltage is permitted to fluctuate only 5%, the generator is designed to provide rated kVA at rated power factor at fluctuated voltage. The armature should not get overheated while supplying larger current at reduced voltage.

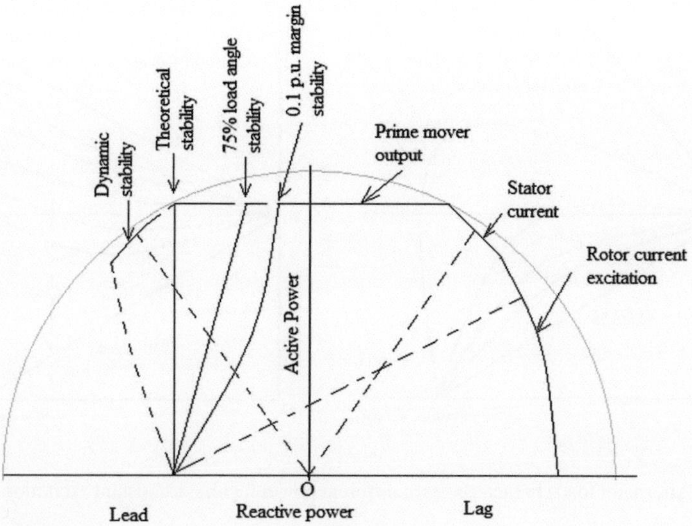

Fig. 6.48 Capability curve of synchronous generator

Practically one cannot load a single synchronous machine under steady-state conditions up to its theoretical stability limit due to different reasons:

(a) Sudden overloading without exceeding the excitation.
(b) Stator heating.
(c) Rotor heating.
(d) Minimum permissible excitation.
(e) Operational limit of the turbine.

The working region of the operation at angle exceeding 90° degree electrical is the region of dynamic stability. The theoretical stability limit is based on 90° load angle with 0.1 per unit margin.

Capability curve or compounding curve of synchronous machine is the operational chart of synchronous machine which gives the limits of its operation to provide active and reactive power for constant voltage. The capability curve is shown in Fig. 6.48.

6.11.7 Alternator Load Characteristics

The alternator load characteristics are shown in Fig. 6.49. Figure 6.49a shows the variation of load current with respect to supply voltage for different power factors at constant excitation. Figure 6.49b shows the variation of excitation with respect to stator current for different power factor keeping the supply voltage constant.

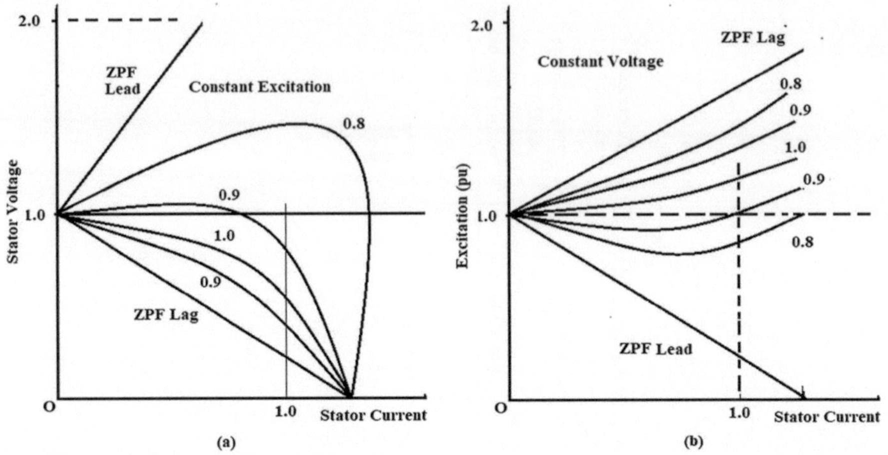

Fig. 6.49 Alternator load characteristics at different power factors: **a** constant excitation, **b** constant voltage

6.12 V-Curve

Figures 6.50 and 6.51 show constant power ($E \sin\delta$) phasor diagram for cylindrical rotor alternator and synchronous motor respectively for different power factors. We know that the power is given by [11]

$$P = P_\mathrm{e} = P_\mathrm{m} = \frac{3VE}{Z_\mathrm{s}} \sin\delta = 3VI\cos\theta$$

Thus, for different power factors, the power will be constant if ($E \sin\delta$) or $I\cos\theta$ is constant.

Fig. 6.50 Phasor diagram of alternator for constant power and different power factor

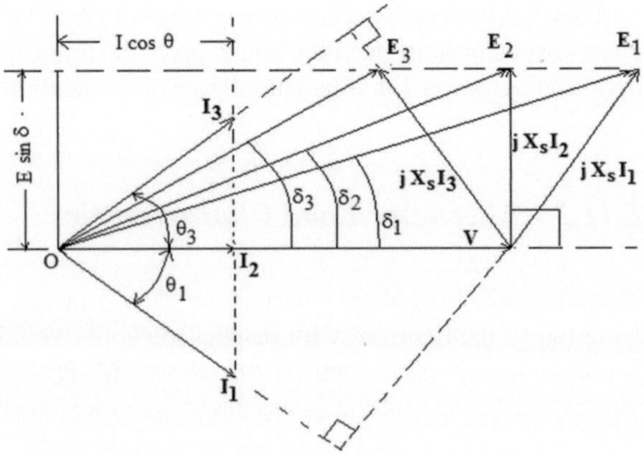

Fig. 6.51 Phasor diagram of cylindrical rotor synchronous motor for constant power and different power factor

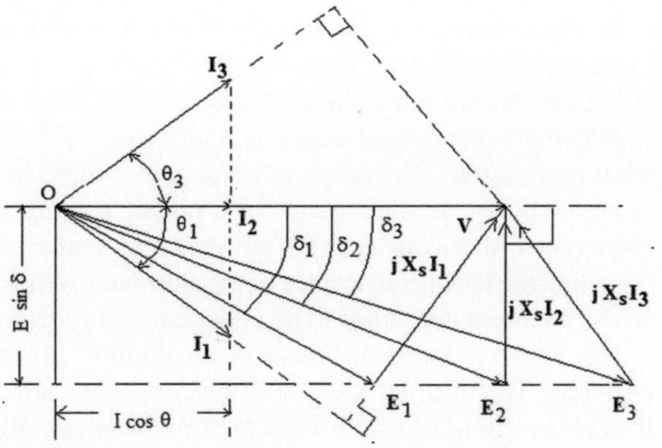

In case of alternator, as the power factor is changed from lagging power factor to leading, the magnitude of current first decreases to minimum at unity power factor and then increases for leading power factor. Thus, for two similar magnitudes of the current, the power factor may be either lagging or leading. This is also true for synchronous motor.

In both Figs. 6.50 and 6.51, I_1 is a lagging current, $\mathbf{I}_2$ is at unity power factor, and $\mathbf{I}_3$ is at leading power factor, such that for constant power

$$I_1 \cos\theta_1 = I_2 = I_3 \cos\theta_3$$

For alternator in Fig. 6.50, we can write

$$E_1 \cos\delta_1 > E_2 \cos\delta_2 > E_3 \cos\delta_3$$

For synchronous motor, in Fig. 6.51, we can write

$$E_3 \cos\delta_3 > E_2 \cos\delta_2 > E_1 \cos\delta_1$$

When machine has the following induced voltages and voltage relationship it is said to be excited accordingly, as shown in Table 6.1.

Table 6.1 Effect of excitation on induced voltage

E and V	Excitation	Power factors	
		Alternator	Synchronous motor
$E \cos\delta < V$	Under-excited	Leading	Lagging
$E \cos\delta = V$	Normally excited	Unity	Unity
$E \cos\delta > V$	Overexcited	Lagging	Leading

If a synchronous motor is overexcited, its power factor is leading and it will act like a capacitive load on the power system. This property of a synchronous motor is utilized in overall power factor correction of a large installation.

When a synchronous motor is under-excited, it works as an inductive load on the power system. The property is sometimes used in obtaining ZPF characteristics of an alternator. In the absence of a proper lagging reactive load in laboratory, an overexcited alternator supplies power to an under-excited synchronous motor for obtaining the leakage reactance of the alternator. When a capacitive load is connected across an alternator, it has to be under-excited for rated voltage.

The above property of synchronous motor is represented in V-curve, as shown in Fig. 6.52. The inverted V-curve is the curve between power factor and field current. If field current of a synchronous motor is very low, the operating point may shift in the unstable zone. In this region, the machine exhibits the *hunting phenomenon.* The speed and current start oscillating due to induced currents in the damper bars. The excessive oscillations in current either blow the fuses or trip the supply. These oscillations in speed can be experimentally visualized using stroboscopic RPM meter—stroboscope. Do you know the principle of operation of stroboscope? It is possible for you to record the oscillations in voltage and current using Hall effect voltage and current transducers respectively? The phasor diagram for constant torque angle of a synchronous motor is shown in Fig. 6.53. Here, the leading current required is higher than the lagging current.

Fig. 6.52 V-curve of alternator and synchronous motor

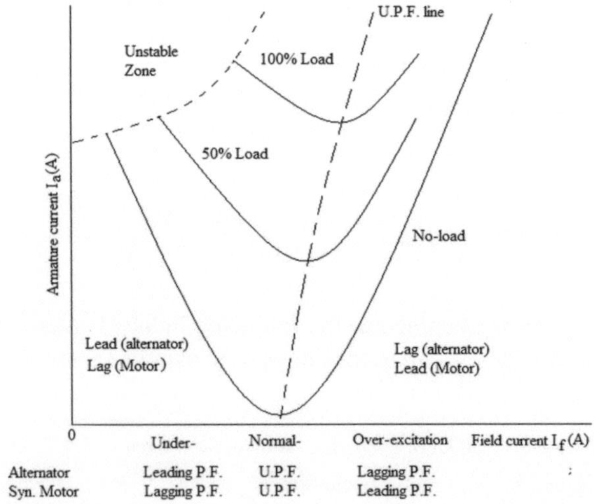

Fig. 6.53 Phasor diagram of synchronous motor for constant torque angle

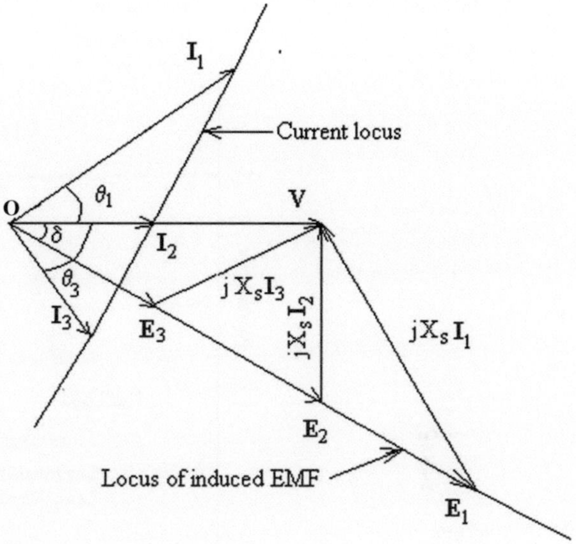

6.12.1 Inverted V-Curve

The inverted V-curves of synchronous machine are the power factor (lagging and leading) versus excitation current plot. We know that an under-excited alternator can deliver power to a leading capacitive load at rated voltage, on the other hand when it is overexcited it delivers power to a lagging load at rated voltage.

In case of synchronous motor, it draws lagging power from source when under-excited and draws leading power when it is overexcited. An under-excited synchronous motor acts as an inductive load on power system whereas an overexcited synchronous motors acts as a capacitive load to the power system.

The inverted V-curve for motor and alternator has been shown in Fig. 6.54.

Example 6.17 A 440 V, two-pole, 50 Hz, three-phase star-connected synchronous motor has a synchronous reactance of 8.0Ω/phase. The armature resistance is negligible and the induced excitation EMF per phase is 200 Volt. The power angle between V and E_f is 36°. Figure 6.55

(a) Calculate the line current and power factor, power, and torque.
(b) Calculate the excitation EMF and power angle necessary to make the power factor unity for the same input power. Also, calculate power and torque.

Solution

Given that $V = 440/\sqrt{3} = 254$ Volt; $E = 200$ Volt; Power angle $\delta = 36°$; $X_s = 8.0$ Ω, synchronous speed, $\omega = 2\pi f = 314$ rad/s (Fig. 6.55).

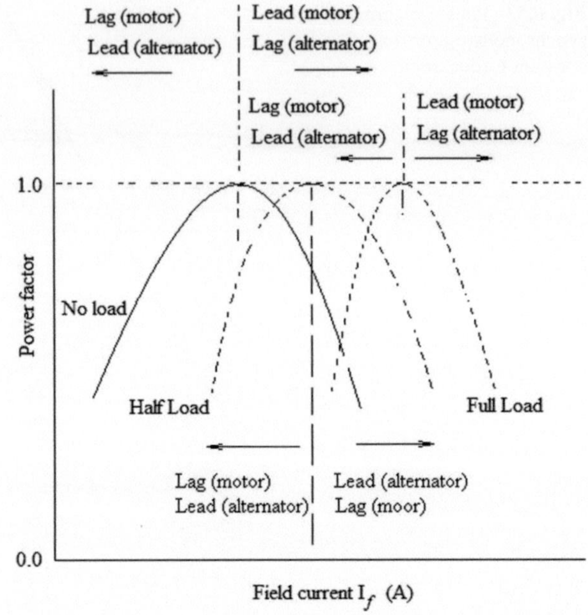

Fig. 6.54 Inverted V-curve

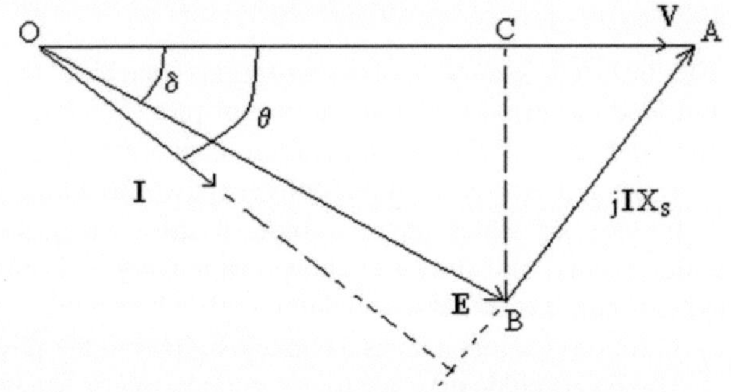

Fig. 6.55 Phasor diagram of synchronous motor-lagging power factor

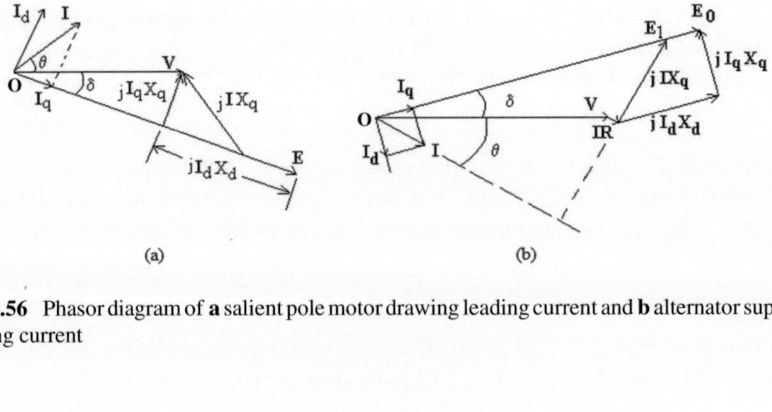

Fig. 6.56 Phasor diagram of **a** salient pole motor drawing leading current and **b** alternator supplying lagging current

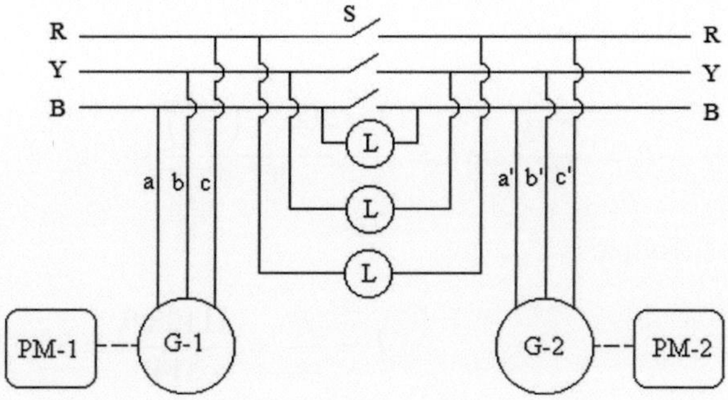

Fig. 6.57 Parallel connection of two alternators—three dark lamp method

(a) Since $E \cos \delta < V$, the power factor is lagging. Let the power factor angle be θ. According to phasor diagram mentioned above in Figs. 6.55, the phase voltage can be given as

$$V = E \cos \delta + IX_s \sin \theta$$

$$254 = 200 \times 0.809 + 8 \times I \sin \theta$$

Thus, we get

$$I \sin \theta = (254 - 161.8)/8 = 11.525$$

Similarly, we have

$$E \sin \delta = IX_s \cos \theta$$

$$I \cos \theta = 200 \times 0.59/8 = 14.7$$

Thus,

$$\tan \theta = 11.525/14.7 = 0.784$$

$$\theta = -38.1^\circ \text{ lagging}$$

Current is

$$I = 14.7/\cos 38.1 = 18.68 \text{ A}$$

Input to the motor is given as

$$P = 3VI \cos \theta = 3 \times 254 \times 14.7 = 11201.4W$$

Also, power

$$P = \frac{3VE_0}{Z_s}\sin\delta = \frac{3\times 200\times\left(\frac{440}{\sqrt{3}}\right)}{8}\sin 36^\circ = 11198.8\,\text{W} = 11.198\,\text{kW}$$

Torque is

$$T = \frac{P}{\omega_s} = \frac{11198.9}{314} = 35.67\,\text{Nm}$$

(b) For unity power factor, let the current be I_2, Thus

$$3VI_2 = 3VI\ \cos\theta$$

Thus

$$I_2 = I\cos\,\theta = 14.7\,\text{A}$$

At unity power factor, let the excitation EMF be E_2

$$E_2\ \cos\ \delta_2 = V$$

$$E_2\sin\ \delta_2 = I_2X_s$$

$$E_2\cos\ \delta_2 = 254$$

or,

$$E_2\sin\ \delta_2 = 14.7\times 8 = 117.6$$

The excitation EMF or no-load EMF is given as

$$E_2 = \sqrt{(254^2 + 117.6^2)} = 279.9\,\text{Volt}$$

$$\delta_2 = \tan^{-1}(117.6/254) = 24.84^\circ$$

Power

$$P = \frac{3VE_0}{Z_s}\sin\delta = \frac{3\times 279\times\left(\frac{440}{\sqrt{3}}\right)}{8}\sin 24.84^\circ = 11199.7\,\text{W} = 11.199\,\text{kW}$$

Torque is given as

$$T = \frac{P}{\omega_s} = \frac{11199.7}{314} = 35.66\,\text{Nm}$$

For a synchronous motor supplying a constant torque load, the power factor can be changed by changing the induced EMF and torque angle.

Example 6.18 A 1100 V, four-pole, three-phase, 50 Hz star-connected cylindrical rotor synchronous motor has a synchronous reactance of 5 Ω per phase. When it delivers 150 HP, the efficiency is 89% and power angle is 9°. Neglecting the armature resistance, determine

(a) excitation EMF per phase
(b) line current
(c) power factor
(d) Power, and
(e) torque.

Solution

Given power angle $\delta = 9°$; per phase voltage $V = 1100/\sqrt{3} = 635.1$ V; Frequency 50 Hz; $X_s = 5.0$ Ω per phase; power = 150 HP; efficiency = 89%

Input power can be given as

$$746 \times 150/0.89 = 125730.34\,\text{W} = \sqrt{3} \times 1100 \times I\cos\theta$$

Synchronous speed, $\omega = 4\pi f/p = 157$ rad/s
Thus, we get

$$I\cos\theta = 125730.34/1100\sqrt{3} = 65.99$$

$$E\sin\delta = IX_s\cos\theta$$

(a) We know, $V = 635$ V

$$E = 5 \times 65.99/\sin 9 = 2109.2\,\text{Volt}$$

$$E\cos 9 = 2109.2\cos 9 = 2109.2 \times 0.9876 = 2083\,\text{V} > 635\,\text{V}$$

Since $E\cos\delta > V$, the machine is overexcited and power factor is leading. Thus, we get

$$V = E\cos\delta - IX_s\sin\theta$$

$$I\sin\theta = (-635.1 + 2109.2\cos 9)/5 = 1447.93/5 = 289.5$$

$$\theta = \tan^{-1}(289.586/65.99) = 77.16^\circ$$

(b) Current I is given as

$$I = 65.99/\cos 77.16^\circ = 296.97\text{ A}$$

(c) Power factor = cos 77.16° = 0.222 leading.
(d) $P = \frac{3VE_0}{Z_s}\sin\delta = \frac{3\times 635.1\times 2109.2}{5}\sin 9^\circ = 125731\text{ W} = 125.731\text{ kW}$
(e) Torque, $T = \frac{P}{\omega_s} = \frac{125731}{157} = 800.83\,Nm$

Example 6.19 A 440 V, 50 Hz, six-pole, three-phase star-connected synchronous motor delivers 10 HP, operating at 0.8 lagging power factor. The total iron, copper, and friction losses are 1000 W. The armature resistance is negligible. If the motor has synchronous reactance of 5 Ω /phase, determine

(a) Efficiency
(b) The induced voltage and torque angle if the motor draws 10A at 0.8 leading power factor
(c) Power
(d) Torque.

Solution
Given per phase voltage $V = 440/\sqrt{3}$ V; synchronous speed, $\omega = 4\pi f/p = 104.66$ rad/s; output power $P_o = 10 \times 746$ W; losses = 1000 W; input power $P_i = 7460 + 1000 = 8460$ W.

(a) Efficiency is

$$\eta = 7460/8460 = 88.12\%$$

(b) Current $I = 10$ A, power factor is 0.8 leading

$$V = E\cos\delta - IX_s\sin\theta$$

$$254 = E\cos\delta - 5\times 10\times 0.6$$

$$E\cos\delta = 254 + 30 = 284\text{ V per phase}$$

Similarly,

$$E\sin\delta = IX_s\cos\theta$$

$$E\sin\delta = 5 \times 10 \times 0.8 = 40$$

Thus torque angle

$$\delta = \tan^{-1}(40/284) = 8.01^\circ$$

The induced EMF

$$E = 40/\sin 8.01^\circ = 287\ \text{Volt per phase}$$

(c) $P = \frac{3VE_0}{Z_s}\sin\delta = \frac{3\times 254.1\times 286}{5}\sin 8.01^\circ = 6095\text{W} = 6.095\text{kW}$

(d) Torque is given as

$$T = \frac{P}{\omega_s} = \frac{6095}{104.66} = 58.23\,\text{Nm}$$

Example 6.20 A 1500 kVA, 11 kV, four-pole, three-phase, 50 Hz, star-connected, synchronous motor has a synchronous reactance of 7 Ω per phase and negligible armature resistance. Calculate the induced voltage, power and torque at

(a) 0.8 power factor lag
(b) 0.8 power factor lead
(c) Unity power factor.

The motor takes 500 kVA in each case.

Solution

Given that the motor is three-phase and star-connected and has phase current 1500/(11 $\sqrt{3}$)= 78.73 A; phase voltage $V = 11000/\sqrt{3} = 6351$ V; synchronous speed, $\omega = 4\pi f/p = 157$ rad/s.

(a) 0.8 power factor lagging

$$E\cos\delta = V - IX_s\sin\theta = 6351 - 78.73 \times 7 \times 0.6 = 6020.33$$

$$E\sin\delta = IX_s\cos\theta = 78.73 \times 7 \times 0.8 = 440.89$$

Thus, we get

$$E = 6036.46\,\text{V}$$

Now,

$$\delta = \tan^{-1}(440.89/6020.33) = 4.19^\circ$$

$$P = \frac{3VE_0}{Z_s}\sin\delta = \frac{3 \times 6351 \times 6036}{7}\sin 4.19^\circ = 1200380.1\text{W} = 1.2\text{MW}$$

Torque is given as

$$T = \frac{P}{\omega_s} = \frac{1200380.1}{157} = 7653\,\text{Nm}$$

(b) 0.8 power factor Leading

$$E\cos\delta = V + IX_s\ \sin\theta = 6351 + 78.73 \times 7 \times 0.6 = 6681.67$$

$$E\sin\delta = IX_s\ \cos\theta = 78.73 \times 7 \times 0.8 = 440.89$$

Thus,

$$E = 6696.2\,\text{V}$$

$$\delta = \tan^{-1}(440.89/6681.67) = 3.78^\circ$$

$$P = \frac{3VE_0}{Z_s}\sin\delta = \frac{3 \times 6351 \times 6696}{7}\sin 3.78^\circ = 1201530.6\text{W} = 1.201\text{MW}$$

Torque,

$$T = \frac{P}{\omega_s} = \frac{1201530.6}{157} = 7653.1\,\text{Nm}$$

(c) UPF

$$E\cos\delta = V = 6351$$

$$E\ \sin\delta = I\,X_s = 78.73 \times 7 = 551.11$$

Thus, we get

$$E = 6374.9\,\text{V}$$

$$\delta = \tan^{-1}(551.11/6351) = 4.96^\circ$$

$$P = \frac{3VE_0}{Z_s}\sin\delta = \frac{3 \times 6351 \times 6374}{7}\sin 4.96^\circ = 1500009.6\text{W} = 1.5\text{MW}$$

Torque

$$T = \frac{P}{\omega_s} = \frac{1500009.6}{157} = 9556\,\text{Nm}$$

Example 6.21 A 440 V, 50 Hz, six-pole, salient pole synchronous motor is drawing 10A, current at 0.8 power factor leading. It has $X_d = 10.0\,\Omega$ and $X_q = 7.0\,\Omega$. Calculate (a) induced EMF (b) torque angle, power, and torque (c) If the machine is running as alternator, calculate current, and power factor for the same magnitude of induced EMF and power angle.

Solution

The phasor diagrams are shown in 6.56 a and b. Given that per phase voltage $V =$ 254 V; current $I = 10.0$A at 0.8 power factor lagging, i.e.,

$$\mathbf{I} = 10(0.8 + j0.6)A$$

Synchronous speed, $\omega = 4\pi f/p = 104.67$ rad/s; $X_d = 10.0\,\Omega$ per phase; $X_q = 7.0$ Ω per phase.

Now, power factor angle $\theta = \cos^{-1}(0.8) = 36.87°$.

For motoring mode, we have (See Fig. 6.56a).

$$\delta = \tan^{-1}\left[\frac{IX_q\cos\theta}{V + IX_q\sin\theta}\right] = \tan^{-1}\left[\frac{10 \times 7.0 \times 0.8}{254 + 10 \times 7.0 \times 0.6}\right] = 10.71°$$

Currents are

$$I_q = I\cos(36.87 + 10.71) = 10\cos(47.58) = 6.75\,\text{A}$$

$$I_d = I\sin(36.87 + 10.71) = 10\sin(47.58) = 7.38\,\text{A}$$

(a) Induced EMF is

$$E_0 = V\cos\delta + I_d X_d = 254 \times \cos 10.71° + 7.38 \times 10 = 323.38\text{V}$$

(b) Torque angle is $\delta = 10.71°$.

Power is given as

$$P = P_e = P_m = \frac{3VE_0}{X_d}\sin\delta + \frac{3V^2}{2}\left(\frac{1}{X_q} - \frac{1}{X_d}\right)\sin 2\delta$$

$$P = \frac{3 \times 254 \times 323.38}{10}\sin 10.71 + \frac{3 \times 254^2}{2}\left(\frac{1}{7} - \frac{1}{10}\right)\sin(2 \times 10.71)$$

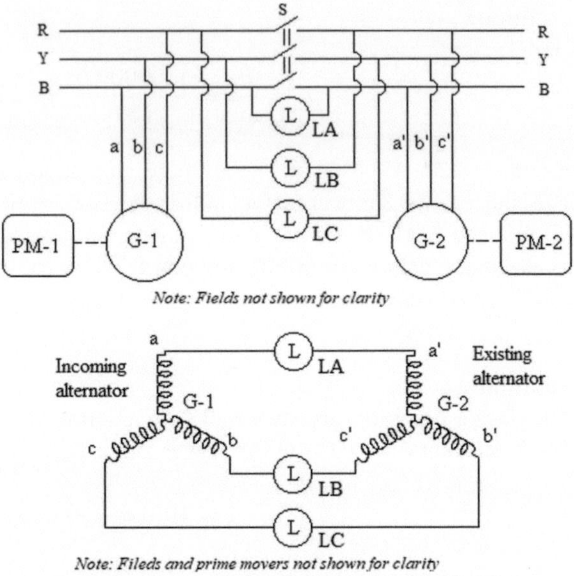

Fig. 6.58 Two dark and one bright lamp method—parallel connections of two alternators—with transposed

$$= 4579 + 1514.66 = 6092.6W$$

$$T = \frac{3 \times 254 \times 323.38}{10 \times 104.67}\sin 10.71 + \frac{3 \times 254^2}{2 \times 104.67}\left(\frac{1}{7} - \frac{1}{10}\right)\sin(2 \times 10.71)$$
$$= 43.74 + 14.47 = 58.2\,\text{Nm}$$

(c) When machine is running as alternator, the magnitude of induced EMF is $E =$ 323.38 V (Fig. 6.56b). Let the new current be $\mathbf{I}$ at lagging power factor angle θ. Now, power angle is $\delta = 10.71°$

From phasor diagram, we get

$$V + I_d X_d \cos\delta - I_q X_q \sin\delta = E_0 \cos\delta$$

Thus, we get

$$254 + 9.825 I_d - 1.3 I_q = 317.75$$
$$9.825 I_d - 1.3 I_q = 63.75$$
$$7.56 I_d - I_q = 49$$

Now, we have

$$I_d X_d \sin\delta + I_q X_q \cos\delta = E_0 \sin\delta$$
$$1.85 I_d + 6.88 I_q = 60.1$$

$$0.27I_d + I_q = 8.74$$

Solving above equations, we get

$$I_d = 7.34\,\text{A and } I_q = 6.76\,\text{A}$$

Current can now be given as

$$I = \sqrt{I_d^2 + I_q^2} = 9.98A$$

We know that torque angle is

$$\tan\delta = \left[\frac{IX_q\cos\theta}{V + IX_q\sin\theta}\right]$$

$$\tan 10.71 = \left[\frac{9.98 \times 7\cos\theta}{254 + 9.98 \times 7\sin\theta}\right]$$

$$254 - 369.7\cos\theta + 69.8\sin\theta = 0$$

Solving above equation, we get, power factor angle $\theta \approx 36.84°$
Power factor cos 36.84° = 0.8 power factor lagging
Check:
Now, let us suppose function

$$\mathrm{f}(\theta) = 254 - 369.7\cos\theta + 69.8\sin\theta$$

$$\mathrm{f}'(\theta) = 369.7\sin\theta + 69.8\cos\theta$$

Use Newton Raphson method for solving for cos θ. Choose some initial angle. Compare the results with analytical results. MSEXEL can also be used for solution.

6.13 Synchronization

In a power system, there are several generators connected to same three-phase line referred to as *grid* supplying power to a widely distributed common load. An isolated alternator needs to be synchronized with the existing power system for sharing the load. The following are the essential conditions for its synchronization:

- Its speed should correspond exactly to the frequency of the power system.
- The voltage should be numerically equal to that of bus-bar voltage.
- The voltage phasor should be exactly coincident with the voltage phasor of the bus bar at the instant of connection.

The prime mover speed of the incoming alternator is to be adjusted that it corresponds exactly to the frequency of the bus bar. In other words, the frequency of the incoming alternator must be identical. The voltage is made numerically equal by adjusting the field current of the incoming alternator G1.

In Fig. 6.57, the 'a-b-c' phases of an alternator are connected to 'a′-b′-c′' phases of another alternator. The lamps are rated to operate at twice the individual phase voltages. If the voltage is higher than 440 V, a potential transformer is required to convert the voltage to a lower level. The switch *S* is connected at the instant when all the lamps are dark. This indicates that the identical phases a & a′, b & b′, and c & c′ are coincident. However, the lamps may remain dark even if there is an appreciable voltage across them. This method is referred as three dark lamp methods. The synchroscope indicates the exact instant of synchronization.

Another method of synchronization is known as *one dark and two bright lamps method*, developed by Siemens and Halske. In this method, the lamp terminals are transposed as shown in Fig. 6.58. In other words, lamps LA, LB, LC are connected across a-a′, b-c′ and c-b′, respectively. These lamps glow in rotation. At synchronism, the lamp LA is dark and the other two lamps LB and LC are equally bright.

Let the frequency of incoming alternator G-1 be f_1 and that of G-2 be f_2. The voltages of each phases are numerically equal, but the incoming generator G1 might be running at near synchronous speed. Figures 6.59, 6.60, and 6.61 show the voltages across lamps LA, LB, LC, when $f_1 > f_2$ and $f_1 < f_2$ respectively.

A schematic representation of the phasor is shown in Fig. 6.59. The lamps LA, LB, and LC experience the difference of their phase voltages, i.e.,

$$\begin{aligned} V_{\mathrm{LA}} &= V_{\mathrm{a}} - V_{\mathrm{a}'} \\ V_{\mathrm{LB}} &= V_{\mathrm{b}} - V_{\mathrm{c}'} \\ V_{\mathrm{LC}} &= V_{\mathrm{c}} - V_{\mathrm{b}'} \end{aligned} \tag{6.96}$$

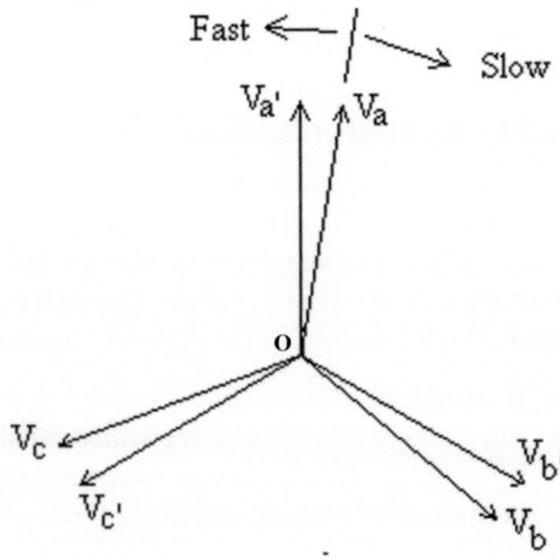

Fig. 6.59 Voltage phasors of alternators G1 and G2 rotation of glow

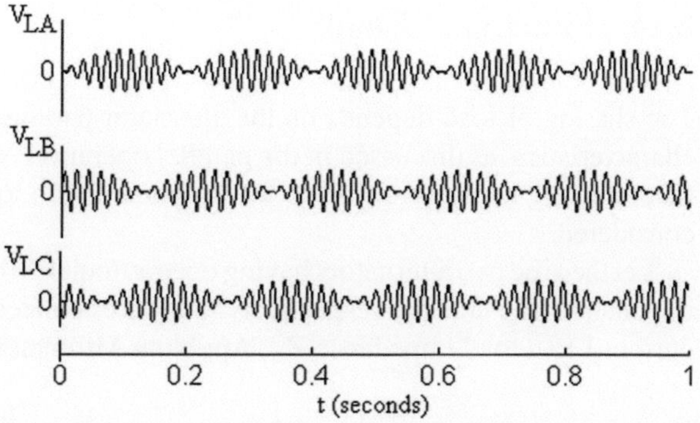

Fig. 6.60 Lamp voltages for $f_1 = 55$ Hz and $f_2 = 50$ Hz (glow sequence LA, LC, LB)

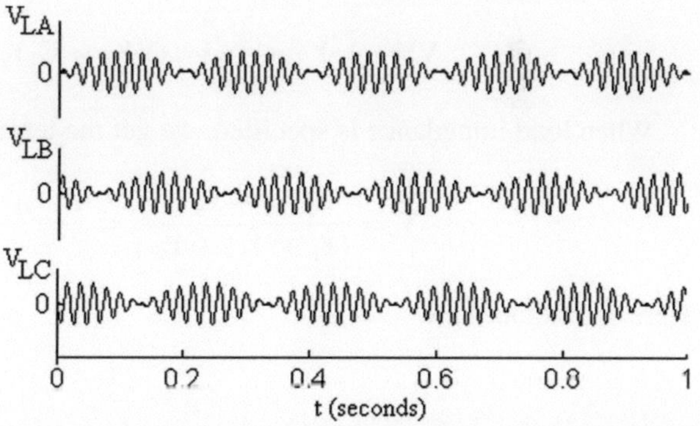

Fig. 6.61 Lamp voltages for $f_1 = 45$ Hz and $f_2 = 50$ Hz (glow sequence LA, LB, LC)

It is assumed that the a′-b′-c′ has the fixed frequency of 50 Hz. The difference in the voltages for $f_1 = 55$ Hz and $f_2 = 50$ is plotted in Fig. 6.60. When incoming a-b-c phasor are fast (55 Hz), the sequence of glow of lamps is LA-LC-LB. If lamp LA is initially dark, the voltage of lamp LC decreases and become dark and then LB does the same. Thus rotation of glow is LA-LC-LB. The peak voltage across each lamp will be $2\sqrt{2}$ time the rms voltage of a phase. As such each lamp must be able to withstand twice the phase voltage.

When the incoming phasor a-b-c is slow ($f_1 = 45$ Hz), the sequence of glow of lamp is LA-LB-LC. If lamp LA is initially dark, the voltage of lamp LB decreases and become dark and then lamp LC follows. Thus, the rotation of glow is LA-LB-LC, as shown in Fig. 6.61. Students are expected to visualize the same on the basis of rotation of phasor a-b-c as shown in Figs. 6.59, 6.60, and 6.61.

When the speed of alternator G1 is so adjusted that frequency is identical to the frequency of bus bar or existing alternator, the voltage across LA is zero, while voltages across LB and LC are identical, i.e., the voltages $V_{LA} = 0$ and $V_{LB} = V_{LC}$. Lamp LA is dark, while the other two lamps are equally bright. At this instant, the switch is closed for joining generator G1 in parallel to generator G2.

If the conditions of synchronization are not satisfied, it results in large transient currents in the system, posing danger to the stability of the power system.

6.14 Sharing of Load

The sharing of load depends on the alternator parameters as well as the governor's characteristics, as discussed in the parallel operation of the DC generator. Millman's theorem may be useful if the speed variation due to loading of the generators is not considered.

Let there be two alternators having open-circuit EMFs $\mathbf{E}_1$ and $\mathbf{E}_2$, and synchronous impedances Z_{s1} and Z_{s2}, respectively. They are connected in parallel supplying a load current $\mathbf{I}$ to a load impedance Z_L. Applying Millman's theorem, we have

$$\mathbf{V}\left[\frac{1}{Z_L}+\frac{1}{Z_{s1}}+\frac{1}{Z_{s2}}\right]=\frac{\mathbf{V}}{Z_0}=\frac{\mathbf{E}_1}{Z_{s1}}+\frac{\mathbf{E}_2}{Z_{s2}}$$

$$\mathbf{V}[Y_L+Y_{s1}+Y_{s2}]=\mathbf{V}Y_0=\mathbf{E}_1Y_{s1}+\mathbf{E}_2Y_{s1} \tag{6.97}$$

When load impedance is specified, we get the terminal voltage as

$$\mathbf{V}=\frac{\mathbf{E}_1Y_{s1}+\mathbf{E}_2Y_{s1}}{[Y_L+Y_{s1}+Y_{s2}]}=\frac{\mathbf{E}_1Y_{s1}+\mathbf{E}_2Y_{s1}}{Y_0}$$

Here admittances are

$$Y_{s1}=\frac{1}{Z_{s1}}$$

$$Y_{s2}=\frac{1}{Z_{s2}}$$

$$Y_L=\frac{1}{Z_L}$$

$$Y_0=Y_L+Y_{s1}+Y_{s2}$$

The current can be calculated as

$$\mathbf{I}_1=\frac{\mathbf{E}_1-\mathbf{V}}{Z_{s1}} \quad \text{and} \quad \mathbf{I}_2=\frac{\mathbf{E}_2-\mathbf{V}}{Z_{s2}}$$

When load current is specified, we can write

$$\mathbf{V}=\frac{(\mathbf{E}_1Y_{s1}+\mathbf{E}_2Y_{s1})-\mathbf{I}}{[Y_{s1}+Y_{s2}]} \tag{6.98}$$

Example 6.22 Two identical synchronous generators have EMFs of 1100 V per phase, 50 Hz, and their impedances are $Z_{s1}=(0.15+j2.1)\ \Omega$ and $Z_{s2}=(0.2+j3.3)\Omega$ per phase are connected in parallel. They supply a common load having impedance of $(1.0 + j1.0)\Omega$ per phase. Calculate the terminal voltage, load currents, active power output, and no-load circulating current for a phase divergence of 5° electrical.

Solution
Given,

$$E_1 = (1100 + j0) \quad \text{V per phase}$$
$$E_2 = 1100(\cos 5 - j \sin 5) = (1096 - j96) \quad \text{V per phase}$$

The load impedance is

$$Z_L = 1.0 + 1.0j = 1.4142\angle 45^\circ \quad \Omega \text{ per phase}$$

The synchronous impedances are given as

$$Z_{s1} = (0.15 + j2.1) = 2.105\angle 85.9^\circ \quad \Omega \text{ per phase}$$

$$Z_{s2} = (0.2 + j3.3) = 3.31\angle 86.5^\circ \quad \Omega \text{ per phase}$$

Applying Millman's theorem, we have

$$V\left[\frac{1}{Z_L} + \frac{1}{Z_{s1}} + \frac{1}{Z_{s2}}\right] = \frac{V}{Z_0} = \frac{E_1}{Z_{s1}} + \frac{E_2}{Z_{s2}}$$

$$V = 614.5\angle -21.5^\circ \quad \text{Volt}$$

The individual currents are,

$$I_1 = \frac{E_1 - V}{Z_{s1}} = 273\angle -62.8^\circ \text{ A}$$

$$\mathbf{I}_2 = \frac{\mathbf{E}_2 - \mathbf{V}}{Z_{s2}} = 163.3\angle -72.7^\circ \text{ A}$$

The per phase active power,

$$P_1 = 614.5 \times 273 \times \cos(62.8 - 21.5) = 126 \quad \text{kW}$$

$$P_2 = 614.5 \times 163.3 \times \cos(72.7 - 21.5) = 62.9 \quad \text{kW}$$

The no-load circulating current is given as

$$\mathbf{I}_c = \frac{\mathbf{E}_2 - \mathbf{E}_1}{Z_{s1} + Z_{s2}} = \frac{1096 - j96 - 1100 + 0j}{0.15 + 2.1j + 0.2 + 3.3j} = 25.1\angle - 131.3^\circ \text{ A}$$

6.15 Synchronizing Power and Torque

In a synchronous machine, a momentary change in its angular speed may be due to following facts [12]:

(a) Fault in the system.
(b) Sudden large load.
(c) Transient behavior.

If an alternator is operating in parallel with other alternators, a synchronizing action is developed due to a momentary changes in its angular speed. This tends to restore the proper phase angles. Consider two alternators operating in parallel. Let the terminal voltages be $\mathbf{V}_1$ and $\mathbf{V}_2$ of the alternators running in exact synchronism. These voltages are equal in magnitude and exactly in phase opposition with respect to the local circuit between them. The machine supplies the load according to the excitation voltage and prime mover power. A sudden increase in the load will cause momentarily slow down both the machines. Due to differences in their governor characteristics, one of the machine will share more load than the other, and as a result, it slows down. The resultant voltage $\mathbf{V}_c$ acts in the local circuit and sets up a current lagging behind the $\mathbf{V}_c$ by an angle γ_c. this has been shown in Fig. 6.62. The current in the local circuit is

$$\mathbf{I}_c = \frac{\mathbf{V}_c}{(R_1 + R_2) + j(X_{s1} + X_{s2})} = I_c\angle - \gamma_c \tag{6.99}$$

where $\gamma_c = \tan^{-1}\left(\frac{X_{s1}+X_{s2}}{R_1+R_2}\right)$

Here R_1 and R_2 are the per phase resistance, X_{s1} and X_{s2} are the per phase synchronous reactance of the two machines.

This synchronizing current I_c is in addition to the load current supplied by one of the generator having a component in phase with its voltage. A generator gets additional load, which slows it down. The other generator gets a motoring power, which tends to raise its speed. Thus there is a tendency toward proper distribution of the load. The process is referred to as synchronizing action caused by additional

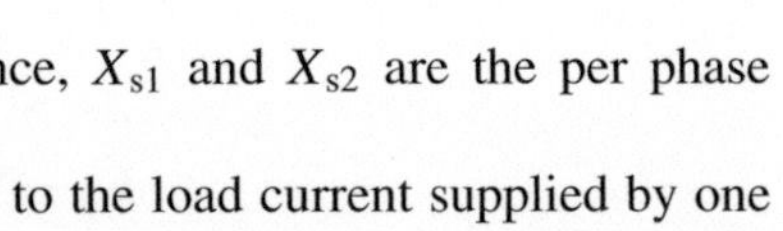

Fig. 6.62 Synchronizing current

power known as *synchronizing power* acting on the machine to bring its normal operation.

The synchronizing torque is defined as the power per electrical radian or degree of deviation of power angle from the normal equilibrium position. On the other hand, the *synchronizing torque* is the synchronizing power divided by synchronous angular speed. Synchronizing power and torque are present in synchronous motors for restoring their transient behavior due to sudden load changes.

Now, the electromagnetic power of a three-phase cylindrical rotor alternator is given as

$$P_{\mathrm{m}} = 3\mathbf{E}_0 \cdot \mathbf{I} = \frac{3E_0^2}{Z_0}\cos\gamma - \frac{3E_0V}{Z_0}\cos(\gamma+\delta)$$

The synchronizing power per electrical radian

$$P_{\mathrm{S}} = \frac{\mathrm{d}P_{\mathrm{m}}}{\mathrm{d}\delta} = \frac{3E_0V}{Z_{\mathrm{s}}}\sin(\gamma+\delta)$$

The synchronizing torque per electrical radians for an alternator is

$$T_{\mathrm{S}} = \frac{P_{\mathrm{S}}}{\omega_{\mathrm{m}}} = \frac{3E_0V}{\omega Z_{\mathrm{s}}}\sin(\gamma+\delta) \tag{6.100}$$

where ω_{m} = mechanical angular velocity in radians per second. If the armature resistance is neglected, then we can write

$$P_{\mathrm{S}} = \frac{\mathrm{d}P_{\mathrm{m}}}{\mathrm{d}\delta} = \frac{3E_0V}{X_{\mathrm{s}}}\cos\delta$$

$$T_{\mathrm{S}} = \frac{3E_0V}{\omega_{\mathrm{m}}X_{\mathrm{s}}}\cos\delta = \frac{P_{\max}}{\omega_{\mathrm{m}}}\cos\delta$$

where $P_{\max} = \frac{3E_0V}{X_{\mathrm{s}}}$. Here $\gamma = 90°$ and $Z_{\mathrm{s}} = X_{\mathrm{s}}$.

Similarly, for a three-phase cylindrical rotor synchronous motors, the synchronizing power per electrical radian

$$P_{\mathrm{S}} = \frac{\mathrm{d}P_{\mathrm{m}}}{\mathrm{d}\delta} = \frac{3E_0V}{Z_{\mathrm{s}}}\sin(\gamma-\delta)$$

The synchronizing torque per electrical radians for a motor is

$$T_{\mathrm{S}} = \frac{P_{\mathrm{S}}}{\omega_{\mathrm{m}}} = \frac{3E_0V}{\omega Z_{\mathrm{s}}}\sin(\gamma-\delta) \tag{6.101}$$

Similarly in a salient pole alternator, if armature resistance is negligible, the power expression can be given as

$$P = P_e = P_m = \frac{3VE_0}{X_d}\sin\delta + \frac{3V^2}{2}\left(\frac{1}{X_q} - \frac{1}{X_d}\right)\sin 2\delta$$

The synchronizing power per electrical radian,

$$P_S = \frac{3VE_0}{X_d}\cos\delta + 3V^2\left(\frac{1}{X_q} - \frac{1}{X_d}\right)\cos 2\delta$$

$$P_S = P_{max}\cos\delta + 2P_{s1}\cos 2\delta \tag{6.102}$$

The synchronizing torque per electrical radians for a salient pole alternator,

$$T_S = \frac{P_{max}}{\omega_m}\cos\delta + \frac{2P_{sl}}{\omega_m}\cos 2\delta$$

where $P_{max} = \frac{3VE_0}{X_d}$ $P_{sl} = \frac{3V^2}{2}\left(\frac{1}{X_q} - \frac{1}{X_d}\right)$

The synchronizing power and torque per electrical degree are be obtained by multiplying the expressions (6.100) to (6.102) with $\frac{\pi}{180}$. For stable operation of a machine, synchronizing power and torque must be positive.

6.16 Swing Equation

It is assumed that the alternator is initially supplying a power P_1, when the prime mover power is changed to supply a total power of P_2. There is a sudden change of power, i.e., $\Delta P = P_2 - P_1$. For a two-pole machine, the angular synchronous velocity is $\omega_s = 2\pi f$, and the accelerating power is equal to $(\Delta P/\Delta\omega)$. This depends on the difference between the actual power delivered at the instant, i.e., P_1 and the desired final value, i.e., P_2. The value of ΔP is not constant. It decreases positively as P_2 is approached. This power accelerates the rotor and the motional equation is given as [12]

$$J\frac{d\omega}{dt} + T_d + T_s = \frac{\Delta P}{\Delta\omega} = T_{sh} \tag{6.103}$$

where

T_{sh} = the shaft torque = $(\Delta P/\Delta\omega)$.
T_d = torque developed due to damper bar current.
T_s = counter electromagnetic torque which changes with power angle.

The energy stored in the rotating mass is given as $\frac{1}{2}J\omega_m^2$, where J is the moment of inertia and ω_m is the angular velocity of rotating mass. Let us suppose that the machine is operating on no load. At this no load, the power angle between **V** and E is zero. The displacement angle of a point on the rotor field structure with respect to the fixed external reference point is θ, then, while in synchronism with the bus system, this angle with time t is

$$\theta = \omega_s t \tag{6.104}$$

Here ω_s is synchronous angular velocity of the bus system and is measured in electrical degrees. Now, the prime mover of the alternator is adjusted to increase the shaft power of the generator. The increased shaft power momentarily increases the angular velocity of the rotor field structure and it may be written as

$$\omega = \frac{d\theta}{dt} \tag{6.105}$$

Equation (6.105) is a general expression of the angular speed of the rotor at any instant of time. The induced EMF phasor **E** has the momentarily angular velocity ω and the bus terminal voltage **V** has the synchronous angular velocity ω_s.

The rotor experiences a momentarily increase in the angular speed. Due to increased shaft power, the difference in angular speeds of phasor **E** and **V** causes induced EMF **E** to advance in phase relative to terminal voltage **V**. The amount of increase is,

$$\delta = \int_0^t (\omega - \omega_s)dt \tag{6.106}$$

Here, integration limit, 0–t denotes the time period during which the generator is temporarily out of synchronism with the bus voltage. The larger is the time t, the greater will be angle δ. The momentarily change in the shaft speed yields a new value of torque or power angle δ, which is in consistent with the increased shaft power input. Once synchronism is reestablished, then we get

$$\theta = \omega_s t + \delta. \tag{6.107}$$

Equation (6.107) differs from the Eq. (6.104), as it represents the total angular displacement of the rotor structure relative to a fixed reference, when alternator is loaded. The instantaneous value of angular velocity is

$$\omega = \frac{d\theta}{dt} = \omega_s + \frac{d\delta}{dt} \tag{6.108}$$

Equations (6.104) and (6.108) imply that when the angular velocity becomes ω_s, then $d\delta/dt = 0$. We can also write the expression for mechanical acceleration

$$\frac{d\omega}{dt} = \frac{d^2\theta}{dt^2} = \frac{d^2\delta}{dt^2} \tag{6.109}$$

For a synchronous alternator having p poles, acceleration in mechanical degree is given as

$$\frac{d^2\theta_m}{dt^2} = \left(\frac{2}{p}\right)\frac{d^2\delta}{dt^2} \tag{6.110}$$

Here θ_m is the mechanical displacement of rotor structure. Thus, in electrical degrees, shaft torque can be given as

$$T_{sh} = \left(\frac{2}{p}\right)J\frac{d^2\delta}{dt^2} + T_d + T_s \tag{6.111}$$

The expression in electrical radians is,

$$T_{sh} = \left(\frac{2\pi}{180p}\right)J\frac{d^2\delta}{dt^2} + T_d + T_s \tag{6.112}$$

The synchronizing torque is $T_s = 3\frac{V \cdot E}{\omega_s \cdot X_s} \sin\delta$. When $T_{sh} = T_s$, the transients decay completely to zero and the machine is synchronized to the bus system.

The damping torque is due to the damper bars currents. The action of the damper bar current is always to produce a torque that opposes the torque which produced it. Thus the damper bar produces a reaction to help restore the generator in equilibrium condition of synchronism with the bus system. The damping torque is given as

$$T_d = K_d\frac{d\delta}{dt} \tag{6.113}$$

Thus, the swing equation is,

$$T_{sh} = \left(\frac{2\pi}{180p}\right)J\frac{d^2\delta}{dt^2} + K_d\frac{d\delta}{dt} + 3\frac{VE}{\omega_s X_s}\sin\delta \tag{6.114}$$

where $\left(\frac{2\pi}{180p}\right)J$ = Inertial torque per unit acceleration, here J is in kg-m^2 or N-m/rad/s^2.

K_d = damping constant.
$3\frac{VE}{\omega_s X_s}$ = Synchronizing torque coefficient.

Example 6.23 A four-pole 50 Hz alternator with constant excitation is delivering a load of 500 kW electrical load at a power angle of 9° electrical. The inertia acceleration coefficient for the combined prime mover –alternator is 1.5 N-m/electrical degree/s^2. The damping torque coefficient is 12 N-m/electrical degrees/second. The input power from prime mover is suddenly increased so that the generator increases its share of the common electrical load to 1000 kW.

(a) Determine the maximum value of power angle as it moves to its new steady state.
(b) How long does it take for the steady state to be reached to be expressed in terms of 1% tolerance band? The % maximum overshoot is 42%.
(c) Develop MATLAB Simulink model for obtaining variation of power angle with respect to time.

Solution

Initial value of power angle $\delta_1 = 9°$ electrical.

Considering a linear variation, the final value

$$\delta_2 = (1000/500)\delta_1 = 18° \text{ electrical}$$

Now, given that $K_J = 1.5$ N-m/electrical degree and $K_d = 12$ N-m/electrical degree /s^2.

At 500 kW, the synchronizing torque is

$$T_s = \frac{500,000}{\omega_s} = \frac{500,000}{\left(\frac{4\pi f}{p}\right)} = \frac{500,000}{\left(\frac{4\pi\times 50}{4}\right)} = 3184.7\,\text{N-m}$$

If the synchronizing torque coefficient is K_s, then

$$K_s = \frac{T_s}{\delta_1} = \frac{3184.7}{9} = 353.86\,\text{N-m/electrical degree}$$

The Laplace transform of swing equation can be written as

$$s^2 + \frac{K_d}{K_J}s + \frac{K_s}{K_J} = 0$$

Or,

$$s^2 + 8s + 235.9 = 0 \quad \text{(a)}$$

Comparing the above equation (a) with

$$s^2 + 2\zeta\omega_n s + \omega_n^2 = 0$$

The natural frequency of oscillation is

$$\omega_n = \sqrt{\frac{353.86}{1.5}} = 15.36\,\text{rad/s}$$

The frequency of natural oscillation in Hz

$$= \frac{15.35}{2\pi} = 2.44\,\text{Hz}$$

The damping ratio is

$$\zeta = \frac{1}{2\omega_n} \times \frac{12}{1.5} = 0.2604$$

The frequency of damped oscillations is

$$\omega_d = \omega_n\sqrt{1-\zeta^2} = 15.36\sqrt{1-(0.2604)^2} = 14.83\,\text{rad/s}$$

The frequency of damped oscillations in Hz

$$= \frac{14.83}{2\pi} = 2.36\,\text{Hz}$$

The settling time is

$$t_s = \frac{5}{\zeta\omega_n} = \frac{5}{0.2604 \times 15.36} = 1.25\,\text{second}$$

(b) The maximum overshoot for damping ratio of 0.2604 is about 42%. The maximum approximate value of overshoot is

$$\delta_{max} = \delta_1 + 1.42(\delta_2 - \delta_1) = 9 + 1.42(18 - 9) = 21.78°\ \text{electrical degree}$$

(c) The swing equation is

$$\frac{d^2\delta}{dt^2} + 8\frac{d\delta}{dt} + 235.9\delta = \frac{1000,000 - 500,000}{\left(\frac{4\pi \times 50}{4}\right)} = 3184.7 \qquad \text{(b)}$$

Figure 6.63 shows the MATLAB Simulink model and the variation of delta with respect to time.

6.17 Equal Area Criterion

Generating power stations are interconnected. If the load on one of the alternator is changed, then the power flows from the interconnected grid must also change. The stability of interconnected system is very vital. Let us discuss a single alternator connected to infinite bus. Let **V** be the terminal voltage and **E** the induced EMF or open-circuit EMF. Consider a balanced system. The per phase power delivered by the machine is given by [12],

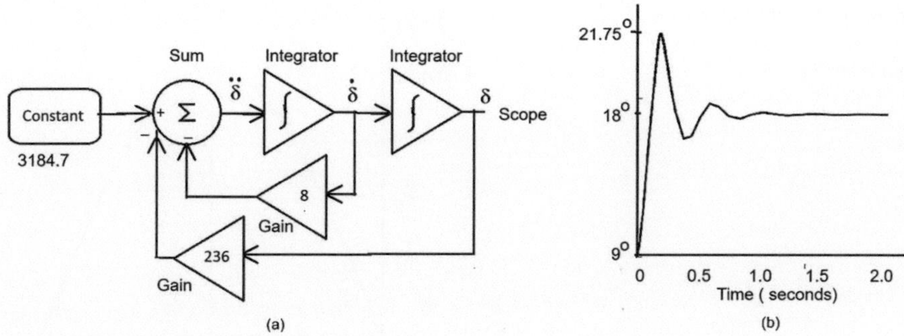

Fig. 6.63 **a** MATLAB Simulink model for solving swing equation **b** response—variation of power angle with sudden load

$$\mathrm{P} = VI \cos\theta$$

From the phasor diagram of a cylindrical rotor alternator, we can write

$$\frac{V}{\sin(90^\circ - \theta - \delta)} = \frac{E}{\sin(90^\circ + \theta)} = \frac{IX_s}{\sin\delta} \tag{6.115}$$

Thus, we can rederive the per phase power angle or P-δ expression as

$$I = \frac{E\sin\delta}{X_s\sin(90^\circ + \theta)} \tag{6.116}$$

$$P = \frac{VE\sin\delta\cos\theta}{X_s\cos\theta} = \frac{VE\sin\delta}{X_s} \tag{6.117}$$

It is assume that both the terminal voltage **V** and synchronous reactance X_s are constant. As a result the induced EMF **E** will also tend to be constant due to the automatic voltage regulators fitted to the alternator. The power angle diagram is shown in Fig. 6.64. The maximum value of power the machine can supply occurs when δ is 90°. Any attempt to extract more power from the machine under these conditions will result in the unit becoming unstable and dropping out of synchronism. This value of $P_{\max}$ is known as *steady-state stability limit* of the machine.

If the power delivered to the bus bar is P_1, the corresponding load angle is δ_1, If the load is increased to P_2, current will increase, as a result, power angle will increase to δ_2. The machine must increase its speed temporarily to achieve the necessary increase in the angle δ. Provided the load is applied **gradually** the rotor will move steadily relative to **V** until the new angle δ_2 is attained, as shown in Fig. 6.64.

If the load on the alternator is increased **suddenly** from P_1 to P_2, as in Figs. 6.65 and 6.66 and 6.67, the rotor will accelerate (relative to **V**) until it reaches its final position given by the power angle δ_2. Since the rotor will have a definite moment of inertia, power must be supplied from the infinite bus bar to achieve this. This

Fig. 6.64 *P-δ* characteristics power is gradually raised from P_1 to P_2

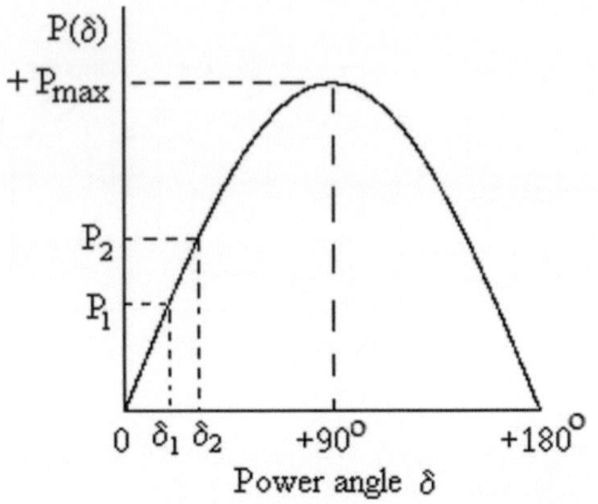

corresponds to the synchronizing power. When machine reaches its new position, the relative forward movement of the rotor will not halt immediately, and its inertia will carry the rotor forward until the damping forces on the system bring it to halt at some power angle given by δ_3. The synchronizing power will then flow in the reverse direction, thereby retarding the rotor. The oscillations will occur about the final position δ_2, until the rotor eventually comes to rest (relative to **V**). Although δ_3 is well below the $\pi/2$—the steady-state stability limit, the oscillations may carry the rotor beyond this into an unstable region, and this latter condition is known as *transient stability*. The amount of load which can be applied suddenly to the machine without the transient stability limit being exceeded is an important factor in the operation of the power system. It should be noted that this maximum suddenly applied load depends on the following factors

(a) Voltage level.
(b) Reactance.
(c) Initial load P_1.

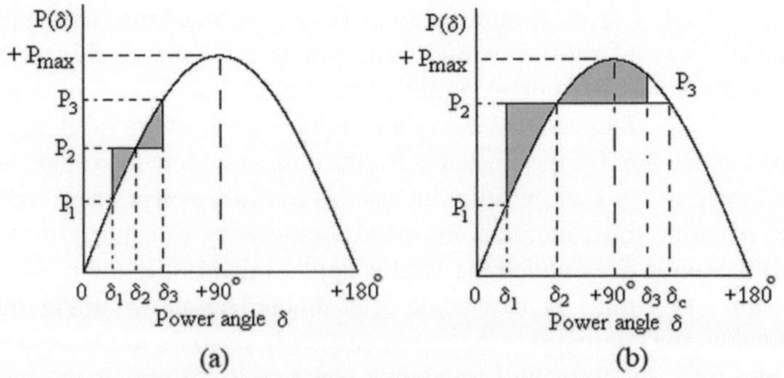

Fig. 6.65 **a** *P-δ* diagram when ΔP is small—power is suddenly raised from P_1 to P_2 (b) *P-δ* diagram when ΔP is large—power is suddenly raised from P_1 to P_2

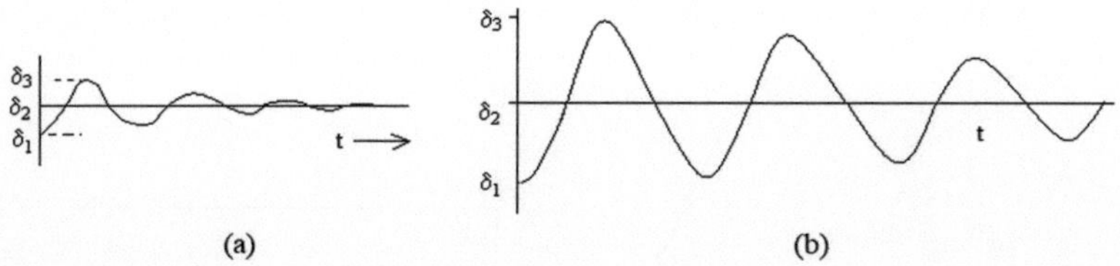

Fig. 6.66 **a** Variation of power angle w.r.t. time—with damper bars, **b** variation of power angle w.r.t. time—with damper bars

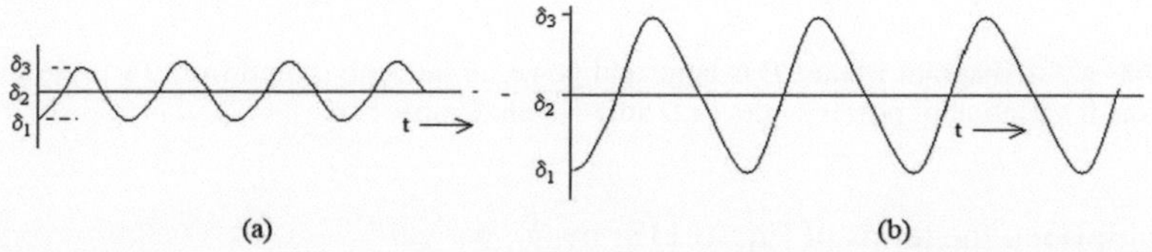

Fig. 6.67 **a** Variation of power angle w.r.t. time –without damper bars, **b** variation of power angle w.r.t. time—without damper bars

The above phenomenon of oscillation of power angle with respect to time, when load is suddenly applied is illustrated in Fig. 6.65a, b, where the machine is initially supplying a load P_1, which is suddenly increased to P_2. The accelerating power supplied is $\Delta P = P_2 - P_1$ and as soon as the power angle corresponding to P_2, i.c., δ_2, is reached, ΔP changes its sign and the rotor begins to decelerate. If however, the overshoot is such that the rotor is still moving forward relative to the angular position of **V**, when power angle becomes δ_3, then ΔP will again change sign and the rotor will begin to accelerate once more. The system is then in an unstable state.

If a synchronous alternator running at power P_1 is suddenly loaded with power P_2, we can define the following:

- The area between δ_1 to δ_2 as A_1.
- The area between δ_2 to δ_c, where critical power angle is $\delta_c = \pi\text{-}\delta_2$, as A_2.

Mark areas A_1 and A_2 in Fig. 6.67a, b. Depending on powers P_1 and P_2, the following conditions may arise:

(a) When area $A_1 < A_2$, the machine is stable, as in Fig. 6.65a, 6.66a and 6.67a.
(b) When area $A_1 = A_2$, the machine is critically stable, as in Fig. 6.65b, 6.66b and 6.67b.
(c) When $A_1 > A_2$, the machine is unstable, as in Fig. 6.68a, b.

The dynamics of machine is given by swing equation. Let the electromagnetic torque (or power) is T (or P) when the new torque is T_2 (or power P_2), the swing equation can be written as

$$T - T_2 = -\frac{J}{p} \cdot \frac{\mathrm{d}^2\delta}{\mathrm{d}t^2} \tag{6.118}$$

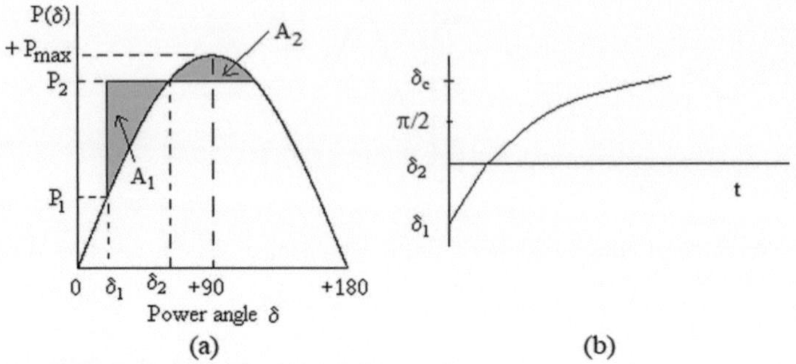

Fig. 6.68 **a** P-δ diagram when ΔP is large and power is suddenly raised from P_1 to P_2-unstable operation, **b** variation of power angle w.r.t. time—with damper bars

Multiplying the terms of Eq. (6.118) by $\frac{d\delta}{dt}$, we get

$$(T - T_2)\frac{d\delta}{dt} = \frac{J}{p} \times \frac{d^2\delta}{dt^2} \times \frac{d\delta}{dt} = \frac{J}{2p} \times \frac{d}{dt}\left[\left\{\frac{d\delta}{dt}\right\}^2\right] \tag{6.119}$$

$$\int_{\delta_1}^{\delta_3} (T - T_2)d\delta = \frac{J}{2p}\int_{\delta_1}^{\delta_3} d\left[\left\{\frac{d\delta}{dt}\right\}^2\right] \tag{6.120}$$

If the dynamic stability is to be ensured, when $\delta = \delta_3$, the rotor speed will be equal to the synchronous speed and

$$\frac{d\delta}{dt} = p(\omega_s - \omega) = 0 \tag{6.121}$$

Thus, we get

$$\int_{\delta_1}^{\delta_3} (T - T_2)d\delta = 0 \tag{6.122}$$

Or,

$$\int_{\delta_1}^{\delta_2} (T - T_2)d\delta - \int_{\delta_2}^{\delta_3} (T_2 - T)d\delta = 0 \tag{6.123}$$

$$\text{Area}\,A_1 - \text{Area}\,A_2 = 0$$

Here, for cylindrical rotor machine torque $T = \frac{P_{\max}}{\omega_s} \sin \delta$. Thus, by integrating the above equation, we get

$$T_2\delta_1 + T_{\max} \cos \delta_1 = T_2\delta_3 + T_{\max} \cos \delta_3 \quad (6.124)$$

Or,

$$P_2\delta_1 + P_{\max} \cos \delta_1 = P_2\delta_3 + P_{\max} \cos \delta_3 \quad (6.125)$$

The minimum value of δ_2 so that the machine does not loose stability can be determined from the above Eq. (6.125). In that case, $\delta_3 = \delta_c = \pi - \delta_2$ for that the right-hand side of equation is minimum. The above Eq. (6.125) is valid for synchronous motor as well as alternator and can be solved using Newton Raphson method. If the system parameters are known, the solution of swing equations can also be obtained for plotting the exact variation of angle $\delta_3 = \delta_c = \pi - \delta_2$ with respect to time, using MATLAB Simulink.

Example 6.24 A cylindrical rotor synchronous motor running at 0.1 $P_{\max}$. Determine the maximum shaft load which can be applied, when the shaft load is 0.1 $P_{\max}$. Use equal area criterion.

Solution

Given $P_1 = 0.1\, P_{\max}$ Thus, we get

$$\delta_1 = \sin^{-1}(0.1) = 5.739^\circ = 0.1001 \text{ Radians}$$

Now

$P_2 = P_{\max} \sin \delta_2$ and $\delta_c = \delta_3 = \pi - \delta_2.$

We know that

$$P_2\delta_1 + P_{\max} \cos \delta_1 = P_2\delta_3 + P_{\max} \cos \delta_3$$

$$\delta_1 \sin \delta_2 + \cos \delta_1 = (\pi - \delta_2) \sin \delta_2 + \cos(\pi - \delta_2)$$

$$\sin \delta_2 \times 0.1001 + \cos 5.739^\circ = \sin \delta_2 \times (\pi - \delta_2) + \cos (\pi - \delta_2)$$

$$0.1001 \sin \delta_2 + 0.9949 + \cos \delta_2 \text{-} (\pi - \delta_2) \times \sin \delta_2 = 0$$

Let function

$$f(\delta_2) = 0.9949 + \cos \delta_2 - (\pi - \delta_2 - 0.1001) \times \sin \delta_2$$

Derivative of $f(\delta_2)$ with respect to δ_2

$$f'(\delta_2) = (\delta_2 + 0.1001 - \pi) \times \cos\,\delta_2$$

Solving the above equation using Newton Raphson method, the initial value of δ_2 comes out to be 0.1001. Now putting the value of δ_2 in functions $f(\delta_2)$ and $f'(\delta_2)$, we get.

1st Iteration

$$f(\delta_2) = 0.9949 + \cos\,0.1001 - (\pi - 0.1001 - 0.1001) \times\ \sin\ 0.1001 = 1.6961$$

$$f'(\delta_2) = (0.1001 + 0.1001 - \pi) \times \cos\ 0.1001 = -2.9250$$

Increment in $\Delta\delta_2 = -\frac{f(\delta_2)}{f'(\delta_2)}$.
Thus new value of δ_2 = Old value of $\delta_2 + \Delta\delta_2$.
Thus new value of $\delta_2 = 0.1001 + (1.6961/2.9250) = 0.6789$.

2nd Iteration

$$f(\delta_2) = 0.9949 + \cos 0.6789 - (\pi - 0.6789 - 0.1001) \times \sin\ 0.6789 = 0.2897$$

$$f'(\delta_2) = (0.6789 + 0.1001 - \pi) \times \cos\ 0.6789 = -1.8390$$

Thus new value of δ_2 = Old value of $\delta_2 + \Delta\delta_2$.
Thus new value of $\delta_2 = 0.6789 + (0.2897/1.8390) = 0.8364$.

3th Iteration

$$f(\delta_2) = 0.9949 + \cos\ 0.8364 - (\pi - 0.8364 - 0.1001) \times 0.8364 = 0.0284$$

$$f'(\delta_2) = (0.6789 + 0.1001 - \pi) \times \cos\ 0.6789 = -1.8390$$

Thus new value of $\delta_2 = 0.8364 + (0.0284/1.4778) = 0.8556$.

4th Iteration

$$f(\delta_2) = 0.9949 + \cos 0.8556 - (\pi - 0.8556 - 0.1001) \times \sin 0.8556 = 0.0005$$

$$f(\delta_2) = (0.8556 + 0.1001 - \pi) \times \cos 0.8556 = -1.433$$

Thus new value of $\delta_2 = 0.8556 + (0.0005/1.433) \cong 0.8559$. Thus the final value of $\delta_2 = 0.8559$ radians. Now, the maximum torque which can be suddenly applied on the shaft of the machine is corresponding to $\delta_2 = 0.8559$ rad, or power $P = P_{\max}$ $\sin\,\delta_2 = 0.755\ P_{\max}$.

6.18 Synchronous Motor

Synchronous machine can operate either as a generator or as a motor. When operating as a motor it converts electrical power into mechanical power. These motors are used for operating large capacity pumps, compressors, conveyers, rollers, spinning mills, etc. At fixed frequency operation, it cannot operate as brake because torque is produced if rotor is rotating at synchronous speed. When synchronous motor is operated using variable frequency drive (VFD) control, for halting or braking the synchronous motor applied voltage and frequency, both are gradually decreased.

A synchronous motor has no natural starting torque. Assuming that the three-phase windings of machine are connected to a three-phase source. The currents in the three-phase windings produces rotating magnetic fields which is rotating in space at an angular speed

$$n = \frac{\omega}{p}$$

Here p is pair of poles. Let the rotor winding is connected to a DC source and is rotating at an angular speed n', not equal to n. The flux linkages produced by rotating excitation field through the turns of stator phase windings is

$$\lambda(t) = \lambda_{\mathrm{m}} \cos p\psi = \lambda_{\mathrm{m}} \cos pn't$$

where $\psi = n't$ is the instantaneous value of the angle between N pole-axis and the axis of symmetry of coil of phase windings at a given moment t. The phase winding is having a current of

$$i(t) = I\sqrt{2}\cos(\omega t - \varphi).$$

The energy of interaction between the rotor and the phase winding is,

$$W = \lambda(t) \cdot i(t) = \lambda_{\mathrm{m}} I\sqrt{2}\cos(\omega t - \varphi)\, \cos pn't$$

The torque acting on the stator according to the principle of action and reaction will have an equal but opposite sense with respect to torque on rotor. The torque can be obtained as

$$T = \left(\frac{\partial W}{\partial \psi}\right)_{i=C} = -p\lambda_{\mathrm{m}} I\sqrt{2}\sin pn't\cos(\omega t - \varphi) \tag{6.126}$$

This torque is a pulsating torque having zero average value when speed $n' \neq n$. When field winding is running at other than the speed of rotating magnetic field or synchronous speed, the machine produces zero average torque. When speed of field winding is zero, $n' = 0$, the torque produced is also zero. As such machine has zero natural starting torque.

On the contrary, when $n' = n$, the machine develops a torque with finite average value, and a positive torque with respect to rotor is produced. If the rotor rotates in the same direction and at same angular speed of the rotating magnetic field of the stator, the synchronous motor develops an electromagnetic torque.

It follows that although a synchronous motor does not develop natural starting torque, it can by any means be brought up to synchronous speed, then it can develop an active electromagnetic torque.

6.18.1 Steady-State Equation of Synchronous Motor

For developing equivalent circuit of synchronous motor, only fundamental harmonics are being considered. The EMF equation has already been discussed in Sect. 5.4. It is assumed that there is no saturation in magnetic parts, iron losses are negligible, machines have perfect electrical and magnetic symmetries, the air-gap is uniform, slotting effect is ignored and machine is running at synchronous speed. The magnetic field in air-gap due to rotating of field winding in a two-pole machine can be given as

$$B_f(\theta, t) = B_{fm}\cos(\omega t - \theta) \tag{6.127}$$

The peak value of flux density is decided by the number of conductors in the field winding and the DC current in it. Since, three-phase winding of stator is connected to a balanced three-phase network having angular frequency ω, the phase currents will be

$$\begin{aligned} i_1 &= \sqrt{2}I\cos\left(\omega t + \frac{\pi}{2} + \psi\right) \\ i_2 &= \sqrt{2}I\cos\left(\omega t + \frac{\pi}{2} + \psi - \frac{2\pi}{3}\right) \\ i_3 &= \sqrt{2}I\cos\left(\omega t + \frac{\pi}{2} + \psi - \frac{4\pi}{3}\right) \end{aligned} \tag{6.128}$$

The angle ψ is an unknown angle. The three-phase current produces a rotating magnetic field which is rotating in space at synchronous speed.

$$B_a(\theta, t) = B_{am}\cos\left(\omega t - \theta + \frac{\pi}{2} + \psi\right) \tag{6.129}$$

The peak value of rotating magnetic fields depends on the number of conductors and armature currents and number of poles. The rotating magnetic field is a reaction field being rotated at same speed of synchronous speed at which the rotor field is rotating in the same direction. But this is being displaced by angle $\frac{\pi}{2} + \psi$ from reference current i_1. These two fields together produce the useful resultant field; this resultant field leads the excitation field by an angle δ, which is called internal angle of machine.

$$B_r(\theta, t) = B_{rm}\cos(\omega t - \theta + \delta)$$

We know that in case of induction motor and transformer, the magnetizing current in the stator reference winding '1' is

$$i_{m1} = \sqrt{2}I_m\cos(\omega t + \delta)$$

The magnetizing current in all the three-phase windings of rotating machine together forms a three-phase symmetrical balanced current system which produces a rotating magnetic field with all the features of resultant magnetic field. Similarly, a referred value of rotor excitation currents flow in the balanced three-phase windings having RMS I_f' referred to stator side. For the reference phase '1', the phase current is given as

$$i_{f1}' = I_f'\sqrt{2}\cos\omega t$$

The current I_f' is called the excitation current or field current referred to stator. The stator current phasor is denoted by $\overline{I}$. By doing so, the rotating magnetic fields are replaced by the current phasor, and all the currents have same angular frequency and reciprocal phase shifts identical with the field displacements.

Thus, in phasor notation, we can write

$$\overline{I} + \overline{I}_f' = \overline{I}_r$$

This is similar to Eq. (6.3) of Sect. 6.4.2. If the component of current corresponding to iron loss is I_w, then the above equation is modified as

$$\overline{I} + \overline{I}_f' = \overline{I}_r + \overline{I}_w \tag{6.130}$$

The resultant field induces the system of resultant EMF in the respective windings,

$$\overline{E}_m = -jX_m\overline{I}_r$$

Here quantity X_m is the magnetizing reactance.

The induced EMF in stator winding by leakage field is introduced, and is given by,

$$\overline{E}_l = -jX_l\overline{I}$$

Here quantity X_l is the leakage reactance. The following equations are obtained for cylindrical rotor synchronous motor

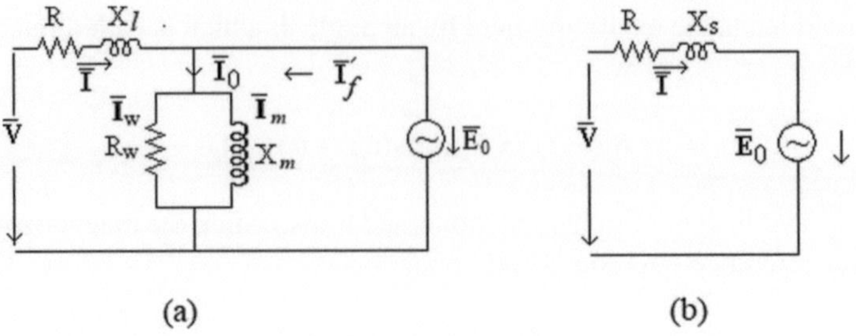

Fig. 6.69 Equivalent circuit of synchronous motor: **a** exact, **b** approximate

$$\overline{V} = R\overline{I} + jX_l\overline{I} - \overline{E}_m$$
$$\overline{E}_m = -jX_m\overline{I}_m = -R_w\overline{I}_w$$
$$\overline{I} + \overline{I}'_f = \overline{I}_m + \overline{I}_w \tag{6.131}$$

Here R_w is representing the equivalent resistance corresponding to iron losses in the stator. The equivalent circuit of synchronous motor (non-salient pole is shown in Fig. 6.69a.

If the losses are ignored, the applied voltage can be given as

$$\overline{V} = -\overline{E}_m = -\overline{E}_0$$

If the mechanical friction is ignored the electromagnetic torque is given as

$$T = -\frac{3E_mI\cos(E,I)}{\omega} = \frac{3E_mI\cos(-E,I)}{\omega} \tag{6.132}$$

The developed torque must be equal to the load torque. For all practical purposes one can take

$$I\cos(-E,I) = I\cos\phi$$

where Ø is the phase shift between V and I.

The voltage equation of synchronous motor can be written as

$$\overline{V} = R\overline{I} + jX_s\overline{I} - \overline{E}_0 \tag{6.133}$$

This corresponds to the equivalent circuit shown in Fig. 6.69b.

6.18.2 Losses in Synchronous Machines

Synchronous machines are doubly fed machines having DC as well as AC windings. The losses in synchronous machines are:

(a) Ohmic losses in excitation or field circuit
(b) Ohmic losses in armature circuit
(c) Core loss in magnetic circuit of alternator
(d) Frictional and windage loss.

The ohmic losses in damper bars appear only during starting, hunting period and disturbances. Large capacity synchronous machines require special arrangements of cooling. The power required for cooling is taken as loss and this is not mentioned in the above list.

In synchronous motor, the input power, $P_{\text{in}} = 3V_{\text{ph}}I_{\text{ph}} \cos\theta$ is three-phase electrical power supplied to its armature and DC source is connected to the excitation windings on the rotor. The output power $P_{\text{m}} = T\omega_{\text{s}}$ is the mechanical power available on the shaft at synchronous speed ω_{s}. This has been shown in Fig. 6.70a.

In case of alternator, the excitation winding is energized by DC source and the rotor is driven by a prime mover at speed ω_{s}. The prime mover develops an active torque T_{a}. The excitation winding of rotor produces an inducing rotating magnetic field having synchronous speed. The input power is given by, $P_{\text{in}} = T_{\text{a}}\omega_{\text{s}}$. Due to inducing field on rotor, a sinusoidal EMF is induced in the stationary armature winding; the frequency of generation is decided by rotor speed and number of poles.

If the three-phase winding of stator is connected to a symmetrical three-phase impedance, it will carry currents forming a symmetrical three-phase system. An active power $P_{\text{out}} = 3V_{\text{ph}}I_{\text{ph}} \cos\theta$ is transferred from the prime mover. Here power factor angle is 'θ'. The alternator develops a counter torque which opposes the active driving torque of the prime mover. The losses in an alternator have been shown in Fig. 6.70b.

6.18.3 Starting of Synchronous Motor

A synchronous motor does not possess a natural starting torque; however, once started it is capable of developing torque at synchronous speed. Synchronous motor is normally started:

(a) Using some auxiliary (pony-motor) motor coupled to the shaft of unloaded/unexcited synchronous motor. Such synchronous motors may not have damper bars. The DC excitation to the field winding and three-phase power to the armature winding are connected when rotor is running closer to the synchronous speed. The three-phase terminals of motor are connected to three-phase supply terminals, such that the direction of rotating magnetic fields is same as the direction of rotation of rotor.

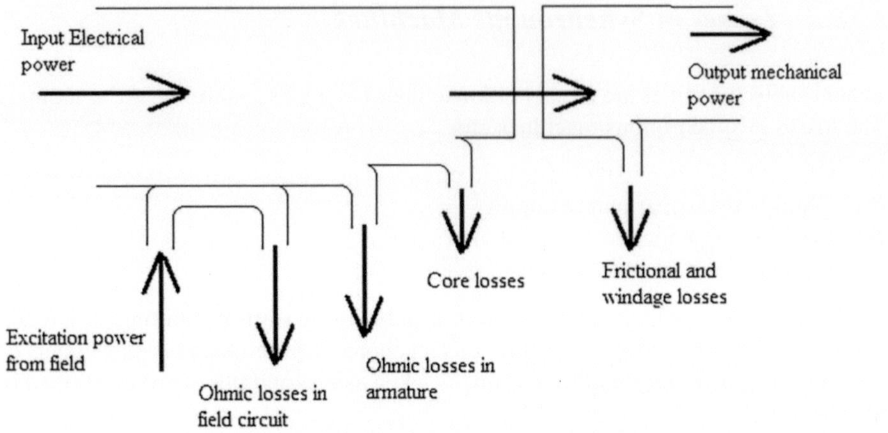

(a) Losses in Synchronous motor

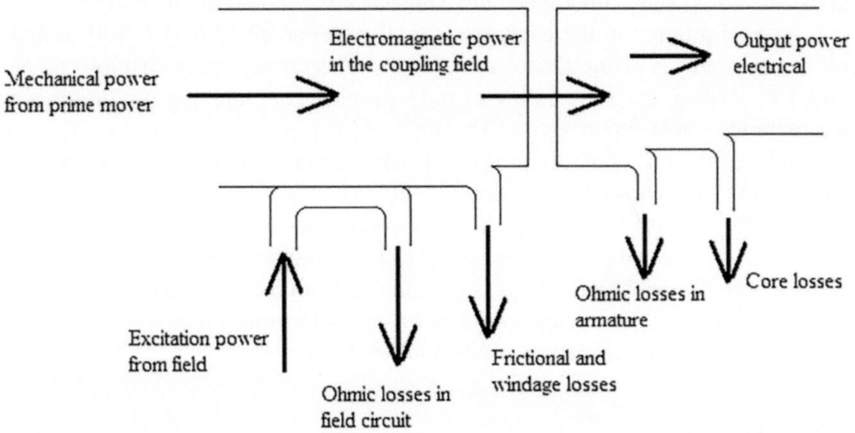

(b) Losses in Alternator

Fig. 6.70 **a** Losses in synchronous motors **b** losses in an alternator

(b) Using induction synchronous motor: Starting of synchronous motor as an induction motor is quite common. The methods of starting are similar to that of slip-ring induction motors. The starting of slip-ring induction motor using star/delta starter in conjunction with rotor resistance starter and running as synchronous motor- induction synchronous motor has already been explained in Sect. 5.18.

(c) Using damper bars: The part squirrel cage winding on the pole face develops a starting torque due to induction principle. The three-phase balanced supply connected to three-phase winding of the stator creates a rotating magnetic field, which rotates at synchronous speed with respect to the stator surface. This rotating field interlinks with the damper winding which behaves just like the squirrel cage rotor of induction motor. The motor starts as an induction motor in the direction of rotating magnetic field and comes to the speed slightly less than the synchronous speed. The field circuit, which was short-circuited through a high resistance, is now connected to a DC supply and high resistance is removed from the field circuit. The polarity produced due to rotating magnetic field magnetizes the rotor poles, resulting in two opposite poles to interlock together.

(d) Using a salient pole motor with damper bars: This increases the starting torque due to saliency. In absence of damper bars in salient pole motor, the machine can be started because of torque due to saliency. Such a machine is referred as reluctances synchronous motor (RSM) and has been discussed in Sect. 6.22.

The field winding has a large number of turns. At standstill condition, if it is left open, an induction of high voltage may damage the insulation of the field winding. The EMF induced in the field winding is less, when the speed is near about synchronous speed. As a precautionary measure, during starting period, the field winding is closed through high resistance. This resistance is removed from the field circuit after connecting the field winding to DC excitation.

For limiting the three-phase armature currents during starting, the schemes used in induction motor starting are used. These are

(a) Star-delta starting.
(b) Autotransformer starting.
(c) VVVF starting.
(d) Rotor resistance starting.

6.18.4 Torque-Speed Characteristics

The torque-speed characteristics of synchronous motor is shown in Fig. 6.71. If the machine is out of synchronism, the torque automatically drops to zero (provided there is no damper bars). The natural torque-speed characteristic of synchronous motor is a hard torque-speed characteristic. The motor will run in steady-state operation at synchronous speed (decided by supply frequency and number of poles), if the developed torque is higher than the load torque. Synchronous motor produces torque only at synchronous speed.

Figure 6.72 shows the choice of induction and synchronous motors in industries for supply frequencies of either 50 Hz or 60 Hz. For low speed high-power drive, synchronous motors are preferred owing to their power factor correction ability. Large power induction motors are energy efficient motor having better efficiency, more than 90%.

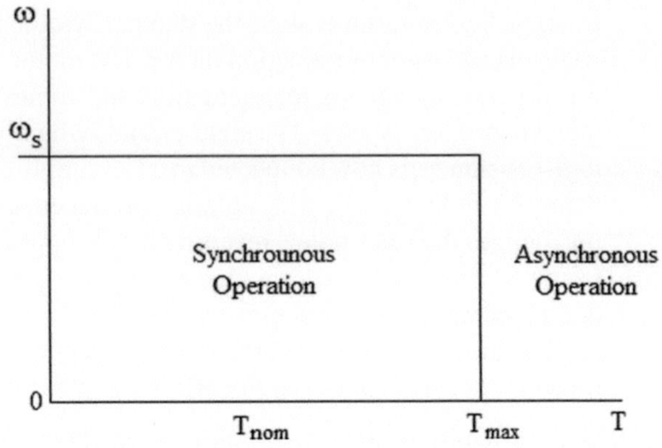

Fig. 6.71 Synchronous motor torque-speed characteristics

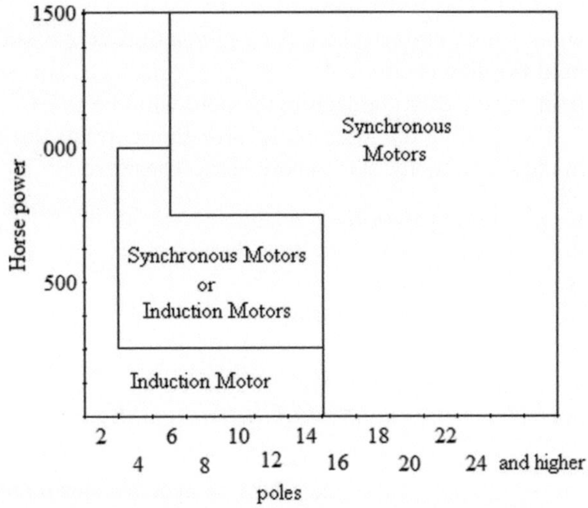

Fig. 6.72 Comparison of applications of synchronous motor and induction motor at power frequency of 50–60 Hz

6.19 Synchronous Condenser

Synchronous motors have the capability of power factor correction. An overexcited synchronous motor running on no load is used in large industries for power factor correction. This is referred as *synchronous condenser*. In a large industry, there are many lagging power factor loads such as induction motors, arc furnaces, arc welding machines, induction furnace, choke coils and reactors, etc. The overall power factor of the industry should be within limit as prescribed by the power undertaking company. A penalty is levied, once the overall power factor of the industry falls below the prescribed value.

The lagging power factor can be corrected using Switched capacitors or synchronous condenser. The latter can control the instantaneous power factor by controlling the field current of synchronous condenser.

6.20 Sudden Short-Circuit Test

Sudden short-circuit test is not conducted on alternator (actual machine) with rated voltage, because the current will be enormously large, and the counter torque produced by the machine may damage the mechanical coupling and foundation of the machine. This test is normally conducted on (scaled) micromachine or on actual machines at lower voltage, thereby limiting the peak current to the rated current. A typical short-circuit current of one phase of an alternator having damper bars with symmetrical three- phase sudden short circuit is shown in Fig. 6.73. When an alternator is suddenly short-circuited, the following conditions arise:

- A sub-transient current for few cycle due to sub-transient reactance.
- A transient current for few seconds due to transient reactance.
- A transient DC component. (not shown here, present in other two phases)
- Steady-state short-circuit current when all transient current have vanished.

Here we define the following:

(a) X_d'' = the *d*-axis sub-transient reactance offered by the machine during sub-transient period.
(b) X_d' = the *d*-axis transient reactance offered by the machine during transient period.
(c) $X_d = X_s$ = *d*-axis reactance during steady-state condition obtained from OCC and SCC.

For a typical cylindrical rotor machine, we have

$$X_d'' \approx 0.5X_d' \approx 0.1X_d$$

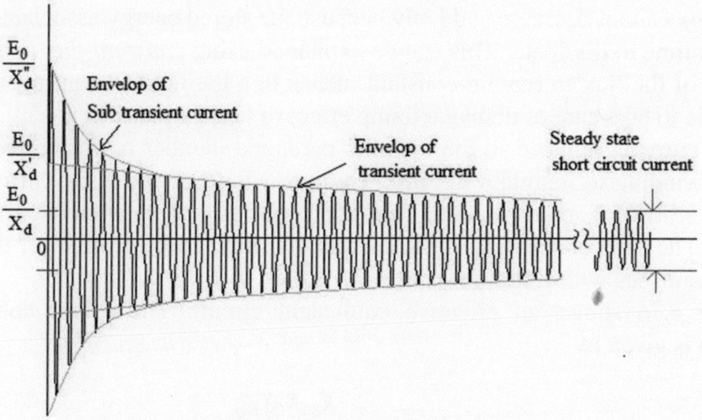

Fig. 6.73 Profile of short-circuit current during sudden short circuit of an alternator in one phase-showing sub-transient, transient, and steady-state short-circuit current

$$X_{d''} \approx 0.5X_{d'} \approx 0.1X_d$$

For a typical salient pole machine

$$X_d = 1.0 \, \text{per unit}$$

$$X_d'' \approx 0.4\,X_d' \approx 0.25X_d$$

$$X_q'' \approx 0.4\,X_q' \approx 0.4X_q$$

and

$$X_q = (2/3)X_d$$

The machine having no damper bars does not exhibit the sub-transient period. It is assumed that the alternator is running at constant speed. When all the terminals of the alternator are suddenly short-circuited, a large short-circuit current flows in the armature winding. The sudden change causes the d- and q-axis MMF to change. A change in d-axis MMF causes the following:

- The induced EMF in the field winding resists the change in the flux linked to main field. The field winding behaves as a transformer due to large alternating flux of armature. A large EMF is induced in the field winding.
- Transient current in the short-circuited damper winding, if present.
- Eddy currents in the pole face and adjacent metal parts.

The initial short-circuit current lags the voltage by 90°, it creates a large demagnetizing MMF in the d-axis tending to reduce the pole flux from its initial value. But the flux cannot decrease suddenly because the stored energy associated with the flux takes time to dissipate. This can be explained using *constant-flux theorem.* The tendency of the flux to remain constant means that the field current must rise to a large value to take care of demagnetizing effect of large armature current. Similarly, the eddy currents induced in the metallic parts and damper bars increase to assist the field winding to maintain the flux. Damper windings and eddy currents in the other metallic parts, pole face, etc. produce large resistances associated with these currents. The time constant of these currents is small and the effect of these eddy currents vanishes within first few cycles.

Figure 6.76 shows an effective equivalent circuit. The *d*-axis sub-transient reactance is given by

$$X_d'' = X_l + \frac{X_{ad}X_f X_{KD}}{X_{ad}X_f + X_f X_{KD} + X_{ad}X_{KD}} \tag{6.134}$$

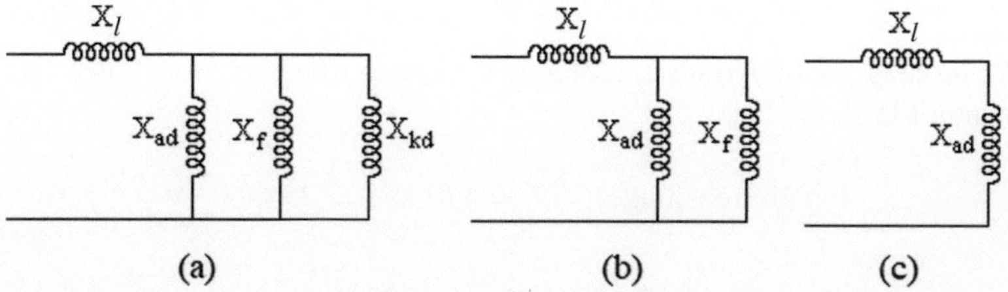

Fig. 6.74 Approximate equivalent circuit of alternator during **a** sub-transient period, **b** transient period, **c** steady-state short circuit

After a few cycles, the equivalent circuit is shown in Fig. 6.76b, which represents the transient operation. The damper bar currents have vanished but since the field winding has lesser effective resistance than damper bars and larger time constant; the effect of field current is present in the transient duration. It lasts for few seconds. The *d*-axis transient reactance is given by

$$X'_{\mathrm{d}} = X_l + \frac{X_{\mathrm{ad}} X_{\mathrm{f}}}{X_{\mathrm{ad}} + X_{\mathrm{f}}} \tag{6.135}$$

After the transient currents are decayed, the steady-state short-circuit currents flows. The d-axis steady-state reactance is given by

$$X_{\mathrm{d}} = X_l + X_{\mathrm{ad}} \tag{6.136}$$

The equivalent circuits during sub-transient, transient, and steady-state short-circuit conditions are shown in Fig. 6.74.

The positive sequence, negative sequence, and zero sequence reactance of alternator are denoted by X_1, X_2 and X_0.

Where

$$X_1 = X_{\mathrm{s}} \quad ; \quad X_2 = \frac{X''_{\mathrm{d}} + X''_{\mathrm{q}}}{2} \quad ; \quad X_1 > X_2 > X_0$$

A brief explanation has been given in Chap. 7.

Example 6.25 A three-phase star-connected load takes 100 A at 0.8 p.f. lagging at 440 V, 50 Hz line. A three-phase star-connected cylindrical rotor synchronous motor having a synchronous impedance of 2-Ω per phase is connected in parallel with the supply and load. The power developed by the motor is 50 kW at a power angle of 20°. Calculate

(a) the reactive kVA of the motor,
(b) the overall power factor of the load and the motor and resultant current.

Solution
The phasor diagram of cylindrical rotor synchronous motor in overexcited condition is shown in Fig. 6.75.

$$\text{Per phase voltage,} \quad V = 440/\sqrt{3} = 254.03 \quad \text{V}$$

$$\text{Power angle} \quad \delta = 20^\circ$$

Let the current be I_m per phase, the per phase power

$$P = 50000/3 = 254.03 \times E_0 \times \sin 20/2 = 254.03 \times E_0 \times 0.342/2$$

Thus, we get

$$E_0 = 383.84 \quad \text{V per phase}$$

From phasor diagram shown in Fig. 6.75, we get

$$\mathrm{E_0} \cos \delta = \mathrm{V} + \mathrm{I_m X_s} \sin \theta_\mathrm{m}$$

$$I_m X_s \cos \theta_m = E_0 \sin \delta$$

Here θ_m is the power factor angle of synchronous motor

$$I_m \sin \theta_m = 53.35$$

$$I_m \cos \theta_m = 65.60$$

Thus, motor current is

$$I = \sqrt{[65.602 + 53.352]} = 84.68\,\text{A}$$

Motor power factor angle

$$\theta_m = \tan^{-1}(53.35/65.60) = 39.13^\circ \text{ leading}$$

Now, reactive kVA of the motor

$$= \sqrt{3} \times 254.03 \times 84.56 \times \sin 39.13^\circ$$
$$= 23472\,\text{VAR} = 23.472 \quad \text{kVAR}$$

(b) Load current $I_L = 100$ A at 0.8 power factor lagging.

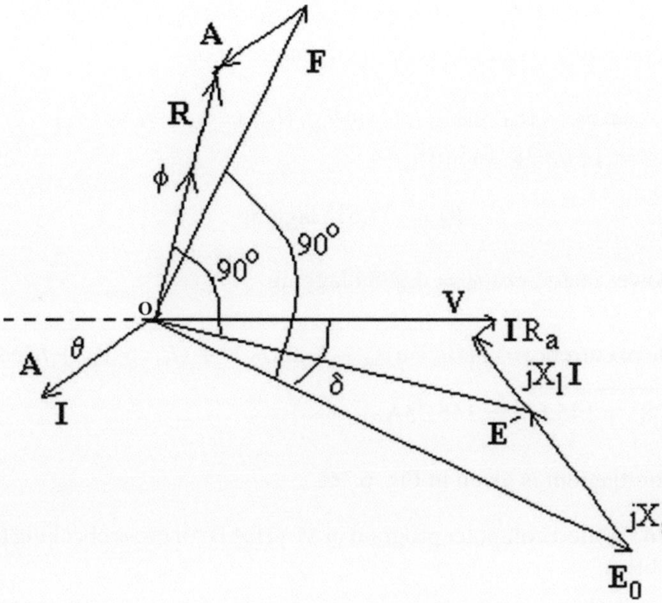

Fig. 6.75 Phasor diagram of cylindrical rotor overexcited synchronous motor

Fig. 6.76 Effect of overexcited synchronous motor for power factor correction

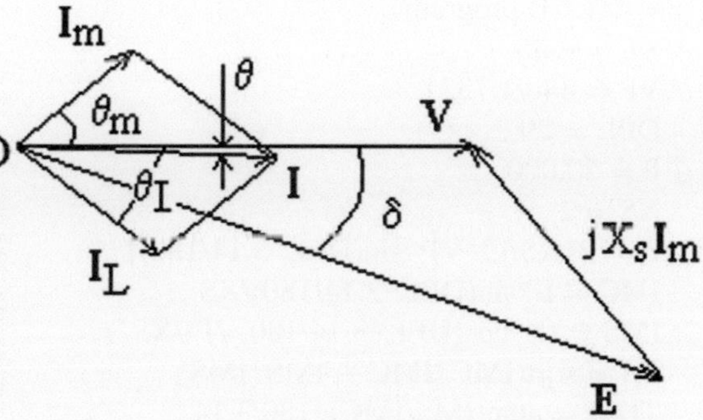

Fig. 6.77 Improvement of power factor of an induction motor

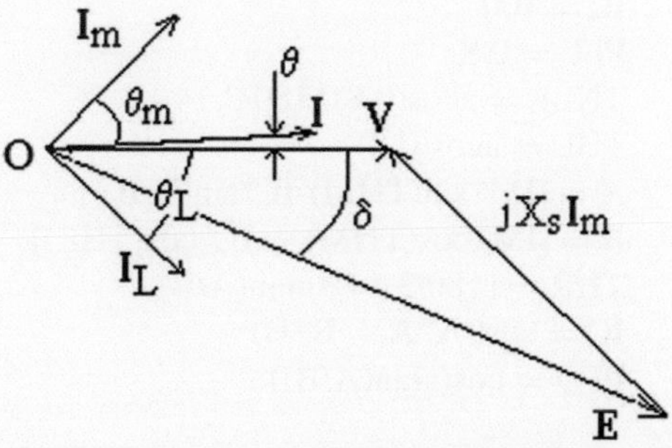

Power factor angle $\theta_L = \cos^{-1}(0.8) = 36.869°$.
Let us suppose that resultant power factor angle be θ_R,

$$\tan\theta_R = [I_m \sin\theta_m - I_L \sin\theta_L]/[I_m \cos\theta_m + I_L \cos\theta_L]$$
$$= [-6.638/145.6]$$

$$\theta_R = -2.61° \text{ lagging}$$

Overall power factor, $cos\theta_R = 0.9989$ lagging

$$\text{Magnitude of current} = \sqrt{\left[(I_m \sin\theta_m - I_L \sin\theta_L)^2 + (I_m \cos\theta_m + I_L \cos\theta_L)^2\right]}$$

$$= \sqrt{\left[6.638^2 + 145.6^2\right]} = 145.75\text{A}.$$

The phasor diagram is given in Fig. 6.76.

Example 6.26 Write a computer program in MATLAB for cross checking the results of above problem.

Solution

```
% MATLAB program
   VL = 440
   VP = 440/1.7321
   DEL = 20
   P = 50,000
   XS = 2
   E = P*XS/[3*VP*sin(DEL*3.14/180)]
   IMC = E*sin(DEL*3.14/180)/XS
   IMS = (E*cos(DEL*3.14/180)VP)/XS
   IM = sqrt(IMC*IMC + IMS*IMS)
   THM = atan(IMS/IMC)*180/3.14
   KVAR = 1.732*VP*IM*sin(THM*3.14/180)
   IL = 100
   PFL = 0.8
   THLL = acos(0.8)*180/3.14
   THL = acos(0.8)
   A = [IM*sin(THM)-IL*sin(THL)]
   B = [IM*cos(THM) + IL*cos(THL)]
   THR = (180/3.14)*atan(A/B)
   IO = sqrt(A*A + B*B)
   PFO = cos(atan(A/B))
```

Example 6.27 A three-phase 440 V, 50 Hz, star-connected induction motor is operating at 0.03 slip. The per phase parameters referred to stator side are $R_1 = R_2' =$

0.1Ω and $X_1 = X_2' = 1.55\Omega$. A three-phase cylindrical rotor synchronous motor having an induced EMF per phase of 500V is developing a power of 30 kW. The motor is having a synchronous reactance of $P = 3EV\sin\delta/X_s$ per phase is operated in overexcited mode in the same supply. Calculate the individual machines' and overall current and power factor.

Solution

The per phase voltage

$$V = 440s/\sqrt{3} = 254.03 \quad \text{V}$$

The per phase induction motor current is given as

$$I_\text{I} = 254.03/\left[(0.1 + 0.1/0.03)^2 + (1.55 + 1.55)^2\right] = 54.92 \quad \text{A}$$

Power factor of induction motor is given as θ = atan {(1.55+1.55)/(0.1+(0.1/0.03))}=42.1 degree lagging. The power factor of induction motor - 0.7462 lagging. The induction motor current is = (40.75 - j 36.82) A The power factor of synchronous motor can be calculated as

$$E\sin\delta = I_\text{m}X_\text{s}\cos\theta_\text{m}$$

Let the suppose that synchronous motor is drawing a current of I_m per phase at a leading power factor of θ_m. Induced EMF E is 300 V. per phase. Developed power can be given as.

$$P = 3EV\sin\delta/X_\text{s}$$

$$\sin\delta = 5 \times 30000/[3 \times 254.04 \times 500]$$

Thus, $\delta = 23.18°$.

$$E\sin\delta = I_\text{m}X_\text{s}\cos\theta_\text{m}$$

From phasor diagram shown in Fig. 6.77, we get

$$I_\text{m}\cos\theta_\text{m} = 500 \times \sin 23.18°/5 = 39.37$$

$$E_0\cos\delta = V + \text{I}_\text{m}X_\text{s}\sin\theta_\text{m}$$

$$I_\text{m}\sin\theta_\text{m} = \left[500 \times \cos 23.18° V\right]/X_\text{s} = 41.12$$

Motor current

$$I_{\text{m}} = \sqrt{\left[39.37^2 + 41.12^2\right]} = 56.93 \quad \text{A}$$

Motor Power factor angle is

$$\theta_{\text{m}} = a\tan(41.12/39.37) = 46.25^\circ \text{ leading}$$

$\text{KVAR} = \sqrt{3} \times 254.04 \times 56.93 \times \sin 46.25^\circ = 18092\,\text{VAR} = 18.092\,\text{kVAR}.$
The current in the power line are:
Induction motor,

$$P_{\text{h}} = 800 - 10 = 790 \quad \text{W}$$

Synchronous motor,

$$\mathbf{I}_{\text{m}} = 56.93\angle + 46.25^\circ = 39.37 + j41.12\,\text{A}$$

Total per phase current

$$\mathbf{I}_{\text{I}} + \mathbf{I}_{\text{m}} = 80.12 + j4.3 = 80.24/\angle + 3.07^\circ\text{A}$$

Overall power factor cos 3.07 = 0.9985 lead. The phasor diagram is given in Fig. 6.77.

6.21 Alternator Connected to Infinite Bus

In power system, an *infinite bus* represents a system of large capacity whose three-phase voltage and frequency remain unaltered even if there is variation of excitation or power of a synchronous machine connected to it. The input mechanical power of prime mover is controlled by controlling the governor settings. The power factor of the current supplied to the bus is controlled by controlling the field excitation [12].

Circle diagram of alternator is a useful idea for predicting its behavior due to different load and excitation. In case of alternator, the electromagnetic power is the prime mover power minus stray loss is given as

$$P_{\text{d}} = VI\cos\theta + I^2R$$

Rearranging we get,

$$I^2 + 2 \times \frac{V}{2 \times R} I\cos\theta + \left(\frac{V}{2R}\right)^2 = \left\{\left(\frac{P_{\text{d}}}{R}\right) + \left(\frac{V}{2R}\right)^2\right\} = R_{\text{d}}^2 \qquad (6.137)$$

The above equation can be represented by a power circle of an alternator, shown in Fig. 6.78a. Here C is the center of power circle and OC is the radius of zero power circle. If we take voltage V as reference, then

$$\bar{I} = \frac{E}{Z_s}\angle(\delta - \gamma) - \frac{V}{Z_s}\angle - \gamma$$

If we draw another circle with center at O and radius V/Z_s, which intersects the zero power circle at point B, then phasor BO is V/Z_s. Excitation circles can be drawn for radii E/Z_s with center at point B. The intersection between power circles and excitation arc determines current I for a given power and excitation. Neglecting

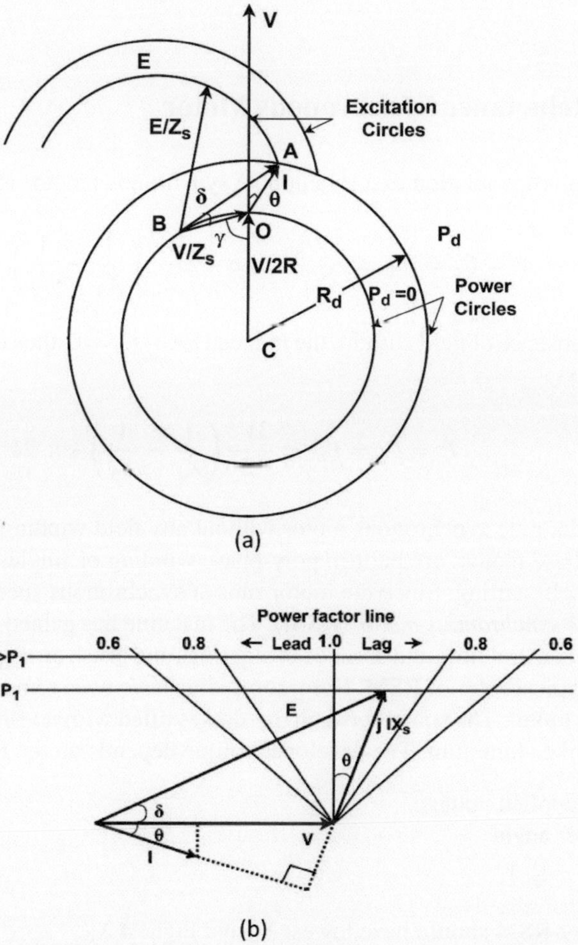

Fig. 6.78 **a** Circle diagram of a cylindrical rotor synchronous alternator, **b** performance characteristics of cylindrical rotor alternator ignoring armature resistance

resistance, the electromagnetic power is electrical output power

$$\overline{E} = \overline{V} + \overline{I}Z_s = \mathrm{V} + (\mathrm{I}\ \cos\theta - j\mathrm{I}\sin\theta)jX_s$$

$$\overline{E} = V + jI\cos\theta X_s + jI\sin\theta X_s \tag{6.138}$$

Here θ is the angle by which the current I lags the terminal voltage V. The operating characteristic of alternator ignoring armature resistance is shown in Fig. 6.78b.

Here θ is the angle by which the current I lags the terminal voltage V. The operating characteristics of alternator ignoring armature resistance are shown in Fig. 6.78b. Develop MATLAB program for plotting 'O' curves of synchronous motor and alternator.

6.22 Reluctance Synchronous Motor

The power angle relation of a salient pole synchronous motor is [6]

$$P = P_e = P_m = \frac{3VE_0}{X_d}\sin\delta + \frac{3V^2}{2}\left(\frac{1}{X_q} - \frac{1}{X_d}\right)\sin 2\delta$$

In the absence of field current, the induced EMF $E_0 = 0$, thus the above expression reduces to

$$P = P_e = P_m = \frac{3V^2}{2}\left(\frac{1}{X_q} - \frac{1}{X_d}\right)\sin 2\delta \tag{6.139}$$

A salient pole synchronous motor without any field winding has a finite starting torque. These motors are built-in poly-phase winding or single-phase winding. The motor is self-starting. Since the motor runs at synchronous speed, it is also called a *reluctance synchronous motor* (RSM). The machine has gained importance because of simple construction, and absence of slip rings and brush arrangements. Figure 6.79 shows a typical rotor of RSM. Here, some slots have been removed from rotor of an induction motor. The space between the slots is filled with non-magnetic conducting material like aluminum. The developed torque depends on the following:

(a) The applied voltage
(b) Power angle
(c) $\left(\frac{1}{X_q} - \frac{1}{X_d}\right)$.

A better RSM should have lowest X_q and highest X_d.

A reduced voltage is initially applied on the stator windings during starting. When the rotor is locked or synchronized with the rotating magnetic field of the stator, full voltage is applied. These are constant speed motors.

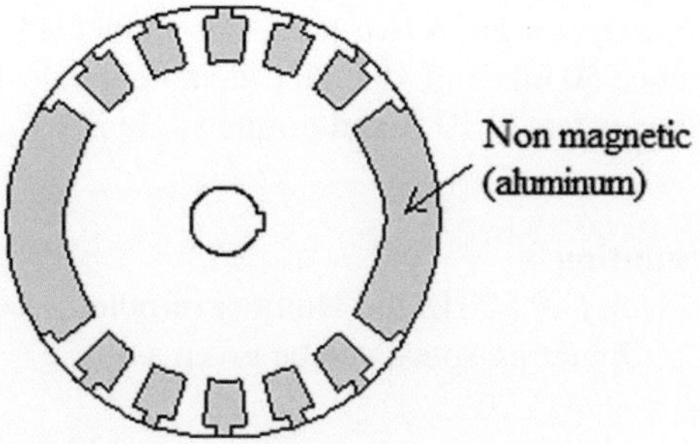

Fig. 6.79 Rotor of two-pole reluctance synchronous motor

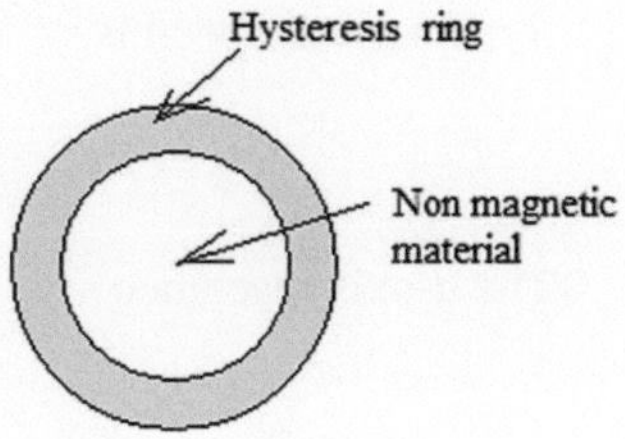

Fig. 6.80 Rotor of hysteresis motor

Example 6.28 A two-pole, three-phase, 50 Hz RSM has d-axis and q-axis reactance of 12 Ω and 5 Ω respectively. Calculate the developed torque and operating speed when the supply voltage is 400 V and torque angle is 15°.

Solution

Given $f = 50$ Hz, number of poles $p = 2$.

Operating speed can be given as

$$N_s = 120f/p = 120 \times 50/2 = 3000\,\text{RPM}$$

Synchronous speed is

$$\omega_s = 2\pi f/(p/2) = 2 \times 3.14 \times 50 = 314\,\text{rad/s}$$

Torque is

$$\begin{aligned} \text{T} &= \frac{3V^2}{2\omega_s}\left(\frac{1}{X_q} - \frac{1}{X_d}\right)\sin 2\delta \\ &= \frac{3}{2 \times 314} \times \left(\frac{400 \times 400}{3}\right)\left(\frac{1}{5} - \frac{1}{12}\right)\sin 30 \\ &= \frac{3}{2 \times 314} \times \left(\frac{400 \times 400}{3}\right)\left(\frac{1}{5} - \frac{1}{12}\right)\sin 30 \\ &= 14.86\,\text{Nm} \end{aligned}$$

Example 6.29 A two-pole, star-connected, 400 Hz RSM has d-axis and q-axis reactance 50 mH and 15 mH. Calculate the developed torque and operating speed when the current is 10 A and torque angle is 15°.

Solution

Given $f = 50$ Hz and Number of poles $p = 2$.

Operating speed can be given as

$$N_s = 120f/p = 120 \times 400/2 = 24000 \,\text{RPM}$$

Synchronous speed is

$$\omega_s = 2\pi f/(p/2) = 2 \times 3.14 \times 400 = 2512 \,\text{rad/s}$$

The d-axis reactance

$$X_d = 2\pi f L_d = 2 \times 3.14 \times 400 \times 50/1000 = 12.56 \,\Omega$$

The q-axis reactance

$$X_q = 2\pi f \, L_q = 2 \times 3.14 \times 400 \times 15/1000 = 3.768 \,\text{W}$$

For a salient pole motor, we have

$$\mathbf{I} = \frac{\mathbf{V} - \mathbf{E}_1}{Z_q} \approx \frac{\mathbf{V} - \mathbf{E}_1}{X_q}$$

When field winding current is zero, EMF $\mathbf{E}_1$ is also zero. Thus, applied voltage per phase is

$$V = X_q I = 3.768 \times 10 = 36.68 \,\text{V}$$

Torque is

$$\begin{aligned} T &= \frac{3V^2}{2\omega_s}\left(\frac{1}{X_q} - \frac{1}{X_d}\right) \sin 2\delta \\ &= \frac{3}{2 \times 2512} \times (37.68 \times 37.68)\left(\frac{1}{3.768} - \frac{1}{12.56}\right) \sin 30 \\ &= 0.07875 \,\text{Nm} \end{aligned}$$

6.23 Hysteresis Motor

Hysteresis motors are a constant speed motors, with built-in fractional horse power. These motors are built with either poly-phase windings or capacitor start single-phase windings. They have hardened cylindrical steel rotor having large retentivity. Figure 6.80 shows the rotor of a hysteresis motor. It has a hysteresis ring on a non-magnetic (aluminum) arbor. The B-H curve of rotor material has been shown in Fig. 6.81. The rotor is smooth rotor without any windings. The rotating magnetic field due to the stator windings cuts the rotor, and results in hysteresis and eddy current losses. Hysteresis motors are started as an induction motor. The rotating magnetic field set-up by the stator currents and the rotor magnetic-axis has an angle of lag δ, as shown in Fig. 6.82. Torque is produced due to interaction between rotor magnetic field and the stator MMF [6].

We know that in an induction motor, the rotor has certain resistance. The power dissipated in this resistance determines the electromagnetic torque (Fig. 6.83) developed in the machines. The eddy current loss can be expressed as

$$P_e = K_e f_r^2 B_{max}^2$$

Fig. 6.81 B-H curve of the hysteresis ring of the rotor

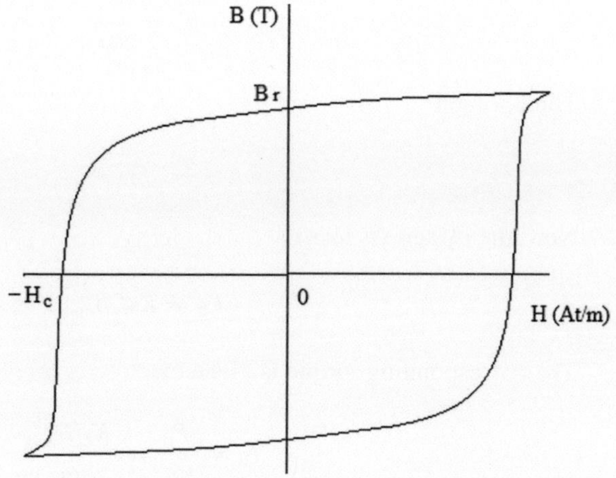

Fig. 6.82 Stator rotating MMF and rotor-induced flux density

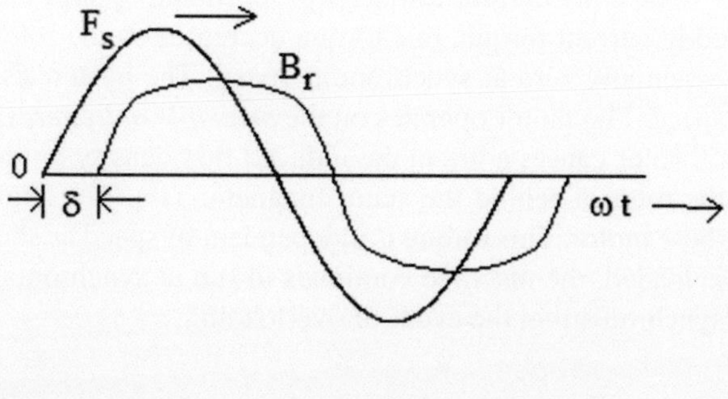

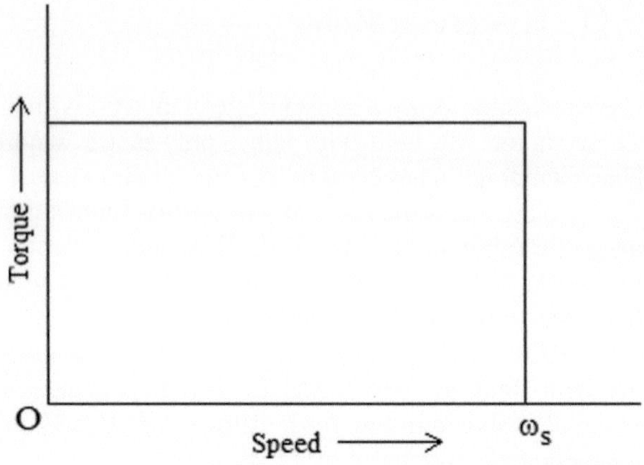

Fig. 6.83 Torque-speed characteristics

where K_e is a constant f_r is the frequency of eddy currents. If machine is running at slip s and the supply frequency is f then $f_r = sf$.

Then the eddy current loss is $P_e = K_e s^2 f^2 B_{max}^2$.The torque related to eddy current is,

$$T_e = \frac{P_e}{s\omega_s} \tag{6.140}$$

Thus, we get

$$T_e \;\alpha\; s$$

Now, the hysteresis loss is

$$P_h = K_h f_r B_{max}^x$$

The corresponding torque is given by

$$T_h = \frac{P_h}{s\omega_s} = \frac{K_h f B_{max}^x}{\omega_s} \tag{6.141}$$

The eddy current torque is proportional to slip; the motor is self starting due to eddy current torque, this torque decreases as the rotor picks up the speed and it is essentially zero at synchronous speed. The hysteresis torque is constant for all the speed. The motor operates on the *principle of hysteresis*. The rotating magnetic field of stator causes a lag in the induced flux density on the rotor. The magnetic-axis of the rotor is behind the stator magnetic-axis. The angle of lag causes the torque in these motor. This torque is independent of speed as shown in Fig. 6.83. As the motor is loaded, the machine continues to run at synchronous speed. The machine looses synchronism in the event of overloading.

The motor was used in paper tape drive and magnetic tape drive of second generation computer, gramophone, recorders, etc. Currently, these motors are being replaced by PMBLDC motor. PMBLDC motors are discussed in Chap. 7.

Example 6.30 The hysteresis loss in the material of a two-pole, 50 Hz hysteresis motor is 800 W. If the rotational losses are 10 W, calculate the power and torque available at the shaft.

Solution

Synchronous angular speed, $\omega_s = \frac{4\pi f}{p} = 315 \quad \text{rad/s}.$

Hysteresis loss, $P_h = 800 - 10 = 790$ W

Torque, $T_h = \frac{\text{Loss}}{\omega_s} = \frac{790}{314} = 2.52$ Nm

Problems

1 Describe the effects produced by the armature MMF on the air-gap flux produced by the field MMF of an alternator when the electrical power is delivered to

(a) lagging load
(b) leading load and
(c) resistive loads

2. Is a synchronous motor self-starting? Explain the procedure of starting a synchronous motor.
3. Why is EMF method referred to as optimistic method and MMF method as pessimistic method?
4. Which method is equally suited for both the salient pole alternator and cylindrical rotor alternator for determination of % regulation?
5. Why is d-axis synchronous reactance higher than q-axis synchronous reactance?
6. A synchronous motor is operated at half load. An increase in the field current causes a decrease in the armature current. Is power factor leading or lagging?
7. Why is field winding of a synchronous motor short-circuited using high resistance at the instant of starting?
8. What are the adjustments required in a synchronous motor to change the power factor of the current drawn?
9. In brushless excitation of synchronous machines, why is the frequency of supply connected to the primary of rotating converter around 400 Hz?

10 Define the following terms:

(a) Normal excitation
(b) overexcitation
(c) under-excitation

11. Draw V-curves of synchronous motor and alternator mentioning power factors and excitation.
12. What is reluctance power?
13. Why are the synchronous motors rarely used in small horsepower rating?

14. Explain in brief, 'A synchronous motor is operating at no load. The field circuit is accidentally opened'.
15. Explain in brief, 'An alternator is connected to bus bar. The field circuit is accidentally opened'.
16. Explain in brief, 'A synchronous motor running at no load, one of the three-phase line is suddenly disconnected and reconnected after a short duration'.
17. What is the role of damper bars in

    (a) synchronous motor
    (b) alternator

18 Show that the power angle of cylindrical rotor alternator is given by

    (a) For lagging power factor $\delta = \tan^{-1}\frac{IX_s\cos\theta}{V+IX_s\sin\theta}$.
    (b) For leading power factor $\delta = \tan^{-1}\frac{IX_s\cos\theta}{V-IX_s\sin\theta}$.
    Assume resistance to be negligible.

19 Show that the per phase-induced EMF of a cylindrical rotor alternator is given by

    (a) For lagging power factor

$$E = \sqrt{(V + IR\cos\theta + IX_s\sin\theta)^2 + (IX_s\cos\theta - IR\sin\theta)^2}$$

    (b) For leading power factor

$$E = \sqrt{(V + IR\cos\theta - IX_s\sin\theta)^2 + (IX_s\cos\theta + IR\sin\theta)^2}$$

20 Show that the power angle of salient pole alternator is given by

    (a) For lagging power factor $\delta = \tan^{-1}\frac{IX_q\cos\theta}{V+IX_q\sin\theta}$.
    (b) For leading power factor $\delta = \tan^{-1}\frac{IX_q\cos\theta}{V-IX_q\sin\theta}$.
    Assume resistance to be negligible.

21 Show that the induced EMF of a salient pole alternator is given by

    (a) Lagging power factor

$$E_1 = \sqrt{(V + IX_q\sin\theta)^2 + (IX_q\cos\theta)^2}$$

$$E_0 = \sqrt{(V + I_dX_d\cos\delta)^2 + (I_dX_d\sin\delta + I_qX_q\cos\delta)^2}$$

    (b) Leading power factor,

$$E_1 = \sqrt{(V - IX_q\sin\theta)^2 + (IX_q\cos\theta)^2}$$

$$E_0 = \sqrt{(V - I_d X_d \cos\delta)^2 + \left(I_q X_q \cos\delta - I_d X_d \sin\delta\right)^2}$$

Assume resistance to be negligible.

22. A two-pole induction motor is running at 3% slip at 1455 RPM. It is supplied by a six-pole alternator. Calculate its speed.
23. Is it possible to connect cylindrical rotor and salient pole alternators in parallel? Discuss in brief.
24. The rated apparent power of a 50 Hz, six-pole, star-connected three-phase synchronous motor with non-salient poles is 500 kVA at 6.6 kV. The synchronous reactance is 40 Ω and the resistance is negligible. Calculate
    (a) stator current, if motor is running at unity power factor
    (b) torque
    (c) if the electromagnetic torque is constant but motor is operating at 0.8 leading power factor, calculate the current and induced EMF.

25. A 50 kVA, 440 V three-phase, star-connected salient pole alternator runs at 75% full load at 0.8 leading power factor. The per phase d- and q-axis reactance are 1.5 Ω and 0.9 Ω, respectively. Neglecting the armature resistance, calculate

    (a) the power angle
    (b) regulation and
    (c) the developed power.

26 A three-phase, two-pole, 100 Hz, star-connected RSM has d-axis and q-axis reactance 16 Ω and 3 Ω respectively. Calculate the developed torque and operating speed

    (a) when the supply voltage is 380 V and torque angle is 14°
    (b) the motor takes 5.0 A and torque angle is 14°.

27. An overexcited star connected, 11 kV synchronous motor is connected across a 100 kVA inductive load of 0.8 lagging power factor. The motor takes 15 kW when running at no load. Calculate the kVA rating of the motor if it is desired to bring the overall power factor the motor inductive load combination to unity.
28. A 100 HP, 440 V 50 Hz delta connected synchronous motor is operating at 75% output at torque angle of 10° and drawing leading current. The per phase synchronous reactance is 2.1 Ω. If the torque angle to remain same, calculate the current when machine is drawing unity power factor and 0.8 lagging power factor currents. What are the output powers? Per phase resistance is assumed to be 0.4Ω.
29. Following are the test data of 11 kV, 1.0 MVA three-phase, star-connected alternator having resistance of 2 Ω per phase.

V_{OC} (kV)	5.8	7	12.5	13.75	15
I_f(A)	40	50	110	140	180
V_{ZPF}	0	1.5	8.5	10.5	12.5

Determine the induced EMF at rated current 0.8 power factor lagging and leading using

(a) EMF method
(b) MMF method
(c) ASA method

30. A three-phase, 2.0 MVA, 6.6 kV, 50 Hz star-connected cylindrical rotor alternator has following open-circuit characteristics:

V_{OC} p.u.	0.26	0.52	0.8	1.0	1.16	1.19
I_f p.u.	0.25	0.50	0.75	1.0	1.25	1.50

In short-circuit test, the 0.6-p.u. field current is required to produce rated load current. In ZPF test the field current required is 1.7 p.u. Determine

(a) the per unit Synchronous reactance
(b) per unit Leakage reactance
(c) Induced EMF, power angle and regulation for 0.8 lagging power factor using ASA method.

31. Two identical synchronous generators have EMF of 220V per phase, 50 Hz. And their impedances are $Z_{s1} = (1+10.0j)\ \Omega$ and $Z_{s2}=(1+11.0j)\ \Omega$ per phase are connected in parallel. They supply a common load having impedance of $(5.0+j1.0)\ \Omega$ per phase. Calculate the terminal voltages, load currents, active power output and no-load circulating current for a phase divergence of 9° electrical.
32. A two-pole, star-connected, 50 Hz RSM has d-axis and q-axis reactance 9.0 Ω and 2.0 Ω respectively. Calculate the developed torque and operating speed when the supply voltage is 400 V and torque angle is 10°.
33. A two-pole, star-connected 400 Hz, RSM has d-axis and q-axis reactance 70 mH and 20 mH. Calculate the developed torque and operating speed when motor takes a current is 15A and torque angle is 10°.
34. A three-phase, star-connected 6.6 kV, 2.0 MVA, 50 Hz cylindrical rotor alternator has the following characteristics:

V (kV)	4	5	6	7	8	8.25
I_f (A)	13	16.7	21.4	28	40	45

The required field current during short-circuit test is 10 A and in ZPF test is 40 A. Obtain induced EMF, power angle, power and percentage regulation using ASA method at full load 0.8 lagging, 0.8 leading and unity power factors.

35. The test data for three-phase, star-connected 10,000 kVA, 11 KV alternator are as follows:

V_{LL} (kV)	0	5	8	10	11	12	13
I_f (A)	0	38.5	65	89	109	128	154
ZPF I_f (A)		0			185		

Determine percentage regulation at full load 0.8 lagging power factor using EMF and ASA method.

36 A cylindrical rotor synchronous motor running at no load. Determine the maximum shaft load which can be applied. Use equal area criterion.

37. Two three-phase star-connected alternators are running in parallel sharing a common load of (3+j) Ω per phase. The synchronous impedance of the alternators are (1+j6) Ω per phase each. The open circuits EMFs are 440 V and 450 V. Find the terminal voltage per phase and power output of each machine. Also, calculate circulating current at no load.

38. If the machine in Problem 6.37 supplies a common load current of 30 A per phase at 0.8 power factor lagging, find the terminal voltage per phase and power output of each machine.

39 Determine which of the following cases of a cylindrical rotor synchronous machine will be stable, critically stable and unstable. Use equal area criterion

(a) $P_1 = 0\ P_2 = 0.9\ P_{max}$
(b) $P_1 = 0.9\ P_{max}\ P_2 = 1.0\ P_{max}$
(c) $P_1 = 0.1\ P_{max}\ P_2 = 0.9\ P_{max}$
(d) $P_1 = 0.5\ P_{max}\ P_2 = 0.96\ P_{max}$

40. A three-phase cylindrical pole synchronous motor of rating 500 kW is running at no load. Determine the maximum power of the load, which can be suddenly applied on the shaft without the motor losing its dynamic stability. Ignore losses. Also, determine what will be maximum suddenly applied load if the motor initially develops a power of 100 kW?

41. A two pole, 50 Hz alternator with constant excitation is delivering 200 kW at power angle of 2° electrical degrees. The inertia acceleration coefficient for the combined prime mover Alternator is 1.6 N-m/electrical degree/s^2. The damping torque coefficient is 10 N-m/electrical degrees/second. The input power from prime mover is suddenly increased so that the generator increases its share of the common electrical load to 1000 kW. Determine

(a) the natural frequency of oscillation
(b) damped frequency of oscillation
(c) damping coefficient.

42. A three-phase star-connected load takes 80 A at 0.74 power factor lagging at 400 V, 50 Hz line. A three-phase, star-connected cylindrical rotor synchronous motor having a synchronous impedance of 1.2 Ω per phase is connected in

parallel with the supply and load. The power developed by the motor is 33 kW at a power angle of 15°. Calculate

(a) the reactive kVA of the motor
(b) the overall power factor of the load and the motor and resultant current.

43. A three-phase 400 V, 50 Hz, star-connected induction motor is operating at 0.05 slip. The per phase parameters referred to stator side are $R_1 = R'_2 = 0.05\ \Omega$ and $X_1 = X'_2 = 0.9\ \Omega$. A three-phase cylindrical rotor synchronous motor having an induced EMF per phase of 450 V is developing a power of 10 kW. The motor is having a synchronous reactance of 8.0 Ω per phase, and is operated in overexcited mode in the same supply. Calculate the individual machines' and overall current and power factor.
44 The hysteresis loss in the material of a two-pole, 400 Hz, two-pole hysteresis motor is 1.0 kW. If the rotational losses are 1%, calculate the power and torque and operating speed.
45. A two pole, 50 Hz hysteresis motor exhibits an eddy current loss and hysteresis loss of 300 W and 400 W respectively. Determine the starting and running torque of the motor. What is the operating speed?
46. A star-connected, three-phase 50 Hz, 11,000 V synchronous motor has an armature resistance of 1.0Ω/phase and synchronous reactance of 30-Ω/phase. When the motor delivers 1000 HP, the efficiency is 90% and the field current is adjusted such that the motor takes a leading current of 90 A. Calculate

(a) power factor
(b) excitation EMF
(c) the mechanical power developed
(d) if load is removed describe how the magnitude and power factor of the resulting armature current compare with the original current.

47. A 20 MVA, three-phase Y-connected, 13.2 KV, 50 Hz salient pole alternator has $X_d = 1.1$ p.u. and $X_q = 0.8$ p.u. Neglect resistance. When the machine is carrying a load of 15 MVA at 0.95 p.u. lag at rated voltage. Calculate the excitation voltage, the power angle, the direct and quadrature-axis currents and regulation. Give values in p.u. for actual voltage and ampere.
48. A 2200 V three-phase 50 Hz four-pole Y-connected synchronous motor has per phase synchronous reactance of 5.7 Ω and an armature resistance of 0.19 Ω per phase. The excitation is so adjusted that the induced voltage is 2200 V (line to line) if the line current is 200 A at a certain load. Calculate

(a) the input power
(b) the developed torque and
(c) the power angle.

49. A 400 V, three-phase, Y-connected synchronous motor delivers 15 HP at the shaft and operates at 0.8 lagging power factor. The total iron, friction and field copper losses are 1.0 kW. If the armature resistance is 0.83 Ω per phase determine efficiency of the motor.

50. The direct and quadrature-axis reactance of a three-phase, 4-pole, star-connected reluctance synchronous motor (RSM) are 15.0 Ω and 3.8 Ω per phase, respectively. The motor operates at a power angle of 11°. Calculate the torque developed by the motor when

(a) motor is drawing a current of 20 A
(b) The motor is supplied by a source of 400 V. Neglect winding resistance. Choose the supply frequency is 60 Hz.

Objective-Type Questions

1. Machines having slip rings and brushes are

(a) Squirrel cage induction motor and synchronous motor
(b) Slip-ring induction motor and synchronous motor
(c) Slip-ring induction motor and DC motor
(d) BLDC motor, slip-ring induction motor and synchronous motor

2. A four-pole, 440 V, 50 Hz synchronous motor is running at 400 V, 60 Hz supply. The synchronous speed is

(a) 1500 RPM
(b) 1800 RPM
(c) 1500 to 1800 RPM
(d) 1650 RPM

3. The power angle diagram of an alternator is shown. It will be stable, when

(a) Area $A1 < A2$
(b) Area $A1 > A2$
(c) Area $A1 = A2$
(d) None

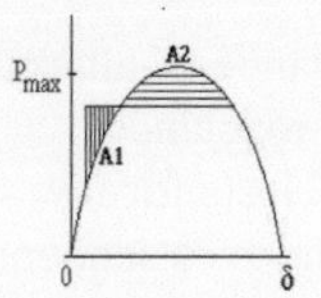

4. In which of the following case, the transient current due to re-switching will be at maximum

(a) Squirrel cage induction motor
(b) Slip-ring induction motor
(c) Synchronous motor
(d) Reluctance synchronous motor

5. Sub-transient region in sudden short-circuit test of an alternator is absent, if
   (a) Machine is running at synchronous speed
   (b) Damper bars are present
   (c) Damper bars are not present
   (d) Machine is running at a slightly reduced speed.
6. In brushless excitation of two-pole, 50 Hz synchronous machines, 400 Hz supply is produced using
   (a) 24 pole PMSG
   (b) 20 pole PMSG
   (c) 16 pole PMSG
   (d) 8 pole PMSG
7. When a synchronous motor is started, its field winding is
   (a) Short-circuited
   (b) Open circuited
   (c) Short-circuited with a high resistance
   (d) Short-circuited with low resistance
8. A salient pole synchronous machine without any damper bars and without any field current is excited with reduced three-phase source
   (a) Machine fails to start
   (b) Machine is started in the direction of rotating magnetic field as a reluctance synchronous motor due to reluctance torque
   (c) Machine is started in the direction of opposite to direction of rotating magnetic field.
   (d) Machine acts as a claw-pole motor
9. Which of the following statement is not true?
   (a) $X_{\mathrm{d}}'' < X_{\mathrm{d}}' < X_{\mathrm{d}}$
   (b) Negative sequence reactance of alternator can be expressed in terms of d and q-axis sub-transient reactance
   (c) The sub-transient period lasts for few seconds
   (d) The positive, negative and zero sequence reactance of an alternator can be calculated using fault tests on alternator
10. Georges's phenomenon can affect the starting of
    (a) DC motor
    (b) Induction start synchronous run motor
    (c) Reluctance motor
    (d) PMBLDC motor
11. The overall power factor of an industry can be corrected to unity using,
    (a) Under-excited synchronous motor
    (b) Overexcited synchronous motor
    (c) Overexcited synchronous generator

(d) Normally excited synchronous motor

12. Hunting occurs in a synchronous machines having damper bars, when

    (a) Sudden load is applied
    (b) Fault conditions
    (c) Both (a) and (b)
    (d) None

References

1. Philip L. Alger, Induction Machines, Gordon & Breach Science Pub. N.Y. 2nd Ed. 1965
2. M.G.Say, Alternating current Machines, 4th ELBS edition, 1976
3. E. Openshaw Taylor: The Performance and design of A.C. commutator motors, Wheeler Publications, India, 11th Print, 1990.
4. A. Fransua, R. Magureanu, Electrical Machines and Drive Systems, TP Technical Press, Oxford, 1984
5. V Del Toro, Electric machines and power systems, PHI, 1980
6. Bhattacharyya, M, Electrical Machines Modelling and Analysis, PHI, Delhi, 2016
7. A.S. Langsdorf, Theory of alternating current machinery, TMH, 2001
8. A. Draper, Electrical Machines, ELBS, 1st Ed, 1956
9. Stephen J Chapman, Electric Machinery Fundamentals, Indian Edition, 4e, Mcgraw hill education (Indi) Pvt Ltd, 2005
10. Siskind, Electrical Machines, 2nd edition International student edition, McGraw hill international book company, 12th printing, 1983
11. Nagrath and Kothari , Electrical Machines, 4/e, 2010, TMH
12. P. K. Mukherjee, S. Chakravorti, Electrical Machines, Dhanpat Rai & sons, 1993

Chapter 7
Special Motors and Generalized Machine Theory

Abstract This chapter discusses the special motors like servo motor-printed circuit motor and AC servo motor used in control system, commutator AC motors, stepper motors, PMBLDC motor, Switched Reluctance motor (SRM), linear motors, written pole motor-generator (WPM, WPG) and eddy current devices. The generalized machine theory and symmetrical components have been briefly discussed.

Keywords Servo motor · Printed circuit motor · PMBLDC motor · SRM · Linear motor · Levitator · WPM · WPG · Eddy current coupling and brake · Generalized machine theory · d-axis · q-axis · Symmetrical components

7.1 Introduction

This chapter discusses special machines used in control system, traction systems and newer machines such as BLDC motor, SRM, WPM and eddy current coupling and eddy current brake. Generalized machine theory and symmetrical components have also been discussed briefly.

7.2 Servo Motors

In control system applications, it is desired to achieve the position of certain machine parts by variation of control signals. In case of a servo-stabilizer, the output voltage needs to be readjusted by changing number of turns of an autotransformer. This is achieved by a servo motor which works on certain control voltage derived by comparing the actual output voltage and the reference voltage. Similarly, the direction of a gun-barrel can be changed according to a control-signal defining the stationary or moving targets. The end-effector of a robotic arm is controlled by certain known closed loop control signal. There are many other automatic industrial applications

R. K. Srivastava, *Electrical Machines*, https://doi.org/10.1007/978-981-96-5039-2_7

where servo motors are used. These servo motors can be operated by either AC or DC supply.

The conventional separately excited DC motor having large axial length and lesser diameter can be used as a DC servo-motor. It is essential that a servo motor must have lowest moment of inertia, which minimizes both the acceleration and deceleration time. This helps in responding more quickly to a control signal. We will now discuss DC servo motor–printed circuit type, and two-phase AC servo motors.

7.2.1 DC Servo Motor-Conventional DC Motor and Printed Circuit Motor

We have already discussed that a conventional DC motor with lowest moment of inertia can be used for servo application. Such motors have large axial length and lesser diameter. These are higher capacity motors frequently used in larger power automatically controlled process industries such as steel mills, aluminum sheet metal industry, etc. The printed circuit motor has the lowest moment of inertia and is an excellent choice for the servo applications.

Printed circuit motor was developed in the late 1950s in France. In these motors, the field is provided using sector shape permanent magnets and the armature winding is in the form of printed circuit (disk). The winding and commutator are in the form of printed circuit, similar to that of a PCB (Printed circuit board) of an electronics circuit. The permanent magnets are of sector-shaped having axial magnetization. The magnetic circuit is completed through the flux return steel plate which also supports the brush holders. The armature is of disk shape and contains no ferromagnetic material. The armature is made of copper plated epoxy–resin impregnated glass fabric sheet known as Printed Circuit Board (PCB), in which copper is plated on both the sides of the sheet. The conductors are etched on both the sides of the plate by printed circuit technique. Connections from side-to-side are made through the holes. Silver-graphite brushes bear directly on the armature conductors which serve as a commutator. The large capacity disk motor has wire wound armature with face or barrel commutator. The resin is finally given a completion cure for strengthening the winding. The supply voltage is normally 12 V battery supply. The armature is placed in between the sector-shape magnets having axial magnetization. The ratings of motor are from FHP and to three HP capacity. It exhibits the shunt type characteristics. The constructional details of printed circuit motor have been described in Fig. 7.1. Figure 7.2a shows a printed circuit motor with single PM disk stator, in which flux return path is through end cover. Figure 7.2b shows printed circuit motor with dual PM disk stators, in which flux return path is through permanent magnets. These are hetero-polar machines in which the number of magnetic poles will be even number [3–7, 12].

There is no cogging in these motors. Both the armature inductance and armature time constant are the lowest. As a results, electrical time constant is also lowest.

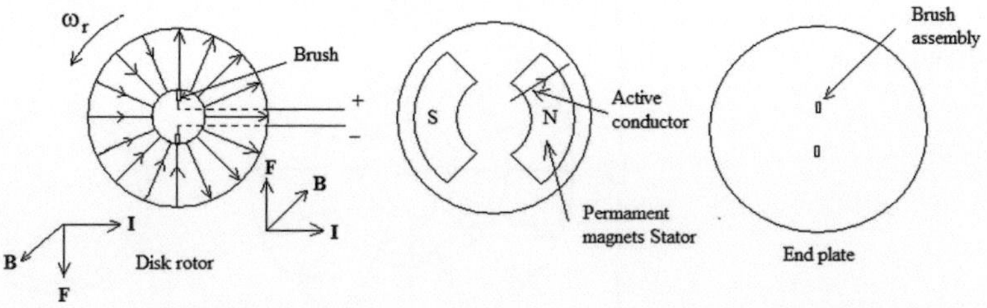

Fig. 7.1 Typical representation assembly of PCB motor showing disk rotor, stator and end plate

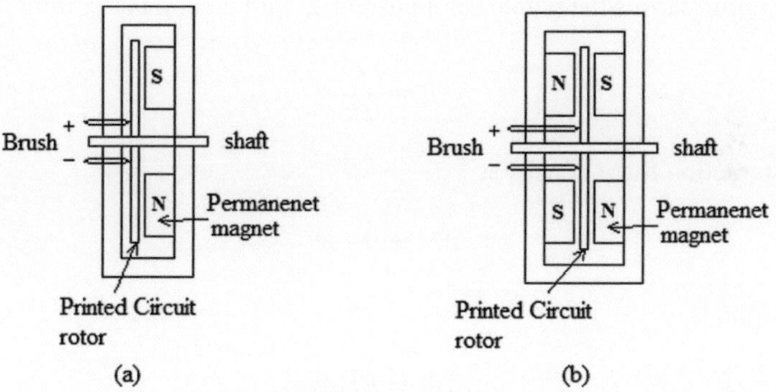

Fig. 7.2 Printed circuit motor: **a** single magnetic stator and **b** double magnetic stators

There are no heavy rotating parts as such moment of inertia and mechanical time constant both are the lowest. The motor quickly responds to the control signal, without any delay in acceleration. Low inherent back EMF in these motors is increased by using stronger rare earth magnets like NdFeB. The torque irregularities due to finite number of conductors are virtually eliminated by skewing the conductors. The motor has excellent heat dissipation and can work up to a temperature of 120 °C.

Let us consider the conduction of current from radius 0.5d to 0.5D in a homopolar machine as shown in Fig. 7.3. The disk is assumed to carry a uniform distribution of radial current **I** in uniform axial magnetic flux density **B**. The disk is assumed to have an angular speed ω_r. Consider an elemental annulus of radius x and radial differential length of dx. It moves at a peripheral speed $x\omega_r$. The motional EMF is

$$\mathrm{d}e_r = Bx\omega_r \mathrm{d}x \tag{7.1}$$

The total disk EMF between d and D is therefore

$$E_r = B\omega_r \int\limits_{0.5d}^{0.5D} x\mathrm{d}x = \frac{1}{8}(D^2 - d^2)B\omega_r \tag{7.2}$$

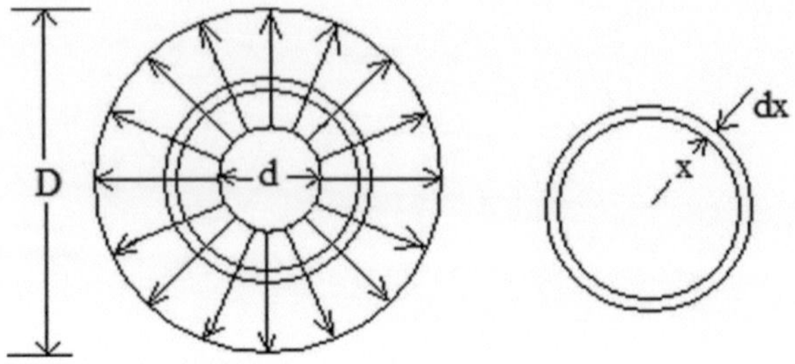

Fig. 7.3 Homo-polar disk rotor carrying radial current

The annulus carries the whole rotor current **I**, with an angular current density,

$$\delta = \frac{I}{2\pi} \tag{7.3}$$

The interaction force density is

$$\mathrm{d}F_e = BI\mathrm{d}x$$

The torque is

$$\mathrm{d}T_e = BIx\mathrm{d}x$$

$$T_e = \frac{1}{8}\left(D^2 - d^2\right)BI \tag{7.4}$$

The output power is

$$E_r I = T_e \omega_r$$

The power developed is

$$P = \frac{1}{8}\left(D^2 - d^2\right)BI\omega_r = \frac{1}{4}\pi\left(D^2 - d^2\right)B\delta\omega_r = \frac{1}{2}\pi^2\left(D^2 - d^2\right)B\delta n \tag{7.5}$$

Here n is the rotation per second. If outer diameter D is increased, then the power, torque, rotational EMF will also increase. But outer diameter is limited by the rotational stresses, flexibility of the disk and the requirement of clearance between stator and rotor.

The above expressions (7.1) to (7.5) are also useful for a hetero-polar disk machine. Let the average gap field density be **B**. The EMF in a single rotor conductor is same as in case of homo-polar machine. After one pole pitch, the direction of current is changed due to the commutator action. Further, the direction of magnetic flux is also changed. The net torque is in the same direction. Let there be N turns on the armature, carrying current **I**. If N conductors are in series between brushes, the angular current density is

$$\delta = \frac{NI}{2\pi} \tag{7.6}$$

Thus, the rotational EMF is

$$E_r = \frac{1}{8}\left(D^2 - d^2\right)NB\omega_r \tag{7.7}$$

The torque is

$$T_e = \frac{1}{8}\left(D^2 - d^2\right)NBI \tag{7.8}$$

The power is

$$P = \frac{1}{4}\pi\left(D^2 - d^2\right)B\delta\omega_r = \frac{1}{2}\pi^2\left(D^2 - d^2\right)B\delta n \tag{7.9}$$

The geometry of disk machine is more suitable for six, eight or more number of poles even in very small machine. The magnetic poles are light, and cheap. They have high coercive field intensity H_c. The poles are arranged to be fixed on stator disk. The return flux path is provided by another magnetic structure of similarly mounted poles of opposite polarity or a steel disk. The depth of magnet is around 20 times the gap length. The magnets are having shape of a sector. The magnetic flux is more effective where the peripheral speed is greater. The optimum ratio of diameter is $D/d = \sqrt{3}$ for axial flux machines.

Figure 7.4a shows armature windings on the disk of a printed circuit motor. The conductors on the front sides are electrically jointed with those on the rear side. The armature is placed in between the sector-shaped magnets having axial magnetization, as shown in Fig. 7.4b.

The wave winding is preferred. The conductors are carrying large current; the larger resistive loss increases the temperature. The temperature is normally limited

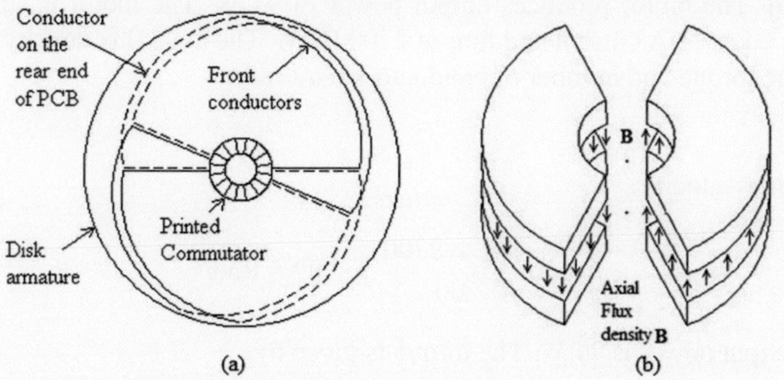

Fig. 7.4 **a** Printed armature (part winding shown for clarity) and **b** sector shape magnet with axial magnetization

up to 150 °C. The ventilation of motor is achieved using circulation of water or oil. The severe sparking is observed when the speed rises. The mechanical structure of motor cannot withstand a high increase in speed. The centrifugal forces and vibration causes mechanical imbalance under dynamic conditions.

Printed circuit DC servo motors do not have hysteresis effect or iron loss. Their size is very compact and can easily fit into small space. However, these motors are costly as compared to other options. The following are the applications of these motors:

(a) Paper tape reader.
(b) Positioning servo motor.
(c) Velocity servo motor.
(d) Industrial instrumentation.
(e) Low cost applications such as opening and closing of automobile doors or fans etc.
(f) DC Tacho-generator.

Example 7.1 A conducting disk of outer and inner diameters of 35 cm and 10 cm, respectively, is mounted on a rotating shaft running at 900 RPM in an axial magnetic field of 1.0 T. Determine the induced EMF in the outer and inner rim of the disk.

Solution

Rotational angular speed

$$\omega = \frac{2\pi \times 900}{60} = 94.2 \text{ Rad/s}$$

The EMF induced is

$$E_r == \frac{1}{8}(D^2 - d^2)B\omega_r = \frac{1}{8}\left(0.35^2 - 0.10^2\right) \times 1 \times 94.2 = 1.3246 \text{ V}$$

Example 7.2 A printed circuit motor has outer radius of 120 mm and inner radius of 60 mm. The motor produces output power of 94 W. The motor is supplied by 12 V and takes 12 A current and runs at 2700 RPM. The axial flux density is 0.70 T. Determine torque and number of conductors required.

Solution

The angular velocity

$$\omega = \frac{2\pi \times 2700}{60} = 282.6 \text{ Rad/s}$$

The output power is 94 W. The torque is given by,

$$T = \frac{P}{\omega} = \frac{94}{282.6} = 0.33 \text{ Nm}$$

The torque is given by

$$T = \frac{1}{8}(D^2 - d^2)NBI$$

Thus the total number of conductors is

$$N = \frac{8T}{(D^2 - d^2)BI} = \frac{8 \times 0.33}{(0.12^2 - 0.06^2)0.70 \times 12} = 29.1 \approx 29 \text{ conductors}$$

7.2.2 Permanent Magnet DC Machines

In Permanent magnet DC machines, the field flux is provided by permanent magnets having radial magnetization. A typical PMDC motor has been shown in Fig. 7.5. In recent years, due to advances in the metallurgical engineering, the powerful rare earth permanent magnets like NdFeB are commercially available. These motors have gained more importance. Like conventional DC machine, the armature winding and commutator are there on the inner side of the machine and carbon or graphite alloy brushes are used. Printed circuit motor is also PM DC motor and has been explained earlier. The PMDC motors have similar characteristics as that of separately excited motor. These motors are used in control systems, toys, e-bikes, automobiles and electric power steering (EPS), etc. Permanent magnet DC generator exhibits characteristics similar to that of separately excited generator. These generators have found applications in battery charging using wind energy and also as tacho-generator. The demagnetization of permanent magnets is possible, if machine is used under overloaded condition. This is prevented by suitable protective measures.

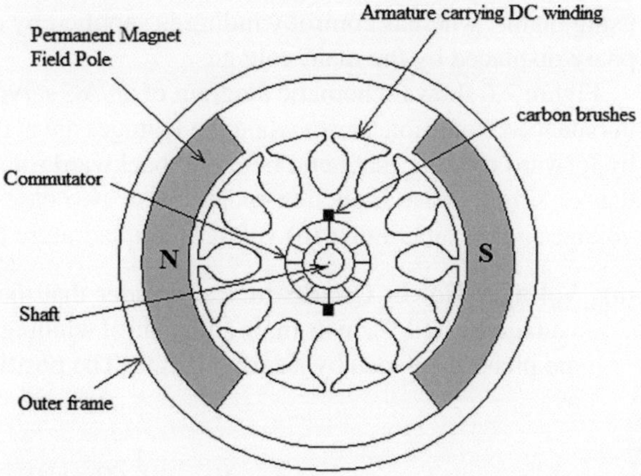

Fig. 7.5 Permanent magnet DC motor (*armature windings not shown for clarity*)

The availability of high power permanent magnet has led to development of several machines such as PMBL, PMSM and PMSG. High speed and high torque motors may have ferrite and NdFeB magnets, respectively. The PMBLDC motor has been discussed in brief in later section. These motors, PMBLDC and PMSM, require the use of Hall Effect sensor or resolver for rotor position sensing. The coil switching is synchronized with the rotor position using solid state electronic circuit. PMSG giving AC supply, is widely used in brushless excitation of synchronous machines, wind power generation and aircraft power supply etc. The discussions of PMSM and PMSG are beyond the scope of this book.

7.2.3 AC Servo Motor

Most of the motors used in control systems are AC and are 2-phase induction motor type. Their mechanical power output varies from 0.5 W to few hundreds Watt. Two-phase servo motors are constructed in several different ways. They all have two-phase distributed windings in the stator. The rotor is either squirrel cage type or drag cup type. The rotor has high resistance, which gives maximum starting torque. Most common servo motors are squirrel cage type except that diameter is made very small to reduce the moment of inertia. This improves the acceleration characteristics.

Drag cup motor has hollow cup-rotor made of copper, aluminum or alloy. The outer stator has the windings while another inner stationary ferromagnetic stator provides a flux return path. The stator winding has high resistance to reduce the possibility of single phasing due to high current. This improves the stability of the control system. The rotor is in form of a conducting-cup, which rotates in between the two stators.

The AC servo motor is basically a two-phase induction motor having larger axial length and lesser diameter so as to keep the moment of inertia at its lowest. The stator winding is a two-phase 90° space displaced windings. The main winding is energized using mains, whereas control winding is supplied by a control voltage which is 90° phase displaced by the main voltage.

Figure 7.6 shows schematic diagram of an AC servo motor. The motor operates in unbalanced conditions, even when the voltages are at quadrature. This is represented by forward rotating magnetic field and backward rotating magnetic field, similar to that of single phase induction motor. Here two cases have been discussed (a) the voltages not in quadrature (b) voltages in quadrature [6].

(a) **Voltages Not in Quadrature**: Consider that the servo motor is energized by voltage $\boldsymbol{V}_a$ and $\boldsymbol{V}_b$ in its main and control windings, respectively. These voltages are phase displaced by a certain angle. The positive sequence voltage is related as

$$\boldsymbol{V}_{a1} = \frac{1}{2}(\boldsymbol{V}_a + j\boldsymbol{V}_b) \tag{7.10}$$

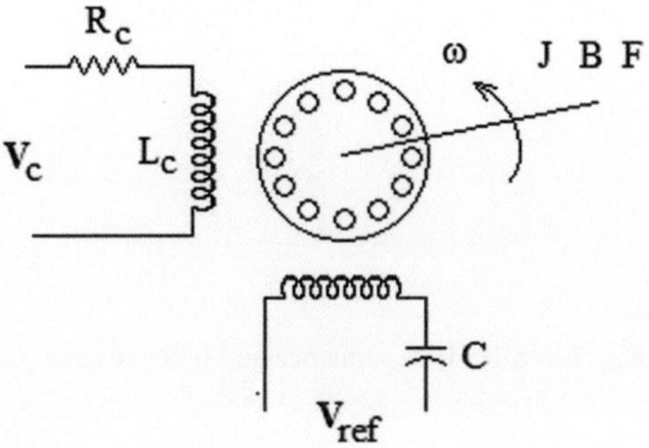

Fig. 7.6 Schematic representation of AC servo motor

The negative sequence voltage is,

$$V_{a2} = \frac{1}{2}(V_a - jV_b) \tag{7.11}$$

The zero sequence voltage is,

$$V_{a0} = \frac{1}{4}(V_a - V_b - V_a + V_b) \tag{7.12}$$

Thus, the unbalance two-phase system can be completely represented by a balanced positive sequence component and a balanced negative sequence component. The two-phase voltages can be expressed as

$$V_a = V_{a1} + V_{a2} \tag{7.13}$$

$$V_b = aV_{a1} + a^3V_{a2} = -jV_{a1} + jV_{a2}, \tag{7.14}$$

where operator $a = -j$.

The operation of unbalanced two-phase servo motor is regarded in terms of positive sequence and negative sequence voltages, as shown in Fig. 7.7.

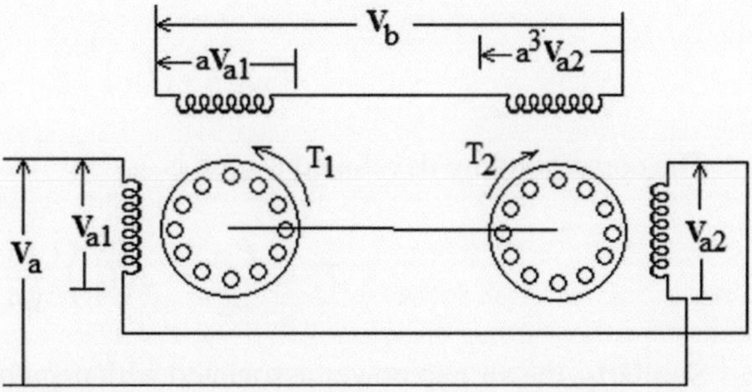

Fig. 7.7 Operation of AC servomotor on two-phase unbalanced supply

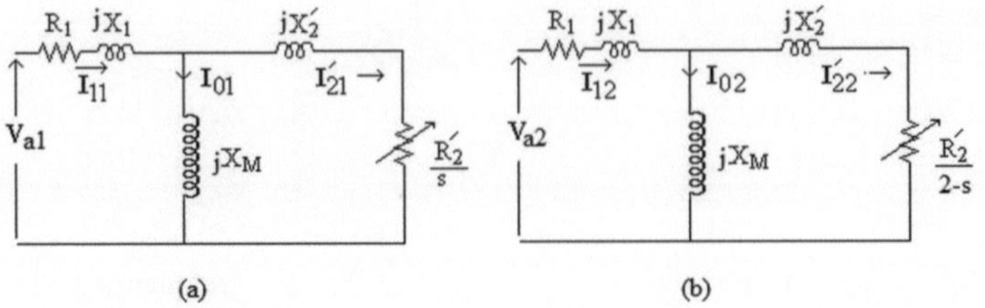

Fig. 7.8 **a** Positive sequence and **b** negative sequence equivalent circuits of AC servo motor

Figure 7.8 shows equivalent circuit for positive sequence and negative sequence components. The motor is running in forward direction at speed N_r, the forward slip is given as

$$s_+ = \frac{N_s - N_r}{N_s} = s$$

The negative sequence voltage set establishes a field that revolves in the opposite direction to the forward field. The slip due to negative sequence set of voltages is

$$s_- = \frac{N_s - (-N_r)}{N_s} = 1 + \frac{N_r}{N_s} = 2 - s$$

Figure 7.8 shows the positive sequence and negative sequence equivalent circuit of an AC servo motor. Here, the rotor currents referred to primary for both the sequence voltages are as follows:

$\boldsymbol{V}_{a1}$ = Positive sequence set of voltage.
$\boldsymbol{V}_{a2}$ = Negative sequence set of voltage.
$I_{21}^{/}$ = Positive sequence rotor current referred to stator.
$I_{22}^{/}$ = Negative sequence rotor current referred to stator.

The positive sequence and negative sequence rotor currents referred to primary side can be obtained by normal circuit analysis. The air gap power associated with positive sequence rotating field,

$$P_{g1} = 2\left(I_{21}^{/}\right)^2 \frac{R_2^{/}}{s}$$

The corresponding developed torque is,

$$T_1 = \frac{P_{g1}}{\omega_s} = 2\left(I_{21}^{/}\right)^2 \frac{R_2^{/}}{s\omega_s} \tag{7.15}$$

Similarly, the air gap power associated with negative sequence rotating magnetic field,

$$P_{g2} = 2\left(I_{22}^{/}\right)^2 \frac{R_2^{/}}{2 - s}$$

The corresponding developed torque is,

$$T_2 = \frac{P_{g2}}{\omega_s} = 2\left(I_{22}^{/}\right)^2 \frac{R_2^{/}}{(2 - s)\omega_s} \tag{7.16}$$

The resultant torque,

$$T = T_1 - T_2 = \frac{1}{\omega_s}\left(P_{g1} - P_{g2}\right)$$

The resultant line currents are

$$\boldsymbol{I}_a = \boldsymbol{I}_{a1} + \boldsymbol{I}_{a2} \quad \text{and} \quad \boldsymbol{I}_b = -j\boldsymbol{I}_{a1} + j\boldsymbol{I}_{a2} \tag{7.17}$$

(b) **Voltages in Quadrature**: If $\boldsymbol{V}_a$ is the main winding voltage and $\boldsymbol{V}_b$ is the control winding voltage which is always 90° apart in phase, we can define the ratio of control winding and main winding voltages as

$$\rho = \frac{V_b}{V_a}$$

The control winding voltage is in quadrature. Therefore, we have

$$\boldsymbol{V}_b = -j\rho \boldsymbol{V}_a \tag{7.18}$$

Thus, the positive sequence can be given as

$$\boldsymbol{V}_{a1} = \frac{\boldsymbol{V}_a}{2}(1 + \rho) \tag{7.19}$$

And, the negative sequence voltage is

$$\boldsymbol{V}_{a2} = \frac{\boldsymbol{V}_a}{2}(1 - \rho) \tag{7.20}$$

The air gap power and torque associated with positive sequence and negative sequence voltage can be obtained as discussed earlier Sect. 7.2.3(a).

Example 7.3 A 220 V, 50 Hz, two-phase servo motor has the following parameters:

$$R_1 = 250\Omega \quad R_2^{/} = 750\Omega \quad X_1 = 50\Omega \quad X_2^{/} = 50\Omega \quad X_m = 1000\Omega.$$

When operating at a slip of 0.6, determine:

(a) resultant torque in synchronous Watts.
(b) stator phase currents.

Assume that the servo motor operates at unbalanced two-phase voltages:

$$\boldsymbol{V}_a = 220\angle 0^\circ \quad \text{and} \quad \boldsymbol{V}_b = 150\angle -60^\circ \text{ V}$$

Solution

The voltages are

$$\boldsymbol{V}_a = 220\angle 0^\circ = 220 + j0 \text{ V}$$

$$\boldsymbol{V}_b = 150\angle -60^\circ = 150(\cos 60 - j\sin 60) = 75 - j129.9 \text{ V}$$

The positive sequence voltage is

$$\boldsymbol{V}_{a1} = \frac{1}{2}(\boldsymbol{V}_a + j\boldsymbol{V}_b) = 0.5(220 + 75j + 129.9)$$
$$= 175 + 37.5j = 179\angle 12.1^\circ \text{ V}$$

The negative sequence voltage is

$$\boldsymbol{V}_{a2} = \frac{1}{2}(\boldsymbol{V}_a - j\boldsymbol{V}_b) = 0.5(220 - 75j - 129.9)$$
$$= 45.05 - 37.5j = 58.6\angle -39.8^\circ \text{ V}$$

Positive sequence impedance

$$Z_1 = R_1 + jX_1 + jX_m \left\| \left(\frac{R_2^/}{s} + jX_2^/ \right) \right. = 250 + 50j + 1000j \| (1250 + 50j)$$
$$= 250 + 50j + 469 + 606j = 973.3\angle 42.4^\circ \ \Omega$$

Thus, the positive sequence current

$$I_{a1} = \frac{V_{a1}}{Z_1} = \frac{179\angle 13.44^\circ}{973.7\angle 42.4^\circ} = 0.1839\angle -28.96^\circ \text{ A}$$

$$I_{12} = \frac{I_{a1} \times 766.3\angle 43.3^\circ}{1250 + 50j} = 0.11\angle 17.34^\circ \text{ A}$$

Positive sequence torque is

$$T_1 = 2\left(I_{21}^{/}\right)^2 \frac{R_2^{/}}{s} = 2 \times 0.11^2 \times \frac{750}{0.6} = 30.25 \text{ Nm}$$

Negative sequence impedance is

$$Z_2 = R_1 + jX_1 + jX_m \left\| \left(\frac{R_2^{/}}{2-s} + jX_2^{/} \right) = 250 + 50j + 1000j \right\| (535.7 + 50j)$$

$$= 250 + 50j + 388.4 + j240.1 = 701.2\angle 24.4^\circ \ \Omega$$

Thus, the negative sequence current

$$I_{a2} = \frac{V_{a2}}{Z_2} = \frac{58.6\angle - 44.2^\circ}{701.2\angle 24.4^\circ} = 0.084\angle - 68.6^\circ \text{ A}$$

$$I_{22} = \frac{I_{a2} \times 700.4\angle 24.82^\circ}{535.7 + 50j} \approx 0.1098\angle - 44.2^\circ \text{ A}$$

Negative sequence torque is

$$T_2 = 2\left(I_{22}^{/}\right)^2 \frac{R_2^{/}}{2-s} = 2 \times 0.1098^2 \times \frac{750}{1.4} = 12.91 \text{ Nm}$$

The resultant torque can be given as

$$T = T_1 - T_2 = 30.25 - 12.91 = 17.34 \text{ Nm}$$

The resultant line currents are

$$\begin{aligned} \boldsymbol{I}_a &= \boldsymbol{I}_{a1} + \boldsymbol{I}_{a2} = 0.1804\angle - 25.96^\circ + 0.084\angle - 69^\circ \\ &= 0.21 - 0.16j \text{ A} \end{aligned}$$

$$\boldsymbol{I}_b = -j\boldsymbol{I}_{a1} + j\boldsymbol{I}_{a2} = 0.001 - 0.13j = 0.13\angle 89.6^\circ \text{ A}$$

Example 7.4 A two-phase, two pole AC drag-cup servo motor has the following parameters:

$$R_1 = 350\Omega \quad R_2^{/} = 250\Omega \quad X_1 = 60\Omega \quad X_2^{/} = 50\Omega \quad X_m = 900\Omega$$

The main winding and control winding are identical, and each is rated for 220 V. The voltages are in exact quadrature. When operating at a slip of 0.3, determine the resultant torque in synchronous Watt when control voltage is 80% of main winding voltage.

Solution

The ratio of control winding and main winding voltage is $\rho = 0.8$. Thus the positive sequence and negative sequence voltages are

$$\boldsymbol{V}_{a1} = \frac{\boldsymbol{V}_a}{2}(1+\rho) = \frac{220}{2}(1+0.8) = 198 \text{ V}$$

The negative sequence voltage is

$$\boldsymbol{V}_{a2} = \frac{\boldsymbol{V}_a}{2}(1-\rho) = \frac{220}{2}(1-0.8) = 22 \text{ V}$$

Positive sequence impedance is

$$Z_1 = R_1 + jX_1 + jX_m \left\| \left(\frac{R_2^/}{s} + jX_2^/ \right) \right. = 350 + 60j + 900j \| (833.33 + 50j)$$

$$= 350 + 60j + 594.6\angle 45.03^\circ = 907.9\angle 32^\circ \ \Omega$$

Thus, the positive sequence current can be given as

$$I_{a1} = \frac{V_{a1}}{Z_1} = \frac{198}{907.9\angle 32^\circ} = 0.2180\angle - 32^\circ \text{ A}$$

$$I_{12} = \frac{I_{a1} \times 594.6\angle 45.3^\circ}{833.33 + 50j} = 0.1553\angle 9.6^\circ \text{ A}$$

Positive sequence torque is

$$T_1 = 2\left(I_{21}^/\right)^2 \frac{R_2^/}{s} = 2 \times 0.1543^2 \times \frac{250}{0.3} = 39.68 \text{ Nm}$$

Negative sequence impedance is

$$Z_2 = R_1 + jX_1 + jX_m \left\| \left(\frac{R_2^/}{2-s} + jX_2^/ \right) \right. = 350 + 60j + 900j \| (147 + 50j)$$

$$= 350 + 60j + 145.4\angle 27.6^\circ = 478.85 + 127.4j = 495.4\angle 14.89^\circ \ \Omega$$

Thus the negative sequence current can be given as

$$I_{a2} = \frac{V_{a2}}{Z_2} = \frac{22}{495.4\angle 14.89^\circ} = 0.044\angle - 14.89^\circ \text{ A}$$

$$I_{22} = \frac{I_{a2} \times 145.4\angle 27.6^\circ}{147 + 50j} = \frac{0.044\angle - 14.89^\circ \times 145.4\angle 27.6^\circ}{147} 0.043\angle 9.71^\circ \text{ A}$$

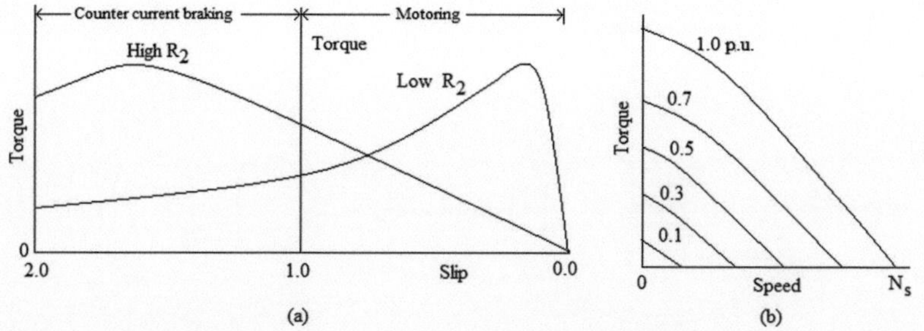

Fig. 7.9 **a** Torque slip characteristics of low and high resistance rotor and **b** torque speed characteristics of servo motor at different control voltages

Negative sequence torque is

$$T_2 = 2\left(I_{22}^{/}\right)^2 \frac{R_2^{/}}{2-s} = 2 \times 0.043^2 \times \frac{250}{1.7} = 0.54 \text{ Nm}$$

The resultant torque is

$$T = T_1 - T_2 = 39.68 - 0.54 = 39.14 \text{ Nm}$$

It can be shown that the resultant torque in two-phase servo motor supplied by control voltage which in quadrature to the main voltage is

$$T = \left(\frac{V_{a1}}{V_a}\right)^2 T_{B1} - \left(\frac{V_{a2}}{V_a}\right)^2 T_{B2}$$

where T_{B1} is the positive sequence motoring torque at slip s, when voltages are balanced. Similarly, T_{B2} is the negative sequence or braking torque at slip 2-s, when voltages are balanced.

Figure 7.9a shows the torque-slip characteristics of low rotor resistance and high rotor resistance of two-phase induction motor supplied by balanced two-phase voltages. The effect of different control voltages on torque speed characteristics of servo motor is been shown in Fig. 7.9b. The motoring slip is from 1.0 to 0.0 whereas counter current braking slip is from 2.0 to 1.0. High rotor resistance torque speed characteristics have no unstable zone of operation and torque is linear.

7.2.4 Synchro

These are alternating current electromagnetic rotating devices used in control applications. These devices are widely used in servo applications. They are capable of remotely providing the signals of position of a moving shaft. These are of following types [6]:

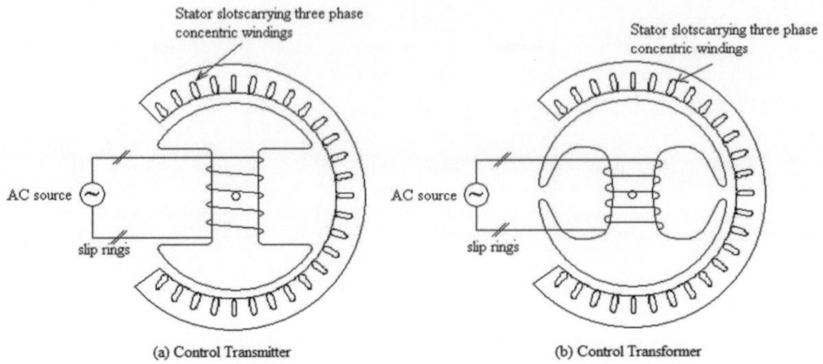

Fig. 7.10 **a** Control transmitter and receiver, **b** control transformer

(a) Control transmitters (CX).
(b) Control transformer (CT).
(c) Control receiver (CR).
(d) Control differential (CD).

Similar to electrical machines, it consists of a stator and a rotor. The stator is having slots on inner periphery which is carrying a three-phase winding. The three phase windings are made either with distributed windings or with concentric windings. The rotor is of dumbbell shape, as shown in Fig. 7.10a, carrying several turns on it, the end terminals are connected to a pair of slip rings mounted on the shaft.

An alternating current (AC) voltage is connected to the rotor windings, and the flow of excitation current on the rotor produces the necessary excitation flux. The electrical zero is defined when the rotor winding is aligned toward stator winding S2 giving maximum coupling between the rotor and the stator S2 winding. Maximum voltage is induced in the S2 winding. The induced voltage in the stator winding changes when the rotor axis is displaced by an angle. The change in induced EMF of control transmitter is an indication of exact position of the rotor. The rotor windings of CX and CT are excited by single phase. The field flux distribution due to the single phase AC current in rotor interacts with the ampere conductor distribution of the stator for producing torque and hence rotation.

The construction of synchro is similar to salient pole synchronous machines. However, the following are essential difference between the two: (i) the synchro rotor is excited by alternating voltage whereas in alternator it is excited by dc voltage (ii) A synchro is moved for a finite angle whereas an alternator serves its purpose when continuously rotating at synchronous speed.

A control transformer has got a salient pole construction, but the air gap is made almost uniform. The rotor has salient pole construction, but its shape looks like an umbrella, as shown in Fig. 7.10b. Such a shape minimizes the air gap which helps in reducing the magnetizing current. It has higher impedance per phase as compared to the transmitter. A differential synchro has uniform air gap. It has balanced three phase distributed windings both on stator and rotor.

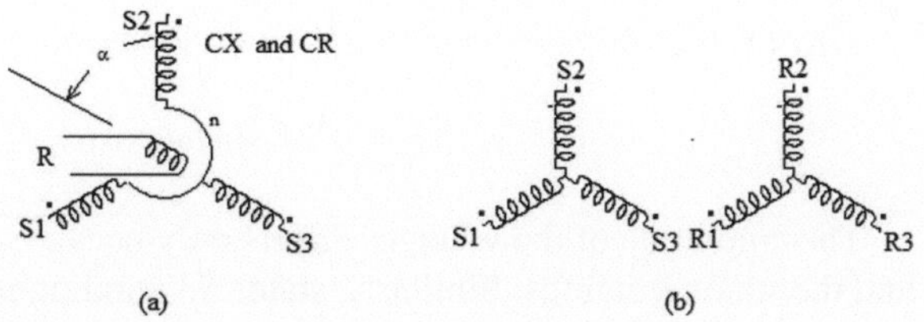

Fig. 7.11 Schematic representation of **a** synchro transmitter and receiver, **b** differential synchro

A synchro receiver is similar to that of control transmitter. It has mechanical viscous damping on its shaft. The damper permits the rotor to respond to the changing stator ampere conductor distribution without causing the rotor to overshoot.

Voltage Relationship: The schematic representation of control transmitter/receiver and control transformer are shown in Fig. 7.10a, b, respectively. In synchro, the voltage varies according to the rotor position. The rotor winding is excited by the alternating current source. Due to transformer action the voltages are induced in the stator windings. The value of induced EMF depends on the coupling. The schematic representation of synchro has been shown in Fig. 7.11. The electrical zero corresponds to position of maximum coupling between the rotor winding and the S2 on stator winding. Let us assume that the rotor is at an angle α in anti-clockwise direction from electrical zero.

The turn ratio is,

$$a - \frac{\text{Effetive turns on the stator}}{\text{Effective turns on the rotor}} \tag{7.21}$$

The RMS value of induced EMF in rotor winding is $\overline{E_r}$, and the RMS value of induced EMF in stator winding is $\overline{E} = a\overline{E_r}$, when the rotor winding is at an angle α from winding S2 which yields the induced EMF on the stator is

$$\overline{E}_{2n} = \overline{E}\cos\alpha \tag{7.22}$$

When angle α is zero, rotor is fully linked to the stator winding $S2$ which yields maximum induced voltage in the stator winding. The instantaneous voltage is given as

$$e_{2n}(t) = \sqrt{2}E\sin\omega t\cos\alpha \tag{7.23}$$

Here ω is the rotational frequency of supply. In the above expression, the term sin ωt is due to the transformer action and the term cosα is due to the coupling between the rotor winding and the stator $S2$ winding.

Since the axis of $S1$ winding is located 120° ahead of the axis of $S2$ winding, the value of RMS voltage induced in $S1$ stator winding is given by

$$\overline{E}_{1n} = \overline{E}\cos(\alpha - 120^\circ) \tag{7.24}$$

The instantaneous voltage of $S1$ winding is given as

$$e_{1n}(t) = \sqrt{2}E \sin \omega t \cos(\alpha - 120^\circ) \tag{7.25}$$

The amplitude of the voltages differs only because of coupling between the rotor and the stator windings. Similarly, stator $S3$ winding is located 120° behind the axis of $S2$, the EMF induced in $S3$ winding is

$$\overline{E}_{3n} = \overline{E}\cos(\alpha + 120^\circ) \tag{7.26}$$

The instantaneous voltage of $S3$ winding is given as

$$e_{3n}(t) = \sqrt{2}E \sin \omega t \cos(\alpha + 120^\circ) \tag{7.27}$$

The above equation describes the induced voltages on the stator phase windings. The terminals voltage between 1 and 2 is given as

$$\overline{E}_{12} = \overline{E}_{1n} + \overline{E}_{n2} = \overline{E}_{1n} - \overline{E}_{2n} \tag{7.28}$$

$$\overline{E}_{12} = \left\{-\frac{1}{2}\cos\alpha + \frac{\sqrt{3}}{2}\sin\alpha\right\}\overline{E} = \sqrt{3}\overline{E}\cos(\alpha - 150^\circ) \tag{7.29}$$

Similarly, we can show that line voltages are

$$\overline{E}_{23} = \left\{\frac{1}{2}\cos\alpha + \frac{\sqrt{3}}{2}\sin\alpha\right\}\overline{E} = \sqrt{3}\overline{E}\cos(\alpha - 30^\circ)$$

$$\overline{E}_{31} = \sqrt{3}\overline{E}\cos(\alpha + 90^\circ) \tag{7.30}$$

We know that the arithmetic sum of any two-line voltage is equal to the third. The variation of line voltages is affected by the displacement of the rotor. If these voltages are fed to the stator of a control transformer, a received or a differential synchro, it produces necessary changes in the end devices.

Applications: A typical application of remotely sensing the rotor position is shown in Fig. 7.12. It utilizes a control transmitter connected to a shaft, whose position is to be sensed, and a control receiver. The control receiver is remotely located at a large distance from the main shaft. However, its stator is electrically connected to the control transmitter. The rotor of control receiver aligns itself accordingly to the EMF induced in the control transmitter. The rotor of control transmitter is excited by an AC source. The rotor of control receiver moves to an angle until electrical null is attained. The EMF induced in the rotor of control receiver is $\overline{E}_r \sin\alpha$, where α is the angle of alignment.

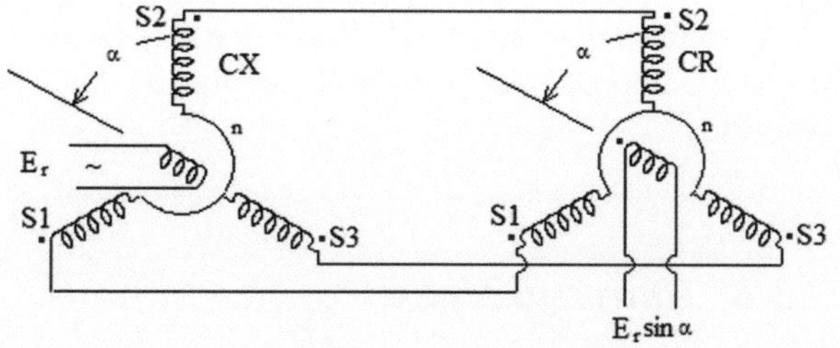

Fig. 7.12 Remote sensing of main shaft position using control transmitter (CX) and control receiver (CR)

The applications of synchro are in control system area. Some of the applications are as follows:

(a) Remotely rotor position indication.
(b) Servo mechanism–error detection.
(c) Torque transmitter.
(d) Addition of angle signal to toque transmitting system.
(e) Error generation for difference in two shaft positions.

These are control machines used in industries for controlling various processes remotely. For example, an operator will know remotely the exact position of different shafts and accordingly decides the next control step, either manually or electronically. These machines were also used for torque transmitters (say), for example, control of robotic arms, process control and power steering of a large vehicle. Nowadays, the application has been replaced by BLDC motors. Similarly, for controlling the radar we require large torque; but by using torque transmitter, we can control it with little efforts.

7.3 AC Commutator Motors

In the past, AC commutator motors were widely used in variety of industrial and transport applications. These motors were the prime source of large motive power in train propulsion and ship propulsion. However, due to the advent of solid state inverter driven induction motors, these motors are not being currently preferred for newer applications, owing to the higher cost of commutator, sparking and cost of maintenance of DC machines. The industrial applications working on these motors are continue to run as usual till they are finally replaced by newer system. In the drive applications, the vector controlled or VVVF control of squirrel cage induction motors are preferred, where total cost of the system is not a criterion of selection. The basic concept of certain solid state induction motor like injection of slip frequency EMF in slip ring induction motor- relies on the theoretical back ground slip frequency EMF generation in the Schrage motor. Therefore, it is essential to discuss AC commutator motors. Following are some important AC commutator motors [2–4, 7, 12]:

(a) Universal AC/DC motor.
(b) Schrage motor.

Apart from these motors, the machine such as rotary converter converts the three-phase AC into DC and vice versa may find its application in unconventional energy resourses.

7.3.1 Universal Motor

Consider the operation of a DC shunt motor across an AC supply. In case of a DC shunt motor, the inductance of the field winding is large as compared to that of armature winding. This results in difference in the phase angles between armature and field winding currents. Here we assume that there is no magnetic saturation. Thus, when the alternating field flux is at its peak value, the alternating armature current may not be at its peak value. The torque produced is very feeble in spite of large AC current. The DC shunt motor is not suitable to work on AC supply. However, when a DC series motor is connected across an AC supply, both the armature and field windings have similar circuit parameters and carry similar current. As such, when alternating pole-flux is at its peak value, the current through the armature conductors also reaches its peak value. When the alternating pole-flux reverses, the armature current also reverses after each half cycle of AC supply. The torque produced is always unidirectional, and is comparable with the DC operation. In the case of DC operation, there is no inductive coupling between the two windings. However in AC operation, the resultant current is smaller, the power factor is very poor and, there are excessive losses in the solid poles and yoke.

In order to limit the core loss, the magnetic circuit of the AC/DC motors is laminated. The reactance of stator and rotor windings are reduced to minimum. The armature inductance is reduced with the help of compensating windings. There two types of compensating windings, namely, *conductively coupled* and *inductively coupled*. The conductively coupled compensating winding can act with both AC & DC supply, whereas the inductive coupled compensating winding is effective only in AC supply. Figures 7.13 and 7.14 show AC series motors with conductively and inductively compensating windings, respectively. The function of compensating winding is to oppose the flux produced in the armature winding due to loading. The inductively coupled compensating winding acts as a short-circuited secondary winding of a transformer.

The effect of compensating winding is to create the flux in q-axis ϕ_c for neutralizing the armature flux ϕ_q, produced due the armature current. This minimizes the armature leakage flux (proportional to $\phi_q - \phi_c$) and, hence armature inductance. The net effect is the increase in the speed, and rotational-induced EMF $\boldsymbol{E}_r$, improvement of power factor and efficiency.

A fractional horse power AC series motor is a simple DC series motor without any compensating winding. These motors can run on AC as well as DC supply, and is known as Universal motor. A plain DC series motor has been shown in Fig. 7.15. They are built in sizes of up to 0.75 kW and are mainly used in domestic appliances, such as food processors, mixer, juicer, grinder, hair drier and vacuum cleaner.

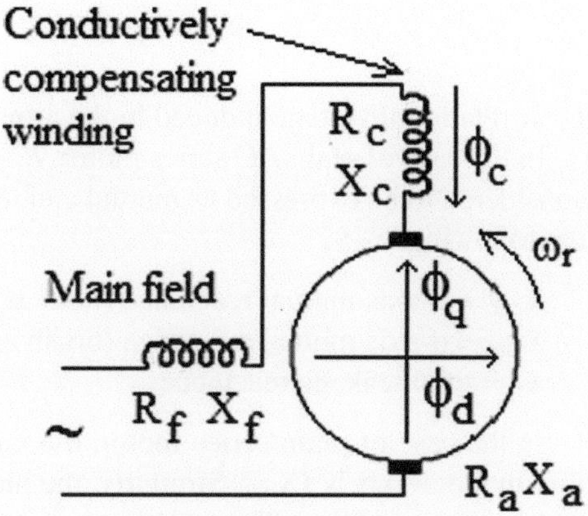

Fig. 7.13 Conductively compensated AC series motor

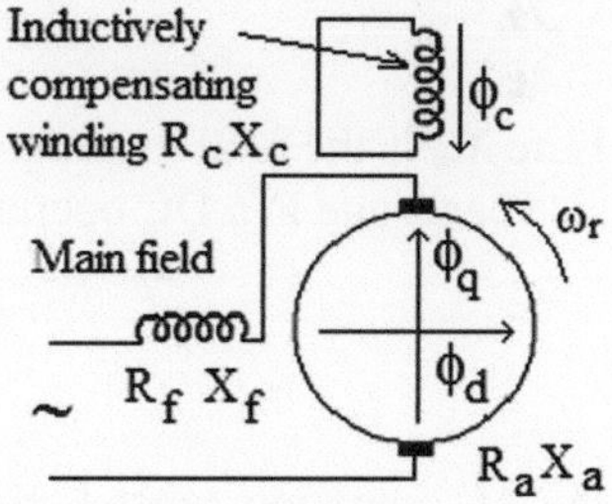

Fig. 7.14 Inductively compensated AC series motor

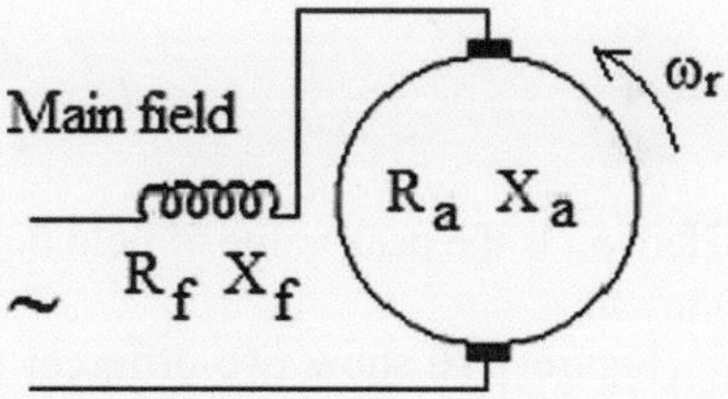

Fig. 7.15 Plain AC series motor

Let R_f and X_f are the resistance and reactance of the field winding, respectively. R_a and X_a are the resistance and leakage reactance of the armature winding, respectively, while R_c and X_c are the resistance and leakage reactance of the compensating winding referred to armature side, respectively. Now, the voltage equation is

$$V = IR + jIX + E_r, \tag{7.31}$$

where

$$X = X_f + X_a + X_c$$

$$R = R_f + R_a + R_c$$

E_r = rotational voltage induced in the armature winding.

In the case of plain AC series motor, $R_c = 0$ and $X_c = 0$. The total reactance X is considered to be expressed as mutual and leakage reactance. Here, we further define the following:

X_{md} = d-axis mutual reactance (field axis).
X_{mq} = q-axis mutual reactance (brush axis).
X_{l} = total leakage reactance.

In the case of plain series motor, the induced voltage in the field due to air gap flux in the d-axis is $\mathbf{I}X_{\text{md}}$. Similarly, the induced voltage in armature due to air gap flux in q-axis is $\mathbf{I}X_{\text{mq}}$. The total reactance can be given as

$$X = X_{\text{md}} + X_{\text{mq}} + X_l = X_m + X_l$$

Here X_m is the total mutual reactance. For fully compensated motor the $X_{\text{mq}} = 0.0$.

The torque in a DC operation is given as

$$T = \frac{\phi ZI}{2\pi} \times \frac{p}{a} \tag{7.32}$$

For AC operation, when both the current and field flux are alternating at a frequency f Hz, the average torque is,

$$T = \frac{\phi_m ZI}{2\sqrt{2}\pi} \times \frac{p}{a} \times \cos\varphi \tag{7.33}$$

Here ϕ_m is the peak value of field flux and φ is the phase difference between flux and current.

Figure 7.16 show two different fluxes due to field poles ϕ and due to compensating winding $\phi^{/}$. These are responsible, respectively, for rotational induced EMF and transformer-induced EMF in AC commutator machines. The phasor diagram corresponding to Eq. (7.31) is shown in Fig. 7.17 and 7.18, respectively, without and with core loss.

Rotational voltage in the DC machine running at speed of N RPM is

$$E = \frac{\phi ZN}{60} \times \frac{p}{a} \tag{7.34}$$

In case of AC operation, the rotational voltage is

$$E_r = \frac{\phi_m ZN}{60\sqrt{2}} \times \frac{p}{a} \tag{7.35}$$

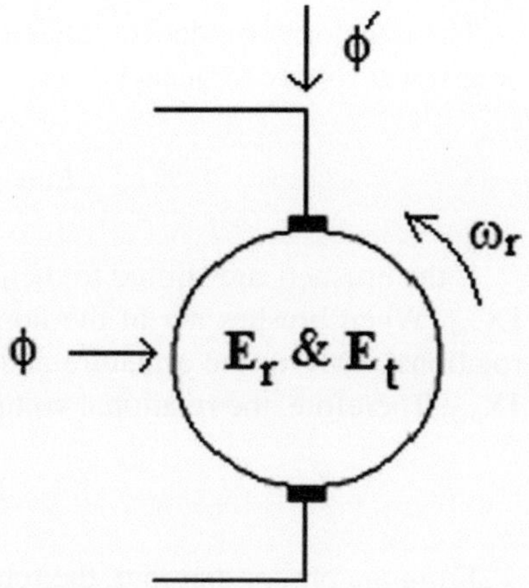

Fig. 7.16 Rotational and transformer-induced voltage in armature

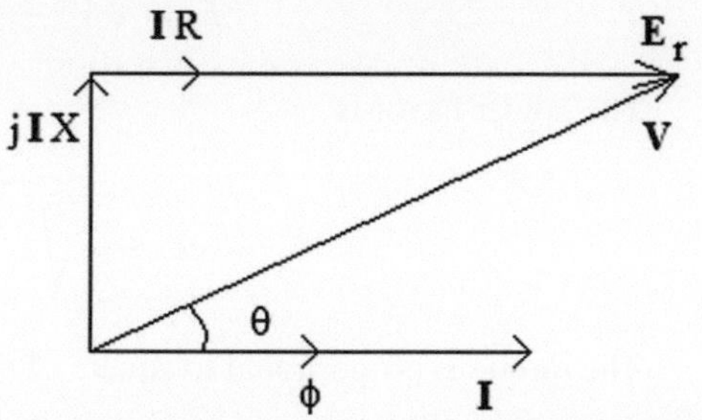

Fig. 7.17 Phasor diagram of AC series motor

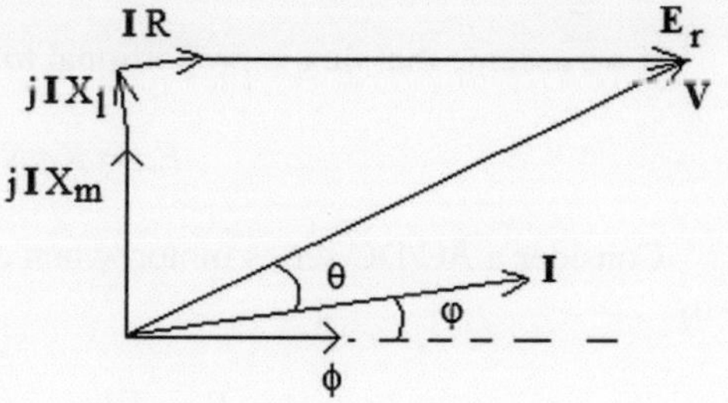

Fig. 7.18 Phasor diagram of AC series motor with core loss

The rotational frequency is defined as

$$f_r = \frac{Np}{120}$$

Thus, the rotational voltage in AC is,

$$E_r = \phi_m f_r \times \frac{Z}{a} \times \sqrt{2} \tag{7.36}$$

The transformer-induced voltage in the armature conductors due to flux $\phi_m^/$ can be given at supply frequency f as

$$E_t = \phi_m^/ f \times \frac{Z}{a} \times \sqrt{2} \tag{7.37}$$

If the brushers are shifted to the pole axis, the transformer-induced EMF will be $\mathbf{IX}_{\text{md}}$. When brushes are in the normal position, the same pole flux produces the rotational EMF in the armature and its magnitude will be $S = \frac{f_r}{f} = \frac{N}{N_s}$ times the $\mathbf{IX}_{\text{md}}$. Therefore, the rotational voltage at different speeds is,

$$E_r = S\text{IX}_{\text{md}} \tag{7.38}$$

From the phasor diagram, the rotational voltage is

$$E_r = \sqrt{V^2 - I^2X^2} - \text{IR} \tag{7.39}$$

The power factor is

$$\cos\theta = \sqrt{1 - \left(\frac{\text{IX}}{V}\right)^2} \tag{7.40}$$

The torque is proportional to square of current, i.e.,

$$T = K_T I^2 \tag{7.41}$$

If we assume that flux is proportional to current I, then, we get

$$E_r = K\phi N = KIN$$

Consider a AC/DC series motor, when excited by AC source, the speed is given by

$$N = \frac{E_r - \text{IR}}{\text{KI}} = \frac{\sqrt{V^2 - I^2X^2} - \text{IR}}{\text{KI}} \tag{7.42}$$

The series compensated motor can run on AC as well as DC. Let the motor runs at speed N_{DC}, the induced EMF will be

$$E_{\text{DC}} = \frac{\phi Z N_{\text{DC}}}{60} \times \frac{p}{a}$$

Similarly, when the motor runs at speed N_{AC} with AC voltage, the induced EMF will be

$$E_{\mathrm{AC}} = \frac{\phi_m Z N_{\mathrm{AC}}}{60\sqrt{2}} \times \frac{p}{a}$$

Here, the DC flux ϕ is produced by current I. Similarly the peak value of AC flux ϕ_m is produced by peak current $I\sqrt{2}$. Thus

$$\frac{E_{\mathrm{AC}}}{E_{\mathrm{DC}}} = \frac{N_{\mathrm{AC}}}{N_{\mathrm{DC}}} \tag{7.43}$$

$$\frac{E_{\mathrm{AC}}}{E_{\mathrm{DC}}} = \frac{N_{\mathrm{AC}}}{N_{\mathrm{DC}}} = \frac{V\cos\theta - IR}{V - IR} = \frac{\cos\theta - \left(\frac{IR}{V}\right)}{1 - \left(\frac{IR}{V}\right)} \approx \cos\theta \tag{7.44}$$

Since, the ratio $\left(\frac{IR}{V}\right)$ is very small. Thus the speed during AC operation will be lesser than DC operation.

The mechanical power developed is $E_r I$. If the rotational losses are known, then we have

$$\text{mechanical output power} = E_r I - \text{Rotational loss}$$

The torque speed characteristics for AC/DC operation have been given in Fig. 7.19. Figure 7.20 shows the characteristic of AC series motor. The power factor and efficiency drop upon loading.

FHP plain AC/DC series motors are having salient poles similar to that of DC machines. However, the stator and rotor are made using laminated stampings. The speed is controlled by tapped field control.

The large size AC/DC series motors with compensating windings are having cylindrical stator stamping in which there are regularly distributed slots are separately

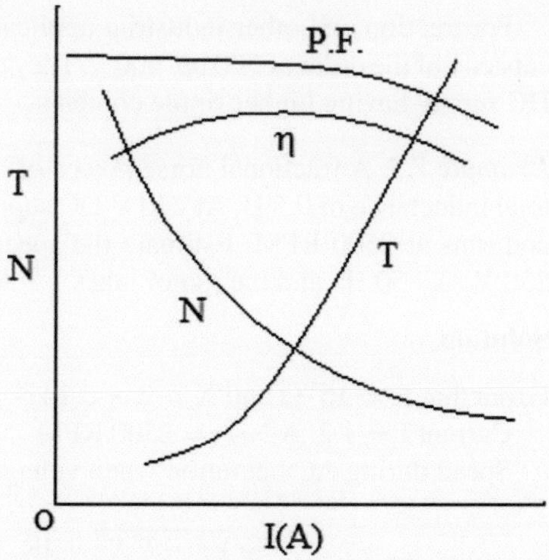

Fig. 7.19 Characteristics of AC series motor

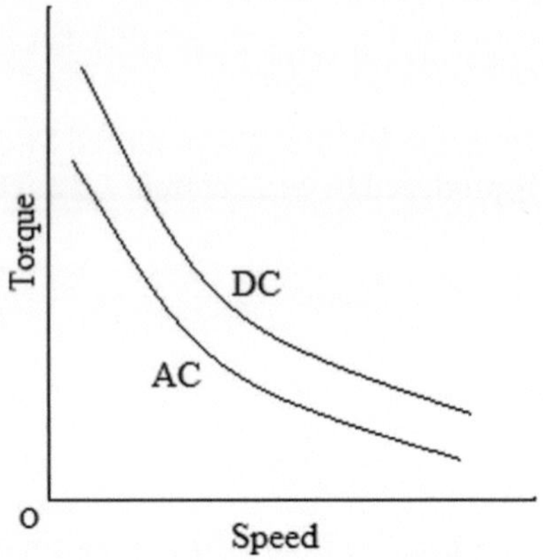

Fig. 7.20 Torque speed characteristics of Universal motor on AC & DC operation

provided for field winding and for compensating windings. The speed control is done using either by voltage control or by shunting the field windings.

Following are the advantages of AC/DC series motors:

(a) They have high operation speed, ranging from 3000 to 25,000 RPM. This makes them suitable for compressors, vacuum cleaners, portable drilling tools and industrial sewing machine.
(b) They provide high torque at lower speed.
(c) These machines have high starting torque suitable for traction applications.

However, these machines have demerits as well, some of which are as follows:

(a) They require High maintenance.
(b) They produce higher noise level at high-speed operation.
(c) The have RI or Radio interference, interfering with other electronic equipment

For traction and other industrial applications, these machines are made in large capacity of the order of 500 hp, 600 V. They can operate on AC/DC supply or rectified DC supply having higher ripple contents.

Example 7.5 A fractional horse power AC series motor has resistance of 15 Ω and total inductance of 0.5 H. At 230 V DC supply voltage the motor takes 1.2 A current and runs at 2500 RPM. Estimate the speed, power factor when supply voltage is 230 V AC, 50 Hz and the motor takes 1.2 A of current.

Solution

Given that R = 15 Ω and X = 2 × 3.14 × 50 × 0.5 = 157 Ω.

Current I = 1.2 A N_{DC} = 2500 RPM.

Speed during AC operation when voltage and current are same

$$N_{AC} = N_{DC}\frac{\sqrt{V^2 - I^2X^2} - \mathrm{IR}}{V - \mathrm{IR}}$$

$$= 2500 \frac{\sqrt{230^2 - 1.2^2 \times 157^2} - 1.2 \times 15}{230 - 1.2 \times 15} = 1344 \text{ RPM}$$

Power factor is

$$\cos\theta = \sqrt{1 - \left(\frac{\text{IX}}{V}\right)^2} = \sqrt{1 - \left(\frac{1.2 \times 157}{230}\right)^2} = 0.5736 \text{ lagging}$$

Example 7.6 A 115 V, 250 W, 5000 RPM, 50 Hz AC series motor has the total resistance of 1.4 Ω and total reactance of 12 Ω. The rotational losses are 18W. Determine (a) Current and power factor and developed torque at rated condition. (b) Speed, power factor, developed torque and efficiency when the load current is reduced to half.

Solution

(a) The mechanical power developed is

$$P_d = 250 + 18 = 268 \text{ W}$$

We know that $P_d = E_r I$, thus the rotational voltage is

$$E_r = \frac{P_d}{I} = \frac{268}{I}$$

From the phasor diagram shown in Fig. 7.17, we have

$$V^2 = (E_r + IR)^2 + (IX)^2$$

$$115^2 = \left(\frac{268}{I} + 1.41I\right)^2 + (12I)^2$$

$$13225 \times I^2 = 71824 + 1.9881 \times I^4 - 755.76 \times I^2 + 144 \times I^2$$

Solving this we get,

$$1.9881 \times I^4 - 13980.76 \times I^2 + 71824 = 0$$

$$I^4 - 7032 \times I^2 + 36127 = 0$$

Thus I = 2.28 A or 83.8 A. Ignoring the larger value of current.
Power factor is

$$\cos\theta = \sqrt{1 - \left(\frac{IX}{V}\right)^2} = \sqrt{1 - \left(\frac{2.28 \times 12}{115}\right)^2} = 0.971 \text{ lagging}$$

The rotational voltage is

$$E_r = \sqrt{V^2 - I^2X^2} - IR = \sqrt{115^2 - 2.26^2 \times 12^2} - 2.26 \times 1.4 = 108.5\ \text{V}$$

Developed Torque is

$$T = \frac{E_r \times I}{(2\pi N/60)} = \frac{108.5 \times 2.28}{(2\pi \times 5000/60)} = 0.47\ \text{Nm}$$

(b) When current is halved, I = 1.14A. The new rotational voltage is

$$E_r = \sqrt{V^2 - I^2X^2} - \text{IR}$$
$$= \sqrt{115^2 - 1.14^2 \times 12^2} - 1.14 \times 1.4 = 112.58\ \text{V}$$

The rotational voltage is proportional to product of speed and current, i.e.,

$$\frac{E_{r2}}{E_{r1}} = \frac{I_2 N_2}{I_1 N_1}$$

Thus new speed is

$$N_2 = N_1 \frac{E_{r2}}{E_{r1}} \left(\frac{I_2}{I_1}\right) = 5000 \frac{112.58}{108.5} \times \frac{2.28}{1.14} = 10376\ \text{RPM}$$

Power factor is

$$\cos\theta = \sqrt{1 - \left(\frac{IX}{V}\right)^2} = \sqrt{1 - \left(\frac{1.14 \times 12}{115}\right)^2} = 0.9929\ \text{lagging}$$

Developed torque is

$$T = \frac{E_r \times I}{(2\pi N/60)} = \frac{108.5 \times 1.14}{(2\pi \times 10376/60)} = 0.12\ \text{Nm}$$

Efficiency is given as

$$\eta = \frac{\text{Output}}{\text{Input}} = \frac{E_r I}{VI\cos\theta} = \frac{112.58 \times 1.14}{115 \times 1.14 \times 0.9929} = 98.6\%$$

7.3.2 Commutator as Frequency Changer

In two-pole DC machine, the coil voltage goes through a complete cycle each time the coil makes one complete revolution with respect to the brushes. The magnitude of voltage at the brushes is proportional to the angular velocity of the coil ω_r and the flux density $B(\theta)$ produced by stationary field poles. In other words, the frequency of variation of voltage at brushes depends on the velocity of coil respective to the brushes. The magnitude of this voltage depends on the velocity of the rotor coil respective to the air gap field. If the air gap field is traveling in space at a speed of ω_s, the magnitude of brush voltage V_b can be expressed as

$$|V_b| = k(\omega_r - \omega_s) \tag{7.45}$$

where k is a constant and ω_r is the mechanical velocity of the rotor coil. If we assume the brushes to be stationary, the frequency of brush voltage is,

$$\omega = (\omega_r - \omega_s) \tag{7.46}$$

A commutator is a kind of frequency changer. Let us suppose that the rotor is excited by two-phase (or three-phase) currents through slip rings. If the frequency of these currents is ω_s, the resulting air gap field is rotating magnetic field having velocity ω_s relative to stationary rotor. If the rotor runs with a velocity ω_r, the resultant velocity of the rotating field as seen from the brushes is $(\omega_s - \omega_r)$. Thus, the frequency of brush voltage is $(\omega_s - \omega_r)$, which corresponds to slip frequency. By changing the rotor speed, the frequency across the brushes can be changed.

The scheme is used in induction motors requiring slip frequency EMF injection, like Schrage motor. Due to advent of solid state devices, it is possible to develop a three phase voltage and inject required EMF in the rotor circuit of slip-ring induction motor for controlling its speed.

7.3.3 Schrage Motor

In Schrage motor, the primary winding is on rotor, which carries the three phase mush winding in its slots. The rotating primary winding receives three phase power by slip ring and brush arrangements and runs opposite to the direction of rotation of rotating magnetic field. This is shown in Fig. 7.21. The slots on the rotor also have commutator winding, connected to the commutator on the other end of the shaft. A rotor slot of Schrage motor is shown in Fig. 7.22. The stator carries three secondary windings which collect the slip frequency injected EMF from three-pairs of brushes rocking on the commutator [1–4].

When the commutator brush pair is on the same commutator segment as shown in Fig. 7.23a, the injected EMF is zero and the secondary windings are short-circuited.

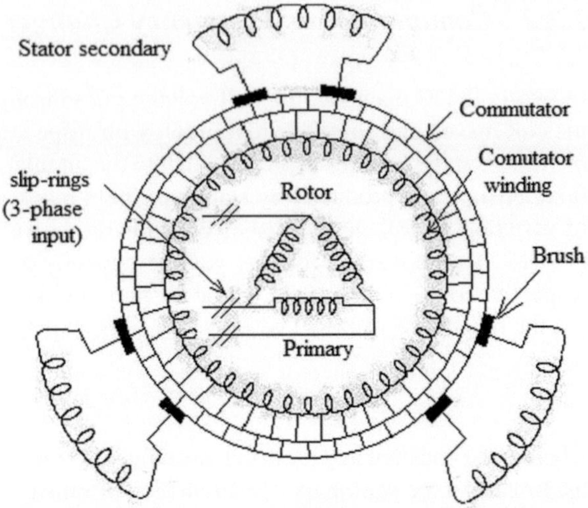

Fig. 7.21 Schematic diagram of Schrage motor

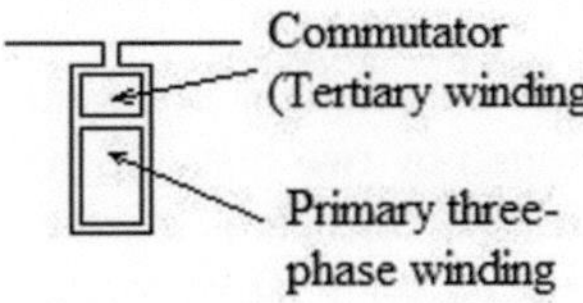

Fig. 7.22 Rotor slot of Schrage motor carrying primary winding and commutator winding conductors

The machine then works as an inverted three phase induction motor. The three phase currents in the primary windings set up a resultant rotating magnetic field rotating at speed N_s relative to the primary or rotor conductors. An induction motor runs in such a direction so as to tend to reduce the frequency of induced EMF set up in the secondary conductors. Accordingly, a rotor excited three phase induction motor runs in a direction opposite to the rotating magnetic field at speed N_r. The speed of rotating magnetic field with respect to stator surface is slip speed, given as N_s-N_r. This results in slip frequency induced EMF in the stator windings.

The rotating magnetic field pasts the brushes at slip frequency. If the brushes are separated, the EMF appearing on the brushes will always be of slip frequency, which is suitable for injection into the stator windings.

In Fig. 7.23a, when injected EMF is zero, the machines runs just below the synchronous speed and the ohmic loss due to current flowing in commutator winding is zero. Therefore, the machine gives highest efficiency in synchronous (like induction motor) operation of Schrage motor. The phasor $s\boldsymbol{E}_2$ is cophasal with $\boldsymbol{E}_2$ or secondary induced EMF at stationary condition.

In Fig. 7.23b, the brushes are separated symmetrically, the injected EMF is obtained from the section of commutator winding. The secondary induced EMF and injected EMF are in phase opposition, or $s\boldsymbol{E}_2 - \boldsymbol{E}_j = 0$. This gives a positive slip. This results in the sub-synchronous operation of Schrage motor.

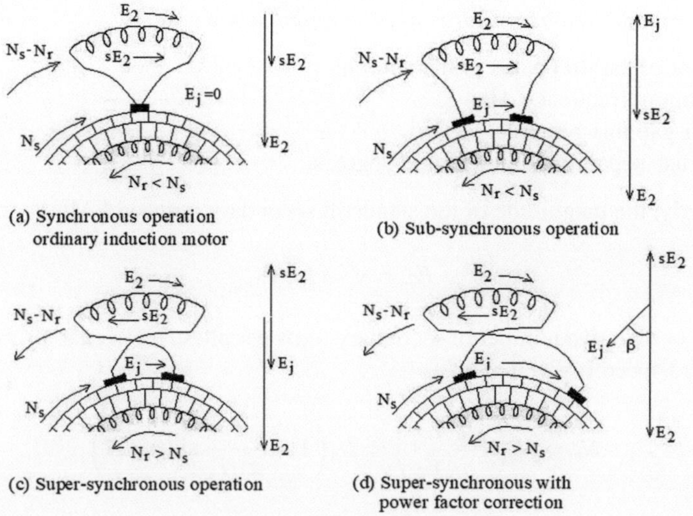

Fig. 7.23 Operation of Schrage motor for different injected EMF

In Fig. 7.23c, when the brushes are moved in opposite direction, the injected EMF and induced EMF are in phase addition. Thus $s\boldsymbol{E}_2 + \boldsymbol{E}_j = 0$, this gives a negative slip. This results in super-synchronous operation of Schrage motor. When the center line of brush axis has moved away from the axis of stator windings as shown in Fig. 7.23d. It results in super-synchronous speed, with power factor correction improvement.

The machine can be made to run in:

- Sub synchronous speed.
- Synchronous speed like ordinary Induction motor.
- Super-synchronous speed.
- Power factor correction and speed control.

The no-load speed is expressed as

$$N_r = N_s(1 - s) = N_s\left(1 - \frac{E_J}{E_2}\right)$$

The no-load slip is

$$s = \pm\frac{E_J}{E_2} + \frac{I_2 Z_{2s}}{E_2} \tag{7.47}$$

The injected EMF is

$$E_j = \sqrt{2}\left(\frac{T_a}{a}\right) f_1 \phi \sin(\theta/2), \tag{7.48}$$

where T_a = total number of turns in the commutator windings.

a = pair of parallel paths in the winding.
f_1 = supply frequency, Hz.
ϕ = air gap flux per pole.
θ = brush separation in electrical degrees.

Similarly, the magnitude of the standstill secondary induced EMF is given as

$$E_2 = \sqrt{2}\pi f_1 \phi T_2 \tag{7.49}$$

where T_2 = effective number of secondary turns per phase referred to primary. Thus the no-load speed is given as

$$N_r = N_s\left(1 - \frac{E_J}{E_2}\right) = N_s\left(1 - \frac{T_a}{\pi a T_2}\sin(\theta/2)\right)$$

However, T_2 and T_a are constants for a given motor, thus

$$N_r = N_s(1 - C\sin(\theta/2)), \tag{7.50}$$

where C is a constant. Thus by varying the brush separation angle, the shape of no-load speed curve with respect to brush separation angle can be changed. This has been shown in Fig. 7.24.

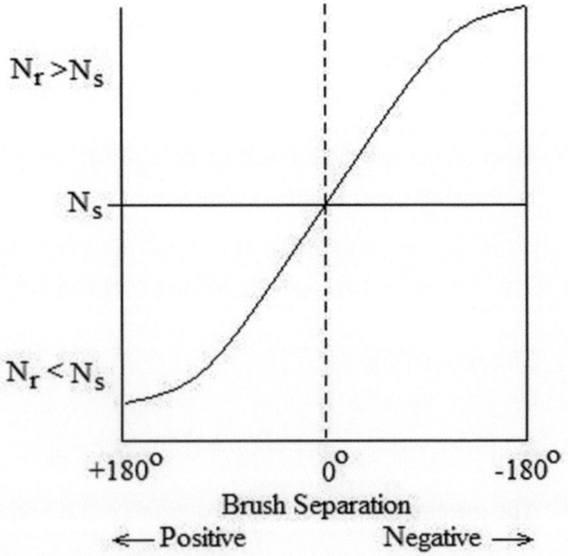

Fig. 7.24 Variation of no-load speed with respect to brush separation in Schrage motor

For a given brush separation or injected EMF E_j, at no-load the resultant voltage is zero. Thus, $s_0\boldsymbol{E}_2 = \boldsymbol{E}_j$ where s_0 is the no-load slip. At any speed the torque produced by motor is proportional to followings:

(a) Secondary current.
(b) Resultant secondary voltage.
(c) $s\boldsymbol{E}_2 - \boldsymbol{E}_j$, where s is the slip corresponding to loaded speed of the motor.

But we know that injected EMF is

$$\boldsymbol{E}_j = s_0\boldsymbol{E}_2$$

Thus, the torque is

$$T \;\; \alpha \;\; sE_2 - s_0E_2 \;\; \alpha \;\; (s - s_0) \tag{7.51}$$

In other words, the drop in speed of a Schrage motor is proportional to the load torque and is independent of the no-load speed corresponding to brush position. Thus, for a given brush separation, speed–torque characteristics are drooping characteristics, similar to that of DC shunt motor. The speed torque curves for different brush position will be almost parallel lines as shown in Fig. 7.25.

In case of ordinary induction motor, the (no-load) power factor is poor when machine is running on no-load. The power factor rises with increase in the load torque. This is due to the fact that with the increase in the load current, the secondary current I_2 increases. Therefore, the corresponding primary component of the load current I_2' also increases, and becomes more and more dominant over the no-load component of primary current I_0. The primary current lags behind the supply voltage

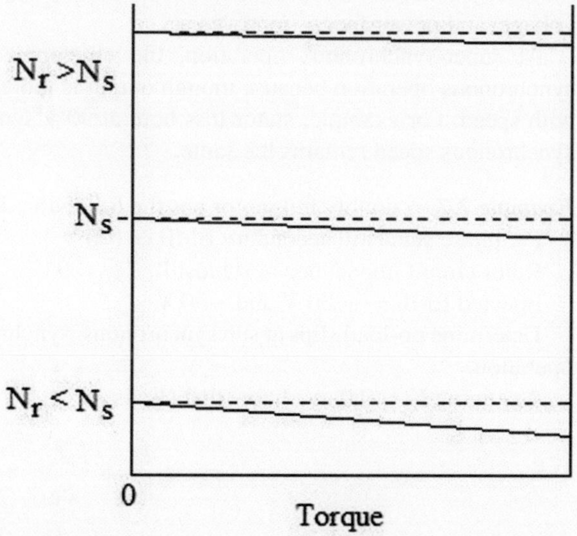

Fig. 7.25 Torque speed characteristics of Schrage motor for sub-synchronous, synchronous, and super-synchronous

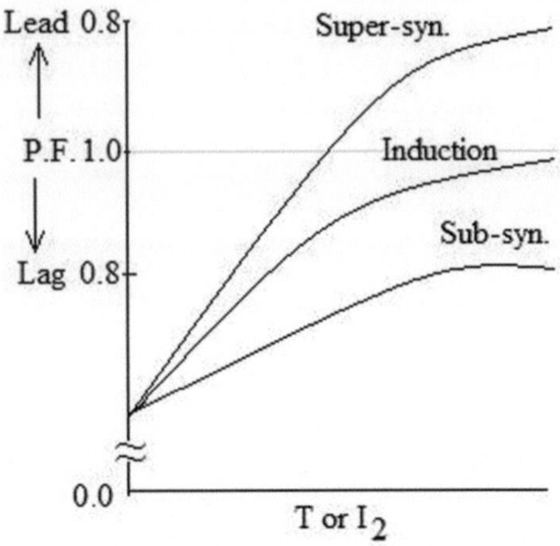

Fig. 7.26 Variation of power factor w.r.t to torque of a Schrage motor for different brush positions

mainly due the lagging component of I_0. At higher load, the no load current becomes more and more negligible in comparison to $I_2^{/}$. The overall power factor of the Schrage motor, therefore, increases with the load for a given speed of operation (or for a given angle of separation of brush). However, the speed of operation has a marked influence on the power factor of the motor, as can be visualized in Fig. 7.26.

At lower speed, the slip is higher and therefore the secondary reactance drop is larger. This results in poor power factor. At synchronous speed or ordinary induction motor operation, the slip is zero the secondary reactance vanishes. The secondary current is in phase with the resultant secondary voltage I_2Z_{2S}. As a result, the primary power factor slightly improves.

At super-synchronous speed of operation, the slip is negative. Therefore, the secondary reactance drop reverses. As a result, I_2 leads the resultant secondary voltage, and power factor improves.

At super-synchronous operation, the efficiency is slightly higher than sub synchronous operation because though output is more, certain losses do not change with speed. For example, stator loss both at 90% synchronous speed and at 110% synchronous speed remains the same.

Example 7.7 A doubly fed motor has the following details:

Per phase standstill secondary EMF = 100 V.

Rotor circuit impedance at standstill, $Z_{2S} = 0.025 + j0.08\ \Omega$.

Injected EMF = + 50 V and −50 V.

Determine no-load slips at sub synchronous, synchronous and super-synchronous operation.

Assuming $I_2 = 0.0$, we know that slip is

$$s = \pm\frac{E_j}{E_2} + \frac{I_2 Z_{2s}}{E_2}$$

When injected EMF is opposite to E_2

$$s = +\frac{50}{100} = 0.5$$

When injected EMF is in phase with E_2

$$s = -\frac{50}{100} = -0.5$$

It is possible to run the Schrage motor at 50% and 150% of synchronous speed, i.e., sub-synchronous and super speed zone, respectively.

Due to advancement in the solid state devices, it is possible to develop the three phase slip frequency EMF using solid state circuit and inject in the rotor circuit of a slip ring induction motor for variable speed operation in motoring and generating region.

7.4 Linear Motor

In 1905, Westinghouse Corporation of America developed the concept of a unique transportation system, in which they laid a three-phase flat winding with magnetic core between railways track. The lower part of bogie of the moving train was made of an aluminum plate facing the stationary linear winding between the railway track. For a continuous transport, the entire track was required to be laid with the linear induction motor having three-phase windings. However, it is not cost-effective. During the first half of the twentieth century, many patents came for applying Linear Induction Motor (LIM) in textile industries. In 1956, Prof E. R. Laithewaite of Imperial College, London suggested the modification of the experiment of Westinghouse Corporation of America. He proposed that the transportation using flat-type of motor is possible if the bogie of the train carried LIM primary, and the aluminum plates backed by iron was laid over the entire length of railways line. The technology to provide the electrical power to moving vehicle and also for providing on-board generation of power was also developed by that time. The research of Laithewaite gave a new dimension. The train can be made to run at very high speed, if the train is lifted in air with the help of electromagnetic levitation and run using LIM. The speed control required merely the variation of supply frequency, because once the moving vehicle is lifted in air, the friction between rail and wheel no longer limits the speed. In levitated transport system, wheels and rails are required only for primary suspension, so that in case of failure of levitation, the train can travel with primary suspension. Many countries, like U.S.A., U.K., Canada, Japan, Germany, France, China etc. are now actively engaged in LIM research. In Japan, the newer trains (tube-railways) are based on LIM technology as it reduces the overall height, thereby reducing the total cost of tunneling in underground railways as well as cost of ventilation and air-conditioning of the tunnel.

Every type of rotating machine may take a form of linear motor. The following are the types of linear motors [2, 7, 8]:

(a) Linear Induction motor.
(b) Linear Synchronous motor.
(c) Linear Reluctance motor.
(d) Linear DC motor.
(e) Linear Stepper motor.
(f) Linear Pulse motor.
(g) Linear Oscillating motor.

These motors are basically application -specific. Here, we will discuss only linear induction motors. The LIM may be of the following different types:

(a) Longitudinal flux LIM:
    (i) Single-sided LIM (SLIM).
    (ii) Double-sided LIM (DSLIM).
(b) Transverse Flux LIM- made using E-core or U-core stampings:
    (i) Single-sided TFLIM (STFLIM).
    (ii) Double-sided TFLIM (DTFLIM).
(c) Tubular LIM.
(d) Electromagnetic Levitators:
    (i) Attraction type.
    (ii) Repulsion type.

7.4.1 Linear Induction Motor

Topologically, LIM can be derived from rotary induction motor. The stator of rotary induction motor is cut by an imaginary plane and unrolled it in flat form of LIM primary. The rotor is replaced by a conducting plate as shown in Fig. 7.27a, b. A typical Single–sided linear induction motor (SLIM) (*windings not shown for clarity*) is shown in Fig. 7.28 in 2D. Thin aluminum sheet faces the active surface of SLIM. The aluminum sheet is backed by ferromagnetic back iron, this improves its performance. The width of aluminum sheet is larger than that of core. The transverse secondary (aluminum sheet) overhang provides path for return of secondary eddy current. A 3D picture of SLIM is shown in Fig. 7.29. If we re-roll the SLIM, as in Fig. 7.28 in the transverse (z-) direction, it results in a tubular motor as shown Fig. 7.30. The primary winding is in form of circular coils embedded in magnetic core and the secondary is in form of a conducting aluminum tube on an iron rod of smaller diameter. The secondary of tubular motor moves along its axis.

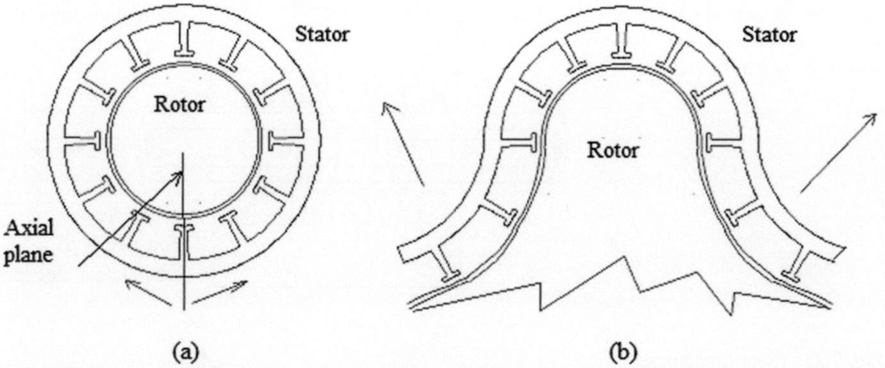

Fig. 7.27 **a** Conventional sheet-rotor induction motor being cut by an imaginary axial plane and **b** unrolling of sheet-rotor induction motor in flat form

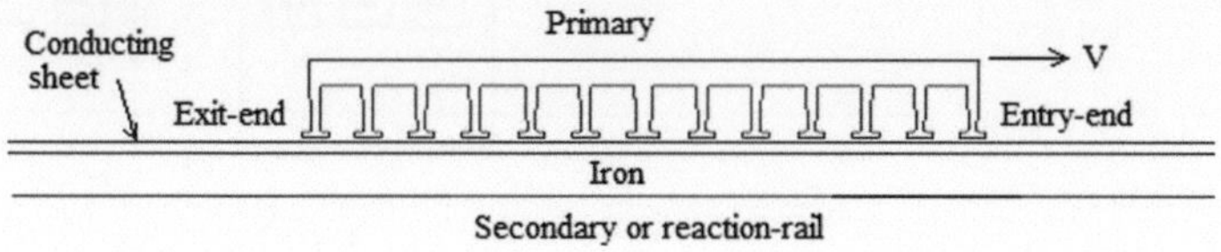

Fig. 7.28 A typical single-sided linear induction motor (SLIM) (*windings not shown for clarity*)

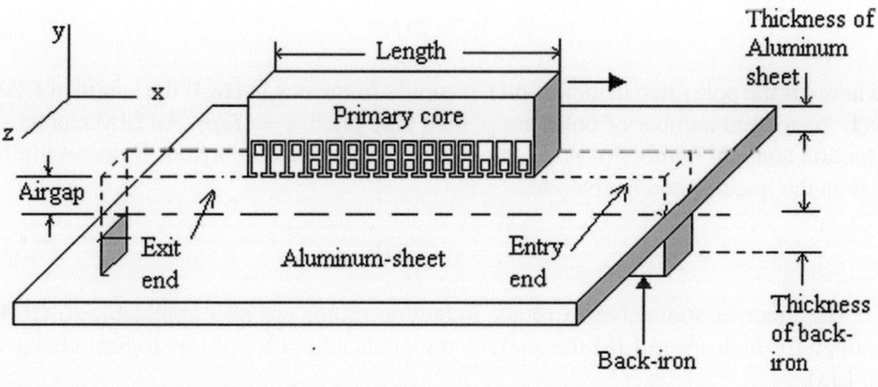

Fig. 7.29 A typical single-sided linear induction motor (SLIM) showing 3D geometry (windings are not shown for clarity)

If we imagine that in a tubular motor, one end is expanded and radially stretched forcing the core/coils on the ground, we get an electromagnetic levitator, as shown in Fig. 7.31. The coils are concentric over a flat magnetic core and secondary takes a form of a conducting plate, which will repel a conducting plate away from primary of levitator.

Linear synchronous speed of a LIM is

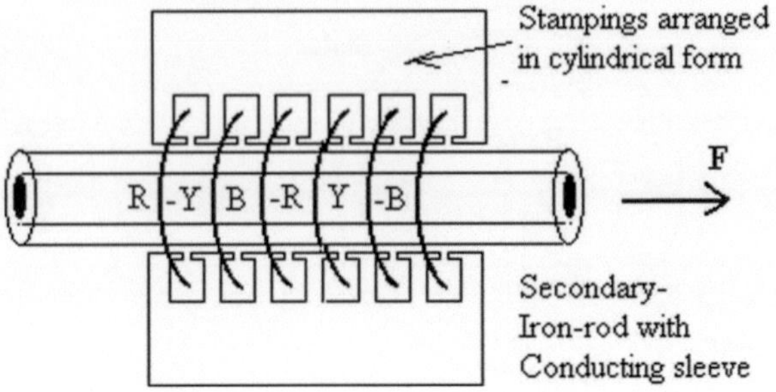

Fig. 7.30 A typical tubular motor (*windings are not shown for clarity*)

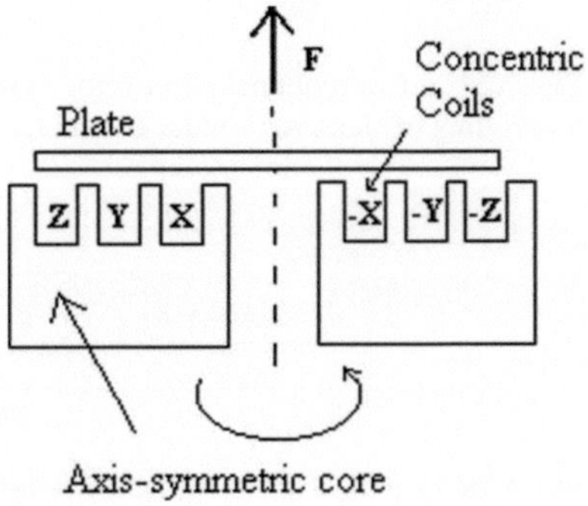

Fig. 7.31 Electromagnetic levitator

$$V_s = 2\tau f \tag{7.52}$$

where τ is the pole pitch in meter and f is supply frequency in Hz. If the length of LIM is L meter and number of poles are p, then pole pitch, $\tau = (L/p)$. An LIM can have fraction and odd number of poles depending on the winding layout. If linear-slip is s then the linear speed is given as,

$$V = (1 - s)V_s \tag{7.53}$$

The theories applicable to rotary induction motor are also applicable to LIM, though for high speed LIM the analytical models takes care of topological changes in LIM.

7.4.2 Topological Changes in LIM

The following are certain topological changes in LIM.

(a) **Static End-Effect**: The inductances of the coil sides closer to either entry-end or exit-end are slightly lesser than those in the mid-portion of the LIM. This causes unbalanced currents even when a balanced set of three-phase voltage is applied in a single LIM. This effect is present even at standstill condition. We know that in a three-phase rotary induction motor, the unbalanced currents give rise to inferior performance of the induction motor. In LIM also, the developed thrust is slightly deteriorated. The static end effect is nullified in an application where three LIM stators having common secondary and the windings of these three LIM stators are transposed to yield equal impedances in all the three phases.

(b) **End Effect**: As compared to the rotary induction motor, an LIM has both an entry-end and exit-end. The reaction rail or secondary entering into the active portion of LIM primary does not have the eddy current induced on it. As a result, certain amount of energy is continuously required to set-up the eddy currents in the reaction plate. Similarly, the portion of reaction rail continuously going away from the LIM exit-end has decaying eddy current, a sizable amount of magnetic energy is continuously being wasted from the exit end. These two effects deteriorate the performance of LIM up to certain extent at high speed. For minimizing these effects, compensating windings are required. For large number of poles (>24), these effects are negligible.

(c) **Transverse–Edge Effect**: This effect arises due to difference in secondary overhang. Under dynamic conditions, it produces destabilizing forces in transverse direction. In wheel-on-rail transport system, the stress is taken by wheel and rail. However, in levitated transport system and recently proposed hyperloop system of transport suitable measures are taken to mitigate this effect. In rotary motors also, similar forces are present in axial direction. But the bearings in end covers do not allow the axial movement of shaft in rotary motors.

(d) **Normal Force**: Strong magnetizing forces between stator and rotor are present in rotary motors, which are nullified because of cylindrical structure of rotary motor. UMP (unbalanced magnetic pull) is present in rotary motor if the air gap is non-uniform. In SLIM having composite reaction rail, these forces are very large and vary with speed. Its effect is to increase the deadweight of the moving primary at high speed in wheel on rail system of transport. This is minimized by increasing the clearance. In normal induction motor the air clearance is of the order of fraction of a millimeter, whereas in case of LIM the clearance is of the order of 10 mm. The increased air clearance increases the leakage reactance of LIM and decreases the net thrust. SLIM propelled overhead suspended Hyper loop pod can utilize the normal force to levitate it against gravity.

Figure 7.32, respectively, shows a typical thrust and normal force versus slip characteristics of LIM at different. Normal force is maximum at synchronous speed when thrust is minimum.

7.4.3 Comparison with Rotary Induction Motor

Table 7.1 compares the LIM with rotary induction motor.

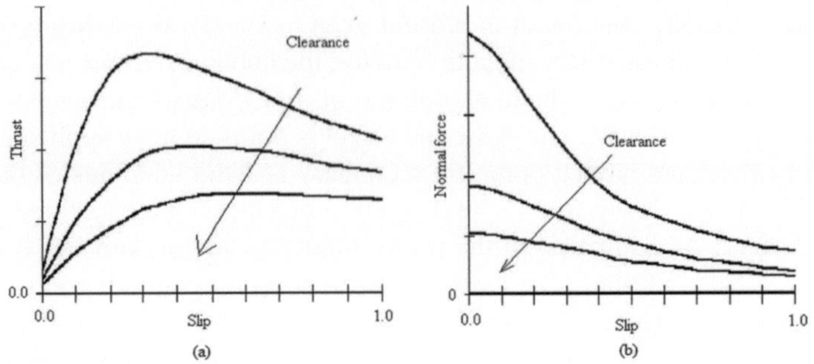

Fig. 7.32 **a** Effect of variation of clearance on thrust slip characteristics and **b** effect of variation of clearance on normal force slip characteristics

Table 7.1 Comparison with rotary induction motor

	Linear induction motor	Rotary induction motor
1	Develops thrust, N	Develops torque, Nm
2	Linear speed in meter per second	Revolution in rotation per minutes
3	Clearance can be easily changed	Clearance between stator and rotor is fixed and cannot be changed
4	Clearance may vary up to 10 mm for lowering the normal force	Clearance is of the order of fraction of a millimeter
5	Normal force between primary and secondary is present	Normal force between stator and rotor gets canceled
6	Application specific	Widely used induction motor
7	Performance deteriorates at high speed due to end effects	There is no such effect in rotary machine
8	In levitated transport system, highest speed is not limited by friction	The highest speed is limited by mechanical constraints such as bearing friction
9	Thrust transferred in a non-contact manner	Thrust is transferred mechanically

These machines have become popular because of the specific advantages offered by them. Nowadays, these motors are finding their application in widely diverse areas.

(a) Product distribution systems, transport of material, parts and products.
(b) Wheel on rail transportation and tube railways
(c) Electromagnetic Aircraft Launchers (EMALS).
(d) Hyperloop System of transport.
(e) Delivery systems in space and hazardous environment.
(f) Levitated transportation.
(g) Transportation switch using x–y LIM.
(h) Under water car.

(i) Office automations and Factory automation.
(j) Inspecting machines.
(k) PV water pumping, draft machine driver.
(l) Vehicle test facility.
(m) Super conducting electromagnetic thrust ship.
(n) Electromagnetic-gun.
(o) Control rod drive (Linear step motor).
(p) Elevator drives.
(q) Fuel pump drive in artificial satellite.
(r) Door / curtain operation.
(s) Postal transfer from one post office to another post office via tube / tunnels.
(t) Linear fork.
(u) Metal strip coil telescopeness control.
(v) Entertainment arcade.
(w) Agitators for melting nonferrous metals.
(x) High-speed mail sorter.
(y) Surface control of molten metal.
(z) Show business.

7.4.4 Analysis of LIM

Due to the topological changes in LIM, it is quite cumbersome to analyze the LIM using techniques used in analyzing rotary induction motor like equivalent circuit analysis and dq0 model. The analysis of LIM is normally done by solving the Maxwell's equation in and around the geometry of the machine. These methods are listed here.

(a) Fourier series model.
(b) Fourier transform model.
(c) Finite element method-ANSYS software.
(d) Boundary element method-BEM software.

The discussions of these methods are beyond the scope of this book. The finite element based analysis is done using commercially available FEM code ANSYS Multiphysics. These software are quite common now-a-days in engineering institutions. The conventional equivalent circuit is also sometimes used for analyzing low speed LIM.

Example 7.8 Calculate the synchronous linear speed of a 1.0 m length, 60-slot, 8 pole linear induction motor operating on 50 Hz supply. The equivalent circuit of a star connected single sided linear induction motor is given in Fig. 7.33. The machine is operating on 440-V, three phase 50 Hz supply. Calculate the per phase current when machine is running at 11.5 m/s.

Solution

The synchronous linear speed is

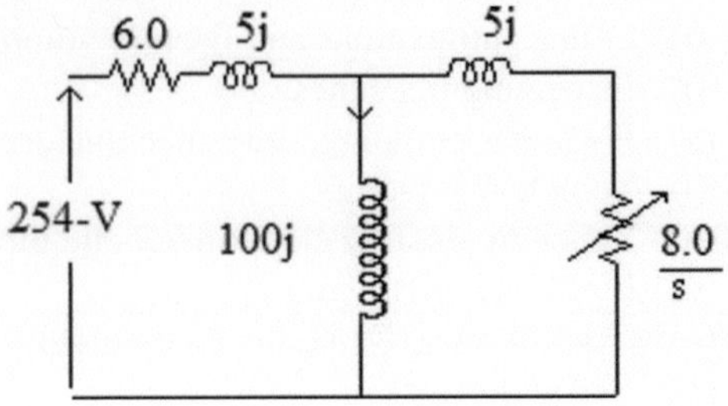

Fig. 7.33 Per phase equivalent circuit of linear induction motor

$$V_s = 2\tau f = 2 \times \frac{L}{p} \times f = 2 \times \frac{1.0}{8} \times 50 = 12.5 \text{ m/s}$$

Linear slip,

$$s = \frac{12.5 - 11.5}{12.5} = 0.08$$

The phase voltage $V_{\text{ph}} = 440/1.7321 = 254$ V.
The total impedance seen on the terminal at slip 0.08 is,

$$Z = 6 + 5j + 100j || \left(\frac{8.0}{0.08} + 5j \right) = 76.76\angle 45.78^\circ\ \Omega$$

Per phase current,

$$I = \frac{V_{ph}}{Z} = \frac{254\angle 0^\circ}{76.76\angle 45.78^\circ} = 3.31\angle - 45.78^\circ \text{ A}$$

Power factor $\cos 45.78^\circ = 0.6974$ lagging.

In linear machine, the clearance between primary and reaction rail is large of the order of 10.0 mm. The leakage reactance is also large. This gives a low power factor. Suitable measures are adopted for improving the overall power factor.

7.5 Stepper Motor

Stepper motors move in steps. They are manufactured to have step sizes as small as 0.72° or as large as 90°. The common step sizes are 1.8°, 7.5° and 15°. A PM stepper motor of step size 1.8° has 40 poles on stator and 50 poles on rotor. The following are some types of stepper motors [2, 4, 6, 12]:

(a) Reluctance type.
(b) PM stepper motor.

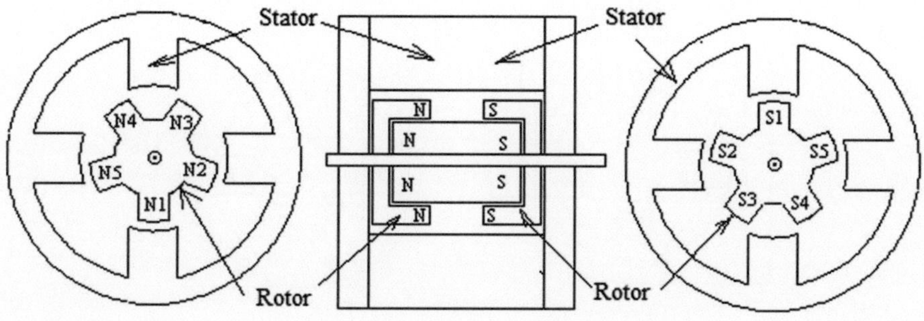

Fig. 7.34 Permanent magnet stepper motor (*coils not shown for clarity*)

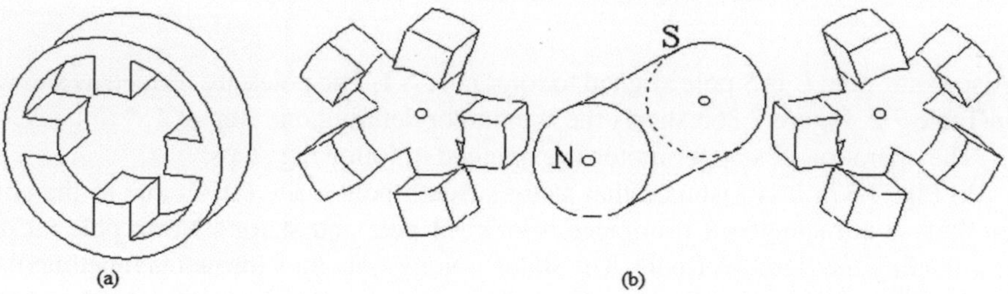

Fig. 7.35 Parts of PM stepper motor (**a**) stator (**b**) rotor projections and permanent magnet

7.5.1 Construction Details

The construction details and operation of reluctance-type stepper motors are almost similar to Switched Reluctance motor (SRM). A stepper motor can operate in open loop whereas a SRM requires the position-based switching of the coils. Both the stator and rotor have saliency. Figure 7.34 shows the construction details of 18° step size PM stepper motor, while Fig. 7.35 shows various parts of stepper motor. The shaft carries a permanent magnet of cylindrical shape. It is assumed that there are five projections on the rotor. The projected soft iron pieces are assembled as shown in Fig. 7.34. The stator poles have either simple winding or bifilar winding.

7.5.2 Operation

The stator poles have either simple winding or bifilar winding. The operation of PM stepper motor is based on the interaction of soft iron magnetized pieces with stator flux. Figure 7.36(a) shows coils *X*-*X*′ and *Y*-*Y*′ on the stator poles *A*-*B*-*C*-*D*. The terminals *X*-*X*′, *Y*-*Y*′ are connected to required polarity of the DC supply. When *X* is positive and *X*′ is negative, stator pole *A* is *N* pole to which rotor pole ***S*1** is aligned.

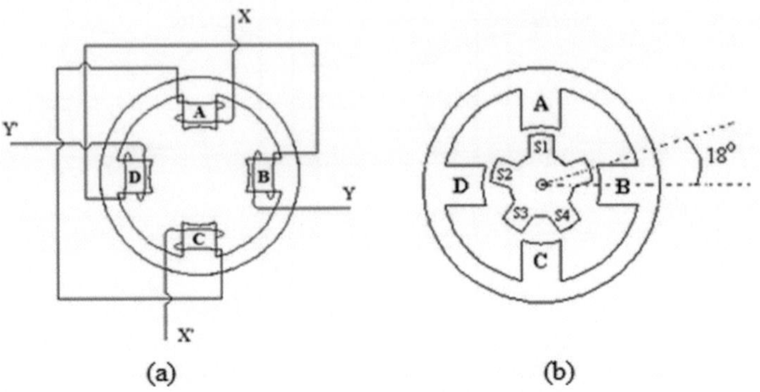

Fig. 7.36 **a** Coils in stepper motor and **b** stepper motor defining one step

Also stator pole *C* is *S* pole aligned to rotor pole ***N*1**. The poles are formed as shown in Table 7.2. Figure 7.36b shows the step motor defining one step of 18°.

The operation of stepper motor is explained in following steps:

In Fig. 7.37a it is assumed that stator's South pole made (at *C*) due to the coil currents is attracting soft iron piece rotor's ***N*1** pole and stator's North pole (at *A*) is attracting the rotor's ***S*1** pole. The stator's North pole thus formed is repelling the rotor's ***N*3** and ***N*4** poles of rotor. Similarly, stator's South pole repels the ***S*3** and ***S*4** rotor poles.

In Fig. 7.37b, the coil *X*-*X*′ is switched OFF and another coil *Y*-*Y*′ is switched ON. The stator's South pole (at *D*) thus formed attracts nearby rotor's ***N*5** pole. Simultaneously the stator North pole (at *B*) attracts the rotor's ***S*5** pole. The rotor moves by 18° in clockwise direction from its original position, and completes one rotation in 20 steps (360°/18° = 20). The rotor moves in the direction of net positive force and acquires a position when net torque is zero. In a similar manner the rotation can be explained as shown in Figs. 7.37, 7.38, 7.39, 7.40 and 7.41.

Tables 7.3 and Table 7.4, respectively, show truth table for clockwise and anti-clockwise directions.

Thrust is zero when the rotor pole is aligned toward a stator pole. If the rotor pole is deviated from stator pole by an angle θ, the force of attraction between opposing poles

Table 7.2 Formation of poles

Stator coils				Stator poles			
X	*X*′	*Y*	*Y*′	*A*	*B*	*C*	*D*
+	–			N		S	
		+	–		N		S
–	+			S		N	
		–	+		S		N

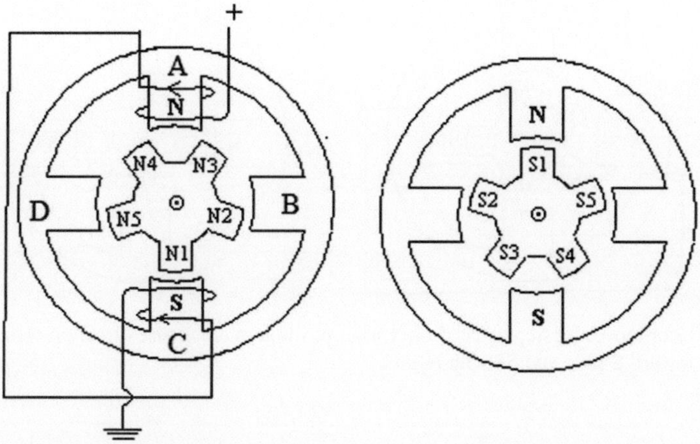

Fig. 7.37 Switching of PM stepper motor for clockwise rotation–rotor projection $N1$ aligned toward S-pole and $S1$ with N-pole

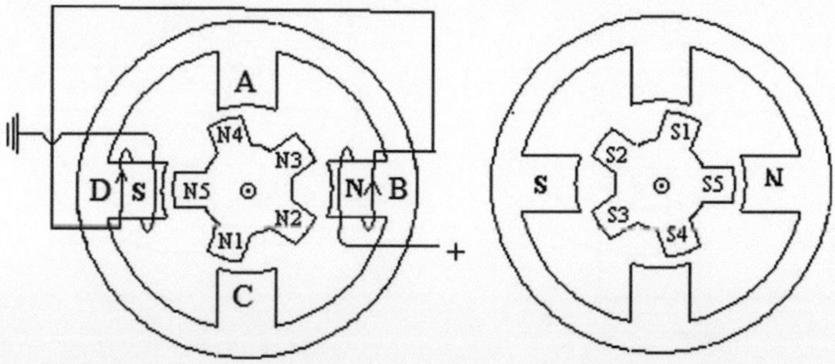

Fig. 7.38 Rotor moves by 18° from original position to clockwise direction-rotor projection $N5$ aligned toward S-pole and $S5$ with N-pole

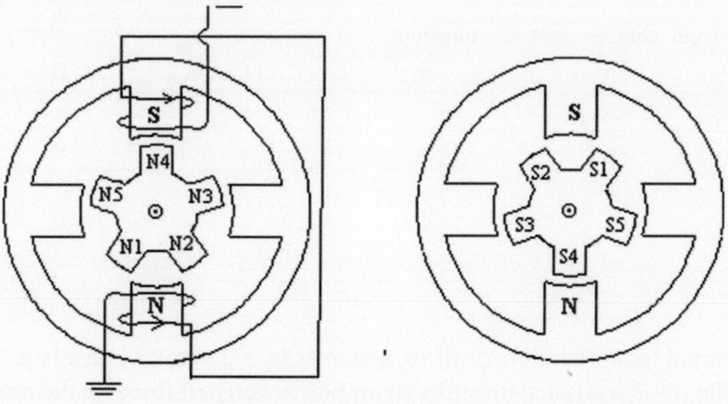

Fig. 7.39 Rotor moves by step of 18° from earlier position in clockwise direction-rotor projection $N4$ aligned toward S-pole and $S4$ with N-pole

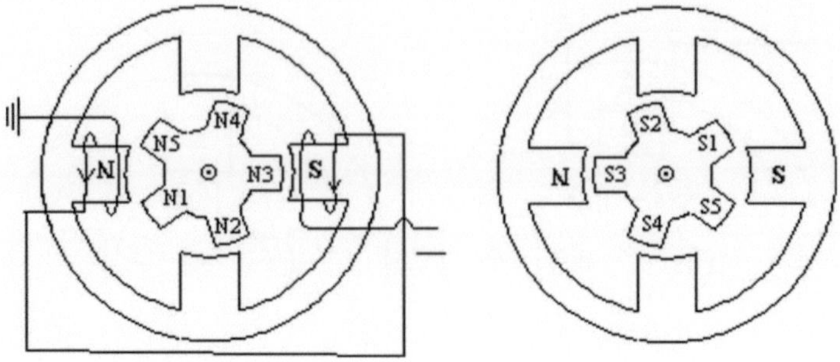

Fig. 7.40 Rotor moves by step of 18° from earlier position in clockwise direction-rotor projection *N*3 aligned toward S-pole and *S*3 with N-pole

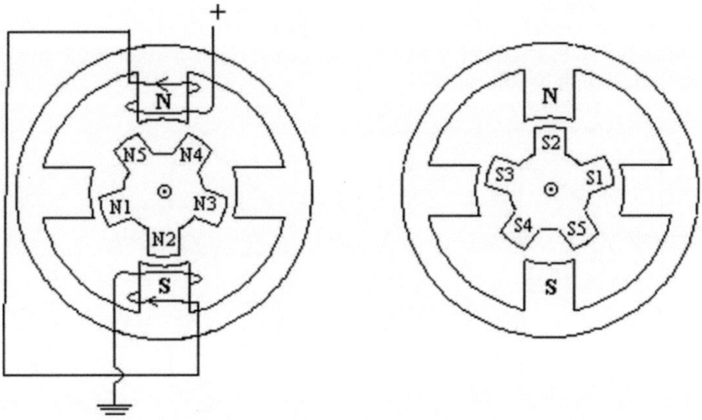

Fig. 7.41 Rotor moves by step of 18° from earlier position in clockwise direction rotor projection N2 aligned toward S-pole and S2 with N-pole (*windings are shown only once for clarity*)

Table 7.3 Truth table for clockwise direction

Step	A	B	C	D
1	N	–	S	–
2	–	N	–	S
3	S	–	N	–
4	–	S	–	N
5	N	–	S	–

is proportional to sinθ and force of repulsion between similar poles is proportional to cosθ. The rotor is aligned toward a stator pole when nett force on the rotor is zero.

When the rotor is aligned as in Fig. 7.37 and the coils are switched as in Fig. 7.38. The forces on rotor are:

Table 7.4 Truth table for anti-clockwise direction

Step	A	B	C	D
1	N	–	S	–
2	–	S	–	N
3	S	–	N	–
4	–	N	–	S
5	N		S	–

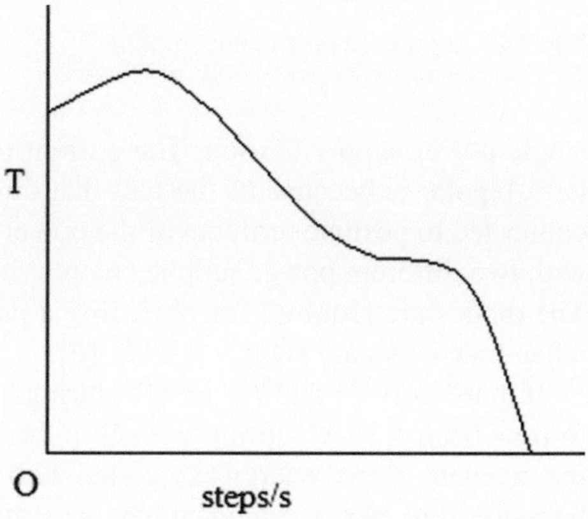

Fig. 7.42 Torque versus Steps per second curve of a PM stepper motor

(a) The South pole in D attracts the $N5$ (18°), $N1$ (90°) and $-N4$ (54°).
(b) The North pole thus formed on B attracts the $S5$ (18°), $S1$ (90°) and $-S4$ (54).

Let the maximum torque be T_{max}, the total torque in Fig. 7.37, when coils are switched ON according to Fig. 7.38 is

$$T = T_{\text{max}}(\sin 18° + \sin 90° - \sin 54° + \cos 18° + \cos 90° - \cos 54°) = 0.863\ T_{\text{max}} \tag{7.54}$$

Torque can be calculated in a similar manner for different positions. Figure 7.42 shows torque speed (steps per second) characteristics of a PM stepper motor. Let us suppose a stepper motor S steps per seconds. If we run it less with than around 2.5% of these steps per seconds the torque drops to zero.

7.5.3 Power Supply for Stepper Motor

Figure 7.43 shows power circuit for bipolar stepper motor. Here the respective coils X-X' and Y-Y' are switched ON/OFF by means of transistors, this is bipolar type amplifier. A total of four transistors are needed for each phase coil whenever a

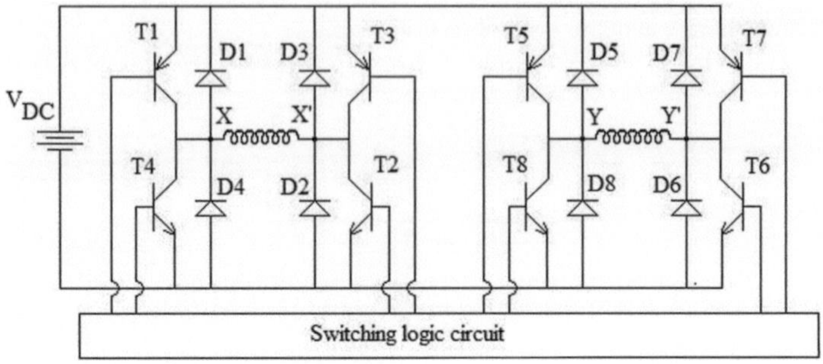

Fig. 7.43 Bipolar-stepper motor amplifier

single power supply is used. The current reversal in the phase coil is desired. The term bipolar is because of the fact that circuit permits either end of the coil to be connected to positive terminal of the power supply. Similar circuit can be developed with two different power supplies employing only or two transistors per phase coil. The diodes are required for providing a path for the inductive coil current to flow when transistors are switched OFF [6].

If transistors $T1$ and $T2$ are ON during a switching cycle thus causing the current to flow from X to X', during transition these transistors are to switched OFF. Since the magnetic stored energy associated with the current in the coil, the current cannot be reduced to zero instantaneously. Any attempt to do so results in a large induced EMF in the coil with terminal X' positive and X as negative. This induced voltage puts the diodes $D1$ and $D2$ in forward bias thus furnishing the path of the switching OFF current in the coil X- X'. The current rapidly decays to zero because of small time constant.

The coils are energized for counter clockwise rotations as shown in Table 7.5

Here ON+ means X or Y terminal is positive and ON- means the X' or Y' terminal is positive. The transistors are switched ON/OFF according to truth table given in Table 7.6.

The cycle is repeated. Can you develop the truth table for clock-wise rotation? In a simpler circuit the coils are made with a center tapped bifilar winding. The current flows in only one section of the winding. A unipolar bifilar winding stepper motor drive amplifier is shown in Fig. 7.44. The circuit requires lesser number of transistor or switching devices and diodes. However, the disadvantage is that the current flows in only half the coil. It is referred as unipolar circuit, because the end

Table 7.5 Switching of coils

Coil X-X'	Coil Y-Y'
ON+	OFF
OFF	ON+
ON-	OFF
OFF	ON-

Table 7.6 Switching of transistor's/IGBTs

Coil *X-X′*				Coil *Y-Y′*			
T1	T2	T3	T4	T5	T6	T7	T8
1	1	0	0	0	0	0	0
0	0	0	0	1	1	0	0
0	0	1	1	0	0	0	0
0	0	0	0	0	0	1	1

of both coil sections stay connected to the negative side of the power supply. The logic for unipolar circuit is given in Table 7.7.

The diodes $D1$, $D2$, $D3$, $D4$ provide the switching current path. Let initially $T1$ is ON. When the current starts flowing through left half of coil *X-X′*. Now suppose that $T1$ is switched OFF. We know that the inductive current cannot be brought to zero instantaneously. Due to transformer action, the second section of coil allows the current in the first section to drop to zero. As a result, the current in the second section instantaneously rises to the level of current in the first section. The latter current flows through the path provided by diode $D2$. The transfer of current from left section to the right section is allowed as there is no change in the stored magnetic energy.

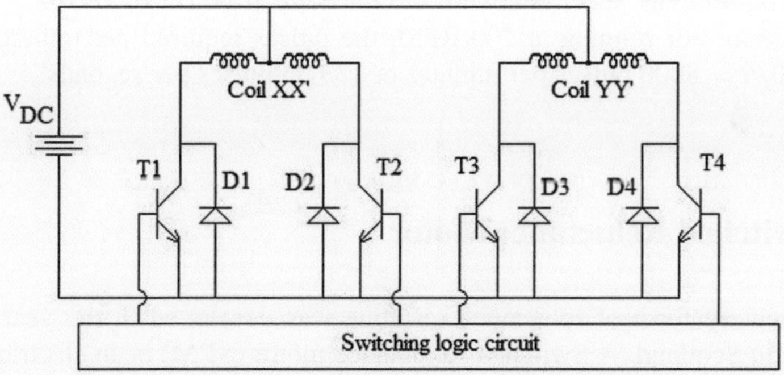

Fig. 7.44 Switching circuit for bifilar winding with unipolar power circuit

Table 7.7 Logic for unipolar circuit

Coil *X-X′*		Coil *Y-Y′*	
T1	T2	T3	T4
1	0	0	0
0	0	1	0
0	1	0	0
0	0	0	1

Consider that initially only coil X-X' is energized, if it is switched OFF and another coil Y-Y' is energized the rotor moves by one step in either direction depending on the polarity of the coil connections.

Let us assume that the initial coil is not switched OFF and another coil is also ON, the rotor moves by half step 9° due to magnetic symmetric conditions. The rotor is aligned to the position so as give zero torque. The precision of the stepper motor is improved by half stepping. The coil switching and truth table can be easily developed.

Personal computer or micro controller are also used for running a stepper motor. A personal computer requires an add-on-card with several input and output ports. There are several commercially available stepper motor controller chips or boards such as L297 and L298, DMC-100 and DMC-201. The control can also be done using dSpace, DSP drive, FPGA drive or microcontroller.

Stepper motors have applications in such a diverse fields as instrumentation, Computer peripherals (dot matrix, printer, Floppy disk drive, Paper tape drive), numerical control (NC) of machines tools like CNC machines, robotics, medical applications, digital X–Y plotter, electronic-type writer etc.

Example 7.9 A stepper motor has 9° per step and must rotate at 200 RPM. What is input pulse rate in pulses per second required?

Solution

A full revolution has 360°, thus with 9° per steps, it will take $360/9 = 40$ steps per revolution. For running at 200 RPM, the pulses required per minutes will be $200 \times 360/9 = 8000$ pulses per minutes or 133.33 pulses per seconds.

7.6 Switched Reluctance Motor

The elementary form of reluctance machine was developed in the year 1838 by Davidson in Scotland. A Switched Reluctance motor (SRM) is an electric machine in which torque is developed by the tendency of ferromagnetic rotor poles to align in a minimum reluctance position (or maximum inductance position) [4, 9].

The stator or primary winding consists of coils in stator poles. These coils are grouped together (phase) which are switched ON or OFF simultaneously depending on the rotor position. A phase is excited when the coil inductance is increasing and switched OFF when coil inductance is decreasing. The reverse is true for generating and braking operations. Since there are several phases, once a particular phase is switched OFF, the next is ON so as to produce a torque. A particular phase or coil group is switched ON / OFF depending on the position of rotor.

The machine has salient poles both on the stator and on the rotor. A 8/6 SRM has eight poles on the stator and six poles on the rotor. The coil groups or phases are switched ON/OFF electronically, a process referred to as electronic commutation. In these motors, normally the angular position of the shaft is sensed and the commutations of the different phases are synchronized with the shaft position. These motors are designed up to 300 kW capacity. These machines are single excited and have no brushes. However, both the stator and rotor are having saliency.

If the rotor and stator poles are symmetric, the unidirectional torque will be produced over one half of the rotor pole pitch, as in case of Variable Reluctance Motor (VRM). If the total number of rotor poles is different than stator poles, a unidirectional torque in all the positions of the rotor is possible. A typical 6/4 SRM has been shown in Fig. 7.45. The torque zone between stator and rotor poles has been described in Fig. 7.46. A rotor pole is not in the torque zone if it is not in the proximity to stator pole. When one of the stator coil is switched ON, the torque is given by

$$T = \left.\frac{\partial W^{/}}{\partial \theta}\right|_{I=\text{Constant}} \tag{7.55}$$

As already explained, the stored energy is

$$W^{/} = \int_{0}^{i} \lambda di \tag{7.56}$$

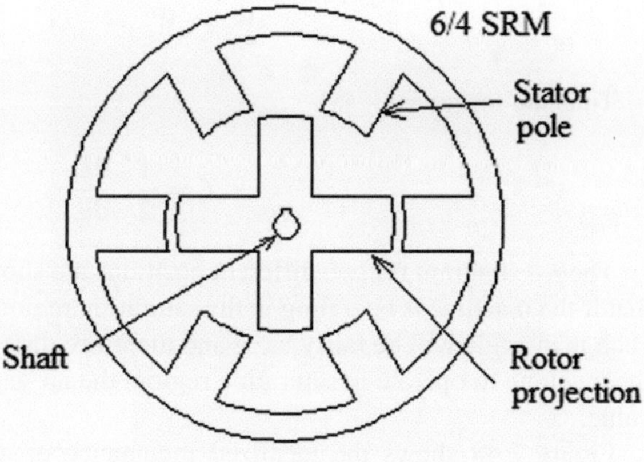

Fig. 7.45 Switched reluctance motor (*stator coils not shown for clarity*)

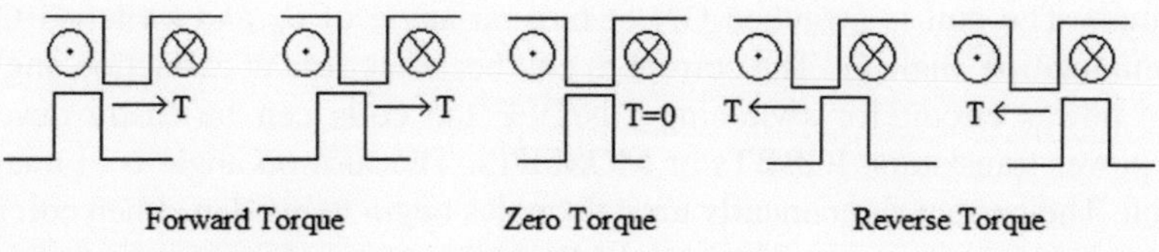

Fig. 7.46 Torque zone in SRM

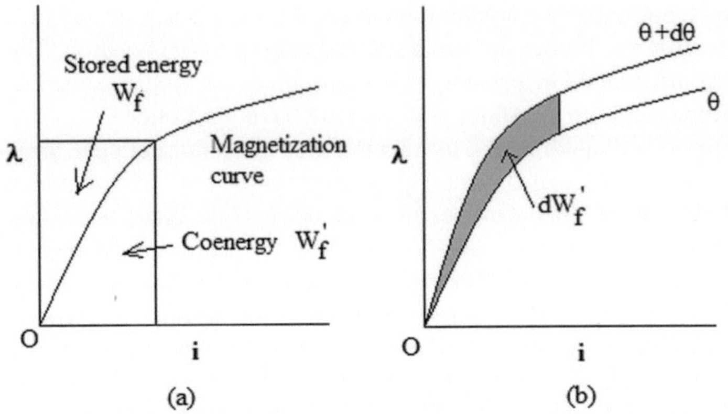

Fig. 7.47 **a** Flux linkage-current diagram and **b** change in co-energy with rotor position

Figure 7.47a shows flux linkage versus current diagram. When the rotor moves by an angle dθ, the λ–I curve also changes. Suppose the rotor is moved by dθ, the change in the stored co-energy is shown in Fig. 7.47b. The torque is accordingly calculated. Since stator and rotor both are having saliency, the conventional analytical technique, such as equivalent circuit approach and generalized machine theory cannot be used in SRM. In a linear magnetic circuit, the stored energy and co-energy are given as

$$W_f = W_f^{/} = \frac{1}{2}Li^2 \tag{7.57}$$

Thus the torque is

$$T = \frac{1}{2}i^2\frac{\mathrm{d}L}{\mathrm{d}\theta} \tag{7.58}$$

The λ–I diagram for two different positions are shown in Fig. 7.47b. It is apparent that if the machine is operating in the saturation region the change in co-energy for a change in angle will be fairly large and therefore, large torque will be produced. For the machine to operate in saturation region, the air gap is kept at minimum possible value.

Figure 7.48 shows the idealized inductance of a stator coil, when rotor pole approaches it. When the two poles overlap each other, the coil inductance is maximum. The coil is switched ON at turn-on angle of θ_o and switched OFF at the commutation angle θ_c. The current in coil becomes zero at extinction angle θ_q.

The power circuit for switching ON/OFF the coils can be easily developed using power transistors, IGBBTs or MOSFETs. The turn-on angle is at unaligned position. The current rises linearly until the poles begin to overlap, when coil inductance tends to be maximum. The back EMF rises because of increase in inductance. At peak current, the supply voltage and back EMF are equal. Subsequently, the back EMF becomes more than supply voltage because the flux linkage continues to rise

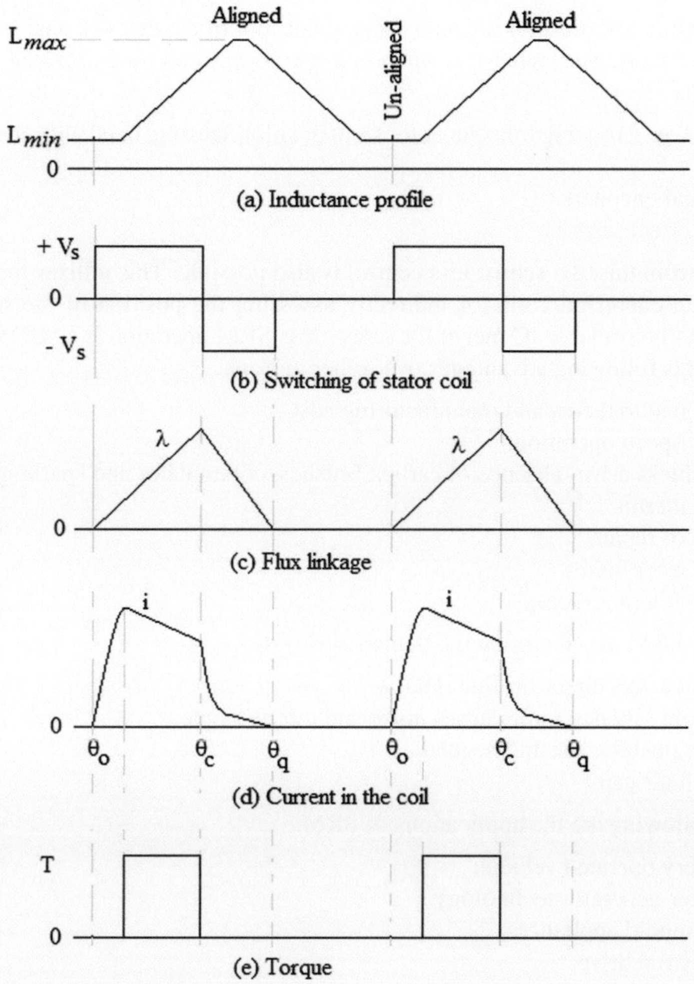

Fig. 7.48 Operation of switched reluctance motor

even when the speed is constant. Now, the excess of applied forward voltage becomes a deficit and the current begins to decrease. At the point of commutation, the applied voltage is reversed and there is a sharp rise in the rate of change of current, which forces the current to be zero at angle θ_q. In single pulse commutation, the coil is switched ON before the poles begins to overlap. When the device is switched ON, the current varies widely from the turn-on angle to the quenching angle. Current chopping technique is also utilized. In current chopping, during ON period of coil instead of providing a continuous voltage, the applied voltage is in a form of train of high frequency pulses.

The SRM switching is based on the rotor position sensing. The switching of different stator coils or phases are synchronized with the rotor position. The inductive pick-up is not preferred for position sensing because of the delayed position signals due to hysteresis. The signal corresponding to rotor positions are obtained using the following:

(a) Resolver-a rotating transducer for shaft position sensing, fixed with shaft on one end.
(b) Optical encoders.
(c) Hall effect sensors.

Apart from this, the sensor less control is also possible. This utilizes the induced EMF in un-energized coils for indirectly assessing the position of the rotor. The commercially available IC meant for sensor less SRM operation is LMB1008.

SRM has following advantages over other motors:

(a) Low material cost and manufacturing cost.
(b) High speed operation.
(c) Brushless drive- absence of carbon brushes, commutator and sparking.
(d) Low inertia.
(e) Ease of repair.
(f) Low rotor losses.
(g) Fault tolerant motor.

However, SRM has the following demerits as well:

(a) Cannot start direct-on-line (DOL).
(b) Double saliency rotor causes noise and torque ripple.
(c) Shaft position sensing needed.
(d) Small air gap.

The following are the applications of SRM:

(a) Battery operated vehicle.
(b) Starter-generator technology.
(c) Household appliances.
(d) Industrial drive.

Example 7.10 A primitive 4/3 SRM is shown in Fig. 7.49. Develop switching sequences for running the rotor in clockwise and anti-clockwise directions.

Solution

Let us assume that coil **A** is energized with rotor pole **a** is aligned toward stator pole **A**. Rotor is free to rotate in either direction. Now, if coil **A** is switched OFF and coil **B** is energized, rotor will move in counter clockwise direction. This moves the rotor by 30° such that rotor pole **b** is aligned toward stator pole **B**.

When coil **B** is switched OFF and coil **C** is energized the rotor will again move in counter clockwise direction by another 30°. Rotor pole **c** is aligned toward stator pole **C**. When coil **C** is switched OFF and coil **D** is energized, the rotor will again move in counter clockwise direction by another 30°. Rotor pole **a** aligned toward

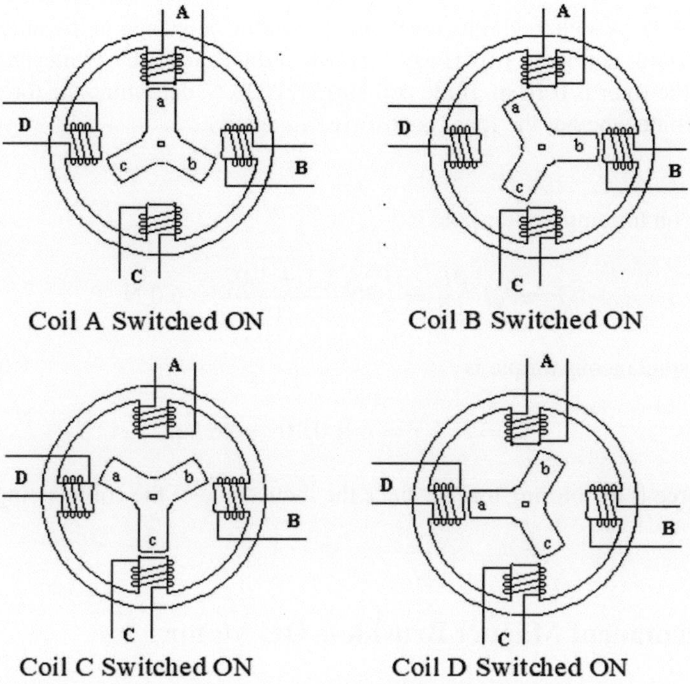

Fig. 7.49 Switching of switched reluctance motor having 4 stator and 3 rotor poles

pole **D**. In one cycle of switching, four coils in sequence makes the rotor to move by 90°. The truth table for forward running and backwards running are shown in Table 7.8 and 7.9, respectively

Table 7.8 Counter-clockwise direction

Steps	A	B	C	D
1	1	0	0	0
2	0	1	0	0
3	0	0	1	0
4	0	0	0	1

Table 7.9 Clockwise direction

Steps	A	B	C	D
1	1	0	0	0
2	0	0	0	1
3	0	0	1	0
4	0	1	0	0

Example 7.11 A switched reluctance motor exhibits a change in the inductance of its one of stator coil from 1.07 H to 1.1 H, when rotor moves by 1° mechanical. The radius of the rotor is 0.10 m. If the coil current is 20 A, determine the torque at that position. State whether the force is positive or negative?

Solution

The force on the single rotor pole is

$$F = \frac{1}{2} I^2 \frac{\partial L}{\partial \theta} = \frac{1}{2} 20^2 \frac{1.1 - 1.07}{1} = 6.0 \text{ N}$$

The instantaneous torque is

$$T = F \times r = 6 \times 0.10 = 0.6 \text{ Nm}$$

The force is a motoring torque, since the inductance of the coil is rising.

7.7 Permanent Magnet Brushless DC Motor

A stepper motor is a brushless motor. Similarly, the motors such as SRM, Reluctance synchronous motor, squirrel cage induction motor and PMBLDC motor and PMSM are also brushless motors. Such motors are becoming popular as they have no mechanical commutator and absence of sparking. Commutation in some of these motors is achieved using electronic circuits [10, 11].

Figure 7.50 shows a permanent magnet brushless DC motor (PMBLDC or BLDC). These motors are being preferred because of their larger power density. The rotor is normally a smooth rotor having even number of alternate N-S pole. These magnetic poles are in a form of flexible sheet and are pasted on the inner surface of outer hub rotor. Such motors are ubiquitous as fan drive in personal computers and laptop. The commutations of the coils are performed using electronic circuit. The torque is produced due to interaction of dissimilar poles. The Hall Effect sensor and latch IC is there in small PMBLDC motors. Here, the inner stator is having four projecting poles carrying bifilar winding. Bifillar winding of poles made using two parallel insulated (SWG) wires. Once coil is made, one common end of parallel wire is grounded whereas other two ends may be connected to +ve and –ve polarity as the case may be. The winding has three terminals. The mid- terminals being at zero potential. By suitable connecting the other two terminals, polarity (N or S) of the poles formed on the stator is altered for producing a forward torque. In case of failure of the Hall Effect probe and latch circuit the operation of the machine is affected.

Low power PMBLDC motors are used in existing computer peripherals, like cooling fan drive of SMPS and CPU, CDROM drive and HDD drives. The cooling fan drive of SMPS is four-pole PMBL motor having bifilar winding. The stationary

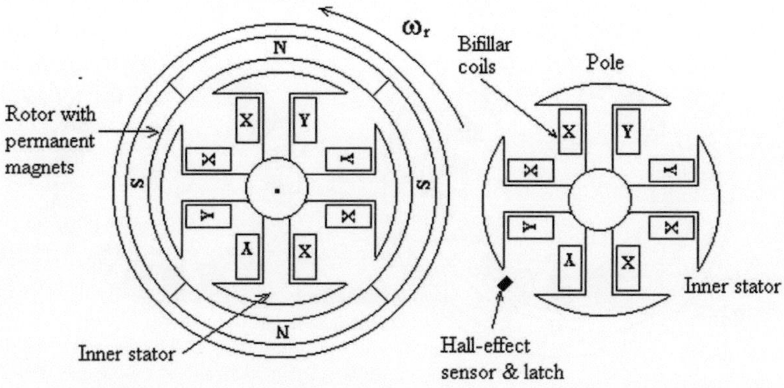

Fig. 7.50 PMBLDC motor with bifilar winding

primary is on the inner side and PM rotor is on the outside. The windings are alternatively switched ON/OFF depending on the rotor position. Hall-effect bi-polar sensor and latch IC senses the position of rotating magnets, and the coils are accordingly switched ON/OFF.

The drive motor in CDROM drive is a tiny 24-pole three-phase PMBL motor. The rotor consists of twenty-four tiny magnetic poles on the inner ferrite sleeve. These magnets are formed with *in-situ* magnetization. The drive circuit consists of three tiny Hall-effect sensors and switching devices. The electronic circuit essentially needs Hall Effect sensor for its operation.

BLDC motors are also built with pancake armature as shown in Fig. 7.51. This has six sector-shape magnets having axial magnetization and six sector-shape flat coils with magnetic core. The position of permanent magnet rotor is sensed using Hall-effect devices. The switching of coils is synchronized with the position of rotor.

In three-phase PMBLDC motors, there are having three distinct winding groups which are regularly distributed on the slots (Fig. 7.52). These windings are alternatively switched ON /OFF with forward current and reverse currents depending on the proximity of North or South pole. A PMBLDC motor may have more than three phases. The power and mechanical torque produced by multiphase motors is expected to be high. These motors have replaced conventional DC motors at many places. The motors are also fault tolerant. In the absence of one phase the motor can run.

Schematic diagram for bipolar Hall-Effect probe and latch- IC is given in Fig. 7.53a. The switch 1 is ON, when North pole is in proximity and is OFF when

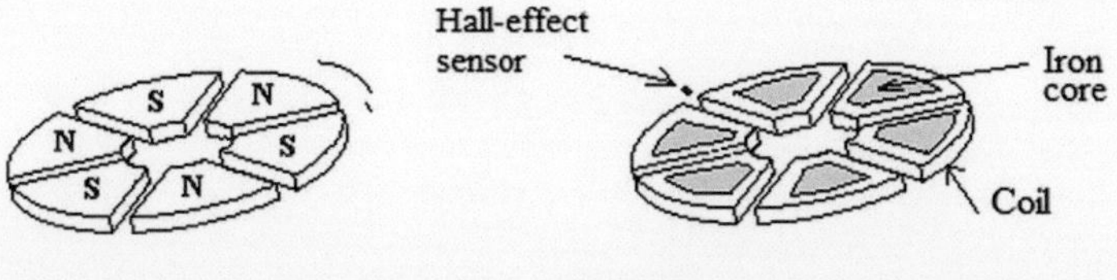

Fig. 7.51 Sector magnet for six-pole axial flux PMBLDC machine with pancake armature

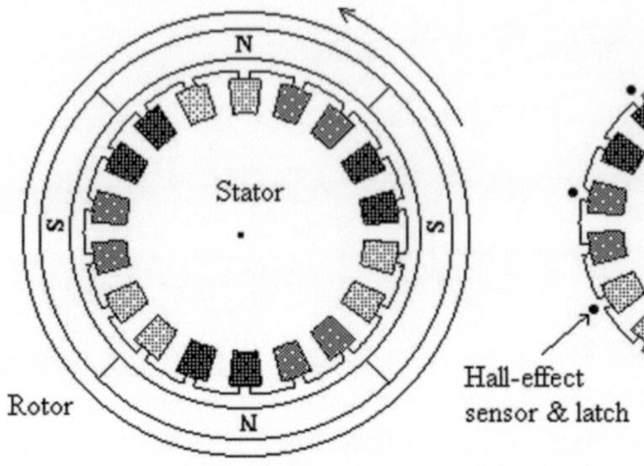

Fig. 7.52 PMBLDC motor with three-phase winding

the magnetic polarity reverses. Similarly, when South pole is in proximity the switch 2 is ON and is OFF when the magnetic polarity reverses. They exhibit the torque speed characteristics similar to that of DC shunt motor as shown in Fig. 7.53b.

The Hall Effect bipolar sensor and latch (IC) switch will remain ON even when magnetic pole is removed and will remain ON until a reverse magnetic polarity is applied. In smaller capacity machine of few Watts, the winding is directly connected to the Hall-effect sensor and latch IC. However, in large capacity PMBLDC motor, the output of Hall-effect sensor and latch is used to drive a power conditioning circuit, which switches ON/OFF the respective winding. PMBLDC motors are used in battery operated cycles, e-bikes, EV, EHV, computer peripherals, CD/DVD players etc.

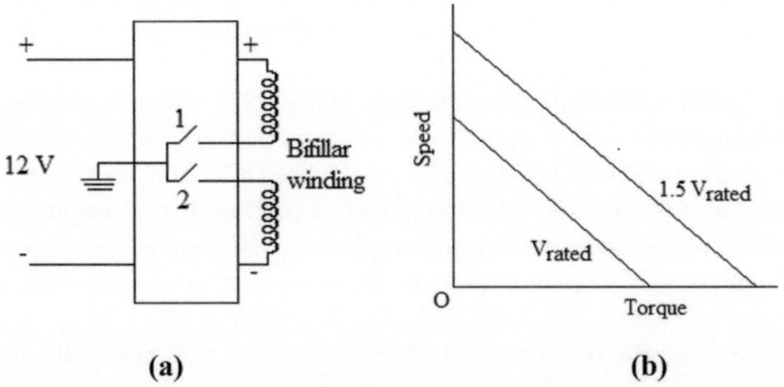

Fig. 7.53 **a** Schematic for hall effect probe and latch H1-IC with bifilar winding, **b** torque speed characteristics of BLDC motor

7.8 Written Pole Motor Generator

Written pole motor technology was invented around 1980 in United States of America. In these machines the squirrel cage rotor is having a magnetizable layer such as ferrite. The inner stator has distributed slots along with a writing coil. The writing coil is carrying a writing signal of frequency 60–50 Hz. The machine is built to operate on single-phase and three-phase supply. It is basically a synchronous machine in which the poles on the magnetizable layer are re-written in the event of any disturbance. The rotor may be either outside or inside. Figure 7.54 shows the details of a WPM with outside rotor.

The following are some of the salient features of written pole motor and generator.

(a) Low starting current WPM (Written Pole Motor).
(b) High starting torque WPM.
(c) Cool operation.
(d) Ability to synchronize and run a large inertia for WPM.
(e) Ability to ride through momentary power blackout for WPG (Written Pole Generator).
(f) Load induced voltage sag and surge for WPG.
(g) Fault clearing induced power interruption of several cycles up to several seconds duration in WPG.

The synchronous motor–alternator set has been built with rotor outside. The large moment of inertia enables the synchronous motor alternator set to ride through the transient conditions and even black out. In event of power failure, the writing process starts as such power is available for more than 15 seconds. If the power supply is not yet restored, it automatically starts an alternative source of power.

The ferrite sleeve has its original poles written due to its earlier use. In the event of disturbance, the poles are re-written by the writing coil in a similar manner as we record the music with the tape recorder. Single phase written pole motors of 20–40 hp capacity have found application in agriculture and farming. These machines have

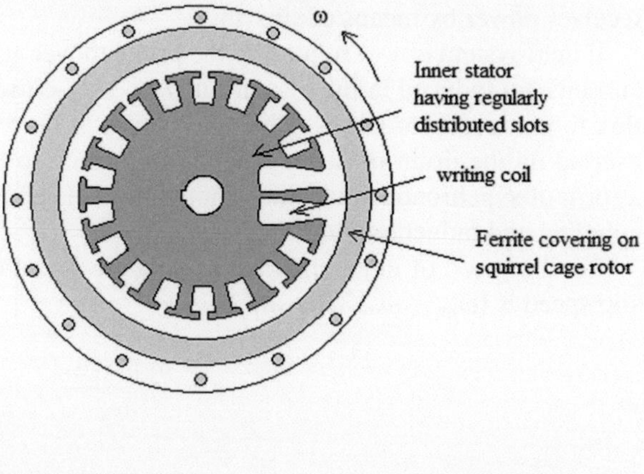

Fig. 7.54 Written pole motor–generator

gained enormous importance in large power conditioning and in power sector. They are also important for supplying power to critical instruments.

7.9 Eddy Current Devices

There are basically two types of eddy current devices such as eddy current coupling and eddy current brake. These have been described here.

7.9.1 Eddy Current Coupling

In certain applications, it is sometimes required to run the shaft at controlled low speed. In recent year the development of solid state induction drive has enable one to run induction motor at different operating characteristics and speed, using VVVF technique. Historically, the induction motors were used for low speed operations using cascade connections and gear motor, prior to availability of solid state drives. The development in power electronics and drives has completely revolutionized the electrical drive sector [2, 5].

Although conventional machines can be used for low speed operation, but when the drive unit is an engine (diesel, petrol or gas) for on-board power generation of 400 Hz in an aircraft, the engine speed can be controlled using Eddy current coupler. Similarly, in textile mills the rotational speed of bobbins can be precisely controlled. In wind power generation subjected to gush of wind, eddy current coupler can provide quality power by perfect speed control. Eddy current coupling have been in use since many years for a large variety of industrial applications rating from fractional horse power to few MW.

An eddy current coupling is a torque transmitter and not a torque converter. It has a DC field structure which is connected to a rotating shaft and the loss drum is connected to another shaft, as shown in Fig. 7.55a. The rotating coil structure receives power by means of slip rings.

If field system is now rotated by the prime mover at certain higher speed, the eddy currents are induced in the loss-drum. The eddy-current field interacts with air-gap flux to produce a torque, which is a function of current and slip speed. The torque exerted on the drum must be equal and opposite to that exerted on the rotor. It is a form of synchronous machine. Though there is an analogy between eddy current coupling and induction motor.

Let the speed of input shaft be ω_{in} and the speed of output shaft is ω_{o}. thus the slip speed is $(\omega_{\text{in}} - \omega_{\text{o}})$. The slip is

$$s = \frac{\omega_{\text{in}} - \omega_o}{\omega_{\text{in}}}$$

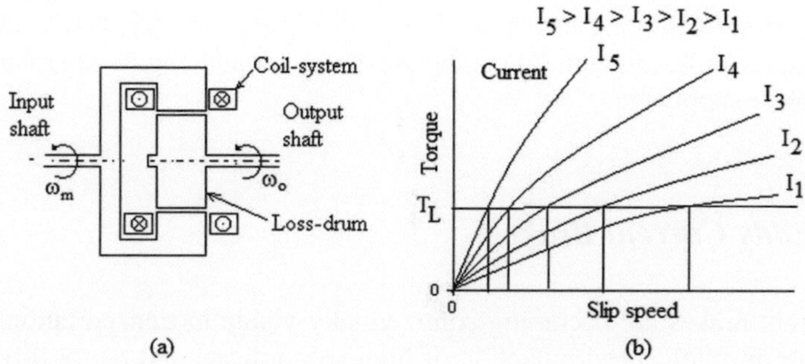

Fig. 7.55 **a** Eddy current coupling; **b** Torque speed characteristics

Here, we should have

$$T_{in} = T_o = T$$

The Input power, $P_m = T\omega_{\text{in}}$.

The Output power, $P_o = T(1 - s)\omega_{\text{in}}$.

Neglecting frictional and windage loss slip power $= T(\omega_{\text{in}} - \omega_o) = sT\omega_{\text{in}}$.

Variable speed of output shaft is obtained simply by varying the slip power dissipation in the drum, which is achieved by controlling the DC excitation of the field system. The output speed can be varied from zero to the full speed of the driving prime mover. The speed range in practice is limited by the heat dissipation capacity of the loss-drum. The torque speed characteristic is identical to the induction motor having higher resistance or solid iron rotor. The developed torque is,

$$T \ \alpha \ \phi^2 \ \alpha \ I_f^2 \tag{7.59}$$

Figure 7.55b shows torque versus slip speed curve for different field currents. The higher is the excitation current, the lesser is the slip speed. By controlling the excitation current, the speed of output shaft can be controlled. An eddy current coupling is normally rated approximately 50% of its maximum torque and is designed to deliver this torque at 4–8% slip. The losses are due to friction and windage loss, $R_f I_f^2$ loss and slip power. The cooling is either by air or water. The following are the advantages of eddy current coupling:

(a) Low control power.
(b) Closed loop speed control.
(c) Less running cost.
(d) Wide speed range.
(e) Inherent reliability.
(f) Ease of control.
(g) Low cost.
(h) Simplicity.

Eddy-Current Coupling has applications in process-industry, textile-mill, coal-crusher, paper-mills, large diesel engine-generator set and engine-alternator set for aircraft power supply etc.

7.9.2 *Eddy Current Brake*

Eddy current brakes are becoming commercially viable in transportation sector—for braking or quick speed reduction of high speed diesel and petrol driven four-, two-wheelers, EV (Electric Vehicle) and EHV (Electric Hybrid Vehicle). In these devices, the DC excitation in the stationary primary coil system produces speed induced eddy current in the rotating loss drum or disk made of conducting material. The construction details are similar to that of eddy current coupling except that one of the shaft having field coils is stationary. The interaction of eddy currents with the flux pattern produces a braking force on the drum or the disk. These machines have also been proposed for closed loop speed control of the driven unit and also for over speed protection of wind turbines. They provide good reliability, low cost and involve no maintenance. Further, a controlled braking torque can also be provided over a wide range of speed. These brakes are being preferred over conventional frictional brakes, because of their superior performances in high speed zone. A combination of frictional brake and eddy current brake is being demanded for low speed and high-speed braking in EV and EHV.

7.10 Generalized Machine Theory

For understanding generalized machines, let us assume a two-pole machine, though analysis is applicable to multi-pole machines as well. It is assumed that saliency is present only on one of its member. This theory is not applicable to the machines having saliency on both of its members like switched reluctance motor and stepper motor. The polar axis is regarded as the direct or d-axis (Fig. 7.56). The three-phase machines in which the MMF produced by the poly phase balanced three-phase windings, have current i_1, i_2 and i_3 is shown in Fig. 6.35. The d-axis salient pole has field winding energized by DC. The configuration is similar to that of synchronous machine having salient pole. It is assumed that there are no damper bars [1, 2].

Since the armature windings are balanced, mathematically the windings can be represented in two phase. The field poles have two axis of symmetry, because of salient pole. (refer Sec 6.8 Two axis theory).

In a three-phase system, the currents are described as,

$$i_1 = I_m \cos \omega t$$
$$i_2 = I_m \cos(\omega t - 120^\circ)$$

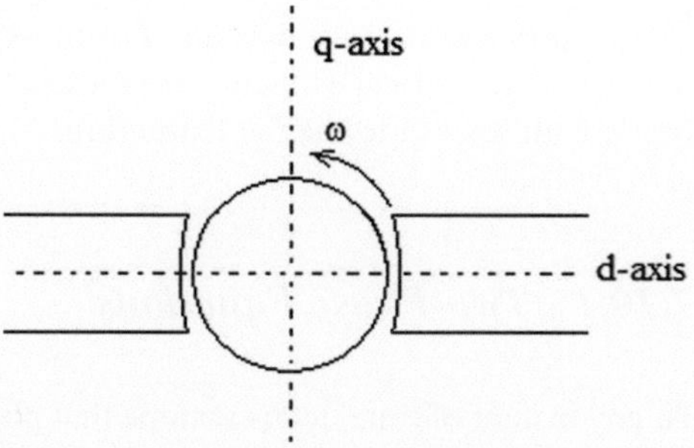

Fig. 7.56 Generalized model of machine showing d- and q-axes

$$i_3 = I_m \cos(\omega t - 240°) \quad (7.60)$$

The machine can be re-winded in two-phase system. Mathematically it is assumed that phase-a is coplanar to i_1. Thus, we get

$$i_1 = K_1 i_a$$

$$K_1 i_b = I_m \cos(90 + \theta) = -I_m \sin \theta$$

The equations are

$$\begin{aligned} i_1 &= K_1 i_a \\ i_2 &= I_m\left[\cos \omega t \cos 120° + \sin \omega t \sin 120°\right] = K_1\left[i_a \cos 120° - i_b \sin 120°\right] \\ i_3 &= I_m\left[\cos \omega t \cos 240° + \sin \omega t \sin 240°\right] = K_1\left[i_a \cos 240° - i_b \sin 240°\right] \end{aligned} \quad (7.61)$$

Similarly, a three-phase voltage system may have an equivalent two-phase system. Thus, it can be shown that

$$\begin{aligned} v_1 &= K_2 v_a \\ v_2 &= V_m\left[\cos \omega t \cos 120° + \sin \omega t \sin 120°\right] = K_2\left[v_a \cos 120° - v_b \sin 120°\right] \\ v_3 &= V_m\left[\cos \omega t \cos 240° + \sin \omega t \sin 240°\right] = K_2\left[v_a \cos 240° - v_b \sin 240°\right] \end{aligned} \quad (7.62)$$

It is assumed that each phase winding is having single turn coil. These results are set up for steady state analysis. The performance of a multiphase system can be matched with that of a two-phase system. In practice, a three-phase machine can be re-winded into two-phase system. The difference in the performance may be only due to imperfect distribution of windings in the slots. If the conductors are distributed in an ideal way, the two-phase machine will have identical performance as that of three-phase machine.

Any n-phase machine may also be represented into two-phase machine in a similar manner. Here, the MMF or current phasors not only vary with time but also physically rotate with respect to the field structure.

7.10.1 Two-Phase Equations

At any instant of time, let us assume that phase-a current and field flux have an angle θ. The voltage equations are

$$v = ri + \frac{\mathrm{d}}{\mathrm{d}t}\lambda$$

$$v = ri + p\lambda, \tag{7.63}$$

where v is the instantaneous value of applied voltage, and operator $p = \mathrm{d}/\mathrm{d}t$. Current i is the instantaneous current and λ is the flux linkage and r is the resistance of the winding. Thus we can write,

$$\begin{bmatrix} v_f \\ v_a \\ v_b \end{bmatrix} = \begin{bmatrix} r_f & 0 & 0 \\ 0 & r_a & 0 \\ 0 & 0 & r_b \end{bmatrix} \begin{bmatrix} i_f \\ i_a \\ i_b \end{bmatrix} + p \begin{bmatrix} \lambda_f \\ \lambda_a \\ \lambda_b \end{bmatrix} \tag{7.64}$$

The phase-a voltage may be written as

$$v_a = r_a i_a + p\lambda_a$$

The flux linkage of phase-a is

$$\lambda_a = L_a i_a + L_{ab} i_b + L_{af} i_f$$

Thus we can write for the machine,

$$\begin{bmatrix} \lambda_f \\ \lambda_a \\ \lambda_b \end{bmatrix} = \begin{bmatrix} L_f & L_{fa} & L_{fb} \\ L_{af} & L_a & L_{ab} \\ L_{bf} & L_{ba} & L_b \end{bmatrix} \begin{bmatrix} i_f \\ i_a \\ i_b \end{bmatrix} \tag{7.65}$$

where

L_a, L_b, L_f = self-inductance of the phase-a, phase-b and field windings, respectively.
L_{ab}, L_{ba} = Mutual inductance between a and b phase windings.
L_{af}, L_{bf} = Mutual inductance between a-, b- phase windings with field winding, respectively.

These inductances are not the functions of time, but since there is a relative motion between the windings, they are functions of time. Thus, operator p applies to the inductance as well as the current.

7.10.2 Self-Inductance

When phase-a is in the d-axis, the inductance is maximum for angle $\theta = 0°$ and 180°. Similarly, when phase-a is in q-axis (maximum air gap) the inductance of coil is minimum at angle $\theta = 90°$ and 270°. The variation of self-inductance of phase-a and phase-b with respect to θ are shown in Fig. 7.57a, b, respectively.

The self-inductance of coil-a can be written as

$$L_a = A + B\cos 2\theta$$

Now,

$$A + B\cos 2\theta = L_d \quad \text{for} \quad \theta = 0° \quad \text{and} \quad \theta = 180°$$

$$A + B\cos 2\theta = L_q \quad \text{for} \quad \theta = 90° \quad \text{and} \quad \theta = 270°$$

Solving we get,

$$A = \frac{1}{2}\left(L_d + L_q\right) \quad \text{and} \quad B = \frac{1}{2}\left(L_d - L_q\right)$$

Thus,

$$L_a = \frac{1}{2}\left(L_d + L_q\right) + \frac{1}{2}\left(L_d - L_q\right)\cos 2\theta = L_d\cos^2\theta + L_q\sin^2\theta \tag{7.66}$$

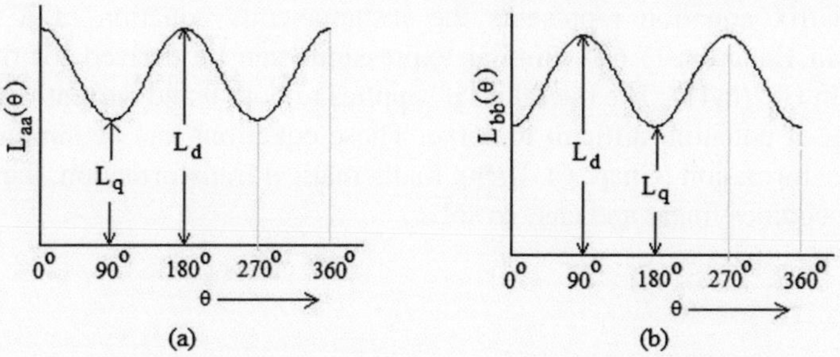

Fig. 7.57 Variation of self-inductance of **a** phase-a coil and **b** phase-b coil

Putting $\theta = 90° + \theta$, we get for phase-b.

$$L_b = L_d \cos^2(90^o + \theta) + L_q \sin^2(90^o + \theta) = L_d \sin^2\theta + L_q \cos^2\theta \qquad (7.67)$$

7.10.3 Mutual Inductance

Regarding mutual inductance, when pole is in between a and b phase winding (maximum iron minimum air gap position), the mutual inductance between the two coils is maximum. Thus, we can write that

$$\begin{aligned} L_{ab} &= (L_d - L_q)\cos\theta \sin\theta \\ L_{af} &= L_{ad}\cos\theta \\ L_{bf} &= L_{ad}\cos(90 + \theta) = -L_{ad}\sin\theta \end{aligned} \qquad (7.68)$$

Putting these values in earlier equation, we get

$$v_a = r_a i_a + p\left[\left(L_d \cos^2\theta + L_q \sin^2\theta\right)i_a + \left(L_d - L_q\right)\sin\theta\cos\theta i_b + L_{ad}\cos\theta i_f\right]$$

Similarly,

$$v_b = r_b i_b + p\left[\left(L_d \sin^2\theta + L_q \cos^2\theta\right)i_b - \left(L_d - L_q\right)\cos\theta \sin\theta i_a - L_{ad}\sin\theta i_f\right]$$

$$v_f = r_f i_f + p\left[L_f . i_f + L_{ad} . \cos\theta . i_a - L_{ad} . \sin\theta . i_b\right]$$

$$\begin{bmatrix} v_f \\ v_a \\ v_b \end{bmatrix} = \begin{bmatrix} r_f + pL_f & pL_{ad}\cos\theta & -pL_{ad}\sin\theta \\ pL_{ad}\cos\theta & r_a + p(L_d\cos^2\theta + L_q\sin^2\theta) & p(L_d - L_q)\sin\theta\cos\theta \\ -pL_{ad}\sin\theta & -p(L_d - L_q)\cos\theta\sin\theta & r_b + p(L_d\sin^2\theta + L_q\cos^2\theta) \end{bmatrix}\begin{bmatrix} i_f \\ i_a \\ i_b \end{bmatrix} \qquad (7.69)$$

The matrix equation represents the instantaneous equation of a two-phase machine, in Equation (7.69). Similar expressions can be derived for three phase as shown in Eq. (6.41). The operator "p" applies to both θ and current which makes the system of equation difficult to solve. These equations can be simplified using Park's transformation (Chap. 6). Using mathematical transformation, the system of equation becomes linear and easy to solve.

7.11 Symmetrical Components—Fortescue's Theorem

An unbalanced three-phase voltage/ current may be expressed in terms of positive sequence, negative sequence and zero sequence voltage/current. Fortescu's Theorem states the following for three phase system:

(a) Positive sequence components consisting of three phasor equal in magnitude displaced from each other by 120° in phase, and having the same phase sequence as the original phasor. Positive sequence components are denoted by subscript 1.
(b) Negative sequence components consisting of three phasor equal in magnitude displaced from each other by 120° in phase, and having the opposite phase sequence as the original phasor. Negative sequence components are denoted by subscript 2.
(c) Zero sequence components consisting of three phasor equal in magnitude displaced from each other by 0° in phase. Zero sequence components are denoted by subscript 0.

Figure 7.58 shows the positive sequence, negative sequence and zero sequence voltage of an unbalanced voltage. These are used to study the behavior of unbalanced poly-phase alternating current machines.

The voltage and current phasor are

$$\begin{aligned} \boldsymbol{V}_a &= \boldsymbol{V}_{a1} + \boldsymbol{V}_{a2} + \boldsymbol{V}_{a0}\ \boldsymbol{I}_a = \boldsymbol{I}_{a1} + \boldsymbol{I}_{a2} + \boldsymbol{I}_{a0} \\ \boldsymbol{V}_b &= \boldsymbol{V}_{b1} + \boldsymbol{V}_{b2} + \boldsymbol{V}_{b0}\ \boldsymbol{I}_b = \boldsymbol{I}_{b1} + \boldsymbol{I}_{b2} + \boldsymbol{I}_{b0} \\ \boldsymbol{V}_c &= \boldsymbol{V}_{c1} + \boldsymbol{V}_{c2} + \boldsymbol{V}_{c0}\ \boldsymbol{I}_c = \boldsymbol{I}_{c1} + \boldsymbol{I}_{c2} + \boldsymbol{I}_{c0} \end{aligned} \tag{7.70}$$

In three-phase system, we define the operators

$$\begin{aligned} a &= 1.0\angle 120^o = e^{j2\pi/3} = -0.5 + 0.866j \\ a^2 &= 1.0\angle 240^o = e^{j4\pi/3} = -0.5 - 0.866j \\ a^3 &= 1.0\angle 0^o = e^{j0} = 1.0 \end{aligned} \tag{7.71}$$

The symmetrical components of unsymmetrical phasor are

$$\begin{aligned} &\boldsymbol{V}_{a1} = a^3\boldsymbol{V}_{a1} = \boldsymbol{V}_{a1};\ \boldsymbol{V}_{b1} = a^2\boldsymbol{V}_{a1};\ \boldsymbol{V}_{c1} = a\boldsymbol{V}_{a1} \\ &\boldsymbol{V}_{a2} = a^3\boldsymbol{V}_{a2} = \boldsymbol{V}_{a2};\ \boldsymbol{V}_{b2} = a\boldsymbol{V}_{a2};\ \boldsymbol{V}_{c2} = a^2\boldsymbol{V}_{a2} \\ &\boldsymbol{V}_{a0} = \boldsymbol{V}_{b0} = \boldsymbol{V}_{c0} \end{aligned} \tag{7.72}$$

Similarly, the positive sequence, negative sequence and zero sequence components are calculated. Thus the respective phase voltages and currents are

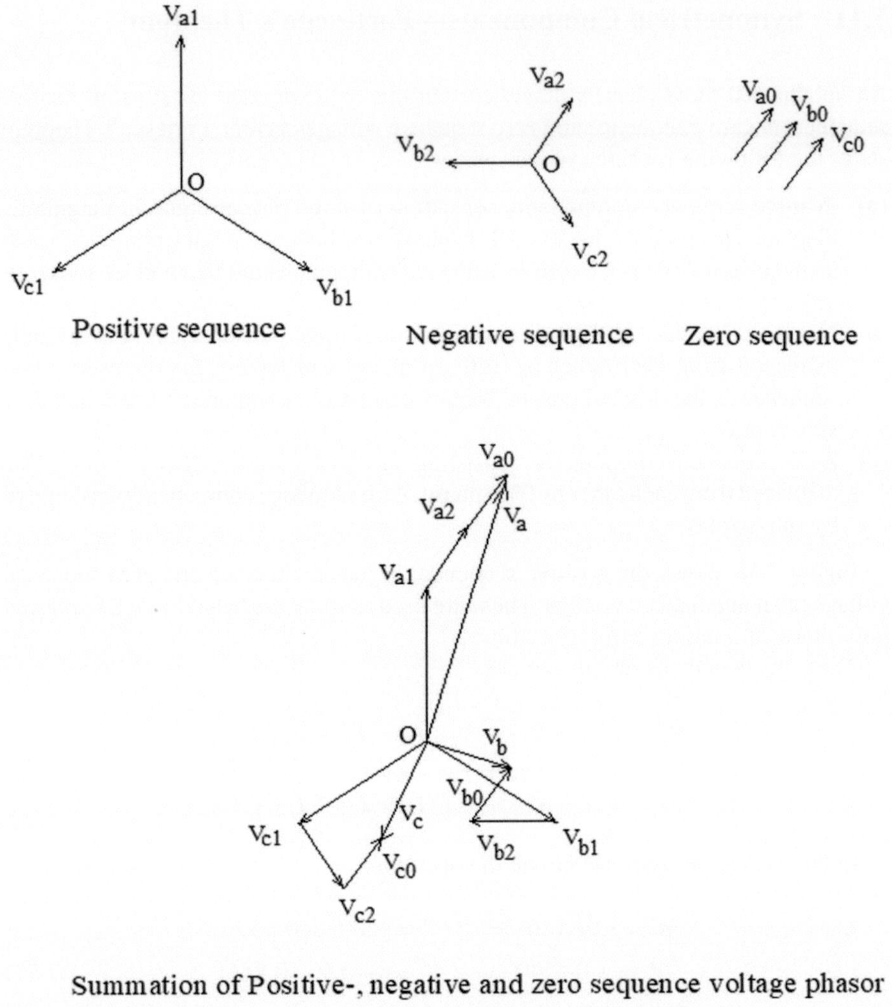

Fig. 7.58 Summation of positive-, negative and zero sequence voltage phasors

$$\begin{aligned} &\boldsymbol{V}_a = \boldsymbol{V}_{a1} + \boldsymbol{V}_{a2} + \boldsymbol{V}_{a0} \qquad \boldsymbol{I}_a = \boldsymbol{I}_{a1} + \boldsymbol{I}_{a2} + \boldsymbol{I}_{a0} \\ &\boldsymbol{V}_b = a^2\boldsymbol{V}_{a1} + a\boldsymbol{V}_{a2} + \boldsymbol{V}_{a0}\ \boldsymbol{I}_b = a^2\boldsymbol{I}_{a1} + a\boldsymbol{I}_{a2} + \boldsymbol{I}_{a0} \\ &\boldsymbol{V}_c = a\boldsymbol{V}_{a1} + a^2\boldsymbol{V}_{a2} + \boldsymbol{V}_{a0}\ \boldsymbol{I}_c = a\boldsymbol{I}_{a1} + a^2\boldsymbol{I}_{a2} + \boldsymbol{I}_{a0} \end{aligned} \tag{7.73}$$

The positive sequence negative sequence and zero sequence components of voltage and currents are:

$$\begin{aligned} &\boldsymbol{V}_{a0} = \tfrac{1}{3}(\boldsymbol{V}_a + \boldsymbol{V}_b + \boldsymbol{V}_c) \qquad \boldsymbol{I}_{a0} = \tfrac{1}{3}(\boldsymbol{I}_a + \boldsymbol{I}_b + \boldsymbol{I}_c) \\ &\boldsymbol{V}_{a1} = \tfrac{1}{3}\left(\boldsymbol{V}_a + a\boldsymbol{V}_b + a^2\boldsymbol{V}_c\right)\ \boldsymbol{I}_{a1} = \tfrac{1}{3}\left(\boldsymbol{I}_a + a\boldsymbol{I}_b + a^2\boldsymbol{I}_c\right) \\ &\boldsymbol{V}_{a2} = \tfrac{1}{3}\left(\boldsymbol{V}_a + a^2\boldsymbol{V}_b + a\boldsymbol{V}_c\right)\ \boldsymbol{I}_{a2} = \tfrac{1}{3}\left(\boldsymbol{I}_a + a^2\boldsymbol{I}_b + a\boldsymbol{I}_c\right) \end{aligned}$$

In a three-phase system, we have

$$\boldsymbol{I}_a + \boldsymbol{I}_b + \boldsymbol{I}_c = \boldsymbol{I}_n = 3\boldsymbol{I}_{a0} \tag{7.74}$$

Power in three-phase system is

$$S = P + jQ = \boldsymbol{V}_a\boldsymbol{I}_a^* + \boldsymbol{V}_b\boldsymbol{I}_b^* + \boldsymbol{V}_c\boldsymbol{I}_c^*$$

$$S = \begin{bmatrix} \boldsymbol{V}_a \ \boldsymbol{V}_b \ \boldsymbol{V}_c \end{bmatrix} \begin{bmatrix} \boldsymbol{I}_a \\ \boldsymbol{I}_b \\ \boldsymbol{I}_c \end{bmatrix}^* = \begin{bmatrix} \boldsymbol{V}_a \\ \boldsymbol{V}_b \\ \boldsymbol{V}_c \end{bmatrix}^T \begin{bmatrix} \boldsymbol{I}_a \\ \boldsymbol{I}_b \\ \boldsymbol{I}_c \end{bmatrix}^*$$

$$S = 3\begin{bmatrix} \boldsymbol{V}_{a0} \ \boldsymbol{V}_{a1} \ \boldsymbol{V}_{a2} \end{bmatrix} \begin{bmatrix} \boldsymbol{I}_{a0} \\ \boldsymbol{I}_{a1} \\ \boldsymbol{I}_{a2} \end{bmatrix}^*$$

Thus, we get

$$S = P + jQ = \boldsymbol{V}_a\boldsymbol{I}_a^* + \boldsymbol{V}_b\boldsymbol{I}_b^* + \boldsymbol{V}_c\boldsymbol{I}_c^* = 3\boldsymbol{V}_0\boldsymbol{I}_0^* + 3\boldsymbol{V}_1\boldsymbol{I}_1^* + 3\boldsymbol{V}_2\boldsymbol{I}_2^* \tag{7.75}$$

Let us apply it for different fault condition in an alternator. The behaviour of alternator during fault conditions are explained using positive sequence, negative sequence and zero sequence voltages, currents and impedances. A positive sequence impedance of an alternator Z_1 interacts with the positive sequence voltage and results in positive sequence current. Similarly, a negative sequence impedance of an alternator Z_2 interacts with the negative sequence voltage and results in negative sequence current. Further, a zero sequence impedance of an alternator Z_0 interacts with the zero sequence voltage and results in zero sequence current.

The negative sequence reactance is related with d- & q-axis sub-transient reactance of alternator as

$$X_2 = \frac{X_d'' + X_q''}{2} \tag{7.76}$$

The negative sequence currents are at double the supply frequency or due to counter rotating magnetic field. For an electrical machine $X_0 < X_2 < X_1$. The positive sequence reactance (Synchronous Impedance, X_s) is same as derived by OCC and SCC or alternator as discussed in Chap. 6.

Under fault condition the voltages as well as currents in alternating current machines are unbalance. In case of alternator, we can write that,

$$\begin{bmatrix} \boldsymbol{V}_{a0} \\ \boldsymbol{V}_{a1} \\ \boldsymbol{V}_{a2} \end{bmatrix} = \begin{bmatrix} 0 \\ \boldsymbol{E}_a \\ 0 \end{bmatrix} - \begin{bmatrix} Z_0 & 0 & 0 \\ 0 & Z_1 & 0 \\ 0 & 0 & Z_2 \end{bmatrix} \begin{bmatrix} \boldsymbol{I}_{a0} \\ \boldsymbol{I}_{a1} \\ \boldsymbol{I}_{a2} \end{bmatrix} \tag{7.77}$$

The matrix is used for analysing unsymmetrical faults in machine. For line to ground fault, we can choose that $\boldsymbol{I}_b = 0; \boldsymbol{I}_c = 0; \boldsymbol{V}_a = 0$.

This gives $\boldsymbol{I}_{a1} = \boldsymbol{I}_{a2} = \boldsymbol{I}_{a0}$.

In line-to-line fault, positive, negative and zero sequence currents are present. Thus

$$\boldsymbol{I}_{a1} = \frac{\boldsymbol{E}_a}{Z_1 + Z_2 + Z_0} \tag{7.78}$$

In Line to ground fault, $\boldsymbol{V}_b = \boldsymbol{V}_c; \boldsymbol{I}_a = 0$ and $\boldsymbol{I}_b = -\boldsymbol{I}_c$.

In line-to-ground fault, only positive and negative sequence currents are present. This gives

$$\boldsymbol{V}_{a1} = \boldsymbol{V}_{a2}; \boldsymbol{V}_{a0} = 0; \boldsymbol{I}_{a0} = 0; \boldsymbol{I}_{a2} = -\boldsymbol{I}_{a1}$$

$$\boldsymbol{I}_{a1} = \frac{\boldsymbol{E}_a}{Z_1 + Z_2} \tag{7.79}$$

In line to line fault, zero sequence current is absent. Symmetrical components are used to determine the performance of electrical machines under fault conditions.

Example 7.12 The three-phase voltage are 220∠0° V, 230∠-115° and 250∠-245° Volt. Determine positive sequence, negative sequence and zero sequence voltages.

Solution

Here

$$\begin{aligned}
V_a &= 220 + j_0\ V \\
V_b &= 230(\cos 115^\circ - j \sin 115^\circ) = (-97.2 - 208.45j)V \\
V_c &= 250(\cos 245^\circ - j \sin 245^\circ) = (-105.65 + 226.57j)V
\end{aligned}$$

The operators are

$$\begin{aligned}
a &= -0.5 + 0.866j \\
a^2 &= -0.5 - 0.866j
\end{aligned}$$

Zero sequence voltage is

$$\begin{aligned}
\boldsymbol{V}_{a0} &= \frac{1}{3}(\boldsymbol{V}_a + \boldsymbol{V}_b + \boldsymbol{V}_c) = \frac{1}{3}(220 + j0 - 97.2 - 208.45j - 105.65 + 226.57j) \\
&= 5.72 + 6.04j = 8.32\angle 46.55^o\text{V}
\end{aligned}$$

Positive sequence voltage is

$$\begin{aligned}\boldsymbol{V}_{a1} &= \frac{1}{3}\left(\boldsymbol{V}_a + a\boldsymbol{V}_b + a^2\boldsymbol{V}_c\right)\\ &= \frac{1}{3}(220 + (-0.5 + 0.866j)(-97.2 - 208.45j) + (-0.5 - 0.866j)(-105.65 + 226.57j))\\ &= 232.7 - 0.58j = 232.7\angle - 0.14^{\circ}V\end{aligned}$$

Negative sequence voltage is

$$\begin{aligned}\boldsymbol{V}_{a2} &= \frac{1}{3}\left(\boldsymbol{V}_a + a^2\boldsymbol{V}_b + a\boldsymbol{V}_c\right)\\ &= \frac{1}{3}(220 + (-0.5 - 0.866j)(-97.2 - 208.45j) + (-0.5 + 0.866j)(-105.65 + 226.57j))\\ &= -18.43 - 5.46j = 19.23\angle - 163.5^{\circ}V\end{aligned}$$

Problems

1. A conducting disk of outer and inner diameters of 1.0 m and 0.10 m, respectively, is mounted on a rotating shaft running at a 1000 RPM in an axial magnetic field of 1.0 T. Determine the induced EMF in between the outer and inner rim of the disk.
2. A printed circuit motor has outer radius of 100 mm and inner radius of 40 mm. The motor produces output power of 100 W at an efficiency of 75%. It is supplied by 12 V and takes 15A current and runs at 3000 RPM. The axial flux density is 1.0 T. Determine torque and number of conductors required.
3. A 220 V, 50 Hz, two-phase, four-pole servo motor has the following parameters:

$$R_1 = 325\,\Omega;\, X_1 = 75\,\Omega;\, R_2^{/} = 800\,\Omega;\, X_2^{/} = 75\,\Omega;\, X_m = 850\,\Omega$$

It is operating at a slip of 0.4. Determine.

(a) the resultant torque in synchronous Watts
(b) the stator phase currents.

Assume that the servo motor operates at unbalanced two-phase voltages:

$$V_a = 220\angle 0^o \quad and \quad V_b = 150 Volts$$

4. A two-phase, two pole drag-cup servo motor has the following parameters:

$$R_1 = 200\,\Omega;\, X_1 = 90\,\Omega;\, R_2^{/} = 200\,\Omega;\, X_2^{/} = 90\,\Omega;\, X_m = 875\,\Omega$$

The main winding and control winding are identical and each are rated for 120 V. The voltages are in exact quadrature, when operating at a slip of 0.3 with control voltage being 80% of main winding voltage. Determine the resultant torque.

5. A fractional horse power AC series motor has resistance of 12.5 Ω and total inductance of 0.75 H. At 220 V DC supply voltage the motor takes 1.5 A current and runs at 2700 RPM. Estimate the speed, power factor when supply voltage is 220 V AC, 50 Hz and the motor takes 1.8 A current.
6. A 220 V, 750W, 22,000 RPM, 50 Hz AC series motor has the total resistance of 0.6 Ω and total reactance of 5 Ω. The rotational losses are 15 W. Determine

   (a) Current and power factor and developed torque at rated condition
   (b) Speed, power factor, developed torque and efficiency when the load current is reduced to half.

7. Calculate the linear synchronous speed of a 1.0 m length, 60 slot, four-pole linear induction motor operating on 60-Hz supply. The equivalent circuit of a star connected single sided linear induction motor is as follows:

$$R_1 = 5.0\,\Omega;\, X_1 = 7.0\,\Omega;\, R_2^{/} = 6.0\,\Omega;\, X_2^{/} = 7.0\,\Omega;\, X_m = 150\,\Omega$$

The machine is operating on 380 V, three-phase 60 Hz supply. Calculate the per phase current, power factor and force developed, when machine is running at 21.5 m/s.

8. A stepper motor has 90° per step and must rotate at 50 RPM. What input pulse rate in pulses per second is required?
9. The three-phase voltage are $220\angle 0°$ V, $220\angle -115°$ and $220\angle -235°$ Volt. Determine positive sequence, negative sequence and zero sequence voltages.
10. The three-phase voltage are $220\angle 0°$ V, $-220\angle +120°$ and $220\angle -240°$ Volt. Determine positive sequence, negative sequence and zero sequence voltages.

Objective-Type Questions

1. The eddy current coupler and eddy current brake are work on

   (a) AC excitation
   (b) DC excitation
   (c) AC or DC excitation
   (d) Both (a) & (b)

2. The motor driving cooling fan of a personal computer is

   (a) Reluctance Synchronous motor
   (b) Switched Reluctance motor
   (c) PMBLDC motor
   (d) Stepper motor

3. In PMBLDC motor, permanent magnet rotor position is judged using

   (a) Inductive proximity sensor
   (b) Capacitive proximity sensor

(c) Hall effect sensor
(d) Resistive proximity sensor

4. The motor driving the CD in CDROM drive of a personal computer is
   (a) Reluctance Synchronous motor
   (b) Switched Reluctance motor
   (c) PMBLDC motor
   (d) Stepper motor

5. The linear synchronous speed of a 380 V, 50 Hz, eight-poles, 1.0 m long linear induction motor is
   (a) 12.5 m/s
   (b) 25 m/s
   (c) 15.5 m/s
   (d) 18.5 m/s

6. An eight pole, 2 m length SLIM is excited by 75 Hz, the linear synchronous speed is
   (a) 12.5 m/s
   (b) 37.5 m/s
   (c) 25 m/s
   (d) 40.5 m/s

7. Isolated Signals corresponding to DC current & voltage can be generated using
   (a) Instrument transformer
   (b) Hall effect transducers
   (c) Ferrite transformers
   (d) accelerometer

8. The purpose of compensating winding is AC series motor is to
   (a) Minimization of gap MMF
   (b) Minimization of armature reactance
   (c) Minimization of field reactance
   (d) Maximization of armature reactance

9. The purpose of compensating winding in cross field machine is
   (a) Minimize the reactance
   (b) Cancel the flux due to load current
   (c) check sparking
   (d) Improve commutation

10. Electrical motors having lowest moment of inertia is
    (a) Servo motor
    (b) Printed circuit DC motor
    (c) Claw pole motor

(d) Synchronous motor

11. Axial flux motors are

    (a) Conventional DC motor with field winding and DC motor with pancake armature
    (b) PMBLDC motor and printed circuit DC motor
    (c) Printed circuit DC motor and DC motor with pancake armature
    (d) Conventional DC motor with permanent magnetic and DC motor with pancake armature

12. In servo motor, the moment of inertia is kept to minimum at design stage by keeping

    (a) Large Diameter and large length
    (b) Large diameter and short length
    (c) Less diameter and large length
    (d) Less diameter and short length

13. Which of following statement is correct?

    (a) AC servo motor is a two-phase induction motor with smaller rotor resistance
    (b) AC servo motor is a two-phase induction motor with large rotor resistance
    (c) AC servo motor is a single phase induction motor with smaller rotor reactance
    (d) AC servo motor is a two-phase induction motor with smaller rotor reactance

14. For a DC series motor with compensating windings, the incorrect input may be,

    (a) DC supply
    (b) AC supply
    (c) DC supply having 35% ripple contents
    (d) Pulsed power supply

15. In a Schrage motor, following statement is not correct,

    (a) It in an inverted three-phase slip ring induction motor
    (b) The rotor rotates in direction opposite to the direction of rotating field
    (c) The rotor rotates in the direction of rotating field
    (d) It can run at sub- and super-synchronous speeds.

16. The linear synchronous speed of a 1.0 m, 150 Hz, 2 -pole linear induction motor is

    (a) 150 m/s
    (b) 75 m/s
    (c) 37.5 m/s
    (d) 18.75 m/s

17. Which of the following motor / motors requires electronic circuits for their operation
    (a) SRM, Stepper motor, BLDC motor, PMSM
    (b) SRM, Stepper motor, BLDC motor
    (c) SRM, Stepper motor
    (d) SRM
18. The direction of rotation of a stepper motor is reversed by
    (a) Reversing the supply terminals
    (b) Reversing the bias of transistor of control circuit
    (c) Changing the switching sequence
    (d) Using another control circuit
19. A composite secondary Linear Induction Motor will exhibit larger thrust when clearance between them is
    (a) very large
    (b) minimum possible
    (c) Any clearance
    (d) Optimum
20. Following machines require Hall Effect Position sensing for their operation
    (a) PMBLDC Motor, DC motor
    (b) Pancake DC motor and PMBLDC motor
    (c) PMBLDC motor and PMSG
    (d) Pancake DC motor and printed circuit DC motor
21. The order of air gap in linear induction motor used in transportation may be
    (a) 0.10 mm
    (b) 1.00 mm
    (c) 10.0 mm
    (d) 100.0 mm

Bibliography

1. Philip L. Alger, Indian Machines, Gordon & Breach Science Pub. N.Y. 2nd Ed. 1965
2. M.G.Say, Alternating current Machines, 4th ELBS edition, 1976
3. E. Openshaw Taylor: The Performance and design of A.C. commutator motors, Wheeler Publications, India, 11th Print, 1990.
4. A. Fransua, R. Magureanu, Electrical Machines and Drive Systems, TP Technical Press, Oxford, 1984
5. M.G. Say & E.O. Taylor, Direct current machines, ELBS, 1980
6. V Del Toro, Electric machines and power systems, PHI, 1980
7. S.A.Nasar, Electromagnetic energy conversion & Systems, Prentice-Hall, Inc., N.J., 1970

8. E.R.Laithwaite, Transport without wheels, Elek Science, London, 1977
9. TJE Miller, Switched Reluctance Motors & their control, Clarendon Press, Oxford, 1993
10. TJE Miller, Reluctance Synchronous motors, Clarendon Press, Oxford, 1993
11. J. R. Hendershot Jr. & THE Miller, Design of Permanent-Magnet Motors, Clarendon Press, Oxford, 1994
12. Veinot, C.G., Fractional Horse power machine

MIX
Papier aus verantwortungsvollen Quellen
Paper from responsible sources
FSC® C105338

If you have any concerns about our products,
you can contact us on
ProductSafety@springernature.com

In case Publisher is established outside the EU,
the EU authorized representative is:
Springer Nature Customer Service Center GmbH
Europaplatz 3, 69115 Heidelberg, Germany

Printed by Libri Plureos GmbH
in Hamburg, Germany